BASIC ELECTRONICS

Prepared by

THE U. S. NAVY

(BUREAU OF NAVAL PERSONNEL)

Revised Edition

DOVER PUBLICATIONS, INC.

NEW YORK

This Dover edition, first published in 1973, is an unabridged and unaltered republication of the fourth edition of *Basic Electronics Volume I,* as published by the United States Navy Training Publications Center in 1980. The text is designated as Rate Training Manual NAVPERS 10087-C.

International Standard Book Number:
ISBN-13: 978-0-486-21076-6
ISBN-10: 0-486-21076-6

Library of Congress Catalog Card Number: 73-87174

Manufactured in the United States by RR Donnelley
21076630 2015
www.doverpublications.com

PREFACE

Basic Electronics, is written for men of the U. S. Navy and Naval Reserve whose duties require them to have a knowledge of the fundamentals of electronics. Electronics concerns itself with the emission, behavior, and effect of electrons in vacuums, gases, and semiconductors. Technically speaking, electronics is a broad term extending into many fields of endeavor. Today, electronics projects itself into Navy life at every turn. It facilitates a means of rapid communications, navigates ships, helps control engineering plants, aims guns, drops bombs, and perform logistic functions. It is, therefore, important to become well informed in all areas of basic electronics in order to be able to qualify for any of the many applicable rates or ratings.

The coverage of basic electronics principles is presented in two volumes of which this is Volume I. The opening chapters treat general information concerning Naval electronics equipments, testing devices, and safety procedures. This is followed by a study of fundamental electronics devices such as transistors and electron tubes, and some of the basic circuit arrangements in which they are frequently found.

You will note that a broad area of this manual treats communications theory. We have seen fit to do this since the concept of transmission and reception is fundamental in the electronics field. The coverage includes a detailed discussion of basic receiving and transmitting circuits for conventional amplitude-modulated, frequency-modulated, and single-sideband systems. Some troubleshooting techniques of these circuits are also presented.

The concluding chapters present a discussion of transmission lines, antennas, and wave propagation. This information will be helpful to the reader in understanding the relationship between transmitting and receiving equipments.

Basic Electronics, Volume II,* NavPers 10087-C treats pulse shaping circuits, microwave applications, basic logic circuits, and some fundamentals of synchro and servo systems. The reader should understand the concepts treated in Volume I before any attempt is made to study the more advanced subjects presented in Volume II.

As one of the basic Navy training manuals, this book was prepared by the Training Publications Division of the Naval Personnel Program Support Activity, Washington, D. C., which is a field activity of the Bureau of Naval Personnel. Technical assistance was provided by the Electronics Technician (Class A) Schools at Great Lakes, Illinois and Treasure Island, California, and the Naval Training Publications Center, Memphis, Tennessee.

**Basic Electronics, Volume II* has been reprinted by Dover Publications, Inc. under the title *Second-level Basic Electronics* (22841-X).

CONTENTS

CHAPTER 1

INTRODUCTION TO ELECTRONIC EQUIPMENTS

The purpose of this chapter is to explain the functional (as opposed to operational) characteristics of various Navy electronic equipments, and tell how these equipments are used in the performance of Navy functions.

The equipments discussed and illustrated in this chapter are selected as representative of the many models and types of electronic equipments used in the Navy today. No attempt is made to cover all of the equipments in use.

Equipments to be covered can be broken down into six broad categories: Communications, Navigational, Radar, Identification Friend or Foe (IFF), Countermeasures, and Sonar. Included in communications are radio, radio teletype, multiplex, facsimile, and digital data equipment. Radar equipment can be divided into distinct types, such as surface search, air search, altitude determining, airborne search and fire control.

Modern Navy equipments must be of rugged construction for long service life. Due to limited space, vibrations, and stresses found aboard ships and aircraft, compactness and ruggedness must be important design considerations for these equipments.

FREQUENCY SPECTRUM

Before actually describing the various equipments, it will be advantageous to cover the frequency spectrum, which generally applies to all equipments.

Electronic equipments operate on frequencies ranging from 10,000 hertz through frequencies in the visible light range (see fig. 1-1). These frequencies are generally divided into eight bands as shown in table 1-1.

The Navy normally uses the VLF and LF bands for shore station transmissions. The commercial broadcast band extends from about 550 kHz to 1700 kHz, thereby limiting naval use to the upper and lower ends of the MF band. The HF band lends itself well for long-range communications, therefore most shipboard radio communications are conducted in the HF band.

A large portion of the lower end of the VHF band is assigned to the commerical television industry and is used by the Armed Forces only in special instances. The upper portion of the VHF band (225 MHz to 300 MHz) and the lower portion of the UHF band (300 MHz to 400 MHz) are used extensively by the Navy for its UHF

67.190

Table 1-1.—Bands of frequencies

Abbreviation	Frequency band	Frequency range
VLF	Very low frequency	below 30 kHz
LF	Low frequency	30-300 kHz
MF	Medium frequency	300-3000 kHz
HF	High frequency	3000-30,000 kHz
VHF	Very high frequency	30-300 MHz
UHF	Ultrahigh frequency	300-3000 MHz
SHF	Superhigh frequency	3000-30,000 MHz
EHF	Extremely high frequency	30-300 GHz

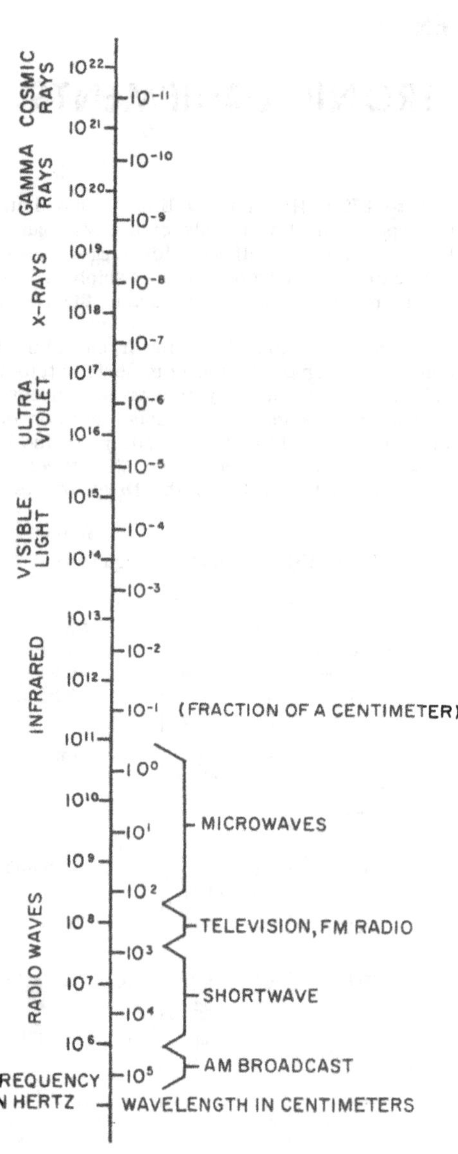

25.96(20C)
Figure 1-1. — Frequency spectrum.

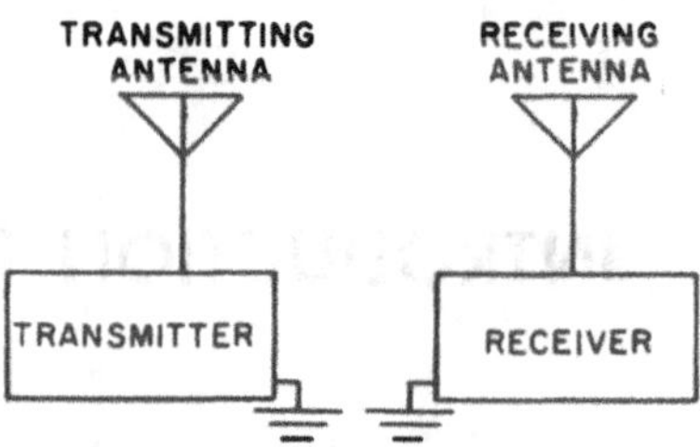

72.50
Figure 1-2. — Basic radio communication system.

communications. The frequencies above 400 MHz in the UHF band through the SHF and EHF bands are normally used for radar and special equipment.

COMMUNICATIONS EQUIPMENT

In the Navy the ships, planes and shore bases do not act independently of one another but operate as a team working together to accomplish a specific task. Radio equipment is used to coordinate the activities of the many units in the Fleet by linking them with each other and with shore stations.

The word "radio" can be defined briefly as the transmission of signals through space by means of electromagnetic waves. Usually, the term is used in referring to the transmission of intelligence code and sound signals, although television (picture signals) and radar (pulse signals) also depend on electromagnetic waves.

Radio equipment can be divided into two broad categories, transmitting and receiving equipment.

A basic radio communication system consists of a transmitter and a receiver (fig. 1-2). The electromagnetic variations are propagated through space from the transmitting antenna to the receiving antenna. The receiver converts these signals into the intelligence that is contained in the transmission.

Of the several methods of radio communications available, those utilized most commonly by the Navy are radiotelegraphy, radiotelephony, radioteletype, radiofacsimile, and digital data.

VERY HIGH OR ULTRA HIGH FREQUENCY

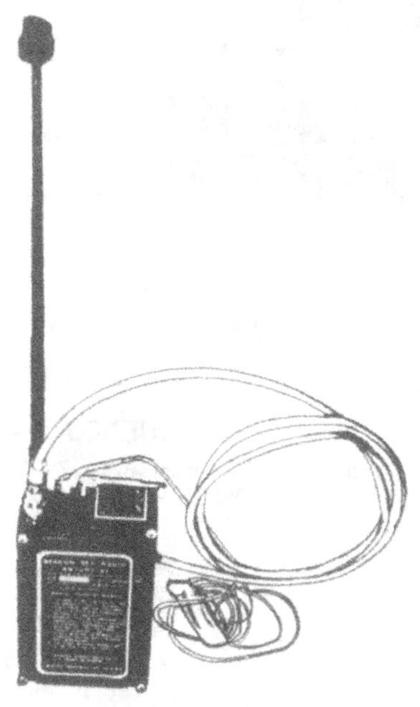

ULTRA HIGH FREQUENCY (RESCUE BEACON)

(A) 32.278(76) (B) 120.65 (C) 76.32 (D) 32.40 (E) 120.5

Figure 1-3.—Representative Navy radio transmitters.

MEDIUM FREQUENCY

(SIZE RELATIONSHIP IS APPROXIMATELY ONE-HALF ACTUAL)

HIGH FREQUENCY

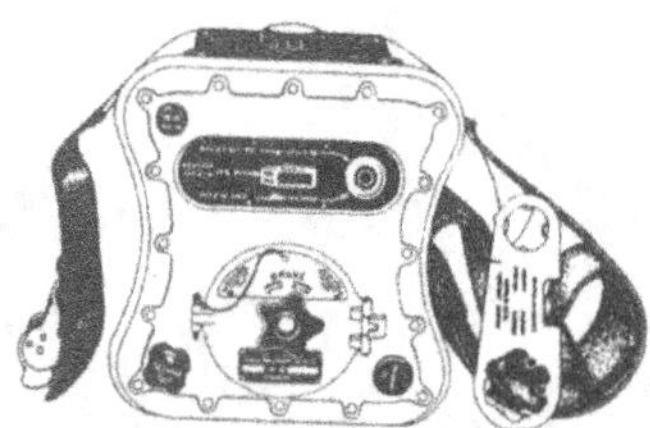

HIGH FREQUENCY (EMERGENCY LIFEBOAT)

(A) 32.278(76) (B) 120.65 (C) 76.32 (D) 32.40 (E) 120.5

Figure 1-3.—Representative Navy radio transmitters—continued.

TRANSMITTERS

A transmitter is a device for converting intelligence, such as voice or code, into electrical impulses for transmission either on closed lines, or through space from a radiating antenna. Transmitters take many forms, have varying degrees of complexity, develop various levels of power, and employ numerous methods for sending the desired information or energy component from one point to another.

Transmitters operating in the medium- and high-frequency bands of the frequency spectrum are used chiefly for communication at medium and long ranges. Some transmitters in these bands, however, are designed for short-range communication. In most instances, short-range transmitters have a lower output power than those designed for communication at medium and long ranges.

The term "short range" (or "distance") means a measurement less than 200 miles; "medium range" is between 200 and 1500 miles; and "long range" exceeds 1500 miles. These values are approximate, because the range of a given equipment varies considerably according to terrain, atmospheric conditions, frequencies, and time of day, month, and year.

Transmissions in the VHF and UHF ranges normally are restricted to line-of-sight distances. Under unusual atmospheric conditions, they have been received at considerably longer distances—500 miles or more.

Shipboard installations of VHF equipments are retained for emergency communications, and for communication with allied forces that have not yet converted to UHF equipments. The VHF equipment is also being used as a backup to UHF equipment.

Most UHF radio transmitters (and receivers) used by the Navy operate in the 225- to 400-MHz frequency range. Actually, this range of frequencies covers portions of both the VHF band and the UHF band. For convenience, however, radio equipments operating within this frequency range are considered to be UHF equipments.

Some of the various types of transmitters used in the Navy are illustrated in figure 1-3.

RECEIVERS

A receiver is a device which converts the electrical impulses picked up by the receiving

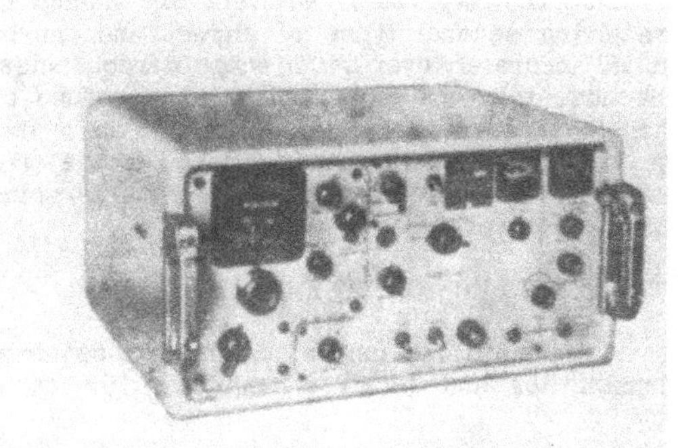

VERY LOW / LOW FREQUENCY

VERY HIGH OR ULTRA HIGH FREQUENCY

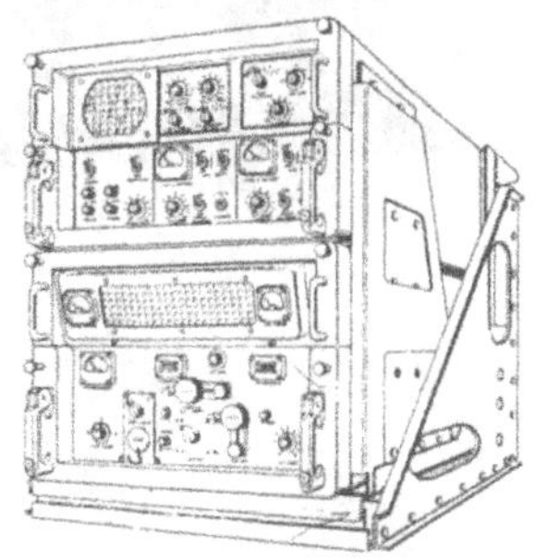

MEDIUM / HIGH FREQUENCY

(A) 1.157 (B) 50.40 (C) 32.42

Figure 1-4.—Representative Navy radio receivers.

antenna back to the original transmitted intelligence.

Modern Navy radio receivers are capable of receiving several types of signals and can be tuned accurately over a wide range of frequencies. Because they are not required to produce or handle large currents and voltages, their size is relatively small when compared to the size of most transmitters. Figure 1-4 shows some of the receivers used by the Navy.

TRANSMITTER-RECEIVERS

A transmitter-receiver comprises a separate transmitter and receiver mounted in the same rack or cabinet. The same antenna may be used for the transmitter-receiver arrangement. When so used, the capability for simultaneous operation of both the transmitting and receiving equipments does not exist. The equipments may be operated independently, using separate antennas. An example of Navy type transmitter-receivers is displayed in figure 1-5.

TRANSCEIVERS

A transceiver is a combined transmitter and receiver in one unit which uses switching arrangements in order to utilize parts of the same electronic circuitry for both transmitting and

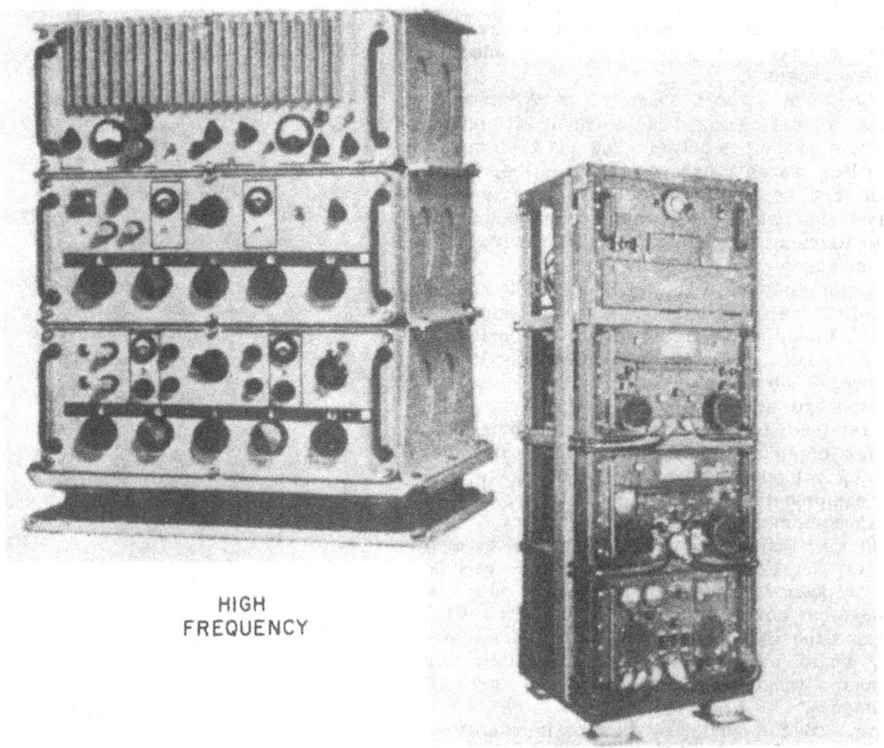

Figure 1-5.—Representative Navy transmitter-receivers.

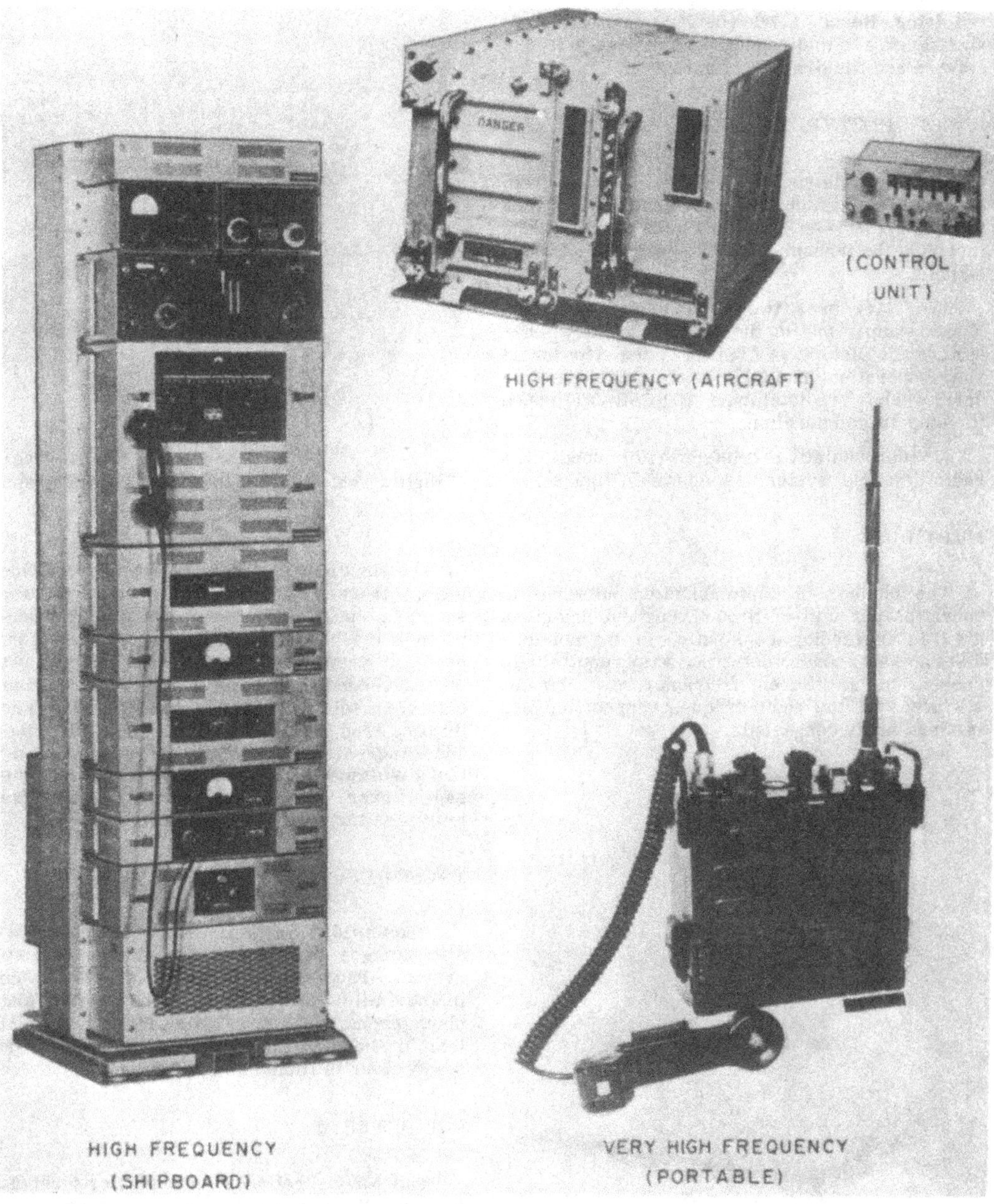

32.135:.120.5
Figure 1-6.—Representative Navy transceivers.

receiving. Hence, a transceiver cannot transmit and receive simultaneously. Some Navy transceivers are illustrated in figure 1-6.

RADIO TELETYPE

A radio teletype system is a series (loop) connected telegraphic communications network operating between two or more points, and involving the transmission and reception of coded signals.

The Navy uses two basic teletype systems aboard ship. One is the audio frequency tone-shift radioteletype (AFTSRATT) used for short range operation. The other is the radiofrequency carrier-shift radioteletype (RFCSRATT) used for long range operation.

A representative teletypewriter used in a radio teletype system is shown in figure 1-7.

120.26(120C)
Figure 1-8.—Representative Navy telegraph multiplex terminal.

MULTIPLEX

The number of communication networks in operation per unit of time throughout any given area is increasing constantly. In the not-too-distant past, each network was required to operate on a different frequency. As a result, all areas of the radio-frequency spectrum had become highly congested.

1.361
Figure 1-7.—Representative Navy teletypewriter.

The maximum permissible number of intelligible transmissions taking place in the radio spectrum per unit of time can be increased through the use of multiplexing. The main purpose of a multiplex system is to increase the message-handling capacity of radio communication, or teletypewriter channels and the transmitters and receivers associated with them. This increase in capacity is accomplished by the simultaneous transmission of several messages over a common channel. A telegraph multiplex terminal is shown in figure 1-8.

FACSIMILE

Facsimile (FAX) is a method for transmitting still images over an electrical communication system. The images, called pictures or copy in facsimile terminology, may be weather maps, photographs, sketches, typewritten or printed text, or handwriting. A representative facsimile transceiver is illustrated in figure 1-9.

COMPUTERS

The Navy's reliance on computer technology has greatly increased in recent years because of the ability of computing systems to provide fast and accurate analysis of logistical and tactical operations. Computing systems are highly

13.70
Figure 1-9.—Representative Navy facsimile transceiver.

reliable and can perform several operations simultaneously.

Computers are divided into two general types, analog computers and digital computers. The analog computer is one which solves problems by translating physical conditions; such as flow, temperature, pressure, angular position, and size of displacement into equivalent electrical quantities. These electrical measurements are continually being updated as the physical values change.

The gasoline gage on an automobile is one type of analog computer. The liquid in the gasoline tank operates a float which, in turn, physically controls the amount of battery current flowing through a rheostat mounted at the tank. The amount of current is proportional to the amount of gasoline in the tank and positions the gasoline gage to a position on the scale indicator which indicates the approximate amount of gasoline in the tank. Thus an analog computer gives an approximate solution in a continuous form, whereas a digital computer gives exact data solutions of discrete values.

The digital computer is one which solves problems by repeated high speed use of the fundamental arithmetic process of addition, subtraction, multiplication, or division in binary, decimal or any pre-determined notation. The digital computer uses increments (digits) to express distinct quantities.

An example of a digital computer is a cash register. Certain specific digits are entered at the console and stored, and upon request, a digital output representing the sum (or difference) is printed out. Other digital computers include the abacus, desk calculator, punchcard machine, and the modern electronic computer.

Digital electronic computers are classified as special-purpose or general-purpose computers. The special-purpose computer is designed to handle a specific type of data processing task as exacting and as efficiently as possible. A general-purpose computer is designed to handle a variety of data processing tasks in which its adaptability, storage capacity and speed are adequate.

Some of the more common places where computers are used (mostly ashore but some afloat) are in command activities, operation centers, communications, finance, medical, weather, supply, maintenance, oceanography, weapon systems, and Naval Tactical Data Systems. Figure 1-10 shows a representative digital computer and a representative analog computer.

NAVIGATIONAL EQUIPMENT

Basically, piloting is a form of navigation. Piloting is that branch of navigation in

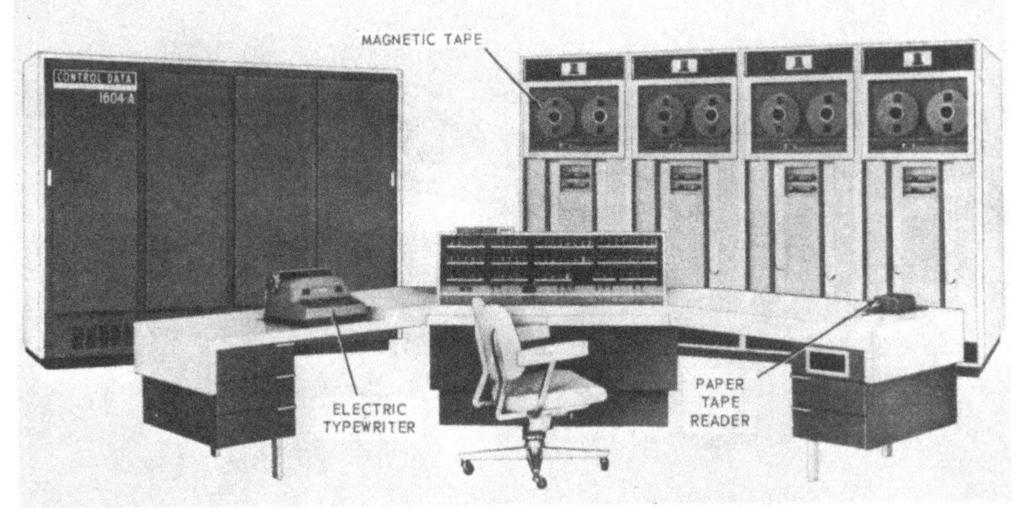

A. DIGITAL

B. ANALOG

Figure 1-10.—Representative Navy computers. 124.1X

which a ship's position is obtained by referring to visible objects on the earth whose locations are known. This reference usually consists of bearing and distance of a single object, cross bearings on two or more objects, or two bearings on the same object with an interval between them.

Position in electronic navigation is determined in practically the same way that it is in piloting. There is one important difference, however. The objects by which the ship's position is determined need not be visible from the ship. Instead, their bearings (and in most instances their ranges) are obtained by electronic means—usually radio.

The advantages of piloting by radio are obvious. A ship's position may be fixed electronically in fog or thick weather that otherwise would make it impossible to obtain visual bearings. Moreover, it may be determined from stations located far beyond the range of even clear-weather visibility.

RADIO DIRECTION FINDERS

Essentially, the radio direction finder (RDF) is a sensitive receiver to which a directional antenna is connected. Radio direction finding is used primarily in air navigation. It is also used in the location of personnel afloat in liferafts or lifeboats, provided these craft are equipped with a radio transmitter, and that the RDF equipment is capable of receiving the transmitted frequencies.

74.130.1
Figure 1-11.—Representative Navy radio direction finder indicator unit.

The RDF indicator unit, illustrated in figure 1-11, produces a visual indication of the bearing of a received signal by means of an oscilloscopic display of the output from a receiver.

Radio Compass

Radio compass equipment finds its greatest present-day usage in aircraft. With a network of radio beacons covering much of the earth's surface, the radio compass is essentially a radio direction finder that automatically indicates the plane's bearing at all times and thus helps the pilot to maintain his course and to locate his position. Where the beamed energy from the beacons cross, it is possible for the pilot or navigator to fix his position with considerable accuracy.

Ground-station radio beacons transmit either continuously or at automatically scheduled times; the pilot tunes his compass receiver to the frequency of the stations listed in the area through which he is passing. Indicators automatically show relative bearing information with respect to the station being received.

TACAN

Tactical air navigation (tacan), is an electronic polar coordinate system that enables an aircraft pilot to read—instantaneously and continuously—the distance and bearing of a radio beacon transmitter installed on a ship or at a ground station. In aircraft equipped with tacan receiving equipment, an azimuth indicator shows the position of the transmitting sources in degrees of magnetic bearing from the aircraft. Also, the distance in nautical miles to the same reference point is registered as a numerical indication, similar to that of an automobile odometer. The transmitter, receiver, and control units of a shipboard and an aircraft TACAN system are illustrated in figure 1-12.

LORAN

The long-range navigation (loran) system provides a means of obtaining accurate navigational fixes from pulsed radio signals radiated

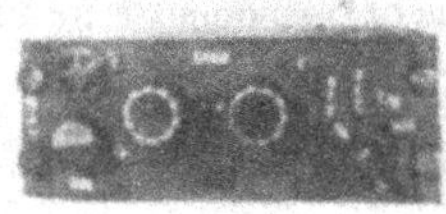

179.1

Figure 1-12. — Representative Navy tacan equipment.

by shore-based transmitters. Depending on the mode of loran operation and the time of day or night, fixes are possible at distances up to 3000 nautical miles from the transmitting stations.

A loran set aboard ship (fig. 1-13A) or airborne (fig. 1-13B) is a receiving set and indicator that displays the pulses from loran transmitting stations ashore.

RADAR EQUIPMENT

Radar (from the words radio detection and ranging) is one of the greatest scientific developments that emerged from World War II. It makes possible the detection and range determination of such objects as ships and airplanes

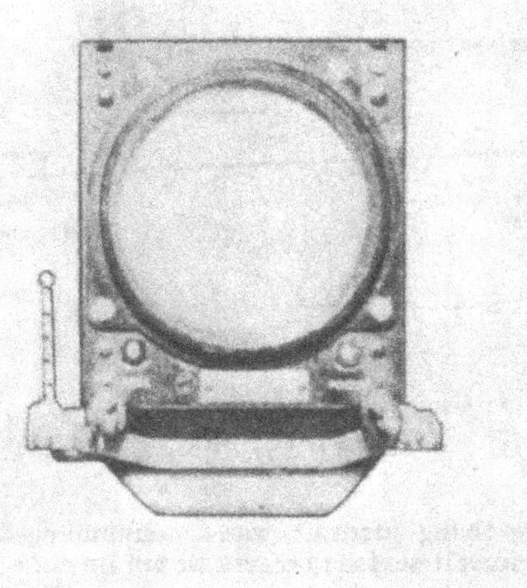

B. AIRCRAFT

120.48

Figure 1-13.—Representative Navy loran equipment.

over long distances. The range of radar is unaffected by darkness, but it often is affected by various weather conditions; for example, heavy fog or violent storms.

No single radar set has yet been developed to perform all the combined functions of air-search, surface-search, altitude-determination, and fire control because of size, weight, power requirements, frequency band limitations, and so on. As a result, individual sets have been developed to perform each function separately. The function of the set is usually most readily identified by the antenna associated with it (fig. 1-14). The information obtained from the radar sets is presented as visual displays on repeaters in CIC (Combat Information Center), on the bridge, and in spaces deemed necessary for efficient operation of the ship.

SURFACE-SEARCH

The principal function of surface-search radars is the detection and determination of accurate range and bearing of surface targets

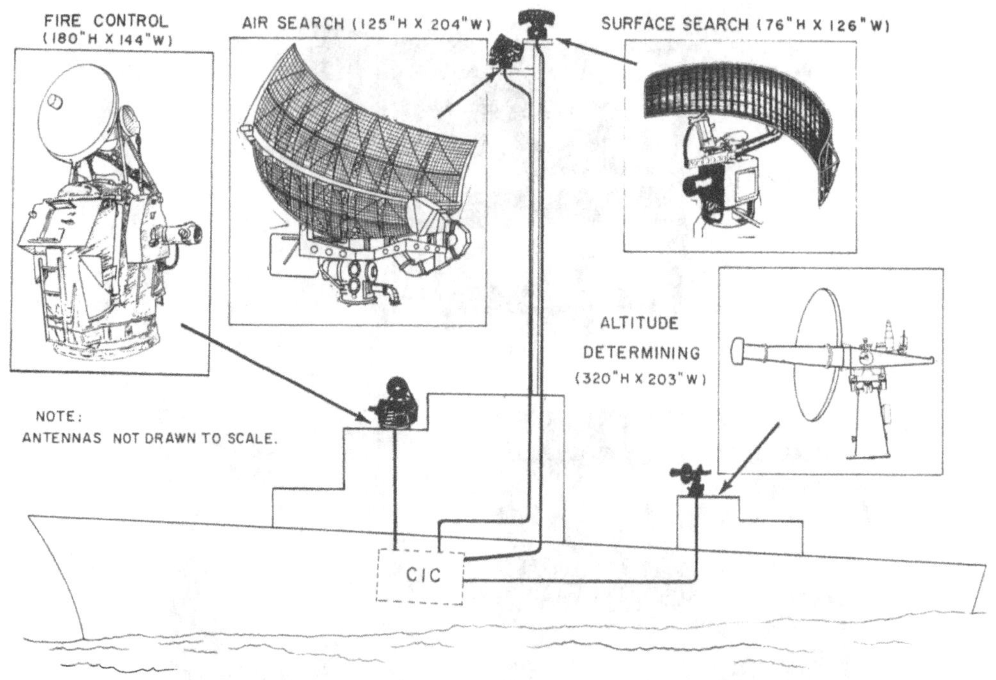

179.2
Figure 1-14.—Navy radar type identification by representative antenna.

and low-flying aircraft while maintaining 360° search for all surface targets within line of sight distance of the radar antenna. A representative surface search radar antenna is shown in figure 1-14.

AIR-SEARCH

The chief function of an air-search radar is the detection and determination of ranges and bearings of aircraft targets at long ranges (greater than 50 miles), while maintaining complete 360° azimuth search. Figure 1-14 contains an illustration of a representative air-search radar antenna.

ALTITUDE-DETERMINING

The function of the altitude-determining radar (antenna shown in fig. 1-14) is to find the accurate range, bearing, and altitude of aircraft targets detected by air-search radar.

FIRE CONTROL

A representative fire-control radar antenna is pictured in figure 1-14. The principal function of fire control radar is the acquisition of targets originally detected and designated from search radars. The determination of extremely accurate ranges, bearings, and position angles of targets, and the automatic tracking of a target provides continuous information for the solution of fire control problems and generation of gun/launcher orders.

AIRBORNE SEARCH

An aircraft with a representative search radar antenna is shown in figure 1-15.

Search radar, the basic type used in most radar-equipped aircraft, shows the range and azimuth of targets such as ships, islands, coastlines, and other aircraft appearing within its beam.

179.3

Figure 1-15.—Aircraft with representative Navy search radar antenna.

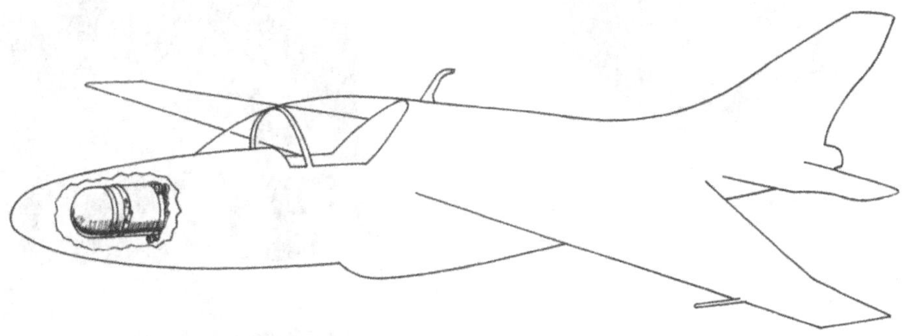

179.4

Figure 1-16.—Aircraft with representative Navy fire-control radar antenna.

AIRCRAFT FIRE CONTROL

The purpose of an aircraft fire control system is to solve the fire control problem so as to accurately direct an aircraft's fire power. An accurate means of range measurement is required for the solution of the fire control problem. Also, a means of aiding the pilot in searching out a target is desirable. The radar set antenna (located in the nose of the aircraft) is shown in figure 1-16.

REPEATERS (INDICATORS)

As the tactics of warfare became more sophisticated, there was more and more evidence that the information obtained from radar would have to be displayed at any one of several physically separated stations. The size and weight of the relatively bulky and complex radar console made it unsuitable for remote installations. The need was for a smaller and lighter general-purpose unit, capable of accepting inputs from more than one type of radar. To fulfill this need,

(A) 120.34 (B) 120.34 (C) 120.86

Figure 1-17.—Representative Navy radar repeaters.

the present-day remote indicator (repeater) was developed.

Three types of radar repeaters currently installed on Navy ships are shown in figure 1-17.

IFF EQUIPMENTS

An Identification Friend or Foe (IFF) system is employed with radar to permit a friendly craft to identify itself automatically before approaching near enough to threaten the security of other naval units. It consists of a pair of special transmitter-receiver units. One set is aboard the friendly ship (fig. 1-18A); the other is aboard the friendly unit (ship or aircraft). Because space and weight aboard aircraft are limited, the airborne system (fig. 1-18B) is smaller, lighter, and requires less power than the shipboard transmitter-receiver. The airborne equipments are automatic, and operate only when triggered by a signal from a shipboard unit.

COUNTERMEASURES

Electronic countermeasures (ECM) is the use of devices or techniques intended to impair the operational effectiveness of enemy electronic equipments or to detect the presence of enemy counter activities. Electronic countermeasures

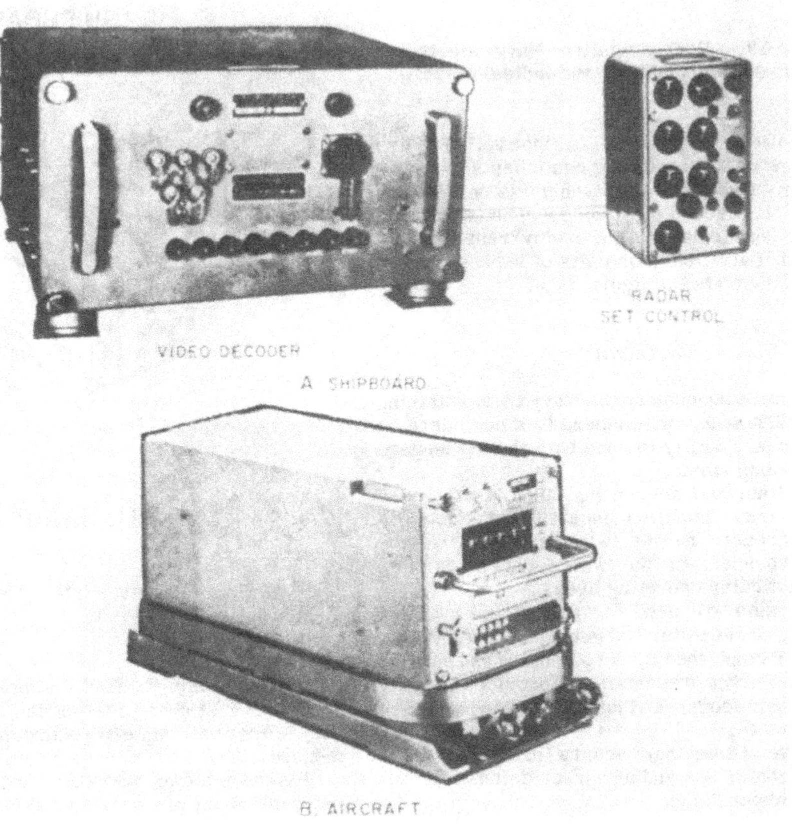

120.39

Figure 1-18.—Representative Navy IFF equipment.

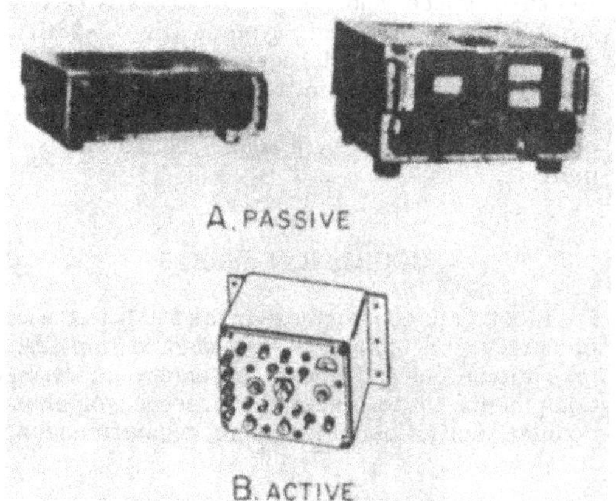

32.1
Figure 1-19.—Representative Navy electronic countermeasures control and indicator units.

are classified as "passive" or "active." Passive ECM is the use of receiving equipments to intercept enemy radar or radio transmissions. Active ECM is the use of electronic or nonelectronic means to jam or deceive the enemy transmission. The control and indicator units of representative sets are illustrated in figure 1-19.

SONAR

The future success of the Navy in maintaining control of the seas will depend to a considerable extent on her ability to cope with the high-speed nuclear submarines.

Immediate and compelling, then, is the need for submarine detection capabilities at significantly increased ranges, with reliable performance independent of the water characteristics of any particular operating area.

Sonar sets are used for detecting, tracking, displaying underwater targets, and navigation. This is accomplished by echo ranging and passive listening. Target presentation is provided visually on indicator scopes and audibly by loudspeakers or headphones.

Passive (listening) sonars are used more aboard submarines and at harbor defense activities than aboard ships.

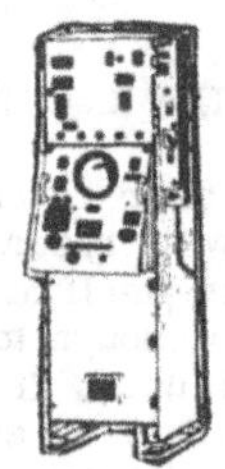

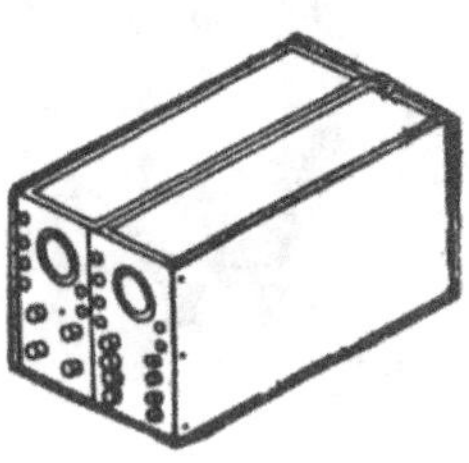

51.52(20C)
Figure 1-20.—Representative Navy sonar control units.

Aircraft using sonar equipment can quickly search a vast area of the ocean, and represent a very effective countermeasure to the submarine menace.

Representative shipboard and airborne sonar set control units are illustrated in figure 1-20.

CHAPTER 2

SAFETY

A technician will install, maintain, and repair electrical and electronic equipment in which dangerously high voltages are present. This work is often done in confined spaces. Among the hazards of this work are electric shock, electrical fires, harmful gases which are sometimes generated by faulty electrical and electronic devices, and injuries which may be caused by improper use of tools.

Because of these dangers, one should formulate safe and intelligent work habits since these are fully as important as knowledge of electronic equipment. One primary objective should be to learn to recognize and correct dangerous conditions and to avoid unsafe acts. A technician must also know the authorized methods for dealing with fires of electrical origin, for treating burns, and for giving artificial respiration to persons suffering from electric shock. In some cases, artificial respiration may have to be accompanied by external cardiac massage to restore the heartbeat.

ELECTRIC SHOCK

Electric shock may cause burns of varying degree, cessation of breathing and unconsciousness, ventricular fibrillation or cardiac arrest, and death. If a 60-hertz alternating current is passed through a person from hand to hand or from hand to foot, the effects when current is gradually increased from zero are as follows:

1. At about 1 milliampere (0.001 ampere) the shock will be felt.
2. At about 10 milliamperes (0.01 ampere) the shock is severe enough to paralyze muscles and a person may be unable to release the conductor.
3. At about 100 milliamperes (0.1 ampere) the shock is usually fatal if it lasts for one second or more. IT IS IMPORTANT TO REMEMBER THAT FUNDAMENTALLY, CURRENT, RATHER THAN VOLTAGE, IS THE CRITERION OF SHOCK INTENSITY.

It should be clearly understood that resistance of body will vary. That is, if the skin is dry and unbroken, body resistance will be quite high, on the order of 300,000 to 500,000 ohms. However, if the skin becomes moist or broken, body resistance may drop to as low as 300 ohms. Thus, a potential as low as 30 volts could cause a fatal current flow. Therefore, any circuit with a potential in excess of this value must be considered dangerous.

IT SHOULD BE NOTED AT THIS POINT THAT THE TAKING OF AN INTENTIONAL SHOCK IS PROHIBITED BY NAVY REGULATION.

CARE OF SHOCK VICTIMS

Electric shock is a jarring, shaking sensation resulting from contact with electric circuits or from the effects of lightning. The victim usually experiences the sensation of a sudden blow, and if the voltage is sufficiently high, unconsciousness. Severe burns may appear on the skin at the place of contact; muscular spasm can occur, causing a person to clasp the apparatus or wire which caused the shock and be unable to turn loose. Electric shock can kill its victim by stopping the heart or by stopping breathing, or both. It may sometimes damage nerve tissue and result in a slow wasting away of muscles that may not become apparent until several weeks or months after the shock is received.

The following procedure is recommended for rescue and care of shock victims.

1. Remove the victim from electrical contact at once, but DO NOT ENDANGER YOURSELF. This can be done by: (1) throwing the switch, if it is nearby; (2) using a dry stick, rope, leather belt, coat, blanket, or any other nonconductor of electricity; or (3) cutting the cable or wires

to the apparatus, using a damage control ax while taking care to protect your eyes from the flash when the wires are severed. Even the victim's shoes, if dry, can be used to pull him clear.

2. Determine whether the victim is breathing. If so, keep him lying down in a comfortable position. Loosen the clothing about his neck, chest, and abdomen so that he can breathe freely. Protect him from exposure to cold, and watch him carefully.

3. Keep him from moving about. After shock, the heart is very weak, and any sudden muscular effort or activity on the part of the patient may result in heart failure.

4. Do not give stimulants or opiates. Send for a medical officer at once and do not leave the patient until he has adequate medical care.

5. If the victim is not breathing, it will be necessary to apply artificial respiration without delay, even though he may appear to be lifeless.

Artificial Respiration

Artificial respiration, the mechanical promotion of breathing, is used to resuscitate persons whose breathing has stopped as a result of electric shock, drowning, asphyxiation, strangling, or the presence of a foreign body in the throat. To revive persons suffering from electrocution resulting in stoppage of breathing, begin artificial respiration immediately. Send another person for competent medical aid. If there is any serious bleeding, stop it first, but do not waste time on anything else. Seconds count, the longer you wait to begin, the less are the chances of saving the victim.

4.224
Figure 2-1.—Direct breathing airway.

4.224
Figure 2-2.—Clearing the throat.

The approved method of artificial respiration is direct breathing airway resuscitation. An airway (fig. 2-1) will be found in all electronic spaces and should be used for giving artificial respiration. The following is the step-by-step procedure for direct breathing airway resuscitation.

Step 1. Clean the throat (fig. 2-2) if necessary otherwise start with Step 2. Place the victim on his back (face up). If foreign matter is visible at his mouth, turn his head to one side, force the mouth open and wipe the throat clean with your fingers or a piece of cloth.

Step 2. Insertion of the Airway. Take up a position at the top of his head (fig. 2-3). With the head tilted back, insert the Airway OVER tongue until the flange rests upon the lips. If the mouth is tightly closed, wedge it open with your index finger inserted between the cheek and teeth, behind the wisdom teeth. Make certain the tongue is not pushed back into the throat—if necessary, hold it forward with your fingers during insertion of the tube. Note that the airway has two tube sizes. Insert the long end of the tube for an adult, insert the short end for a child.

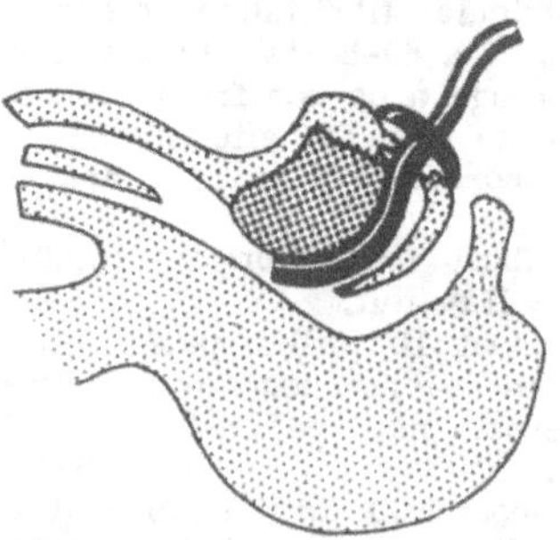

4.224
Figure 2-3.—Insertion of airway.

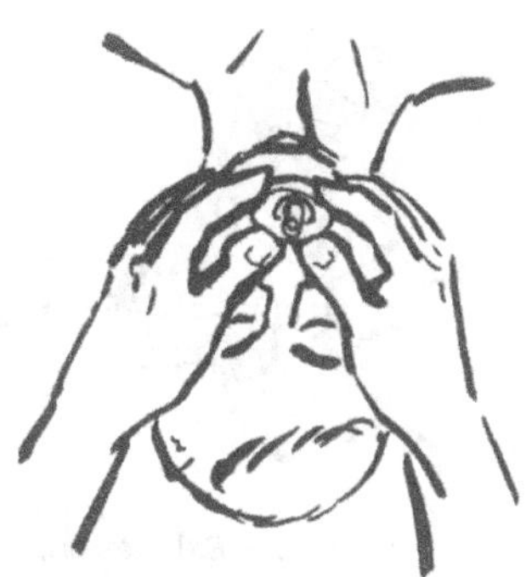

4.224
Figure 2-4.—Prevention of air leakage.

Step 3. Hold the chin upward. Keep head tilted back (front of neck stretched). Pinch victim's nose with your thumbs and press the flange over the lips with your index fingers to prevent air leakage (fig. 2-4). Hold chin upward and toward yourself with remaining fingers. NEVER LET THE CHIN SAG!

Step 4. Blow into the tube (fig. 2-5) as follows: Take a deep breath and blow forcefully into the mouthpiece of the tube. Watch the victim's chest. When it moves, take your mouth off the tube and let the victim exhale passively. When his exhalation is finished, blow in the next deep breath. The first few breaths you blow must be deep and at a rapid rate. Thereafter, about one breath every 3 to 4 seconds is adequate. If the chest does not move, increase the "chin-up" position, improve the position of your fingers, and blow more forcefully. If the chest still does not move, readjust the position of the AIRWAY, which may have been inserted too deeply or not deeply enough. Make sure the flange is resting firmly on the lips.

4.224
Figure 2-5.—Blowing into tube.

To assist shallow natural breathing, blow in at the moment the victim inhales and take your mouth off quickly when he exhales. CONTINUE ARTIFICIAL RESPIRATION UNTIL THE VICTIM BREATHES NATURALLY OR PROFESSIONAL MEDICAL AID ARRIVES!

If an Airway is not immediately available, initiate artificial respiration, using the mouth-to-mouth technique. This method is as follows:

Step 1. Clean mouth of all foreign material.

Step 2. Place victim on his back (face up) (fig. 2-6). Lift his neck and place a folded coat, blanket, etc. under his shoulders. Tilt his head as far back as possible.

Step 3. Assume a position at the side of victim's head. Grasp the jaw with the thumbs in one side of the mouth and pull it forward (fig. 2-7). Maintain this position to keep air passage open.

Step 4. Pinch victim's nostrils shut with the thumb and forefinger (fig. 2-8). Take a deep breath and place your mouth over his mouth and your thumb, creating a tight seal. Blow into the victim's mouth until you see his chest rise.

Step 5. Remove your mouth and let victim exhale passively (fig. 2-9). Inflate the lungs at a rate of about 12 times per minute for adults and 20 times per minute for children. Continue until the victim begins to breathe normally.

- Keep the victim sufficiently covered so that he will not suffer from exposure.

- Do not allow bystanders to crowd around the victim and interfere with the first aid treatment.

4.224
Figure 2-6.—Placement of victim.

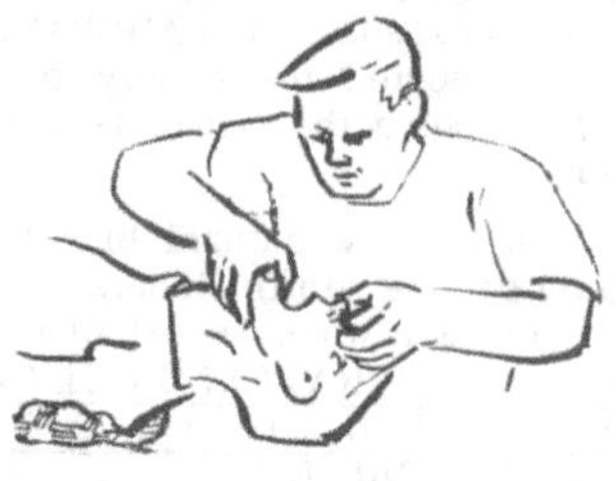

4.224
Figure 2-7.—Opening of air passage.

4.224
Figure 2-8.—Inflation.

- Do not attempt to give the victim anything to drink while he is unconscious, since this may cause choking.

- Artificial respiration must be continued for at least 4 hours unless the natural breathing is restored before that time or a medical officer declares the person dead. Some people have been saved after as much as 8 hours of artificial respiration.

There are other methods of artificial resuscitation with which one should become familiar and be able to master. These are described in Standard First Aid Training Course, NavPers 10081-B.

Cardiac Arrest (Loss of Heartbeat)

If the victim of an electrical shock has no heartbeat, (no detectable pulse at wrist or neck, and heart cannot be heard) he has then suffered a cardiac arrest. In this case, severe brain damage may occur if circulation is not restored within four minutes. Closed cardiac massage is the recommended procedure to be used

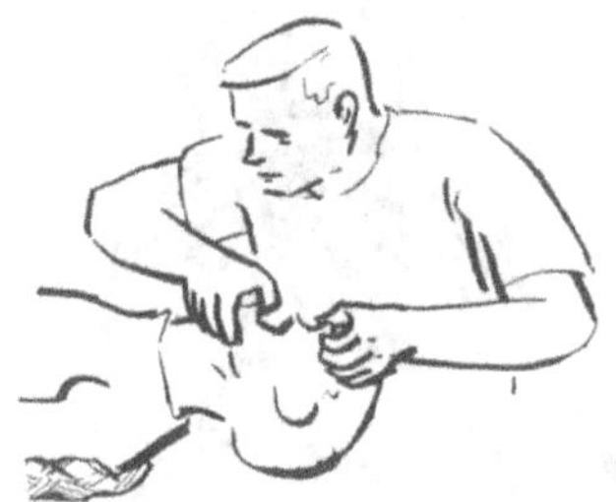

4.224
Figure 2-9.—Exhalation.

in such cases. The procedure is outlined in chapter 9600 of the NavShips Technical Manual NS 0901-000-0001, and in various first aid books and other technical manuals. However, a person should not attempt to perform closed cardiac massage, unless he has been instructed in the proper procedure by competent personnel (a doctor or hospital corpsman) and then only as a last resort, since serious or even fatal injury may result if improperly applied.

WORKING ON ENERGIZED CIRCUITS

Insofar as is practical, a technician should NOT undertake repair work on energized circuits and equipment. However, if it should become necessary, as when making operational adjustments, then one should carefully observe the following safety precautions.

1. Ensure that you have adequate illumination. You must be able to see clearly, if you are to safely and properly perform the job.
2. Ensure that you are insulated from ground by an approved type rubber mat, or layers of dry canvas and/or wood.
3. Where practical, use only one hand, keeping the other either behind you or in your pocket.
4. If the system voltage exceeds 150 volts, rubber gloves should be worn.
5. An assistant should be stationed near the main switch or circuit breaker, so that the equipment may be immediately deenergized in case of an emergency.
6. A man qualified in first aid for electrical shock should be standing by during the entire operation.
7. DO NOT WORK ALONE.

8. DO NOT work on any type of electrical apparatus when wearing wet clothing, or if the hands are wet.

9. DO NOT wear loose or flapping clothing.

10. The use of thin-soled shoes and shoes with metal plates or hob nails is prohibited.

11. Flammable articles, such as celluloid cap visors should not be worn.

12. All rings, wristwatches, bracelets, and similar metal items should be removed before working on the equipment. Also ensure that clothing does not contain exposed metal fasteners such as zippers, snaps, buttons, and pins.

13. Do not tamper with interlock switches, that is, do not defeat their purpose by shorting them or blocking them open.

14. Ensure that equipment is properly grounded before energizing.

15. Deenergize equipment before attaching alligator clips to any circuit.

16. Use only approved meters and other indicating devices to check for the presence of voltage.

17. Observe the following procedures when measuring voltages in excess of 300 volts.

- Turn off the equipment power.
- Short circuit or ground the terminals of all components capable of retaining a charge.
- Connect the meter leads to the points to be measured.
- Remove any terminal grounds previously connected.
- Turn on the power and observe the voltage reading.
- Turn off the power.
- Short circuit or ground all components capable of retaining a charge.
- Disconnect the meter leads.

18. On all circuits where the voltage is in excess of 30 volts and where the deck, bulkheads, or work benches are of metallic construction, the worker should be insulated from accidental ground by use of approved insulating material. The insulating material should have the following qualities:

- It should be dry, without holes, and should not contain conducting materials.

- The voltage rating for which it is made should be clearly marked on the material, and the proper material should be used so that adequate protection from the voltage can be supplied.

- Dry wood may be used, or as an alternative, several layers of dry canvas, sheets of phenolic insulating material, or suitable rubber mats.

- Care should be exercised to ensure that moisture, dust, metal chips, etc. which may collect on insulating material is removed at once. Small deposits of such materials can become electrical hazards.

- All insulating materials on machinery and in the area should be kept free of oil, grease, carbon dust, etc., since such deposits destroy insulation.

SAFETY SHORTING PROBE

ALWAYS ASSUME THAT THERE IS A VOLTAGE PRESENT when working with circuits having high capacitance even when the circuit has been disconnected from its power source. Therefore, capacitors in such circuits should be discharged individually, using an approved type shorting probe. High capacity capacitors may retain their charge for a considerable period of time after having been disconnected from the power source. An approved type shorting probe is shown in figure 2-10.

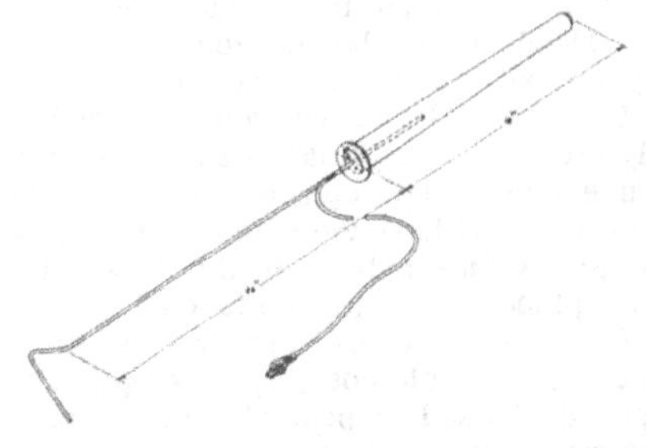

1.1

Figure 2-10.—An approved type shorting probe.

When using the safety shorting probe, always be sure to first connect the test clip to a good ground (if necessary, scrape the paint of the grounding metal to make a good contact). Then hold the safety shorting probe by the insulated handle and touch the probe end of the shorting rod to the point to be shorted out. The probe end is fashioned so that it can be hooked over the part or terminal to provide a constant connection by the weight of the handle alone. Always take care not to touch any of the metal parts of the safety shorting probe while touching the probe to the exposed "hot" terminal. It pays to be safe; use the safety shorting probe with care.

Some equipments are provided with walk-around shorting (such as fixed grounding studs or permanently attached grounding rods). When this is the case, these should be used rather than the safety shorting probe described previously.

WORKING ON DEENERGIZED CIRCUITS

When any electronic equipment is to be repaired or overhauled, certain general safety precautions should be observed. They are as follows:

1. Remember that electrical and electronic circuits often have more than one source of power. Take time to study the schematics or wiring diagrams of the entire system to ensure that all sources of power have been disconnected.
2. If pertinent, inform the remote station regarding the circuit on which work will be performed.
3. Use one hand when turning switches on or off.
4. Safety devices such as interlocks, overload relays, and fuses should never be altered or disconnected except for replacement. In addition, they should never be changed or modified in any way without specific authorization.
5. Fuses should be removed and replaced only after the circuit has been deenergized. When a fuse blows, the replacement should be of the same type and have the same current and voltage ratings. A fuse puller should be used to remove and replace cartridge type fuses.
6. All circuit breakers and switches from which power could possibly be supplied should be secured (locked if possible) in the open or off (safety) position and tagged. Figure 2-11 shows one type of tag used for this purpose.
7. After the work has been completed, the tag (or tags) should be removed ONLY by the same person who signed it when the work began.

Figure 2-11.—Warning tag for marking open position of switches and cut-out circuits.

8. Keep clothing, hands, and feet dry if at all possible. When it is necessary to work in wet or damp locations, use a dry platform or wooden stool to sit or stand on, and place a rubber mat or other nonconductive material on top of the wood. Use insulated tools and insulated flashlights of the molded type when required to work on exposed parts.

ELECTRICAL FIRES

Fire is a hazard. The three general classes of fires are A, B, and C. Class A fires involve wood, paper, cotton and wool fabrics, rubbish, and the like. Class B fires involve oil, grease, gasoline and aircraft fuels, paints, and oil-soaked materials. Class C fires involve insulation and other combustible materials in electrical and electronic equipment.

Electrical or electronic equipment fires result from overheating, short circuits (parts failure), friction (static electricity), or radio-frequency arcs. Also, equipment may be ignited by exposure to nearby Class A or B fires. Since Class C fires involve electrical circuits, electrical shock is an added hazardous condition. Thus, whenever possible, any electrical equipments exposed to a Class A or Class B fire, or actually ignited by such a fire, should be deenergized immediately. If the equipment cannot be deenergized completely, protective measures must be enforced to guard against electrical shock. In addition, extinguishing agents other than gases will contaminate delicate instruments, contacts, and similar electrical devices. Therefore, carbon dioxide (CO_2) is the preferred extinguishing agent for electrical fires because it does not conduct electricity and rapidly evaporates leaving little or no residue. Thus, its use reduces the possibility of electrical shock to personnel and damage to delicate equipment as a result of contamination.

A dry chemical type of extinguishing agent, composed chiefly of potassium bicarbonate (Purple-K), is suitable for electrical fires because it also is a nonconductor and, therefore, protects against electrical shock. However, damage to electrical or electronic parts may result from the use of this agent. The dry chemical extinguisher is similar in appearance to the CO_2 extinguisher.

A SOLID STREAM OF WATER MUST NEVER BE USED TO EXTINGUISH ELECTRICAL FIRES IN ENERGIZED EQUIPMENT because water usually contains minerals which make it conductive; the conductivity of sea water is many times greater than that of fresh water. (Pure distilled water is not a good electrical conductor and may be used in an emergency on small electrical fires.) If circumstances demand the use of fresh or sea water, fog produced by a special hose nozzle (fog head or tip) may be used in electrical or electronic equipment spaces. However, even though the fog is a fine diffusion or mist of water particles, with very little conductivity, there is still some danger of electric shock, unless the equipment is completely deenergized. Also, fog condensation on electrical equipment frequently damages the components, and this damage must be corrected after the fire. Exercise care when using the fire hose. Pressure at the fireplug may be as much as 100 psi and an unrestrained fire hose may result in whiplash.

Foam is not recommended for electrical fires because of equipment damage and possible shock hazard to personnel; however, if necessary, foam may be used only on DEENERGIZED circuits. When a blanket of foam is applied to a burning substance, the foam smothers the fire; that is, it cuts off the air supply to the burning substance. Thus the supply of oxygen necessary to support combustion is isolated from the substance, and the fire will be extinguished.

The following general procedure is used for fighting an electrical fire:

1. Promptly deenergize the circuit or equipment affected.
2. Sound an alarm in accordance with station regulations or the ship's fire bill. When ashore, notify the fire department; if afloat, notify the Officer of the Deck. Give the fire location and state what is burning. If possible, report the extent of the fire, that is, what its effects are upon the surrounding area.
3. Secure ventilation by closing compartment air vents or windows.
4. Control or extinguish the fire using a CO_2 fire extinguisher.
5. Avoid prolonged exposure to high concentrations of carbon dioxide in confined spaces since there is danger of suffocation unless special breathing apparatus is available.
6. Administer artificial respiration and oxygen to personnel overcome by carbon dioxide fumes and keep the patient warm.

Fire aboard a Navy vessel at sea under normal conditions sometimes is more fatal and damaging to both personnel and the ship itself, than that resulting from battle. It is extremely important that all personnel know and understand the dangers of fire. They should know the type and location of all firefighting equipment and apparatus in the immediate working and berthing spaces, and throughout the ship. It is too late after a fire has started; the time to become familiar with this equipment is now.

WARNING SIGNS, PLATES, AND TAGS

Warning signs and suitable guards shall be provided to prevent personnel from coming into accidental contact with dangerous voltages, to warn personnel of possible presence of explosive vapors, RF radiation hazards, poisonous effects of stack gases, and of other dangers which may cause injury or death. Equipment installation should not be considered completed until assurance has been made that appropriate warning signs have been posted in full view of operating personnel.

Certain types of standard electronics warning signs are available for procurement from the Commander, Philadelphia Naval Shipyard. A list of signs that are available has been distributed to all ships, commands, and shore activities. Any warning signs not listed should be ordered on a separate requesting document.

Some of the more commonly used warning signs are discussed below, most of these were designed by the Naval Ships Systems Command.

Figure 2-12 is a High Voltage Warning Sign (NavShips Drawing No. RE 10 B 608B). This sign is to be displayed at all locations where danger to personnel exists, either through direct contact with high voltage or through high voltage arc-over. Appropriate guards should also be installed at these locations.

Warning Sign for Personnel Working Aloft in Way of Smoke Pipe Gases (NavShips Drawing No. RE 10 AA 529A) is to be displayed at the bottom and top of access ladders to electronic equipment in the way of smoke pipe gases (fig. 2-13).

RF Radiation Hazard Warning Signs (NavShips Drawing No. RE 10 D 2282) are of four types (fig. 2-14).

Warning Plates for Electronic Equipment Installed in Small Craft (NavShips Drawing No. RE 10 A 589): Although the drawing title indicates this warning plate is to be installed in small craft, it may and should be displayed in all spaces where there is a possibility of the accumulation of explosive vapors (fig. 2-15).

40.67(31)
Figure 2-12.—High voltage warning sign.

DANGER
PERSONNEL ARE CAUTIONED TO GUARD AGAINST POISONOUS EFFECTS OF SMOKE PIPE GASES WHILE SERVICING EQUIPMENT ALOFT.
WHEN SERVICING EQUIPMENT IN THE WAY OF SMOKE PIPE GASES USE OXYGEN BREATHING APPARATUS AND A TELEPHONE CHEST OR THROAT MICROPHONE SET FOR COMMUNICATION WITH OTHERS IN WORKING PARTY.
OBTAIN NECESSARY EQUIPMENT BEFORE GOING ALOFT.

40.67(26D)
Figure 2-13.—Smoke pipe gases warning sign.

Warning Tags for Marking Open Position of Switches and Cutout Circuits (NavShips 3950 (3-63)-GPO: 1963-0-674658 (on reverse side of tag): This tag is used to indicate a switch which must be left in the OFF or OPEN (safe) position during repairs. These tags (fig. 2-11) are available for ship and shore personnel through normal supply channels.

PRECAUTIONS WHEN WORKING ALOFT

When radio or radar antennas are energized by transmitters, workmen must not go aloft unless advance tests show that little or no danger exists. A casualty can occur from even a small spark drawn from a charged piece of metal or rigging. Although the spark itself may be harmless, the "surprise" may cause the man to let go his grasp involuntarily. There is also shock hazard if nearby antennas are energized, such as those on stations ashore or aboard a ship moored alongside or across a pier.

Danger also exists that rotating antennas might cause men working aloft to fall by knocking them from their perch. Motor safety switches controlling the motion of rotating antennas must

40.67(76)

Figure 2-14.—R.F. radiation hazard warning signs.

be tagged and locked open before anyone is allowed aloft near such antennas.

When working near a stack, draw and wear the recommended oxygen breathing apparatus. Among other toxic substances, stack gas contains carbon monoxide. Carbon monoxide is too unstable to build up to a high concentration in the open, but prolonged exposure to even small quantities is dangerous.

Observe these safety precautions when going aloft:

1. Obtain permission from the CIC watch officer, communication watch officer (CWO), and OOD.
2. Ensure that the boiler safety valves are not being set by checking with the engineer officer.
3. Procure the assistance of another man along with a ship's Boatswain's Mate qualified in rigging.

WARNING
DO NOT ENERGIZE ELECTRONIC EQUIPMENT UNTIL VENTILATION BLOWERS HAVE BEEN OPERATING A MINIMUM OF FIVE MINUTES TO EXPEL EXPLOSIVE VAPORS

40.67(140)
Figure 2-15.—Warning plate for electronic equipment installed in small craft.

4. Wear a safety belt. To be of any benefit, the belt must be fastened securely as soon as the work place is reached. Some men have complained on occasion that a belt is clumsy and interferes with movement. It is true the job may take a few minutes longer, but it is also true that a fall from the vicinity of an antenna is usually fatal.

5. Do not attempt to climb loaded with tools. Keep both hands free for climbing. Tools can be raised by an assistant below. Tools should be secured with preventer lines to avoid dropping them.

6. Ensure good footing and grasp at all times.

7. Remember the nautical expression of old seafarers: HOLD FAST.

RADIOACTIVE ELECTRON TUBES

Electron tubes containing radioactive material are now commonly used. These tubes are known as TR, ATR, PRE-TR, spark-gap, voltage-regulator, gas-switching, and cold-cathode gas-rectifier tubes. Some of these tubes contain radioactive material and have intensity levels which are dangerous; they are so marked in accordance with Military Specifications.

So long as these electron tubes remain intact and are not broken, no great hazard exists. However, if these tubes are broken and the radioactive material is exposed, or escapes from the confines of the electron tube, the radioactive material becomes a potential hazard. The concentration of radioactivity in a normal collection of electron tubes in a maintenance shop does not approach a dangerous level, and the hazards of injury from exposure are slight. However, at major supply points, the storage of large quantities of radioactive electron tubes in a relatively small area may create a hazard. For this reason, personnel working with equipments employing electron tubes containing radioactive material, or in areas where a large quantity of radioactive tubes is stored, should read and become thoroughly familiar with the safety practices contained in Radiation, Health, and Protection Manual, NavMed P-5055. Strict compliance with the prescribed safety precautions and procedures of this manual will help to avoid preventable accidents, and to maintain a work environment which is conducive to good health.

The following precautions should be taken to ensure proper handling of radioactive electron tubes and safety of personnel:

1. Radioactive tubes should not be removed from cartons until immediately prior to actual installation.
2. When a tube containing a radioactive material is removed from equipment, it should be placed in an appropriate carton to prevent possible breakage.
3. A radioactive tube should never be carried in ones pocket, or elsewhere on a person in such a manner that breakage can occur.
4. If breakage does occur during handling or removing of a radioactive electron tube, notify the cognizant authority and obtain the services of qualified radiological personnel immediately.
5. Isolate the immediate area of exposure to protect other personnel from possible contamination and exposure.
6. Follow the established procedures set forth in NavMed P-5055.
7. Do not permit contaminated material to come in contact with any part of a person's body.
8. Take care to avoid breathing any vapor or dust which may be released by tube breakage.
9. Wear rubber or plastic gloves at all times during cleanup and decontamination procedures.
10. Use forceps for the removal of large fragments of a broken radioactive tube. The remaining small particles can be removed with a vacuum cleaner, using an approved disposal collection bag. If a vacuum cleaner is not available, use a wet cloth to wipe the affected area. In this case, be sure to make one stroke at a time. DO NOT use a back-and-forth motion. After each stroke, fold the cloth in half, always holding one clean side and using the other for the new stroke. Dispose of the cloth in the manner stated later.
11. No food or drink should be brought into the contaminated area or near any radioactive material.

12. Immediately after leaving a contaminated area, personnel who have handled radioactive material in any way should remove any clothing found to be contaminated. They should also thoroughly wash their hands and arms with soap and water, and rinse with clean water.

13. Immediately notify a medical officer if a wound is sustained from a sharp radioactive object. If a medical officer cannot reach the scene immediately, mild bleeding should be stimulated by pressure about the wound and the use of suction bulbs. DO NOT USE THE MOUTH, if the wound is of the puncture type, or the opening is small, make an incision to promote free bleeding, and to facilitate cleaning and flushing of the wound.

14. When cleaning a contaminated area, seal all debris, cleaning cloths, and collection bags in a container such as a plastic bag, heavy wax paper, or glass jar, and place in a steel can until disposed of in accordance with existing instruction. Decontaminate all tools and implements used to remove a radioactive substance, using soap and water. Monitor the tools and implements for radiation with an authorized radiac set; they should emit less than 0.1 MR/HR at the surface.

CATHODE-RAY TUBES (CRTs)

Cathode-ray tubes should always be handled with extreme caution. The glass envelope encloses a high vacuum and, because of its large surface area, is subject to considerable force caused by atmospheric pressure. (The total force on the surface of a 10-inch CRT is 3750 pounds, or nearly two tons; over 1000 pounds is exerted on its face alone.) Proper handling and disposal instructions for CRTs are as follows:

1. Avoid scratching or striking the surface.
2. Do not use excessive force when removing or replacing the CRT in its deflection yoke or its socket.
3. Do not try to remove an electromagnetic type CRT from its yoke until the high voltage has been discharged from its anode connector (hole).
4. Never hold the CRT by its neck.
5. Always set a CRT with its face down on a thick piece of felt, rubber, or smooth cloth.
6. Always handle the CRT gently. Rough handling or a sharp blow on the service bench can displace the electrodes within the tube, causing faulty operation.
7. Safety glasses and gloves should be worn when handling CRT's.
8. Before a CRT is discarded, it should be made harmless by breaking the vacuum glass seal. To accomplish this, proceed as follows:

- Place the tube that is to be discarded in an empty carton, with its face down and protection over its sides and back.

- Carefully break off the locating pin from its base (fig. 2-16) to remove the vacuum. If this does not break the vacuum use the alternate method below.

WARNING

> The chemical phosphor coating of the CRT face is extremely toxic. When disposing of a broken tube, be careful not to come into contact with this compound.

An alternate method of rendering a CRT harmless is to place it in a carton. Then, using a long, thin rod, pierce through the carton and the side of the CRT.

REPLACEMENT OF ELECTRON TUBES

Do NOT use bare hands to remove hot tubes from their sockets; use a tube puller or asbestos gloves. Before replacing high voltage tubes, ensure that the plate (anode) cap or the lead terminal (on CRT's) has been properly discharged. When replacing (or working close to) radioactive

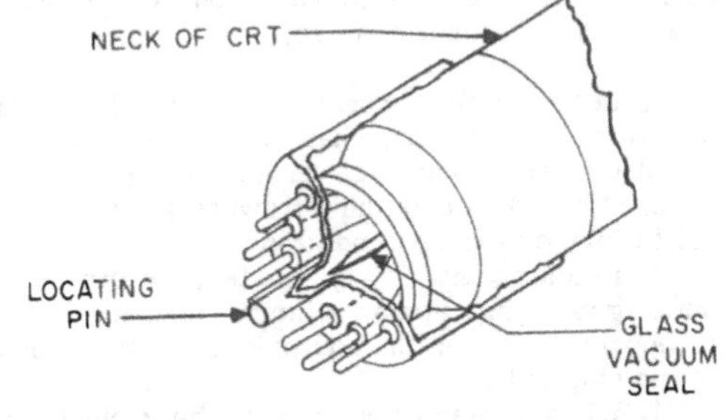

Figure 2-16.—Cathode-ray tube base structure.

tubes, ensure that special precautions (discussed previously) are observed.

GROUNDING OF POWER TOOLS AND EQUIPMENT

The possibility of electrical shock can be reduced by ensuring that all motor and generator frames, metal bases, and other structural parts of electrical and electronic equipment are at ground potential.

Normally, on steel-hull vessels, such grounds are inherently provided because the metal cases or frames of the equipment are in contact with one another and with the metal structure of the vessel. In some instances where such inherent grounding is not provided by the mounting arrangements, such as equipment supported on shock mounts, suitable ground connections must be provided.

The conductors employed for this purpose generally are composed of flexible material (copper or aluminum) that provides sufficient current-carrying capacity to ensure an effective ground. In this manner, equipment cases and frames which are not intended to be above ground potential are effectively grounded, and the possibility of electrical shock to personnel coming in contact with metal parts of the equipment is minimized. The secondary function of grounds is to improve the operation and continuity of service of all equipments.

In all instances where equipment grounding is provided, certain general precautions and preventive maintenance measures must be taken to ensure that all bonding surfaces (connection points or metallic junctions) are securely fastened and free of paint, grease, or other foreign matter that could interfere with the positive metal-to-metal contact at the ground connection point. A few of the precautions are:

1. Periodically clean all strap-and-clamp type connectors to ensure that all direct metal-to-metal contacts are free from foreign matter.
2. Check all mounting hardware for mechanical failure or loose connections.
3. Replace any faulty, rusted, or otherwise unfit grounding strap, clamp, connection, or component between the equipment and the ground to the ship hull.
4. When replacing a part of the ground connection, make certain that the metallic contact surfaces are clean, and that electrical continuity is reestablished.
5. After the foregoing steps have been completed, recheck to be sure that the connection is securely fastened with the correct mounting hardware, and paint the ground strap and hardware in accordance with currently accepted procedures.

Because of the electrical shock hazards that could be encountered aboard ship, plugs and convenience outlets for use with portable test equipment and power tools normally are the standard three-prong type. These are so designed that the plug must be in the correct position before it can be inserted into the receptacle. To ensure that the safety factors incorporated in these devices are in serviceable condition and are safe for use, the following precautions and inspections must be performed.

1. Inspect the pins of the plug to see that they are firmly in place and are not bent or damaged.
2. Check the wiring terminals and connections of the plug. Loose connections and frayed wires on the plug surface must be corrected, and any foreign matter removed, before the plug is inserted into the receptacle.
3. Use a meter to ensure that the ground pin has a resistance of less than 1 ohm to equipment ground.
4. Do not attempt to insert a grounded-type plug into a grounded receptacle without first aligning it properly.
5. Remember, NEVER USE A POWER TOOL OR A PIECE OF PORTABLE TEST EQUIPMENT UNLESS YOU ARE ABSOLUTELY SURE THAT IT IS EQUIPPED WITH A PROPERLY GROUNDED CONDUCTOR.

CLEANING SOLVENT

The Navy does not permit the use of gasoline, benzine, ether, or like substances for cleaning purposes. Only nonvolatile solvents should be used to clean electrical or electronic apparatus.

In addition to the potential hazard of accidental fires, many cleaning solvents are capable of damaging the human respiratory system in cases of prolonged inhalation. The following is a list of "Do Nots" when using cleaning solvents.

1. DO NOT work alone in a poorly ventilated compartment.
2. DO NOT breathe directly the vapor of any cleaning solvent for a long time.

3. DO NOT spray cleaning solvents on electrical windings or insulation.

4. DO NOT apply solvents to warm or hot equipment, since this increases the toxicity hazard.

The following reminders are positive safety steps to be taken when cleaning operations are underway.

1. Use a blower or canvas wind chute to blow air into a compartment in which a cleaning solvent is being used.
2. Open all usable portholes, and place wind scoops in them.
3. Place a fire extinguisher close by, ready for use.
4. Use water compounds in lieu of other solvents where feasible.
5. Wear rubber gloves to prevent direct contact.
6. Use goggles when a solvent is being sprayed.
7. Hold the nozzle close to the object being sprayed.

Inhibited methyl chloroform (1,1,1 trichloroethane) should be used where water compounds are not feasible. Methyl chloroform is toxic and should be used with care since concentrations of the vapor can be fatal. Methyl chloroform is an effective cleaner and safe when reasonable care is exercised. Care requires plenty of ventilation and observance of fire precautions.

For additional information on the safety precautions to be observed when using solvents, see chapter 9670 of the Naval Ships Technical Manual.

AEROSOL DISPENSERS

Deviation from prescribed procedures in the selection, application, storage, or disposal of aerosol dispensers containing industrial sprays has resulted in serious injury to personnel because of toxic effects, fire, explosion, and so on. Specific instructions concerning the precautions and procedures that must be observed to prevent physical injury cannot be given because there are so many types of industrial sprays available. However, all personnel concerned with the handling of aerosol dispensers containing volatile substances should clearly understand the hazards involved and the need to use all protective measures required to prevent personal injury. Strict compliance with the instructions printed on the aerosol dispenser will prevent many of the accidents which result from improper application, mishandling, or improper storage of industrial sprays used in the Naval service for electrical and electronic equipment.

SAFETY IS YOUR JOB—DO IT PROPERLY!

CHAPTER 3

BASIC TEST EQUIPMENT

In the field of electricity and electronics, as in all the other physical sciences, accurate quantitative measurements are essential. This involves two important items — numbers and units. Simple arithmetic is used in most cases and the units are well defined and easily understood. The standard units of current, voltage, and resistance, as well as other units are defined by the National Bureau of Standards. At the factory various instruments are calibrated by comparing them with established standards.

The technician commonly works with ammeters, voltmeters, and ohmmeters; but he may also have many occasions to use wattmeters and various other types of measuring devices.

Electrical equipments are designed to operate at certain efficiency levels. To aid the technician in maintaining the equipment, technical instruction books and sheets containing optimum performance data, such as voltages and resistances, are prepared for Navy equipment.

To the technician, a good understanding of the functional design and operation of electrical instruments is important. In electrical service work, one or more of the following methods are commonly used to determine if circuits are operating properly.

1. Use an ammeter to measure the intensity of current flow in a circuit.
2. Use a voltmeter to determine the voltage or potential existing between two points in a circuit.
3. Use an ohmmeter or megger (megohmmeter) to measure circuit continuity and total or partial circuit resistance.

The technician may also find it necessary to employ a wattmeter to determine the POWER being consumed or emitted by certain equipments. If he wishes to measure the TOTAL ENERGY consumed by certain equipments or certain circuits, a watt-hour or kilowatt-hour meter is used.

For measuring other quantities the technician must employ the appropriate instrument. In order to employ the proper instrument the technician must be aware of the capabilities of various instruments. In each case the instrument indicates the value of the quantity measured, and the technician interprets the information in a manner that will help him understand the way the circuit is operating. A thorough understanding of the construction, operation and limitations of the basic types of measuring instruments, coupled with the theory of circuit operation, is most essential in selecting the proper instrument and in servicing and maintaining electrical equipment.

BASIC NONELECTRONIC METERS

All measuring instruments must have some form of indicating device in order to be of any use to the technician. The most basic indicating device, used in instruments measuring current and voltage, is called a GALVANOMETER. This device (and many other indicating devices) operates by virtue of the magnetic field associated with current flow. Therefore, it would be well for the reader to review the properties of magnetism and electro-magnetism found in Basic Electricity, NavPers 10086-B, before continuing.

BASIC METER MOVEMENT

The stationary permanent-magnet moving coil meter is the basic movement used in most measuring instruments for servicing electrical equipment. The basic movement consists of a stationary permanent magnet and a movable coil. When current flows through the coil the resulting magnetic field reacts with the magnetic field of the permanent magnet and causes the coil to rotate. The greater the intensity of current flow through the coil the stronger the magnetic field produced, and the stronger the magnetic field the greater the rotation of the coil.

Galvanometer

A simplified diagram of one type of stationary permanent-magnet moving coil instrument is shown in figure 3-1. Such an instrument is commonly called a GALVANOMETER. The galvanometer indicates very small amounts (or the relative amounts) of current or voltage, and is distinguished from other instruments used for the same purpose in that the movable coil is suspended by means of metal ribbons instead of by means of a shaft and jewel bearings.

The movable coil (bobbin) of the galvanometer in figure 3-1 is suspended between the poles of the magnet by means of thin flat ribbons of phosphor bronze. These ribbons provide the conducting path for the current between the circuit under test and the movable coil. They also provide the restoring force for the coil. The restoring force, exerted against the driving force of the coil's magnetic field (to be described later) is balanced in order to obtain measurement of the current intensity. The ribbons thus tend to oppose the motion of the coil, and will twist through an angle that is proportional to the force applied to the coil by the action of the coil's magnetic field against the permanent magnet's field. The ribbons thus restrain or provide a counter force for the magnetic force acting on the coil. When the driving force of the coil current is removed, the restoring force returns the coil to its zero position. In order to determine the amount of current flow a means must be provided to indicate the amount of coil rotation. Either of two methods may be used—(1) the pointer arrangement, or (2) the light and mirror arrangement. In the pointer arrangement, the end of the pointer is fastened to the rotation coil and as the coil turns the pointer also turns. The other end of the pointer moves across a graduated scale and indicates the amount of current flow. An advantage of this arrangement is that it permits overall simplicity. A disadvantage of the pointer arrangement is that it introduces the problem of coil balance, especially if the pointer is long.

The use of a mirror and a beam of light simplifies the problem of coil balance. When this arrangement is used to measure the turning of the coil, a small mirror is mounted on the supporting ribbon (fig. 3-1). An internal light source is directed to the mirror, and then reflected to the scale of the meter. As the moving coil turns, so does the mirror, causing the light reflection to move over the scale of the meter. The movement of the reflection is proportional to the movement of the coil; thus, the intensity of current being measured by the meter is indicated.

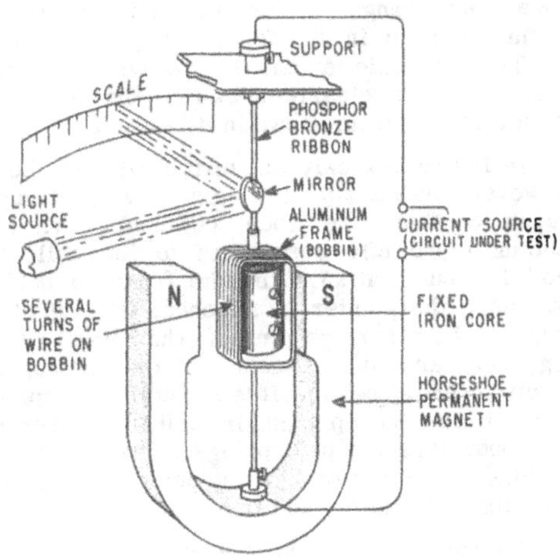

Figure 3-1.—Simplified diagram of a galvanometer.

If a beam of light and mirrors are used, the beam of light is swept to the right or left across a central-zero translucent screen (scaled) having uniform divisions. If a pointer is used, the pointer is moved in a horizontal plane to the right or left across a center-zero scale having uniform divisions. The direction in which the beam of light or the pointer moves depends on the direction of current through the coil.

This instrument is used to measure minute currents as, for example, in bridge circuits. In modified form, the galvanometer has the highest sensitivity of any of the various types of meters in use today.

D'Arsonval Meter Movement

Most d.c. instruments use meters based on some form of the D'Arsonval meter movement. In the D'Arsonval type meter, the length of the conductor is fixed and the strength of the field between the poles of the magnet is fixed. Therefore, any change in current causes a proportionate change in the force acting on the coil.

The principle of the D'Arsonval movement may be more clearly shown by the use of the simplified diagram shown in figure 3-2.

In figure 3-2 only one turn of wire is shown; however, in an actual meter movement many turns of fine wire would be used, each turn adding more effective length to the coil. The coil is wound on an aluminum frame or bobbin, to which the pointer is attached. OPPOSITELY WOUND hair springs (one of which is shown in fig. 3-2) are also attached to the bobbin, one at either end. The circuit to the coil is completed through the hairsprings. In addition to serving as conductors the hairsprings serve as the restoring force that returns the pointer to the zero position when no current flows.

As has been stated, the deflecting force is proportional to the current flowing through the coil. The deflecting force tends to rotate the coil against the restraining force of the hairspring. The angle of rotation is proportional to the deflection force. When the deflecting force and the restraining force are equal, the coil and the pointer cease to move. Since the deflecting force is proportional to the current in the coil and the angle of rotation is proportional to the deflecting force, then the angle of rotation is proportional to the current through the coil. When current ceases to flow through the coil, the driving force ceases and the restoring force of the springs return the pointer to the zero position.

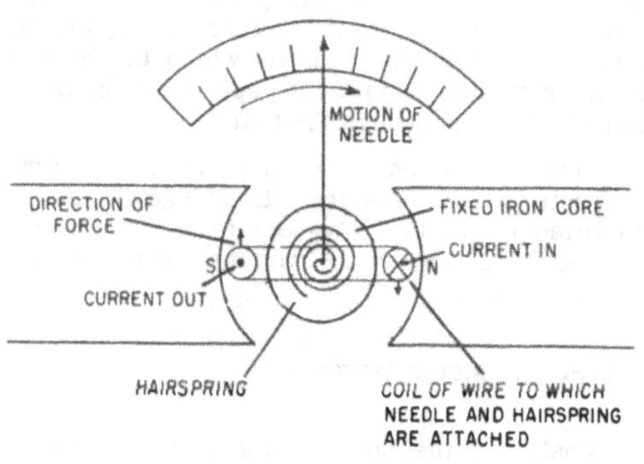

73.36.2
Figure 3-2.—D'Arsonval meter movement.

If the current through the single turn of wire is in the direction indicated (away from the observer on the right-hand side and toward the observer on the left-hand side), the direction of force, by the application of the right-hand motor rule, is upward on the left hand side and downward on the right hand side. The direction of motion of the coil and pointer is clockwise. If the current is reversed in the wire, the direction of motion of the coil and pointer is reversed.

A detailed view of the basic D'Arsonal movement, as commonly used in ammeters and voltmeters is shown in figure 3-3. The principle of operation is the same as that of the simplified version discussed previously. The iron core is rigidly supported between the pole pieces and serves to concentrate the flux in the narrow space between the iron core and the pole piece—in other words in the space through which the coil and bobbin move. Current flows into one hairspring, through the coil, and out of the other hairspring. The restoring forces of the spiral springs return the pointer to the normal, or zero, position when the current through the coil is interrupted. Conductors connect the hairsprings with the outside terminals of the meter. If the instrument is not DAMPED, that is if some type of loss is not introduced to absorb the energy of the moving element, the pointer will oscillate for a long time about its final position before coming to rest. This action makes it nearly impossible to obtain a reading and some form of damping is necessary to make the meter practicable. Damping is accomplished in many D'Arsonval movements by means of the motion of the aluminum bobbin upon which the coil is wound. As the bobbin oscillates in the magnetic field, an electro motive force (EMF) is induced in it because it cuts through the lines of force. Therefore, induced currents flow in the bobbin in such a direction to oppose the motion, and the bobbin quickly comes to rest in the final position after going beyond it only once. In addition to factors such as increasing the flux density in the air gap, the overall sensitivity of the meter can be increased by the use of a light-weight rotating assembly (bobbin, coil, and pointer) and by the use of jewel bearings as shown.

It is noted that the pole pieces (figs. 3-2 and 3-3) have curved faces. The advantage of this type of construction can be seen if it is remembered that lines of force enter and leave a magnetic field in the air gap at right angles to the coil regardless of the coil's angular position.

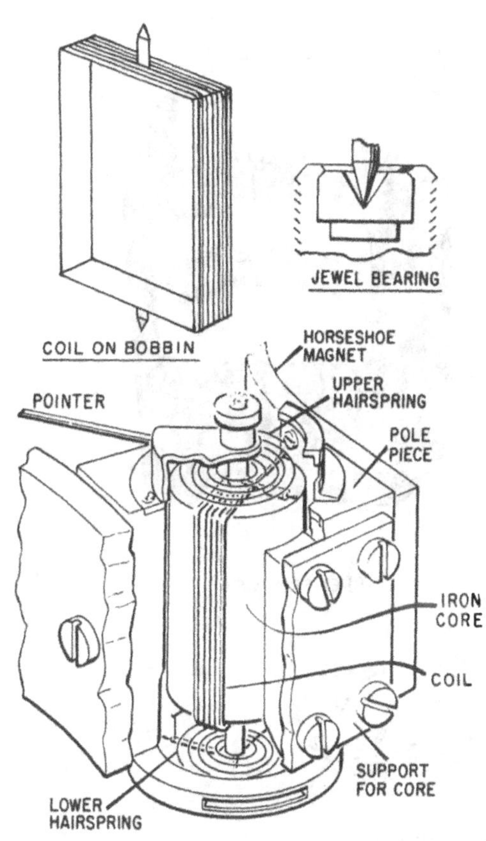

73.36.3

Figure 3-3.—Detailed view of basic D'Arsonval meter movement.

This type of construction makes possible a more linear scale than if the pole faces were flat.

D.C. AMMETER

The small size of the wire with which an ammeter's movable coil is wound places limits on the current that may be passed through the coil. Consequently, the basic D'Arsonval movement discussed thus far may be used to indicate or measure only very small currents.

In order to measure a larger current, a shunt must be used with the meter. A SHUNT is a physically large, low-resistance, conductor which is connected in parallel (shunt) with the meter terminals. It is used to carry the majority of the load current. This shunt has the correct amount of resistance so that only a small portion of the total current will flow through the meter coil. The meter current will be proportional to the total load current. If the shunt is of such a value that the meter is calibrated in milliamperes, the instrument is called a MILLIAMMETER. If the shunt has such a value so that the meter must be calibrated in terms of amperes, it is called an AMMETER.

A single standardized meter movement is normally used in all ammeters, no matter what the range is for a particular meter. For example, meters with working ranges of zero to 10 amperes, zero to 5 amperes, or zero to 1 ampere will all use the same meter movement. The various ranges are achieved through the use of different values of shunt resistance with the same meter movement. The designer of the ammeter simply calculates the correct shunt resistance required to extend the range of the meter movement to measure any desired value of current. This shunt is then connected across the meter terminals. Shunts may be located inside the meter case (internal shunts) with the proper switching arrangements for changing them, or they may be located outside (external shunts) the meter case with the necessary leads to connect them to the meter.

External Shunts

An external shunt is shown in figure 3-4A. The physical appearance of some typical external shunts is shown in figure 3-4C. Figure 3-4B shows a meter movement mounted in a case which provides protection against breakage, magnetic shielding in some cases, portability, etc.

The shunt strips are usually made of manganin. This is an alloy which has an almost zero temperature coefficient. The zero temperature coefficient is desirable as the heavy currents which often flow through the shunts produce heat. Most materials would increase their resistance as they are heated up. This would cause the shunt to carry less current; the additional current would flow through the meter movement, and damage to the meter might occur. Using shunts which are constructed with zero temperature coefficient materials eliminates this problem.

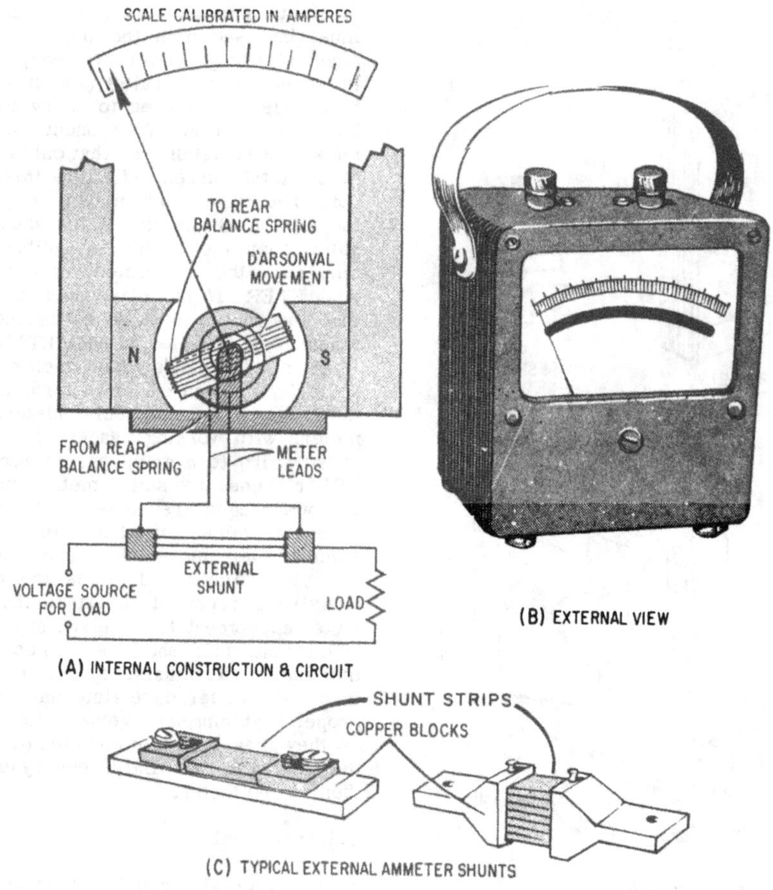

Figure 3-4.—D.c. ammeter utilizing D'Arsonval movement with external shunts.

The ends of the shunt strips are embedded in heavy copper blocks which are attached to the meter coil leads and the line terminals. To insure accurate readings, the meter leads for a particular ammeter should not be used interchangeably with those for a meter of a different range. Slight changes in lead length and size may vary the resistance of the meter circuit and thus its current, and may cause incorrect meter readings. External shunts are generally used where currents greater than 50 amperes must be measured.

It is important to select a suitable shunt when using an external shunt ammeter so that the scale deflection is easily read. For example, if the scale has 150 divisions and the load current to be measured is known to be between 50 and 100 amperes, a 150 ampere shunt is suitable. Under these conditions each division of the scale will represent 1 ampere. In other words, a full scale deflection of the pointer would rest on the 150th division mark indicating that 150 amperes of load current is flowing. At half scale deflection the pointer would rest on the 75th division mark indicating that 75 amperes of load current is flowing.

A shunt having exactly the same current rating as the estimated normal load current should never be selected. If such a shunt were to be selected, then it is possible for higher than normal load currents to drive the pointer off scale and damage the meter movement. A good choice of shunt values will bring the needle somewhere near the midscale indication, when the load current is normal. For example, assume that the meter scale is divided into 100 equal divisions and it is desired to measure a current of 60 amperes. The shunt to use would be a 100 ampere shunt as this would make each division of the scale equal to 1 ampere and the meter indication would fall on the 60th division showing that 60 amperes of load current is flowing. This leaves room (40 amperes) for unexpected surge currents.

Internal Shunts

For current ranges below 50 amperes, internal shunts are most often employed. In this manner the range of the meter may be easily changed by means of a switching arrangement which will select the correct internal shunt having the necessary current rating and resistance. Before the required resistance of the shunt for each range can be calculated, the total resistance of the meter movement must be known.

For example, suppose it is desired to use a 100 microampere D'Arsonval meter having an internal coil resistance of 100 ohms to measure line currents up to 1 ampere. The meter will deflect to its full-scale position when the current through the deflection coil is 100 microamperes. As the coil resistance is 100 ohms the coil's voltage is easily calculated by Ohm's law, $E = I \times R = .0001 \times 100 = .01$ volts, where E is voltage in volts, I is current in amperes, and R is resistance in ohms. When the pointer is deflected to full scale there will be 100 microamperes of current flow through the coil and .01 volts dropped across it. It should be remembered that 100 microamps is the maximum safe current for the meter movement. More than this value of current will damage the meter. It is therefore required that the shunt carry any additional load current.

There will be a .01 volt drop across the meter coil, and because the shunt and coil are in parallel, the shunt must also have a voltage drop of .01 volts. The current that flows through the shunt must be the difference between the full-scale meter current and the line current being fed in. In this case the meter current is 100 microamperes. It is desired that this current cause full scale deflection only when the total current is 1 ampere. Therefore, the shunt current must equal 1 ampere minus 100 microamperes or .9999 amperes. Ohm's law now provides the value of required shunt resistance. $R = E/I = .01/.9999 = .01$ ohm (approximately).

The range of the 100 microampere meter can be increased to 1 ampere (full scale deflection) by placing a .01 ohm shunt in parallel with the meter movement.

The 100 microampere instrument may also be converted to a 10 ampere meter by the use of a proper shunt. For full-scale deflection of the meter, the voltage drop, E, across the coil (and the shunt) is still .01 volt. The meter current is still 100 microamperes. The shunt current must therefore be 9.9999 amperes under full-scale deflection conditions. The required shunt resistance is $R = E/I = .01/9.9999 = .001$ ohm. Again the shunt resistance value is an approximation.

The same instrument can likewise be converted to a 50 ampere meter by the use of a proper shunt. Calculating: $R = E/I = .01/49.9999 = .0002$ ohms (approximately).

The above method of computing the shunt resistance is satisfactory in most cases when the line current is in the ampere range and the meter current is relatively small compared to the load current. In such cases it is allowable to use an approximate value of resistance for the shunt as was done above. However, when the total current is in the milliampere range and the coil current becomes an appreciable percentage of the line current, a more accurate calculation must be made. EXAMPLE: It is desired to use a meter movement which has a full-scale deflection of 1 milliampere and a coil resistance of 50 ohms to measure currents up to 10 milliamperes. Using Ohm's law the voltage across the meter coil (and the shunt) at full-scale deflection is: $E = I \times R = .001 \times 50 = 50$ mv. The current that flows through the shunt is the difference between the line current and the meter current. $I = 10$ ma. - 1 ma. = 9 ma. The shunt resistance must be $R = E/I = .05/.009 = 5.55$ ohms. Notice that in this case the exact value of shunt resistance has been used rather than an approximation.

A formula for determining the resistance of the shunt is given by $R_s = I_m / I_s \times R_m$ where R_s is the shunt resistance in ohms, I_m is the meter current at full-scale deflection, I_s is the shunt current at full-scale deflection, and R_m is the resistance of the meter coil. If the values given in the previous example are used in this equation it will yield 5.55 ohms, the value previously calculated.

Various values of shunt resistance may be used, by means of a suitable switching arrangement, to increase the number of current ranges that may be covered by the meter. Two switching arrangements are shown in figure 3-5.

Part A is the simpler of the two arrangements from the point of view of calculating the value of the shunt resistors when a number of shunts are used. However, it has two disadvantages:

1. When the switch is moved from one shunt resistor to another the shunt is momentarily removed from the meter and the line current then flows through the meter coil. Even a momentary surge of current could easily damage the coil.

2. The contact resistance—that is, the resistance between the blades of the switch when they are in contact—is in series with the shunt but not with the meter coil. In shunts that must pass high currents the contact resistance becomes an appreciable part of the total shunt resistance. Because the contact resistance is of a variable nature, the ammeter indication may not be accurate.

A more generally accepted method of range switching is shown in figure 3-5B. Although only two ranges are shown, as many ranges as needed can be used. In this type of circuit the range selector switch contact resistance is external to the shunt and meter in each range position and therefore has no effect on the accuracy of the current measurement.

Ammeter Connections

CURRENT MEASURING INSTRUMENTS MUST ALWAYS BE CONNECTED IN SERIES WITH A CIRCUIT, NEVER IN PARALLEL WITH IT. If an ammeter were connected across a constant potential source of appreciable voltage, the low internal resistance of the meter bypasses the circuit resistance. This results in source voltage (or a good portion of it) being applied directly to the meter terminals and the resulting excessive current burns up the meter.

If the approximate value of current in a circuit is not known, it is best to start with the HIGHEST RANGE of the ammeter and switch to progressively lower ranges until a suitable reading is obtained.

Most ammeter needles indicate the magnitude of the current being deflected from left to right. If the meter is connected with reversed polarity, the needle will be deflected backwards, and this action may damage the movement. Hence the proper polarity should be observed in connecting the meter in the circuit. That is, the meter should always be connected so that the current flow will be into the negative terminal and out of the positive terminal. Figure 3-6 shows various circuit arrangements and the proper ammeter connection methods to measure current in various portions of the circuits.

Ammeter Sensitivity

METER SENSITIVITY is determined by the amount of current required by the meter coil to

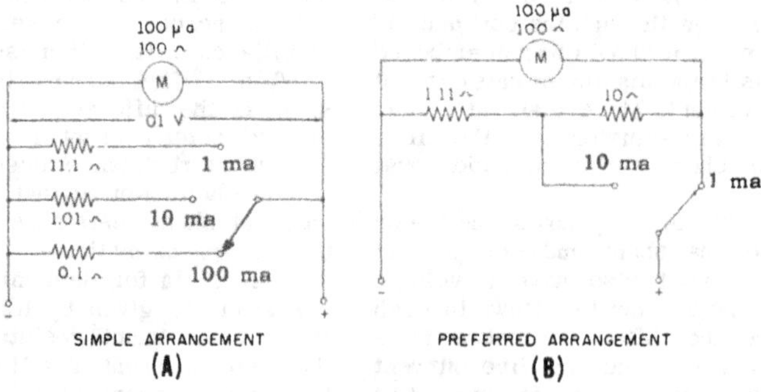

Figure 3-5.—Ways of connecting internal meter shunts.

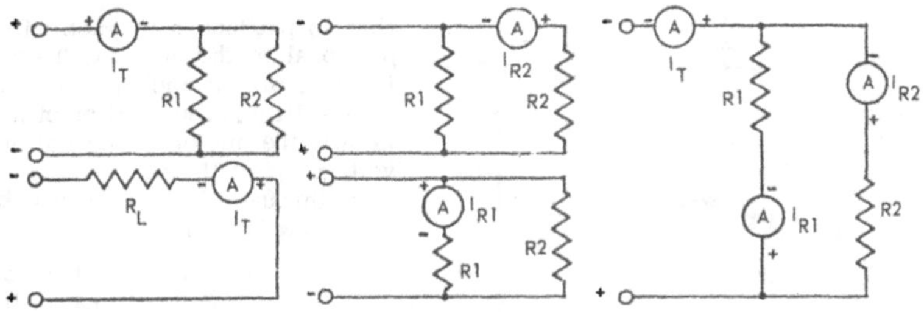

179.6
Figure 3-6.—Proper ammeter connections.

produce full scale deflection of the pointer. The smaller the current required to produce this deflection the better the sensitivity of the meter. Thus, a meter movement which required only 100 microamperes for full scale deflection would have a better sensitivity than a meter movement which requires 1 milliampere for the same deflection.

Good meter sensitivity is especially important in ammeters which are to be used in circuits where small currents are flowing. As the meter is connected in series with the load device, the current flows through the meter. If the internal resistance of the meter is an appreciable portion of the load resistance a METER LOADING EFFECT will occur. Meter loading is defined as "the condition which exists when the insertion of a meter into a circuit changes the operation of that circuit." This condition is not desired. When a meter is inserted into a circuit its use is intended to allow the measurement of circuit current in the normal operating condition. If the meter alters the circuit operation and changes the amount of current flow, the reading obtained must be erroneous. An example of this is shown in figure 3-7.

In figure 3-7A the circuit to be tested is shown. There is an applied voltage of 100 millivolts, and a resistance of 100 ohms. The current normally flowing in this circuit would then be 1 milliampere. In figure 3-7B we insert an ammeter which requires 1 milliampere for full scale deflection and has an internal resistance of 100 ohms. As there is 1 milliampere of current flow in figure 3-7A there is the tendency to say that with the meter inserted a full scale deflection will occur and the 1 milliampere of circuit current will be measured. THIS DOES NOT HAPPEN. In figure 3-7B, with the ammeter inserted into the circuit, the total resistance of the circuit now becomes 200 ohms. With an applied voltage of 100 millivolts Ohm's law yields the current as $I = E/R = 100 \times 10^{-3}/200 = .5 \times 10^{-3}$ $= .0005$ amperes or .5 milliamperes instead of the desired 1 milliampere. Since the meter will

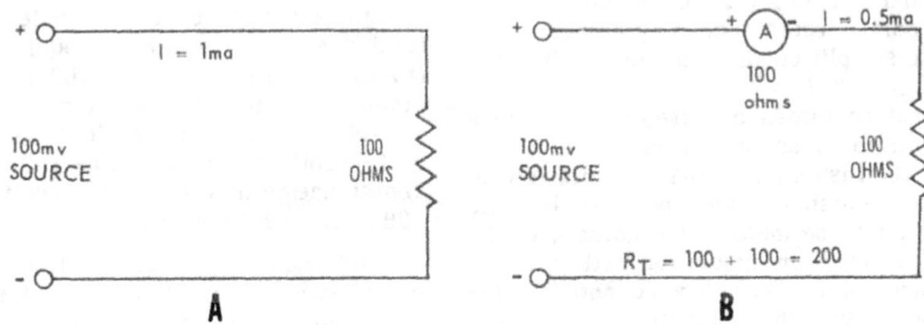

179.7
Figure 3-7.—Ammeter loading effect.

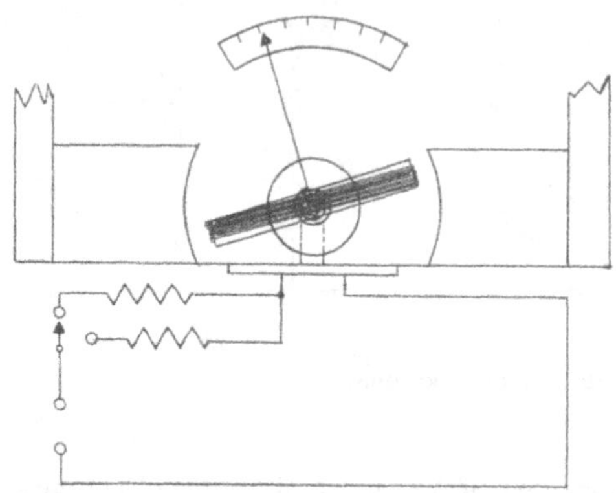

5.104
Figure 3-8.—Internal construction and circuit of simplified voltmeter.

read .5 milliamperes instead of the normal value of current the meter has a definite loading effect. In such cases as this it is desirable to use ammeters which have a better current sensitivity and a lower internal resistance.

D.C. VOLTMETERS

The 100 microampere D'Arsonval movement used as the basic meter for the ammeter may also be used to measure voltage if a high resistance (multiplier) is placed in series with the moving coil of the meter. For low-range instruments this resistance is mounted inside the case with the D'Arsonval movement and typically contains a wire wound resistance having a low temperature coefficient and wound on either spools or card frames. For the higher voltage ranges, the series resistance may be connected externally. A simplified diagram of a voltmeter is shown in figure 3-8.

Wire wound resistors are treated in such a way that a minimum amount of moisture will be absorbed by the insulation. Moisture will reduce the insulation resistance and increase leakage currents, which cause incorrect readings. Leakage currents through the insulation will increase with the length of resistance wire and become a factor that limits the magnitude of voltage that can be measured.

It must be remembered that the D'Arsonval meter movement operates by use of a current flow to produce a magnetic field which is proportional to the current. This meter movement is therefore an indicator of current flow rather than voltage. The addition of a series resistance allows the meter to be calibrated in terms of voltage. It still operates because of the current flow through the meter but the scale may be marked off in volts.

The meter movement being used in figure 3-9, for an example, has an internal resistance of 100 ohms, requires 100 microamperes for full scale deflection, and will have a voltage drop of 10 millivolts when full scale deflection is achieved. If this meter were placed directly across a 10 volt source, an excessive current would flow ($I = E/R = 10/100 = 100$ milliamperes or 99.9 milliamperes too many), and the meter movement would be destroyed.

Since the normal meter voltage drop is 10 millivolts for full scale deflection condition, some means is required to drop 9.99 volts without applying it directly to the meter. This is the function of the multiplier resistor shown in figure 3-9.

Extending Voltmeter Ranges

The value of the necessary series resistance is determined by the current required for full scale deflection of the meter and by the range of voltages to be measured. Because the current through the meter circuits is directly proportional to the applied voltage, the meter scale can be calibrated directly in volts for a fixed value of series resistance.

For example, assume that the basic meter is to be made into a voltmeter with a full scale reading of 1 volt. The coil resistance of the basic meter is 100 ohms, and 100 microamperes of current causes a full scale deflection. The total resistance which is required of the circuit (to limit total current to 100 microamperes) is given by Ohm's law as $R = E/I = 1\text{ v}/100\text{ ua} = 10$ kohms (for a full scale 1 volt range). As the meter coil already contains 100 ohms the series resistance is therefore required to be 10 kohms - 100 ohms = 9.9 kohms.

Multirange voltmeters utilize one meter movement with the required resistances connected in series with the meter by a conventional switching arrangement. A schematic diagram of a multirange voltmeter with three ranges is shown in figure 3-10. The total resistances for

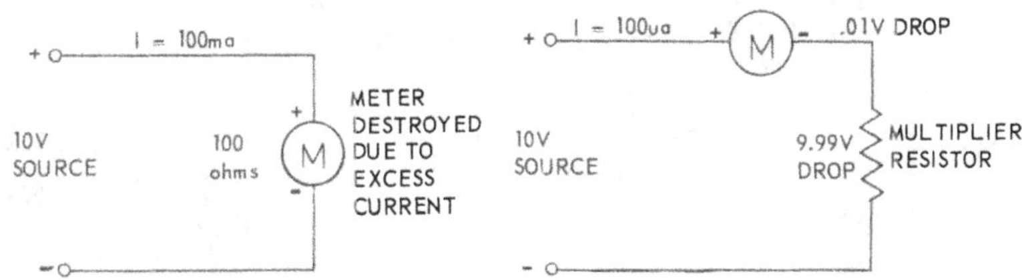

179.8

Figure 3-9.—Usage of multiplier resistors with a D'Arsonval meter movement.

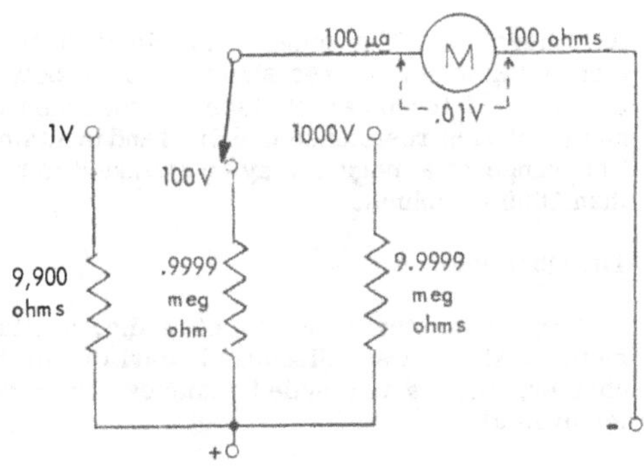

179.9

Figure 3-10.—Multirange voltmeter.

each of the three ranges, beginning with the 1 volt range, is given by Ohm's law as:

For 1 volt range - R = 1/.0001 = 10 kohm
For 100 volt range - R = 100/.0001 = 1 megohm
For 1000 volt range - R = 1000/.0001 = 10 megohms

The multiplying series resistor (R_m) for each of these circuits is 100 ohms less than the total resistance.

Voltmeter Connection

VOLTAGE MEASURING INSTRUMENTS ARE CONNECTED ACROSS (IN PARALLEL WITH) A CIRCUIT. If the voltage to be measured is not known, it is best to start with the highest range of the voltmeter and progressively lower the range until a suitable reading is obtained.

In many cases, the voltmeter is not a center-zero indicating instrument. Thus it is necessary to observe the proper polarity when connecting the instrument to the circuit. As the case in connecting the circuit to the d.c. ammeter previously covered, the voltmeter is connected so that current will flow into the negative terminal of the meter.

Influence In a Circuit

The function of a voltmeter is to indicate the potential difference between two points in a circuit. When the voltmeter is connected across a circuit, it shunts the circuit. If the voltmeter has a low resistance it will draw an appreciable amount of current. The effective resistance of the circuit will be lowered and the voltage reading will consequently be changed. When voltage measurements are made in high-resistance circuits, it is necessary to use a high-resistance voltmeter to prevent this shunting action of the meter. The effect is less noticeable in low-resistance circuits because the shunting effect is less. The problem of voltmeter shunting is illustrated in figure 3-11.

In part A of figure 3-11 a source voltage of 150 volts is applied to a series circuit consisting of two 10-kohm resistors. As shown in figure 3-11A the voltage drop across each resistor is 75 volts. It is now assumed that in the 150 volt range the voltmeter to be used has a total internal resistance of 10 kohms. Figure 3-11B shows the voltmeter connected. It is in parallel with R2. The parallel combination of R2 and the meter now exhibit a total resistance of 5 kohms. As a result the voltage drops change to 100 volts across R1 and 50 volts across R2, which is in parallel with the meter. It is easily seen that this is not the normal voltage drop across R2.

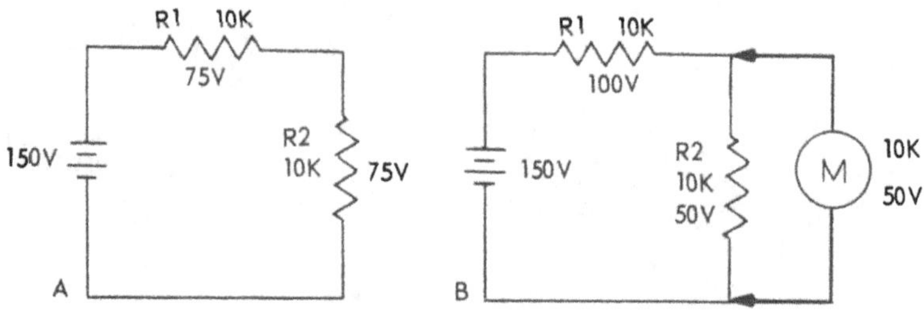

179.10
Figure 3-11.—Shunting action caused by a voltmeter.

Sensitivity

The sensitivity of a voltmeter is given in ohms per volt, (Ω/E), and may be determined by dividing the resistance, R_m, of the meter plus the series resistance R_s, by the full scale reading in volts. Thus, in equation form, sensitivity = $(R_m + R_s)/E$. This is the same as saying that the sensitivity is equal to the reciprocal of the full scale deflection current. In equation form then,

$$\text{Sensitivity} = \frac{\text{ohms}}{\text{volts}} = \frac{1}{\frac{\text{volts}}{\text{ohms}}} = \frac{1}{\text{amperes}}$$

Thus, the sensitivity of a 100 microampere movement is the reciprocal of .0001 ampere, or 10,000 ohms per volt.

Accuracy

The accuracy of a meter is generally expressed in percent. For example, a meter that has an accuracy of 1 percent will indicate a value that is within 1 percent of the correct value. The statement means that if the correct value is 100 units, the meter indication may be anywhere within the range of 99 to 101 units.

METERS USED FOR MEASURING RESISTANCE

The two instruments most commonly used to check the continuity, or to measure the resistance of a circuit or circuit element, are the OHMMETER and the MEGGER (Megohmmeter). The ohmmeter is widely used to measure resistance and check the continuity of electrical circuits and devices. Its range usually extends to only a few megohms. The megger is widely used for measuring insulation resistance, such as between a wire and the outer surface of its insulation, and insulation resistance of cables and insulators. The range of a megger may be extended to more than 1000 megohms.

The Ohmmeter

The ohmmeter consists of a d.c. milliammeter, which was discussed earlier in this chapter, with a few added features. The added features are:

1. A d.c. source of potential.
2. One or more resistors (one of which is variable).

A simple ohmmeter circuit is shown in figure 3-12.

The ohmmeters pointer deflection is controlled by the amount of battery current passing through the moving coil. Before measuring the resistance of an unknown resistor or electrical circuit, the ohmmeter must be calibrated. If the value of resistance to be measured can be estimated within reasonable limits, a range is selected which will give approximately half scale deflection when this resistance is inserted between the probes. If the resistance is unknown the selector switch is set on the highest scale. Whatever range is selected the meter must be calibrated to read zero before the unknown resistance is measured.

Calibration is accomplished by first shorting the test leads together as shown in figure 3-12. With the test leads shorted, there will be a complete series circuit consisting of the 3 volt source, the resistance of the meter coil, R_m, the resistance of the zero-adjust rheostat,

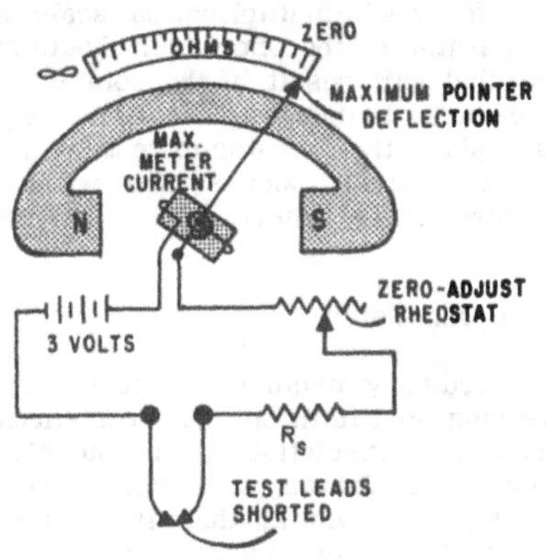

179.11

Figure 3-12.—Simple ohmmeter circuit.

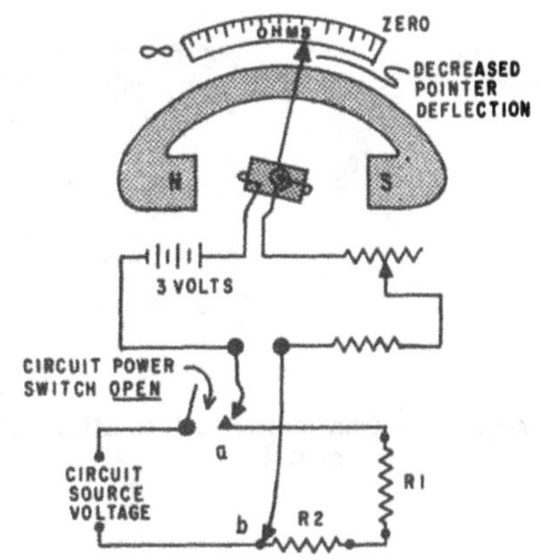

179.12

Figure 3-13.—Measuring circuit resistance with an ohmmeter.

and the series multiplying resistor, R_s. Current will flow and the meter pointer will be deflected. The zero point on the ohmmeter scale (as opposed to the zero for voltage and current), is located at the extreme right side of the scale. With the test leads shorted, the zero-adjust potentiometer is set so that the pointer rests on the zero mark. Therefore, full scale deflection indicates zero resistance between the test leads.

If the range is changed the meter must be "zeroed" again to obtain an accurate reading. When the test leads of an ohmmeter are separated, the pointer of the meter will return to the left side of the scale, due to the spring tension acting on the movable coil assembly. This reading indicates an infinite resistance.

After the ohmmeter is adjusted for zero reading, it is ready to be connected in a circuit to measure resistance. A typical circuit and ohmmeter arrangement is shown in figure 3-13. The power switch of the circuit to be measured should always be in the OFF position. This prevents the circuit's source voltage from being applied across the meter, which could cause damage to the meter movement.

As indicated above the ohmmeter is an open circuit when the test leads are separated. In order to be capable of taking a resistance reading the path for current produced by the meter's battery must be completed. In figure 3-13 this is accomplished by connecting the meter at points A and B, (putting the resistors R1 and R2 in series with the resistance of the meter coil, zero-adjust rheostat, and the series multiplying resistor). Since the meter has been preadjusted (zeroed) the amount of coil movement now depends solely on the resistance of R1 and R2. The inclusion of R1 and R2 raised the total series resistance, decreased the current, and thus decreased the pointer deflection. The pointer will now come to rest at a scale figure indicating the combined resistance of R1 and R2. If R1 or R2, or both, were replaced with a resistor having a larger ohmic value, the current flow in the moving coil of the meter would be decreased still more. The deflection would also be further decreased and the scale indication would read a still higher circuit resistance. Movement of the moving coil is proportional to the intensity of current flow. The scale reading of the meter, in ohms, is inversely proportional to current flow in the moving coil.

The amount of circuit resistance to be measured may vary over a wide range. In some cases it may be only a few ohms, and in other cases it may be as great as 1 megohm. To enable the meter to indicate any value being measured, with the least error, scale multiplication features are incorporated in most ohmmeters. Most ohmmeters are equipped with a selector switch for selecting the multiplication scale desired. For example, a typical meter may have a six position switch, marked as follows: RX1, RX10, RX100, RX1000, RX10,000, and RX100,000. This is shown in figure 3-14.

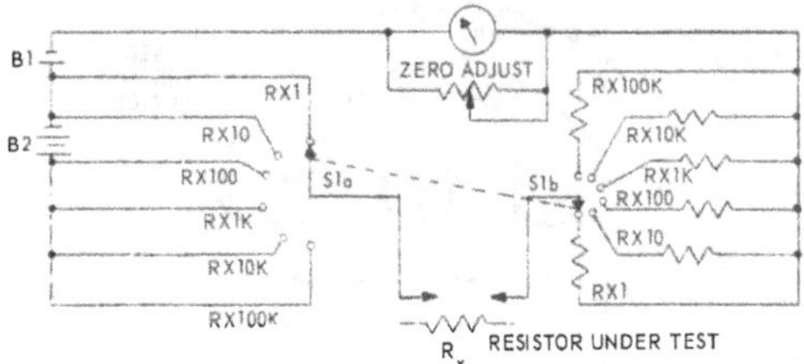

179.13
Figure 3-14.—Ohmmeter with multiplication switch.

The range to be used in measuring any particular unknown resistance (R_x in fig. 3-14) depends on the approximate ohmic value of the unknown resistance. For instance, assume the ohmmeter scale of figure 3-14 is calibrated in divisions from zero to 1000. If R_x is greater than 1000 ohms, and the R x 1 range is being used, the ohmmeter cannot measure it. This occurs because the combined series resistor R x 1 and R_x is too great to allow sufficient battery current to flow to deflect the pointer away from infinity (∞). The switch would have to be turned to the next range R x 10.

When this is done, assume the pointer deflects to indicate 375 ohms. This would indicate that R_x has 375 x 10 = 3,750 ohms of resistance. The change of range caused the deflection because resistor R x 10 has only one tenth the resistance of resistor R x 1. Thus, selecting the smaller series resistance allowed a battery current of sufficient value to cause a useful pointer deflection. If the R x 100 range were used to measure the same 3,750 ohm resistor, the pointer would deflect still further, to the 37.5-ohm position. This increased deflection would occur because resistor R x 100 has only one tenth the resistance of resistor R x 10.

The foregoing circuit arrangement allows the same amount of current to flow through the meter moving coil whether the meter measures 10,000 ohms on the R x 1 scale, or 100,000 ohms on the R x 10 scale, or 1 megohm on the R x 100 scale.

It always takes the same amount of current to deflect the pointer to a certain position on the scale (midscale position for example), regardless of the multiplication factor being used. Since the multiplier resistors are of different values, it is necessary to ALWAYS "zero" adjust the meter for each multiplication scale selected. The operator of the ohmmeter should select the range that will result in the pointer coming to rest as near midpoint of the scale as possible. This enables the operator to read the resistance more accurately, because the scale readings are more easily interpreted at or near midpoint.

The Megohmmeter

An ordinary ohmmeter cannot be used for measuring multimillion ohm resistances, such as conductor insulation. To adequately test for insulation breakdown, it is necessary to use a much higher potential than is furnished by an ohmmeter's battery. This potential is placed between the conductor and the outside of the insulation.

An instrument called a MEGGER (megohmmeter) is used for these tests. The megger (fig. 3-15) is a portable instrument consisting of two primary elements: (1) a hand driven d.c. generator which supplies the necessary voltage for making the measurement, and (2) the instrument portion, which indicates the value of the resistance being measured. The instrument portion is of the opposed coil type as shown in figure 3-15A. Coils A and B are mounted on the movable member C with a fixed angular relationship to each other, and are free to turn as a unit in a magnetic field. Coil B tends to move the pointer counterclockwise, and coil A clockwise.

Coil A is connected in series with R3 and the unknown resistance, R_x, to be measured. The combination of coil A, R3, and R_x form a direct series path between the + and - brushes of the d.c. generator. Coil B is connected in series with R2 and this combination is also connected across the generator. There are no restraining springs on the movable member of the instrument portion of the megger. Therefore, when the generator is not operated, the pointer floats freely and may come to rest at any position on the scale.

The guard ring intercepts leakage current. Any leakage currents intercepted are shunted to the negative side of the generator. They do not flow through coil A; therefore, they do not affect the meter reading.

If the test leads are open-circuited, no current flows in coil A. However, current flows internally through coil B, and deflects the pointer to infinity, which indicates a resistance too large to measure. When a resistance such as R_x is

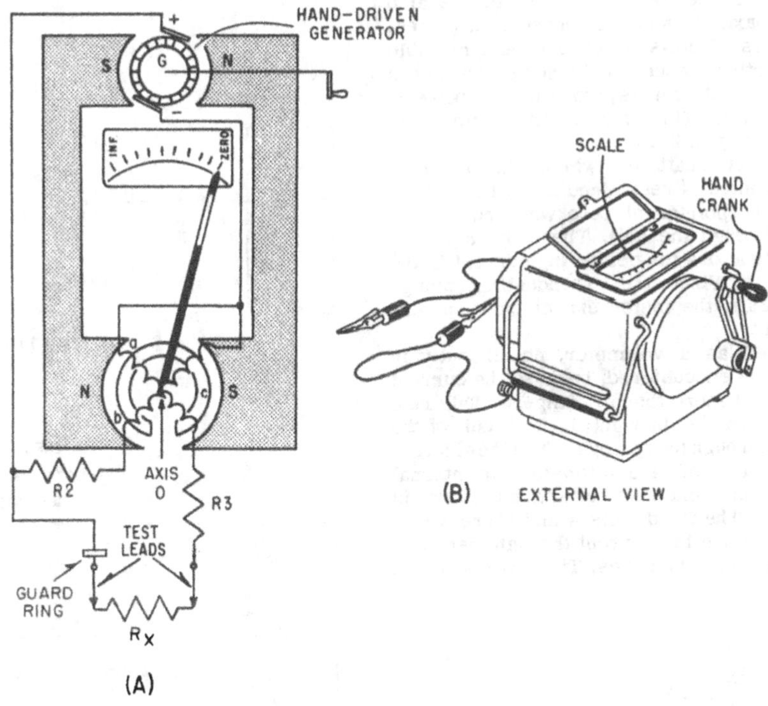

1.55

Figure 3-15.—Megger internal circuit and external view.

connected between the test leads, current also flows in coil A, tending to move the pointer clockwise. At the same time, coil B still tends to move the pointer counterclockwise. Therefore, the moving element, composed of both coils and the pointer, comes to rest at a position at which the two forces are balanced. This position depends upon the value of the external resistance, which controls the relative magnitude of current in coil A. Because changes in voltage affect both coil A and coil B in the same proportion, the position of the moving system is independent of the voltage. If the test leads are short-circuited, the pointer rests at zero because the current in coil A is relatively large. The instrument is not injured under these circumstances because the current is limited by R3. The external view of one type of megger is shown in figure 3-15B.

Meggers provided aboard ship usually are rated at 500 volts. To avoid excessive test voltages most meggers are equipped with friction clutches. When the generator is cranked faster than its rated speed, the clutch slips and the generator speed and output voltages are not permitted to exceed their rated values. For extended ranges, a 1,000 volt generator is available. When extremely high resistance such as 10,000 megohms or more are to be measured, a high voltage is needed to cause sufficient current flow to actuate the meter movement.

ELECTRODYNAMOMETER TYPE METER

The electrodynamometer type meter differs from the galvanometer type in that no permanent magnet is used. Instead, two fixed coils are utilized to produce the magnetic field. Two movable coils are also used in this type meter.

The two fixed coils are connected in series and positioned coaxially (in line), with a space between them. The two movable coils are also positioned coaxially, and are connected in series. The two pairs of coils (fixed pair and movable pair) are further connected in series with each other. The movable coil is pivot-mounted between the fixed coils. This meter arrangement is illustrated in figure 3-16.

The central shaft on which the movable coils are mounted is restrained by spiral springs which hold the pointer at zero when no current is flowing through the coil. These springs also serve as conductors for delivering current to the movable coils. Since these conducting springs are very small, the meter cannot carry a very heavy current.

When used as a voltmeter, no difficulty in construction is encountered, because the current required is not more than 0.1 ampere. This rate of current can be brought in and out of the moving coil through the springs. When the electrodynamometer is used as a voltmeter, its internal connections and construction are as shown in figure 3-17A. The fixed coils A and B are wound of fine wire, since the current through them will be no more than 0.1 amperes. They are connected

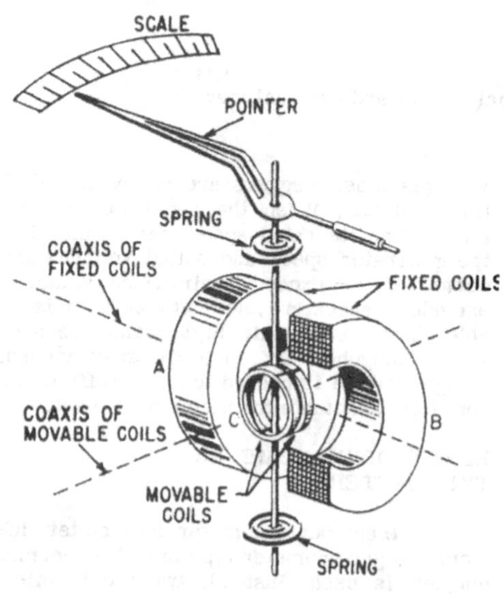

5.109
Figure 3-16.—Inside construction of electrodynamometer.

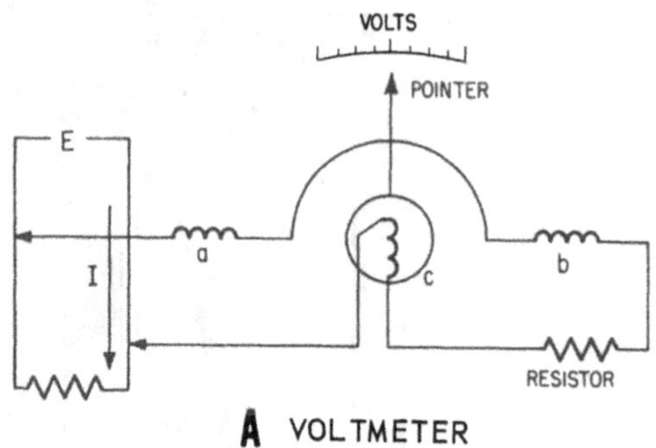

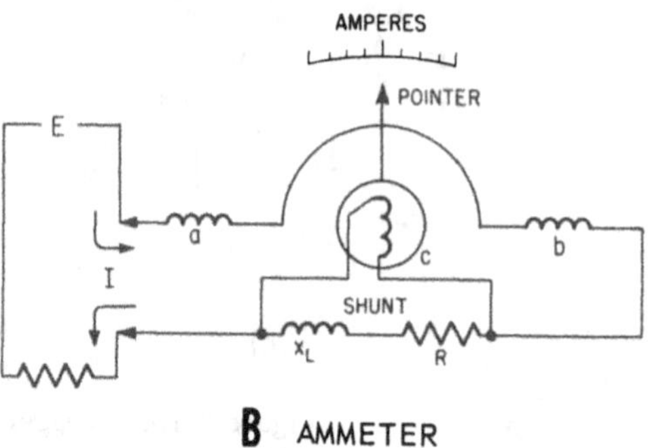

179.14
Figure 3-17.—Circuit arrangement of electrodynamometer for voltmeter and ammeter.

directly in series with the movable coil C and the series current-limited resistance. For ammeter applications, however, a special type construction must be used, because the large currents that flow through the meter cannot be carried through the moving coils.

In the ammeter, the stationary coils A and B (fig. 3-17B) are generally wound of heavier wire, to carry up to five amperes. In parallel with the moving coils is an inductive shunt, which permits only a small part of the total current to flow through the moving coil. This current through the moving coil is directly proportional to the total current through the instrument. The shunt has the same ratio of reactance to resistance as has the moving coil, thus the instrument will be reasonably correct at all frequencies of usage if a.c. currents are to be measured.

The meter is mechanically damped by means of aluminum vanes that move in enclosed air chambers. Although electrodynamometer type meters are very accurate, they do not have the sensitivity of the D'Arsonval type meter movement. For this reason they are not widely used outside the laboratory. The big advantage of the electrodynamometer type meter movement is that it can be used to measure alternating current as well as direct current.

If alternating current is applied to the standard galvanometer type meter, it will produce no reading. Instead, the meter will vibrate about the zero reading. On one half cycle of the a.c., the meter will be deflected to the left and on the other half cycle to the right. As the frequency being measured is usually 60 Hz or greater, the meter is incapable of mechanically responding at this speed. The result is simply a vibration around the zero point and no useful reading of voltage or current is obtained. In order to use the D'Arsonval type meter movement to measure a.c. currents and voltages, some means must be obtained to change the a.c. to d.c. so that a unidirectional current will be applied to the meter.

It should be noted that this problem does not exist with the electrodynamometer type movement. The current flow through the stationary coils sets up a magnetic field. The current flow through the moving coils sets up an opposing magnetic field. With two magnetic fields which are opposing, the pointer deflects to the right. If the current now reverses direction, the magnetic fields of BOTH sets of coils are reversed. By reversing both fields, they still oppose, and the pointer still deflects to the right. Consequently, no rectifying devices are required for the electrodynamometer type meter movement. It will work equally well with a d.c. current or an a.c. current.

WATTMETER

Electric power is measured by means of a wattmeter. This instrument is of the electrodynamometer type. It consists of a pair of fixed coils, known as current coils, and a moving coil, known as the voltage (potential) coil (fig. 3-18). The fixed coils are made up of a few turns of comparatively large conductor. The potential coil consists of many turns of fine wire; it is mounted on a shaft, carried in jeweled bearings, so that

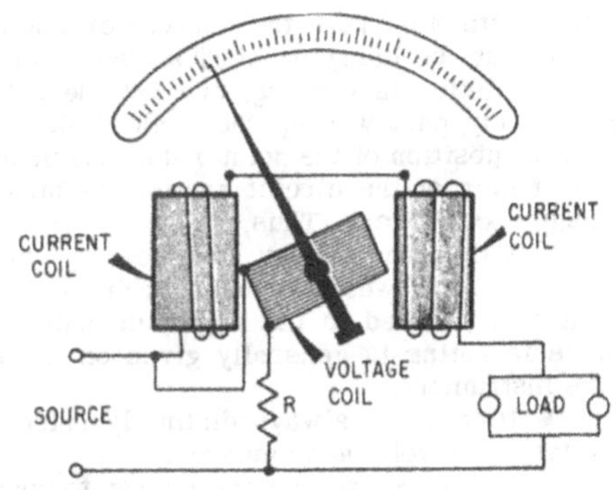

179.15

Figure 3-18.—Simplified electrodynamometer wattmeter circuit.

it may turn inside the stationary coils. The movable coil carries a needle which moves over a suitably graduated scale. Flat coil springs hold the needle to a zero position.

The current coil (stationary coil) of the wattmeter is connected in series with the circuit (load), and the potential coil (movable coil) is connected across the line. When line current flows through the current coil of a wattmeter, a field is set up around the coil. The strength of this field is proportional to the line current and in phase with it. The potential coil of the wattmeter generally has a high resistance resistor connected in series with it. This is for the purpose of making the potential-coil circuit of the meter as purely resistive as possible. As a result, current in the potential circuit is practically in phase with line voltage. Therefore, when voltage is impressed on the potential circuit, current is proportional to and in phase with the line voltage.

The actuating force of a wattmeter is derived from the interaction of the field of its current coil and the field of its potential coil. The force acting on the movable coil at any instant (tending to turn it) is proportional to the product of the instantaneous values of line current and voltage.

The wattmeter consists of two circuits, either of which will be damaged if too much current is passed through them. This fact is to be especially emphasized, because the reading of the instrument does not serve to tell the user that the coils are being overheated. If an ammeter or voltmeter is overloaded, the pointer will be indicating beyond the upper limit of its scale.

In the wattmeter, both the current or potential circuit may be carrying such an overload that their insulation is burning, and yet the pointer may be only part way up the scale. This is because the position of the pointer depends upon the power factor of the circuit as well as upon the voltage and current. Thus, a low power factor circuit will give a very low reading on the wattmeter even when the current and potential circuits are loaded to the maximum safe limit. This safe rating is generally given on the face of the instrument.

A wattmeter is always distinctly rated, not in watts but in volts and amperes.

Figure 3-19 shows the proper way to connect a wattmeter into a circuit.

AUDIO SIGNAL GENERATORS

Audio signal generators produce stable frequency signals from about 20 hertz to 200,000 hertz. They are used primarily for testing audio sections of equipment. The major components of an audio signal generator are an oscillator (or oscillators), one or more amplifiers, an output control and a power supply. In addition, voltage regulator circuits are necessary to ensure stability of the oscillator, because the a.c. line voltage sources may vary.

A representative audio frequency signal generator is shown in figure 3-20. This unit is intended primarily for bench testing of electronic equipment. It operates from line voltage (115 v.a.c.) and produces the output frequencies over a continuous range in conjunction with a four position multiplier.

Any frequency from 20 to 200,000 Hz may be selected by setting the main tuning dial and the range switch so that the two readings, when multiplied together, equal the desired frequency. For example, to select an output frequency of 52,000 Hz, set the main tuning dial to "52" and the range switch to "X1000". Voltages from zero to 10 volts may be selected by using the "OUTPUT LEVEL" control in conjunction with the attenuator switch. The attenuator is calibrated in seven decade steps so that with the output meter set to 1.0, output voltages of 10 volts to 10 microvolts can be obtained by simply switching the attenuator. For intermediate values of output voltage, the "OUTPUT LEVEL" control is varied so that the output meter reads the desired voltage. The attenuator switch is then set so that its value, multiplied by the output meter reading, gives the desired output voltage level. For example, to obtain an output voltage of 0.04 volts set the meter by means of the "OUTPUT LEVEL" control to read .4 and set the "ATTENUATOR" switch to the "X.1" position. The output voltage will then be the meter reading multiplied by the attenuator setting, or 0.04 volts.

Oscillator frequency calibration may be checked, during operation, on the first two bands, by means of the built-in frequency meter. To check the operation of the oscillator at 60 Hz, set the main tuning dial to "60" and the "RANGE" switch to "X1." This sets the frequency of the oscillator at 60 hertz. Turn the "FREQUENCY METER" switch on and move the main dial back and forth slightly until the reed vibrates with maximum amplitude. This point is rather critical, and care must be used to see that the point of maximum response is located. The main tuning dial should indicate "60" within one division, if the frequency calibration is correct. Similarly, the output frequency may be checked at 400 hertz, by means of the 400 hertz reed in the frequency meter.

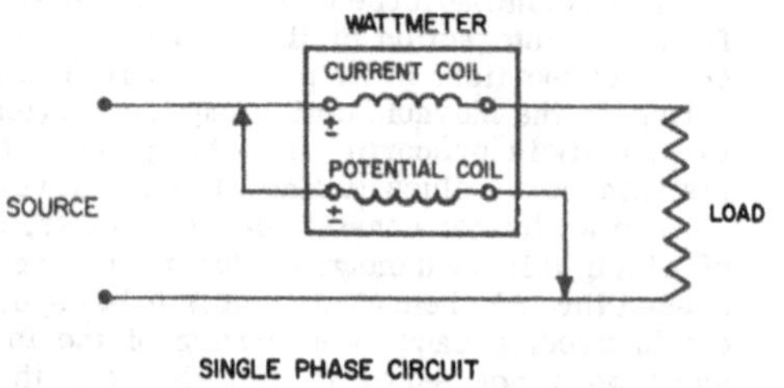

Figure 3-19.—Wattmeter connection.

RADIO FREQUENCY SIGNAL GENERATORS

Signal generators that cover the frequency range from about 10 kHz to 4.5 GHz are classified as RF signal generators. One signal generator alone does not cover this range. A number of RF frequency generators are required to do so. As an example; one may cover 10 kHz to 50 MHz, another 100 kHz to 110 MHz, a third 4 to 405 MHz, a fourth 800 to 2400 MHz, and still another 1.8 to 4.5 GHz.

As an example in this text we will use a representative RF signal generator which covers

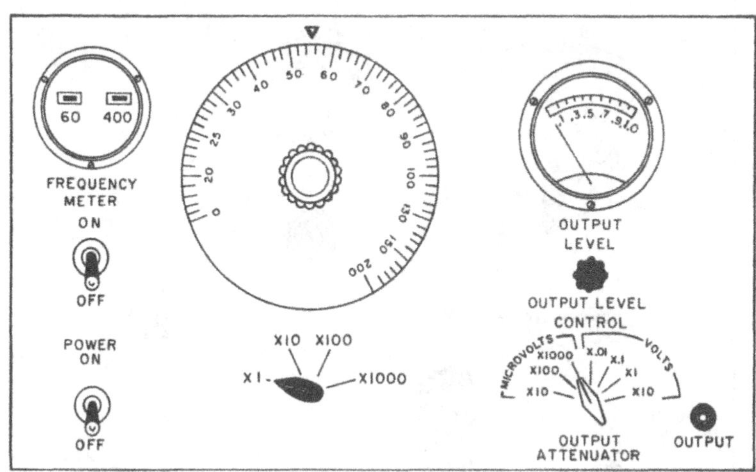

179.17
Figure 3-20.—Representative audio signal generator.

the 10 kHz to 50 MHz range (fig. 3-21). This unit is intended primarily for bench testing of electronic equipment. It operates from line voltage (115 v.a.c.) and produces frequencies over a continuous range from 10,000 to 50,000,000 hertz in eight bands. The band selector control is located on the front panel.

The RF output for the entire frequency range is continuously variable from 0.1 to 100,000 microvolts at the RF OUTPUT X-MULT jack, while approximately 2 volts is available at the RF OUTPUT X200K jack. The voltage at the RF OUTPUT X-MULT jack is varied by the RF MULTIPLIER switch and the MICROVOLT control, whereas the voltage at the RF OUTPUT X200K jack is varied only by the MICROVOLT control. The RF MULTIPLIER is a six step attentuator in multiples of 10. The RF voltage at the jack is determined by multiplying the meter reading by the multiplier factor on the front panel. For instance, if the meter reads 5 and the attenuator is set to the 100 position, the output will be 500 microvolts.

An audio signal of 400 or 1,000 hertz, adjustable from 0 to approximately 3 volts by means of the % MOD AUDIOOUT-XTALCALOUT control, is available at the AUDIOOUT-XTALCALOUT-EXTMODIN jack. This audio voltage is used mainly to modulate the RF carrier. Modulation is adjusted by the % MOD-AUDIOOUT-XTALCALOUT control from 0 to 50 percent. Percentage modulation is read directly on the lower scale of the same meter used to measure the RF voltage.

CATHODE-RAY OSCILLOSCOPE

The most versatile piece of test equipment that a technician has at his disposal is the Cathode-Ray Oscilloscope (CRO). This device gives a visual presentation of any voltage waveform present in a circuit.

The oscilloscope permits many characteristics of a circuit to be observed and measured. Some of these are: (1) frequency of operation; (2) duration (or time) of occurrence of one or more cycles; (3) phase relationships between voltage waveforms appearing at different points in the circuit; (4) shape of the waveform; and (5) amplitude of the waveform.

The oscilloscope is used for alignment of radio and radar receiving and transmitting equipment, frequency comparison, percentage of modulation checks, and many other applications.

BASIC OSCILLOSCOPE

The principal components of a basic oscilloscope include a cathode-ray tube (CRT); a sweep

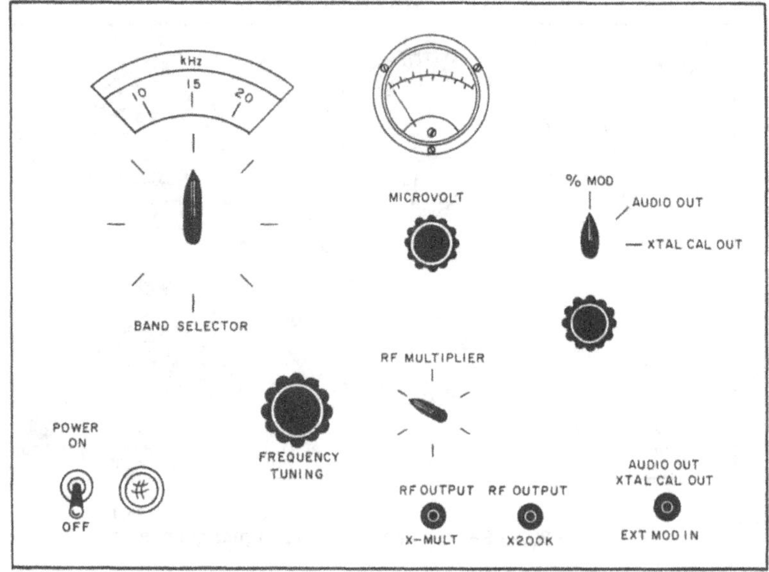

179.18

Figure 3-21.—Representative RF signal generator.

generator; deflection amplifiers (horizontal and vertical); a power supply; and suitable controls, switches, and input connectors for proper operation. Figure 3-22 shows the block diagram of a basic oscilloscope.

The horizontal and vertical deflection amplifiers increase amplitude of the input voltage to the level necessary for deflection of the beam. The sweep generator supplies a sawtooth voltage to the input of the horizontal amplifier. The power supply provides d.c. voltages for operation of the amplifiers and CRT.

The cathode-ray tube is the heart of the oscilloscope. The CRT contains an electron gun, a deflection system, and a screen inside an evacuated glass envelope. The electron gun is an arrangement of tube elements to introduce electrons into the envelope, accelerate them, and focus them into a narrow beam. The deflection system controls position of the narrow beam of electrons arriving at the screen. The screen is a chemical phosphor that converts the energy of electrons into light. Figure 3-23 shows construction of a cathode-ray tube.

Deflection

The deflection system, electrostatic in this type of cathode-ray tube, consists of two pairs of metal plates, mounted at right angles to each other, and placed inside the neck of the CRT. Figure 3-23 shows the location of the plates. When a difference of potential is applied to the plates, an electrostatic field is developed between the plates. This field acts upon the beam of energy and controls the point at which it will arrive at the screen. Figure 3-24 shows a simplified arrangement of the front screen and deflection plates for explanatory purposes.

In figure 3-24A, all plates are at the same potential and no field exists to act upon the beam, so the beam arrives at the center of the screen. In figure 3-24B, a positive potential is applied to the top plate, resulting in beam moving towards the top of screen. The greater the amplitude of positive potential, the farther the beam moves away from the center toward the positive potential.

In figure 3-24C, a positive potential is applied to the bottom plate and the beam moves towards the bottom of the screen. In figure 3-24D, the

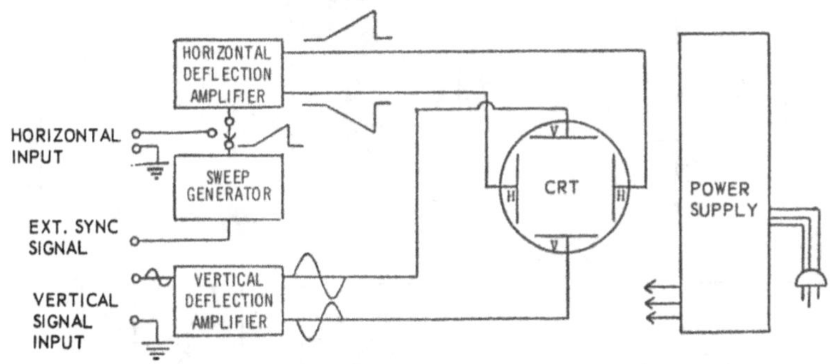

20.324
Figure 3-22.—Block diagram of a basic oscilloscope.

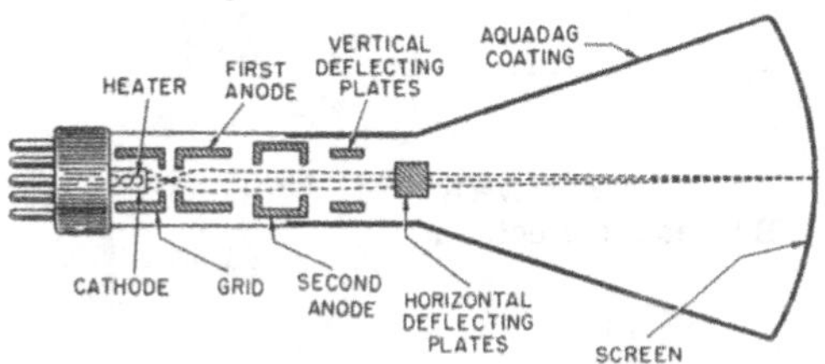

20.320
Figure 3-23.—Construction of a cathode ray tube.

potential is applied to the right horizontal deflection plate, and the beam moves correspondingly. In figure 3-24E, the potential is applied to the left horizontal deflection plate and the beam moves left. In figure 3-24F, a positive potential is applied to the left horizontal deflection plate and the top vertical deflection plate causing beam movement to the upper left. By controlling the amplitude of voltage on one plate with respect to the others, the beam can be positioned.

In a practical oscilloscope deflection amplifiers replace batteries. In many circuits, the voltage waveform to be examined does not have sufficient amplitude to cause beam movement across the screen. Deflection amplifiers increase the voltage waveform amplitude, without changing its shape, to a level necessary for deflection.

In figure 3-25, with a sine wave applied to the vertical signal input terminal, the vertical deflection amplifiers amplify this signal and deliver two sine waves of opposite polarities to the vertical deflection plates. (How two sine waves were developed will be explained later.) This will result in beam movement up and down at a rate determined by the sine wave frequency. Usually, this beam movement is so rapid that the eye cannot follow it, and instead the eye sees just a bright vertical line.

Since the beam has negligible mass, it responds to peak-to-peak excursions of the sine wave. By marking a faceplate in inches or centimeters, the oscilloscope can be calibrated with a known voltage before applying the sine wave, and then with the sine wave applied, its amplitude can be calculated.

Figure 3-26 shows a CRT with a vertical line drawn 8 centimeters long with 1 centimeter graduations. Assuming a 1 volt peak-to-peak signal moved the beam 1 cm, the amplitude of the vertical trace shown would be 3 volts peak-to-peak.

Since a voltage waveform is a graphical presentation of a voltage varying with respect to time, it is desirable that the oscilloscope gives this type of presentation. To accomplish this, a voltage is applied to the horizontal deflection plates and varied in such a manner as to move the beam at a linear rate from left to right, and then return the beam back to the left side as rapidly as possible. A sawtooth waveform is used for this purpose, and it is developed in the sweep generator section of the oscilloscope. Figure 3-27 shows a sawtooth waveform. The horizontal axis represents time, in seconds, while the vertical axis represents amplitude in volts. The values assigned were arbitrarily chosen to demonstrate a linear rate of change.

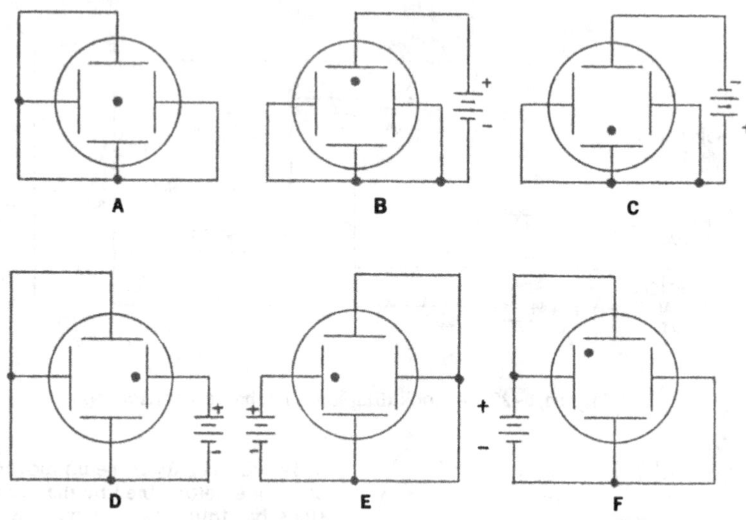

Figure 3-24.—Simplified CRT beam deflection.

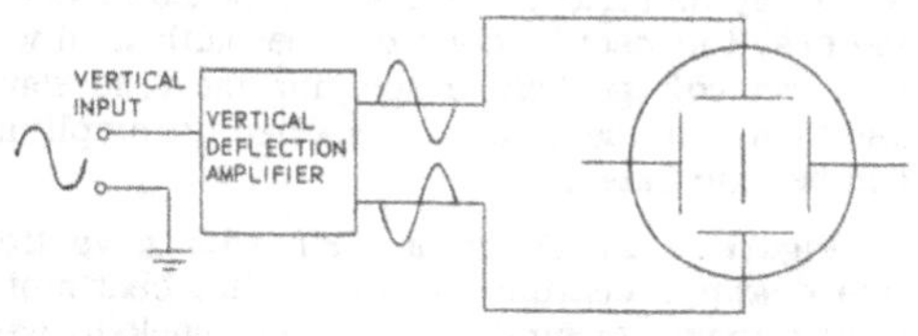

Figure 3-25.—Vertical deflection.

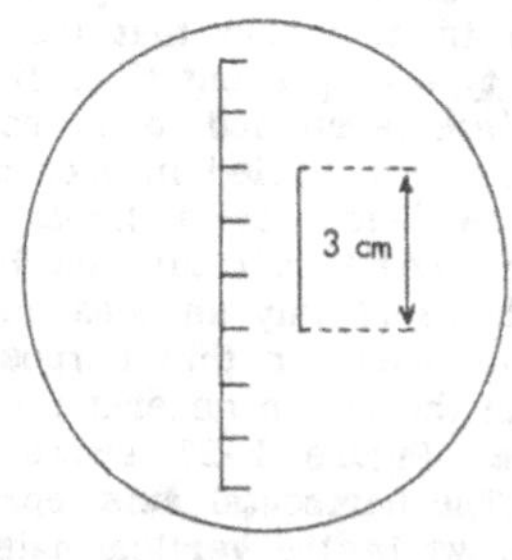

Figure 3-26.—Calculating voltage amplitude with calibrated faceplate.

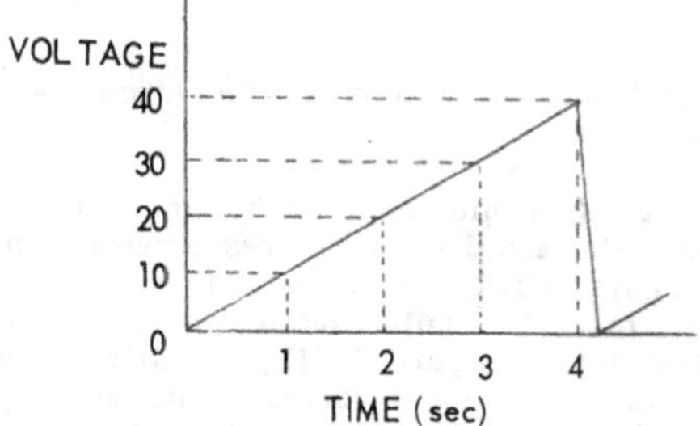

Figure 3-27.—Sawtooth waveform.

At the end of 1 second, the amplitude has risen to 10 volts. In this example, for a linear rate of change to occur, the amplitude of the sawtooth should increase 10 volts for each 1 second increase in time. Thus, at the end of 2 seconds, the amplitude should be 20 volts, and at the end of 3 seconds, the amplitude should be 30 volts.

Figure 3-28 shows the results of applying a sawtooth waveform to the horizontal deflection plates. At the start of the sweep, the beam is at the left side of the screen. As sawtooth

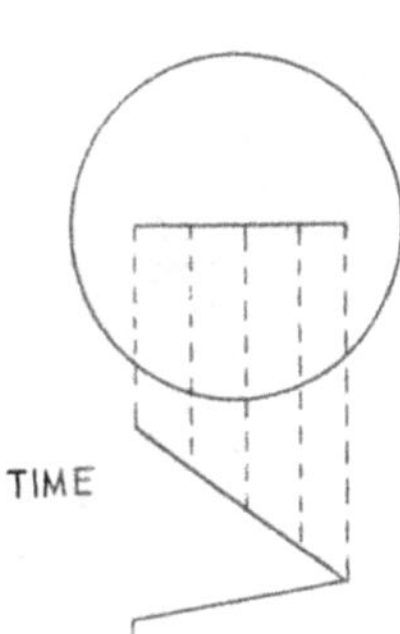

179.23
Figure 3-28.—Horizontal deflection.

amplitude increases, the beam moves to the right until the sawtooth reaches its maximum amplitude. At this time, the sweep is ended, sawtooth amplitude drops down rapidly, pulling the beam back to the left. This action continues to repeat at a frequency determined by the sweep generator. This sweep, or base line, is referred to as the "time base."

During the brief period that the sawtooth is pulling the beam back to the left side of the screen, called the retrace or flyback time, no useful information is displayed. Therefore, the oscilloscope contains a circuit (the blanking circuit) that turns the beam off during flyback time so that this portion of the cycle is not seen.

Figure 3-29 shows results of applying a sawtooth to the horizontal deflection plates and a sine wave to the vertical deflection plates. As the sawtooth moves the beam from left to right, the sine wave simultaneously acts upon the beam to pull it up and down. The combined result of these two actions is the visual reproduction of the sinewave, with the vertical axis representing the amplitude of the sinewave, and the horizontal axis representing the duration, or time, of the sinewave.

Notice the necessity of applying a sawtooth waveform to the horizontal deflection plates. If any other waveform were used, such as a sinewave, the beam would not move at a constant or linear rate from left to right, and the waveform applied to the vertical deflection plates would not be faithfully reproduced.

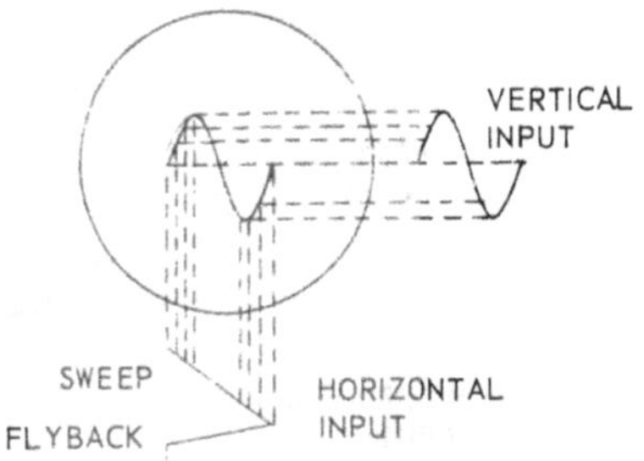

179.24
Figure 3-29.—Sinewave reproduction.

There is a definite relationship between the frequency of the signal applied to the vertical deflection plates and the frequency of the sawtooth. Figure 3-30 illustrates this relationship. Figure 3-30A shows the sinewave and the sawtooth operating at the same frequency. Recall that during retrace the beam is cut off (blanked), therefore a full cycle of the sinewave cannot be observed. Figure 3-30B shows the results when the sinewave is twice the frequency of the sawtooth. Now one full cycle can be observed. In operating the oscilloscope, it is usually best to set the operating controls so that a minimum of two cycles of the input waveform may be viewed. This is done by setting the sweep generator (time-base) frequency to a lower frequency than the vertical input signal.

Just as the vertical axis can be calibrated to measure amplitude, the horizontal axis can be calibrated to measure time. In figure 3-31, the horizontal axis is graduated in centimeters. By knowing the frequency of the sweep generator, the time of one cycle of the input waveform can be calculated. Suppose the frequency of the sweep generator is 1 Hz, and the horizontal scale is 10 cm wide with 1-cm graduations. The time required for one sawtooth to move the beam from left to right (10 cm) is:

$$t = \frac{1}{f} = \frac{1}{1} = 1 \text{ second.}$$

Each centimeter then, in time, is 1/10 = 0.1 second. Since one cycle of the input sinewave shown in figure 3-31 occupies 5 cm, the time is: 5 x 0.1 = 0.5 seconds; and the frequency is: f = 1/t = 1/0.5 = 2 Hz.

Rather than mark the faceplate with graduations, oscilloscope manufactures provide separate

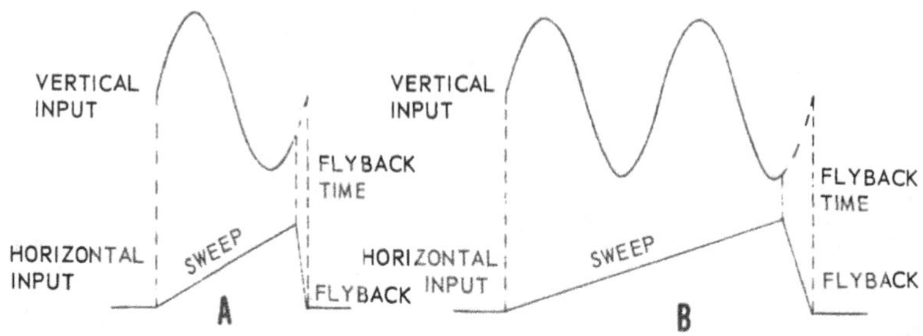

179.25
Figure 3-30.—Relationship between horizontal and vertical signal frequencies.

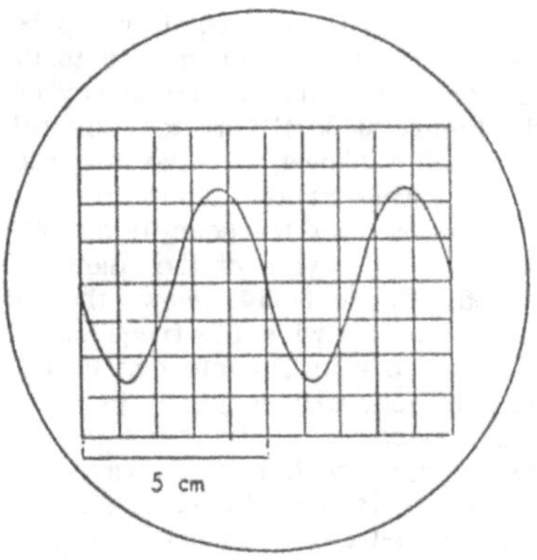

179.26
Figure 3-31.—Calculating frequency with graticule.

plates of glass or plastic that are accurately marked and mounted on the oscilloscope in front of the CRT. These plates are called graticules.

Controls

Figure 3-32 is a drawing of the front panel of a general purpose oscilloscope. Oscilloscopes vary greatly in the number of controls and connectors. Usually, the more controls and connectors, the more versatile is the instrument. Regardless of the number, all oscilloscopes have similar controls and connectors, and once the operation of these are learned, the technician can move with ease from one model oscilloscope to another. Occasionally identical controls will be differently labeled from one model to another, but most controls are logically grouped, and the names are indicative of their function.

The POWER switch may be a toggle, slide, or rotary switch, and it may be mounted separately or on a shaft that is common to another control, such as the intensity control. Its function is to apply the line voltage to the power supply.

The INTENSITY (sometimes called brightness or brilliance) control adjusts the brightness of the beam on the CRT, and is a rotary control. The control is turned clockwise to increase the intensity of the beam and should be adjusted to a minimum brightness level for comfortable viewing.

The FOCUS control is a rotary control that adjusts the spot (beam) size. Figure 3-33 shows the in-focus and out-of-focus extremes. In figure 3-33A, no deflection is applied and the beam is centered, but out of focus. The focus control is adjusted to give a small, clearly defined, circular dot, as shown in figure 3-33B. In figure 3-33C, horizontal deflection is applied with the focus control misadjusted. The focus control is adjusted to give a thin, sharp line as in figure 3-33D.

The HORIZONTAL POSITION and VERTICAL POSITION controls are rotary controls used to position the trace. Since the graticule is often drawn to represent a graph, some oscilloscope have the positioning controls labeled to correspond to the "X" and "Y" axis of the graph. The "X" axis represents the horizontal movement while the "Y" axis represents the vertical movement. Figure 3-34 shows the effects of the positioning controls on the trace.

In figure 3-34A, the horizontal control has been adjusted to move the trace too far to the right while in figure 3-34B the trace has been moved too far to the left. In figure 3-34C the

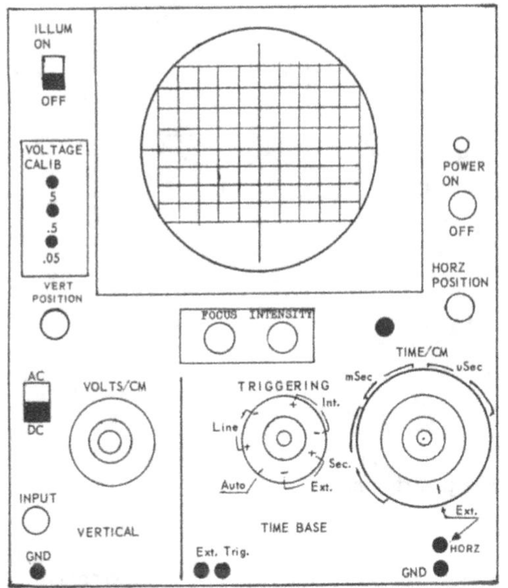

1.86
Figure 3-32.—Typical basic oscilloscope.

vertical positioning control has been adjusted to move the trace to close to the top while in figure 3-34D the trace has been moved too close to the bottom. Figure 3-34E shows the trace properly positioned.

The vertical INPUT (or "Y" INPUT or SIGNAL INPUT) jack connects the desired signal to be examined to the vertical deflection amplifiers. On some oscilloscopes, there may be two input jacks, one labeled AC and the other labeled DC. Other models may have a single input jack with an associated switch to select the AC or DC connection. In the DC position, the signal is connected directly to the vertical deflection amplifiers while in the AC position the signal is fed through a capacitor first. Figure 3-35 shows the schematic of one arrangement.

The deflection amplifiers increase the amplitude of the input signal level required for the deflection of the CRT beam. These amplifiers must not have any other effect on the signal, such as changing the shape (called distortion). Figure 3-36 shows the results of distortion occuring in a deflection amplifier.

An amplifier can handle only a limited range of input amplitudes before it begins to distort the signal. To prevent this, oscilloscopes incorporate circuitry to permit adjustment of the input signal amplitude to a level that will prevent distortion from occurring. This adjustment is the ATTENUATOR control. This control extends the usefulness of the oscilloscope by enabling it to handle a wide range of signal amplitudes.

The attenuator usually consists of two controls. One is a multi-position switch and the other is a potentiometer (fig. 3-37). Each position of the switch may be marked; (1) as to the amount of voltage required to deflect the beam a unit distance, such as volts/cm, or (2) as to the amount of attenuation given to the signal, such as 100, 10, or 1.

In the first case, suppose the .05 volt/cm position were selected. In this position, the beam will be deflected vertically 1 centimeter for every .05 volts of applied signal. If a sinewave occupied 4 cm peak-to-peak, its amplitude would be 4 x .05 = .2 volts pk/pk (refer to fig. 3-38).

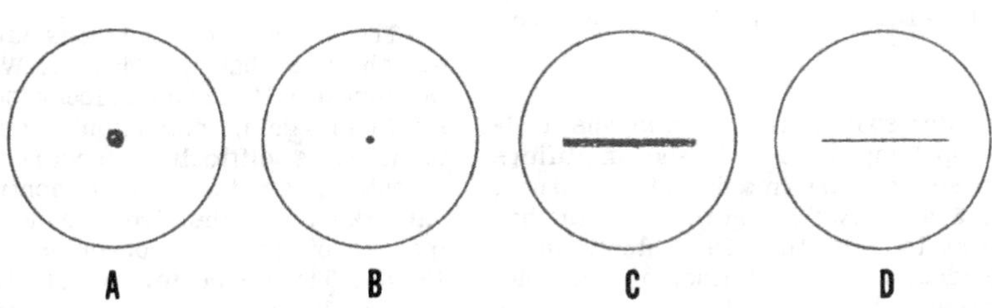

179.27
Figure 3-33.—Focus extremes.

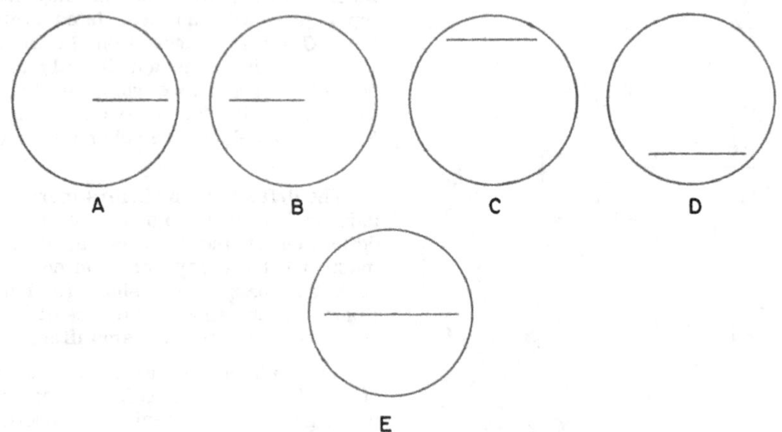

179.28
Figure 3-34.—Trace positioning.

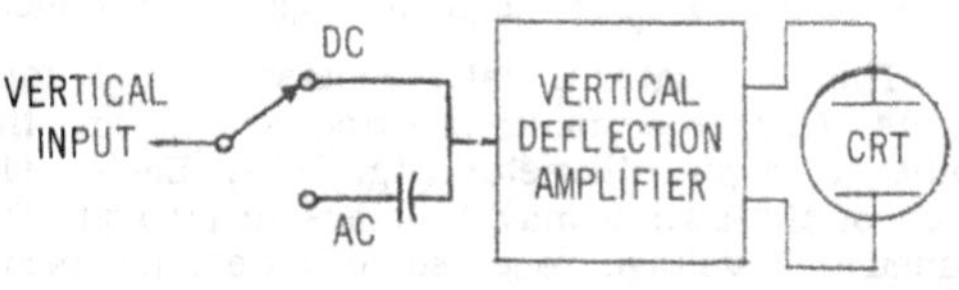

179.29
Figure 3-35.—Vertical input arrangement.

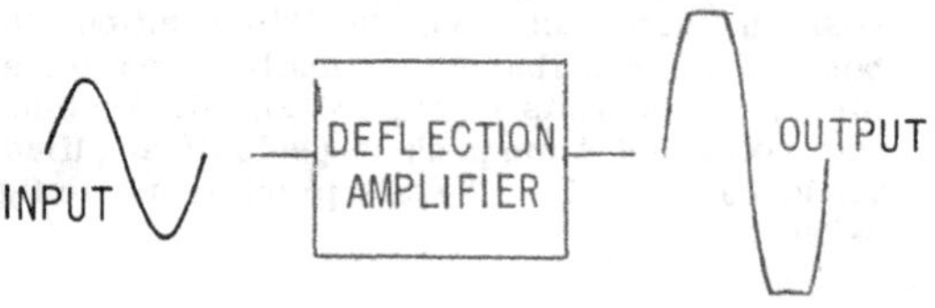

179.30
Figure 3-36.—Deflection amplifier distortion.

The attenuator switch provides a means of adjusting the input signal level to the amplifiers by steps. These steps are in a definite sequence, such as 1-2-5 as shown in figure 3-37. Another sequence used is 1-10-100. The potentiometer control provides a means of fine, or variable, control between steps. This control may be mounted separately or it may be mounted on the attenuator switch. When the control is mounted

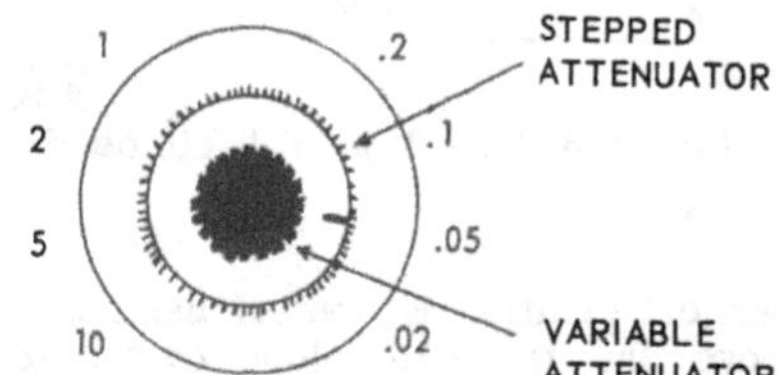

179.31
Figure 3-37.—Attenuator control.

separately, it is often marked as "Fine Gain" or simply "Gain." When mounted on the attenuator switch, it is usually marked "Variable," with the panel markings and knob of a different color for ease of identification.

The variable control adds attenuation to the switch step that is selected. When the control is turned fully counterclockwise, it is at the minimum gain, maximum attenuation setting. Since it is difficult to accurately calibrate a potentiometer, the variable control is either left unmarked, or the front panel is arbitrarily marked off in some convenient units (e.g., 1-10, 1-100). The attenuator switch, however, can be accurately calibrated to the front panel designations. To do this, the variable control is turned fully clockwise to cut it out of the attenuator

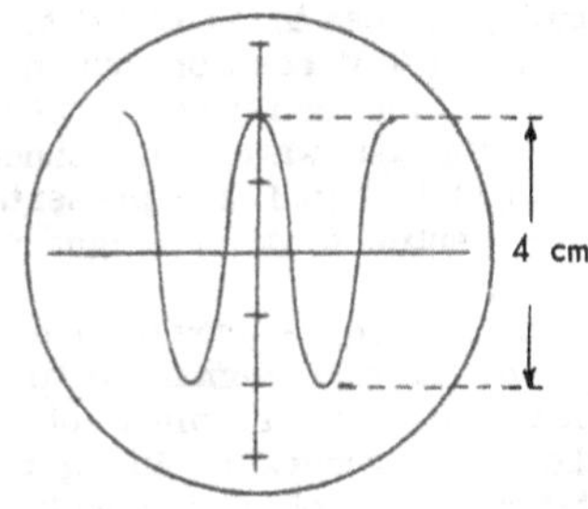

Figure 3-38.—Sinewave attenuation.

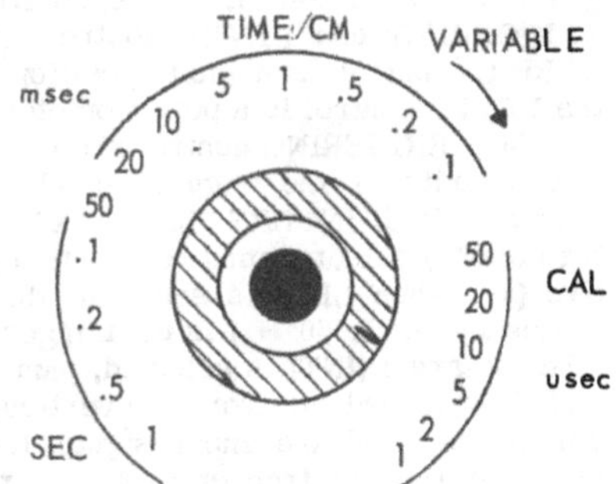

Figure 3-39.—Time base control.

circuit, and this position is usually marked CAL (calibrate) on the panel.

As mentioned previously, the sweep generator develops the sawtooth waveform that is applied to the horizontal deflection plates of the CRT. This sawtooth causes the beam to move at a linear rate from the left side of the screen to the right side. This trace, or sweep, is the TIME BASE of the oscilloscope. To enable the oscilloscope to accept a wide range of input frequencies, the frequency of the time base is variable. Again, two controls are used. One is a multiposition switch that changes the frequency of the sweep generator in steps, and the second control is a potentiometer that varies the frequency between steps (fig. 3-39). The switch has each step calibrated and the front panel is marked TIME/CM. A 1-2-5 sequence is used for numbering the switch positions and the front panel has markings that group the numbers into microseconds (μs), milliseconds (ms), and seconds (SEC).

The potentiometer is labeled variable and the panel is marked with an arrow indicating the direction to turn the pot to the calibrated (CAL) position. When it is desired to accurately measure the time of one cycle of an input signal, the variable control is turned to the CAL position and the TIME/CM switch is turned to select the appropriate time base. Suppose the 5 μsec position was chosen and two cycles of an input signal were being displayed as shown in figure 3-40. One cycle occupies 5 centimeters along the horizontal axis. Each cm is worth 5 μsec in time, so the time of one cycle equals 5 x 5 = 25 μsecs, Frequency may be found by using f = 1/t.

In selecting a time base, remember that it should be lower in frequency than the input signal. If the input signal requires 5 μsec to

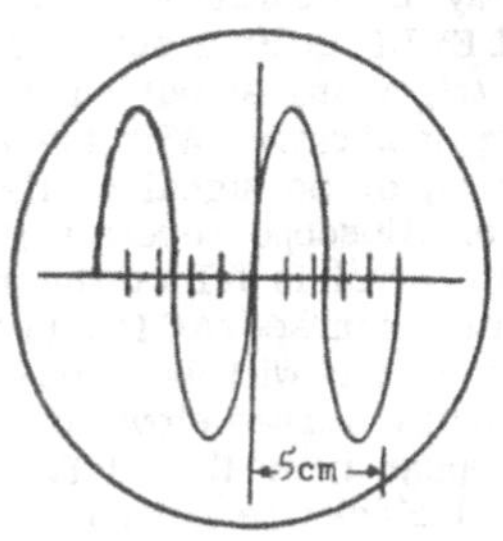

Figure 3-40.—Time measurement.

complete one cycle, and the sawtooth is set for 0.5 μsec/cm with a 10 cm wide graticule, approximately one cycle will be displayed. If the time base is set for 1 μsec/cm, approximately two cycles will be displayed. If the time base is set at a frequency higher than the input frequency, only a part of the input signal will be displayed.

In the basic oscilloscope, the sweep generator runs continuously (free-running) while in the more elaborate oscilloscopes, it is normally turned off. In the oscilloscope shown in figure 3-32, the sweep generator requires a signal from some source to start (trigger) it. This type of oscilloscope is called a "Triggered Oscilloscope." The triggered oscilloscope permits more accurate time measurements to be made and gives a more stable presentation.

Front panel controls are provided to permit the selection of the source and polarity of the trigger signal and the amplitude. In addition, provision is usually made to adjust the sweep generator to free-run. These controls are the TRIGGERING and LEVEL controls. The TRIGGERING control is a multi-position switch and the LEVEL control is a potentiometer.

The TRIGGERING control chooses the source and polarity of the trigger signal. The source may be LINE, INTERNAL, or EXTERNAL, and the polarity selected may be negative (-) or positive (+). When LINE is selected, the power line frequency (e.g., 60 Hz) is the trigger frequency. When internal (INT) is selected, part of the input signal is tapped off from the vertical deflection amplifiers, and the input signal frequency is also the trigger frequency. In external (EXT), a front panel binding post is connected through the switch to the sweep generator circuit, and a signal from any external source that is appropriate may be connected to the binding post.

The LEVEL control sets the amplitude level that the triggering signal must exceed to start the sweep generator. With the stability control misadjusted, or no signal available for triggering, the oscilloscope screen will be blank. When the level control is fully counterclockwise, it is in a position marked AUTO. In this position the sweep generator will be free-running. As the level control is turned slowly clockwise, the trace will disappear from the CRT. Continue turning the control clockwise until the trace reappears, and then stop turning. At this point, there will be a stable presentation on the screen.

The triggering and level controls are used to synchronize the sweep generator with the input signal. This gives a stationary waveform display. If they are unsynchronized, the pattern tends to jitter or move across the screen making observation difficult.

Most oscilloscopes provide a test signal to a front panel connector. This signal may be a few volts a.c. tapped from the power transformer, or it may be an accurately calibrated square wave. There may be just one panel connector with only one amplitude of voltage available, or there may be several connectors, each providing a different amplitude signal. Some models provide one connector with a switch to select any one of a wide range of amplitudes. The connector, or connectors may be labeled TEST SIGNAL, LINE, or VOLTAGE CALIBRATOR. The oscilloscope in figure 3-32 uses three jacks labeled VOLTAGE CALIB, with the output amplitude of a square wave marked over each jack. The voltages available are 5 volts, 0.5 volts, and 0.05 volts peak-to-peak. The voltage calibrator provides a known reference voltage for checking the calibration of the VOLTS/CM control. As an example, suppose the VOLTS/CM control is set to the 1 VOLT/CM position and the VARIABLE control is in the calibrated position. With the 5 volt calibrated waveform connected to the vertical input terminal, the presentation should occupy 5 centimeters in height as shown in figure 3-41.

Most oscilloscopes provide a means of connecting an external signal to the horizontal deflection amplifiers in place of the sawtooth from the sweep generator. In figure 3-32, when the TIME/CM control is rotated fully clockwise to the position marked EXT (external), the sweep generator will be disconnected from the horizontal deflection amplifiers, and a front panel binding post (HORIZ.) will be connected to the input of the amplifiers. A signal may now be connected to this binding post. The VARIABLE TIME/CM control becomes an amplitude (gain) control to provide a means of controlling the width of the trace when a signal is applied. With no signal applied to the vertical or horizontal input connectors, a small, stationary dot will be present on the CRT. CAUTION: this dot should not be allowed to remain on the CRT. Either a signal should be applied to one of the inputs, or the INTENSITY control should be turned counterclockwise until the dot disappears. Once a signal is applied, the intensity may be turned back up to view the trace. If the dot is allowed to remain on the CRT, it will damage the chemical coating and necessitate the replacement of the CRT.

The horizontal input provides a means of connecting a second signal for the purpose of comparing the phase or frequency difference

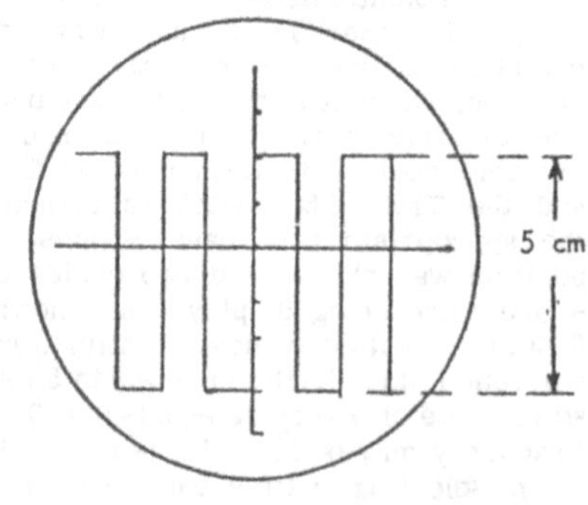

Figure 3-41.—Calibration check.

with respect to a signal applied to the vertical input. The resultant display on the CRT is called a LISSAJOUS figure.

Figure 3-42 shows the resultant pattern, and how it was formed, when two sine waves of the same frequency, but differing in phase by 90 degrees, are applied to the vertical and horizontal inputs. Notice that the pattern is a circle. Figure 3-43 shows the patterns that will result from various sine waves differing in phase. For accurate measurements, both signals should be the same amplitude on the screen. This is accomplished by applying only one signal at a time and adjusting the respective gain control.

Figure 3-44 shows the pattern that will result from applying a sine wave to the vertical input at twice the frequency of a sine wave applied to the horizontal input. To determine the actual frequency, one of the input frequencies must be known, otherwise the only information that will be obtained is the ratio of the two frequencies.

Using lissajous figures to determine an unknown frequency is accomplished by first establishing a ratio. This is done by counting the number of times the pattern touches a horizontal line, A in figure 3-44, and the number of times the pattern touches a vertical line, B in figure 3-44. In this example, the ratio is 2:1, and represents the ratio of the vertical frequency to the horizontal frequency. If the horizontal frequency was known to be 1 kHz, then the vertical frequency would be 2 kHz. The number of patterns that may be obtained is quite varied, and range from the simple to the complex.

The ILLUM (illuminate) control turns on several small light bulbs mounted around the edge of the graticule making the lines visible so that measurements of the displayed signal can be made easily.

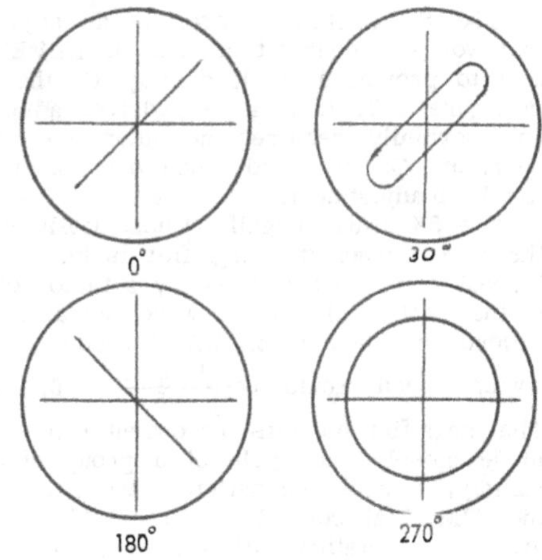

179.37
Figure 3-43.—Phase difference lissajous patterns.

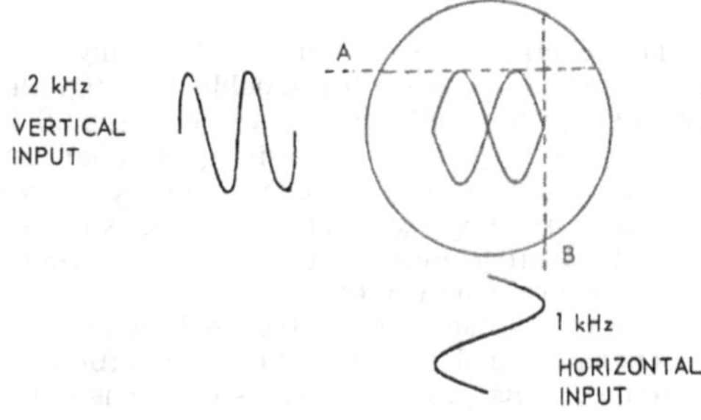

179.37
Figure 3-44.—Frequency ratio lissajous pattern.

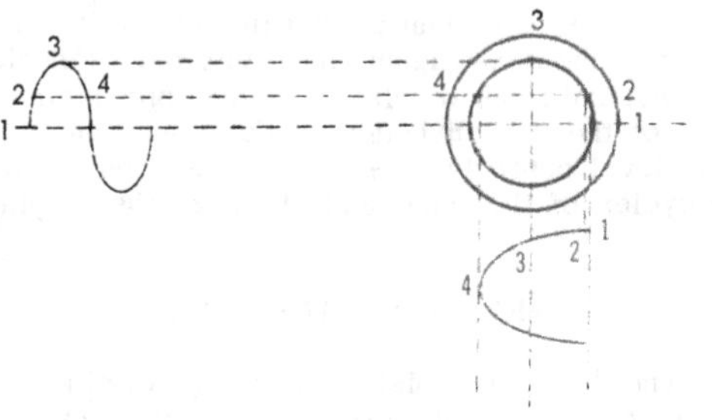

179.36
Figure 3-42.—Forming a lissajous pattern.

The DC BAL (balance) control is a screwdriver adjustment to prevent the whole trace from shifting vertically as the VARIABLE VOLTS/CM control is turned through its range. This control may be mounted on the front panel or it may be a service adjustment located inside the oscilloscope. This control, once set, only requires adjustment at periodic intervals, not every time the oscilloscope is used.

The STABILITY control is an adjustment that works in conjunction with the LEVEL control to provide a stable display. On the model in figure 3-32 it is a screwdriver adjustment and normally requires no adjustment by the operator. On more advanced oscilloscopes, it is a knob adjustment.

The 5X MAG (magnifier) essentially expands the sweep. When the magnifier is turned on, it increases the sweep speed by a factor of five. If the TIME/CM control were set at 1 millisecond, and then the 5X MAG turned on, the new sweep speed would be: $\frac{1 \text{ msec}}{5} = 0.2$ msec. The magnifier permits the operator to examine more closely one cycle of a group of cycles rapidly, without necessitating the resetting of the TIME/CM control. Figure 3-45 shows the visual presentation with the magnifier off and then turned on. On the oscilloscope in figure 3-32 the 5X MAG is turned on by pulling the horizontal positioning control straight out.

Operation

The oscilloscope is one of the many tools that a technician uses for troubleshooting. Before using the oscilloscope, a check should be made to verify that it is in good operating condition. If the oscilloscope is faulty, it will give false readings and the technician will find himself wasting many hours trying to find a trouble where none exists.

The technician should also be familiar with the operation and limitations of the oscilloscope available. Misadjusting controls will cause false readings to be obtained. Attempting to take measurements in a circuit operating beyond the capabilities of the oscilloscope will give false readings, and may even result in the instrument being damaged. When using an oscilloscope for the first time, consult the technical manual for that particular equipment to determine the correct operating procedure and any limitations.

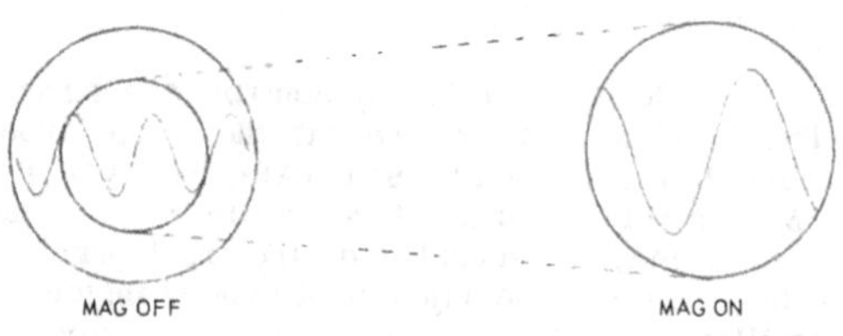

Figure 3-45.—5 X MAG presentation.

Since the oscilloscope provides a visual display which is affected by the manupulation of the controls, and usually has its own built in test signal source, its operating condition can be checked rapidly.

For the initial setup of the oscilloscope in figure 3-32, the controls should be set as listed below:

POSITION, FOCUS, AND INTENSITY controls	:	in the center of their range of rotation
TRIGGERING	:	INT +
LEVEL	:	AUTO
TIME/CM	:	1 msec
VARIABLE	:	CAL
POWER	:	ON

Allow approximately two minutes for the instrument to warm up. After the warm up period, a trace should be visible. Turn the positioning controls, one at a time, to see if the trace can be moved to the edges of the CRT face, and then center the trace. Adjust the intensity control for comfortable viewing, and the focus control for a thin, well defined line.

Turn the VOLTS/CM control to the 0.02 position and the VARIABLE control to CAL. Connect the 0.05 volt output from the voltage calibrator to the vertical input. Turn the LEVEL control from the AUTO position. As this is done, the trace will disappear. Continue turning until the trace reappears. The display should be a stable display of several cycles of a square wave, 2.5 centimeters in height. Disconnect the lead from the voltage calibrator. The equipment is now ready for use.

Before connecting a signal to the input terminal, turn the input attenuator (VOLTS/CM) to the highest attenuation setting (10 volts/cm). Connect the input signal and rotate the attenuator control until the display occupies approximately 2/3 of the screen height. Adjust the time base and level controls to select the desired number of cycles of the input and stabilize the display.

BASIC TRANSISTOR TESTER

The basic transistor tester (shown in fig. 3-46) consists of a bias source, a meter circuit, switching arrangements and a 1-kHz oscillator.

Transistors can be tested in-circuit with either of the two test leads provided. Or they

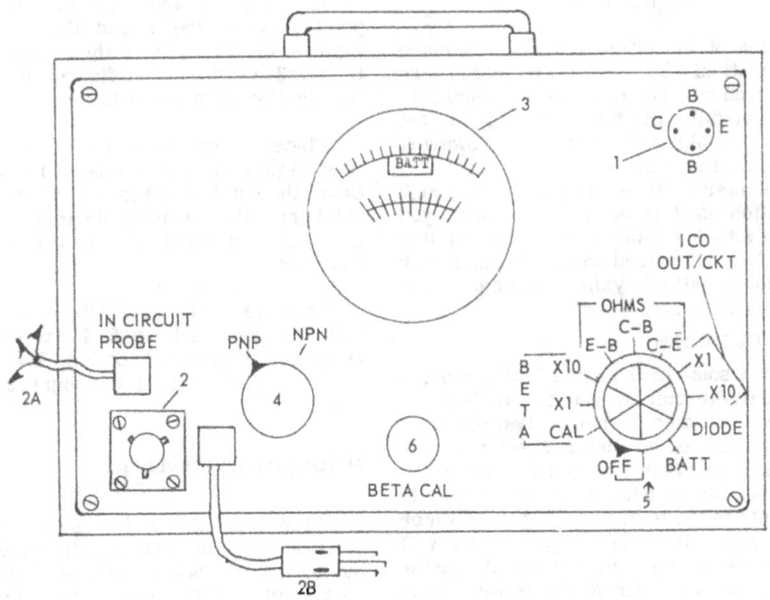

179.39
Figure 3-46.—Basic transistor tester.

can be tested out-of-circuit with either the alligator-clip test lead or the transistor socket on the front panel. If a transistor being tested out-of-circuit has solder on its leads, do not use the transistor socket. Excessive solder on the leads may damage the transistor socket.

The function of the basic transistor tester is to measure (1) current gain (Beta), (2) emitter-base, collector-base and collector-emitter resistance and (3) I_{CO} (which can be measured only when the transistor is out of the circuit).

Controls and Functions

The controls (as numbered in figure 3-46) have the following functions:

1. Transistor socket — Used for testing transistors out of the circuit. To be used only for transistors that have no solder on their leads.

2. In-circuit probe — Allows the use of alligator test leads (2A) or the "backscratcher" test prod (2B). Note: alligator test leads (2A) are used whenever doing an out-of-circuit test on a transistor that has solder on its leads.

3. Meter — Has four scales including one each for battery check, beta, I_{CO}, and resistance.

4. PNP/NPN — Internally sets up proper bias for transistor under test.

5. Function switch — Selects the particular test and range desired.

6. Beta Cal — Establishes a reference level of collector current.

MATH REVIEW

This section of the chapter deals with some of the mathematical skills necessary for understanding electronics. The mathematical subjects covered are not difficult; they are taught in the first semester of high school algebra. However, they are important as a review.

Do not be misled by the simplicity. Although the mathematics used is not very advanced, you will find many topics later in this manual that are very hard to understand unless you can work easily with mathematical symbols and can follow mathematical arguments.

APPLYING FORMULAS

You will find some of the numerical problems in this manual difficult if you do not learn to use formulas. In some instances two different formulas are required to obtain a solution to a problem. It is important that you be able to do this with ease, for in addition to being able to apply a formula to the text, you must know which formula to apply. In later chapters you will constantly be called upon to use simple mathematical formulas in order to understand electronics.

EXPONENT NOTATION

You should be skilled enough in using mathematical symbols to solve simple algebra problems at a glance. You should also be able to work with ease in exponent notation of whole numbers and fractions. (Exponent notation is frequently used in electronics. It is a method of expressing very large quantities in a small space.) In exponent notation, 6,600,000 is written 6.6×10^6 and 0.000043 is written 4.3×10^{-5}. This notation is very handy in electronics. For instance, it would be very awkward to write the equation for determining the resonant frequency of a series LCR circuit in the long form when the value of the capacitor (C) is 159 uf. by placing nine zeroes to the right of the decimal point. The long form (0.000000000159) can be quickly expressed in exponent notation as 1.59×10^{-10}. The following discussion of exponent notation will be of help if you are unfamiliar with this method, or need review.

Whole Numbers

Place a decimal point to the right of the first figure in the number. Count the number of figures to the right of this decimal point. This is the exponent of 10 in the exponent notation.

Example: 64,230,000,000. Place a decimal point between the 6 and the 4. There are 10 figures to the right of this. Thus, this number is 6.423×10^{10}. The figures 6.423 are called the significant digits of this number.

Place a decimal point to the right of the first figure in the number that is not a zero. Count the number of figures between this decimal point and the original decimal point. This is again the exponent of 10, but it is a negative exponent.

Example: 0.000257. Place a decimal point between the 2 and the 5. There are four figures between the decimal points. Thus this number is 2.57×10^{-4}. Here the significant digits are 2.57.

Multiplying and Dividing

When multiplying two numbers in exponent notation, multiply the significant digits and add the exponents. When dividing, divide the significant digits of the numerator (dividend) by the significant digits of the divisor, and subtract the exponent of the denominator from the exponent of the numerator. Study the following examples until you understand this thoroughly.

$$(2\times10^4) \times (3\times10^7) = (2\times3) \times (10^4\times10^7) = (2\times3) \times 10^{(4+7)} = 6\times10^{11}$$

$$(2\times10^9) \times (3\times10^{-11}) = (2\times3) \times (10^9\times10^{-11}) = (2\times3) \times 10^{9+(-11)} = 6\times10^{-2} = 0.06$$

$$\frac{6\times10^9}{3\times10^5} = \frac{6}{3} \times \frac{10^9}{10^5} = \frac{6}{3} \times 10^{9-5} = 2\times10^4$$

$$\frac{6\times10^4}{3\times10^{-7}} = \frac{6}{3} \times \frac{10^4}{10^{-7}} = \frac{6}{3} \times 10^{4-(-7)} = 2\times10^{11}$$

$$\frac{6\times10^{-15}}{3\times10^{-27}} = \frac{6}{3} \times \frac{10^{-15}}{10^{-27}} = \frac{6}{3} \times 10^{-15-(-27)} = 2\times10^{12}$$

The following examples illustrate the rule that when the decimal point of the significant digits is moved to the left, then the exponent of the multiplicand is raised by the same number of places that the decimal point was moved. Conversely, when the decimal point of the significant digits is moved to the right, then the exponent of the multiplicand is lowered by the same number of places that the decimal point was moved.

$$\frac{4\times10^{13}}{8\times10^{5}}=\frac{4}{8}\times\frac{10^{13}}{10^{5}}=\frac{4}{8}\times10^{13-5}=0.5\times10^{8}=5.0\times10^{7}$$

$$5.4\times10^{7} = 0.54\times10^{8} = 54\times10^{6}$$

Since students of electronics and personnel working with electronics are required to handle cumbersome numbers ranging from the extremely small fractions of electrical units to the very large numbers represented by radio frequencies, you can see that a thorough knowledge of exponent notation and the ability to apply the theory of exponents will greatly assist you in reducing unnecessary work.

Table 3-1 is furnished as an aid to assist you in acquiring the skills necessary to work rapidly and correctly with exponents. Note that table 3-1 provides only some of the values of the powers of 10.

Non-Intergral Exponents

When a root of a number is to be found, the operation may be indicated by the radical sign $a\sqrt{n}$, "a" indicating the desired root. It is frequently more useful to write this operation as the number to the reciprocal power of the desired root. For example:

$$\sqrt[2]{n} = n^{\frac{1}{2}},\ \sqrt[3]{n} = n^{1/3},\ \sqrt[1.2]{n} = n^{1/1.2}$$

Exponents of this type are treated exactly the same as integral exponents. For example:

$$\sqrt{n}\cdot\sqrt{n} = n^{½}\cdot n^{½} = n^{(½ + ½)} = n$$

$$\frac{n}{\sqrt{n}} = \frac{n^1}{n^{\frac{1}{2}}} = n^{1-\frac{1}{2}} = n^{\frac{1}{2}}$$

Raising to Powers

When a number is raised to any power, the exponent of the number is multiplied by the power. This follows from the rule of addition of exponents:

$$\sqrt{n}\cdot\sqrt{n} = n \text{ or } (n^{½})^2 = n^{½}\cdot n^{½} = n^{(½ + ½)} = n$$

$$\sqrt{\sqrt{n}} = (n^{\frac{1}{2}})^{\frac{1}{2}} = n^{\frac{1}{4}}$$

Table 3-1.—Table of powers of 10

NUMERICAL MAGNITUDE	NUMBER OF 10 FACTORS	PREFIX	SYMBOL
1 000 000 000 000 =	10^{12}	TERA	T
1 000 000 000 =	10^{9}	GIGA	G
1 000 000 =	10^{6}	MEGA	M
1 000 =	10^{3}	KILO	k
100 =	10^{2}	HECTO	h
10 =	10	DEKA	dk
0.1 =	10^{-1}	DECI	d
0.01 =	10^{-2}	CENTI	c
0.001 =	10^{-3}	MILLI	m
0.000 001 =	10^{-6}	MICRO	u or μ
0.000 000 001 =	10^{-9}	NANO	n
0.000 000 000 001 =	10^{-12}	PICO	p
0.000 000 000 000 001 =	10^{-15}	FEMTO	f
0.000 000 000 000 000 001 =	10^{-18}	ATTO	a

20.384.1

Note that when a term consisting of more than one factor is raised to a power, each factor is raised to the power. For example:

$$(a\cdot b)^2 = a^2b^2$$

$$(a\cdot10^6)^{\frac{1}{2}} = a^{\frac{1}{2}}\times10^{6/2} = a^{\frac{1}{2}}\times10^3$$

If you have experienced difficulty in understanding the above mathematical review, you are referred to the latest revision of Mathematics, Vol. 1, NavPers 10069-( ); Mathematics, Vol. 2, NavPers 10071-( ); and Mathematics, Vol. 3, NavPers 10073-( ) as sources of further study material on the mathematical processes required in the science of electronics.

Knowledge of the mathematics represented in the preceding discussion is essential to learning much of the material in this manual. Both you and the Navy will profit if you study the manual, even though you do not improve your mathematical abilities further. However, if you desire a complete understanding and mastery of electronics, you will find that knowledge of the type of mathematics discussed here is essential.

Although the use of a slide rule is not required, you will find the ability to use a slide rule of great value in arriving at speedy solutions to some of the problems in subsequent chapters.

CHAPTER 4

CHARACTERISTICS OF MATTER AND ENERGY

A technician will encounter many different types of materials in his association with electrical and electronic equipments. Thus, a basic introduction to the structure of matter and an understanding of its characteristics will enable one to know the basic components of matter and the role of these components in the production of electric current.

MASS AND ENERGY

Matter is defined as anything which has mass and occupies space. It exists as solids, liquids, or gases.

Energy is the ability to do work. Mass is the amount of matter an object possesses. Energy and mass, taken together, are indestructible but the two are interchangeable. This means that in order for mass to go out of existence (as occurs in many sub-atomic particle reactions) new energy must simultaneously become present in some form. Or, some available energy might disappear resulting in mass being "created." This may be otherwise expressed as: the total mass and energy in the universe is constant. Thus, energy and mass are different forms of the same thing. The effect on the surroundings determine the particular form at a given time, but the two forms are simply and inseparably related. The relation is stated simply in Einstein's equation:

$$E = MC^2$$

where E is energy, M is mass, and C^2 represents the speed of light squared. Since C^2 is a constant, the equation states that energy and mass are directly related. Thus, for two stationary particles (or matter) free from external forces, if the one has greater mass, it must also have greater energy, and, if one has greater energy, it must also have greater mass. The mass of a particle (or matter) stationary and free from external forces is known as its "rest mass."

A particle may be stationary yet be under the influence of equal but opposite forces; so if the cancelling forces become unbalanced, the mass would be subjected to some net force and it would be accelerated according to the second law of Newton:

$$F = MA$$

where F is the net force, M is the mass, and A is the resulting acceleration. Under these circumstances, the particle has the possibility of, or potential for, doing some work. Therefore, the particle here possesses potential energy, or energy due to position. A familiar example of potential energy is a book sitting on a table. The mass of the book is stationary because the net force on the book is zero. It is true that gravity tends to force the book down so that the book exerts a force on the table, but at the same time the table exerts an equal but oppositely directed force on the book, so that the net force on the book is zero. The book has potential energy due to its position in the gravitational field.

Another type of energy to be considered is energy of motion, known as kinetic energy. Any moving mass possesses this energy, and the amount of kinetic energy depends on its mass and its velocity. The book on the table has some potential energy but no kinetic energy with respect to the table. If the table were suddenly removed, the possibility of performing work, or potential energy of the book, would become actual work since the book would fall.

Rest energy, potential energy, and kinetic energy depend on the particle's mass. In addition, the kinetic energy depends on the particle's velocity. Energy, mass, and velocity are closely interrelated. The molecules of some substances normally move about with random velocity. This means that at any instant their directions are random and their speeds are random—some high, some low. It is the average of these velocities which in part determines their average kinetic energy and therefore what is commonly known as the temperature of the substance. If heat is added, the average velocity of the molecules increases, their average kinetic energy increases, and the substance's temperature rises.

One example of these energy, mass, and velocity relationships would be two cars, each having an equal mass of 1000 kilograms. If the first car runs on a track at 100 kilometers per hour and the second car runs at 120 kilometers per hour, the second must possess greater kinetic energy. Therefore, in order to stop both in the same length of time, the brakes of the second car would have to dissipate more energy than the first (to reduce the velocity and kinetic energy to zero).

The mass of a particle increases as its velocity increases, although the amount of increase in our relatively slow world goes virtually undetected. When velocity becomes high enough to be near the speed of light, the mass increase becomes appreciable. And, of course, as the mass and velocity change, the energy of the particle will also change.

BASIC PARTICLES OF MATTER

An atom is defined as the smallest particle into which an element can be divided and still retain the chemical properties of that element.

The atom is the smallest part of an element that enters into a chemical change, but it does so in the form of a charged particle. These charged particles are called IONS, and they are of two types—POSITIVE and NEGATIVE. A positive ion may be defined as an atom that has become positively charged. A negative ion may be defined as an atom that has become negatively charged. One of the properties of charged ions is that ions of the same charge tend to repel one another, whereas ions of unlike charge will attract one another. (The term charge has been used loosely. At present, charge will be taken to mean a quantity of electricity which can be one of two kinds, positive or negative.)

MOLECULE

The combination of two or more atoms to form the smallest part of a compound comprises a structure known as a MOLECULE. For example, when the compound water is formed, two atoms of hydrogen and one atom of oxygen combine to form a molecule of water. A single molecule is very small and is not visible to the naked eye. Therefore, a few drops of water may contain as many as a million molecules. A single molecule is the smallest particle into which the compound can be broken down and still be the same substance. Once the last molecule of a compound is divided into atoms, the substance no longer exists.

All atoms of all elements are similar in structure because their components are alike. Atoms are composed of minute particles, the discovery and the characteristics of which will now be discussed. Atoms basically consist of ELECTRONS, PROTONS, and NEUTRONS.

AN ATOMIC MODEL

Once the basic constituents of the atom are known, an attempt can be made to construct a suitable atomic model. This model must accurately represent and be compatible with all of the facts known at the time the model is constructed. Dalton, an English chemist and physicist who lived in the late 18th and early 19th century, viewed the atom as a small indestructible sphere having the ability to become firmly attached to other atomic spheres. Later and more advanced experimentation proved that tiny charged particles could be removed from inside the atom. As a result, Dalton's model could no longer be considered satisfactory.

Joseph Osgood Thompson, an American physicist, advanced the theory that atoms must have a structure since a fundamental particle may be extracted from them. He envisioned the atom as being a sphere in which were contained a sufficient number of positive and negative charges to make the overall charge of the atom neutral. Thompson's idea that the positive and negative charges were evenly distributed throughout a sphere was disproved in an experiment conducted by Sir Ernest Rutherford.

In this experiment, a narrow beam of alpha particles (positive double charged helium ions) was obtained from a sample of radium and directed through a small hole in a lead block toward a thin sheet of gold foil. If the atom were constructed as Thompson visualized, the positive alpha particles should have had their paths deflected by small amounts due to the positive charge distributed evenly through the atom. The results were hardly what was expected. Rutherford found that most of the alpha particles went right through the gold foil without being deflected at all. The remaining particles received large amounts of deflection, some as high as 180°. This could only be explained by assuming that all of the positive charge in the atom was concentrated in one area away from the negative charge. Any alpha particle coming close to this center of charge would be severely deflected, while one passing some distance away would go through the foil undeflected.

From the results of Rutherford's experiments, emerged our present concept of the structure of the atom. The atom is now believed to consist of a group of positive and neutral particles (protons and neutrons) called the NUCLEUS, surrounded by one or more negative orbital electrons. Figure 4-1 shows the arrangement of these particles for an atom of the element boron. This concept of the atom can be likened to our solar system in which the sun is the massive central body, and the planets revolve in orbits at discrete distances from the sun. The nucleus commands a position in the atom similar to the position held by the sun in the solar system. The electrons whirl about the nucleus of the atom much as the planets whirl about the sun. In both the solar system and the atom practically all the matter in the system is contained within the central body.

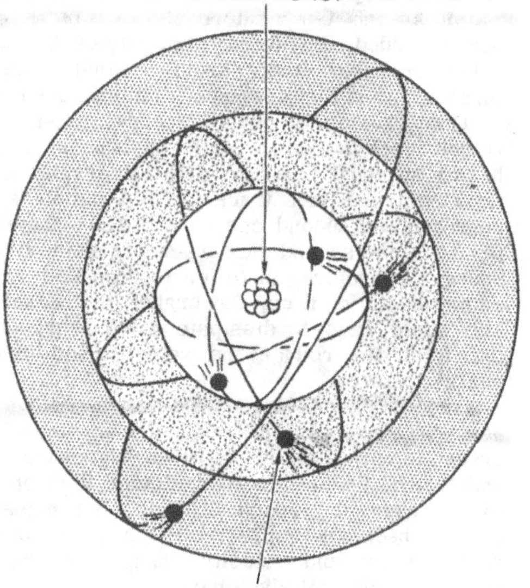

5.37:.38
Figure 4-1.—Boron atom.

THE ATOMIC NUCLEUS

Excluding short lived subatomic particles such as mesons and neutrinos, which are of little importance to electronics, the nucleus of an atom (fig. 4-1) is made up of heavy particles called PROTONS and NEUTRONS. The proton is a tiny charged particle containing the smallest known unit of positive electricity. The neutron has no electrical charge.

In the lighter elements the nucleus contains approximately one neutron for each proton while in the heavier elements there is a tendency for the neutrons to out-number the protons. The nucleus of the helium atom consists of two protons and two neutrons. In contrast an atom of mercury has eighty protons and one hundred and twenty neutrons in its nucleus.

The mass of a proton and a neutron is very nearly the same and is equal to approximately 1.67×10^{-24} gram. This mass is about 1845 times as great as the mass determined for the electron.

To obtain some idea of the relative dimensions of a typical atom, assume the atom to be expanded in size until its outer diameter is equal to twice the length of a football field. The nucleus, positioned in the center, would appear as a sphere having a diameter equal to that of a penny! This example vividly illustrates the vast emptiness which exists within the atom. One can now readily see why most of Rutherford's alpha particles streamed through the thin gold foil with little or no deflection.

THE PLANETARY ELECTRONS

Surrounding the positive nucleus of a typical atom is a cloud of negative charge made up of planetary electrons. Each of these electrons contains one unit of negative electricity equal in amount to the unit of positive electricity contained in the proton. In a normal atom the number of electrons in this cloud is exactly equal to the number of protons in the nucleus. The net charge of a normal atom is therefore zero, since the equal and opposite effects of the positive and negative charges balance one another.

If an external force is applied to an atom one or more of the outermost electrons may be removed. This is possible because the outer electrons are not attracted as strongly to the positive nucleus as are the inner electrons. When atoms combine to form an elemental substance, the outer electrons of one atom will interact with the outer electrons of neighboring atoms to form bonds between the atoms. These atomic bonds constitute the binding force which holds all matter together. When bonding occurs in some substances, each atom retains its full complement of electrons. In other substances one or more outer electrons will be gained or lost as a result of bonding. As indicated by the above statements, the electron configuration of the atom is of great importance. The chemical and electrical properties of a material are almost wholly dependent upon the electron arrangement within its atoms.

As might be expected, the nucleus is well shielded by the electron cloud and does not enter into chemical or electrical processes. To disrupt the nucleus of an atom requires a vast amount of energy such as is released by each atom in the explosion of an atomic bomb.

AN ELEMENTARY ATOM

The structure of an atom is best explained by a detailed analysis of the simplest of all atoms, that of the element hydrogen. The hydrogen atom (fig. 4-2) is composed of a nucleus containing one proton and a single planetary electron. As the electron revolves around the nucleus it is held in this orbit by two counteracting forces. One of these forces is called CENTRIFUGAL FORCE, and is the force which tends to cause the electron to fly outward as it travels in its circular orbit. This is the same force which causes a car to roll off a highway when rounding a curve at too high a speed. The second force acting on the electron is CENTRIPETAL FORCE. This force tends to pull the electron in towards the nucleus and is provided by the mutual attraction between the positive nucleus and negative electron. At some given radius the two forces will exactly balance each other providing a stable path for the electron. For the hydrogen atom this radius is approximately 5.3×10^{-11} meters.

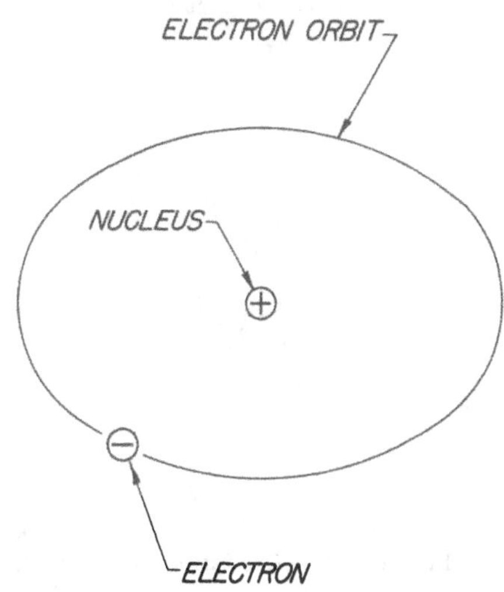

Figure 4-2.—The hydrogen atom.

Energy Levels

Since the electron in the hydrogen atom has both mass and motion it contains two types of energy. By virtue of its motion the electron contains KINETIC energy. Due to its position it also contains POTENTIAL energy. The total energy contained by the electron (kinetic plus potential) is the factor which determines the radius of the electron orbit. The orbit shown in figure 4-2 is the smallest possible orbit the hydrogen electron can have. In order for the electron to remain in this orbit it must neither gain nor lose energy.

Light energy exists in tiny packets or bundles of energy called photons. Each photon contains a definite amount of energy depending on the color (wavelength) of light it represents. Should a photon of sufficient energy collide with the orbital hydrogen electron, the electron will absorb the photon's energy as shown in figure 4-3. The electron which now has a greater than normal amount of energy will jump to a new orbit farther from the nucleus. The first new orbit to which the electron can jump has a radius four times as large as the radius of the original orbit. Had the electron received a greater amount of energy, the next possible orbit to which it could jump would have a radius nine times the

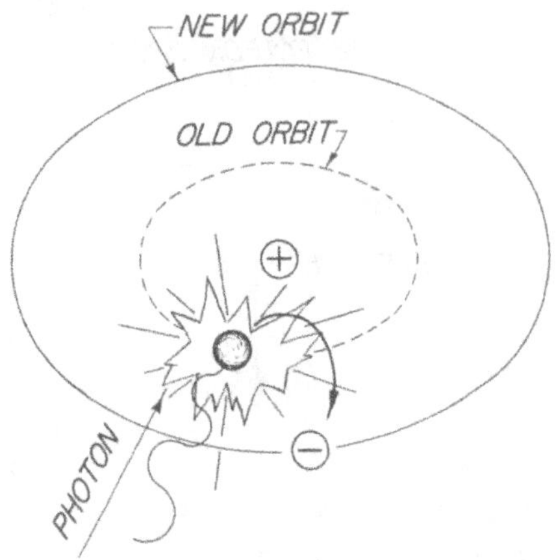

179.40
Figure 4-3.— Excitation by a photon.

original. Thus, each orbit may be considered to represent one of a large number of energy levels that the electron may attain. It must be emphasized that the electron cannot jump to just any ORBIT. The electron will remain in its lowest orbit until a sufficient amount of energy is available, at which time the electron will accept the energy and jump to one of a series of PERMISSIBLE orbits. An electron cannot exist in the space between permissible orbits or energy levels. This indicates that the electron will not accept a photon of energy unless it contains enough energy to elevate the electron to one of the allowed energy levels. Heat energy and collisions with other particles can also cause the electron to jump orbits.

Once the electron has been elevated to an energy level higher than the lowest possible energy level, the atom is said to be in an EXCITED state. The electron will not remain in this excited condition for more than a fraction of a second before it will radiate the excess energy and return to a lower energy orbit. To illustrate this principle assume that a normal electron has just received a photon of energy sufficient to raise it from the first to the third energy level. In a short period of time the electron may jump back to the first level emitting a new photon identical to the one it received.

A second alternative would be for the electron to return to the lower level in two jumps; from the third to the second, and then from the second to the first. In this case the electron would emit two photons, one for each jump. Each of these photons would have less energy than the original photon which excited the electron and would represent a longer wavelength of light.

This principle is used in the fluorescent light where ultraviolet light photons, which are not visible to the human eye, bombard a phosphor coating on the inside of a glass tube. The phosphor electrons in returning to their normal orbits emit photons of light that are of a visible wavelength (longer wavelength). By using the proper chemicals for the phosphor coating any color of light may be obtained, including white (all colors combined). This same principle is also used in lighting up the screen of a television picture tube.

Complex Atoms

Although hydrogen has the simplest of all atoms, the basic principles just developed apply equally well to the atoms of more complex elements. The manner in which the orbits are established in an atom containing more than one electron is somewhat complicated and is part of a science known as quantum mechanics. In an atom containing two or more electrons, the electrons interact with each other and the exact path of any one electron is very difficult to predict. However, each electron will lie in a specific energy band and the above mentioned orbits will be considered as an average of the electrons positions.

Shells and Subshells

The difference between the atoms, insofar as their chemical activity and stability is concerned, is dependent upon the number and position of the particles included within the atom. Atoms range from the simplest, the hydrogen atom containing one proton and one electron, to the very complex atomic structures such as silver containing forty-seven protons and forty-seven electrons. How then are these electrons positioned within the atom? In general, the electrons reside in groups of orbits called shells. These shells are elliptically shaped and are assumed to be located at fixed intervals. Thus, the shells are arranged in steps that correspond to fixed energy levels. The shells, and the number of electrons required to fill them, may be predicted by the employment of PAULI'S EXCLUSION PRINCIPLE. Simply stated, this principle specifies that each shell may contain no more than $2n^2$ electrons, where (n) corresponds to the shell number starting with the one closest to the nucleus. By this principle the second shell for example, would contain $2(2)^2$ or 8 electrons when full. In addition to being numbered, the shells are also given letter designations as pictured in figure 4-4.

Starting with the shell closest to the nucleus and progressing outward, the shells are labeled K, L, M, N, O, P, and Q respectively. The shells are considered to be full or complete when they contain the following quantities of electrons: two in the K shell, eight in the L shell, eighteen in the M shell, and 32 in the N shell. (The formula $2n^2$ can be used to determine the number of electrons only in the four shells closest to the nucleus of an atom. Succeeding shells have, as maximum number of electrons; O shell—18 electrons, P shell—12 electrons, and Q shell—2 electrons.) Each of these shells is a major shell and can be divided into subshells of which there are four labeled s, p, d, and f. A subshell exists at a given energy level (that is, at a given distance from the nucleus). There may be one or more subshells at a specified distance from the nucleus with the electron(s) of each moving in a different direction.

Like the major shells, the subshells are also limited as to the number of electrons which they can contain. Thus, the s subshell is complete when it contains two electrons, the p subshell when it contains six, the d subshell when it contains ten, and the last subshell when it contains fourteen electrons.

Inasmuch as the K shell can contain no more than two electrons, it must have only one subshell, the s subshell. The M shell is composed of three subshells: s, p, and d. If the electrons in the s, p, and d subshells are added, their total is found to be eighteen the exact number required to fill the M shell. This relationship exists between the shells and subshells up to and including the N shell. Notice the difference between the electron configurations for copper and sodium illustrated in figure 4-5.

Atomic Weight and Atomic Number

There are a wide variety of atoms, a different type of atom comprising each of the 104 known elements. Each atom is similar in that all atoms consist of protons and electrons. However, atoms of different elements contain varying numbers of basic particles, thus causing a difference in weight. A classification, based on the ATOMIC WEIGHT AND ATOMIC

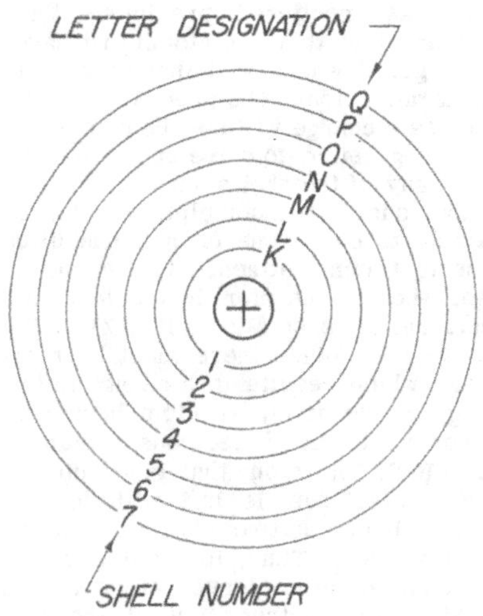

179.41
Figure 4-4.—Shell designation.

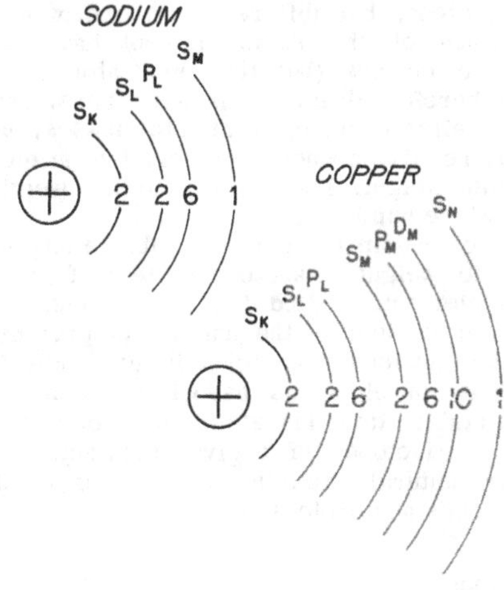

179.42
Figure 4-5.—Copper and sodium atoms.

NUMBER of the atoms, has been devised to differentiate between different atoms.

Although atoms are far too small to be weighed, a system has been set up whereby the weight of one atom is given with reference to a universally accepted standard. This system of weights, called the atomic weight of the elements, uses the element oxygen as a reference. The atomic weight of oxygen is assigned a numerical value of sixteen, and the atomic weights of other elements are determined by comparison with oxygen. No element will have an atomic weight less than one. The lightest element known, hydrogen, has an atomic weight equal to 1.008. The atomic weight of the atoms of different elements is found in the periodic table (shown in the appendix).

As previously stated, the proton and the neutron have very nearly the same mass. If the mass of a proton represents an atomic weight of one, the mass of an electron (1/1845 that of a proton) is negligible. The mass of an atom would therefore be governed by the number of protons and neutrons in its nucleus. An element such as the gas helium, with an atomic weight of 4.003, obtains its mass from two protons and two neutrons in its nucleus.

Two atoms of the same element may not have the same number of neutrons in their nuclei. These atoms are isotopes. Isotopes are defined as atoms whose nuclei have the same number of protons, but different numbers of neutrons. Isotopes of the same element have different atomic masses (but the same atomic number) and therefore different mass numbers. The average weight of isotopes' atomic masses, based on their relative abundance, determines an element's atomic weight. This is why atomic weights are not whole numbers.

Included in the periodic table along with the atomic weight is another system of classifying the elements called the atomic number. The atomic number is the number of protons found in the nucleus, therefore, it also indicates the number of electrons associated with the atom in a stable state. The atomic number is the same for all isotopes of a given element. Atoms in their natural state have an equal number of electrons and protons.

Valence

The number of electrons in the outermost shell determines the VALENCE of the atom. For this reason, the outer shell of an atom is called the VALENCE SHELL; and the electrons contained in this shell are called the valence electrons. The valence of an atom determines its ability to gain or lose an electron, which, in turn, determines the chemical and electrical properties of the atom. An atom that is lacking only one or two electrons from its outer shell will easily gain electrons to complete its shell, but a large amount of energy is required to free any of its electrons. An atom having a relatively small number of electrons in its outer shell in comparison to the number of electrons required to fill the shell will easily lose the valence electrons.

The valence shell always refers to the outermost shell, whether it be a major shell or a subshell. The copper and sodium atoms each have one electron in the outermost shell. Even though the atomic weights and atomic numbers of copper and sodium are quite different, the atoms are similar in that they both contain one valence electron. Since copper (Cu) and sodium (Na) have one valence electron (refer to fig. 4-5), they both appear in group one of the periodic table. Group one designates all elements having one valence electron.

Ions and Ionization

It was mentioned previously that ions do exist and that they are atoms that have assumed a charge. It was stated that there are positive and negative ions. The process whereby an atom acquires a charge will now be discussed.

It is possible to drive one or more electrons out of any of the shells surrounding the nucleus. In the case of incomplete shells, it is also possible to cause one or more additional electrons to become attached to the atom. In either case, whether the atom loses electrons or gains electrons, it is said to be IONIZED. For ionization to take place there must be a transfer of energy which results in a change in the internal energy of the atom. An atom having more than its normal count of electrons is called a NEGATIVE ION. The atom that gives up some of its normal electrons is left with fewer negative charges than positive charges and is called a POSITIVE ION. Thus, ionization is the process by which an atom losses or gains electrons.

To drive electrons out of the shells of an atom requires that the internal energy of the atom be raised. The amount of energy required to free electrons from an individual atom is called the ionization potential.

The ionization potential necessary to free an electron from an inner shell is much greater than that required to free an electron from an outer shell. Also, more energy is required to remove an electron from a complete shell than an unfilled shell.

CRYSTAL STRUCTURE

Now that all matter has been shown to consist of a fundamental unit called an atom, the arrangement of the atoms within a material may be investigated. Practically all of the inorganic (non-living) solids occur in CRYSTALLINE form. Even materials like iron, copper, and aluminum are crystalline in nature. A piece of iron is made of a great number of crystals lying in random positions throughout the material. A substance composed of a large number of crystals is called a POLYCRYSTALLINE material.

Crystal Lattice

If one were to examine common table salt under a magnifying glass the small grains would appear as tiny cubes of salt. Each of these cubes has a precise atomic structure and constitutes a single crystal of salt. The arrangement of atoms in a salt (sodium chloride) crystal is shown in figure 4-6. In a salt crystal the atoms become ionized as the crystal is formed. The lines between the ions of sodium and chloride represent the chemical bonds which hold the crystal together. Due to the way in which the bonds form, every perfect crystal will be like every other crystal. This precise repeating arrangement of atoms within a crystal is called a crystal LATTICE. The physical properties of a material (hardness, tensile strength, etc.) are to a great degree dependent upon the lattice structure of the material.

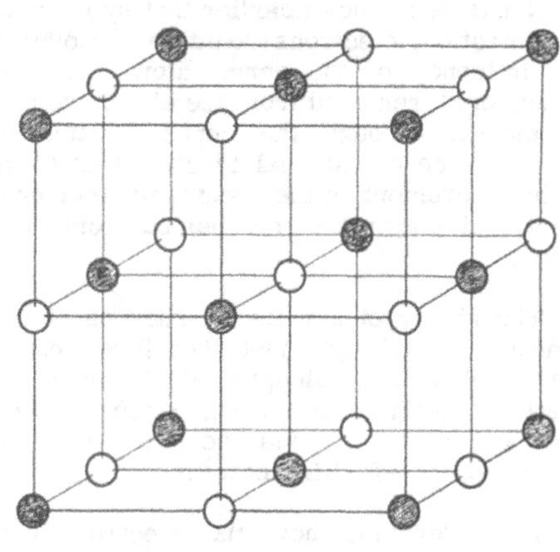

179.43

Figure 4-6.—Atomic lattice structure of salt.

Conductors, Semiconductors and Insulators

In the study of electronics, the association of matter and electricity is of paramount importance. Since every electronic device is constructed of parts made from ordinary matter, the effects of electricity on matter must be well understood. As a means of accomplishing this, all the elements of which matter is made may be placed into one of three categories: CONDUCTORS, SEMICONDUCTORS, and INSULATORS. Conductors for example, are elements such as copper and silver which will conduct a flow of electricity very readily. Due to their good conducting abilities they are formed into wire and used whenever it is desired to transfer electrical energy from one point to another. Insulators (non-conductors) on the other hand, do not conduct electricity to any great degree and are therefore used when it is desirable to prevent a flow of electricity. Substances such as sulphur, rubber, and glass are good insulators. Materials such as germanium and silicon are not good conductors but cannot be used as insulators either, since their electrical characteristics fall between those of conductors and insulators. These in-between materials which do not make good conductors, or good insulators are classified as semiconductors.

The electrical conductivity of matter is ultimately dependent upon the energy levels of the atoms of which the material is constructed. In any solid material such as copper, the atoms which make up the molecular structure are bound together in the crystal lattice. Since the atoms of copper are firmly fixed in position within the lattice structure, they are not free to migrate through the material, and therefore cannot carry the electricity through the conductor without application of external forces as will be explained in the following discussion.

Free Electrons

It has been shown earlier that by the process of ionization, electrons could be removed from the influence of the parent atom. These electrons, once removed from the atom, are capable of moving through the copper lattice under the influence of external forces. It is by virtue of the movement of these charged electrons that electrical energy is transported from place to place.

The ability of a material such as copper to conduct electricity must therefore depend on the number of dislodged electrons normally available within the lattice. Since copper is a good conductor, it must contain vast numbers of dislodged or FREE electrons.

To understand how the electrons become free, it is necessary to refer back to the electron energy levels within the atom. It was previously stated that if precisely the right amount of energy were added to an orbital electron, it would jump to a new orbit located farther from the nucleus. If the energy is sufficiently large, the jump may carry the electron to such a distance from the positive nucleus that the electron becomes free. Once free, the electron constitutes the charge carrier discussed above. The only problem remaining is to determine how the electron in the piece of copper obtains enough energy to become free.

After a moments consideration, a person realizes that the average piece of copper contains some amount of heat energy. In fact, a piece of copper at room temperature (72°F.) is approximately 531° F. above absolute zero! This temperature indicates that the copper, although only warm to the touch, must contain a considerable amount of heat energy. The phonons of heat energy, along with other forms of natural radiation, elevate the electrons to the energy levels where they can become free.

Energy Gaps

From the preceding theories, one might wonder why all materials containing the same amount of heat energy do not conduct electricity equally well. The answer lies in the fact that the electrons in various materials require different amounts of energy to become free. This idea may be best developed by using energy level diagrams like the one in figure 4-7. In this model, the outer shell is depicted as having two energy bands called the valence bond band and the conduction band. Between these two energy bands is an energy gap called the forbidden gap or forbidden band. Electrons residing in the lower band are considered to be firmly attached to the parent atoms and are not available for the conduction of electricity. In order for an electron to become a free electron, it must gain enough energy from external forces to jump the forbidden gap and appear in the

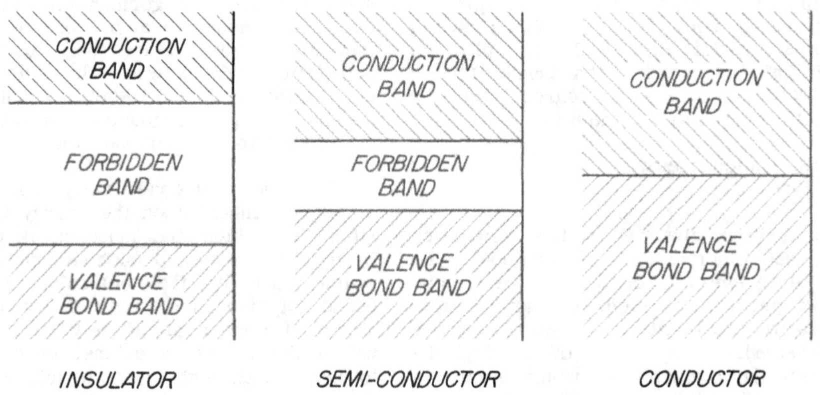

179.44
Figure 4-7.— Energy level diagrams.

conduction band. Once in the conduction band, the electron is free and may be made to move along through the conductor in the form of an electric current. The energy diagram for the insulator shows the insulator to have a very wide energy gap. This means that a large amount of energy must be added to each electron in an insulating material before it can become free. Thus at room temperature sufficient energy is not available to cause electrons to jump to the conduction band and the material has practically no free electrons. In comparing the energy level diagrams for an insulator and a conductor, the conductor is seen to have little or no forbidden gap. Since this is true, under normal conditions the conduction band for a conductor contains a sufficient number of free electrons to make it a good conductor of electricity.

The semiconductor being neither a good conductor nor a good insulator has an energy gap which, on the energy level diagram, has a width between that of a conductor and that of an insulator.

In the following discussion the role of the conductor, semiconductor, and insulator will assume greater and greater importance as the various electronic devices are developed and discussed. In fact, in the final analysis, all electronic phenomena are based on the electrical nature of matter.

TWO KINDS OF CURRENT

The degree of difficulty in dislodging valence electrons from the nucleus of an atom determines whether the element is a conductor, semiconductor, or an insulator. When an electron is freed in a block of pure semiconductor material, it creates a "hole" which acts as a positively charged current carrier. Thus, electron liberation creates two currents, known as electron current and hole current.

Holes and electrons do not necessarily travel at the same rate, and when an electric field is applied, they are accelerated in opposite directions. The life spans (time until recombination) of a hole and a free electron in a given semiconductor sample are not necessarily the same. Hole conduction may be thought of as the unfilled tracks of a moving electron. Because the hole is a region of net positive charge, the apparent motion is like the flow of particles having a positive charge. An analogy of hole motion is the movement of the hole as balls are moved through a tube (fig. 4-8). When ball number 1 is removed from the tube, a space is left. This space is then filled by ball number 2. Ball number 3 then moves into the space left by ball number 2. This action continues until all the balls have moved one space to the left at which time there is a space left by ball number 8 at the right-hand end of the tube.

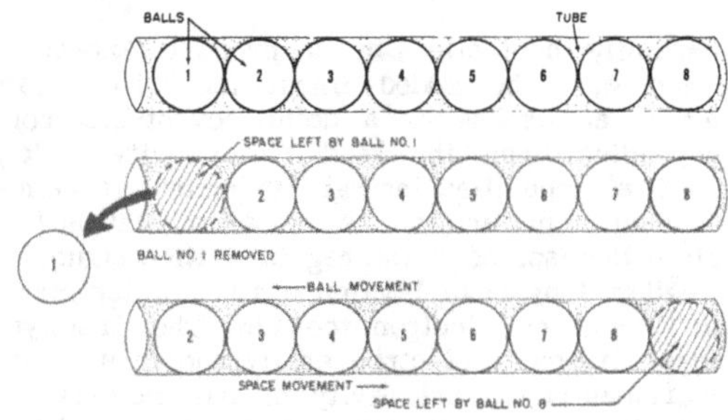

20.21

Figure 4-8.—Analogy of hole movement.

A pure specimen of semiconductor material will have an equal number of free electrons and holes, the number depending on the temperature of the material and the type and size of the specimen. Such a specimen is called an intrinsic semiconductor; and the current, which is borne equally by hole conduction and electron conduction, is called intrinsic conduction.

If a suitable "impurity" is added to the semiconductor, the resulting mixture can be made to have either an excess of electrons, thus causing more electron current, or an excess of holes, thus causing more hole current. An impure specimen of semiconductor material is known as an extrinsic semiconductor.

IMPURITY DONORS AND ACCEPTORS

In the pure form, semiconductor materials are of little use in electronics. When a certain amount of impurity is added however, the material will have more (or less) free electrons than holes depending upon the kind of impurity added. Both forms of conduction will be present, but the "majority carrier" will be dominant. The holes are called positive carriers, and the electrons negative carriers. The one present in the greatest quantity is called the majority carrier; the other is called the minority carrier. The quality and quantity of the impurity are carefully controlled by a process known as "doping." The added impurities will create either an excess or a deficiency of electrons depending upon the kind of impurity added.

The impurities that are important in semiconductor materials are those impurities that aline themselves in the regular lattice structure whether they have 1 valence electron too many, or 1 valence electron too few. The first type loses its extra electron easily and in so doing increases the conductivity of the material by contributing a free electron. This type of impurity has 5 valence electrons and is called a pentavalent impurity. Arsenic, antimony, bismuth, and phosphorous are pentavalent impurities. Because these materials give up or donate 1 electron to the material they are called donor impurities.

The second type of impurity tends to compensate for its deficiency of 1 valence electron by acquiring an electron from its neighbor. Impurities of this type in the lattice structure have only 3 electrons and are called trivalent impurities. Aluminum, indium, gallium, and boron are trivalent impurities. Because these materials accept 1 electron from the material they are called acceptor impurities.

N-TYPE GERMANIUM

When a pentavalent (donor) impurity like arsenic is added to germanium it will form covalent bonds with the germanium atoms. Figure 4-9A illustrates an arsenic atom (As) in a germanium lattice structure. The arsenic atom has 5 valence electrons in its outer shell but uses only 4 of them to form covalent bonds with the germanium atoms, leaving 1 electron relatively free in the crystal structure. Because this type of material conducts by electron movement it is called a negative-carrier type or N-type semiconductor. Pure germanium may be converted into an N-type semiconductor by "doping" it with a donor impurity consisting of any element containing 5 electrons in its outer shell. The amount of the impurity added is very small; it is of the order of 1 atom of impurity in 10 million atoms of germanium.

P-TYPE GERMANIUM

A trivalent (acceptor) impurity element can also be added to pure germanium to "dope" the material. In this case the impurity has 1 less electron than it needs to establish covalent bonds with 4 neighboring atoms. Thus in 1 covalent bond there will be only 1 electron instead of 2. This arrangement leaves a hole in that covalent bond.

Figure 4-9B, shows the germanium lattice structure with the addition of an indium atom (In). The indium atom has 1 electron less than it needs to form covalent bonds with the 4 neighboring atoms and thus creates a hole in the structure. Gallium and boron also exhibit these characteristics. The holes are present

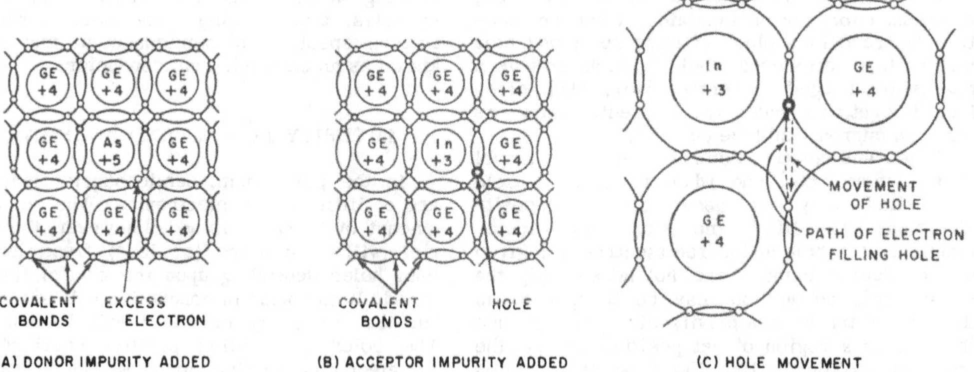

Figure 4-9.—Germanium lattice with impurities added.

only if a trivalent impurity is used. Note that a hole carrier is not created by the removal of an electron from a neutral atom, but is created when a trivalent impurity enters into covalent bonds with a tetravalent (4 valence electrons) crystal structure. Because this semiconductor material conducts by the movement of holes which are positive charges, it is called a positive carrier-type or P-type semiconductor. When an electron fills a hole (fig. 4-9C) the hole appears to move to the spot previously occupied by the electron.

CHARGES IN N AND P TYPE MATERIALS

When a donor material such as arsenic is added to germanium, the fifth electron in the outer ring of the arsenic atom does not become a part of a covalent bond. This extra electron (when acted on by some force) may move away from the arsenic atom to one of the nearby germanium atoms in the N-type material.

The arsenic atom has a positive charge of 5 units on the inner circle, as shown in figure 4-9A, and when the electron moves away from the arsenic atom there will be only 4 electrons to neutralize the positive charge and as a result there will be a region of positive charge around the arsenic atom. Similarly, the excess electron that has moved into the germanium atom outer shell makes a total of 5 electrons instead of 4 electrons for that atom of germanium. Thus, there is a region of negative charge around this atom.

Although there is a region of positive charge around the arsenic atom after the electron has moved away, and a region of negative charge around the germanium atom with the extra electron, the total charge on the N-type crystal remains the same. In other words the total charge is zero. There are exactly enough electrons to neutralize the positive charges on the nuclei of all the atoms in the crystal. However, because some of the electrons may move about in the crystal, there will be regions in the crystal where there are negative charges and other regions where there will be positive charges, even though the net charge on the crystal is zero.

In a P-type material having an impurity such as indium added to it, a similar situation may exist. Indium has only 3 electrons in its outer ring. Three electrons are all that are needed to neutralize the net positive charge of 3 units on the inner circle (fig. 4-9B). However, with only 3 electrons in the outer shell, there is a hole in one of the covalent bonds formed between the indium atom and the 4 adjacent germanium atoms. If an electron moves in to fill this hole (fig. 4-9C) there is one more electron in the indium atom than is needed to neutralize the positive charge of 3 units. Thus there will be a region of negative charge around the indium atom.

Similarly, if one of the germanium atoms gives up an electron to fill the hole in the covalent bond, the germanium atom will be short an electron and there will be a region of positive charge around this atom. While the giving up of an electron by a germanium atom and the acquisition of an electron by the indium atom charges (ionizes) both atoms involved, the net charge on the P-type crystal is still zero. There is one atom that is short an electron and another atom that has one too many. The crystal itself does not acquire any charge.

These ionized atoms produced in both N- and P-type germanium are not concentrated in any one part of the crystal, but instead are spread uniformly throughout the crystal. If any region within the crystal were to have a very large number of positively charged atoms, these atoms would attract free electrons from other parts of the crystal to neutralize part of the charged atoms so that the charge would spread uniformly throughout the crystal. Similarly, if a large number of atoms within a small region had an excess of electrons, these electrons would repel each other and spread throughout the crystal.

As stated previously, both holes and electrons are involved in conduction. In N-type material the electrons are the majority carriers and holes the minority carriers. In P-type material the holes are the majority carriers and electrons the minority carriers.

CURRENT FLOW IN N-TYPE MATERIAL

Current flow through an N-type material is illustrated in figure 4-10. Conduction in this type of semiconductor is similar to conduction in a copper conductor. That is, the application of voltage across the material will cause the loosely bound electron to be released from the impurity

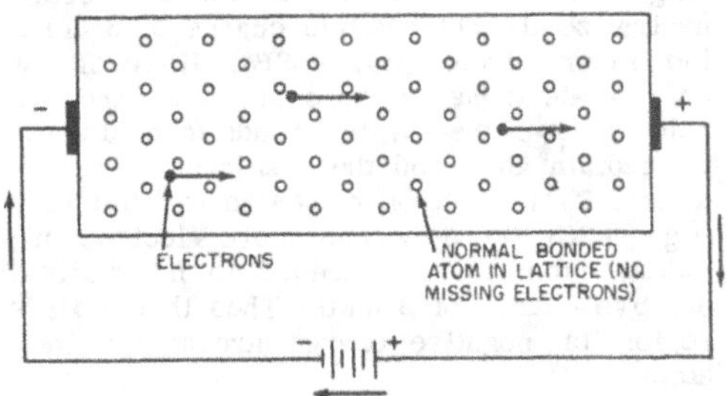

Figure 4-10.—Current flow in N-type material.

atom and move toward the positive potential point.

Certain differences do exist however, between the N-type semiconductor and a copper conductor. For example, the semiconductor resistance decreases with temperature increase, because more carriers are made available at higher temperatures. Increasing the temperature releases electrons from more of the impurity atoms in the lattice causing increased conductivity (decreased resistance). In the copper conductor, increasing the temperature does not increase the number of carriers but increases the thermal agitation or vibration of the structure so as to impede the current flow further (increase the resistance).

CURRENT FLOW IN P-TYPE MATERIAL

Current flow through a P-type material is illustrated in figure 4-11. Conduction in this material is by positive carriers (holes) from the positive to the negative terminal. Electrons from the negative terminal cancel holes in the vicinity of the terminal while at the positive terminal, electrons are being removed from the covalent bonds, thus creating new holes. The new holes then move toward the negative terminal (the electrons shifting to the positive terminal) and are canceled by more electrons emitted from the negative terminal. This process continues as a steady steam of holes (hole current) moving toward the negative terminal.

In both N-type and P-type materials, current flow in the external circuit is out of the negative terminal of the battery and into the positive terminal.

PN JUNCTIONS

Elements whose atoms contain four valence electrons are classified as semiconductors. Examples of such elements are germanium and silicon. These semiconductor materials are of little use in electronics in their pure or intrinsic form. However, when they are tainted or doped with a small amount of impurity material, they form the basis for a myriad of solid state

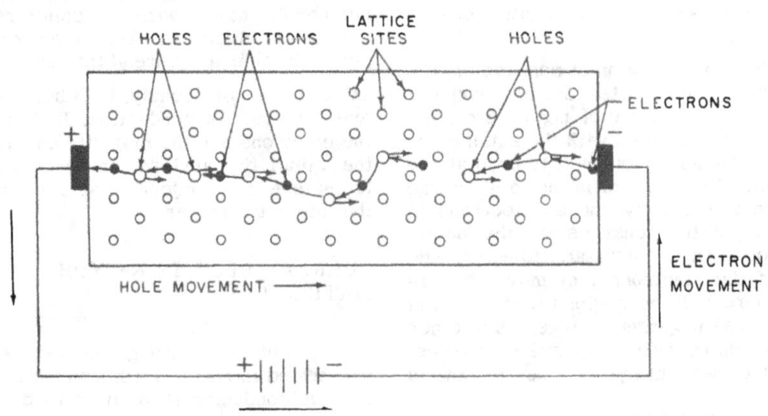

Figure 4-11.— Current flow in P-type material.

electronic devices. The impurity elements, added to semiconductor material, fall into one of two categories: Pentavalent—those with five valence electrons; and Trivalent—those with three valence electrons. When a pentavalent impurity is added to a semiconductor material, the result is called N type material. When a trivalent impurity is added, P type material is formed.

THE BASIC PN JUNCTION

Both N type and P type semiconductor material are electrically neutral. However, a block of intrinsic semiconductor material may be doped with pentavalent and trivalent impurities, so as to make half the crystal N material and the other half P material. A force will then exist across the junction of the N and P material. The force is an electro-chemical attraction by the P material for electrons in the N material. This force exists because the trivalent impurity has caused a deficiency of electrons within the structure of the P material, while the pentavalent impurity has caused an excess of electrons within the structure of the N material.

Due to the above mentioned force, electrons will be caused to leave the N material and enter the P material. This will make the N material in proximity to the junction positive with respect to the remainder of the N material. Also the P material in proximity to junction will become negative with respect to the remaining P material. This is illustrated in figure 4-12.

After the initial movement of charges, further migration of electrons ceases, due to the equalization of electron concentration in the immediate vicinity of the junction. The charged areas on either side of the junction constitutes a potential barrier, or junction barrier, which prevents further current flow. This region is also called a depletion region.

Several facts must be emphasized. The junction barrier exists only for a minute distance on either side of the junction. The formation of a barrier occurs only in a homogeneous crystal which has been properly doped. That is, doping two separate sections of crystal and then placing them in contact will not produce the desired phonomenon. Finally, the barrier is formed at the instant the crystal is manufactured, and the magnitude of the barrier is a function of the particular crystal.

BIASING PN JUNCTIONS

The device described above is a semiconductor junction diode. The schematic symbol for a semiconductor diode is illustrated in figure 4-13. This device allows appreciable current flow in one direction, while restricting current flow to an almost negligible value in the other direction. The N material section of the device is called the cathode, and the P material section is called the anode. The device permits current flow from cathode to anode, and restricts current flow from anode to cathode. This action will be described in the following paragraphs.

Consider the case wherein a potential is placed, externally, across the diode, positive on the anode with respect to the cathode. This is depicted in figure 4-14. The applied voltage, called bias, is in opposition to the junction barrier potential. If this voltage is increased from zero, the junction barrier will be progressively reduced, and current flow through the device will increase. This is depicted in figure 4-15. Eventually, the barrier will be eliminated, and current

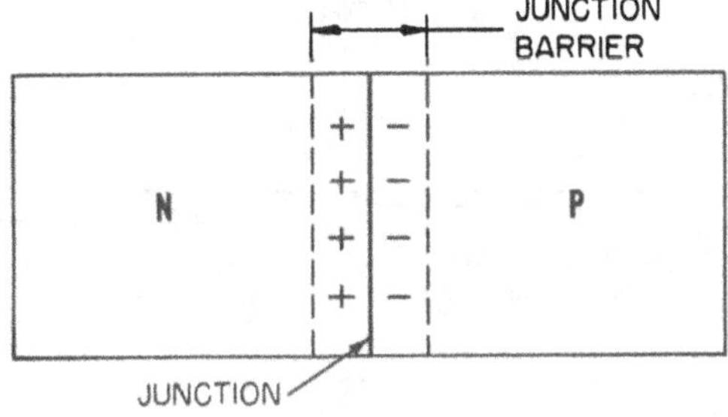

20.23

Figure 4-12.—PN junction.

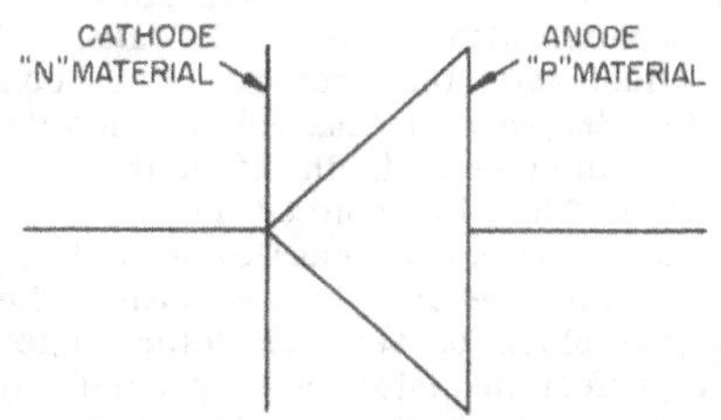

1:9.45

Figure 4-13.—Schematic symbol for a semiconductor diode.

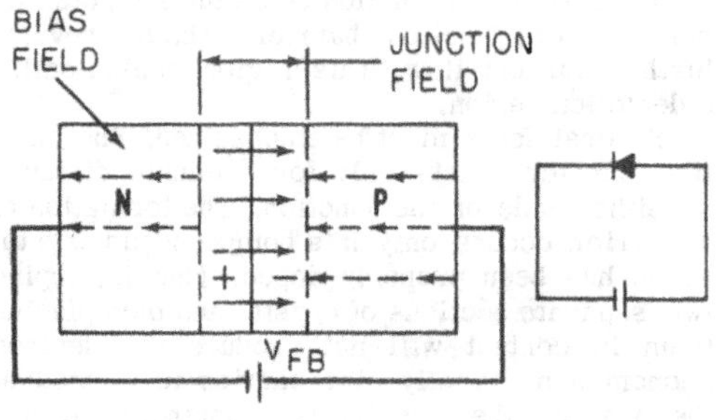

Figure 4-14.—Forward bias.

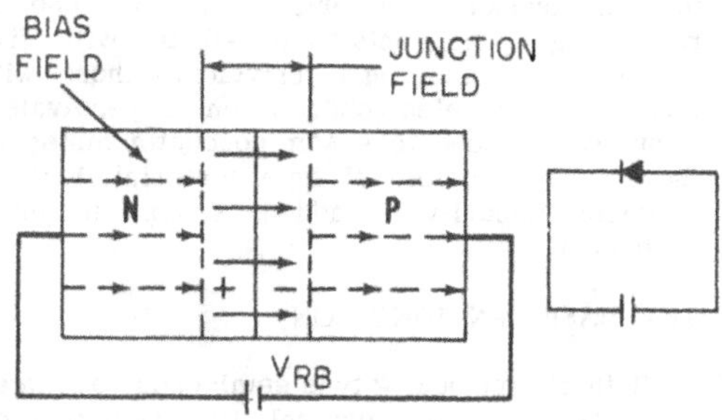

Figure 4-16.—Reverse bias.

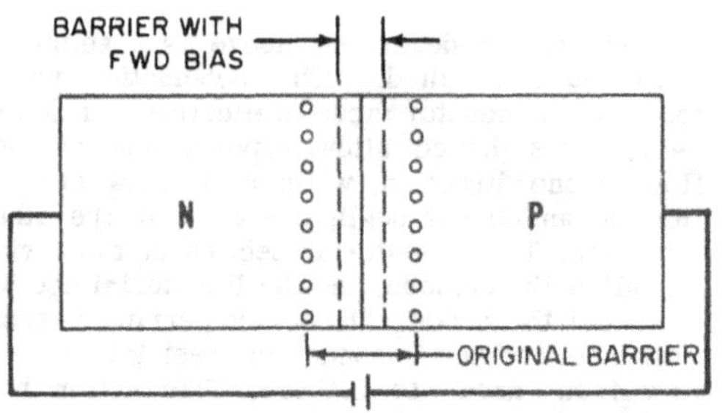

Figure 4-15.—Effect of forward bias on barrier width.

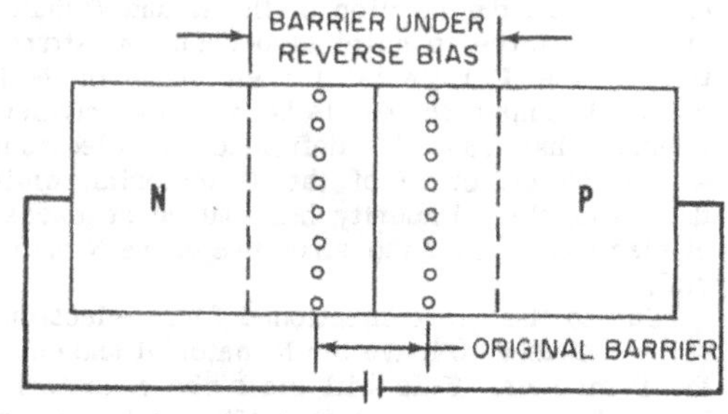

Figure 4-17.—Effect of reverse bias on barrier width.

flow will increase rapidly with an increase in voltage. This polarity of voltage (anode positive with respect to the cathode) is called forward bias, since it causes the device to conduct an appreciable current flow.

Next, consider the case wherein the anode is made negative with respect to the cathode. Figure 4-16 illustrates this reverse bias condition. Note that the reverse bias voltage aids the junction barrier potential. In effect the barrier is increased. This is depicted in figure 4-17. It would seem that no current flow should be possible under this reverse bias condition. However, since the block of semiconductor material is not a perfect insulator, a very small reverse, or leakage current will flow. At normal operating temperatures this current may be neglected. It is noteworthy, however, that leakage current increases with an increase in temperature.

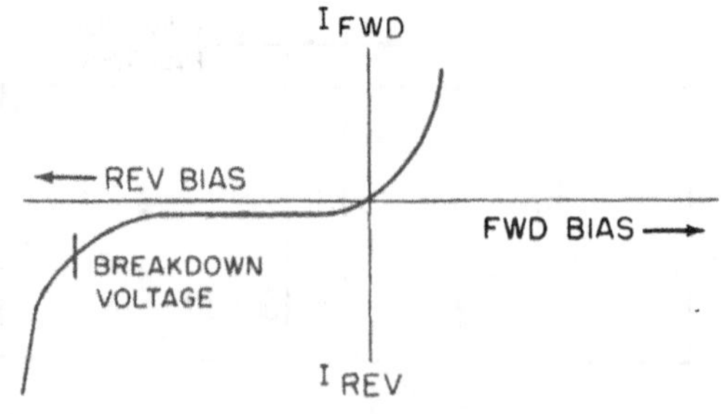

Figure 4-18.—Semiconductor diode characteristic curve.

Figure 4-18 is a graph of the current flow through a semiconductor diode, plotted by values of anode to cathode voltage. Note that the forward current increases slowly at low values of forward bias. As the forward bias is increased, the barrier is neutralized, and current increased rapidly for further increases in applied voltage. It should be noted that excessive forward bias could destroy the device through excessive forward current.

With reverse bias applied, a very small reverse current exists. This reverse current increases minutely, with an increase in reverse current. However, if an excessive reverse voltage is applied, the structure of the semiconductor material may be broken down by the resulting high electric stress, and the device may be terminated. The value of reverse voltage at which this breakdown occurs is called the avalanche or breakdown voltage. At this voltage, current increases rapidly with small increases in reverse bias. This region of rapidly increasing current flow is called the avalanche region. Certain semiconductor devices are designed and doped to operate in the avalanche region without harm. These special devices will be discussed in subsequent chapters.

CHAPTER 5

PN JUNCTION POWER SUPPLIES

The use of PN junction rectifiers in the design of power supplies for electronic equipment is increasing. Reasons for this include these characteristics: no requirement for filament power, immediate operation without need for warm-up time, low internal voltage drop substantially independent of load current, low operating temperature, and generally small physical size. PN junction rectifiers are particularly well-adapted for use in the power supplies of portable and small electronic equipment where weight and space are important considerations.

Semiconductor materials treated to form PN junctions are used extensively in electronic circuitry. Variations in doping agent concentrations and physical size of the substrate produce diodes which are suited for different applications. There are signal diodes, rectifiers, zener diodes, reference diodes, varactors and others.

DIODES

Pictorial representations of various diodes are shown in figure 5-1. This is but a very limited representation of the wide assortment in case design. However, the shape of characteristic curves of these diodes is very similar; primarily, current and voltage limits and relationships are different. Figure 5-2 shows a typical curve of a junction diode. The graph shows two different kinds of bias. Bias in the PN junction is the difference in potential between the anode (P material) and the cathode (N material). Forward bias is the application of a voltage between N and P material, where the P material is positive with respect to the N material. When the P material becomes negative with respect to the N material, the junction is reverse biased. Application of greater and greater amounts of forward bias causes more and more forward current until the power handling capability of the diode is exceeded, unless limited by external circuitry. Small amounts of forward bias cause very little current flow until the internal barrier potential is overcome. The potential difference varies from diode to diode, but is usually no more than a few tenths of a volt. Reverse bias produces a very small amount of reverse current until the breakdown point is reached, then an increase in reverse bias will cause a large increase in reverse current. Therefore, if breakdown is not exceeded, the ratio of forward current to reverse current is large—for example, milliamperes to microamperes or amperes to milliamperes. Changes in temperature may cause alterations in the characteristic curve, such as: slope of curve at any point, breakdown point, amount of reverse current, etc.

DIODE SPECIFICATIONS

There are many specifications listed in various manufacturers' specification sheets and in semiconductor data manuals. Descriptions of various diode types and their more important electrical characteristics follow.

Rectifier Diodes

Rectifier diodes are used primarily in power supplies. These diodes are primarily of the silicon type because of this material's inherent reliability and higher overall performance compared to other materials. Silicon allows higher forward conductance, lower reverse leakage current and operation at higher temperatures compared to other materials.

The major electrical characteristics of rectifier diodes are listed below:

D.C. BLOCKING VOLTAGE (V_R)—maximum reverse d.c. voltage which will not cause breakdown.

AVERAGE FORWARD VOLTAGE DROP (V_F)—average forward voltage drop across

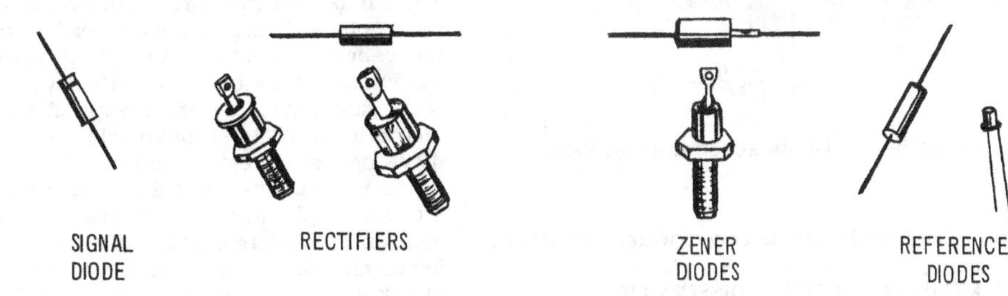

179.51
Figure 5-1.—Junction diodes.

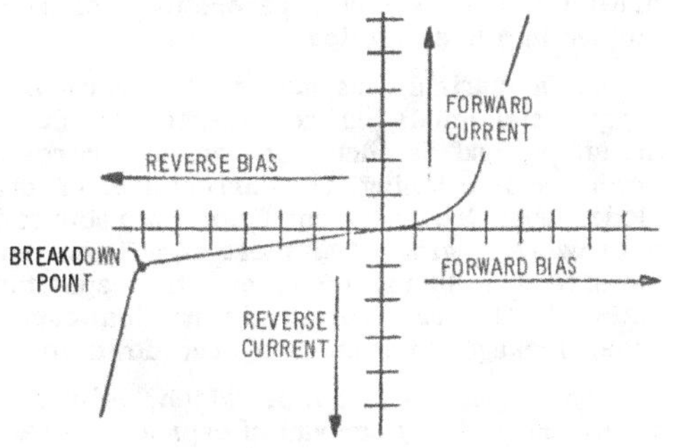

179.52
Figure 5-2.—Diode characteristic curve.

the rectifier given at a specified forward current and temperature, usually specified for rectified forward current at 60 Hz.

AVERAGE RECTIFIER FORWARD CURRENT (I_F)—average rectified forward current at a specified temperature, usually at 60 Hz with a resistive load. The temperature is normally specified for a range, typically -65 to +175 degrees centigrade.

AVERAGE REVERSE CURRENT (I_R)—average reverse current at a specified temperature, usually at 60 Hz.

PEAK SURGE CURRENT (I_{SURGE})—peak current specified for a given number of cycles or portion of a cycle. For example, 1/2 cycle at 60 Hz.

Signal Diodes

Signal diodes fall into various categories, such as general purpose, high-speed switch, parametric amplifiers, etc. These devices are used as mixers, detectors and switches, as well as in many other applications.

Signal diodes major electrical characteristics are:

PEAK REVERSE VOLTAGE (PRV)—maximum reverse voltage which can be applied before reaching the breakdown point.

REVERSE CURRENT (I_R)—small value of direct current that flows when a semiconductor diode has reverse bias.

MAXIMUM FORWARD VOLTAGE DROP AT INDICATED FORWARD CURRENT (V_F @ I_F)—maximum forward voltage drop across the diode at the indicated forward current.

REVERSE RECOVERY TIME (t_{rr})—time required for reverse current to decrease from a value equal to the forward current to a value equal to I_R when a step function of voltage is applied.

The schematic diagram for the rectifier and signal diode is shown in figure 5-3. Forward current flows into the point of the arrow and reverse current is with the arrow.

Zener Diodes

The zener diode is unique compared to other diodes in that it is designed to operate reverse biased in the avalanche or breakdown region. The device is used as a regulator, clipper, coupling device and in other functions.

The major electrical characteristics of zener diodes are:

NOMINAL ZENER BREAKDOWN $V_{Z(NOM)}$ — sometimes a $V_{Z(MAX)}$ and $V_{Z(MIN)}$ are used

179.53
Figure 5-3.—Diode schematic symbol.

to set absolute limits between which breakdown will occur.

MAXIMUM POWER DISSIPATION (P_D)—maximum power the device is capable of handling. Since voltage is a constant, here is a corresponding current maximum (I_{ZM}).

Schematic diagrams of the zener are shown in figure 5-4. Zener current flows in the direction of the arrow. In many schematics a distinction is not made for this diode and a signal diode symbol is used.

Reference Diodes

Reference diodes were developed to replace zener diodes in certain applications because of the zener's temperature instability. Reference diodes provide a constant voltage over a wide temperature range. The important characteristic of this device, besides V_Z, is T_{min} T_{max} which specifies the range over which an indicated temperature co-efficient is applicable. The temperature co-efficient is expressed as a percent of change of reference (V_Z) per degree centigrade change in temperature.

Varactor Diodes

PN junctions exhibit capacitance properties because the depletion area represents a dielectric and the adjacent semiconductor material represents two conductive plates. Increasing reverse bias decreases this capacitance while increasing forward bias increases it. When forward bias is large enough to overcome the barrier potential, high forward conduction destroys the capacitance effect, except at very high frequencies. Therefore, the effective capacitance is a function of external applied voltage. This characteristic is undesirable in conventional diode operation, but is enhanced by special doping in the varactor or variable-capacitance (varicap) diodes. Application categories of the varactor can be divided into two main types, tuning and harmonic generation. Different characteristics are required by the two types but both use the voltage dependent junction capacitance effect. Figure 5-5 shows the voltage capacitance relationships. The use of this diode for frequency multiplication (harmonic generation) will be explained in a later section.

As a variable capacitor, the varactor is rugged and small, is not affected by dust or moisture, and is ideal for remote control and precision fine tuning. The current uses of tuning diodes span the spectrum from AM radio to the microwave region. The most significant parameters of a tuning diode are the capacitance ratio, "Q," series resistance, nominal capacitance, leakage current, and breakdown voltage.

The capacitance ratio, which defines the tuning range, is the amount of capacitance variation over the bias voltage range. It is normally expressed as the ratio of the low voltage capacitance divided by the high voltage capacitance. For example, a typical specification which reads $C_4/C_{60} = 3$ indicates that the capacitance value at 4 volts is 3 times the capacitance value at 60 volts. The high voltage in the ratio is usually the minimum breakdown voltage specification. A 4 volt lower limit is quite common since it describes the approximate lower limit of linear operation for most devices. The capacitance

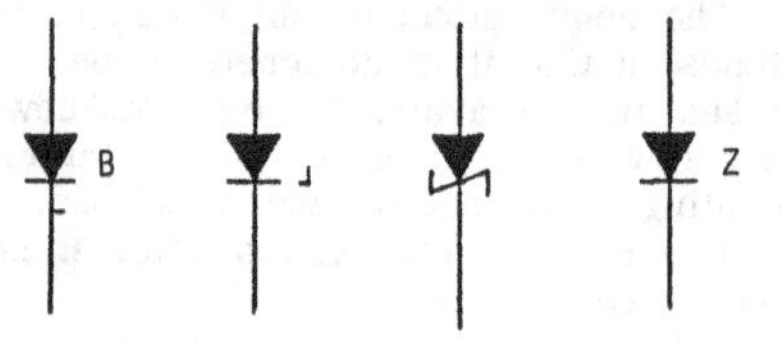

179.54
Figure 5-4.— Zener diode schematic symbols.

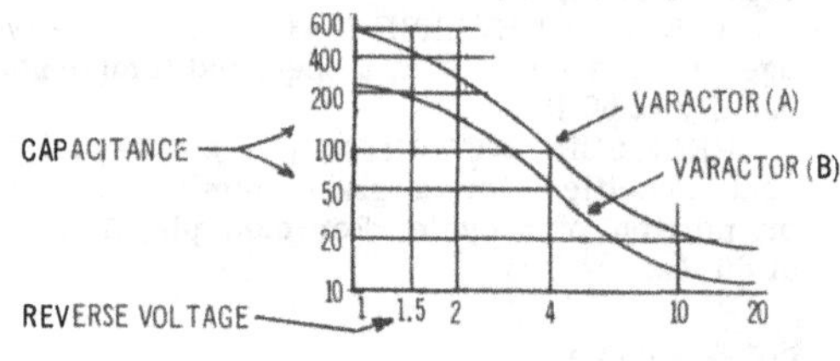

179.55
Figure 5-5.—Typical voltage capacitance relationship in varactors.

ratio of tuning diodes varies in accordance with construction.

"Q" is inversely proportional to frequency, nominal capacitance, and series resistance. Ideally, tuning diodes should have high "Q," low series resistance, low reverse leakage, and high breakdown voltage at any desired capacitance ratio; however, as might be expected, these parameters are not unrelated and improving one degrades another so that often a compromise must be reached. As a rule, diodes with low capacitance values have the highest "Q."

Various schematic symbols are used to designate varactor diodes, as shown in figure 5-6. The application of the varactor as a control device in LC tanks will be described. The resonant frequency of any tank circuit containing L and C is found by the formula $F_o = \frac{1}{2\pi\sqrt{LC}}$. By changing the values of either L or C we change the resonant frequency of the tank. The resonant frequency of a tank can be changed mechanically (i.e., the use of a movable slug in the coil or the movement of plates on a variable capacitor). However, the varactor provides a means of obtaining electronic control of the tuned tank.

Capacitors in parallel add so as to give an increased capacitance value. If one of the capacitors is variable, the range of the combination is also variable. Figure 5-7 shows a varactor that is controlled by a variable voltage supply; the junction capacitance of the device is part of the tank's reactive components. The degree of reverse bias across the varactor will determine the capacitance of the varactor. C2 blocks d.c. current flow through L1. As the varactor's capacitance is varied, the resonant frequency of the tuned tank composed of L1, C1, and the junction capacitance, will change. If C2 is large in comparison to the junction capacitance, C2 will have a minimal effect in determining the resonant frequency. A decrease in the reverse bias on the varactor will cause an increase in its capacitance. The increase in capacitance causes the overall capacitance (C1 and varactor) to increase with a resultant decrease in resonant frequency of the circuit. An increase in reverse bias of the diode causes an opposite effect on resonant frequency.

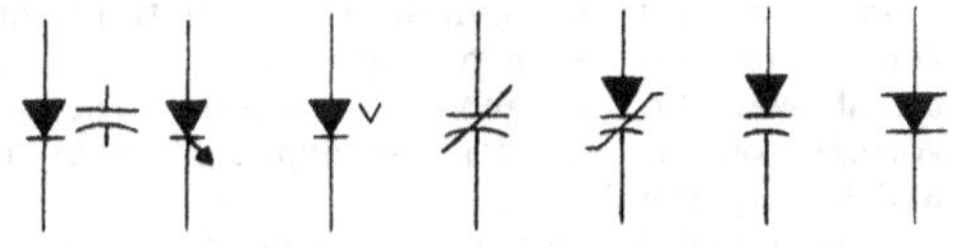

179.56

Figure 5-6.—Schematic symbols of varactors.

PN Junction Diode

Figure 5-8 shows how an ordinary PN junction diode might be used as a switch to control the frequency of an LC tank circuit. Although a switch (SW1) and a battery are used for ease of explanation, the control voltage may be taken from other voltage sources. Therefore, the battery and switch can be replaced by control lines. The combination of C2 and CR1 in parallel with C1 and L1, will be resonant at some specific frequency according to the values of capacitance and inductance present. With SW1 open the value of capacitance presented by the PN junction diode is much lower than it is when the switch is closed. This change in capacitance occurs when the PN junction diode conducts heavily (SW1 closed) due to the forward bias, and is an effective short. The value of capacitance presented in parallel with C1 and L1 is now approximately equal to the value of C2. When the switch is open, the value of capacitance in parallel with C1 and L1 is the series value of C2 and the junction capacitance of CR1. The junction capacitance

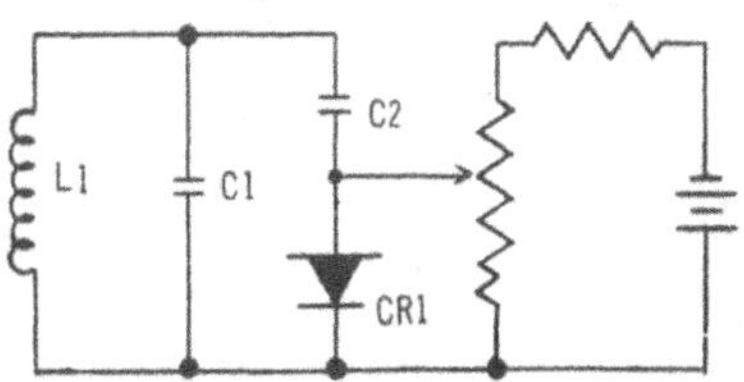

179.57

Figure 5-7.—Changing resonant frequency with a varactor.

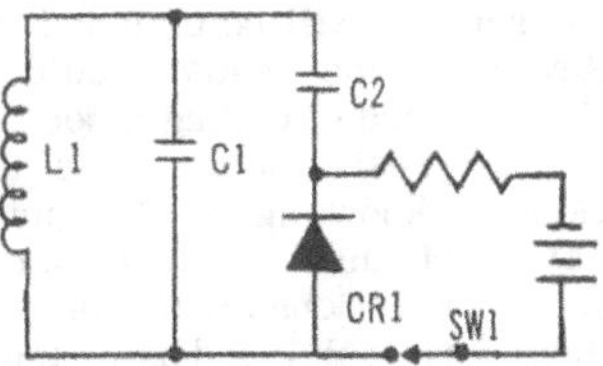

179.58

Figure 5-8.—Changing resonant frequency with a PN junction diode.

of the diode chosen in this application will be inconsequential and the resonant frequency will be determined primarily by L1 and C1. The high resonant frequency condition of this circuit is determined by the non-conducting state of CR1, and the low resonant frequency occurs when CR1 is conducting heavily.

One application of a PN junction diode is as a switch, as was shown above. Remember that a conducting PN junction diode acts as a very low resistance and that a reverse biased PN junction diode presents a very high resistance. This switching characteristic has many other applications. The PN junction diode, as a switch, will be encountered in various equipment.

DIODE APPLICATIONS

So far two applications of the PN junction diode have been presented—one, as a varactor for frequency control and two, as an ordinary junction diode functioning as a switch. There are several other uses such as rectification, detection, and voltage regulation—to name a few. The particular use of a PN junction diode can usually be recognized by the bias on the diode and the circuit in which it is found. Rectifier and detector biasing will be covered in more detail in later sections of this text.

DIODE TESTING

The general condition of a PN junction diode may be tested for a good/bad indication through the use of an ohmmeter. Since an ohmmeter uses a battery for its operation, polarity of the voltage appearing at the leads must be known.

Remember that a forward biased PN junction diode exhibits a very low resistance when the negative (battery) lead of an ohmmeter is connected to the cathode (N material) and the positive (battery) lead of the meter makes contact with the anode (P material), a very low resistance reading will be obtained for a normal diode. Reversing the leads should result in a very high resistance reading since reverse bias causes a high resistance across the junction.

By using a known good PN junction diode, polarity of an ohmmeter's leads can be ascertained as follows: Connect the ohmmeter leads to the diode terminals, and note the resistance reading. If a low resistance reading is obtained, the diode is forward biased, indicating that the negative battery lead is connected to the cathode, and the positive battery lead is connected to the anode. A high resistance reading would indicate opposite polarities.

The ohmmeter being used should be set to the R x 100 scale in order to assure the best readings. A bad diode will be one that either reads extremely high in both directions (indicating an open diode) or extremely low in both directions (indicating a shorted diode). Care must be taken to insure that the battery voltage and series resistance contained in the meter are such that the current and/or voltage rating of the diode under test are not exceeded.

PN JUNCTION DIODE RECTIFIER CIRCUITS

A description of how the PN junction diode may be used as a switch and as a device to automatically change the frequency of a circuit has been given. Now an area of common usage will be presented—rectification.

Rectification is described as the changing of an alternating current (a.c.) to a unidirectional or direct current (d.c.). The normal PN junction diode is well suited for this purpose as it conducts very heavily when forward biased (low resistance direction) and only slightly when reverse biased (high resistance direction).

Figure 5-9 is a block diagram of a power supply showing an a.c. input to and a d.c. output from a block labeled positive power supply and filter network. Although figure 5-9 shows a power supply that provides a unidirectional current which causes a positive voltage output, it might well be designed to furnish a negative voltage output. A detailed explanation of this will be supplied later.

The question of why a change of a.c. to d.c. is necessary probably arises. The answer, simply, is that for proper operation many electronic circuits depend upon d.c. As already pointed out the PN junction diode conducts more easily

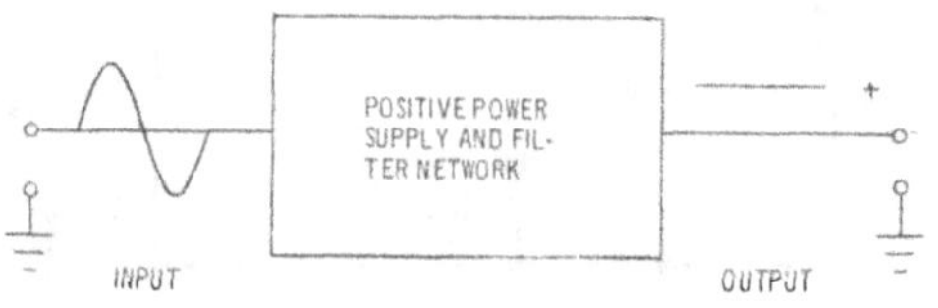

179.59

Figure 5-9.—Positive voltage output from an a.c. input.

in one direction than in the other. Transistors and electron tubes are also unidirectional and a constantly alternating source voltage would be undesirable.

Before describing how an a.c. input is converted into a d.c. output the definition "load" as it applies to power supplies must be understood. "Load" is the current supplied to the power consuming device or devices connected to the power supply. The power consuming device needs voltage and current for proper operation and this voltage and current is supplied by the power supply. The power consuming device may be a simple resistor or one or more electronic circuits using resistors, capacitors, coils, and active devices.

HALF-WAVE RECTIFIER

Figure 5-10 shows the PN junction diode functioning as a half-wave rectifier. A half-wave rectifier is one that uses only half of the input cycle to produce an output.

The induced voltage across L2 (the transformer secondary) will be as shown in figure 5-10. The dots on the transformer indicate points of the same polarity. During that portion of the input cycle which is going positive (solid line), CR1, the PN junction diode, will be forward biased and current will flow through the circuit. L2, acting as the source voltage, will have current flowing from the top to the bottom. This current then flows up through R_L causing a voltage drop across R_L equal to the value of current flowing times the value of R_L. This voltage drop will be positive at the top of R_L, with respect to its other side, and the output will therefore be a positive voltage with respect to ground. It is common practice for the end of a resistor receiving current to be given a sign representing a negative polarity of voltage, and the end of the resistor through which current leaves is assigned a positive polarity of voltage. The voltage drop across R_L, plus the voltage drops across the conducting diode and L2, will equal the applied voltage. Although the output voltage will nearly equal the peak input voltage, it cannot reach this value due to the voltage drops, no matter how small, across CR1 and L2.

The negative half cycle of the input is illustrated by the broken line. When the negative half cycle is felt on CR1, the PN junction diode is reverse biased. The reverse current will be very small—but it will exist. The voltage resulting from the reverse current, as shown below the line in the output is exaggerated in figure 5-10 to bring out the point of its existence. Although only one cycle of input is shown in figure 5-10 it should be realized that the action described above continually repeats itself, so long as there is an input.

By reversing the diode connection in figure 5-10—having the anode on the right instead of the left—the output would now become a negative voltage. The current would be going from the top of R_L toward the bottom—making the output at the top of R_L negative in respect to the bottom or ground side.

The same negative output can be obtained from figure 5-10 if the reference point (ground) is changed from the bottom (where it is shown) to the top, or cathode connected end of the resistor. The bottom of R_L is shown as being negative in respect to the top, and reading the output voltage from the "hot" side of the resistor to ground would result in a negative voltage output.

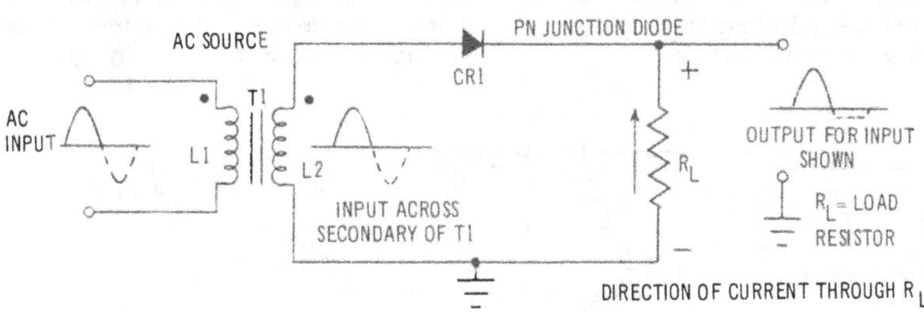

179.60

Figure 5-10.—Positive voltage output half-wave PN junction diode rectifier.

The half-wave rectifier will normally indicate improper functioning in one of two manners—there is no output or the output is low. The NO OUTPUT condition can be caused by no input—the fuse has blown, the transformer primary or secondary winding is open, the PN junction diode is open, or the load is open.

The low output condition might be caused by an aged diode. A check of both forward and reverse resistance of the diode may reveal the condition of the diode. Low output can be the result of an increased forward resistance or a decreased reverse resistance of the diode.

It is necessary to check the a.c. input voltage to see if it is of the correct value—a low input voltage will result in a low output voltage. A check of the transformer secondary voltage should be made to see if it is of the correct value also, as a low secondary voltage will also result in a low output voltage.

By removing the input voltage, the technician can make resistance checks of the components. Is the primary open? The secondary open? Has the diode increased in forward resistance value or decreased in reverse resistance value? Is the diode open? Has the load resistance become shorted? Do the components show signs of excessive heat dissipation—have they become discolored? Does energizing the circuit and putting an ammeter in series with the load make the load current excessive?

All these questions should be answered by the technician when troubleshooting the half-wave rectifier. If a trouble is discovered in the rectifier, one should then determine if the cause is a local one (in the rectifier itself) or due to some changes in the following circuitry, such as the power supply filter components or changing load impedance. While it is important that the trouble be repaired, elimination of the cause of the trouble is of greater importance.

FULL-WAVE RECTIFIER

The PN junction diode works just as well in a full-wave rectifier circuit as shown in figure 5-11. The circuit shown has a negative voltage output, however, it might just as well have a positive voltage output. This can be accomplished by either changing the reference point (ground side of R_L) or by reversing the diodes in the circuit.

The a.c. input is felt across the secondary winding of T1. This winding is center tapped as shown—the center of the secondary is at ground potential. This seems to be a good time to define ground as a reference point which is of no particular polarity. When the polarity is such that the top of T1 secondary is negative, the bottom is positive. At this time, the center tap, as shown, has two polarities, positive with respect to the top half of the winding, and negative with respect to the bottom half of the winding. When the secondary winding is positive at the top, the bottom is negative and the center tap is negative with respect to the top and positive with respect to the bottom. What is the polarity of the reference point (ground)? The answer must be in terms of "with respect to." For each alternation of the input, one of the diodes will be forward biased and the other one reverse biased.

For ease of explanation, the negative alternation will be considered when the rectifier current is initially energized by the a.c. source. CR1 will be forward biased—negative voltage felt on its cathode—and CR2 will be reverse biased—a positive voltage felt on its cathode. Therefore, the top of T1 secondary must be negative with respect to the bottom. When forward bias is applied to CR1, it conducts heavily from cathode to anode (dashed arrow), down through R_L—this current flow creates a voltage drop across R_L—negative at the top with respect to the bottom or ground side of R_L. The current passing through R_L is returned to CR1 by going through the

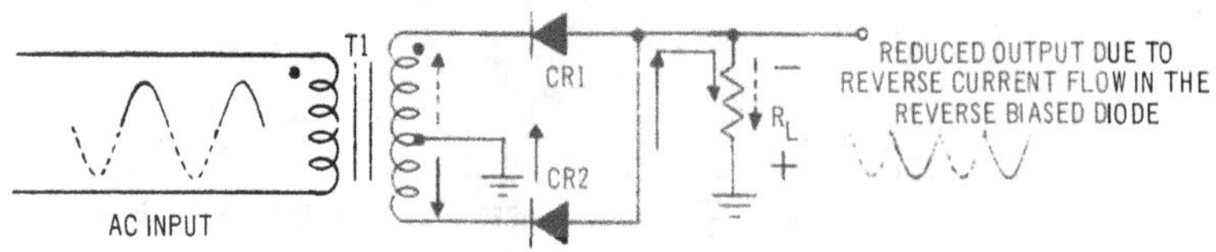

Figure 5-11.—Unfiltered negative output full-wave rectifier.

grounded center tap and up the upper section of the center tapped secondary winding of T1. This completes the first alternation of the input cycle. The second alternation of the input now is of such polarity as to forward bias CR2—a negative voltage at the bottom of T1 secondary winding with respect to ground. CR1 is now reverse biased. CR2 conducts, current moves in the same direction through R_L (solid arrow)—top to bottom—back through the lower half of the center tapped secondary to CR2. One may wonder why current does not flow from the anode of one of the diodes through the anode to cathode of the other diode. The answer is simple—it does—however, current flow through a reverse biased diode is very slight due to the high resistance of the diode when reverse biased. This rectifier has a slightly reduced output—as shown in figure 5-11—because of the reverse current flow.

As can be seen in the output waveform of figure 5-11, there are two pulses of d.c. out for every cycle of a.c. in—this is full-wave rectification. Current flow through R_L is in the same direction, no matter which diode is conducting. The positive going alternation of the input allows one diode to be forward biased and the negative going alternation of the input allows the other diode to be forward biased. The output—for the full-wave rectifier shown—is a negative voltage measured from the top R_L to ground.

As in the half-wave rectifier, there can be two indications of trouble—NO OUTPUT or LOW OUTPUT. No output conditions are indications of no input, shorted load circuits, open primary winding, open or shorted secondary winding, or defective diodes. Low output conditions are possible indications of aging diodes, open diodes, or opens in either half of the secondary winding (allowing the circuit to act as a half-wave rectifier).

The method for troubleshooting the full-wave rectifier is the same as used for the half-wave rectifier. Check voltages of both primary and secondary windings, check current flow, and, when the circuit is deenergized, take resistance measurements. Shorted turns in the secondary windings give a lower voltage output and possibly shorted turns in the primary winding will produce a lower voltage input. (Shorted turns are hard to detect with an ohmmeter—they are more easily detected by taking a voltage reading across various terminals of the energized transformer.)

BRIDGE RECTIFIER

Now, the PN junction diode will be described as it is used in a bridge rectifier circuit. Figure 5-12 shows such a circuit capable of producing a positive output voltage. When the a.c. input is applied across the secondary winding of T1, it will forward bias diodes CR1 and CR3, or CR2 and CR4. When the top of the transformer is positive with respect to the bottom, as illustrated in figure 5-12 by the designation number 1, both CR1 and CR2 will feel this positive voltage. CR1 will have a positive voltage on its cathode, a reverse bias condition, and CR2 will have a positive voltage on its anode, a forward bias condition. At this same time the bottom of the secondary winding will be negative with respect to the top—placing a negative voltage on the anode of CR3 (a reverse bias condition) and on the cathode of CR4 (a forward bias condition).

During the half cycle of the input designated by the number 1 in figure 5-12, we find that CR2 and CR4 are forward biased and will therefore conduct heavily. The conducting path is shown by the solid arrows, from the source (the secondary winding of T1) through CR4 to ground, up through R_L making the top of R_L positive with respect to the grounded end, to the junction of CR2 and CR3. CR2, being forward biased, offers the path of least resistance to current flow and this is the path current will take to get back to the source.

During the alternation designated by the number 2 in figure 5-12, shown by the dashed arrows, the top of the secondary winding is going negative while the bottom is going positive. The negative voltage at the top is felt by both CR1 and CR2, forward biasing CR1 and reverse biasing CR2. The positive voltage on the bottom of T1 secondary is felt by CR3 and CR4, forward biasing CR3 and reverse biasing CR4. Current flow—starting at the source (T1 secondary winding) is through CR1 to ground, up through R_L (this is the same direction as it was when CR2 and CR4 were conducting, making the top of R_L positive with respect to its grounded end) to the junction of CR2 and CR3. This time CR3 is forward biased and offers the least opposition to current flow, and current takes this path to return to its source.

As can be seen, the diodes in the bridge circuit operate in pairs; first one pair—CR1 and CR3—conduct heavily and then the other pair—CR2 and CR4—conduct heavily. As shown in the output waveform we get one pulse out for every half cycle of the input—or two pulses out

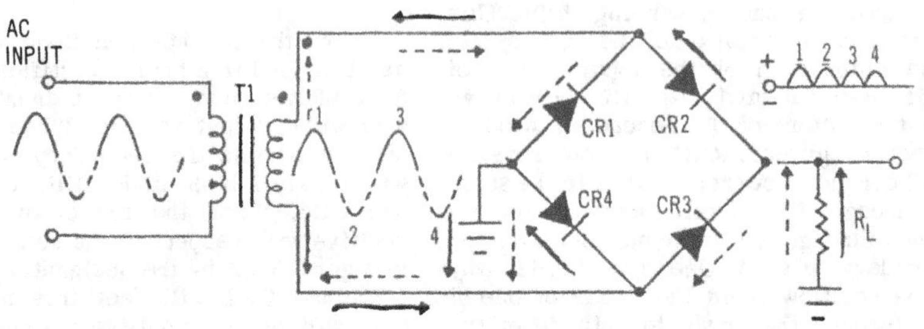

Figure 5-12.— PN junction diode in a bridge rectifier circuit.

for every cycle in. This is the same as for the full-wave rectifier circuit explained previously.

The bridge circuit will also indicate a malfunction in one of two manners—it has no output or a low output. The causes for both conditions are the same as they were for the half- or full-wave rectifier. If any one of the diodes open, the circuit will act as a half-wave rectifier with a resultant lower output voltage.

JUNCTION DIODE CONSIDERATIONS

The junction diode has four important ratings that must be taken into consideration when designing a power supply. They are the maximum

- average forward current
- repetitive reverse voltage
- surge current
- repetitive forward current

These ratings are important to the technician when it becomes necessary to troubleshoot a power supply or when selecting junction diodes for replacement, when the desired one is not readily available.

The maximum average forward current is the maximum amount of average current that can be permitted to flow in the forward direction. This rating is usually given for a specified ambient temperature and should not be exceeded for any length of time as damage to the diode will occur. The maximum repetitive reverse voltage is that value of reverse bias voltage that can be applied to the diode without causing it to break down.

The maximum surge current is that amount of current allowed to flow in the forward direction in non-repetitive pulses. Current should not be allowed to exceed this value at any time and should only equal this value for a period not to exceed one cycle of the input. The maximum repetitive forward current is the maximum value of current that may flow in the forward direction in repetitive pulses.

All of the ratings mentioned above are subject to change with temperature variations. If the temperature increases, the ratings given on the specification sheet should all be lowered or damage to the diode will result.

POWER TRANSFORMERS

Power transformers are used in power supply circuits because of the efficiency and ease with which they transfer energy. The power transformer is capable of receiving a voltage at one level and delivering it at the same level, some higher level, or some lower level. Transformers that convert voltage to a higher level are called step-up transformers; those that convert voltage to a lower level are called step-down transformers; and those that provide the same output voltage level as the applied input level are "one-to-one" transformers. The references in all cases, are to voltage levels.

In power transformers, even though the coils are wound on the same material, the separate windings are each insulated and are therefore electrically isolated from one another. The source of energy for the primary is thus isolated from the secondary and any associated circuitry. The efficiency of power transformers is very high, approximately 90 percent of the input power is usable in the output.

A quick review of voltage, current, and power relationships is in order. The voltage induced into the secondary winding of any given transformer is determined by the ratio of the number of turns in the primary winding to the number of turns in the secondary winding and the amount of voltage applied to the primary. For every volt per turn of the primary winding there will be a volt per turn in the secondary winding. For example, if the primary has 10 windings (turns) and each winding (turn) has 10 volts the entire primary has 100 volts applied.

The secondary winding in this example has only five windings (turns). Since the number of volts per turn must be the same for primary and secondary windings, we find that the secondary winding has only 50 volts—five windings (turns) and 10 volts per winding (turn) = 50 volts. This is an example of a step-down transformer. If, instead of five windings (turns) in the secondary, we had 15 windings, the voltage across the secondary would be 150 volts (15 windings at 10 volts per winding) and the transformer would be a step-up transformer.

From the information given, a simple relationship can be derived. The voltage in the primary (E_p) is equal to the number of volts per turn times the number of turns (N_p) in the primary, and the voltage in the secondary (E_s) is equal to the number of volts per turn times the number of turns (N_s) in the secondary. The voltage induced in the secondary is determined by the ratio $E_p/E_s = N_p/N_s$. It is now a simple matter to calculate any missing value when the other three are given.

Primary power (E_pI_p) would equal secondary power (E_sI_s) in an ideal transformer (the power taken from the source would be equal to the power delivered to the load). However, as mentioned previously, the power transformer is an efficient element, but not 100% efficient. The reasons for the losses will not be reiterated at this time, only the types of losses encountered in power transformers: (1) Copper or I^2R Losses, (2) Eddy Current Losses, and (3) Hysteresis Loss. Due to these losses the efficiency of the power transformer encountered in electronic equipments is, as previously stated, about 90%.

The efficiency of the power transformer is found by the formula % of Efficiency = P_{out}/P_{in} x 100. The current ratio is the reverse of the voltage or turns ratio (I_s/I_p). This means that in a step-up transformer the current flow in the secondary is less than that in the primary. In a step-down transformer, the current flow in the secondary is greater than that in the primary. When the voltage in the primary remains the same—no changes—and the voltage in the secondary increases, the current in the secondary decreases (power out = power in). This is true because the flux lines developed in a transformer core are proportional to the ampere turns of the associated windings—since the flux is the same for both windings the ampere turns must be the same for both—N_pI_p equals N_sI_s. It can be seen that since N_p/N_s equals E_p/E_s it follows that E_pI_p equals E_sI_s. It has been noted that secondary voltage may be higher or lower than the primary voltage. This being true, then E_p/E_s equals I_s/I_p or the primary and secondary currents are inverse to the voltage.

POWER SUPPLY FILTERS

As previously indicated the operation of most electronic circuits is dependent upon a direct current source. It has been illustrated how alternating current can be changed into a pulsating direct current—that is, a current that is always positive or negative with respect to ground although it is not of a steady value; it has "ripple."

Ripple can be defined as the departure of the waveform of a rectifier from pure d.c. It is the amplitude excursions, positive and negative, of a waveform from the pure d.c. value—the alternating component of the rectifier voltage. Ripple contains two factors which must be considered—frequency and amplitude. Ripple frequency, in the rectifiers that have been presented is either the same as line frequency for the half-wave rectifier, or, twice the line frequency for the full-wave rectifiers.

In the half-wave rectifier, one pulse of d.c. output was generated for one cycle of a.c. input; the ripple frequency is the same as the input frequency. In the full-wave rectifiers (center tapped and bridge) two pulses of d.c. output were produced for each cycle of a.c. input—the ripple frequency is twice that of the line frequency. With a 60-hertz input frequency there will be a 60-hertz ripple frequency in the output of the half-wave rectifier and a 120-hertz ripple frequency in the output of the full-wave rectifier.

The amplitude of the ripples in the output of a rectifier circuit will give us a measure of the effectiveness of the filter being used—the ripple factor. The ripple factor is defined as the ratio of the r.m.s. value of the a.c. component to the average d.c. value, $r = \frac{E_{r.m.s.}}{E_{d.c.}}$. The lower the ripple factor the more effective the filter. The term "percent of ripple" may be used. This is different from the ripple factor only because the figure arrived at in the ripple factor formula is multiplied by 100 to give us a percent figure, $\%\ \text{Ripple} - \frac{E_{r.m.s.}}{E_{d.c.}} \times 100.$

In both formulas given, $E_{r.m.s.}$ is the r.m.s. value of ripple voltage and $E_{d.c.}$ is the d.c. value (average value) of the output voltage.

Filter circuits used in power supplies are usually low pass filters. (A low pass filter is a network which passes all frequencies below a specified frequency with little or no loss but is highly discriminate against all higher frequencies.) The filtering is done through the use of resistors or inductors, and capacitors. The purpose of power supply filters is to smooth out the ripple contained in the pulses of d.c. obtained from the rectifier circuit while increasing the average output voltage or current.

Filter circuits used in power supplies are of two general types, capactior input and choke input. There are several combinations that may be used, although they are referred to by different names (Pi, RC, L section, etc.). The closest element electrically to the rectifier determines the basic type of filter being used.

Figure 5-13 depicts the basic types. In A a capacitor shunts the load resistor, therein bypassing the majority of ripple current which passes through the series elements. In B an inductor (choke) in series with the load resistor opposes any change in current in the circuit. The capacitor input filter will keep the output voltage at a higher level compared to a choke input. The choke input will provide a steadier current under changing load conditions. From this it can be seen that a capacitor input filter would be used where voltage is the prime factor and the choke input filter is used where a steady flow of current is required.

CAPACITOR INPUT FILTER

First, an analysis will be made of the simple capacitor input filter depicted in figure 5-14A. The output of the rectifier, without filtering, is shown in B, and the output, after filtering, is illustrated in C. Without the capacitor, the output across R_L will be pulses as previously described. The average value of these pulses would be the $E_{d.c.}$ output of the rectifier.

With the addition of the capacitor the majority of the pulse changes are by-passed through the capacitor and around R_L. As the first pulse appears across the capacitor, changing it from negative to positive, bottom to top, the peak voltage is developed across the capacitor. When the first half cycle has reached its peak and starts its negative going excursion, the capacitor will start to discharge through R_L maintaining the current through R_L in its original direction, thereby holding the voltage across R_L at a higher value than its unfiltered load. Before the capacitor can fully discharge, the positive excursion of the next half cycle is nearing its peak, recharging the capacitor. As the pulse again starts to go negative the capacitor starts to discharge once again. The positive going excursion of the next half cycle comes in and recharges the capacitor; this action continues as long as the circuit is in operation.

The charge path for the capacitor is through the transformer secondary and the conducting diodes, and the discharge path is through the load resistor. The reactance of the capacitor, at the line frequency, is small compared to R_L, which allows the changes to by-pass R_L and, effectively, only pure d.c. appears across R_L.

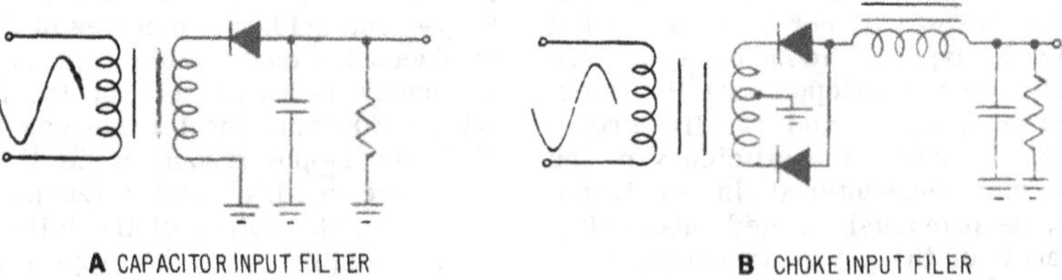

Figure 5-13.—Filter circuits.

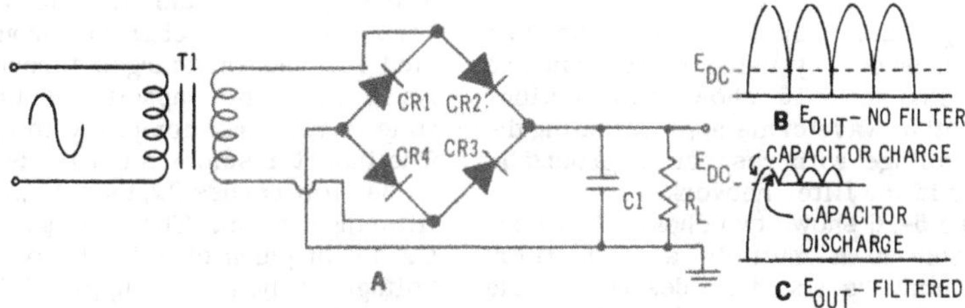

179.64

Figure 5-14.—Low voltage power supply with simple capacitor filter.

This illustrates the use of an RC time constant. If the value of C1 or R_L were such that the discharge time was the same, or less than that of the charge time we would have no filtering action. The larger the values of C1 and R_L, the longer the discharge time constant, and the lower the ripple factor. The charge time of the capacitor is the RC values of the capacitor, the conducting diodes, and the transformer secondary. The impedance offered by these elements is very small when compared to the impedance in the discharge path of the capacitor—the value of R_L. The output voltage is practically the peak value of the input voltage. This circuit provides very good filtering action for low currents, but results in little filtering in higher current power supplies due to the smaller resistance of the load.

CHOKE INPUT FILTER

The next filter to be analyzed is the choke input filter, or the L-section filter. Figure 5-15 shows this filter and the resultant output of the rectifier after filtering has taken place. The series inductor, L (choke), figure 5-15, will oppose rapid changes in current. The output voltage of this filter is less than that of the capacitor input filter since the choke is in series with the output impedance. The parallel combination of R_L and C in connection with L smooths out the peaks of the pulses and results in a steady, although reduced, output.

The inductance chokes off the peaks of the alternating components of the rectified waveform and the d.c. voltage is the average, or d.c. value, of the rectified wave. The choke input filter allows a continuous flow of current from the rectifier diodes rather than the pulsating current flow as seen in the capacitor input filter. The X_L of the choke reduces the ripple voltage by opposing any change in current during either the rapid increases in current during the positive excursions of the pulses or decreases in current during the negative excursions. This keeps a steady current flowing to the load throughout the entire cycle. The voltage developed across the capacitor is maintained at a relatively constant value approaching the average value of input voltage because of this steady current flow.

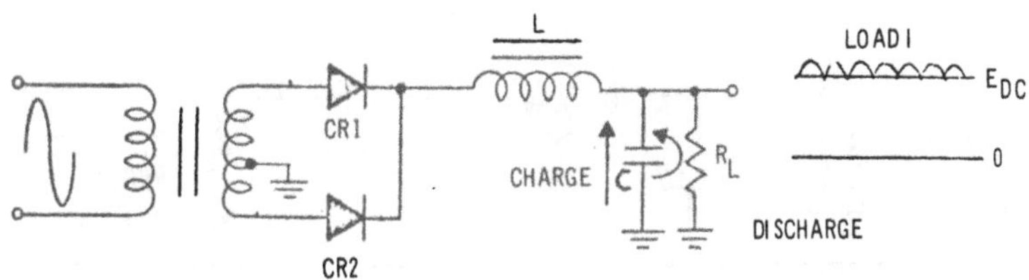

179.65

Figure 5-15.—L-section (choke input) filter showing representative waveforms.

Multiple Section Choke Input Filter

The filtering action provided by the choke input filter can be enhanced by using more than one such section. Figure 5-16 shows two sections with representative waveforms approximating the shape of the voltage with respect to ground at different points in the filter networks.

While figure 5-16 shows two choke input sections being used as a multiple section filter, more sections may be added as desired. While the multiple section filter does reduce the ripple content— and they are found in applications where only a minimum ripple content can be tolerated in the output voltage— they also result in reduced regulation. The additional sections add more resistance in series with the power supply which results in increased voltage variations in the output when the load current varies.

PI FILTER

A filter, called the Pi filter because of its resemblance to the Greek letter Pi, π, is a combination of the simple capacitor input filter and the choke input filter. This filter is shown in figure 5-17.

The resistor, R, is known as a bleeder resistor and is found in practically all power supplies. The purpose of this resistor is two fold: when the equipment has been working and is then turned off it provides a discharge path for the capacitor, preventing a possible shock to maintenance personnel; and it also provides a fixed load, no matter what equipment is connected to the power supply. It is also possible to use this resistor as a voltage dividing network through the use of appropriate taps. More on this will be subsequently discussed.

The Pi filter is basically a capacitor input filter with the addition of an L-section filter. The majority of the filtering action takes place across C1 which charges through the conducting diode(s) and discharges through R, L, and C2. As in the simple capacitor input filter, the charge time is very fast compared to the discharge time. The inductor smooths out the peaks of the current pulses felt across C2, thereby providing additional filtering action. The voltage across C2, since C2 is in parallel with the output, is the output voltage of the power supply. Although the voltage output is lower in this filter than it would be if taken across C1 and the load, the amount of ripple is greatly reduced.

Even though C1 will charge to the peak voltage of the input when the diodes are conducting and discharge through R when they are cut off, the inductor is also in the discharge path and opposes any changes in load current. The voltage dividing action of L and C2 is responsible for the lower output voltage in the Pi filter when compared to the voltage available across C1.

In figure 5-17 the charge path for both C1 and C2 is through the transformer secondary, through the capacitors, and, in the case of C2, through L. Both charge paths are through the conducting diode. However, the discharge path for C1 is through R and L while the discharge path for C2 is through R only. How fast the input capacitor, C1, discharges is mainly determined by the ohmic value of R. The discharge time of the capacitors is directly proportional to the value of R. If C1 has very little chance to discharge, the output voltage will be high. For lower values of R the discharge rate is faster, and the output voltage will decrease. With a lower value of resistance the current will be greater and the capacitor will discharge further. The E output

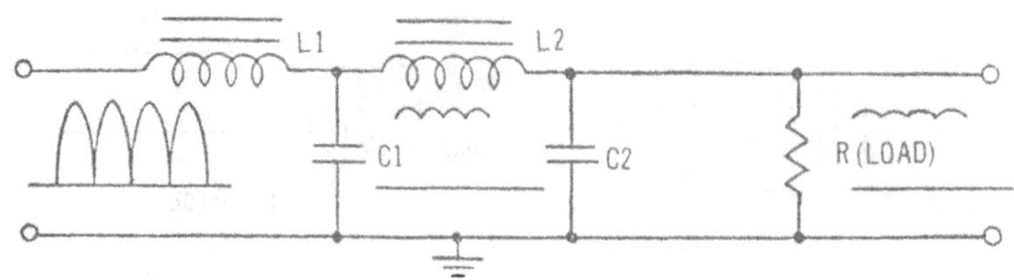

179.66
Figure 5-16.— Multiple section choke input filter with representative waveforms.

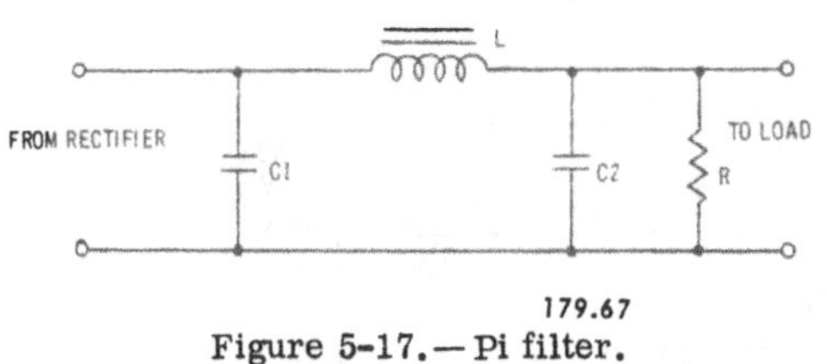

Figure 5-17.—Pi filter.

is the average value of d.c. and the faster the discharge time the lower the average value of d.c. and the lower the E out.

RC Capacitor Input Filter

While the Pi filter previously discussed had an inductor placed between two capacitors, the inductor can be replaced by a resistor, as shown in figure 5-18. The main difference in operation between this Pi filter and the one previously discussed is the reaction of an inductor to a.c. when compared to the resistor. In the former filter the combination of the reactances of L and C2 to a.c. was such as to provide better filtering, giving a relatively smooth d.c. output.

In figure 5-18 both the a.c. and d.c. components of rectified current pass through R1. The output voltage is reduced due to the voltage drop across R1 and the higher the current the greater this voltage drop. This filter is effective in high voltage—low current applications. As in choke input filters, the capacitor input filters shown may be multiplied; i.e., identical sections may be added in series.

The choice of a filter for a particular use is a design problem, but the purpose and operation of filters should be understood by all technicians because of their importance to the proper operation of equipment following the power supply.

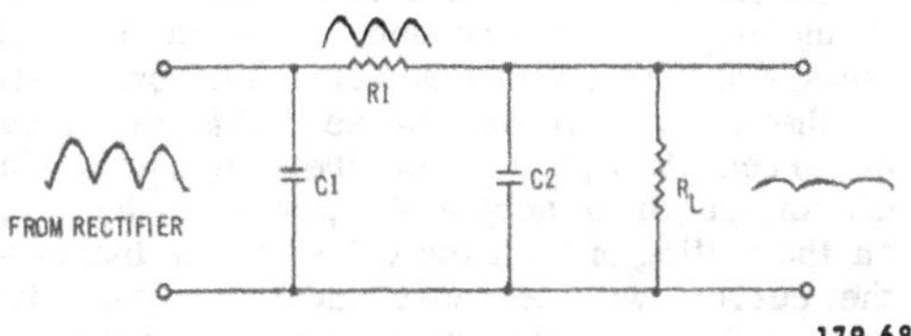

Figure 5-18.—Capacitor input filter and associated waveforms.

BLEEDER RESISTOR VOLTAGE DIVIDER

Figure 5-19 shows how the bleeder resistor may function as a voltage divider network. Terminal 3 is grounded. Current is flowing as indicated, from the bottom to the top, making terminal 4 negative with respect to ground (terminal 3). Terminals 1 and 2, on the other hand, are positive with respect to ground.

Since collector voltage from NPN and PNP transistors used in amplifier circuits needs positive and negative values, respectively, a voltage dividing network, such as that shown in figure 5-19 is typical. At point X, 50 ma. of current enters the junction of R3 and terminal 4 (load A). At point X this current divides, 40 ma. flows through R3 and 10 ma. through load A. They both have a voltage drop of 22.5 volts across them.

At point Y the current again divides into the parallel paths of R2, load B, and load C. As indicated, 40 ma. flows through loads B and C, and 10 ma. flow up through R2. The current through R2 causes a voltage drop of 10v. The top of R2 is positive 10 volts with respect to the bottom, or ground side, of R2. Load B, being in parallel with R2, is also a positive 10 volts, although there are 15 ma. of current flowing through load B. It should be evident that the value of R2 is greater than the impedance of load B by a ratio of 1.5 to 1. If it isn't evident, work it out using the formula $R(Z) = \frac{E}{I}$, since you have the values of both E and I. At point Y' the currents through R2 and load B rejoin at terminal 2 and flow up through R1. The 25 ma. flowing through R1 make the voltage drop across R1 a positive 12.5 volts, with respect to its lower end. The voltage measured from point Z to ground, however, will be 22.5 volts since the voltage drops across R1 (12.5 volts) and R2 (10 volts) are between point Z and ground. (The voltage across load C is in parallel with the series combination of R1 and R2, with load B in parallel with R2.) At point Z the 25 ma. flowing in load C combines with the 25 ma. through R1, which satisfies Kirchhoff's current law.

Figure 5-20 shows this division and joining of current in line form with the arrowheads indicating the direction of current. While it may seem that current is flowing in two directions at one time, if you just think in terms of Kirchhoff's current law—"The algebraic sum of currents entering and leaving a junction of conductors is zero," it will be evident that it is not.

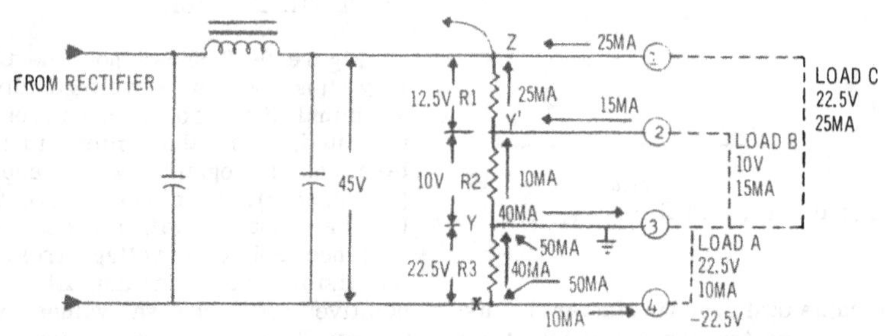

179.69

Figure 5-19.—Voltage divider network with bleeder resistor.

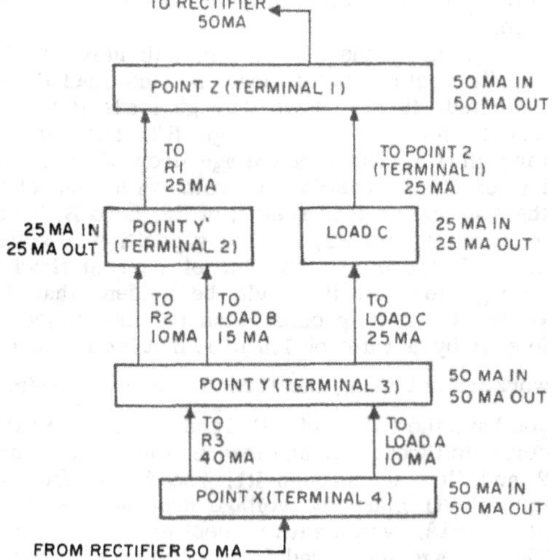

179.70

Figure 5-20.—Flow chart of figure 5-19.

VOLTAGE REGULATORS

It is often required that the output voltage from a power supply be maintained at a constant value regardless of input voltage or load variations. The device used to give us this control is the voltage regulator. Regardless of the specific operating device used, the action is basically the same—that of providing a constant value of output voltage from the power supply irrespective of reasonable variations in input voltage or load current.

A voltage regulator, then, is an electronic device connected in the output of a power supply to maintain the output voltage at its constant rated value. It reacts automatically within its rated limits to any variations in the output voltage. Should the output voltage rise or fall, the voltage regulator automatically compensates for the change and maintains the output voltage at the required value. Although there may be large changes in load current drawn from a power supply, or changes in the applied voltage, the voltage regulator maintains a constant output voltage.

The regulating action of the voltage regulator is, in effect, that of a variable resistor which responds to any changes in the current flowing through it. This "variable resistance" may be in series or in shunt with the load.

SIMPLE SERIES VOLTAGE REGULATOR

A simple series voltage regulator is shown in figure 5-21. Here we see a variable resistor (R) connected in series with a load resistor (R_L).

As in any series circuit the current through R and R_L will depend upon the value of source voltage and the total amount of resistance ($R + R_L$) in the circuit. It can be seen that the amount of current through R_L, and therefore, the amount of voltage drop across R_L, will be dependent on the setting of R. If the value of R is increased the current will decrease and the voltage drop across R_L will also decrease. If the value of R is decreased, the value of total resistance also decreases and current will increase causing a

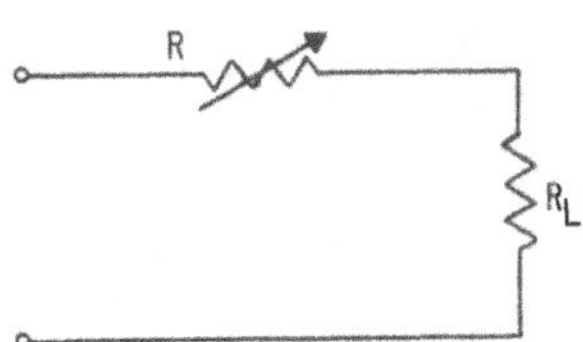

179.71

Figure 5-21.—Simple series voltage regulator.

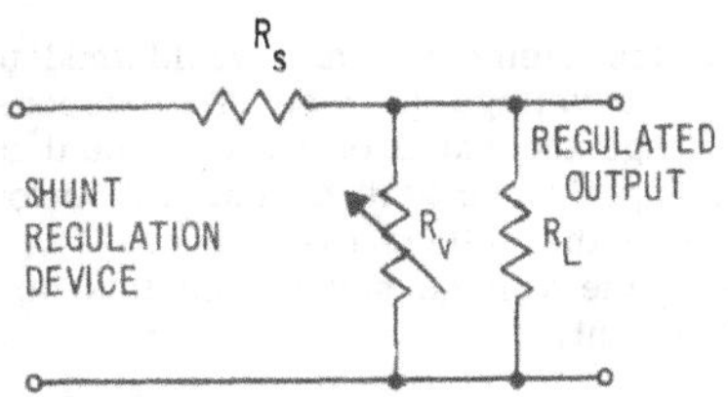

179.72

Figure 5-22.—Simple shunt voltage regulating circuit.

larger voltage drop to be felt across R_L. If a fixed value for the voltage across R_L is set, the regulator will insure that this value remains constant by properly varying the resistor R to compensate for circuit changes.

If the input voltage increases, without changing R, the output voltage would increase. However, if R is increased in value, a larger voltage drop across R results. The increase in the value of R and the voltage drop across it are proportional to the rise in input voltage (a 5v. increase in the input results in a 5v. increase in the drop across R) so that the output voltage (the drop across R_L) remains exactly what it was before the increased input voltage. Conversely, a decrease in the input voltage requires a decrease in the value of R so that the voltage drop across R is proportional to the decreased input voltage (a 5 volt decrease in the input results in a 5 volt decrease in the voltage drop across R) and the output voltage remains constant.

SIMPLE SHUNT VOLTAGE REGULATOR

Figure 5-22 depicts a simple shunt voltage regulator. As in the simple series voltage regulator, a voltage dividing action is used to obtain regulation. The amount of current through R_s is now determined by the shunt regulating device R_v. The larger the current through R_s, the higher its voltage drop, and the lower the voltage across R_L.

The shunt voltage regulator operates in the following manner. R_s (the series resistance) is in series with R_v (the shunt regulating device) and R_L (which represents the load impedance). R_v and R_L, being in parallel, will have the same voltage drop across them. R_s, being in series, will have the total current flowing through R_v and R_L flowing through it. To keep the voltage constant R_v must be adjusted to compensate for changes in the input voltage as well as for changes caused by load impedance changes.

Analysis of an increasing source voltage will exemplify the regulating action. The voltage across R_v and R_L would tend to increase with this increase of source voltage. To compensate for this undesired change, the value of R_v must be decreased. This results in more current flowing through the entire circuit (the total resistance has been decreased) and the increase in current through R_s causes a larger voltage drop across R_s, consequently, the output voltage remains constant.

Conversely, when the source voltage decreases, the voltage across the parallel leg, R_v and R_L, would tend to decrease. To compensate for this change the value of R_v must be increased. This increases the total resistance and decreases total current. Less current through R_s means less voltage drop across R_s, and more voltage available across R_v and R_L. Once again the voltage is held constant. The amount of voltage increase or decrease from the source is the same amount of increase or decrease in the voltage drop across R_s, resulting in the desired amount of voltage for the load.

If the value of R_L were to decrease (the equivalent of adding another resistor in parallel with R_v and R_L) the total resistance would decrease. Current would increase, causing a greater voltage drop across R_s. This would tend to cause the output voltage to decrease. To compensate for this change, the value of R_v must be increased, reducing the amount of current through R_s and bringing the output voltage to the desired amount.

If the value of R_L were to increase (the equivalent of removing a circuit from the system) the total resistance would increase. Total current would decrease, and the decrease in current through R_s would result in a lower voltage drop across the element. This would tend to cause the output voltage to rise. To compensate for this change the value of R_V must be decreased, increasing the amount of current through R_S increasing the voltage drop across R_S, and decreasing the voltage across the load to the desired amount.

Both the voltage regulators discussed rely upon a mechanical adjustment and would require continuous monitoring of the equipment to obtain, at best, barely satisfactory results. These voltage regulators were used for explanation purposes. Electronic devices (discussed next) accomplish the function of the variable resistors electronically and automatically to provide excellent results, immediately.

ZENER VOLTAGE REGULATOR

The breakdown diode, or zener, is an excellent source of variable resistance. Zener diodes come in voltage ratings ranging from 2.4 volts to 200 volts, with tolerances of 5, 10, and 20 percent and with power dissipation ratings as high as 50 watts.

The zener diode will regulate to its rated voltage with changes in load current or input voltage. Referring to the zener diode shunt-type regulator in figure 5-23, the zener diode VR1 is in series with fixed resistor R_S. The voltage across the zener is constant, thus holding the voltage across the parallel load R_L constant. Although the circuit shown depicts a positive voltage output, it is a simple matter to have a negative output voltage—reverse the zener and the polarities shown in figure 5-23.

The value of R_S in figure 5-23 has been fixed so that it can handle the combined currents of the diode and the load and still allow the diode to conduct well within the breakdown region. R_S stabilizes the load voltage by dropping the difference between the diode operating voltage and the unregulated input voltage.

The zener diode is a PN junction that has been specially doped during its manufacture so that when reverse biased it will operate at a specific breakdown voltage level. It operates well within its rated tolerance over a considerable range of reverse current.

Now for the operation of the zener diode shunt regulator as shown in figure 5-23. If input voltage to the regulator circuit decreases, the voltage

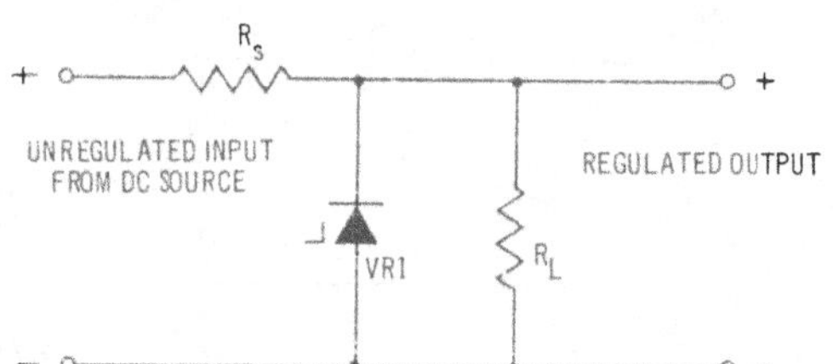

179.73

Figure 5-23.—Simple zener diode shunt type regulator circuit.

across the zener diode must also decrease causing zener current to decrease. The total current in the circuit decreases, much the same as when the value of R_V was increased in the simple shunt regulator of figure 5-22. The current through R_S, having decreased, results in a lower voltage drop across R_S. This results in the voltage drop across the zener and the load returning to the desired voltage. The zener diode has replaced the variable resistor of figure 5-22 and makes the necessary adjustment automatically with the change in input voltage or load current.

Effectively, the variable resistor that required manual adjustment has been replaced with an electronic equivalent—the zener diode.

Zener operation when input voltage decreases has been discussed; now, an analysis of operation will be made for an increase in input voltage. When input voltage increases, the change is immediately felt across the zener. This effectively biases the zener so that there is an increase in zener current. The increase in zener current means an increase in total circuit current. R_S, in series with the source, will have an increase in current through it, resulting in a larger voltage drop across it. With the larger drop across R_S, the voltage across the zener, and, therefore, the load, is reduced to the desired output voltage. The zener, in this instance, has a lower resistance.

For changes in load current the zener makes the adjustment so that source current remains constant, and the output voltage will also be constant. For instance, if the current drawn by the load decreases, the zener current will increase a corresponding amount. The total current remains the same. If the voltage across R_S is the same as it was before the load current decrease, and if source voltage has not changed, then it is logical to infer that the output voltage is at the desired regulated amount. Conversely, if the current drawn by the load increases, the zener current decreases by the same amount and total current in the circuit remains the same. Since total current is the same, the voltage drops across R_S and VR1 (and therefore, R_L also) cannot change because the source voltage has not changed.

Some words of caution about the shunt type voltage regulator are in order. Do not operate this circuit without a load. If a no-load condition exists, the zener must dissipate more power than usual—the load power as well as its normal power. If this condition occurs, it is quite possible that the maximum power rating, or the maximum reverse current rating, of the zener will be exceeded and damage to the zener will result.

If there is a failure in the voltage regulator circuit just discussed, the following checks should help in locating the trouble:

1. Check to find out if there is a load on the regulator circuit. The lack of a load might indicate a damaged zener as the source of trouble.
2. Check the d.c. voltage measurements at the input to determine whether voltage is applied and whether it is within tolerance.
3. Since the operation of this regulator is based upon the voltage divider principle, measurements of the voltages across the output terminals and across R_S might be necessary to determine if the output voltage is within tolerance or if the drop across series resistor R_S is excessive (be sure to observe correct voltage polarity when making these checks).
4. If the load is shorted, or if R_S is open, a voltage measurement across R_S will indicate source voltage. In both cases there will be no output. It will now be necessary to check the value of R_S. (Disconnect R_S from the circuit when making the measurement.)
5. If the zener diode becomes open there will be no regulation and the output voltage should be higher than normal.
6. If the zener diode becomes shorted there will be no output.

In general, if the output voltage is above normal, it is an indication of an open circuit in the shunt elements, either VR1 or the load, or an increase in impedance of these same elements. An output voltage that is below normal is an indication of an increased value of R_S, a low input voltage, or an excessive load current due to a decrease in load impedance.

COMBINATIONAL VOLTAGE SUPPLIES

Power supplies found in modern electronics equipment must supply a great variety of voltages and current. The need for combination power supplies becomes evident when equipment requirements are voltages in ranges of -5v., -15v., -55v., +5v., +120v., +2400v., and -580 volts for a typical oscilloscope, or -90v. at 6 ma., +28v. at 180 ma. and 580 ma., +130v. at 315 ma., and 250 v. at 37 ma. as used in the low voltage power supply for a typical radio transceiver.

The necessity of having a power supply capable of delivering a high voltage-low current output—a low voltage-high current output—and a high voltage-high current output, simultaneously, is the reason for having combinational power supplies.

It is not the purpose of this section to present all possible varieties of combinational power supplies. It has been described how both negative and positive voltages can be obtained from a voltage dividing network connected to a power supply. In the section dealing with voltage multipliers it will be shown how aiding voltages are combined to increase the output. It should be realized that, in addition to the sum of the voltages available, any of the voltages comprising that sum might also be used. Two basic combinations will be presented here—the full-wave-bridge and the full-wave full-wave.

FULL-WAVE BRIDGE

Figure 5-24 is an example of the full-wave-bridge combinational power supply. Part A shows a simplified schematic drawing and part B shows the entire schematic, including the filtering network. It is a typical arrangement of the full-wave bridge combination supply, quite often called the "economy" power supply.

In figure 5-24B, CR1 and CR3 form the full-wave rectifier circuit and C2A, C2B, and L2 form the filter network. R1 is a current limiting resistor used to protect the diodes from surge currents. R4 is the bleeder resistor and also assures that the power supply has a minimum load at all times. CR1 and CR4 form the bridge rectifier circuit with L1, R2, C1A and C1B doing the filtering; R3 is a bleeder resistor and assures that the bridge always has a minimum load. Each circuit, of itself, works in the conventional manner. Troubleshooting will be the same as for the other power supplies covered, entailing the NO OUTPUT—LOW OUTPUT factors.

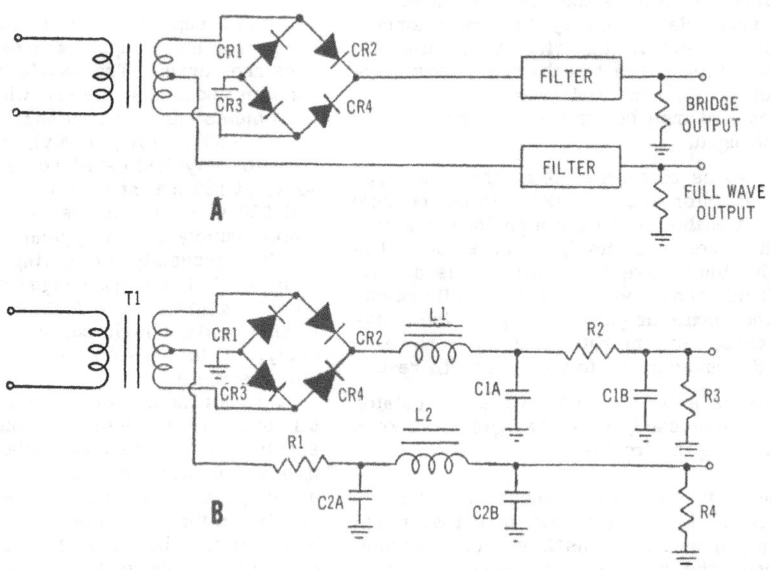

179.74
Figure 5-24.—Full-wave bridge combination power supply.

FULL-WAVE FULL-WAVE

Figure 5-25 illustrates a full-wave full-wave combinational power supply, with positive and negative outputs. It has one primary distinguishing feature compared to a bridge circuit—the center tapped transformer secondary. The components associated with the negative voltage output are CR1, CR3, L2, C2, and R2, while CR2, CR4, L1, C1, and R1 are the components in the positive voltage output. Transformer T1 is a component common to both supplies.

The operation of each full-wave rectifier is identical. As a refresher, the operation will be reiterated. When point A is negative with respect to ground, point B will be positive. This condition causes both CR1 and CR4 to conduct. Both these diodes are associated with different full-wave rectifiers. The negative power supply will be described first—this is the one associated with the conduction of CR1. Current through CR1 flows through L2, C2, and R2 and completes the path via ground to transformer centertap, thence to A. This is the path during the first half cycle while CR1 is conducting. At the same time this action is taking place CR4 is conducting in the positive power supply. Since point B is positive and the lower half of the transformer secondary acts as a source, current from point B to ground flows up through C1, charging it as shown, L1, CR4 and back to the source. When the polarities at point A and point B are reversed, point A now being positive and point B being negative, with respect to ground, the conducting diodes are now CR3 in the negative supply and CR2 in the positive supply.

Again, taking the negative supply first, CR3 conducts and the current path, from point B and back to point B, is through CR3, L2, up through C2 in the same manner as when CR1 was conducting, to ground and through the center tap of the transformer to point B. This completes the full cycle of operation for the negative supply and CR1 and CR3 will be conducting alternately so long as there is an input.

At the time CR3 is conducting in the negative power supply, CR2 is conducting in the positive power supply. Current, from point A back to point A, is going down the upper half of the transformer secondary to ground, up through C1 and L1, through CR2 and back to point A.

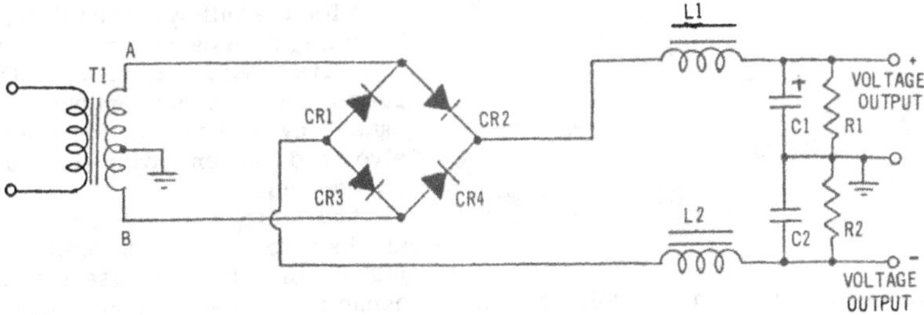

179.75

Figure 5-25.—Full-wave full-wave combinational power supply with positive and negative output voltages.

Again, the charge path for C1 is the same as it was when CR4 was conducting.

While it has been said, that when troubleshooting the combinational power supply, the troubles to look for are of the NO OUTPUT—LOW OUTPUT type, it should be realized that in the full-wave bridge combinational power supply that a low output can occur in either section or the voltage out of the full-wave rectifier can be normal, while the bridge rectifier might indicate a low output. If the load currents are not excessive, and the filter components have checked satisfactorily, the defective diodes, in the first case, would be assumed to be CR1 or CR3 and, in the second case to be CR2 or CR4.

VOLTAGE MULTIPLIERS

Figure 5-26 depicts a simple half-wave rectifier circuit that is capable of delivering a voltage increase (more voltage output than voltage input), providing the current being drawn is low. It is shown that it is possible to get a larger voltage out of a simple half-wave rectifier so long as the current is low. If the current demand increases, the output voltage will decrease. This can best be explained by the use of RC time constant. The charge time for the circuit in figure 5-26 is very fast since the circuit elements in the capacitor's charge path are the diode, CR1, the surge resistor, R, and the secondary of the transformer. These elements combine to form a very low impedance, since CR1 is conducting during the charge time and the value of R is about 20 ohms. In comparison, the discharge path for the capacitor is through the load which offers an impedance several hundred times higher than that of the charge path. The lower the load

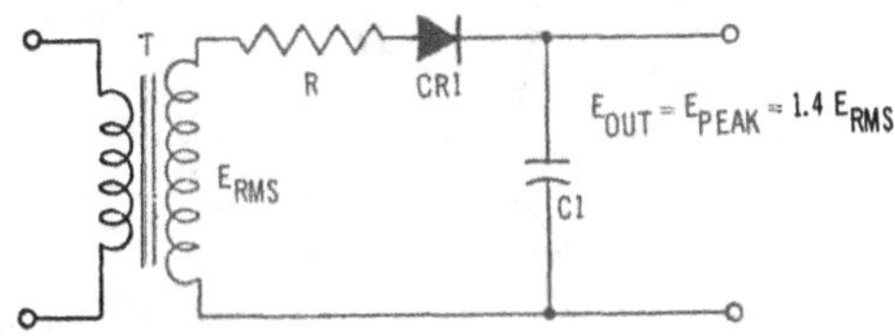

179.76

Figure 5-26.—Simple half-wave rectifier used to deliver an increased voltage output.

impedance the greater the current. If the discharge path for the capacitor offers a lower impedance, the capacitor will discharge further, lowering the output voltage.

Rectifier circuits that can be used to double, triple, and quadruple the input voltage will now be discussed. All these circuits have one thing in common—they use the charge stored on capacitors to increase the output voltage. Figure 5-27 is a block diagram of a voltage multiplier circuit. The input is a.c. and the output is d.c. multiple.

HALF-WAVE VOLTAGE DOUBLER

The first voltage multiplier circuit is the half-wave voltage doubler. As the name implies, this circuit gives a d.c. output that is approximately twice that obtained from the equivalent half-wave rectifier circuit.

Figure 5-28 shows a typical half-wave voltage doubler circuit. While the circuit shown uses a transformer and the output voltage is positive with respect to ground, it could just as well operate as a negative voltage output by reversing

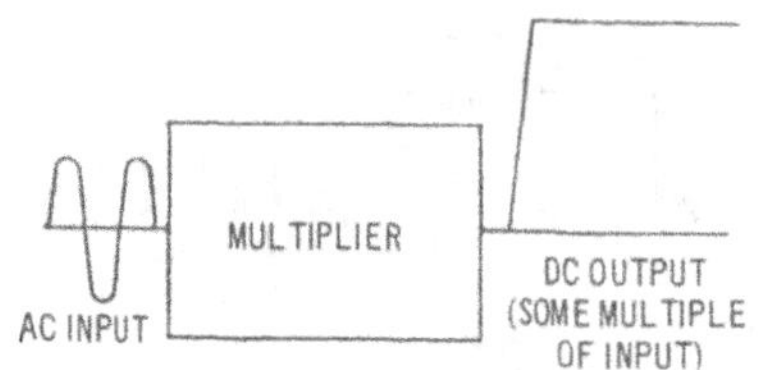

179.77
Figure 5-27.—Block diagram of a voltage multiplier circuit.

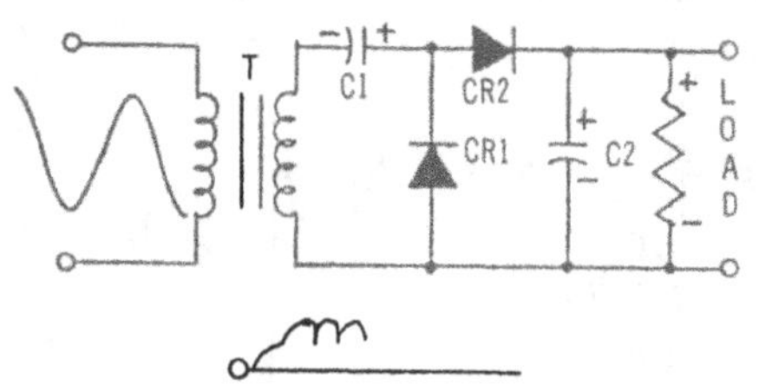

179.78
Figure 5-28.—Half-wave voltage doubler circuit.

the diodes. The transformer, which may be used to step up secondary voltage or as an isolation transformer, may also be eliminated with the proper choice of circuit elements.

When the top of the transformer secondary in figure 5-28 is negative, C1 will charge through conducting CR1 to approximately the peak of the secondary voltage. The direction of charge is indicated by the polarity signs. At this time there is no output. On the next alternation, when the top of the transformer secondary is positive with respect to the bottom, C1 will discharge through the transformer and CR2, which is now conducting. C2, which is also in this discharge path, is charged to approximately twice the peak of the secondary voltage because the charge on C1 is in series with the applied a.c. and therefore, adds to the voltage applied to C2.

Since C2 receives only one charge for every cycle of operation, the ripple frequency, as in the half-wave rectifier, is the same as the input frequency. Also, as in the half-wave rectifier, C2 will discharge slightly between charging cycles so that filtering is required to smooth the output and give us a relatively pure d.c.

The procedures for troubleshooting the half-wave voltage doubler are the same as those used for the half-wave rectifier. No output conditions might be caused by a defective transformer, defective rectifiers, an open C1, or short circuited C2. The low output condition might be caused by a low input, rectifier aging, or excessive load current (caused by a decrease in load impedance).

The voltage multiplier circuits which follow all have one thing in common with the circuit just described—they use the charge stored on a capacitor to increase the output voltage. As the voltage across C1 is added to the input voltage to approximately double the charge applied to C2, so will the charges on other capacitors add to the charge applied to an input capacitor to double, triple or quadruple the output voltage.

FULL-WAVE VOLTAGE DOUBLER

The full-wave rectifier circuit can also be adapted to a voltage doubling circuit. Figure 5-29 depicts a basic full-wave voltage doubler circuit. Depending upon the circuit application it may or may not use a power or isolation transformer. The resistor R_s is a "surge" resistor that is used to limit the charge current and protect the diode; it might not be necessary in some equipments and when used it is placed in series with the a.c. source. Resistors R1 and R2 are not necessary for circuit operation but may be used to act as bleeder resistors to discharge their associated capacitors when the circuit is deenergized. When used they also tend to equalize the voltages across C1 and C2.

The circuit operates much the same as the full-wave rectifier previously discussed with the exception that now two capacitors are employed, each one charging to approximately the

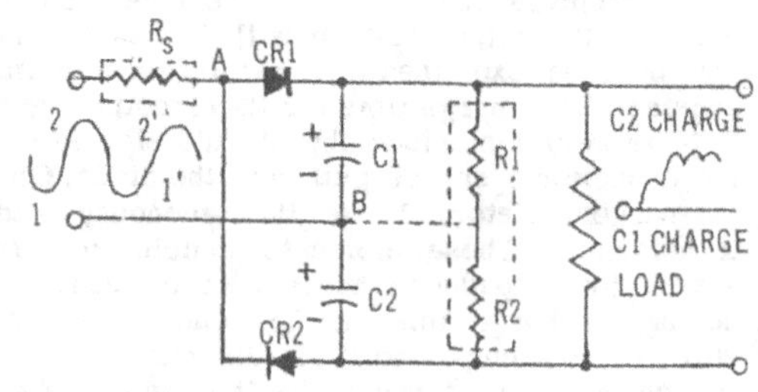

179.79
Figure 5-29.—Basic full-wave voltage doubler.

peak voltage of the input, adding their charges to provide an output. When point A is positive with respect to point B, C1 will charge through the conducting diode, CR1, and the source. It will charge to approximately the peak of the incoming voltage. On the next half cycle of the input point A is now negative with respect to point B, CR2 conducts charging C2 in the direction indicated. The voltage across the load will be the total of the voltages across C1 and C2. C1 and C2 will be equal value capacitors and R1 and R2 will also be equal value resistors. The value of R_s will be small, probably in the 20-500 ohm range.

VOLTAGE TRIPLER

Figure 5-30 depicts a typical voltage tripler circuit with waveforms and circuit operation. Part A shows the complete circuit. In part B, C1 is shown charging as CR2 is conducting. In part C, C3 is illustrated charging as CR3 is conducting. Part D reveals the charge path for C2 while CR1 is conducting. In part E a comparison

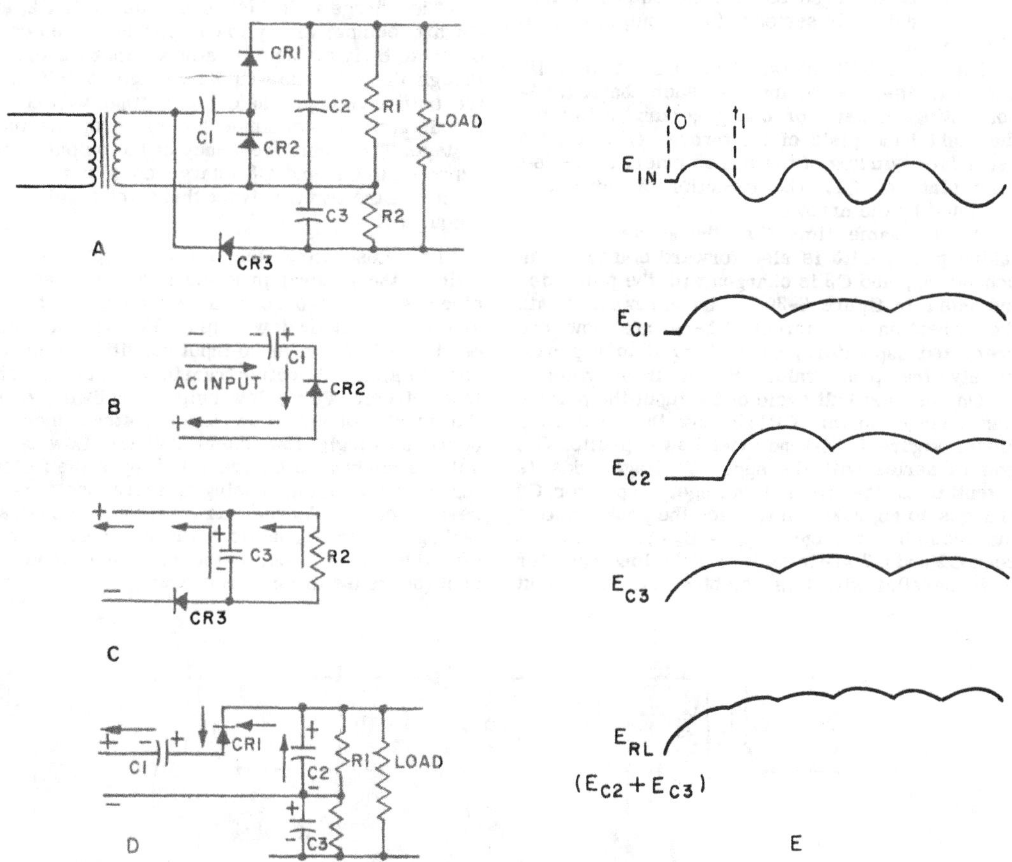

179.80

Figure 5-30.—Typical voltage tripler circuit operation and waveforms.

is made of the input signal and its effects on the voltages felt across C1, C2, C3, and the load. The following explanation uses figure 5-30 as the operating device:

Close inspection of figure 5-30A should reveal that removal of CR3, C3, R2 and the load resistor results in the voltage doubler circuit previously described. The connection of circuit elements CR3, and the parallel network C3 and R2 to the basic doubler circuit is arranged so that they are in series across the load. The combination provides approximately three times more voltage in the output than is felt across the input. Fundamentally, then, this circuit is a combination of a half-wave voltage doubler and a half-wave rectifier circuit arranged so that the output voltage of one circuit is in series with the output voltage of the other.

Figure 5-30B shows how C1 is initially charged. Assume the input is such that CR2 is conducting. A path for charging current is from the right hand plate of C1 through CR2 and the secondary winding of the transformer, to the left hand plate of C1. The direction of current is indicated by the arrows.

At the same time that the above action is taking place, CR3 is also forward biased and is conducting, and C3 is charged with the polarities indicated in figure 5-30C. The arrows indicate the direction of current. There are now two energized capacitors, each charged to approximately the peak value of the input voltage.

On the next half cycle of the input the polarities change so that CR1 is now the conducting diode. Figure 5-30D indicates how capacitor C1, now in series with the applied voltage, adds its potential to the applied voltage. Capacitor C2 charges to approximately twice the peak value of the incoming voltage ($E_{in} + E_{C1}$). As can be seen C2 and C3 are in series and the load resistor is in parallel with this combination. The output voltage then will be the total voltage felt across C2 and C3, or approximately three times the peak voltage of the input.

Figure 5-30E indicates the action taking place using time and the incoming voltage. At time zero (t_0) the a.c. input is starting on its positive excursion. At this time the voltage on C1 and C3 is increasing. When the input starts to go into its negative excursion, t_1, both C1 and C3 start to discharge and the voltage across C2 is increasing. The discharge of C1 adds to the source voltage when charging C2 so that the value of E_{C2} is approximately twice the value of the peak value of the input. Since E_{C2} and E_{C3} are in series across the load resistor, the output is their sum.

The charge paths for capacitors C1, C2, and C3 are comparatively low impedance when compared to their discharge paths and therefore, even though there is some ripple voltage variations in the output voltage, the output voltage will be approximately three times the value of the input voltage. The ripple frequency of the output, since capacitors C2 and C3 charge on alternate half cycles of the input, is twice that of the input ripple frequency.

Troubleshooting the voltage tripler circuit follows the general practice given for rectifier circuits. The two general categories of failure are no output or low output. For the no output condition look for a no input condition, a lack of applied a.c., a defective transformer, or a shorted load circuit. For a low output condition, check the input voltage. Low input voltage means a correspondingly low output voltage. Low output voltage might also be a result of any of the following: rectifier aging causing an increased forward resistance or a decreased reverse resistance; any leakage in the capacitors or decrease in their effective capacitance; an increase in the load current (decrease in load impedance).

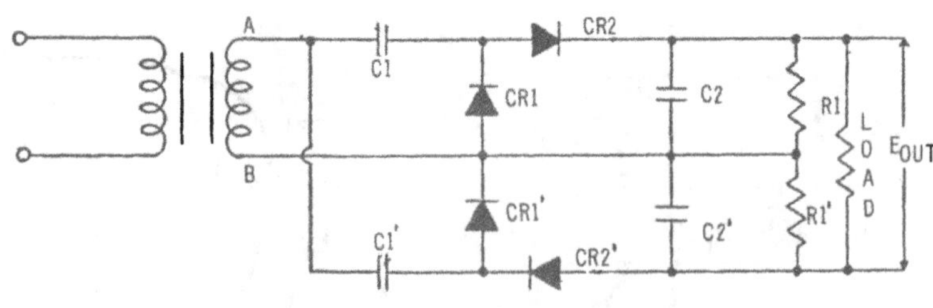

179.81

Figure 5-31.—Voltage quadrupler circuit.

VOLTAGE QUADRUPLER

Figure 5-31 shows a typical voltage quadrupler circuit. Essentially this circuit is two half-wave voltage doubler circuits connected back to back and sharing a common a.c. input. In order to show how the voltage quadrupler works, figure 5-31 has been shown as two voltage doublers. The counter parts of the one circuit are shown as a prime (') in the second circuit. (C1 in the first circuit is the same as C1' in the other circuit; CR1 and CR1' conduct at the same time, etc.) When the circuit is first turned on it will be assumed that the top of the secondary winding, point A, is negative with respect to B. At this instance CR1 is forward biased and conducts allowing C1 to charge. On the next alternation point B is negative with respect to point A. At this time two things are going to occur: (1) C1, which was charged to approximately the peak voltage across the secondary winding, will aid the source and, since CR2 is now conducting, C2 will be charged to approximately twice the incoming voltage; and (2) CR1' will conduct charging C1' to approximately the peak voltage of the input. During the following alternation C1' adds to the input which allows CR2' to conduct, charging C2' to twice the input voltage.

When CR2 conducts, C1 will aid the input voltage in charging C2 to approximately twice the peak voltage of the secondary and CR2'; conducting at the same time charges C1'. On the next alternation CR1' conducts, and since C1 is in series with the input, it aids in charging C2' to twice the peak of the secondary. The voltages across C2 and C2' add to provide four times the peak secondary voltage in the output.

CHAPTER 6

TRANSISTORS

A transistor is a device which utilizes a small change in current to produce a large change in voltage, current, or power. The transistor, therefore, may function as an amplifier or an electronic switch. An amplifier is a device which increases the voltage, current, or power level of a signal applied to that device. The output of an amplifier may or may not be an exact replica of the input waveshape. An electronic switch is a device which utilizes a small voltage or current to turn on or turn off a large current flow, with great rapidity.

ANALYSIS OF TWO JUNCTION TRANSISTORS

There are two fundamental types of junction transistors. These are the NPN and PNP. Their basic construction, or chemical treatment is as implied by their names. That is, either a section of P material is separated, by junction barriers, from two sections of N material; or a section of N material is separated, by junction barriers, from two sections of P material. Simplified cross section diagrams and schematic symbols of the devices are depicted in figure 6-1.

Note that the devices are not simply two junction diodes. The base section is extremely thin compared to the collector and emitter sections. Also, the surface area of the emitter-base junction is less than that of the collector-base junction.

BIASING AND BASIC CURRENT PATHS

Consider the case wherein a reverse bias is applied across the collector-base junction. This is depicted in figure 6-2. The reverse bias will cause almost no current flow. As in the basic PN junction diode, a very small reverse current exists. At normal operating temperatures this reverse current may be considered negligible.

In the case of the transistor, the reverse bias from collector to base is made close to the reverse breakdown voltage of the junction. This

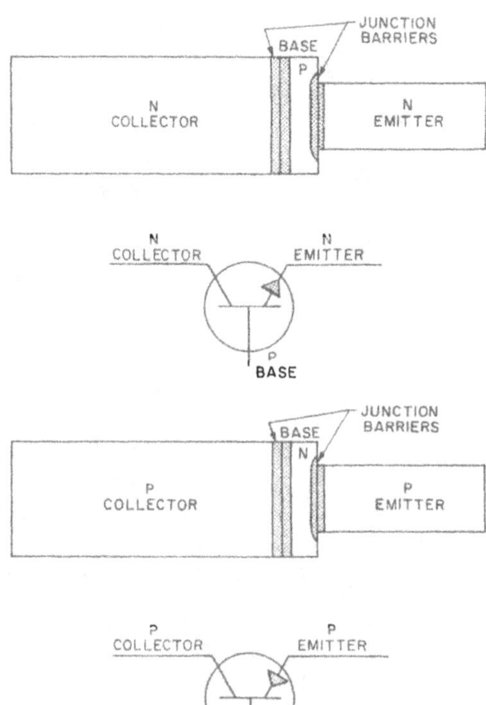

179.82

Figure 6-1.—Transistor cross section diagrams and schematic symbols.

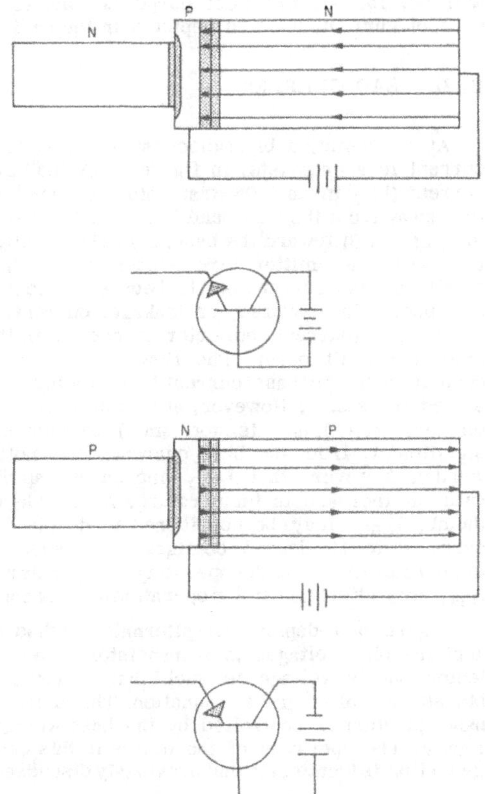

179.83
Figure 6-2.—Reverse bias across collector-base junction.

is depicted in figure 6-3, a graph of the collector-base current versus voltage.

If a small forward bias is now applied to the emitter-base junction, its junction barrier will be eliminated. Also, a concentrated field will now exist between emitter and collector. This is depicted in figure 6-4. The concentrated field, resulting from the emitter-base forward bias, aids the field caused by the collector-base reverse bias and causes the collector-base junction to move into the avalanche (reverse breakdown) region. This is depicted in figure 6-5.

With the collector-base junction in a form of avalanche breakdown, current flow through the

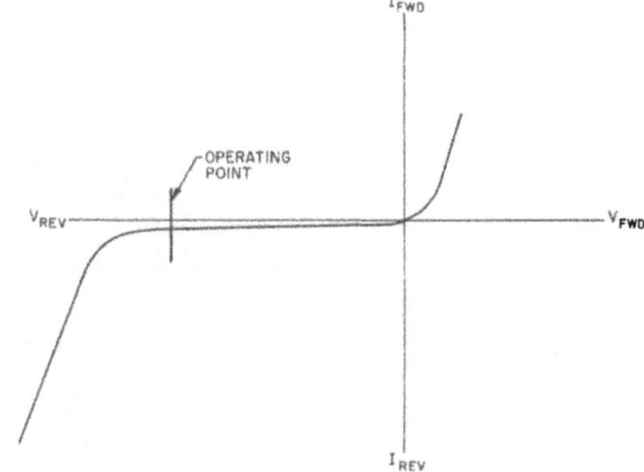

179.84
Figure 6-3.—Graph of collector-base current versus voltage.

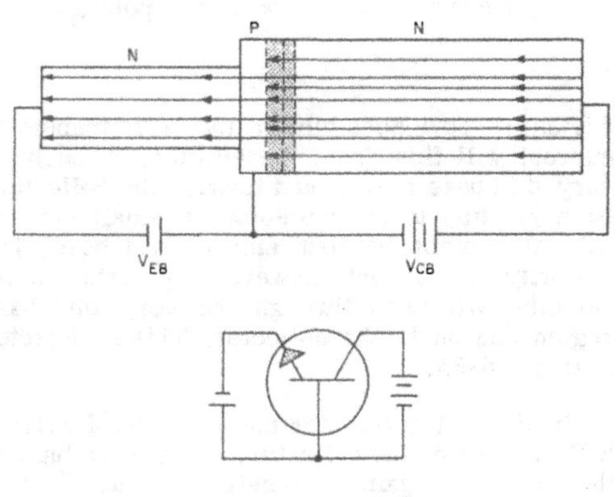

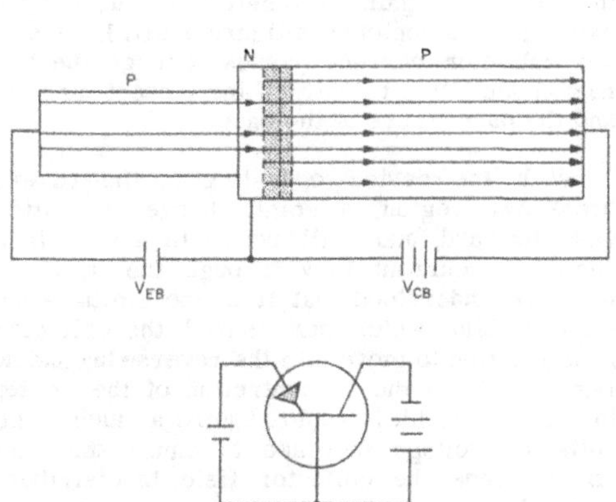

Figure 6-4.—Adding forward bias to emitter-base junction. 179.85

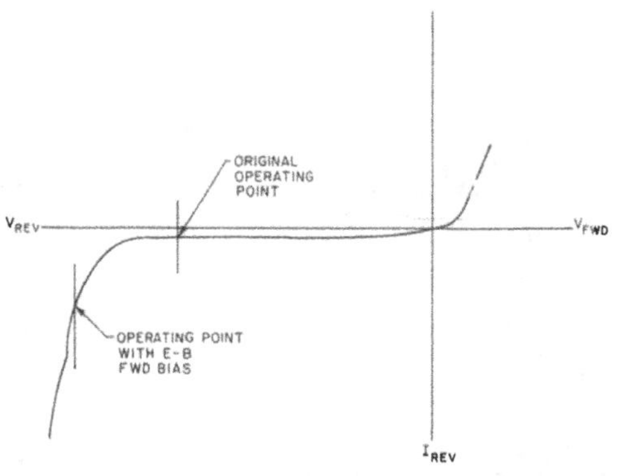

179.86

Figure 6-5.—Characteristic curve showing effect of forward bias on operating point.

device is now possible. In the NPN transistor current will flow from the emitter, through the very thin base region, and through the collector-base junction to the collector. A small current will also exist between emitter and base. The majority of current, however, by virtue of its velocity will pass through the very thin base region and on to the collector. This is depicted in figure 6-6A.

In the PNP transistor the majority of current will flow from the collector, through the base to the emitter. Again, a small base current will exist. This is depicted in figure 6-6B. Here also, the collector current passes through the base region due to the velocity of the current carriers, and the narrowness of the base.

With the device operating in the reverse breakdown region, a small change in emitter-base forward bias will result in a very large change in current flow through the device. It must be understood that it is the emitter-base forward bias which has caused the collector-base junction to move into the reverse breakdown region, due to the concentration of the emitter to collector field. It would require a much larger collector voltage increase to cause the same effect, since the collector field is distributed over a larger area.

If the emitter-base junction is reverse biased, along with the collector-base junction, the device will be, for all practical purposes, unable to conduct current. This is depicted in figure 6-7.

BASE LEAD CURRENT

At this point, a brief discussion of base lead current is appropriate. In figure 6-6A the base current (I_B) in an NPN transistor is shown to flow away from the base, and in the PNP transistor (fig. 6-6B) toward the base. It will be recalled that with the emitter circuit open (fig. 6-2) a small reverse current exists between collector and base. The reverse, or leakage, current is called I_{CBO} (current, collector to base, with the emitter circuit open). The flow of I_{CBO} is in opposition to the base current flow in a forward biased transistor. However, at normal operating temperatures I_{CBO} is too small to subtract significantly from the base current. It is noteworthy, however, that I_{CBO} increases rapidly with an increase in temperature. As a rule of thumb, I_{CBO} may be considered to double for every 8 to 10 degree centigrade increase in temperature. At high operating temperatures I_{CBO} may affect transistor operation unfavorably.

Figure 6-8 depicts an alternate method of applying bias voltages to a transistor. The collector supply voltage is such that it reverse biases the collector-base junction. The emitter-base junction is controlled by the base voltage supply. The operation of the device in this configuration is identical to that previously discussed.

A list of common symbols and abbreviations used in connection with semiconductor devices is included in the appendix. These symbols and abbreviations will be used in the following discussions.

CURRENT GAIN IN THE BASIC TRANSISTOR

Current in the emitter circuit can be varied by changing the emitter-base voltage. The change in the emitter current will cause an almost equal change in the collector current. The change in the collector current is always slightly less than the change in the emitter current because some of the emitter current will leave the transistor as base current. The collector current is the difference between the emitter current and the base current, or:

$$I_C = I_E - I_B$$

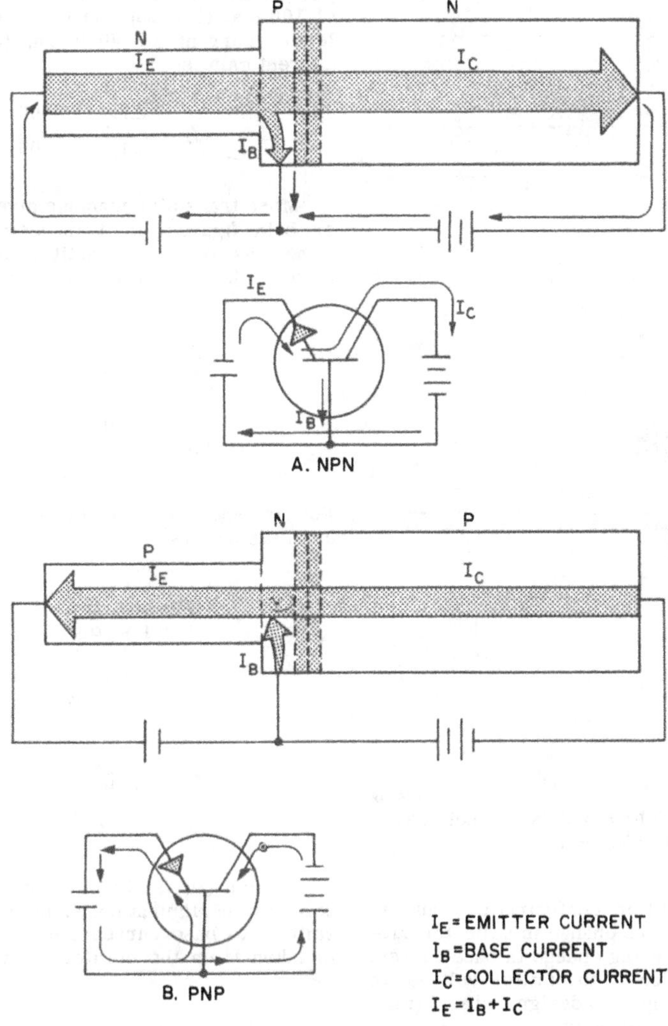

179.87

Figure 6-6.—Current flow in NPN and PNP transistors.

A transistor has either of three current gains depending upon the circuit configuration. Current gain in the common base configuration is the ratio of the change in collector current to the corresponding change in emitter current for a constant collector to base voltage. In this circuit, current gain is designated as alpha $(\alpha)_{ce}$. The equation is:

$$\alpha_{ce} = \frac{\Delta I_C}{\Delta I_E}$$

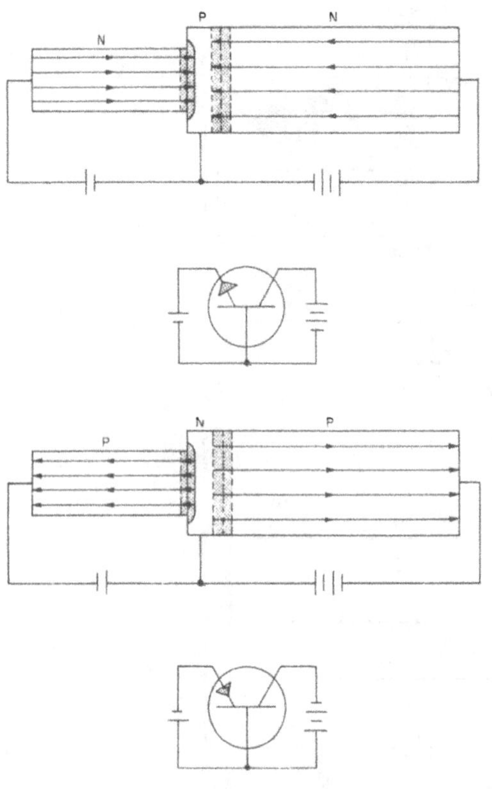

179.88
Figure 6-7.— Transistors with both junctions reverse biased.

In the common emitter configuration, current gain is the ratio of the change in collector current to the corresponding change in base current for a constant collector to emitter voltage. In this circuit, current gain is designated as β (pronounced beta). The equation is:

$$\beta = \frac{\Delta I_C}{\Delta I_B}$$

The method for solving for current gain can best be expressed by use of an example. With collector voltage adjusted to -10 volts and base current equal to 25 microamperes, collector current I_C = 1 ma. or 1000 μa. Increasing I_B to 125 μa. (with constant V_C) increases the collector current to approximately 5 ma. or 5000 μa. Thus, the increase in input base current of 100 μa. (125-25) causes an increase in collector current of 4000 μa. (5000-1000). The current gain is:

$$\beta = \frac{\Delta I_C}{\Delta I_B} = \frac{4000}{100} = 40$$

Most transistor manuals give either the α_{ce} or beta figures. If it is desired to determine either one of these quantities when the other is known, the following equations may be used:

$$\beta = \frac{\alpha_{ce}}{1 - \alpha_{ce}}$$

$$\alpha_{ce} = \frac{\beta}{1 + \beta}$$

For instance, if a transistor has an α_{ce} of 0.97, the beta is:

$$\beta = \frac{\alpha_{ce}}{1 - \alpha_{ce}}$$

$$\beta = \frac{0.97}{1 - 0.97}$$

$$\beta = \frac{0.97}{0.03}$$

$$\beta = 32.33$$

Gamma (γ) is the current gain of a common collector configuration. Gamma is based on the emitter to base current ratio because the output is taken from the emitter circuit. The equation is:

$$\gamma = \frac{\Delta I_E}{\Delta I_B} \text{ with } V_{CC} \text{ constant}$$

The conversion equations from alpha or beta are:

$$\gamma = \frac{1}{1 - \alpha_{ce}}$$

$$\gamma = \beta + 1$$

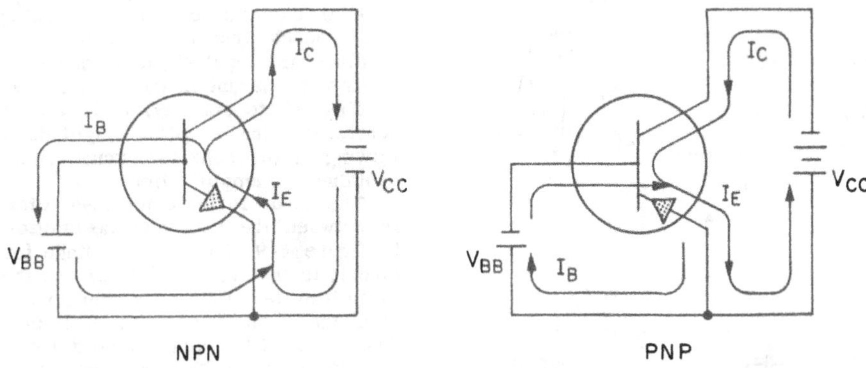

179.89

Figure 6-8.—Alternate method of applying bias to transistors.

BASIC TRANSISTOR AMPLIFIERS

Before going into the basic transistor amplifier, there are three terms to be defined, amplification, amplifier, and classes of operation. Amplification is the process of increasing the strength of a signal. An amplifier is the device used to increase the current, voltage, or power of an input signal. Amplification is the result of controlling a relatively large quantity of current (output circuit) with a small quantity of current (input circuit).

The transistor amplifier may be connected in any one of three basic configurations. These circuits are (1) common emitter, (2) common base, and (3) common collector. The word common, as used here, means the element named is part of (common to) both the input and output circuits.

Transistor amplifiers may be operated class A, class B, class AB, and class C. A class A amplifier is an amplifier in which the emitter-base junction bias and the alternating voltages applied to the emitter-base junction are such that collector current (I_C) in specific transistors flows at all times. It is operated so that the wave shape of the output is the same as that of the signal applied at the input.

A class B amplifier is an amplifier in which the emitter-base bias is approximately equal to the cutoff value, so that I_C is approximately zero when no excitation voltage is applied to the emitter-base junction. I_C flows, in a specific transistor, for approximately one-half of each cycle when an alternating signal voltage is applied to the emitter-base junction.

A class AB amplifier is an amplifier in which the emitter-base bias and the alternating voltages applied to the emitter-base junction are such that I_C in a specific transistor flows for appreciably more than half but less than the entire electrical cycle.

A class C amplifier is an amplifier in which the emitter-base bias is appreciably greater than the cutoff value. I_C in the transistor is zero when no alternating voltage is applied to the emitter-base junction and flows in a specific transistor for appreciably less than one-half of each cycle when an alternating signal voltage is applied to the emitter-base junction.

COMMON EMITTER TRANSISTOR AMPLIFIER

The NPN and PNP common emitter transistor amplifiers are shown in figure 6-9A and B, respectively. Bias (the average difference of potential across the emitter-base junction) of the transistors shown in figure 6-9 is in the forward, or low resistance, direction. This means that direct current is flowing continuously, with or without an input signal, throughout the entire circuit. The bias current, flowing through the emitter-base junction, is supplied by bias battery V_{BB} . The average base voltage, as measured with respect to the emitter, depends upon the magnitude of the bias voltage supply in the emitter-base circuit and the base resistor. The current and the amplitude of the input signal

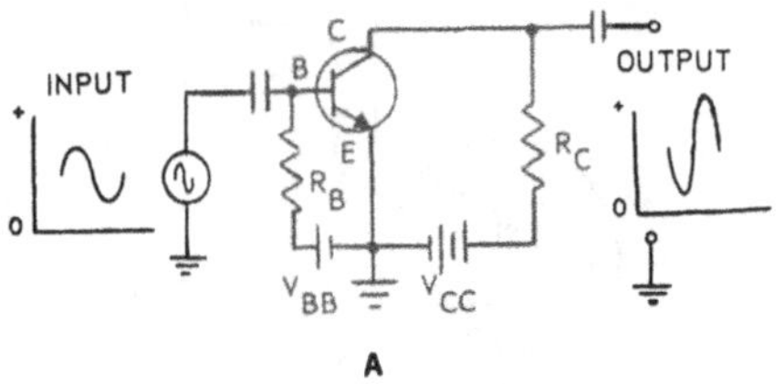

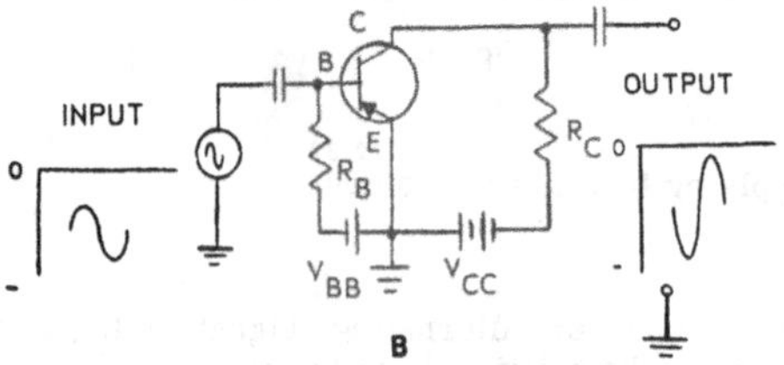

20.413

Figure 6-9.—NPN and PNP common emitter transistor amplifiers.

determine the class of operation of the transistors. (In this explanation the transistor is operating class A.) The average collector voltage depends upon the magnitude of the collector supply voltage, V_{CC}.

When a signal is supplied to the input circuit of the transistor (the base-emitter junction here) the bias current varies about the average, or no signal, value. That portion of the signal that aids the bias causes transistor current to increase, while that portion of the signal that opposes the bias causes transistor current to decrease.

Since the input signal is capable of causing transistor current to increase or decrease, according to its polarity, it is also capable of controlling the greater amplitude of current flowing in the collector circuit. This, as already explained, is amplification.

The input to the common-emitter amplifier is between the base and the emitter as shown in figure 6-9. The output voltage for this type circuit is that voltage felt across the collector to emitter terminals. The emitter is common to both the input and output circuits, hence the designation of the common emitter configuration.

The output voltage will vary with collector current. This is the same as stating that the voltage on the collector will vary with collector current. As collector current increases, collector voltage decreases and vice versa due to the drop across the collector resistor RC.

It is possible to visualize the transistor acting as a variable resistance. When the transistor is allowing more current to flow, its resistance is lower than when it is causing less current to flow. Figure 6-10 will be referred to for further explanation. That portion of the input signal indicated by point 1 in the input circuit effectively aids the bias, thereby reducing resistance of the transistor to current flow. Current flowing in the transistor increases so that current in the entire circuit is increased. This increased current must flow through R_C and thereby increases the voltage drop across R_C. The increased voltage drop across R_C leaves less voltage to be dropped across the transistor. Since the output is taken from the collector terminal to the emitter terminal (ground) it is evident that collector voltage (V_C) is less than it had been at its no signal value of +4 volts. At point 1

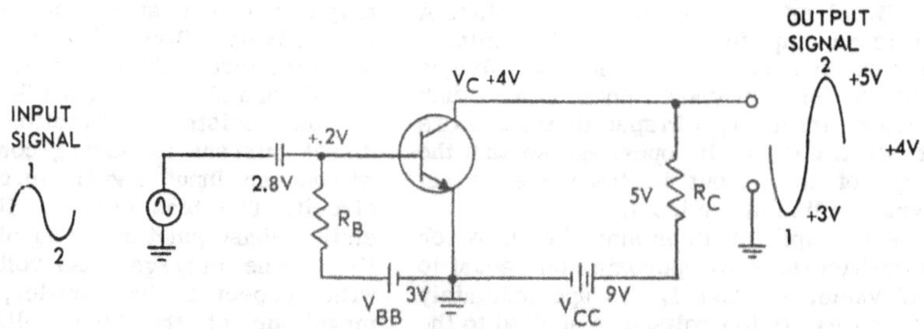

20.413

Figure 6-10.—Basic NPN common emitter transistor amplifier.

of the output waveform (corresponding to point 1 of the input signal) the collector voltage is at +3 volts.

As the input signal goes from point 1 to point 2 the transistor is increasing its resistance until, at point 2, it offers its maximum resistance to current flow. Less current flowing in the transistor means less current is flowing in the circuit. This reduction in current flowing through R_C results in a reduced voltage drop across R_C. The reduction in the voltage drop across R_C means that the collector voltage is higher. This increase is shown as +5 volts at point 2 on the output signal. Again point 2 on the input signal and point 2 on the output signal occur essentially at the same time.

In succeeding cycles, as the input signal aids the forward bias, transistor current increases causing an increased voltage drop across R_C, thereby decreasing the output voltage. As the input signal opposes the forward bias, transistor current decreases. This decrease in current through R_C means a reduced voltage drop across R_C and more voltage available in the output.

In any of the amplifier circuits which are presented, moving the base potential towards the collector voltage (V_C) increases transistor current, and moving the base voltage away from collector voltage decreases transistor current. As shown in figures 6-9 and 6-10, the input and output signals are 180° out of phase. The common emitter amplifier, whether it be an NPN or PNP circuit, provides a phase inversion.

COMMON BASE TRANSISTOR AMPLIFIER

The common base amplifier, using an NPN and a PNP transistor, is shown in figure 6-11. In these amplifier circuits the input signal is placed on the emitter of the transistor and the output signal is taken from the collector, with the base element being common to both.

As noted previously, a PN junction, whether it be in the diode or a transistor, will react to a difference of potential applied across the junction. When this difference of potential is such that the depletion area is reduced, the junction is biased in the forward, or low resistance direction, allowing more current to flow. Conversely, when the difference of potential increases the depletion area, resistance increases and less current flows.

The input signal, being applied to the emitter of the common base amplifier, will cause the emitter-base junction to react in the same manner it did in the common emitter circuit. An input signal that aids the bias will increase the transistor current, and one that opposes the bias will decrease transistor current.

Using figure 6-12, the operation of a common base amplifier will be explained at this time. As the input signal to the emitter is going in the direction of point 1 its magnitude adds to the positive voltage on the emitter. The difference of potential between emitter and base increases, reducing the resistance of this junction and allowing more current to flow in the transistor. This increased magnitude of current flowing through R_C causes an increased voltage drop across R_C — the voltage drop goes from 5 volts to 6 volts. The collector voltage (V_C) which had been at a -4 volts in the quiescent (no signal) stage will now become -3 volts — or more positive than it had been. Again, point 1 on the input signal and point 1 on the output signal represent the same period in time. When the input signal moves toward point 2 its magnitude subtracts from the value of positive voltage on the emitter and the difference of potential across the emitter-

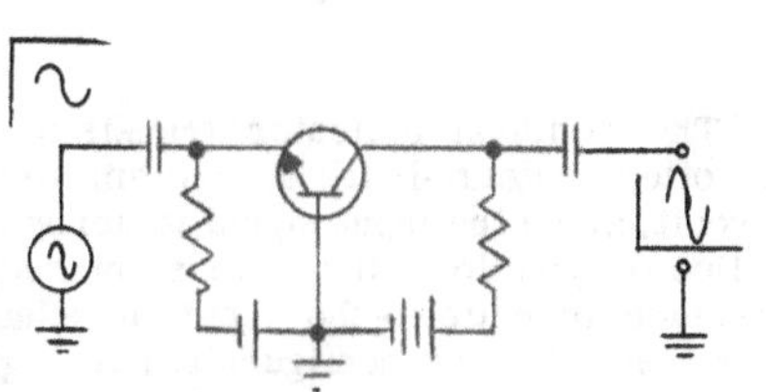

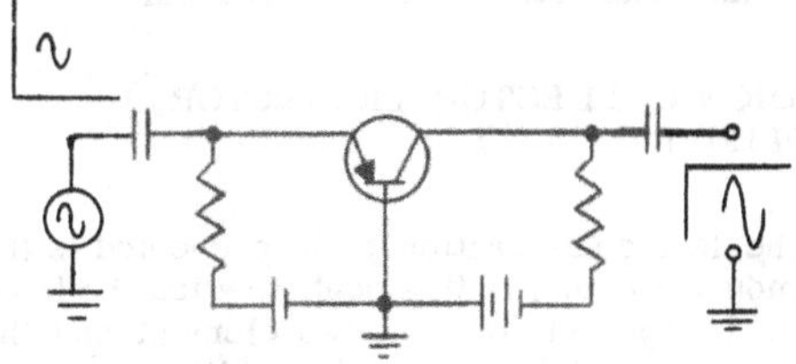

179.90

Figure 6-11. — NPN and PNP common base amplifier circuits.

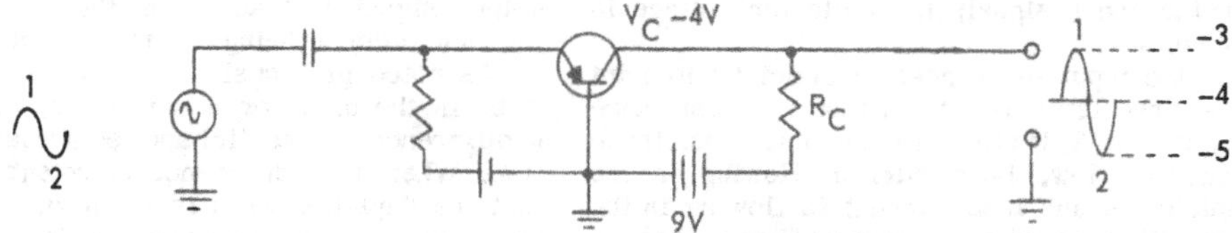

179.91

Figure 6-12.—Basic PNP common base transistor amplifier.

base junction decreases. This decrease in potential across the emitter-base junction results in an increase in the junction resistance. This higher junction resistance decreases the magnitude of current flowing in the transistor. Less current flowing through R_C means a reduction in the voltage drop across R_C —from 5 volts in the no signal condition to -4 volts at point 2 on the input. Collector voltage now becomes a -5 volts or more negative. The output voltage at point 2 occurs at the same time as the input signal is at point 2. Note that the output voltage is taken from the collector to ground, rather than across R_C.

As can be observed from the waveforms, there is no phase inversion in the common base amplifier. The input signal and the output are in phase. Although the input signal in the common base amplifier was applied to the emitter element, and made the emitter more, or less, positive, it had the same effect on the collector as making the base more, or less, negative.

If two points of reference are to be considered such as the emitter and the base, and one is made more positive or negative with respect to the other; the other becomes more negative or positive with respect to the first. When the emitter was made more positive in the previous explanation, the base was moving towards the collector voltage (negative in this case), and transistor current increased.

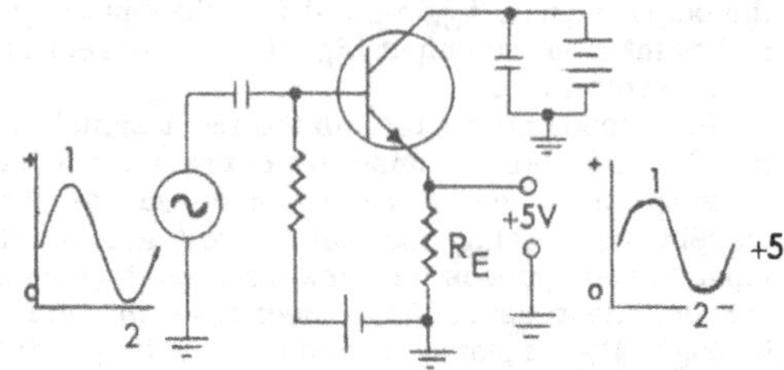

179.92

Figure 6-13.—NPN common collector transistor amplifier.

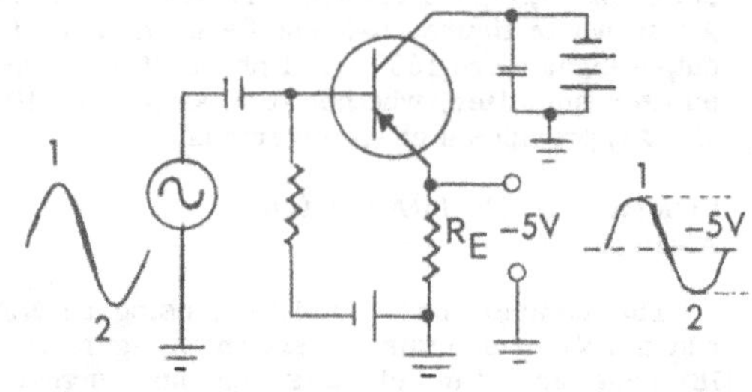

179.93

Figure 6-14.—PNP common collector transistor amplifier.

COMMON COLLECTOR TRANSISTOR AMPLIFIER

The last configuration to be presented is the common collector. In this configuration the input signal is applied to the base element and the output will be taken from the emitter with the collector common to both input and output. The common collector amplifiers are shown in figure 6-13 (NPN) and figure 6-14 (PNP).

The common collector transistor amplifier is often referred to as an emitter follower circuit. Since the input signal is across the base-collector junction—the reverse biased, high resistance direction—the input impedance of the common collector configuration is high. On the other hand, the output impedance of this circuit is low, since it is taken in the forward bias, low resistance direction of the base-emitter junction. The high input impedance and low output impedance of the common collector circuit make it useful as an impedance matching device. Figure 6-14 will be used to explain the circuit operation of the basic common collector transistor amplifier circuit.

As shown in figure 6-14, the transistor is forward biased. When the input signal is moving the base potential away from the collector voltage (point 1 on the input signal), forward bias on the emitter-base junction is opposed and less current flows in the circuit. Less current flowing through R_F means that the negative potential, with respect to ground on the output end of R_F, is something less than its quiescent value of -5 volts. The output signal is moving in a positive direction as shown by point 1 on the output signal. Again points 1 and 2 on the input and output waveforms occur at approximately the same period in time. The input signal at point 2 is of the polarity necessary to increase the difference of potential across the emitter-base junction. This aids the forward bias, causing an increase in transistor current. Increased current flow through R_E results in a larger voltage drop across R_E, and the output voltage becomes more negative as shown by point 2 on the output waveform.

The reason that the common or grounded collector configuration is called an emitter follower is because the output voltage on the emitter is practically a replica of the input voltage. That is, it has the same waveform and amplitude. In a practical circuit, the output amplitude will be slightly less than the magnitude of the input. It was stated that the input is across the base-collector junction, however, a closer examination will reveal that the input is also across the series combination consisting of the base-emitter junction and the output load resistor. Since the input is across two resistors and the output is across only one of these, the output voltage must be less than the input. However, since all currents in the circuit flow through the load resistor and only a small amount of current flows in the input circuit, a considerable current gain may be obtained.

In summary, the transistor behaves like a variable resistance. When it is so biased as to cause decreased resistance, current flow increases and vice versa. A comparison chart for the three configurations is shown in table 6-1.

Table 6-1.—Transistor amplifier comparison chart

AMPLIFIER TYPE	COMMON BASE	COMMON EMITTER	COMMON COLLECTOR
INPUT OUTPUT PHASE RELATIONSHIP	0°	180°	0°
VOLTAGE GAIN	HIGH	MEDIUM	LOW
CURRENT GAIN	LOW (α)	MEDIUM (β)	HIGH (γ)
POWER GAIN	LOW	HIGH	MEDIUM
INPUT RESISTANCE	LOW	MEDIUM	HIGH
OUTPUT RESISTANCE	HIGH	MEDIUM	LOW

179.94

GAIN IN TRANSISTOR CIRCUITS

Gain is the ratio of output voltage, power, or current with respect to an input reading in the same unit of measurement. Each transistor configuration will give a different value of gain even though the same transistor is being used. The configuration used is a matter of design consideration, however the technician is interested in the ratio of output with respect to the input. This is important in determining whether or not the transistor is performing its circuit application function. Transistor specifications usually state the normal current gain for a given transistor, at a particular voltage.

Generally, if the output circuit has a low impedance the transistor provides a large current or power gain. Conversely, a high output impedance is an indication of a voltage gain. Some transistor circuits have a low input impedance, others high input impedances. Some transistor circuits have low output impedances and others high output impedances.

OPERATING LIMITS

Some of the limits imposed on operating transistors are: maximum collector current, maximum collector voltage, maximum collector power dissipation, and collector cutoff (leakage) current. The first three limitations listed are transistor ratings, and the last item is a characteristic of the transistor. A rating is defined as a limiting value which, if exceeded, may result in permanent damage to the device. On the other hand, a characteristic is a measurable property of the device under specific operating conditions.

Collector cutoff current, also called collector-base leakage current, or I_{CBO} for the common base configuration (I_{CEO} for the common emitter configuration), is determined primarily by two factors, temperature and voltage.

The rating of I_{CBO} or I_{CEO} is given as some maximum at a specified collector-to-base or collector-to-emitter voltage and a particular temperature. If either the temperature or the collector voltage increases, cutoff current will increase. The rating must be proportionally increased for voltages or temperatures which are higher than those specified.

The peak collector power dissipation rating P_C is that value of power that can safely be dissipated by the collector. P_C represents the limit of the product of the d.c. quantities V_C and I_C. If the maximum power dissipation rating is 5 watts, the product of collector voltage times collector current flow must not exceed 5 watts.

The maximum collector voltage rating is the maximum d.c. potential that can be applied to the collector with safety. The maximum collector voltage rating is usually given in terms of the collector-emitter voltage (V_{CEO}). The value of this voltage, in part, determines the quantity of reverse bias across the collector-base junction and the quantity of forward bias across the emitter-base junction. For a specified value of I_B both I_C and I_B will remain relatively constant until the maximum value of V_{CEO} is reached. When this value V_{CEO} is exceeded, both I_C and I_B begin increasing. When the breakdown voltage (BV_{CEO}), which is just above the maximum value of V_{CEO}, is reached, I_C and I_B increase very rapidly, and the device breaks down.

The maximum collector current rating is the maximum value of d.c. that can flow through the collector. As may be seen from the above information, permanent damage to the device will occur if this value is exceeded.

BASE LEAD CURRENT

Transistor current is dependent upon many variables. Some of these variables are: (1) the type of bias, i.e., fixed bias, self bias; (2) the magnitude of the input signal; (3) the configuration being used; (4) the type of transistor being used; and (5) the direction of current flow in the base lead.

Of the variables mentioned the direction of current flow in the base lead deserves further examination. Whether the majority of current flow is in the emitter or the collector is determined by this variable. If the base lead current is flowing into the base material in the PNP (fig. 6-15A), or out of the base material into the base lead in an NPN (fig. 6-15B), the majority of current is flowing in the emitter. If the reverse is true (fig. 6-15C and D), then the majority of current is flowing in the collector.

REVERSE CURRENT

Reverse current in transistors as in the PN junction diode, is that portion of electron flow taking place across a reverse biased junction. Normally, the magnitude of this reverse current

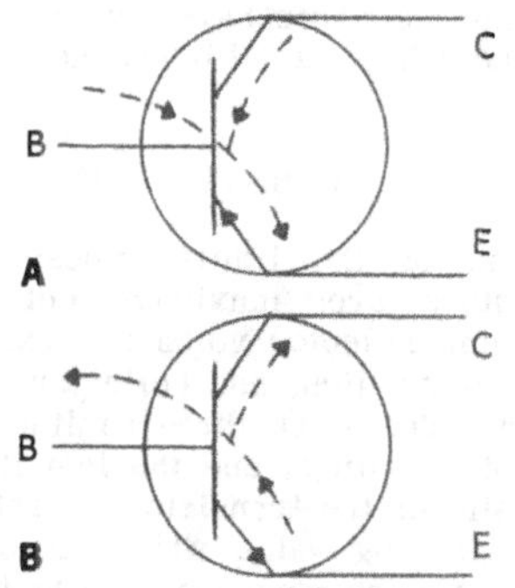

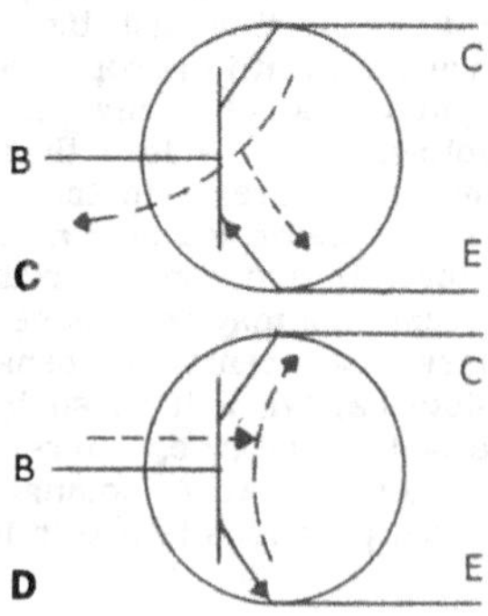

179.95

Figure 6-15.—Base lead current flowing into, or out of, the base material in NPN and PNP transistors.

is very low (in the microampere range) and does not subtract any significant quantity from the base current (current in the base lead or I_B). However, since this reverse current increases rapidly as temperature increases, it might reach a magnitude where damage to the transistor or malfunctioning of the circuit would occur.

In order to explain how this reverse current may cause trouble by reducing the magnitude of I_B, or even reversing the normal flow of I_B, the block diagram of an NPN transistor shown in figure 6-16 will be used. The solid lines indicate the normal flow of current while the dotted lines indicate the direction of reverse current, when the transistor is biased as shown.

In figure 6-16, when S1 is closed, current flow is as indicated by the solid lines: 100% of the current flows in the emitter, some small percentage flows out of the base lead, and the remaining current flows in the collector. With the switch open there is a small amount of current flowing into the base material from the base lead and the collector electrode, continuing through the collector back to the source, V_{CC}.

This is the previously discussed reverse current I_{CBO}, which was defined previously as follows: I-current, C-collector electrode, B-base electrode and O-open. The "open" in this case is the emitter electrode. I_{CBO} is the current flowing in the collector-base elements with the emitter open.

As may be observed in figure 6-16, I_{CBO} flows in a direction that opposes normal base lead current flow. Normal flow is out of the base material into the base lead, and I_{CBO} is into the base material from the base lead. These opposing currents will produce a net magnitude

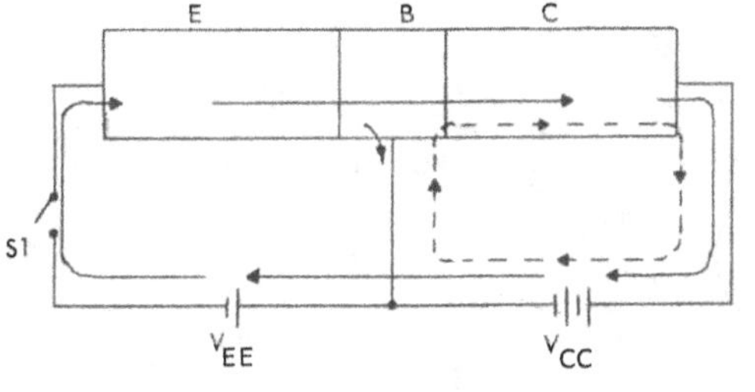

Figure 6-16.—Normal and reverse current in an NPN transistor.

of their difference with the larger magnitude of current prevailing. Normally, in an NPN, the larger current is that flowing out of the base material which is reduced by some amount of I_{CBO}.

I_{CBO} will exist even though the emitter is not actually open. To measure the magnitude of I_{CBO} the opposing current of the emitter-base must be eliminated, and the easiest way is to open the emitter. The reverse current, called I_{CBO}, is flowing, thereby reducing I_B, at all times. When, due to increased ambient temperatures, the collector-base junction resistance reduces to the point where I_{CBO} is greater than normal, it may be found that the transistor is not functioning properly and the circuitry using this transistor is also malfunctioning. Every transistor has a stated value of permissible I_{CBO} that may be found in transistor specifications, in equipment manuals, or in separate manuals. Once the value of I_{CBO} is exceeded the transistor is of little use as it has, at least, started to destroy itself.

While I_{CBO} has been explained in the NPN transistor it must be realized that the same facts hold true for the PNP transistor. To prove it, draw a diagram similar to that in figure 6-16, making the transistor a PNP and reversing the polarities of the batteries. Insert the solid and dotted lines, making sure to observe the correct directions. You should be able to determine if normal current flow is into, or out of, the base material, and what direction in the base lead I_{CBO} flows.

An understanding of I_{CBO} is important not only for the detrimental effects explained here but also because it can be put to beneficial use. Its specific benefits will be explained in the proper place. For now, remember that it is always present and opposing normal base current flow in transistors biased in the forward (low resistance) direction on the emitter-base junction and in the reverse (high resistance) direction on the collector-base junction.

While the transistor amplifiers that have been presented up to now were biased in forward and reverse directions through the use of two batteries, this is very seldom the case. The purpose for using two batteries was to simplify the explanation of transistor operation. Single source biasing, which is covered later in this chapter, will point out the necessity for knowing in which direction base lead current is normally flowing.

TRANSISTOR LEAD IDENTIFICATION

Identification of the leads or terminals of a semiconductor device is necessary before it can be connected into a circuit. As there is no standard method of identifying transistor leads or terminals it would be quite possible to mistake one lead for another.

In figure 6-17 the bases of four sets of transistors are shown. Each set, while similar in appearance, has different elements connected to the leads. In the first set, the leads are emitter, base, collector, reading from left to right, for the top transistor, while in the bottom transistor the leads are emitter, collector, and base. If one of these transistors was connected into a circuit as a replacement for the other, the circuit would not function properly or the transistor might be destroyed. The same general results apply for the other sets of transistors. Note that in the right hand set, the cases can and are used for emitter and/or collector connections.

To further complicate matters, field effect transistors also have the same casing as those shown above. It is emphasized that you CANNOT TRUST to shape in replacing one transistor with another. Be sure the leads are where they should be and that the transistor chosen as a replacement is suitable for performing the circuit application. If there is any doubt, consult the equipment manual or a transistor manual showing the specifications for the transistor being used.

FREQUENCY LIMITATIONS

The range of frequencies over which a transistor performs a useful function in a circuit is limited by inherent parameters. Manufacturer's data sheets usually specify only one or two of these parameters, while the user may need others. Therefore, a clear understanding of these parameters and the relationships between them is of value in evaluating transistor performance.

COMMON BASE CURRENT GAIN

Typical of such parameters is h_{fb}, which denotes the common base a.c. short circuit forward current gain, better known as alpha. Another parameter consists of h_{fbo}, the common base a.c. short circuit forward current gain at a frequency of 1 kHz. As the frequency is increased, h_{fb} remains approximately equal to h_{fbo}. After the upper frequency limit is reached, h_{fb} begins to decrease rapidly.

The frequency at which a significant decrease in h_{fb} occurs provides a basis for comparison of the expected high frequency performance of different transistors. The common base current gain cut-off frequency, f_{ab} is defined as that frequency at which h_{fb} is 3 db below h_{fbo}. Expressed in magnitude, h_{fb} at f_{ab}, is 70.7 percent of h_{fbo}. A curve of h_{fb} versus frequency is shown in figure 6-18A.

The curve in figure 6-18 has the following characteristics: (1) at frequencies below f_{ab}, h_{fb} is nearly constant and approximately equal to h_{fbo}; (2) h_{fb} begins to decrease significantly in the region of f_{ab}, and (3) above f_{ab}, the rate of decrease in h_{fb} with an increasing frequency approaches 6 db per octave.

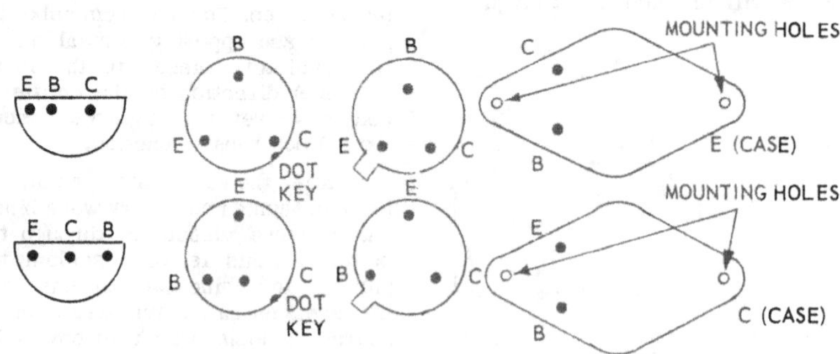

Figure 6-17.—Bottom view of some common transistor cases.

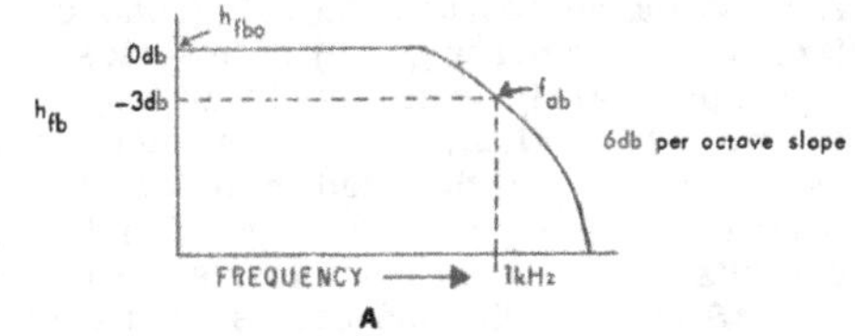

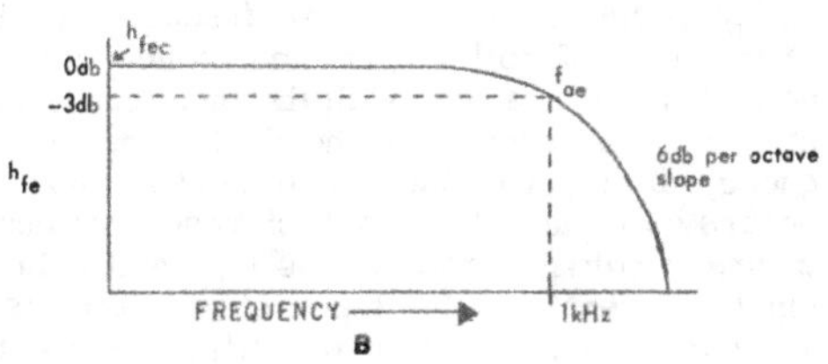

179.98

Figure 6-18.—Common base and emitter current gain plotted against frequency.

COMMON EMITTER CURRENT GAIN

The common emitter parameter which corresponds to f_{ab} is f_{ae}, the common emitter current gain cut-off frequency. By definition, f_{ae} is the frequency at which h_{fe} (the common emitter a.c. short circuit current gain, known as beta) has decreased 3 db below h_{feo} (the value of h_{fe} at 1 kHz). A typical curve of h_{fe} versus frequency is shown in figure 6-18B.

TRANSIT TIME

An important characteristic in establishing the frequency behavior of a transistor is the time required for an electron to travel from the emitter to the collector (transit time). If signals are applied whose frequency changes so rapidly that the electrons are unable to be dislodged rapidly enough to allow travel from the emitter to the collector before another frequency alteration occurs, the frequency limit of the transistor is said to be exceeded.

As mentioned earlier, if the frequency limit of a given transistor is exceeded, the output will be greatly reduced. Manufacturers are very rapidly overcoming this problem by continuously developing newer devices with higher cut-off frequencies. This is mainly accomplished by making the base material as thin as possible without reducing the power gain.

INTERELEMENT CAPACITANCES

Another factor which makes the gain of a transistor dependent on frequency is the interelement capacitances within the transistor. Figure 6-19 shows the interelement capacitances associated with the transistor. The capacitances are shown externally; however, the actual capacitances are produced by the energy depletion regions within the transistors. Bécause the width of these regions vary in accordance with the voltages across them and the energy transferred through them, the capacitance values also vary.

The increase in the width of the energy depletion region between the base and the collector, as the reverse bias voltage is increased, is reflected in lower capacitance values. This phenomenon is equivalent to spreading apart the plates of a capacitor so that lower capacitance results.

An increase in emitter current flow, most of which flows to the collector through the base-collector junction, increases the collector-base capacitance. This increased current flow through the junction may be considered as effectively reducing the width of the energy depletion region. This phenomenon is equivalent to reducing the distance between the plates of a capacitor so that an increased capacitance results.

Effect on High Frequency Gain

In wide band amplifier applications, a uniform gain is required from very low frequencies

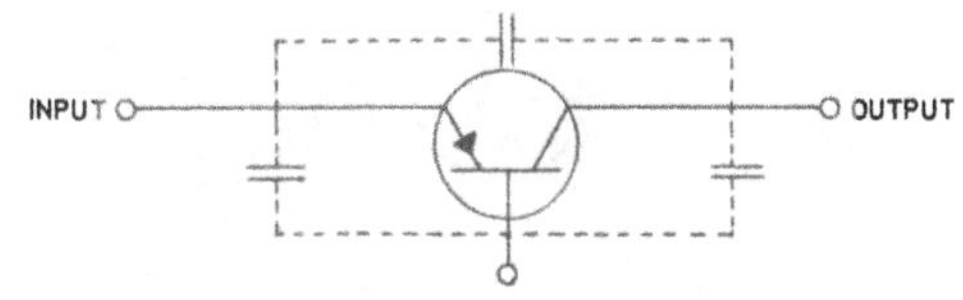

179.99

Figure 6-19.—Effective interelement capacitances of the transistor.

to very high frequencies. These interelement capacitances cause the voltage gain of resistance coupled amplifiers to fall off at high frequencies. The output capacitance of a stage and the input capacitance of the next stage shunt the load resistance of the first stage as shown in figure 6-20. The stray capacitance and the output capacitance for this common emitter circuit is represented by C_{oe} whereas the capacitive component of the input impedance to the next stage is represented by C_{ie}. When the X_c of C_{oe} + C_{ie} equals R_C, the voltage gain drops 3 db. Transistor design can reduce these capacitances somewhat and the resistance of R_C can be reduced. The reduction of R_C reduces overall voltage gain, but the frequency range is extended because the frequency at which X_c of the shunting capacitance equals R_C is raised. In addition to the reduction of R_C, different types of compensating circuits may be added to maintain uniform gain at higher frequencies.

SERIES COMPENSATION.—Series high frequency compensation is illustrated in figure 6-21. It is so called because the coil (commonly referred to as a peaking coil) is in series with the signal path. Analysis of this circuit may be made in two ways. First, the coil isolates the input capacitance from the load resistor. Since the input capacitance is much larger than the output capacitance, there is much less shunting across R_C. Secondly, the inductance of L2 is very small, which causes the inductive reactance to be appreciable only at those frequencies close to the area of roll off. The inductance is chosen so that it will resonate with the input capacitance at some frequency near the desired upper frequency limit. The increased current now that occurs as L2 and C_{ie} approach resonance causes a larger voltage drop across C_{ie} and a larger input current to the input of the transistor. Therefore, the reduction in voltage gain at high frequencies should be offset by the resonant rise in input current of the following stage.

The resonant rise of current is very gradual because the circuit has a low "Q." The low "Q" is due to the resistance of the input junction and the low inherent "Q" of the coil. The proper

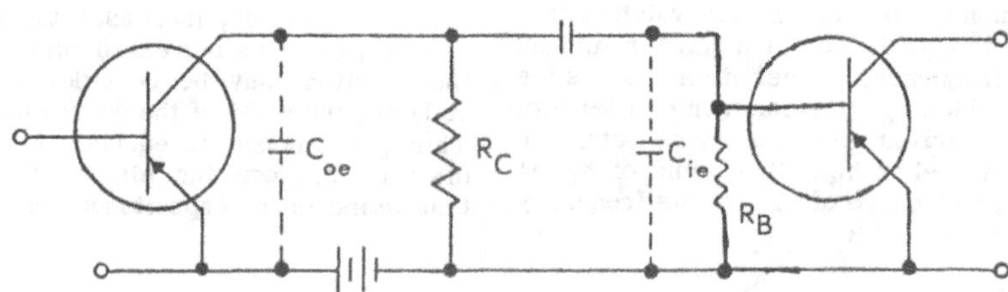

Figure 6-20.—Interelement capacitance in relation to external circuit.

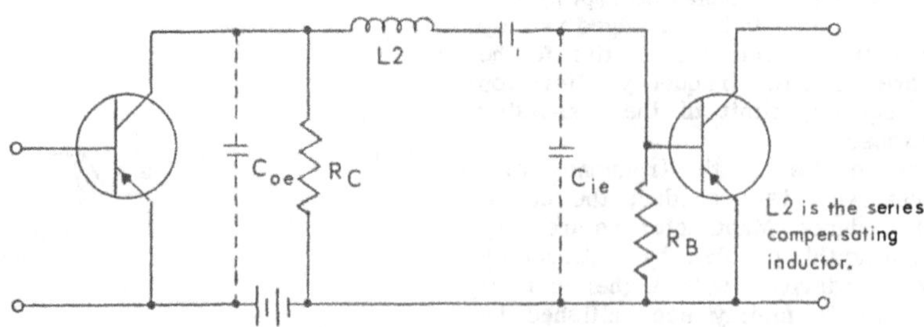

Figure 6-21.—Two stage amplifier employing high frequency series compensation.

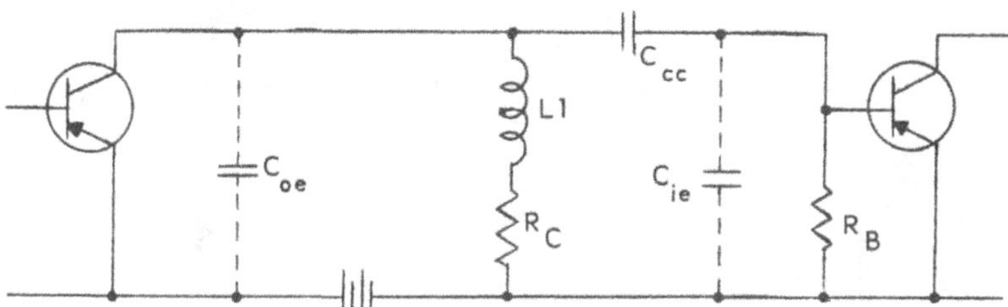

20.422
Figure 6-22. — Two stage amplifier employing shunt compensation for high frequency applications.

value of inductance used in the series circuit will determine the high end of the frequency response curve and will provide a flat response over a wide range of frequencies.

If the output decreases as applied frequency increases, it may be an indication that the compensating component L2 is short circuited. In that case, only a resistance reading across the component will reveal the malfunction. Of course, an open in the inductor would prevent a signal from arriving at the base of the following stage.

SHUNT COMPENSATION. — For shunt compensation (fig. 6-22) inductor L1 is added in series with the load resistor R_C but in shunt with the output signal path. This compensates for the shunting effect of the output capacitance C_{oe} and the input capacitance C_{ie}. These capacitive reactances shunt inductor L1 and resistor R_C at the high end of the frequency range. Capacitor C_{cc}, coupling capacitor, appears as a short circuit at high frequencies. Peaking coil L1, in series with the load resistor, forms a parallel resonant circuit (with capacitive reactances C_{oe} and C_{ie}) which has a very broad frequency response. The resonant peak of this parallel combination maintains a practically uniform gain throughout the high frequency range.

In the uncompensated circuit (fig. 6-20) the high frequency gain was reduced by the capacitive reactances of C_o plus C_i. The impedance of the parallel resonant circuit has a value that is approximately the same as the load resistor R_C. Therefore, when the frequency increases, the decrease in the capacitive reactances of C_o and C_i is exactly compensated by the increase in the inductive reactance of inductor L1. The frequency response is thus increased, extending the flatness of the frequency response curve over a much broader range of frequencies. The high frequency gain of a wide band amplifier using shunt peaking is approximately 50 percent less than if it were using series peaking.

A shorted peaking coil in either compensation network would result in a reduction of high frequency gain. An open inductor would result in an open circuit to the signal, and, consequently, there would be no output possible.

SERIES SHUNT COMPENSATION. — Figure 6-23 illustrates the case where series compensation and shunt compensation are combined; the frequency response is approximately the same as that of a shunt or series network; however, the high frequency gain of this combined setup is approximately 80 percent greater than that of the series peaking coupling circuit.

The graph in figure 6-24 illustrates the comparative gain of the various forms of compensations. It should be noted that the difference in gain characteristics results from the fact that with series compensation a larger value of load resistance may be used and, therefore, more gain from the amplifier may be obtained.

Effect on Low Frequency Gain

On the low end of the frequency response range, the input and the output capacitances of the transistors have no effect on the frequency response. The low frequency response is, however, limited by capacitor C_{cc} (fig. 6-25) and resistor R_B.

COMPENSATION. — The time constant R_B C_{cc} must be large to prevent the low frequency response from falling off and to prevent phase distortion. At low frequencies, the reactance of C_{cc} is very high and thus a good deal of signal voltage is dropped across it. One method of

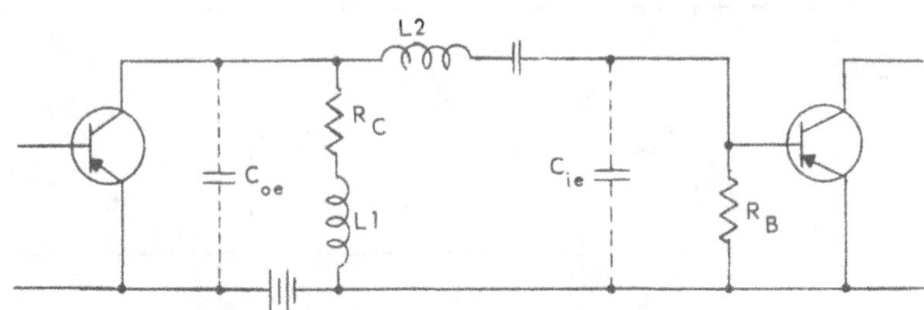

20.422
Figure 6-23.—Series shunt compensation.

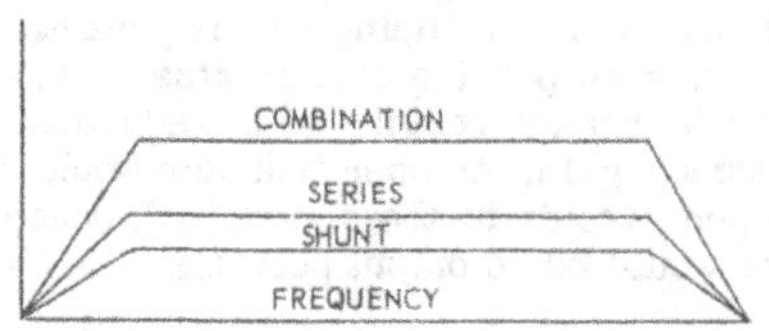

179.101
Figure 6-24.—Graph illustrating gain characteristics with various methods of compensation.

compensating for this drop is to increase the gain of the previous stage by increasing the load impedance of that stage, of course, this increase must occur only at the low end of the frequency range. This is accomplished by the compensating filter, R1 and C1. The filter increases the collector load impedance at low frequencies and compensates for the phase shift and low frequency attenuation produced by capacitor C_{cc} and resistor R_B. As the frequency is increased, the reactance of C1 decreases and so does the collector load impedance, producing less gain from that stage. But as the frequency is decreased, the reactance of C1 increases and with it the collector load impedance resulting in more gain from that stage. This combination extends the frequency response over a much lower range of low frequencies. The gain and phase response become more uniform over the low frequency range. Effectively, this filter is compensating for the voltage drop across the coupling capacitor C_{cc} over the low range of frequencies.

HIGH AND LOW FREQUENCY COMPENSATION COMBINED

A wide band amplifier must be able to produce an undistorted amplified output of a nonsinusoidal

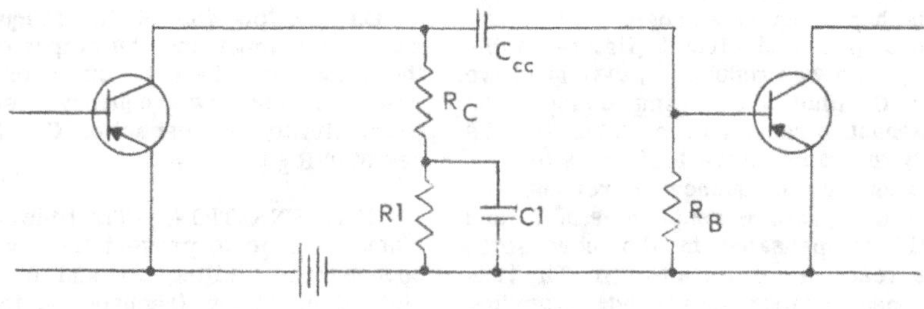

20.423
Figure 6-25.—Low frequency compensation.

signal input. It finds many applications in television, radar, and many other electronic applications. Figure 6-26 shows a typical wide band amplifier (30 hertz to 40 megahertz) using both high and low frequency compensation. The high and low compensating networks operate independently of each other. At low frequencies the series reactance of inductor L2 is very small and has no effect on the input circuit. The reactances of the output impedance capacitance C_{oc} and input impedance capacitance C_{ie} are so large that they have no effect on the signal at low frequencies. This means that the combination peaking circuitry has no effect at low frequencies; however, the low frequency compensating network of C1 and of R1 are increasing the gain of Q1 and compensating for the voltage drop across C_{cc} at the low end of the frequency range. As the frequency is increased the combination compensation network starts to form a resonant circuit with the output and input capacitances. Simultaneously, the reactance of C1 starts to decrease, short circuiting R1, so that this network no longer has any effect at the high end of the frequency range.

Examining the components in figure 6-26, it may be easily determined what malfunction would result from their change in value. For instance, C1 opens: the gain is normal at the low end of the frequency range, but as the frequency is increased, C1 can no longer short out R1, therefore, the collector load impedance remains high and the gain of that stage is abnormally high, since there is a decreasing drop in the reactance of C_{cc}. The overall gain of the amplifier is high at the mid frequency range but it is no longer flat across the band. An open in one of the series components, R1, R_C, L1, or L2 would cause a complete loss of signal in the first three cases, because of loss of collector voltage and in the latter because of an open in the signal path.

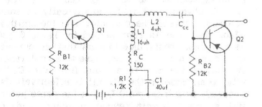

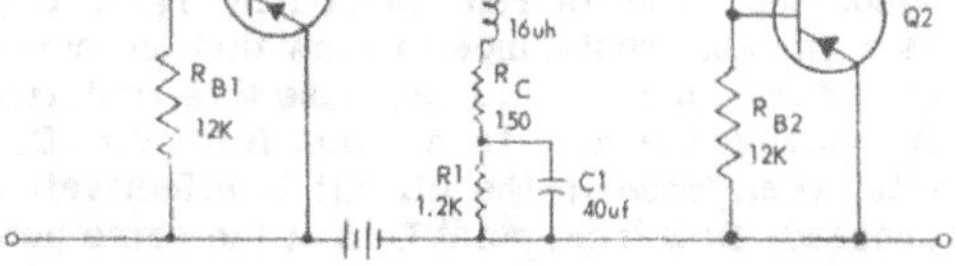

179.102

Figure 6-26.—High and low frequency compensation combined.

INTERNAL FEEDBACK

A unilateral electrical device is one which permits the transfer of energy in one direction only. The transistor is not naturally a unilateral device. Every transistor, due to interelement capacitances, has an internal feedback mechanism which feeds voltage back from the output to the input. In a transistor configuration when the feedback voltage aids the input voltage, it is termed positive feedback—also called regenerative feedback. If the positive feedback voltage is large enough, the amplifier will oscillate. At low frequencies, audio frequencies for example, the voltage feedback is low; therefore, precautions to prevent oscillation are not normally required. High frequency amplifiers, on the other hand, with tuned coupling networks are susceptible to oscillations so that precautions may be taken.

The effect of positive feedback on the input circuit of an electrical device is to alter its input impedance. Normally, both resistive and reactive components of the input impedance are affected. The change in input impedance of a transistor caused by internal feedback can be eliminated by using an external feedback circuit. If the external circuitry cancels the resistive and reactive internal feedback to the input circuit, the transistor is considered to be UNILATERALIZED. If the external feedback circuit cancels only the reactive changes to the input circuit, the transistor amplifier is considered to be NEUTRALIZED. Actually, neutralization is a special case of unilateralization. In either case, the external feedback circuit prevents oscillations in the amplifier.

Besides altering input impedance, this regenerative feedback also increases the gain of the amplifier, since part of the output signal is fed back internally to the input circuit increasing the input signal. This effect can easily snowball and cause the output waveform to become distorted as a result of overdriving the stage.

Figure 6-27 shows a tuned common base amplifier. The internal elements of the transistor that may cause sufficient feedback and oscillations are shown in dashed lines. Resistor R'_b represents the resistance of the base and is referred to as the base spreading resistance. Capacitor C_c represents the capacitance of the collector-base junction. Resistor R_c represents the resistance of the collector-base junction. This resistance is very high in value because of the reverse bias on the collector-base junction.

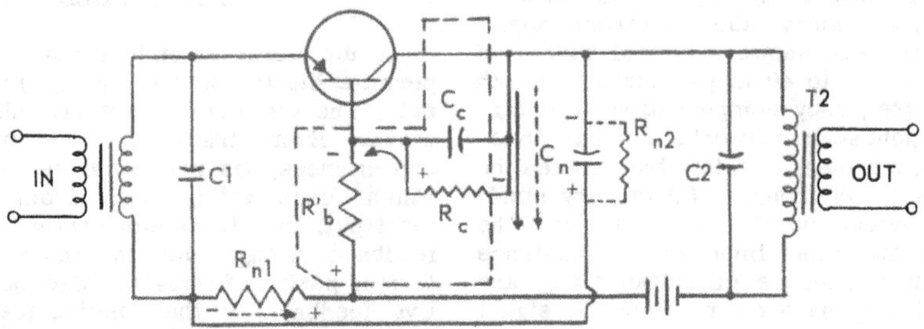

179.103
Figure 6-27.— Properly neutralized (unilateralized) common base amplifier.

At very high frequencies, capacitor C_c effectively shunts resistor R_c.

Assume that the incoming signal aids forward bias, causing the emitter to go more positive with respect to the base. Collector current flow increases in the direction shown (solid arrow). A portion of the collector current passes through capacitor C_c and through resistor R'_b in the direction shown, producing a voltage with the polarities shown. The voltage across R'_b aids the incoming signal and therefore represents a positive or regenerative feedback that may cause oscillations.

To overcome the possibility of oscillations, the external circuit, consisting of resistors R_{n1} and R_{n2} and capacitor C_n, can be added to the circuit, as shown in figure 6-27. Resistors R_{n1} and R_{n2} and capacitor C_n correspond to resistors R'_b and R_c and capacitor C_c respectively. The need of R_{n2} depends on the frequency being amplified; the higher the frequency, the less the need for resistor R_{n2}.

When the incoming signal aids the forward bias, the collector current increases. A portion of the collector current passes through capacitor C_c and through resistor R'_b in the direction shown, producing a voltage with the polarities shown. A portion of the collector current also passes through capacitor C_n and through resistor R_{n1} (dashed arrow), producing a voltage with the indicated polarity. The voltages produced across resistors R'_b and R_{n1} are opposing voltages. If the voltages are equal, no positive feedback from the output circuit to the input circuit occurs. The amplifier is then considered neutralized.

SINGLE SOURCE BIASING

Single source biasing has the advantage of using the existing power supply to satisfy the needs for transistor biasing. This not only eliminates the need for a separate power supply, but also simplifies circuit wiring. The main reason for this simplification is that both the collector and the base require voltages, which possess the same polarity with respect to the emitter, for biasing purposes.

A transistor with its three doped elements may be simplified to three resistances in series with each other. Figure 6-28A shows the schematic diagram of an NPN transistor. It's three elements can be broken down into their equivalent electrical representation (fig. 6-28B). Now, by application of a negative source to the emitter material, that point is made the most negative in the branch, while the collector material is the least negative. In figure 6-28C, point B is less negative than point A, or positive in respect to point A. In effect, this makes the base positive in respect to the emitter, and in an NPN transistor this satisfies the requirements for forward bias.

It should be noted that the base material is situated between the collector and the emitter; it will always find itself electrically at some potential between the collector and the emitter. Also, the base is represented in figure 6-28C as a variable resistance to show that an increase in forward bias should decrease this resistance. By moving the arm from point B to point D, the total resistance in the circuit is effectively decreased; flow from point D is at the same potential as point B with respect to point A, giving more forward bias; and more current is allowed to flow from the collector to the emitter. When

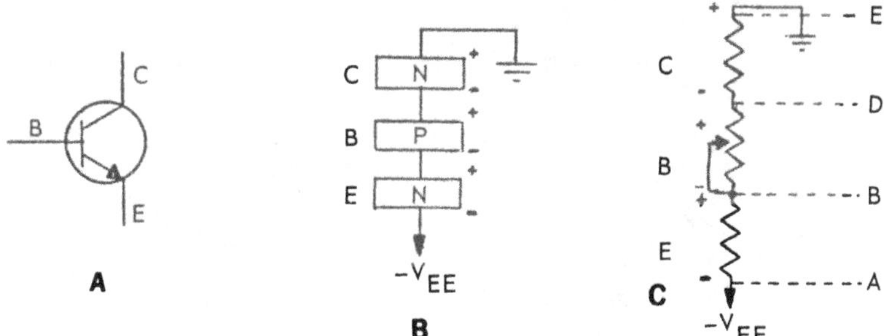

179.104
Figure 6-28.—Schematic diagram of an NPN transistor broken down to equivalent electrical representation and resistance equivalent.

the arm is moved back to point B, the total resistance is increased, decreasing the current flow in the circuit; now point B is less positive than point D with respect to point A, resulting in less forward bias.

Figure 6-29 illustrates the PNP, again three equivalent resistances are used and this time a positive source is applied to the emitter source terminal. It may be noted that point A is the most positive point in the network and that point B is less positive than point A, while point D is less positive than (negative with respect to) point B. In this PNP, the bias is negative with respect to the emitter, which satisfies the forward bias requirement for this type of transistor. Once more, the base is shown as a variable resistance to illustrate that forward bias can be increased or decreased.

The addition of a single resistor from the base to the emitter source terminal simply shunts the base-emitter junction, further decreasing its resistance, permitting more forward bias to be applied. Should the base current flow increase and, consequently, the transistor current flow, with no signal input, you should check the value of R_B; an open base resistor would allow a greater difference of potential to be felt between the base and the emitter and therefore more current through the transistor. The base-emitter resistor reduces the resistance of the base-emitter junction, reducing the amount of forward bias with no signal applied. A direct short circuit from the base to the emitter would of course cause all signal to be lost and an excessive amount of current flow through the device. By considering the transistor as being three resistances in series, one can always easily establish the polarity of any given material in respect to an adjacent material, regardless of the polarity of the source, as shown in figure 6-30.

The NPN transistor may have its collector connected to a positive source as shown in figure 6-31A. In this case, point E of figure

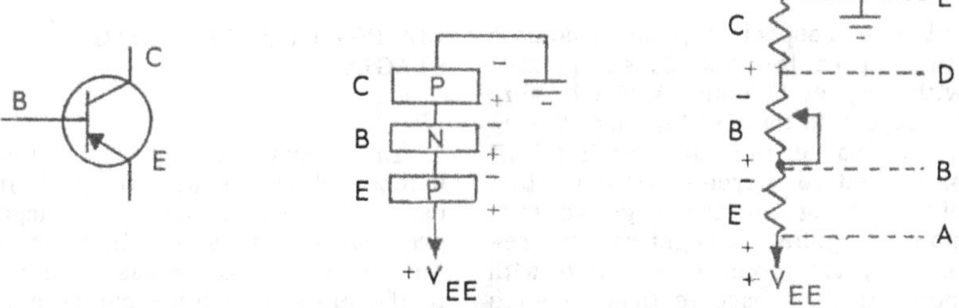

179.105
Figure 6-29.—Schematic diagram of a PNP transistor broken down to equivalent electrical representation and resistance equivalent.

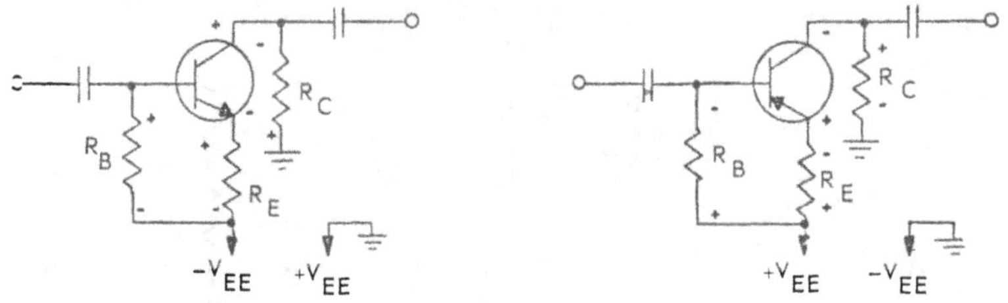

179.106
Figure 6-30.—Establishing transistor polarities.

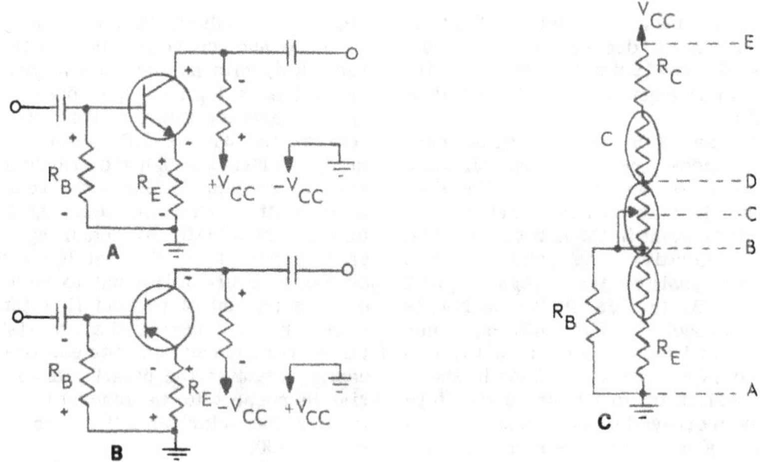

179.107
Figure 6-31.—NPN and PNP transistor equivalent circuit.

6-31C is positive in respect to point D, point D is positive in respect to point C, and point C is positive with respect to point B. The base is positive with respect to the emitter, and, therefore, it is forward biased. In figure 6-31B a PNP transistor has a negative source on its collector, making point E (of fig. 6-31C) negative with respect to point D, point D negative with respect to point C, and point C negative with respect to point B. The base is thus negative with respect to the emitter and in a PNP this is forward bias. In both cases R_B is shunting the base to emitter junction.

BIASING BY SUCCEEDING STAGES

Bias considerations increase as an increasing number of stages are involved. It is possible to use a succeeding stage to change the bias of a previous one as shown in figure 6-32. In Q1, it may be seen that the base is negative in respect to the emitter which permits current flow from the collector source to the emitter terminal. In Q2 the base, due to voltage divider action, is negative with respect to the emitter; this

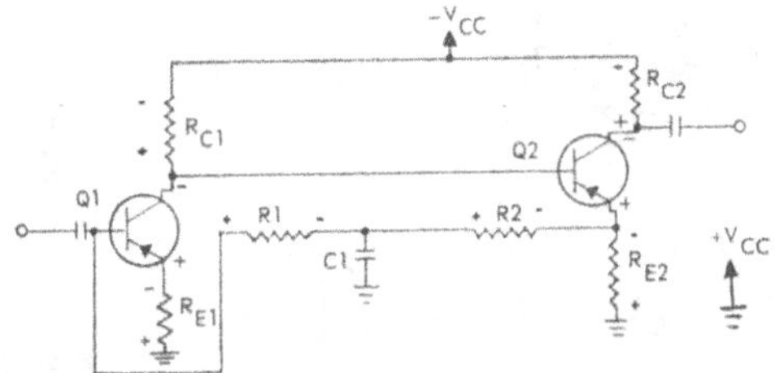

Figure 6-32.—Directly coupled amplifiers with bias stabilization. 179.108

forward bias allows current flow from the collector source through R_{C2} to the emitter source terminal, also, some current flow from the collector terminal of Q1 to the base of Q2 and to the emitter terminal. Lastly, a path for current flow exists from the collector of Q2 through the transistor and from the top of R_{E2} through R2 and R1 to the base of Q1. The purpose of C1 is to isolate Q1 from Q2 by preventing the amplified a.c. signal variations from reaching the base of Q1 while variations in the d.c. voltage level are fed back to the base of Q1. The potential difference from the top of R_{E2} to ground is the source for base-emitter bias of Q1. For instance, let's say the value of R_{EI} decreases allowing more current flow through Q1. This would cause a larger voltage drop across R_{CI}, the collector load, and a decrease in potential difference from collector to ground. Since the collector of Q1 and base of Q2 are directly coupled, this decrease in potential makes the base of Q2 less negative in respect to its emitter and decreases forward bias. This will cause less current flow through Q2 and make the potential at the top of R_{E2} less negative. Since the base of Q1 is tied to this point, this base is made less negative with respect to the emitter of Q1, which decreases the current flow through Q1 minimizing the effects caused by a small change in ohmic value of R_{EI}.

BIASING VIA R_B

Another method of providing proper biasing potentials is to use a resistor from the base to the collector source terminal. It may be seen in figure 6-33A that the base is less negative than V_{CC} but more negative than the emitter, providing forward bias for the PNP transistor. Also, it may be seen in figure 6-33B that the base is less positive than $+V_{CC}$ but more positive than the emitter, providing forward bias for the NPN transistor. The value of R_B must, however, be high for two reasons: (1) to prevent loading of the a.c. signal, and (2) to lower the base-emitter bias sufficiently from the value of V_{CC}.

SHUNT FEEDBACK

The circuits shown in figure 6-34 use straight shunt feedback. The purpose of returning R_B to the collector instead of returning it to the power supply is to stabilize the bias under variations of current flow due to changes in transistors or temperature. If the current flow should increase through a transistor, the voltage at the collector would decrease and thus decrease the

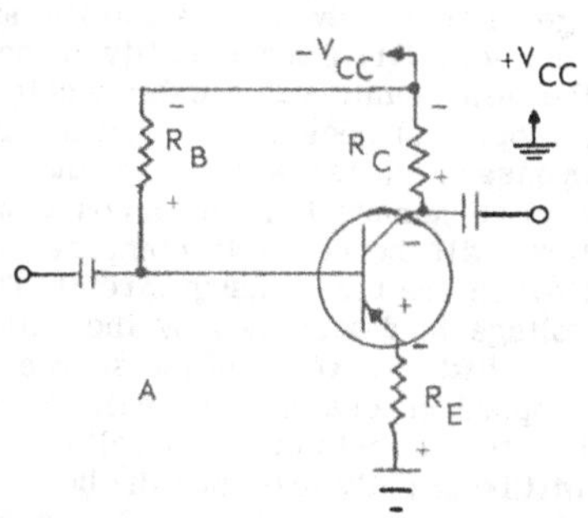

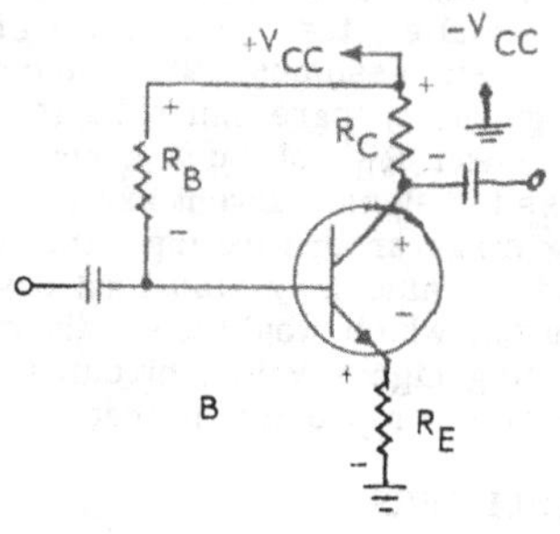

179.109
Figure 6-33.—Shunt biasing.

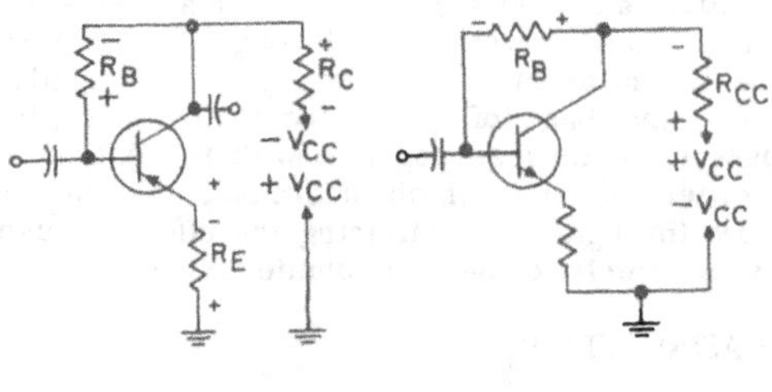

179.110

Figure 6-34.—Shunt feedback circuits.

potential at the base. This would decrease the forward bias on the transistor and reduce the current flow. On the other hand, should the current flow through the transistor decrease, the collector voltage would increase, increasing the potential on the base and increasing the forward bias which will allow more current to flow. This configuration stabilizes the d.c. bias level on the base. This degenerative voltage feedback does, however, decrease the gain of the transistor because a portion of the signal is also fed back. Compensation for this may be made by eliminating the a.c. component from the collector potential. This is done by breaking up R_B into R_{F1} and R_{F2} with a decoupling capacitor between them (fig. 6-35).

A variation of this stabilization circuit is shown in figure 6-36. Here, the base is connected to a voltage divider. This arrangement provides greater stabilization, but the additional resistor does absorb more power from the source. In that sense, this circuit is less efficient; R_{F1} and R_B can assume a wide range of values in achieving the proper voltage division for the base, however; the smaller their value, the greater stability the circuit assumes. With the common emitter configuration, care must be taken not to make R_{F1} and R_B of such a small value thatthey bypass the signal current arriving from the preceding stage around the input impedance. This low ohmic value may also load down the preceding stages, which would lower the overall gain. In selecting higher value, circuit stability is sacrificed, so a compromise is made.

BIASING AMPLIFIERS

A common collector configuration class A amplifier (NPN or PNP) can also be biased by a single positive or negative source. In figure 6-37 it is shown how, through voltage divider

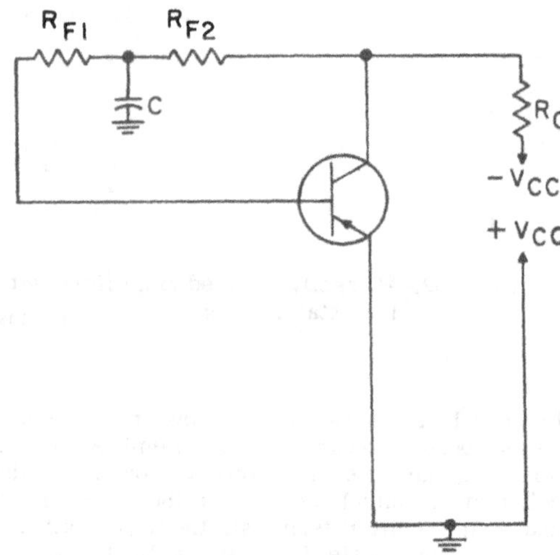

179.111

Figure 6-35.—Compensating for degenerative feedback.

action, the collector is positive with respect to the emitter. Forward bias is therefore established for this NPN transistor, R_B shunts the base-emitter resistance to obtain the precise value of voltage needed for proper forward bias.

A forward bias voltage may also be provided by returning the base resistor to the collector (fig. 6-38). The ohmic value of this resistor sets the voltage level on the base.

A biasing arrangement for a common base class A amplifier is shown in figure 6-39. Base-to-emitter voltage is established by the base voltage divider network, R2 and R3, and the emitter resistor, R1. Good stability is achieved due to the low ohmic value exhibited by the base resistors and because the emitter resistor is not bypassed. This is because the input signal is applied across R1. The amount of base current flow will be neligible compared to the current flow in the base biasing circuit. Therefore, base voltage is determined by the ratio of R2 to R3. Two fifths (6 volts) of the source voltage will be dropped across R2, and will be the base voltage for the transistor. The voltage on the emitter will be slightly less than the base voltage because of the forward voltage drop across the emitter-base junction. A PNP transistor is used in this example, but an NPN may be used by reversing the source voltage polarities.

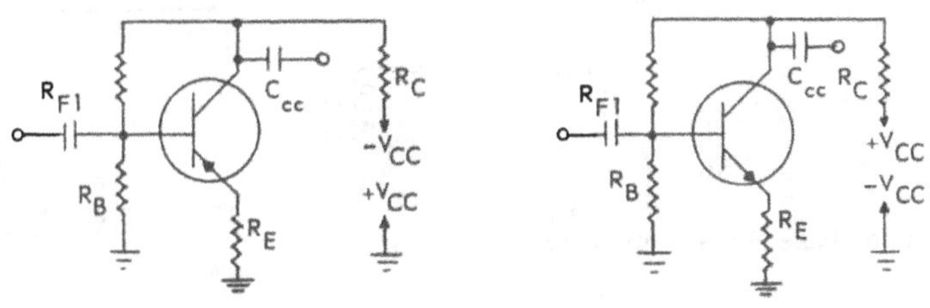

179.112
Figure 6-36.—Stabilization circuits.

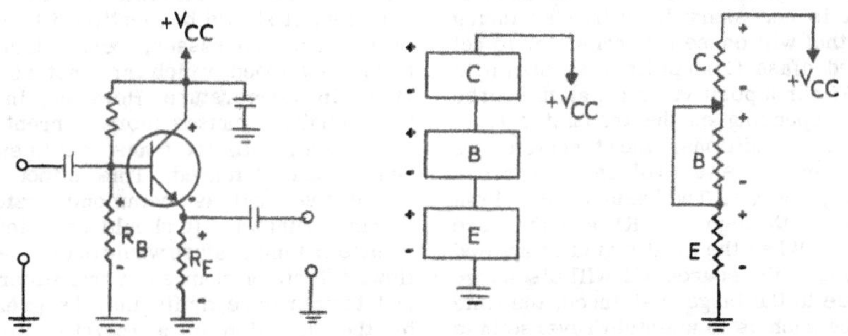

179.113
Figure 6-37.—Forward biased NPN transistor.

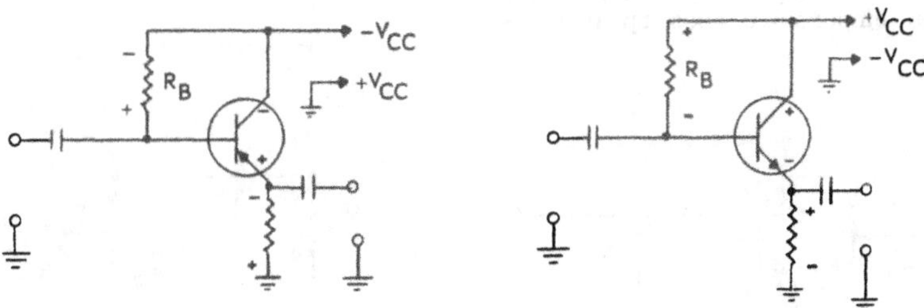

179.114
Figure 6-38.—Forward biased transistors.

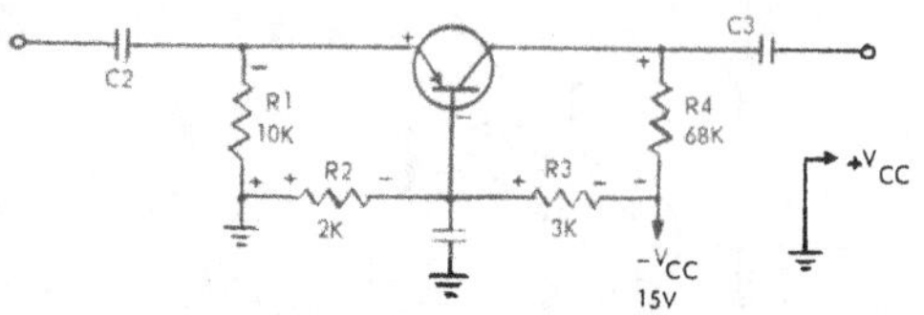

179.115
Figure 6-39.—Common base class A amplifier.

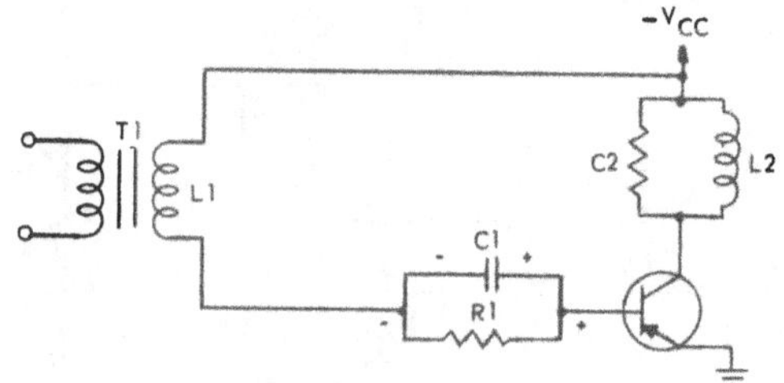

179.117
Figure 6-41.—Class C amplifier.

In the case of class B amplifiers it is desirable to reduce their current drain to practically zero under no signal conditions. As shown in figure 6-40, the base potential is only very slightly more negative than the emitter and only a minute amount of forward bias is developed.

For certain applications, as in the case of oscillators, it is necessary to utilize a biasing arrangement that will cause the transistor to act as a simulated class C amplifier, as shown in figure 6-41. Either a positive or negative source may be used, depending on the transistor type.

Under static conditions, the transistor is forward biased and there is a voltage drop across R1 and a charge on C1. The base current flows from the source through L1, R1 and the base emitter junction. When the input signal is applied so that it opposes the source, C1 will discharge through R1. Due to the large resistance, the time constant will be such as to maintain reverse bias for more than 180° of the input signal. Note that the discharge of C1 places the base at a positive potential in respect to the emitter, establishing a reverse bias on the base-emitter junction.

TEMPERATURE COMPENSATION

Temperature rise is a problem in transistors because as the temperature goes up it tends to increase the flow of electrons in both the P and the N materials. This allows more current flow through the transistor for the same applied voltages. It should be mentioned that as the state of the art progresses, better transistors are being developed which are not as sensitive to rises in temperature. However, in older types this condition causes more current flow, hence more heat, until the transistor is entirely overheated and damaged. This effect tends to be cumulative and is commonly referred to as thermal runaway. It should be noted that a decrease in temperature would decrease the electron flow. Effects of changes in transistor parameters and temperature drifts may be greatly reduced by the insertion of a resistor in the emitter circuit.

As illustrated in figure 6-42 all the collector current will flow through the emitter resistor R_E. The voltage drop across R_E serves to make the emitter negative with respect to ground. However, the base is also at a negative potential with respect to ground. The bias level, therefore,

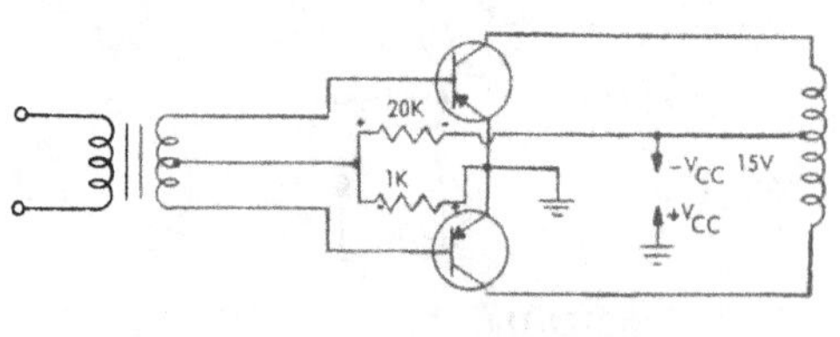

179.116
Figure 6-40.—Class B amplifier.

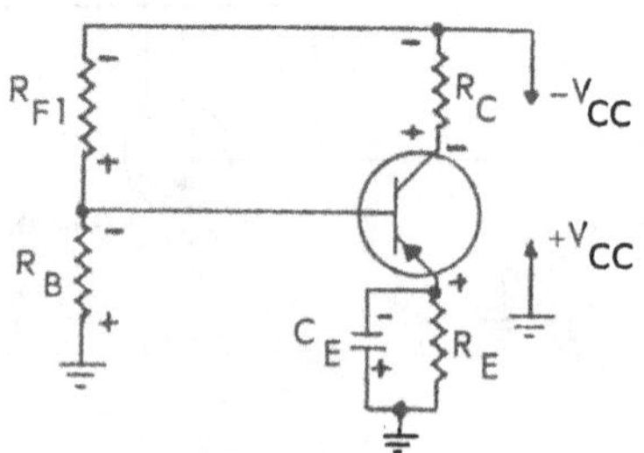

179.118
Figure 6-42.—Emitter biasing a PNP transistor.

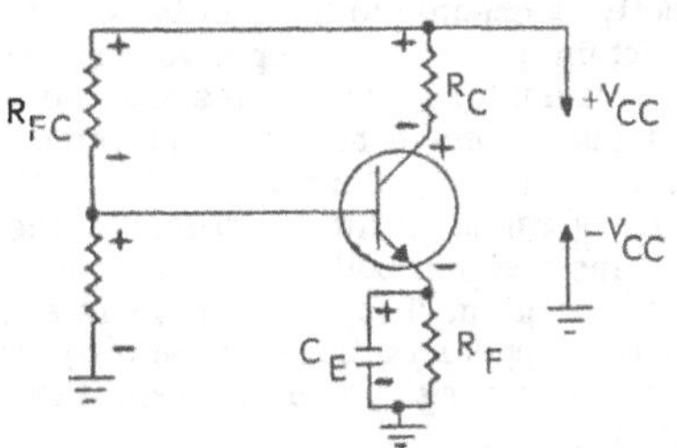

179.119
Figure 6-43.—Emitter biasing an NPN transistor.

will be the difference between the negative base potential and the negative emitter potential. For instance, if the collector current flow should increase due to a rise in temperature, the voltage drop across the emitter resistor will increase due to the increase in current flow. The emitter will now become more negative and this will reduce the forward bias and the current flow through the transistor is reduced. Consequently, collector current flow is reduced by decreasing base-emitter current flow. In figure 6-43, with an NPN transistor, the emitter resistor has the same purpose. The voltage drop across it makes the emitter positive with respect to ground. The base is, also through voltage divider action, positive with respect to ground. If collector current flow increases, the emitter becomes more positive, and the difference between base potential and emitter potential decreases. Forward bias, and collector current flow, is reduced by a decrease in the base-emitter current flow.

This degenerative effort greatly improves the stability of the transistor, yet under dynamic (signal) conditions this causes a reduction in gain. A bypass capacitor, C_E, may be added in parallel to R_E. Temperature drifts would occur at a very slow rate and C_E would not bypass this change in potential; however, signal variations wouldn't affect the emitter potential. As a rule C_E is chosen for a reactance value equal to 1/10 the bypassed resistor at the lowest frequency to be amplified.

Effects of temperature drifts can also be minimized by the use of temperature compensating devices such as thermistors, varistors, and diodes properly located in the circuit. Resistance of a thermistor will decrease as temperature increases. In figure 6-44A, it may be seen that as the temperature rises the base potential is decreased, since there is less resistance to develop a potential. A forward biased diode may also be used, as illustrated in figure 6-44B, as an increase in temperature will lower its voltage drop and thus decrease the base potential, due to the drop in its forward resistance.

COMPOUND-CONNECTED TRANSISTORS

The compound-connected transistor amplifier finds use in the output stage of receivers, public address amplifiers, and modulators, in other words, where large audio power outputs are required. Compound-connected transistors will have at least two active elements. A greater number of transistors may be used in the compound connection but we are going to examine only circuits using two transistors compound-connected. Although two transistors will be used, effectively only three leads will be used: the base lead of one transistor, the emitter lead of the second transistor, and the collector leads

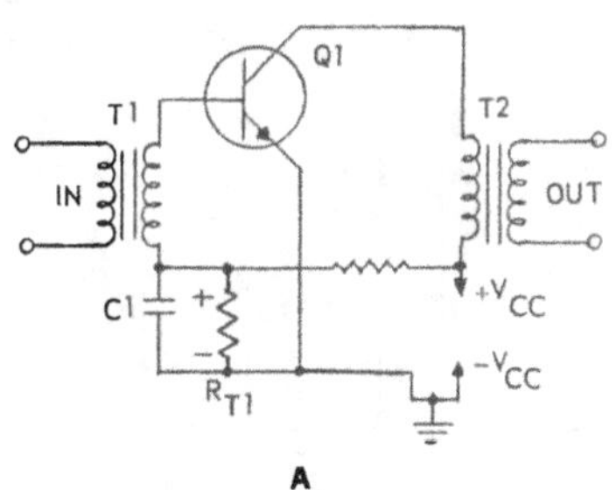

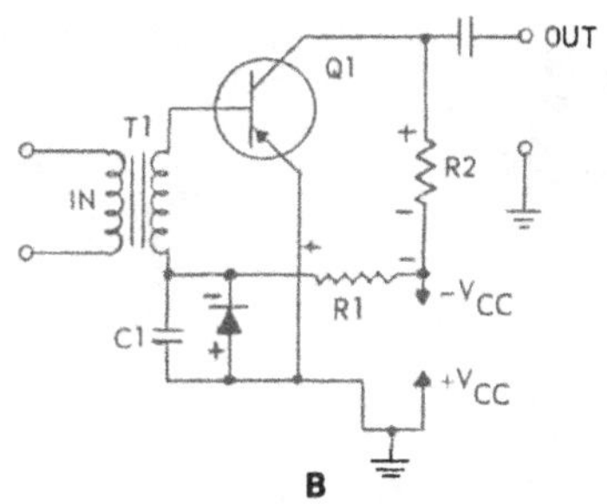

179.120
Figure 6-44. — Temperature compensation.

of both transistors which are connected to a common point.

The input to a compound-connected power amplifier will be a series connection and the output will be a parallel connection. To better explain the foregoing statements, we shall use figure 6-45 to show the operation of a compound-connected common emitter audio amplifier circuit as well as its circuit configuration. That part of the circuit enclosed by the dashed lines can be considered as a single transistor with connection points A, B, and C representing the base, collector, and emitter, respectively, of the equivalent transistor.

Fixed class A bias is supplied to the base of transistor Q1 through voltage divider resistors R1 and R2 connected across the common power supply. The bias of Q2 is supplied by emitter current flow through Q1 and Q2 to ground. Capacitive input coupling is provided by C1. The output load, R_C, is connected in series with both transistors and the common source of power. The collectors of both transistors are connected in parallel and thus share a common load. The output is taken through coupling capacitor C2. If greater power output is desired, R_C and C2 may be replaced by a transformer.

Both transistors are connected in the common emitter configuration. The base of Q2, being directly connected to the emitter of Q1, provides a certain portion of negative feedback. Since both transistors are series connected across the input, a much higher input impedance (100 or more times as large) is offered than with a single common emitter stage. In the absence of an input signal both transistors operate in the class A region. The bias on Q2 is slightly less than that appearing on the base of Q1 because of the small voltage drop between the base and emitter of Q1.

Assuming a negative going input signal at the base of Q1 we find the collector current increasing due to the increase in forward bias. The amount of increase in collector current is determined by the amplitude of the incoming signal and the gain of the transistor. The incoming signal also forward biases Q2, increasing the collector current. The increase in the collector current of Q2 is not as great as that of Q1, since Q2's forward bias is not as large. The collectors, being connected in parallel, have their currents add in the output, since both flow in the same direction through R_C. The increase in current flow through the load will increase the voltage drop across the load and will reduce the voltage on the collectors. Since V_{CC} is a negative voltage in the case of PNP's (being used in fig. 6-45), the output waveform is

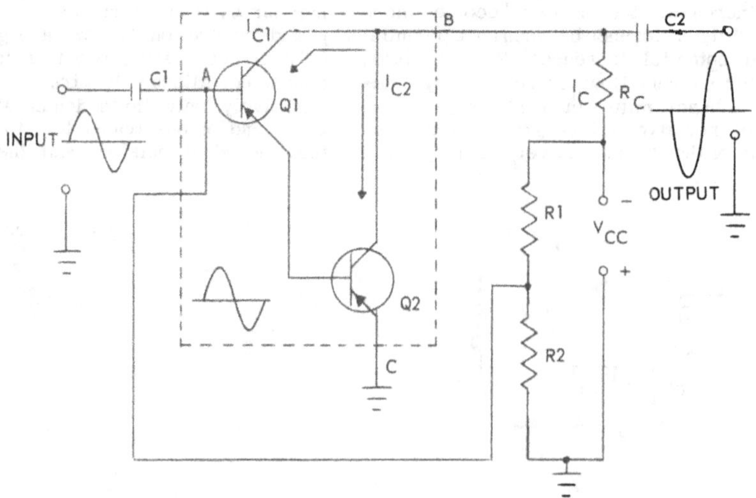

Figure 6-45.—Compound connected common emitter audio amplifier circuit.

going in a positive (less negative) direction. The negative going input signal produces a positive going output signal.

Conversely, when the input signal is going in a positive direction, it opposes forward bias and reduces the current in the circuit. A reduction in collector current means a lower voltage drop across the load resistor, less voltage across R_C and more voltage on the collector. A positive poing input signal produces a negative going output signal. The characteristic of phase reversal, input to output, in common emitter configurations is still present in the compound-connected common emitter amplifier.

Current gain in compound-connected transistors is greater than that of two transistors connected in cascade by approximately 10%. The addition of reactive elements which affect frequency response in the cascaded amplifier circuit is not present in the compound-connected amplifier due to the direct connection of the transistors. Frequency response, therefore, in the compound-connected amplifier is not deteriorated.

Although the input impedance in the compound-connected amplifier is much higher than that of a single transistor amplifier, the drive required for operation is the same in both circuits. Remembering that current in a series circuit is uniform and that a smaller current through a high resistance can produce the same voltage drop as a higher current through a smaller resistance, it can be seen that more drive is not necessary. The additional gain is due to the collector currents adding in the output circuitry. The dynamic output impedance of a compound-connected amplifier is lower than that of a conventional amplifier. This is due to increased currents through the load.

The compound-connected, common collector amplifier uses the same arrangement of transistors as does the CE amplifier. The major change, of course, is in the output circuit. In the CE amplifier we take our output from the collector circuit, and in the CC amplifier the output is taken from the emitter circuit.

Although the voltage gain is less than unity, the current gain is high as is the power gain. Figure 6-46 shows a compound-connected common collector amplifier. Note that this circuit uses four leads in this configuration (B1, B2, C, and E).

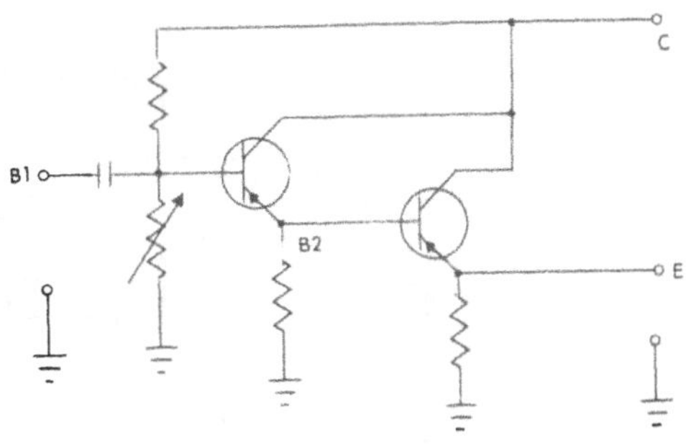

179.122

Figure 6-46.—Compound-connected common collector amplifier.

UNIJUNCTION TRANSISTORS

The unijunction transistor (UJT) is a three terminal semiconductor device. Sometimes referred to as a double base diode, its characteristics are quite different from those of the conventional two junction transistor. Some of its features are: (1) a stable triggering voltage, (2) a very low value of firing current, (3) a negative resistance characteristic, and (4) a high pulse current capability. They find use in oscillators, timing circuits, voltage and current sensing circuits, SCR (to be discussed) trigger circuits, and bistable circuits.

Figure 6-47 is a block diagram of the unijunction transistor showing the leads, types of materials used, and the names of the elements. If terminals 2 and 3 were connected together, the resulting device would have the characteristics of a conventional junction diode.

Figure 6-48A represents the schematic symbol for the unijunction transistor. A simplified

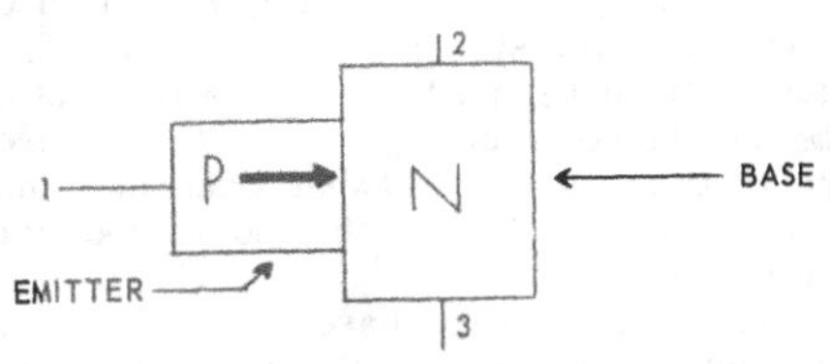

179.123

Figure 6-47.—Block diagram of UJT.

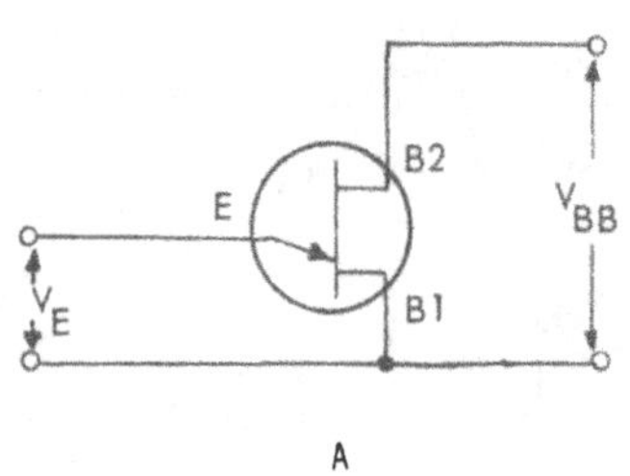

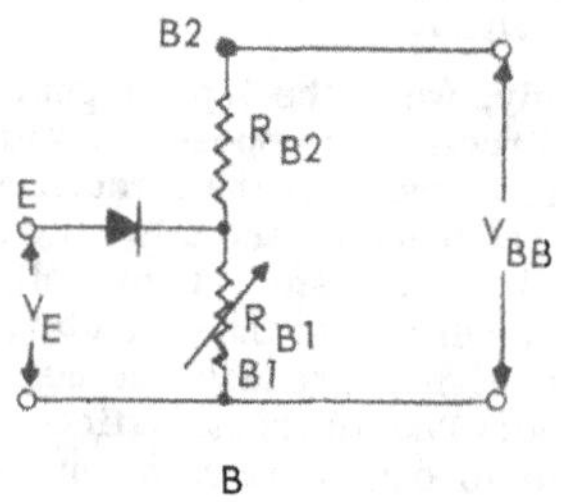

179.124
Figure 6-48.—UJT schematic symbol and simplified equivalent circuit.

equivalent circuit is shown in figure 6-48B. R_{B1} and R_{B2}, in figure 6-48B represent the resistances of the N type silicon bar between the ohmic contacts, called base one (B1) and base two (B2), at opposite ends of the base. A single rectifying contact called the emitter (E) is made between base one and base two. The diode in figure 6-48B represents the unijunction emitter diode. In normal circuit operation base one is grounded and a positive bias voltage (V_{BB}) is applied at base two. For example, if the bias voltage applied at base two is 10 volts positive with respect to base one (R_{B1} is equal to R_{B2}), the voltage drop across each resistor is equal to 5 volts. There is no emitter current flowing, and the silicon bar acts as if it were a simple voltage divider. The emitter, being halfway along the bar, sees 5 volts (with respect to B1) opposite it. So long as the emitter voltage (V_E) remains 5 volts or less, very little emitter current flows, as the emitter will be either in a neutral state or reverse biased. When the emitter voltage (V_E) rises above 5 volts, the junction becomes forward biased, emitter current increases very rapidly, and the resistance (R_{B1}) decreases very rapidly. Current flow at this time is from base one to the emitter.

Referring to figure 6-48B, R_{B1} is shown as a variable resistor and R_{B2} is fixed. Exact values are not important, but the realization that the combination of R_{B1} and R_{B2} represents the total resistance between base one and base two contacts will aid in the understanding of how the UJT functions.

When the emitter-base one contacts are in the reverse biased condition the resistance R_{B1} is at its maximum value, and current flows through the silicon bar from base one to base two. However, with the emitter potential is such that it becomes forward biased the resistance, R_{B1}, decreases to a minimum value and current flow is now from base one to the emitter.

The unijunction transistor exhibits three distinct regions of operation. These are: (1) the cutoff region, (2) the negative resistance region, and (3) the saturation region. Figure 6-49 is a curve depicting these regions. That portion of the emitter current graph shown to the left of the zero reference is where the emitter is either reverse biased or neutral, and this is the cutoff region, which ends at the point marked "peak." At the "peak point" the current between base one and the emitter increases rapidly as now a forward bias condition exists. That part of the resistance of the silicon bar, R_{B1}, in figure 6-48B, between base one and the emitter is now minimal, and emitter current is increasing while emitter voltage is decreasing. This, in effect, is the negative resistance region shown between the peak and valley points of figure 6-49. Beyond the valley point the resistance of the device behaves as a positive resistance; current again

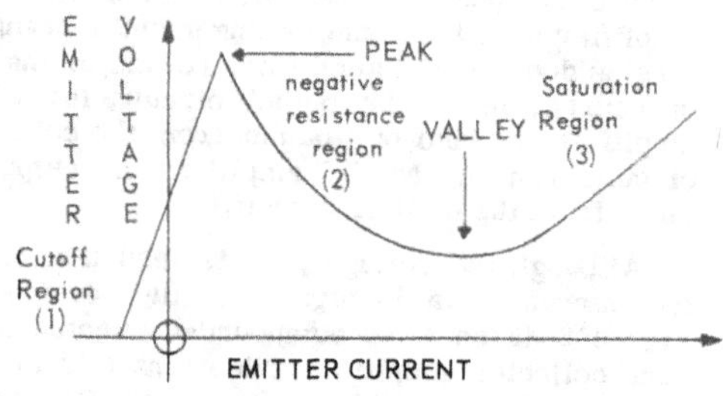

179.125
Figure 6-49.—UJT regions of operation.

increases slowly with voltage. This area of positive resistance is the saturation region of the unijunction transistor.

A typical circuit application of the unijunction transistor is a sawtooth generator. Figure 6-50 shows the schematic diagram of such a circuit with the resultant sawtooth waveform output. Initially, C1 (fig. 6-50) is not charged, and the emitter potential is zero. When power is applied, C1 will start to charge slowly since it is connected to the battery through R1. This charging will continue until the potential on the emitter (the top of C1) becomes positive with respect to base one and forward biases the emitter-base one junction. As soon as this point is reached, current from base one to the emitter rises rapidly and C1 will discharge through base one and the low resistance of the PN junction. When C1 is discharged the cycle repeats. The voltage across C1, then, is a sawtooth waveform possessing a slow rise time (t_1 to t_2, t_3 to t_4, t_5 to t_6) and a rapid descent (t_2 to t_3, t_4 to t_5, t_6 to t_7). When the emitter-base one junction is forward biased, current through the silicon bar also rises. As can be seen, this is a very simple method of obtaining a sawtooth waveform.

FIELD EFFECT TRANSISTORS

The Junction Field Effect Transistor (JFET) and the Insulated Gate Field Effect Transistor (IGFET) operate on a similar principle (current controlled by an electric field) but differ in the method that the control element is made. This difference results in a considerable difference in device characteristics, and, therefore, circuit design variances become a necessity. The first FET to be discussed is the JFET (fig. 6-51).

JFET

Basically, a JFET is a bar of doped silicon, called a channel, that behaves as a resistor.

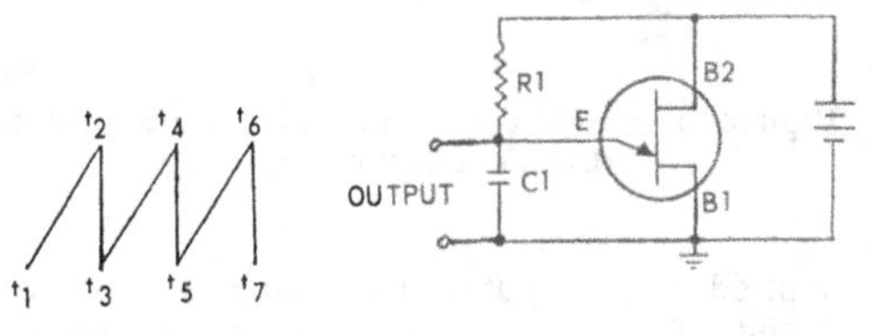

179.126
Figure 6-50.— UJT sawtooth generator.

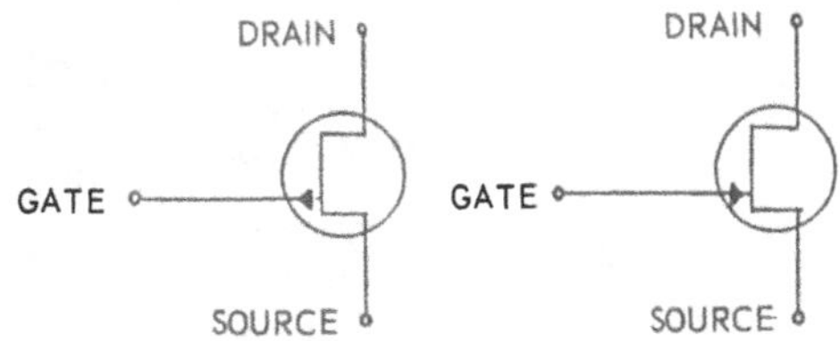

179.127
Figure 6-51.— Schematic symbols for N and P types JFETs.

The doping may be N or P type, creating either an N channel or P channel JFET. There is a terminal at each end of the channel. One terminal is called the source, and the other is called the drain. Current flow between source and drain, for a given drain-source voltage, is dependent upon the resistance of the channel. This resistance is controlled by a gate. The gate consists of P type regions diffused into an N type channel, or N type regions diffused into a P type channel. Cutaway views of both types are shown in figure 6-52.

As with any PN junction, a depletion region, or better, an electric field, surrounds the PN junction when the junctions are reverse biased. As the reverse voltage is increased, the electric fields spread into the channel until they meet, creating an almost infinite resistance between source and drain.

Consider now the case of zero gate voltage (V_G), but with an external voltage applied between source and drain (V_D). Drain current (I_D) flow in the channel sets up a reverse bias, along the surface of the gate parallel to the channel. As the drain source voltage (V_D) increases, the depletion regions increase until they meet. There is now an effective increase in channel resistance that will prevent any further increase in drain current. This V_D that causes this current limiting condition is called the "pinch-off" voltage (V_O). Further increase in drain source voltage produces only a slight increase in drain current. A typical curve showing this action is shown in figure 6-53.

As can be seen in figure 6-53, in the low current region, I_D is linearly related to V_D. As I_D increases, the depletion region between the gate and the silicon bar increases. This causes the "channel" to deplete and the slope of the I_D curve decreases. At the point where the drain source voltage (V_D) is equal to the pinch-off voltage, (V_O), the drain current (I_D)

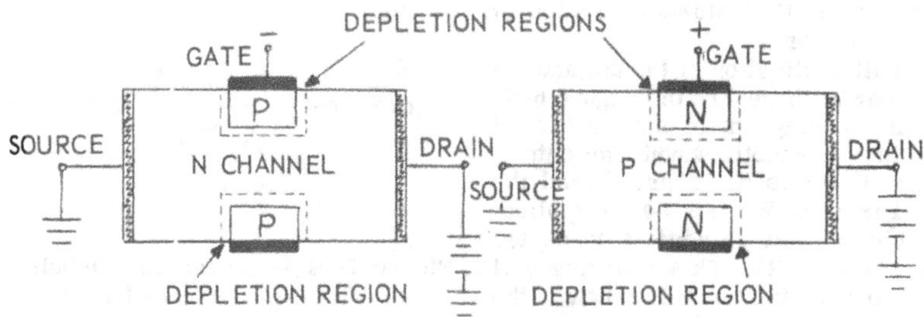

179.128

Figure 6-52.—Cutaway view of N and P channel junction field effect transistors.

saturates and reaches an avalanche voltage point between gate and drain. At this point (V_A), drain current increases rapidly and control of the device is lost.

If a reverse bias (V_{GS}) is applied to the gate channel, "pinch-off" will occur at a lower level of I_D because the depletion area at the gate is larger to begin with and is further increased by drain current flow. The greater the reverse bias applied to the gate, the sooner we reach "pinch-off" and the lower the value of I_D. Figure 6-54 illustrates this point.

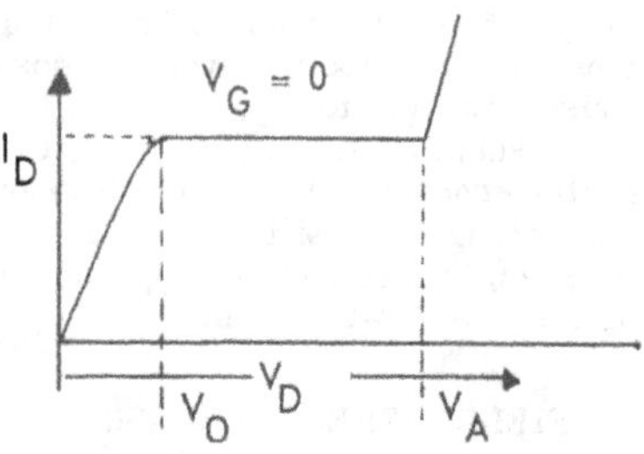

179.129

Figure 6-53.—I_D characteristics with V_G equal to zero.

IGFET

Details of the construction of an N channel depletion mode, IGFET, are shown in figure 6-55. As can be seen, the main body is composed of a P substrate. (In this sense, substrate is the physical material on which a circuit is fabricated. While its primary purpose is to furnish mechanical support, it may also serve a useful thermal or electrical function.) The source and drain material is diffused into the body. In this case it is a low resistivity N type material indicated by N+ in the figure. Between the source and drain materials a moderate resistivity N channel is also diffused into the body. Covering the main body of the IGFET is a layer of insulating oxide. Holes are cut into the oxide layer, allowing metallic contact to the source and drain. The gate metal area is now overlaid in the area between the source and drain contacts on the oxide. There is no physical contact between these metal contact areas.

The N channel between source and drain will allow current flow when a potential is applied between the source and the drain, with a zero gate voltage. When a reverse bias voltage is applied to the gate, drain current will be reduced. The greater the reverse voltage on the gate, the lower the drain current.

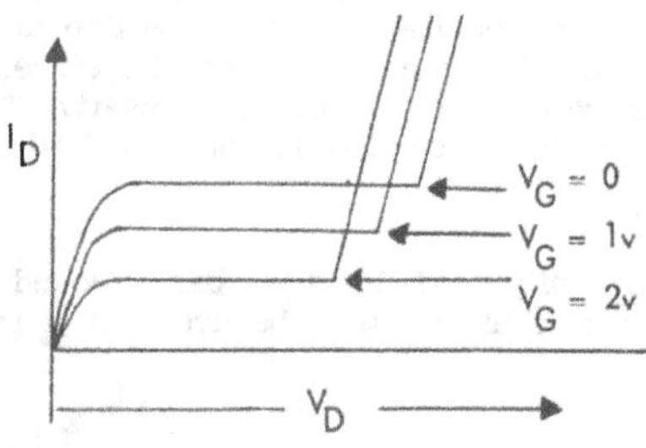

179.130

Figure 6-54.—Drain current characteristics for different values of V_G.

For the N channel IGFET shown in figure 6-55, a negative voltage applied at the gate will

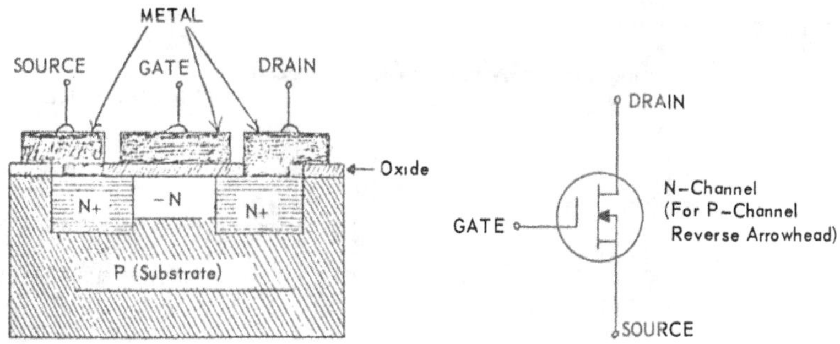

179.131
Figure 6-55.—Construction of depletion mode IGFET and schematic symbol.

cause an increasing in the depletion of the channel, increasing the resitance of the channel, and lowering I_D. Figure 6-56 shows this occurrence.

When the gate voltage on an IGFET is zero, there will be an electric field set up across the oxide insulating material due to drain current flow. The metal contact of the gate forms one side of a capacitor, while the oxide material acts as the dielectric and the N channel is the other side of the capacitor. As the drain current flows from left to right the N channel will exhibit a greater positive potential. The upper plate of our capacitor, the metal of our gate, will be negative with respect to the bottom of the plate, the N channel. We have an electric field that is exhibiting a negative potential at the top of the "capacitor" and a positive potential at the bottom. The increasingly positive potential due to drain current flowing from left to right, makes the field across our capacitor stronger on the right so that the field appears in the shape shown in figure 6-57.

An increase in the reverse bias potential on the gate will increase the strength of the field shown in figure 6-57 and the drain current will be reduced. One of the uses of the IGFET is in electronic voltmeters. You will recall (chapter 3) that the higher the resistance of the voltmeter, the more accurate it is for making measurements in high resistance circuits. Since the input impedance of an IGFET is very high, it is practical to use it in the input of an electronic voltmeter. Remembering that the greater the reverse bias applied to the gate of an IGFET the lower the drain current, and, conversely, the lower the reverse bias applied to the gate, the higher the drain current, it can be seen that

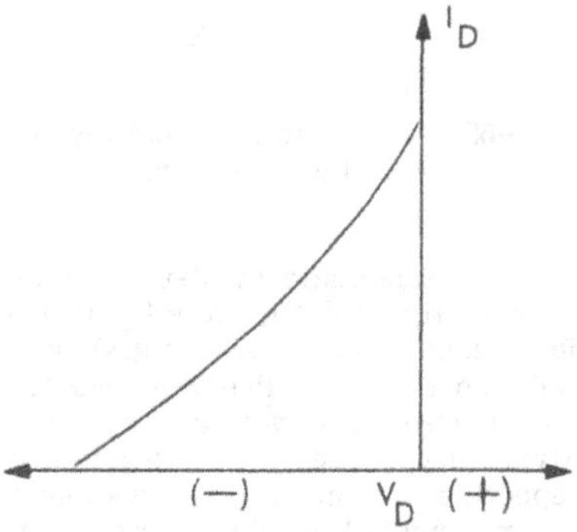

I_D

(—) V_D (+)

179.132
Figure 6-56.—Characteristic of depletion mode IGFET.

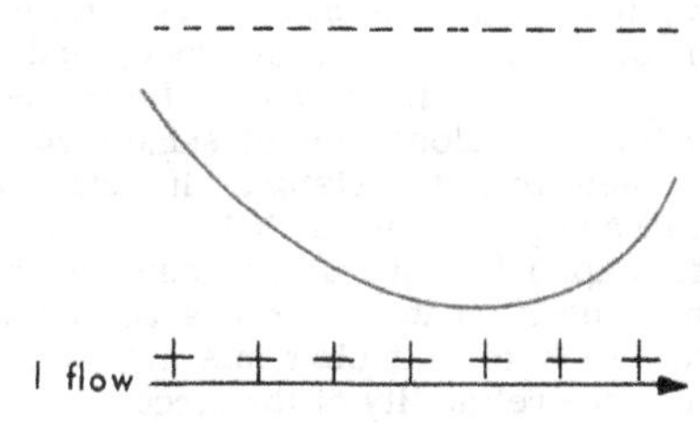

179.133
Figure 6-57.—Potential distribution.

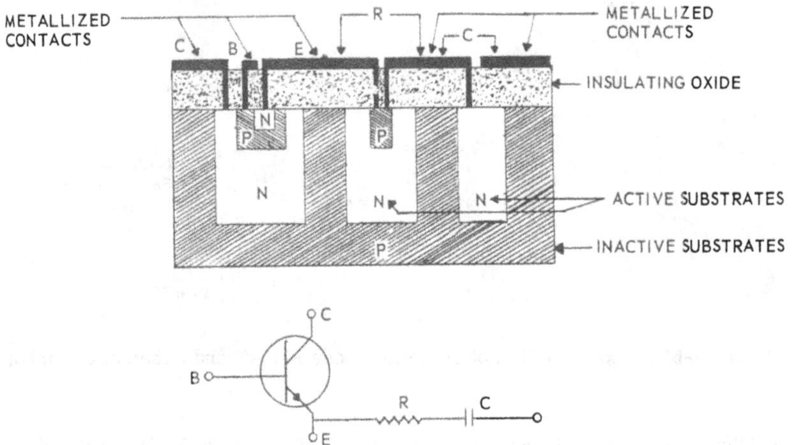

179.134
Figure 6-58.—Schematic and pictorial diagram of a simple IC device.

by connecting the drain terminal to meter circuitry, the meter will respond to changes in drain current.

INTEGRATED CIRCUITS

Another development in solid state electronics is the integrated circuit (IC). As the term integrated implies, several components are contained in one package. One semiconductor chip can contain two or more transistors, several resistors and capacitors, as well as many individual diodes. How this may be accomplished is our next topic. Figure 6-58 is a pictorial and schematic of a simple IC.

As can be seen from figure 6-58 it is easy for one IC device to replace many separate or discrete components with a single chip by using the various properties of semiconductor elements, such as resistance, capacitance, and conductivity. The circuit shown in figure 6-58 will probably be enclosed in the same size case now associated with transistors. In addition to the space saving features of IC's, the fact that they receive almost identical processing enables them to be closely matched in characteristics. The closer such circuit elements are matched, the greater the reliability of the circuit.

IC's come in various packages but the one to be described at this time is the differential amplifier; a schematic symbol for this circuit is shown in figure 6-59, while figure 6-60 is the complete schematic. A differential amplifier is basically an amplifier that responds to the difference between two voltages or currents and effectively suppresses like voltages or currents.

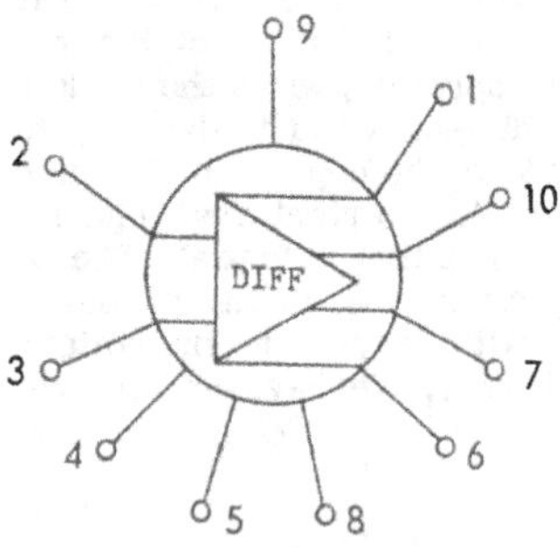

67.159
Figure 6-59.—Schematic symbol for IC differential amplifier.

Inspection of figure 6-60 shows that the signal inputs are applied to the bases of Q1 and Q2 through leads 2 and 3 respectively. These input signals are amplified by Q1 and Q2 and applied to the bases of Q3 and Q4, and the outputs of the emitter followers Q3 and Q4 are available at pins

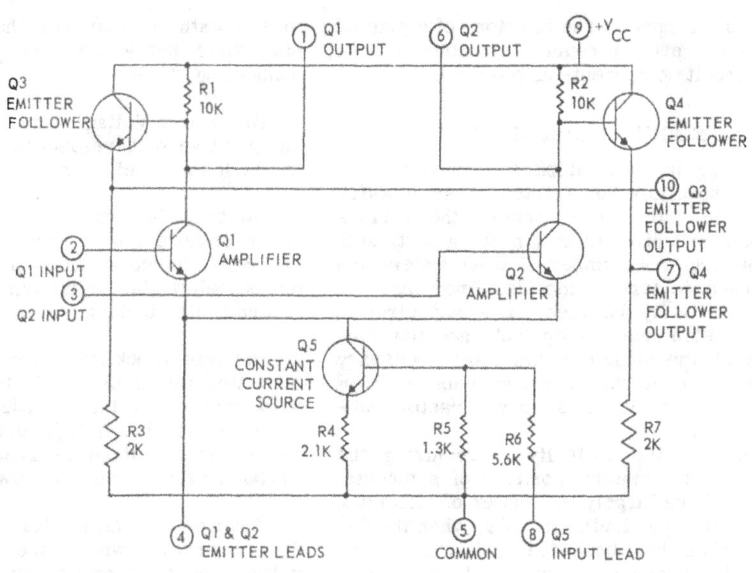

Figure 6-60.—Differential amplifier, integrated circuit device.

10 and 7, respectively, of the integrated circuit. The output signals of Q1 and Q2 (the collector signals) are also available at pins 1 and 6 of the device. The constant current source, Q5, determines the gain of the circuit by determining the emitter currents of Q1 and Q2. Control of the amplifier gain can be accomplished by applying a signal to base of Q5 through pin 8 and R6. If it is desired to bypass the constant current source, a signal may be applied directly to the emitters of Q1 and Q2, at pin 4. Lead 9 is the connecting lead for the positive V_{CC} , and lead 5 is the common connection for the device.

It should be noted that all the components shown in figure 6-60 are contained on the IC chip and that it might be necessary to connect external biasing or coupling components depending upon the required operating conditions. For example, external resistors might be connected from pin 9 to pins 2 and 3 to provide bias for Q1 and Q2. This type of construction, with primary connections, increases the flexibility of the device; less different types are required, and, thus, the cost is less. The values given for figure 6-60 are nominal and are not necessarily those that would be found in any particular IC.

PRINTED CIRCUIT BOARDS

A printed circuit board is an insulator surface upon which conductors and other circuit components have been "printed." To be most effective, the printed circuitry should adhere permanently to the base plate. The printed wiring should have low resistance and be able to pass the desired current without overheating. The other printed components, resistors, capacitors, and inductors, should remain stable under their rated loads.

By using a mold of the desired shape it is possible to produce uniform (as far as physical dimensions are concerned) units. Since printed circuit boards are readily adapted to plug-in units, the elimination of terminal boards, fittings and tie points results in a substantial reduction in the overall size of equipment.

Printed circuit boards come in various sizes depending upon the type of circuitry being used. Common sense dictates that a printed circuit board using transistors will be smaller than one used for the same function utilizing electron tubes. Regardless of the size of the printed circuit board, you can be sure that the same circuit(s) built in the conventional manner would occupy a much larger space. The functions of a printed circuit board are to replace as many of the passive circuit components as possible.

MODULAR CIRCUITRY

The basic building block for modular circuits is a specially constructed ceramic wafer on which is mounted or printed the various components and circuits that make a particular electronic stage. A number of these wafers are then stacked to form a module. Since there is a great similarity between parts and circuits in modern electronic equipment, and the construction of the module permits the necessary variations to conform to the various needs of circuits, the module is a very versatile subassembly unit.

Because of the difficulty in repairing the printed or subminiature portions of a module, servicing will be largely the matter of localizing the defect to a particular module. When the defective module has been found, it will be removed and a replacement module inserted into the circuit.

SILICON CONTROLLED RECTIFIER (SCR)

The Silicon Controlled Rectifier (SCR) is a three junction semiconductor device. It is normally an open circuit until a proper signal is applied to the gate element at which time it rapidly switches to a conducting state. Figure 6-61 depicts the SCR pictorially and schematically.

The following characteristics of the SCR will be used to explain the operation of the device. They are as follows:

Forward Blocking Voltage: That value of voltage (anode terminal positive with respect to the cathode) that puts the SCR in a forward bias state but which does not switch it to the conducting state.

Breakdown Voltage: The voltage (anode terminal positive with respect to cathode) that causes the SCR to switch to the conductive or ON state.

Holding Current: That minimum value of anode-cathode current that is required to keep the SCR in conduction. If the anode current drops below this minimum value the SCR will return to its blocking state.

Reverse Blocking Region: That region of SCR operation where the anode terminal is negative with respect to the cathode (reverse bias) but not in avalanche. A light current will flow until the reverse voltage is great enough to cause avalanche (heavy current flow).

Reverse Avalanche Region: That region of SCR operation where the maximum reverse voltage (negative anode) has been exceeded and avalanche current flows.

Gate Trigger Current: That value of gate current, in the forward direction (gate to cathode) required to trigger an SCR whose anode is positive with respect to its cathode.

Gate Trigger Voltage: That value of voltage across the gate to cathode element of an SCR caused by Gate Trigger Current flow. This voltage is measured prior to the SCR reaching a conduction state.

Inspection of figure 6-62 indicates that the forward blocking region is largest when gate current is zero, it takes a greater value of forward anode-cathode voltage to cause breakover. As gate current is increased the forward

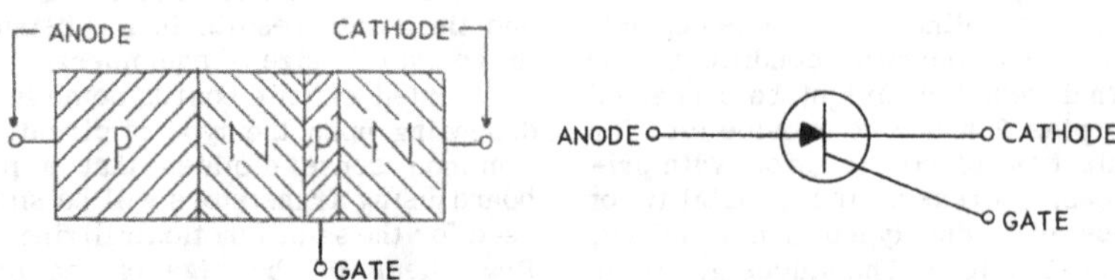

Figure 6-61.—Pictorial and schematic representation of SCR.

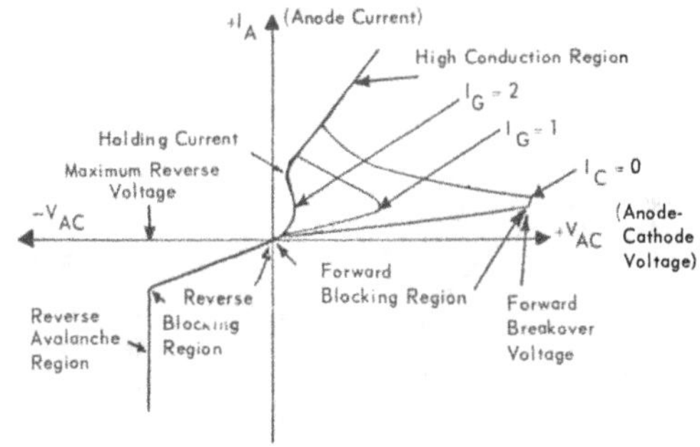

179.136

Figure 6-62.—Voltage-current characteristics of an SCR.

blocking region is decreased and the breakover voltage point occurs sooner. When the breakdown voltage value is reached, the gate region of the SCR is effectively eliminated and the SCR goes into heavy conduction and stays there until anode current drops below the minimum value of holding current at which time the SCR reverts to the forward blocking state.

Gate current is an important characteristic of the SCR. With gate current at zero, a high value of anode-cathode voltage in the forward direction would be necessary to cause the SCR to reach the breakover voltage and go into high conduction. As gate current is increased to increase forward bias, the forward blocking region is reduced and breakover voltage occurs sooner. It is possible to set a value of gate current that would eliminate the forward blocking region entirely at which time the SCR would perform as a conventional PN rectifier.

Once the SCR reaches the high conduction region the gate has no control, the gate area has been effectively eliminated. To regain control, external methods must be applied to the SCR to reduce the flow of anode current. One method is to apply a voltage to the anode that will make the anode negative with respect to the cathode. This will turn the SCR off, putting it back to a forward blocking condition. Another method is to divert enough anode current to some external circuit for a period of time long enough to cause holding current to drop below its minimum value, thereby turning the SCR off. A switch across the anode-cathode terminals, electronically activated or otherwise, will divert the anode current flow. Still another method of turning the SCR off is to cause a heavy reverse current flow from gate to cathode by injecting a reverse voltage which will cause the gate region to increase its opposition to anode current flow and allow the gate to regain control of the device.

Figure 6-63 illustrates how the SCR can be used to control a high power circuit. The breakover voltage with switch 1 open is higher than E source; therefore, the SCR is not conducting and there is no current flow through the load. When switch 1 is closed, the breakover voltage is reduced (the forward blocking region is reduced) because of the voltage applied to the gate. The gate voltage causes the SCR to conduct, and current flows through the load. This arrangement enables a low power line to control a high power circuit.

The SCR may also be used as a phase control device in various power supply applications. "Phase Control" is a rapid on-off switching process used to connect an a.c. supply to a load for a controlled fraction of each cycle. By governing the phase angle of an a.c. wave at which the SCR is allowed to conduct, control of the output of a circuit is accomplished.

A very simple variable resistance half-wave circuit is shown in figure 6-64. This circuit will control the output from essentially zero (full on) to 90 electrical degrees of the anode voltage wave. With the circuit shown (fig. 6-64) it is not possible to prevent the SCR from conducting beyond the 90 degree point, because the supply voltage to the gate and the voltage necessary to trigger the gate allowing the SCR to conduct, are in phase. That is, when E source is offering maximum resistance, the voltage, anode to cathode, is sufficient to cause breakover and

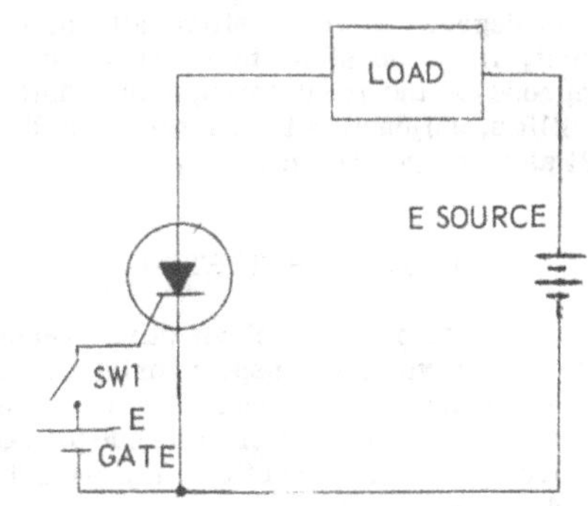

179.137

Figure 6-63.—SCR being used to control high power circuits.

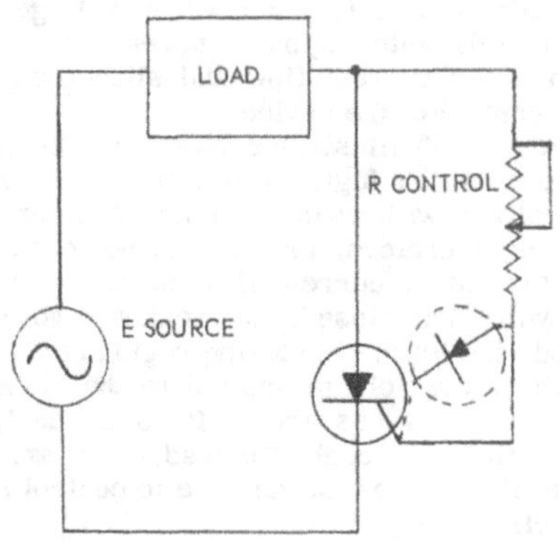

179.138

Figure 6-64.—Simple variable resistance half-wave phase control.

the SCR can be caused to conduct at any point from zero degrees to 90 degrees by adjusting the control resistor, because the SCR will conduct at the point where gate trigger current is reached. Of course, during the entire negative alternation of the source the anode-cathode voltage is reversed and the SCR reverts to a reverse blocking or OFF state. In order to prevent reverse avalanche a diode is placed in series with the control resistor and the gate. This diode is shown in figure 6-64 enclosed in dotted lines. Figure 6-65 depicts the voltage and current characteristics of the circuit when R control is adjusted.

Although we have shown the operation of a single phase control circuit that allows control of up to 90 degrees of the positive alternation of the input, it is possible to control the entire 360 degrees of the input through the addition of other SCR's, unijunction transistors, and diodes, as well as through other methods.

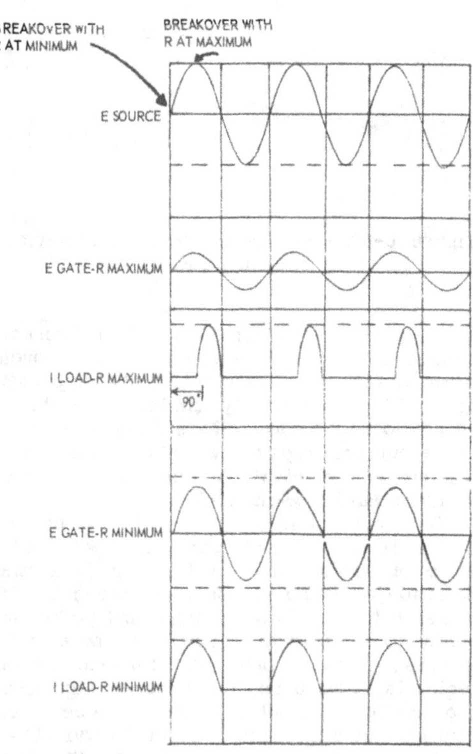

179.139

Figure 6-65.—Characteristics of a simple variable resistor half-wave control circuit.

TRANSISTOR TESTING

The first step in troubleshooting transistor circuits is to visually inspect printed circuit boards for loose connections, short circuits due to poor soldering and open circuits due to cracked foil. Careful visual inspection may save hours of troubleshooting.

Whenever a particular transistor is suspected to be faulty, it may be tested in the circuit if there are no low resistance shunts (such as coils, forward biased diodes, low value resistors, etc.) across any of its leads. A low resistance shunt across any two of the transistor leads can easily cause erroneous indications on the transistor test set.

Erroneous indications can be eliminated by removing the suspected transistor from the circuit. Some manufacturers have simplified the removal and replacement of transistors by using transistor sockets. Whenever a circuit board without transistor sockets is encountered, transistors will have to be carefully unsoldered in order to test them out of circuits.

Any meter used for voltage measurements in transistor circuits should have a sensitivity of greater than 20,000 ohms per volt on all ranges. Meters with a lower sensitivity may draw excessive current from the circuit under test when using the lower voltage ranges. Meters with an input impedance of 100,000 ohms per volt or more are preferred. Precautions to be observed when taking measurements in transistor circuits are:

1. BEFORE making any resistance measurements, make sure that all power to the circuit under test has been disconnected and that all capacitors have been discharged.

2. Do not use an ohmmeter which passes more than one milliampere through the circuit under test.

3. Ensure that all test equipment is isolated from the power line either by the equipment's own power supply transformer or by an external isolation transformer.

4. Always connect a ground lead between COMMON for the circuit under test and COMMON on the test equipment.

5. Do not short circuit any portion of a transistor circuit. Short circuiting individual components or groups of components may allow excessive current to flow, thus damaging components.

6. Do not remove or replace any transistors in an energized circuit.

I_{CO} TEST

Age and exposure to higher than normal temperatures may cause excessive collector leakage current, generally designated as I_{CO}. I_{CO} is greatly dependent on ambient temperature, thus care must be taken to see that temperature is controlled when measuring I_{CO}.

One sign of a defective transistor is instability of I_{CO}. Whenever I_{CO} increases slowly while being measured, it is quite evident that the transistor is defective. Excessive leakage current indicates transistor deterioration and usually is accompanied by lower than normal beta.

Figure 6-66 shows the test circuit for measuring I_{CO}. The transistor must be out of the circuit for this test. If I_{CO} were measured with the transistor in the circuit, the indicated I_{CO} will be higher than the actual I_{CO} due to shunt current paths in the circuit.

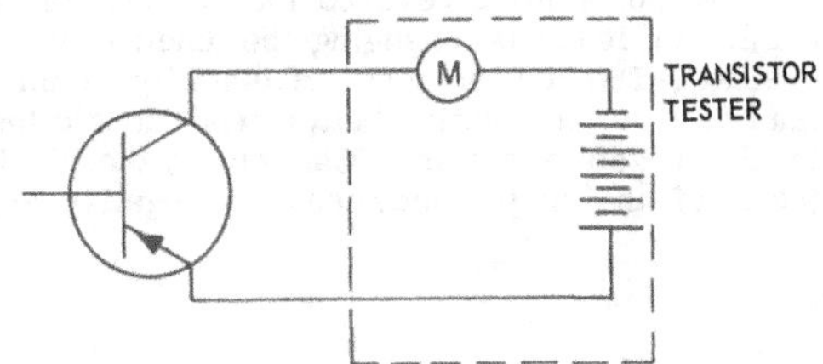

179.140
Figure 6-66.—Measurement of I_{CO}.

TRANSISTOR TESTING USING AN OHMMETER

While an ohmmeter cannot test all the characteristics of transistors, especially transistors used for high frequencies or switching, it is capable of making many simple transistor tests. Since transistor testers might not always be available, and the ohmmeter is usually available, some hints on how to test transistors using an ohmmeter are in order.

Since damage to transistors could occur when using voltages above approximately 6 volts, care must be taken to avoid using resistance scales where the internal voltage of the ohmmeter is greater than 6 volts. This higher potential is usually found on the higher resistance scales. Excess current might also cause damage to the transistor under test. Since the internal current limiting resistance generally increases as the resistance range is increased, the low range of the resistance scale should also be avoided. Basically, if we stay away from the highest resistance range (possible excessive voltage) and the lowest resistance range (possible excessive current), the ohmmeter should present no problems in transistor testing. Generally speaking, the RX10 and RX100 scales may be considered safe.

The polarity of the battery, as well as the voltage value, must be known when using the ohmmeter for transistor testing. Although, in some cases, the ground or common lead (black) is negative and the hot lead (red) is positive, this battery set-up is not always the case.

It is possible to reverse the polarity of the ohmmeter leads by changing the function switch position. This means that the black or common lead is now the positive battery and the red lead is the negative battery. Caution: Although the jacks of the meter may show a negative sign under the common and a positive sign under the other jack, these do not indicate the internal polarity of the battery connected to the jack. In addition, do not use any ohmmeter for transistor testing that draws more than one milliampere.

CHAPTER 7

ELECTRON TUBES

A most important device which performs as an amplifier/rectifier in communications and navigation systems is the electron tube. Electron tubes have been replaced in many applications by solid-state devices. However, innumerable uses for tubes will exist for some time.

ELECTRON TUBE DIODES

Basically, the electron tube diode behaves as a PN junction diode. Preparation for current flow through the two devices is slightly different, however.

Considering first the thermionic or heated cathode electron tube, one discovers that its major components are a cathode (emitter), an anode, and an evacuated envelope (as their environment). Cathodes are formed of compounds which lend themselves to rapid and easily accomplished emission of electrons. Anode material does not readily lend itself to such behavior.

When the cathode is heated, it immediately starts to emit electrons. (This emission of electrons from hot bodies is known as Edison effect.) Although this emission is low the potential between the anode and cathode can be measured (before an external potential difference is applied between them) if adequately sensitive equipment is used. Such an indication is achieved by comparing the stable or reference state of the anode material to the electron emission from the cathode material.

At the instant a difference of potential is applied externally between anode (positive) and cathode (negative), an electric field exists between the electrodes of the diode. Electrons, now under the directed influence of the electric field, flow toward the anode through the source to cathode and in the complete cycle path as long as the field exists.

In electron tube diodes, the intensity of current flow is directly proportional to the level of potential difference between the cathode and anode. This potential difference may also be described as the electric field which exists between the cathode and anode, or, such field's strength and/or density.

Since the material of an electron tube's anode is not conducive to the emission of electrons, and this material is not heated to prepare it for such emission, current flow from the anode to cathode does not normally occur. Hence, the electron tube diode performs as a unidirectional valve or switch.

Should a difference of potential, high in value and reversed in polarity be placed across an electron tube diode, the "dielectric" between anode and cathode may be "broken down," and an excessive quantity of reverse current flow will naturally occur. Electron tube reverse current breakdown may be closely compared to the breakdown of a capacitor's dielectric. Each is due to excessive electric field strength or extraordinarily high difference of potential across device elements. Note that in each case of breakdown there exists some physical means of current conduction between the poles of potential difference. In capacitors this medium is the dielectric; in electron tubes it is also a form of dielectric. Specifically, in electron tubes, the evacuated envelope is not an absolute void or "perfect" vacuum.

OPERATION OF THE ELECTRON TUBE DIODE

As has been stated, the evacuated envelope surrounding the electron tube diode is not a perfect vacuum. Therefore, some matter remains in the existing space. When the cathode is heated, it begins to emit electrons. Upon application of an electric field, these electrons are directed through the material of the "vacuum" to the anode, through the source and back to the cathode.

A question which frequently arises concerning the operation of electron tube diodes is the one of improving its efficiency by not evacuating the envelope. Such thinking results from the fact that residual matter in the evacuated tube provides the current conduction path from cathode to anode. However, increasing this matter by leaving air in the device has two fundamental obstacles to effective operation. One, the oxygen in air would encourage the rapid burn-out of the heater and, two, an abundance of matter in the tube would allow excessive electron dissipation in the forms of ionization and heat, even should problem one be overcome. Ionization, incurred in the process just discussed, develops whenever electron flow occurs. Visible evidence of ionization is manifested when the ionized matter is principally composed of an inert gas or gases.

COLD CATHODE ELECTRON TUBE DIODES

In some cases electron tube diodes are constructed without facilities for heating the cathode. Such diodes are frequently used as voltage regulators as well as conventional rectifiers.

Operation of cold cathode diodes occurs when a difference of potential is placed across the diode anode (positive) and cathode (negative) causing an electric field to exist between these elements. Since cathode material is of a relatively high electron emission level, some of these electrons are started on their path to the anode. Usually, the medium of conduction is an inert gas such as neon, argon, krypton or xenon. When electron flow in these gases occurs, the gases glow in one of several colors and are said to have become ionized.

Since some energy is dissipated in the form of heat while "cold" cathode electron tubes are operating, the cathode temperature does rise. Therefore, the term "cold" cathode is not completely accurate and at best only indicates the lack of a cathode heating element in such electron tubes.

As current flow through an inert gas diode increases more energy is dissipated in ionizing the gas. Within the limits established by the quantity and type of gas, such diodes will maintain a constant voltage drop across their elements, though current flow through the devices varies.

CATHODE CONSTRUCTION

This chapter, for the most part, will be concerned with electron tubes which employ thermionic emission. Thermionic cathodes are heated in

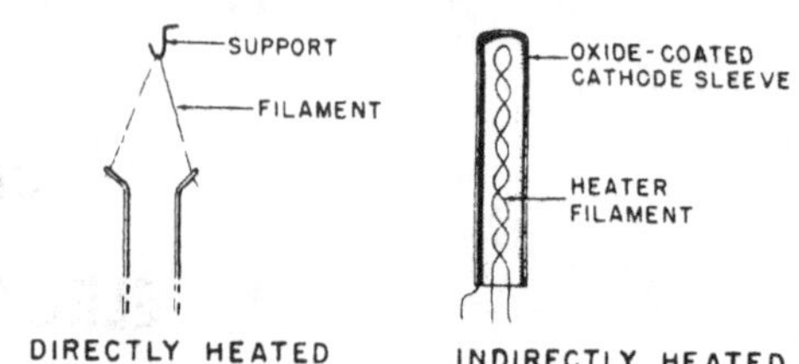

Figure 7-1.—Methods of heating the cathodes of electron tubes.

one of two ways—directly or indirectly. A directly heated cathode is one in which the current used to supply the heat flows directly through the emitting material. In an indirectly heated cathode, the heating current does not flow through the emitting material. Figure 7-1 illustrates the construction of the two types of cathodes.

The directly heated cathode, commonly called a filament, has the advantage of being fairly efficient, and capable of emitting large amounts of energy. However, due to the small mass of the filament wire, the filament temperature fluctuates with changes in current flow. If an a.c. source is used to heat the filament, undesired hum may be introduced into the circuit. This is especially evident in low level signal circuits.

A relatively constant rate of emission under mildly fluctuating current conditions may be obtained with an indirectly heated cathode. This type of cathode is in the form of a cylinder, in the center of which is a twisted, electrically insulated wire called the heater. The cathode is maintained at the correct temperature by heat radiated from the heater. The emitting material in this type of cathode remains at a relatively constant temperature, even with an alternating heater current. Figure 7-2 illustrates the schematic symbols for directly and indirectly heated cathodes.

DIODE CONSTRUCTION

As the name implies, the diode electron tube contains two electrodes or elements (the heater in an indirectly heated cathode type of tube is not considered an element). These electrodes are called the cathode and plate or anode. Physically, the cathode is normally placed in the center of the tube structure, and is surrounded by the plate. Examples of directly and indirectly heated diodes are shown in cut-away views, in figure 7-3.

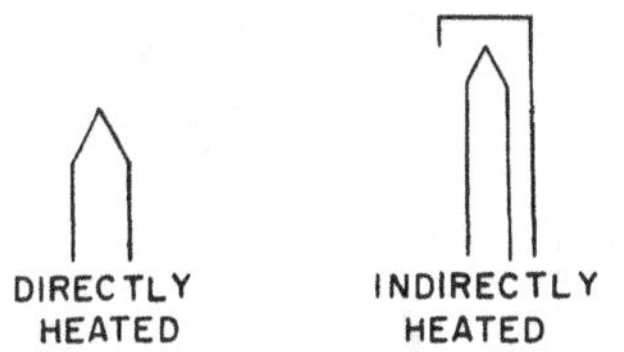

179.141

Figure 7-2.—Schematic symbols for directly and indirectly heated cathodes.

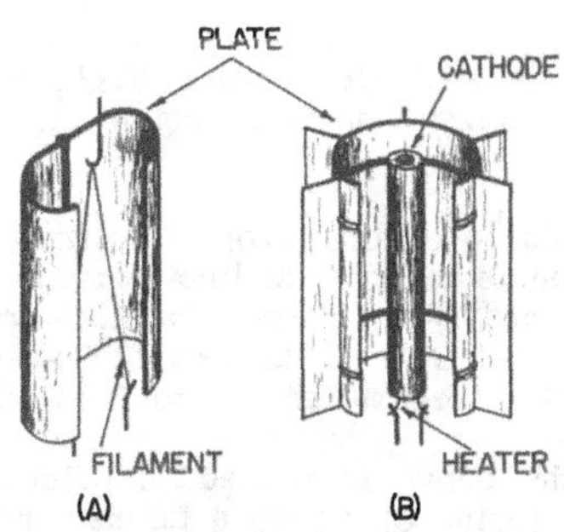

179.142

Figure 7-3.—Cutaway views of directly and indirectly heated diodes.

The plate, as previously stated, is made of a material which is not conducive to the emission of electrons. It must also be capable of dissipating the heat generated by electrons striking the plate.

The schematic symbols used to represent diode tubes are shown in figure 7-4. In many cases, two or more diodes may be included in the same envelope, in order to conserve space. A tube which contains two plates and one or two cathodes is called a duo-diode or twin diode. Figure 7-5 illustrates the construction and schematic symbols of duo-diode electron tubes.

As a rule, electron tubes do not have as long a life expectancy as resistors, capacitors, or other circuit components. This is due, in large part, to wearing out or breaking down of the filament or heater. To provide for easy removal and replacement of the tube, the base of the tube (fig. 7-5) is constructed in the form of a plug, which is inserted into a socket on the chassis. The electrical connections between the tube elements and the circuit are completed through the plug terminals, called pins (fig. 7-5).

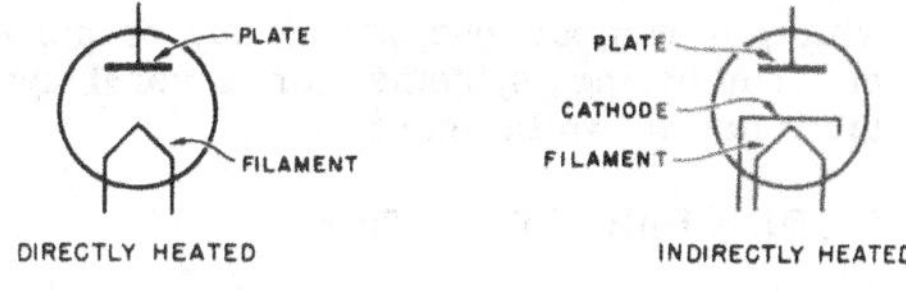

4.140

Figure 7-4.—Schematic symbols used to represent diode tubes.

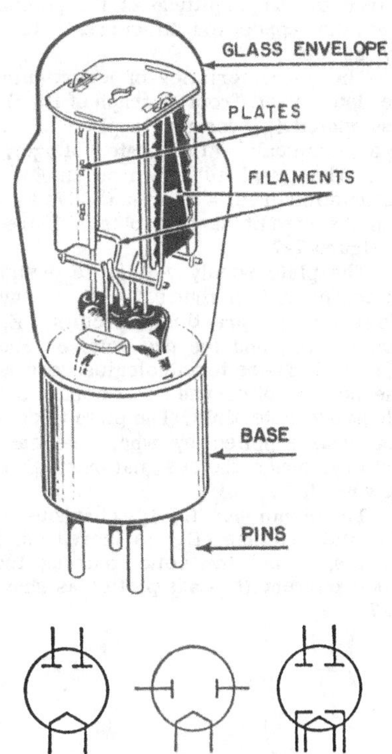

4.140(20C)

Figure 7-5.—Construction and schematic symbols of duo-diodes.

There are various types of tube bases, containing different numbers and sizes of pins. Each type of tube base has some kind of guide or key to prevent the tube from being plugged into the socket improperly. For ease of circuit tracing, the tube pins are assigned numbers. At the BOTTOM of the tube or socket, the pins are numbered in a clockwise direction, beginning with pin number one at the key or guide. The pin numbering systems for several types of tubes are shown in figure 7-6.

DIODE CHARACTERISTICS

As was previously explained, when a difference of potential of the proper polarity (i.e., plate positive—cathode negative) is placed across a diode electron tube, current flows from cathode to plate. The magnitude of this current is determined by the amplitude of the potential applied and the opposition to current flow presented by the tube.

The characteristics of a particular tube may be determined from a graph of its plate current designated I_b, versus its plate to cathode voltage (commonly called plate voltage), designated E_b. The graph is known as an E_b - I_b characteristic curve. A representative E_b - I_b curve and the circuit used to obtain it are illustrated in figure 7-7.

The plate supply voltage is designated E_{bb}. In the circuit in figure 7-7 E_b is equal to E_{bb}. However, in practical circuits, E_b will not equal E_{bb}, and the two voltages should not be confused. These terminologies were derived from the now obsolete use of batteries to supply the electrode potentials. The plate to cathode potential was supplied by what was called a "B" battery, hence the designations E_{bb} and E_b are now used.

To obtain an E_b-I_b characteristic curve, the plate voltage (E_b) is increased, from zero, in steps, and the corresponding increases in plate current (I_b) are plotted as shown in figure 7-7.

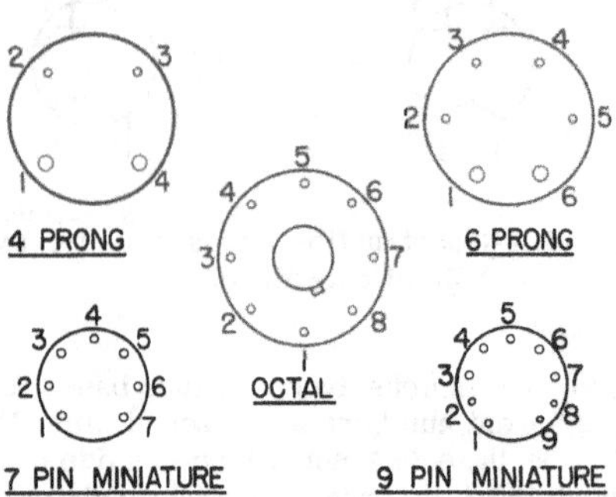

4.4141
Figure 7-6.— Pin numbering systems for several tube types.

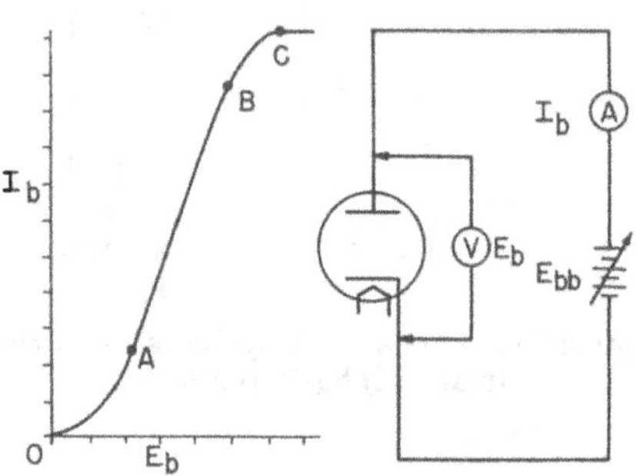

179.143
Figure 7-7.—Representative E_b-I_b curve and circuit used to obtain it.

The curve obtained from the circuit in figure 7-7 consists of three distinct regions; O to A, A to B, and B to C. Note that in the regions of O to A and B to C appreciable non-linearity exists. The region from A to B is essentially linear.

As the curve is plotted, a point is finally reached (point C) where a further increase in plate voltage no longer produces an increase in plate current. At this point the plate is attracting all the electrons the cathode is capable of emitting. This condition is known as saturation, and at this point the upper limit of the tube's conduction capabilities have been reached.

Note that the current flow through a tube is some finite value. This indicates that the tube is effectively offering opposition to current flow. The opposition offered to the flow of direct current, by an electron tube, is called D.C. PLATE RESISTANCE. D.c. plate resistance is measured in ohms and designated as R_P. This resistance is thought of as existing from the cathode to the plate of the tube. R_P may be calculated by the formula

$$R_P = \frac{E_b}{I_b}$$

Where: R_P = d.c. plate resistance in ohms
E_b = the voltage from plate to cathode in volts
I_b = plate current in amperes

The d.c. resistance of a diode is not constant throughout the operating current range of the tube. This can be verified by calculating the R_P at several points along the E_b-I_b characteristic

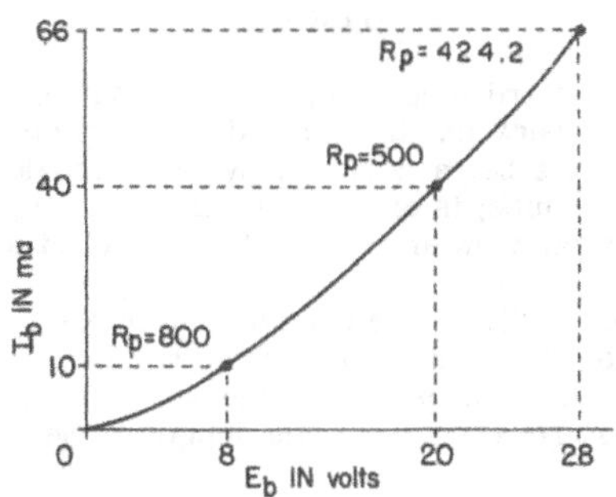

179.144

Figure 7-8.—Calculating R_p from the E_b-I_b characteristic curve.

curve, as shown in figure 7-8. However, it is practically constant throughout the linear portion of the curve.

D.c. plate resistance is a "static" characteristic. A static characteristic is one which is obtained with d.c. potentials applied to the tube. Another characteristic, a.c. plate resistance, is a dynamic characteristic. That is, it is indicative of tube performance under signal, or a.c., conditions.

The term a.c. plate resistance is defined as the opposition offered by an electron tube to the flow of alternating current. The symbol designating a.c. plate resistance is r_p. The r_p of a diode electron tube may be found with the use of the formula

$$r_p = \frac{\Delta e_b}{\Delta i_b}$$

Where: r_p = a.c. plate resistance in ohms
Δe_b = a small change in plate voltage
Δi_b = a small change in plate current

The values required to calculate r_p may be obtained from the E_b-I_b characteristic curve of the tube concerned. This is illustrated in figure 7-9. A.c. plate resistance also varies throughout the operating range of the tube; however, it too is essentially constant in the linear portion of the curve.

DIODE RATINGS

Each diode has certain voltage, current and power values which should not be exceeded in normal operation. These values are called ratings.

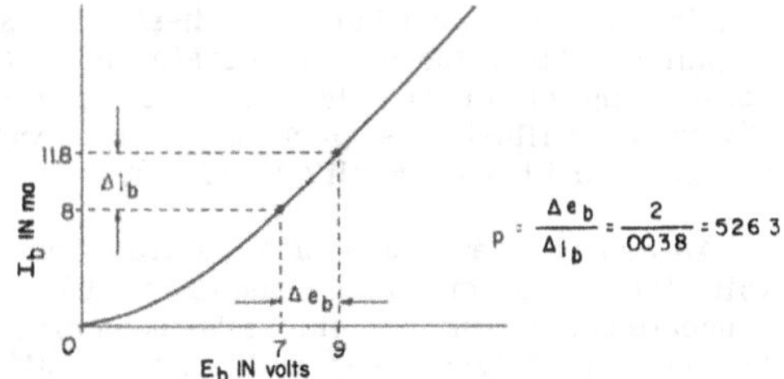

179.145

Figure 7-9.—Calculating r_p from the E_b-I_b characteristic curve.

The following are the most important diode electron tube ratings:

Plate Dissipation is the maximum average power, in the form of heat, which the plate may safely dissipate.

Maximum Average Current is the highest average plate current which may be handled continuously. It is based on the tube's permissible plate dissipation.

Maximum Peak Plate Current is the highest instantaneous plate current that a tube can safely carry, recurrently, in the direction of normal current flow.

Peak Inverse Voltage (PIV) is the highest instantaneous plate voltage which the tube can withstand recurrently, acting in a direction opposite to that in which the tube is designed to pass current (i.e., plate negative—cathode positive).

It should also be noted, that the correct filament or heater voltage and current is required for proper operatpon of an electron tube. If heater current is too low, the cathode will not emit sufficient electrons. This will result in low emission, and the tube will be incapable of proper operation. Excessive heater or filament current may reduce the life of the tube or destroy the heater or filament.

DIODE RECTIFIERS

There is little practical difference between a solid state diode rectifier circuit and an electron tube diode rectifier circuit. It should be mentioned, however, that most electron tube rectifiers require a source of filament or heater voltage (supplies). Also, there is no possibility of reverse current in an electron tube rectifier. Figure 7-10 illustrates electron tube half wave, full wave, and bridge rectifier circuits.

As can be seen, these circuits are identical, with the exception of the heater or filament connections, to their solid state counterparts (chapter 5). Filters used with these rectifier circuits will also be identical to those previously discussed.

TRIODES

If a third electrode, called a control grid, is placed between the cathode and plate of an electron tube, a device known as a triode (three element tube) is created. Figure 7-11 illustrates a cutaway view and schematic symbol of a triode tube.

The addition of a control grid makes possible the use of an electron tube as an amplifier. The grid usually consists of fine wire wound on two support rods, extending the length of the cathode.

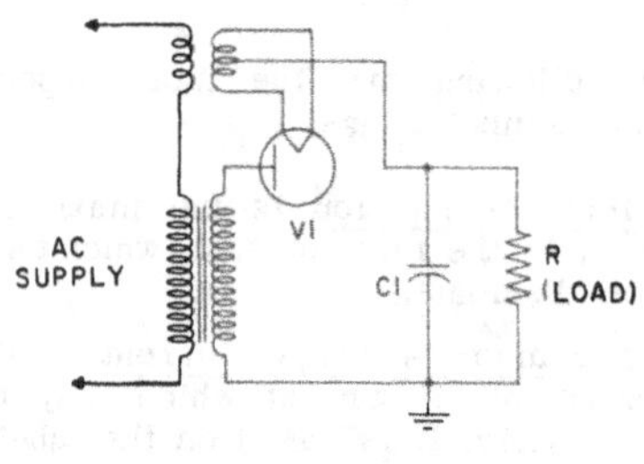

HALF WAVE

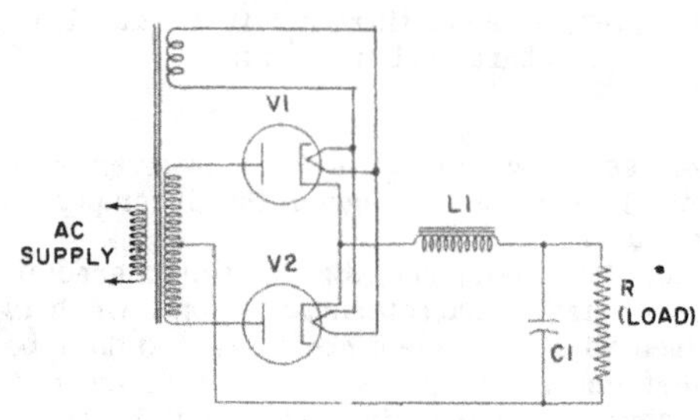

FULL WAVE

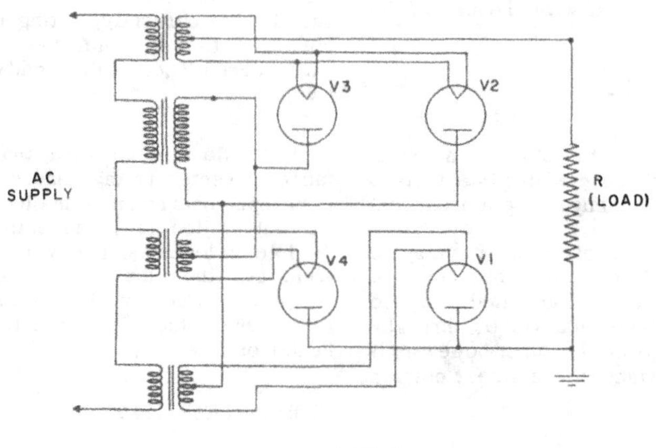

BRIDGE

(A) 20.37 (B) 20.40.1 (C) 20.29.2

Figure 7-10.—Electron tube rectifier circuits.

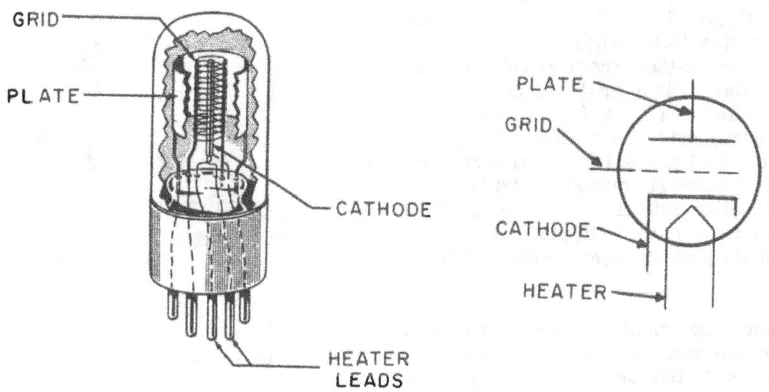

Figure 7-11.—Cutaway view and schematic symbol of a triode tube.

Two types of grid structure are illustrated in figure 7-12.

The grid wire is made of a material which has a low level of electron emission. The spacing between the turns of the grid wire is large compared to the size of the wire. This allows the passage of electrons from the cathode to the plate to be practically unobstructed by the grid structure.

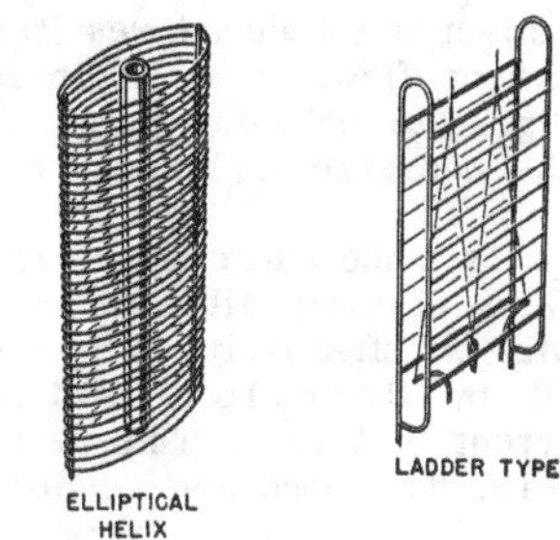

Figure 7-12.—Typical grid structures.

CONTROL OF PLATE CURRENT

The purpose of the control grid is to control the flow of plate current. The grid acts as an electric shield between cathode and plate. Therefore, both grid and plate potentials are effective in controlling plate current. However, as explained later, the grid voltage has much more control over plate current than does the plate voltage. It is this fact which accounts for the triode's ability to amplify.

Basic to an understanding of triode operation, is an understanding of the effect of the control grid on an electric field between plate and cathode. Consider the circuit of figure 7-13. In this circuit, the plate to cathode voltage (E_b) is such that the plate is positive with respect to the cathode (in this case $E_b = E_{bb}$). The grid is connected directly to the cathode. Therefore, the control grid voltage with respect to the cathode, designated E_c, is zero. The control grid and cathode are at the same potential.

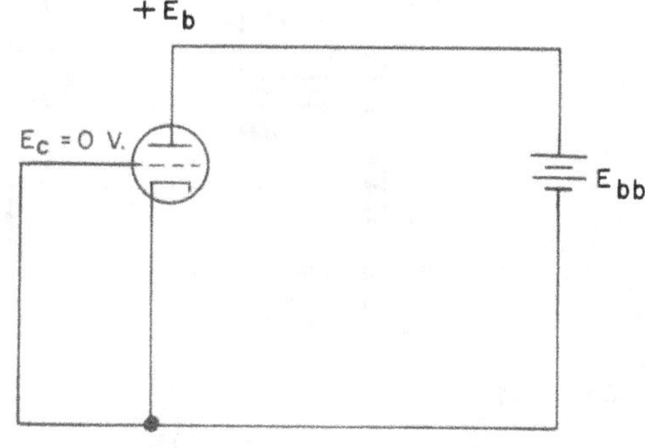

Figure 7-13.—Triode with grid at same potential as cathode.

An electric field exists within the tube, as illustrated in figure 7-14. The lines of force which make up this field originate at the plate and terminate on either the control grid or cathode. Since the control grid and cathode are at the same potential ($E_c = 0$), and the windings of the control grid are spaced wide apart, the majority of the lines of force will terminate on the cathode. Electrons emitted by the cathode, under the directed influence of the electric field, will flow from cathode to plate and return to the cathode through the plate voltage supply, E_{bb}.

Now consider the result of making the grid somewhat negative with respect to the cathode. This is done with the use of a grid voltage supply designated E_{cc} (fig. 7-15A). A greater number of lines of force will now terminate on the grid, and fewer will terminate on the cathode. This is illustrated in figure 7-15B. The corresponding attraction for electrons, by the plate, is effectively reduced, resulting in a lower plate current flow. A small increase in the negative grid to cathode voltage will cause an appreciable decrease in plate current.

If E_c is made sufficiently negative all the electric lines of force will terminate on the grid. There will be effectively no attraction to the plate felt by the cathode electrons, thus no plate current will flow. This is illustrated in figure 7-16. The condition resulting from the

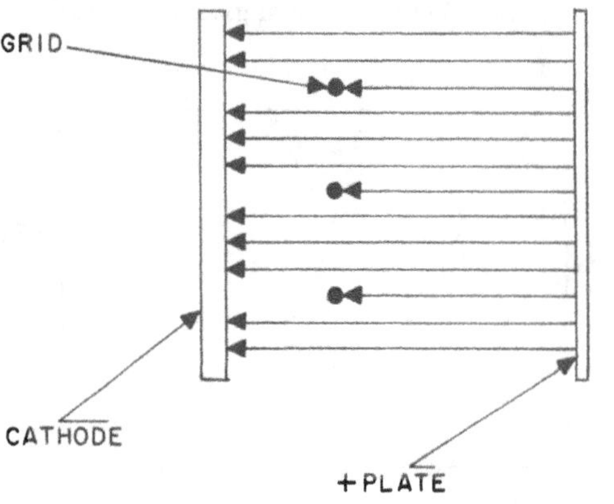

179.148
Figure 7-14.—Electric field with zero grid potential.

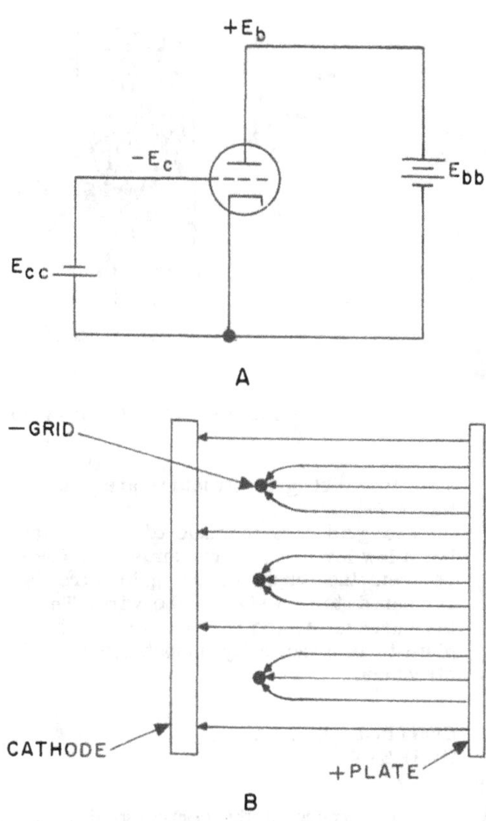

179.149
Figure 7-15.—Circuit and electric field for a triode with a negative grid.

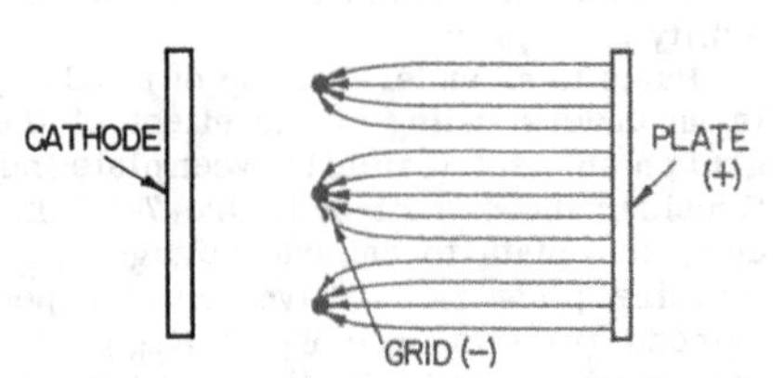

179.150
Figure 7-16.—Electric field at or beyond cutoff.

grid's prevention of current flow is called "cutoff." The negative grid to cathode voltage required to achieve cutoff is called cutoff voltage (E_{co}).

If the grid is made positive with respect to the cathode, a large current will flow. This current will be divided between the plate and control grid. Control grid current is designated I_g. The total current emitted by the cathode is called cathode current and is designated I_k. The sum of plate current and control grid current equals cathode current. That is

$$I_k = I_b + I_g$$

Normally, a tube will not be operated with a positive E_c. However, in many cases the grid will be driven positive with respect to the cathode by the positive peak of an input signal.

Grid current causes heating of the grid. The grid must be capable of dissipating this heat. The grid current and power dissipation ratings of a tube must not be exceeded, or destruction of the grid structure may result.

It should be noted that if E_c is held constant, the magnitude of plate current will be a function of plate voltage, much the same as in a diode. However, a large change in plate voltage is required to cause the same change in plate current as a small change in grid voltage.

In practical triode circuits, a d.c. difference of potential between grid and cathode is used to set the operating level of the tube. This d.c. voltage is called bias. In most electron tube circuits the bias is such as to make the control grid negative with respect to the cathode. Although there are some applications requiring the grid to be biased positive with respect to the cathode, these will be covered separately, when required. For the remainder of this chapter, we will be concerned only with circuits utilizing negative grid to cathode bias.

The bias required for the operation of an electron tube may be obtained by one of two general methods, fixed bias, or self bias. Fixed bias is obtained from a power supply or battery. Self bias is obtained by using the tube's own current to develop a bias voltage. These two methods will be discussed in detail later on. For the present, we will represent bias voltage with a battery (E_{cc}), as was illustrated in figure 7-15A.

The input signal voltage normally applied to the control grid of an electron tube is alternating in character; at least, it is a voltage which varies in amplitude and perhaps in polarity. The input signal (e_g) and bias voltage (E_{cc}) are in series, between the control grid and cathode. This is illustrated in figure 7-17.

The relationship between the input signal, bias, and plate current is illustrated in figure 7-18. Figure 7-18A represents an input signal of 5 volts peak. This is the instantaneous a.c. component of grid to cathode voltage (e_g). It varies between +5 and -5 volts. The grid bias is arbitrarily set at -6 volts, as shown in figure 7-18B. As shown in figure 7-18C, the instantaneous grid to cathode voltage (e_c) will ride on the -6 volt level. As the input signal (e_g) goes positive, the instantaneous grid to cathode voltage (e_c) becomes less negative. This results in more plate current (i_p), as illustrated in figure 7-18D. As e_g goes negative, e_c becomes more negative, resulting in less plate current.

Note that figures 7-18C and 7-18D represent changes in grid to cathode voltage and plate current, respectively. These changes are in phase, and represent a combination of d.c. and a.c. components. The instantaneous a.c. component of plate current is designated i_p.

TUBE CONSTANTS

The behavior of the plate current in a triode, under the influence of various control grid and plate voltages, does not occur at random. It is a design function of the tube, specifically the geometric organization of the tube electrodes. Separation between electrodes, shape and dimensions of the electrodes, and other physical details, are examples of factors which determine tube behavior. All of these factors are expressed by a group of tube constants. The constants may be used to predict tube behavior.

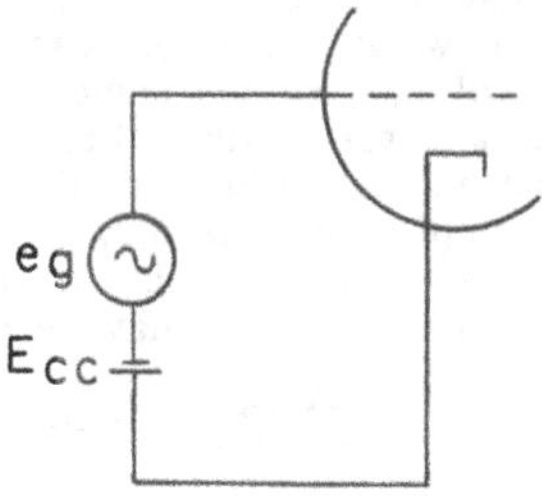

179.151

Figure 7-17.—Input signal and bias voltage representation.

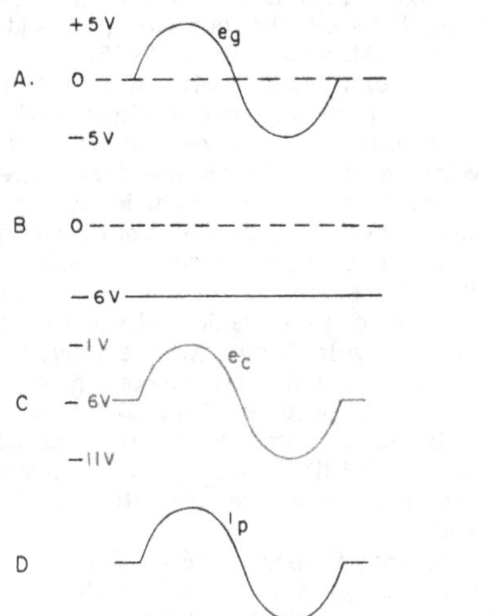

Figure 7-18.—Relationship between input signal, bias, and plate current.

The three primary tube constants are amplification factor, a.c. plate resistance, and transconductance.

Amplification factor is a measure of the effectiveness of the control grid as compared to the effectiveness of the plate in controlling plate current. It may also be stated as the ratio of the change in plate voltage to a change in control grid voltage in the opposite direction, under the condition that plate current remains unchanged. For example, if, when the plate voltage is made 1 volt more positive, the control grid voltage must be made 0.1 volt more negative (with respect to the cathode) to hold plate current unchanged, the amplification factor is 1 divided by 0.1, or 10. In other words, a small variation in grid to cathode voltage has the same effect on plate current as a large plate voltage change—the latter equal to the product of the grid voltage change and amplification factor.

In formula form, amplification factor, designated μ (pronounced mu) may be expressed as

$$\mu = \frac{\Delta e_b}{\Delta e_c} \text{ } (I_b \text{ constant})$$

Triodes may be grouped into three general classes, according to their μ : low-μ triodes, in which the amplification factor is less than 10; medium-μ triodes, in which it is between 10 and 50; and high-μ triodes in which the amplification factor is 50 or more.

The a.c. plate resistance of a triode, designated r_p, is the opposition offered by the tube to the flow of alternating current. It is the quotient of a small change of plate voltage divided by the corresponding change in plate current, with the grid voltage held constant. That is,

$$r_p = \frac{\Delta e_b}{\Delta i_b} (E_c \text{ constant})$$

For example, if a change of 0.1 ma is produced by a plate voltage variation of 1 volt, the plate resistance is 1 divided by 0.1×10^{-3} or 10,000 ohms.

Transconductance, designated g_m, is a tube constant which expresses the specific change in plate current for a unit change in grid voltage, with the plate voltage held constant. In formula form

$$g_m = \frac{\Delta i_b}{\Delta e_c} (E_b \text{ constant})$$

The unit of transconductance is the mho. For convenience, a millionth of a mho, or μmho, is used to express transconductance. Thus, if a grid to cathode change of 0.5 volt causes a plate current change of 1 ma., the transconductance is 1×10^{-3} divided by 0.5 or 2×10^{-3} mhos. This is expressed as 2,000 μmhos.

The transconductance of a tube is an important constant. It is the most commonly used of all the constants when comparing similar tubes. A tube with a higher transconductance is capable of furnishing greater signal output than a tube with a lower transconductance, assuming identical levels of input signal voltage and arrangements in the plate circuits.

The three previously discussed constants are interrelated. The operating voltages applied to each type of tube determine the exact value of each constant, and for any one set of operating potentials, each tube bears these three ratings. The constants vary in magnitude relative to each other in a definite manner.

Amplification factor remains substantially constant. Being a property of the physical geometry of the tube, it varies little with changes in the operating voltages applied to the grid and plate. Plate resistance; r_p, varies greatly with operating voltages, especially through the region of potentials which result in low values of plate current. This occurs at high negative grid bias, low plate voltage, or both. Increasing the physical separation between grad and plate in a triode results in a high amplification factor and also a high a.c. plate resistance. Usually, therefore, the higher the amplification factor, the higher the a.c. plate resistance of a tube.

Transconductance varies inversely with a.c. plate resistance. The higher the a.c. plate resistance of a tube the lower its transconductance, or vice versa.

The three constants are mathematically related as follows:

$$\mu = r_p \quad g_m$$

$$r_p = \frac{\mu}{g_m}$$

$$g_m = \frac{\mu}{r_p}$$

TRIODE AMPLIFYING ACTION

One of the primary functions of a triode electron tube is as a voltage amplifier. Although it has been shown that a signal between the grid and cathode of a tube will cause corresponding variations in plate current, a circuit such as the one illustrated in figure 7-19A cannot produce any plate voltage changes. This is due to the fact that the tube is the only impedance across the source, E_{bb}, and must always have the total voltage dropped across it.

In order to convert plate current variations into plate voltage variations, a resistance (or other impedance) must be placed in the plate circuit, external to the tube, as shown in figure 7-19B. This resistance is called the plate load resistance and is designated R_L.

The plate circuit contains several new terms, since a new action occurs. To understand these new terms, it is imperative to note the location of the plate load resistor. R_L is directly in the path of i_b. The movement of plate current must be through this resistor because the complete plate current path is from the cathode to the plate, through R_L to the positive terminal of the source, E_{bb}, and from the negative terminal of E_{bb} to the cathode.

Under these circumstances, two voltage drops occur in the plate circuit. One of these is across the internal resistance of the tube, or $i_b r_p$, and the other is the drop across R_L, or $i_b R_L$. This leads to three voltages being present in the plate circuit: E_{bb}, the plate supply voltage; e_b, the instantaneous plate to cathode voltage; and the voltage across the load resistance (e_{RL}). Note that e_b plus e_{RL} must equal E_{bb}. Instantaneous designations were used in the above discussion in order to cover all possibilities. Whether e_b and e_{RL} are varying or constant will depend, of course, on the voltage applied between the grid and cathode. In any case the relationships described always hold true.

Consider figure 7-20. Assume the E_{cc} is adjusted to such a high negative value as to cut off plate current. In that event, the voltage across r_p (between plate and cathode) is the same as the plate supply voltage, or $e_b = E_{bb} = 350$ volts. This is true because the tube's internal plate resistance is infinite and plate current is cut off. Under these conditions there will be no voltage drop across R_L.

Now assume that E_{cc} is made less negative, and allows 2 ma. of plate current to flow. With i_b equal to 2 ma., e_{RL} will be

$$e_{RL} = i_b R_L$$

$$e_{RL} = 2 \times 10^{-3} \cdot 25 \times 10^3$$

$$e_{RL} = 50 \text{ volts}$$

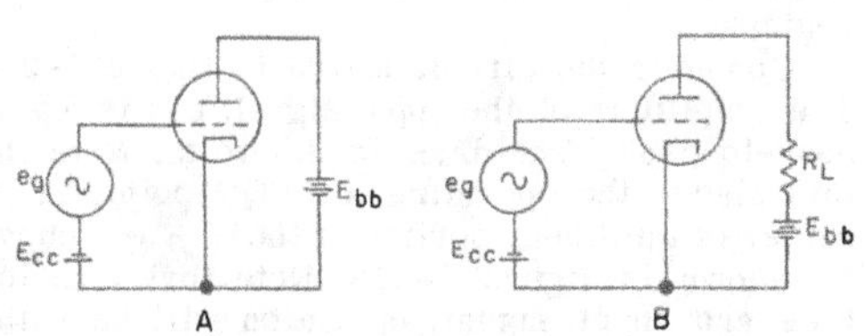

179.153

Figure 7-19.—Addition of plate load resistance.

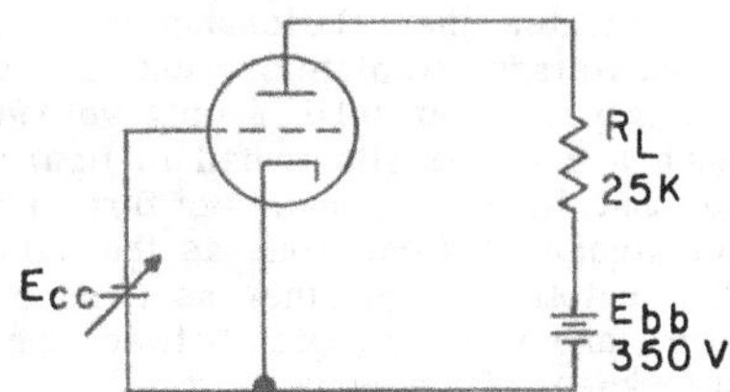

179.154

Figure 7-20.—Controlling e_b through E_{cc}.

Since e_b is the difference between E_{bb} and e_{RL}

$$e_b = E_{bb} - e_{RL}$$

$$e_b = 350 - 50$$

$$e_b = 300 \text{ volts}$$

Most of the voltage drop takes plate across r_p, which is now finite since plate current flows.

Next, assume the grid is made less negative, to a point where a plate current of 12 ma. is allowed to flow. By Ohm's law, the voltage across R_L is

$$e_{RL} = i_b R_L$$

$$e_{RL} = 12 \times 10^{-3} \cdot 25 \times 10^3$$

$$e_{RL} = 300 \text{ volts}$$

Therefore, e_b is

$$e_b = E_{bb} - e_{RL}$$

$$e_b = 350 - 300$$

$$e_b = 50 \text{ volts}$$

Bearing in mind that the plate supply voltage, E_{bb}, is always constant at 350 volts, and that the change in plate current is a function of the control grid to cathode voltage, it is possible to imagine an increase in plate current such that all the voltage drop in the circuit occurs across R_L. If this could occur, e_b would be zero. This is not actually possible, since r_p will never be zero, and therefore e_b, though very low, will not be zero.

In order to study the operation of an actual triode amplifier circuit, under signal or "dynamic" conditions, a curve called a dynamic transfer curve or $i_b - e_c$ curve, is used. Such a curve illustrates the relationship between grid to cathode voltage and plate current, and is valid only for a particular tube, supply voltage, and R_L. Such a curve is illustrated in figure 7-21.

Note that the curve consists of three regions: two non-linear portions, one as the curve approaches cutoff and the other as it approaches saturation, and a linear region between the upper and lower knees of the curve.

The $i_b - e_c$ curve may be used to plot the changes of plate current caused by changes in grid voltage. These plate current variations will

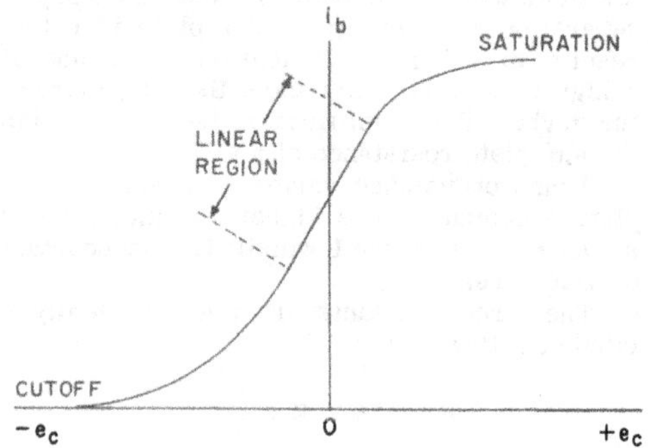

Figure 7-21.—Dynamic transfer (i_b-e_c) curve.

then determine the plate voltage variations caused by the effect of the load resistance. To demonstrate the method of predicting the plate current behavior from the dynamic transfer characteristic, it is necessary first to establish the operating point on the characteristic curve. This point is set by the amount of grid bias applied to the tube. It establishes a steady value of plate current, which prevails for a zero input signal voltage. This value of plate current is called the quiescent value of plate current.

The location of the operating point determines how much change occurs for input signals of different magnitudes. Since it is always desirable to develop a maximum change in plate current for a unit change in grid voltage, the location of the operating point is determined by the maximum amount of signal voltage anticipated. Another factor related to the location of the operating point is the desired behavior of the plate current. If it is desired that the output signal be a faithful reproduction of the input signal, the operating point must be in the center of the linear portion of the curve, and the input signal must not be so large as to drive the tube into the non-linear regions.

Consider the circuit shown in figure 7-22A. The amplitude of the input signal, e_g, is 3 volts peak-to-peak. The bias is 3.5 volts. Note that this places the operating, or "Q" point, in the center of the linear portion of the $i_b - e_c$ curve, as shown in figure 7-22B. Note that with this bias and input signal, operation will be within the linear portion of the curve.

The plate current variations (i_p) resulting from the 3 volt peak-to-peak signal (e_g) applied

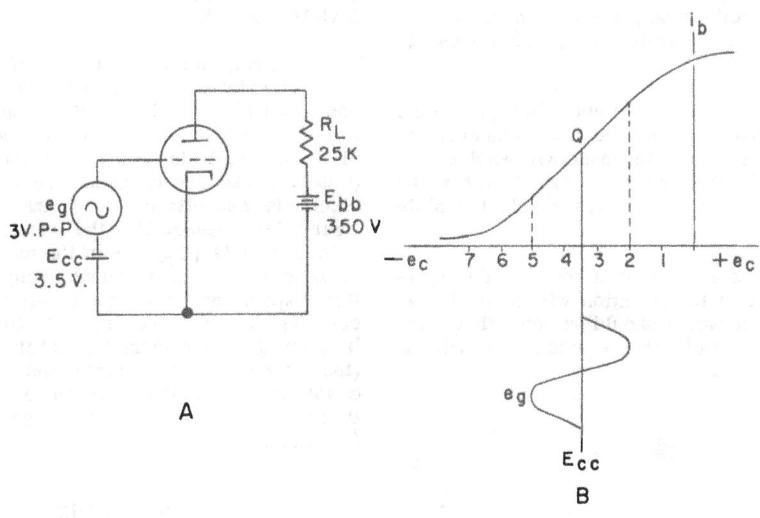

179.156
Figure 7-22.—Setting the Q point.

to the circuit of figure 7-22A, are shown in figure 7-23. Referring to line AB, the input signal voltage waveform is drawn so that its 0-volt point coincides with the bias voltage. All other instantaneous amplitudes of the signal are then projected onto the characteristic curve, and indicate the instantaneous plate currents. Horizontal projections of these intersections are the instantaneous amplitudes of the plate current.

For instance, line AB represents a period during which the signal voltage is zero, producing the period represented by A'B' wherein the quiescent plate current is 7 ma. This current is established by the no signal conditions—that is, the 350 volt supply, the 25,000 ohm load resistance, and the 3.5 volt bias.

As e_g goes positive (points BC), the effective grid to cathode voltage becomes less negative, and i_p increases (B'C'). When e_g goes negative (CE), e_c becomes more negative, and i_p decreases (C'E'). This is shown point by point in figure 7-23.

As was previously explained, variations in plate current cause variations in e_{RL} and e_b. Figure 7-24 shows the relationship between e_g, i_p, e_{RL}, and e_b, of figure 7-22A.

When e_g is zero, i_p is 7 ma. which results in an e_{RL} of 175 volts, and an e_b of 175 volts. When e_g is maximum positive, the effective grid to cathode voltage (e_c) is at its least negative point, and i_p is 10 ma. This results in an e_{RL} of 250 volts and an e_b of 100 volts. When e_g is maximum negative, e_c is at its most negative point, and i_p is 4 ma. This in turn, results in an e_{RL} of 100 volts and an e_b

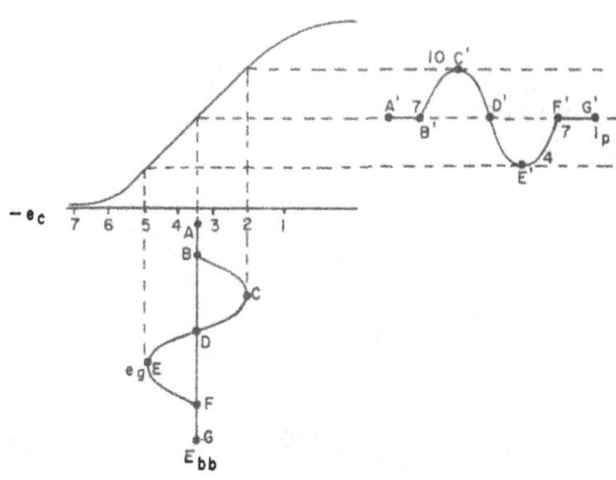

179.157
Figure 7-23.—Variations of i_p (following e_g).

of 250 volts. Note that an e_g of 3 volts peak-to-peak (1.5 volts peak) results in an e_p of 150 peak-to-peak (75 volts peak). This represents a voltage gain of 50.

Several points should be noted in figure 7-24. The input signal e_g and the a.c. component of plate current i_p are in phase with each other, but are out of phase with e_b. Also note that the sum of e_{RL} and e_b always equals the plate supply voltage.

The previous example of a basic triode amplifier was limited to operation within the linear portion of its curve. It should be noted that other "classes of operation" are used and will be discussed later on.

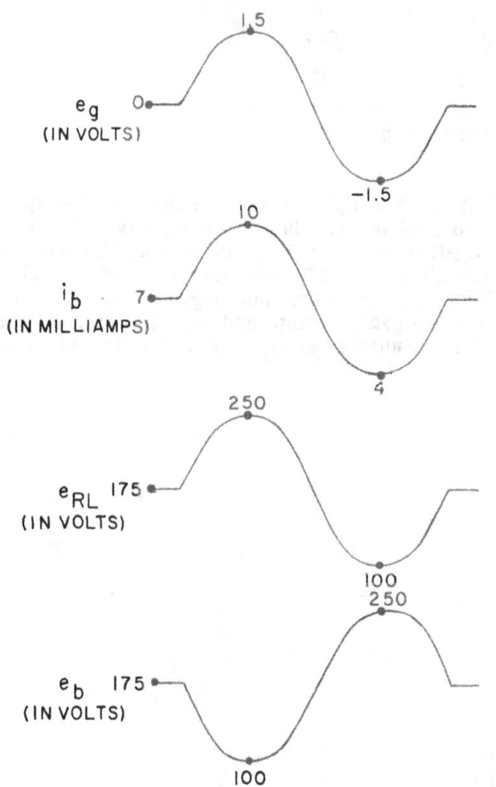

179.158
Figure 7-24.—Relationship between e_g, i_b, e_{RL}, and e_b of figure 7-22A.

INTER-ELECTRODE CAPACITANCE

The grid, plate, and cathode of a triode form an electrostatic system, each electrode acting as one plate of a small capacitor. The capacitances are those existing between grid and plate (C_{gp}), plate and cathode (C_{pk}), and grid and cathode (C_{gk}). These capacitances are known as inter-electrode capacitances and are represented in figure 7-25. Generally, the capacitance between grid and plate (C_{gp}) is of the most importance. In high gain radio frequency amplifier circuits this capacitance may act to produce undesired coupling between the input circuit (the circuit between grid and cathode) and the output circuit (the circuit between plate and cathode). This coupling is undesirable in an amplifier because it may cause instability and unsatisfactory performance.

TETRODES

The capacitance between grid and plate may be diminished by mounting an additional electrode, called the screen grid (grid No. 2), in the tube. With the addition of the screen grid, the tube has four electrodes and is, accordingly, called a tetrode. The screen grid or grid No. 2 is mounted between the control grid (grid No. 1) and the plate, and acts as an electrostatic shield between them, thus reducing the grid to

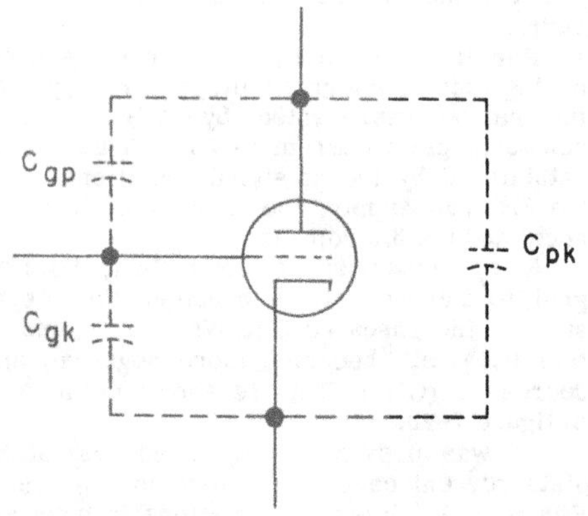

20.8
Figure 7-25.—Schematic representation of inter-electrode capacitance.

plate capacitance. In practice, the grid to plate capacitance is reduced from several picofarads for a triode to 0.01 picofarad or less, for a screen grid tube. The schematic symbol for a tetrode is shown in figure 7-26.

The screen grid has another desirable effect in that it makes plate current practically independent of plate voltage over a certain range. The screen grid is operated at a positive potential with respect to the cathode, and therefore attracts electrons from the cathode. However, because of the comparatively large space between wires of the screen grid, most of the electrons drawn toward the screen grid pass through it, and are attracted by the plate. Hence the screen grid supplies an electrostatic force pulling electrons from the cathode to the plate. At the same time, the screen grid shields the electrons between the cathode and screen grid (from the plate) so that the plate exerts very little electrostatic force on electrons near the cathode.

So long as the plate voltage (E_b) is higher than the effective screen grid to cathode voltage (E_{c2}), plate current in a screen grid tube depends to a great degree on screen grid voltage and very little on plate voltage. The fact that the plate current, in a screen grid tube, is largely independent of plate voltage makes it possible to obtain much higher amplification with a tetrode than with a triode. The low grid to plate capacitance makes it possible to obtain this high amplification without plate to grid feedback and resultant instability.

TETRODE CHARACTERISTICS

In the tetrode with fixed control grid and screen grid potentials, the cathode current (I_k) or total current is practically constant. Hence, that portion of cathode current which is not collected by the plate must be collected by the screen grid. Where the plate current is large, the screen current (i_{c2}) must be small, and vice versa. This is illustrated in figure 7-27.

Although the plate voltage does not affect the total or cathode current to a very great extent, it does determine the division of the cathode current between plate and screen. At zero plate to cathode voltage, few electrons reach the plate, and i_b is low. As the plate to cathode voltage is increased (O to A), a rapid rise in i_b occurs with a corresponding decrease in i_{c2}. However, this increase is followed by a region in which i_b decreases with increasing e_b (A to B). This region is one of negative resistance, since an increasing e_b is accompanied by a decreasing i_b, and is called the negative resistance or dynatron region. When the plate voltage is larger than the screen potential (point B), the plate current increases, and the screen current decreases.

Secondary emission is the action of electrons being knocked free from the plate (due to the bombardment action of high velocity electrons from the cathode). This secondary emission results in the negative resistance of the dynatron region in multielement tubes because loose electrons are between the screen grid and the plate

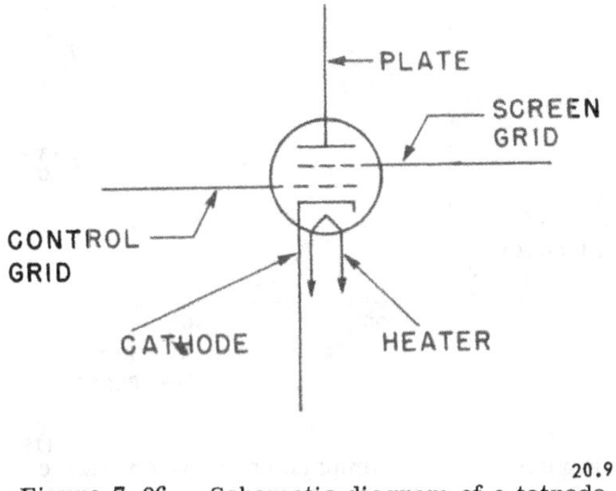

20.9
Figure 7-26.—Schematic diagram of a tetrode.

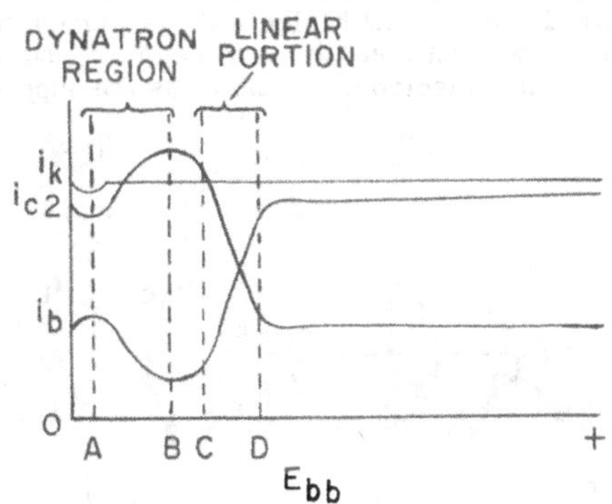

179.159
Figure 7-27.—Comparison of cathode, screen grid, and plate currents in a tetrode.

causing some arriving electrons to be repelled from the plate. In two and three electrode tubes, this secondary emission does not cause trouble because no positive electrode, other than the plate itself, is present to attract electrons.

In the case of screen grid tubes, the proximity of the positive screen grid to the plate offers a strong attraction to the secondary emission energy, and particularly so if the plate voltage is lower than the screen voltage. This effect reduces plate current and limits the useful plate voltage swing for tetrodes.

Although some applications for tetrodes make use of the dynatron region of operation, most tetrode circuits are designed to operate in the linear portion of the curve (points C to D in fig. 7-27). This is usually accomplished by maintaining e_p above E_{c2}.

Figure 7-28 illustrates a basic tetrode amplifier circuit. R_{sg} is called a screen dropping resistor and is used to maintain the screen grid operating potential. C_{sg} is a screen bypass capacitor, and is used to maintain the screen grid at a constant potential.

The following relationships exist in the basic circuit of figure 7-28:

$$E_{R_{sg}} = i_{c2} \cdot R_{sg}$$

$$E_{c2} = E_{bb} - E_{R_{sg}}$$

$$i_k = i_b + i_{c2}$$

PENTODES

The effects of secondary emission are minimized when a fifth electrode is placed, within the tube, between the screen grid and plate. This fifth electrode is known as the suppressor grid (grid No. 3) and is usually connected, internally or externally, to the cathode. Both methods are shown, schematically, in figure 7-29. Because of its negative potential with respect to the plate, the suppressor grid retards the flight of secondary emission electrons and diverts them back to the plate. The main electron stream is not materially affected by the presence of the suppressor grid. Figure 7-30 illustrates a comparison between the $e_b - i_b$ curves for a tetrode and pentode.

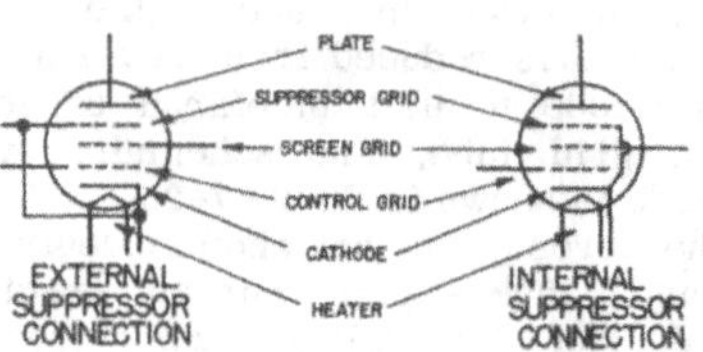

179.161
Figure 7-29.—Schematic representation of pentode tube.

The pentode has replaced the tetrode in RF voltage amplifiers, because it permits a somewhat higher voltage amplification in moderate values of plate potential. Likewise, it permits a greater plate potential excursion without distortion. However, tetrodes are used extensively in high power tuned amplifiers.

Figure 7-31 illustrates a basic pentode amplifier circuit. Since the suppressor grid is at a negative potential with respect to the screen grid or plate, it does not draw current. Therefore,

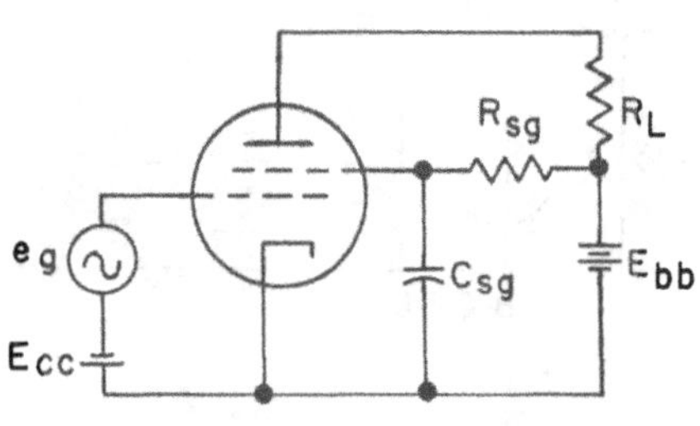

179.160
Figure 7-28.—Basic tetrode amplifier circuit.

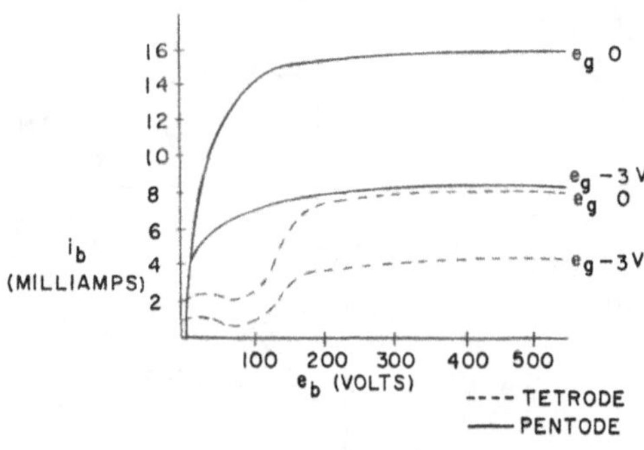

179.162
Figure 7-30.—Comparison between the e_b-i_b curves for a tetrode and pentode.

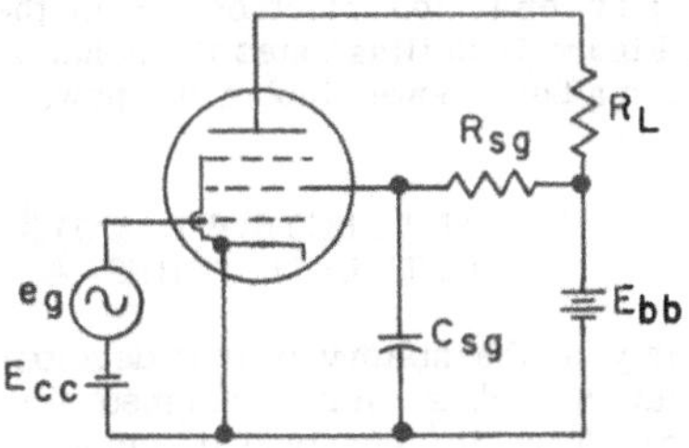

179.163
Figure 7-31.—Basic pentode amplifier circuit.

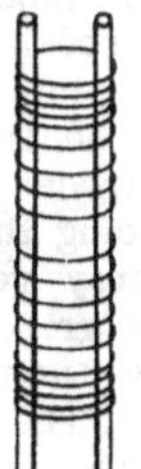

20.14(72)
Figure 7-32.—Variable-mu control grid.

the relationships of voltages and currents are essentially the same as those given for the tetrode. It should be noted that for some specialized applications the suppressor grid may be used as a signal grid. Explanation of these circuits will be given when appropriate.

REMOTE CUTOFF (VARIABLE-MU) TUBES

If, in a pentode, the spacing between grid wires is not uniform along the entire length of the control grid structure, the various portions of the grid will possess different degrees of electrostatic control over the plate current. This type of grid structure is illustrated in figure 7-32.

The remote-cutoff action is due to the structure of the grid, which provides a variation in amplification factor with a change in bias. The control grid is wound with wide spacing at the center and with close spacing at the top and bottom. When weak signals and low grid bias are applied to the tube, the effect of the non-uniform turn spacing of the grid on cathode emission and tube characteristics is essentially the same as for uniform spacing. As the grid bias is made more negative to handle larger input signals, the electron flow from the sections of the cathode enclosed by the ends of the grid is cut off. The plate current and other tube characteristics are then dependent on the electron flow through the open section of the grid. This action reduces the gain of the tube so that large signals may be handled with minimum distortion.

Figure 7-33 illustrates a typical i_b - e_c dynamic transfer curve for remote cutoff and conventional type tubes. It will be noted that while the curves are similar at low grid voltages, the plate current of the variable-mu tube drops off quite slowly with large values of negative grid voltage. This slow change makes it possible to handle large signals satisfactorily. Because remote cutoff tubes can accommodate large and small signals, they are particularly suited for use in equipment having automatic gain control.

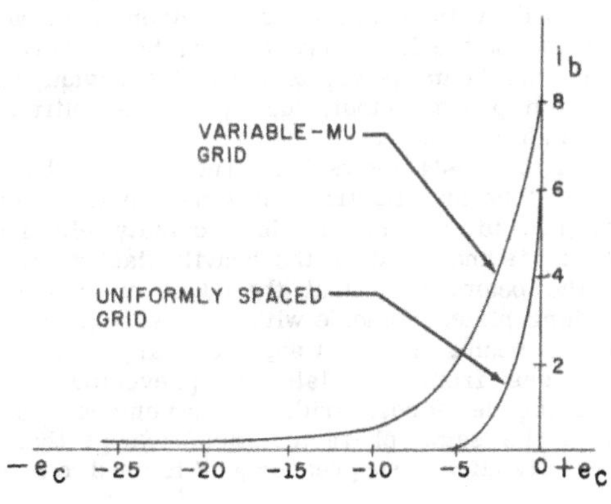

20.14
Figure 7-33.—i_b-e_c curves for remote cutoff and conventional type tubes.

BEAM POWER TUBES

A beam power tube is a tetrode in which directed electron beams are used to substantially increase the power handling capability of the tube. Such a tube contains a cathode, a control grid, a screen grid, a plate, and beam forming electrodes.

A beam power tube is designed with the beam forming electrodes so placed that secondary emission from the plate is repelled back to the plate by the high density electron beam. The beam forming electrodes are at cathode potential, to assist in producing the desired beam effects and to prevent stray reflected electrons from reaching the screen grid.

A feature of a beam power tube is its low screen grid current. The screen grid and control grid are spiral wires, wound so that each turn of the screen grid is shaded from the cathode by a grid turn. This alignment of the screen grid and control grid causes the electrons to travel in sheets between the turns of the screen grid so that few electrons are absorbed by the screen grid. Because of the effective suppressor action provided by the concentrated electron beam, and because of the low current drawn by the screen grid, the beam power tube has the advantages of high power output, high power sensitivity, and high efficiency.

Figure 7-34 shows the structure of a beam power tube and illustrates how the electrons are confined to beams. The high density electron region is indicated by the heavily dashed lines in the beam. Note that the edges of the beam forming plates coincide with the dashed portions of the beam. In this way, secondary emission electrons from the plate are prevented from reaching the screen grid. Power pentodes have much the same characteristics, except that a specially aligned suppressor grid is used to keep secondary emission electrons from the screen grid. Figure 7-35 illustrates the schematic symbols for a beam power tube and a power pentode.

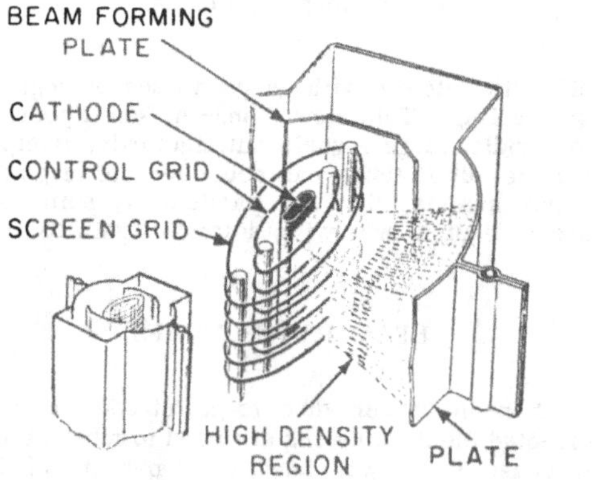

BEAM FORMING PLATE
CATHODE
CONTROL GRID
SCREEN GRID
HIGH DENSITY REGION
PLATE

20.13

Figure 7-34.—Internal structure of a beam power tube.

MULTI-ELECTRODE AND MULTI-UNIT TUBES

Early in the history of tube development and application, tubes were designed for general service. That is, one type of tube, the triode, was used for many varied applications. Obviously, this one tube could not meet all requirements to the best advantage.

Later and present trends of tube design are the development of ''specialty'' type tubes. These tubes are designed either to give optimum performance in a particular application, or to combine in one envelope functions which formerly required two or more tubes.

One class of ''specialty'' tubes is multi-electrode tubes. Types of this class generally require more than three electrodes to obtain the desired characteristics. In addition to tetrodes and pentodes (previously illustrated), this class of tubes includes tubes with up to seven electrodes (fig. 7-36A). The operation of these types of tubes will be discussed when required.

Another class is multi-unit tubes which contain two or more tubes in one element (fig. 7-36B). Examples are twin diode triodes, triode pentodes, twin triodes, and tubes containing dissimilar triodes.

A third class of tubes combines features of each of the other two classes. Tubes in this class are similar to the multi-electrode types in that they may have up to seven electrodes, all of which affect the electron stream; and they are similar to the multi-unit tubes in that they perform, simultaneously, more than one function,

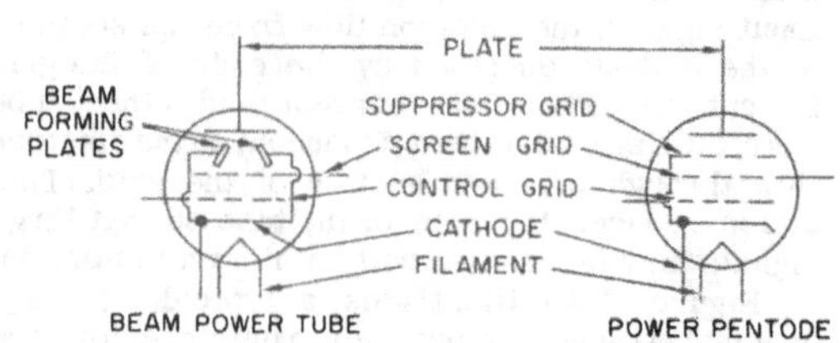

179.164

Figure 7-35.—Schematic symbols for beam power tube and power pentode.

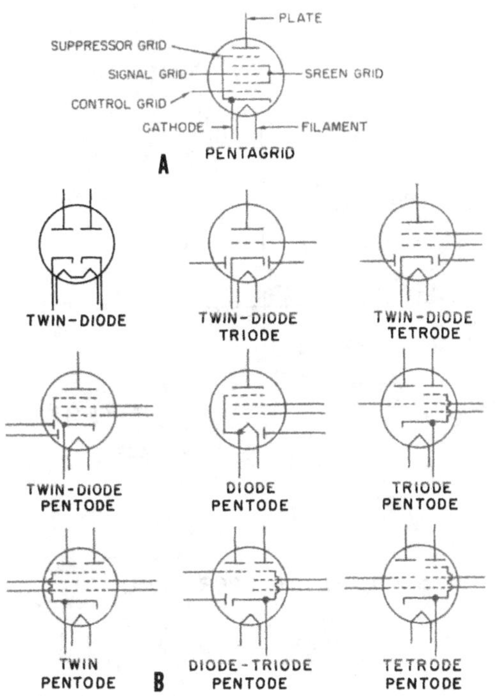

Figure 7-36.—Schematic diagrams of multi-electrode and multi-unit tubes.

e.g., oscillator and mixer (converter). Their schematic appearance is identical to multi-electrode tubes.

CLASSIFICATION OF AMPLIFIERS

Amplifiers may be classified according to their principal applications. There are two principal groups, voltage amplifiers and power amplifiers.

Voltage amplifier tubes have a relatively high amplification factor and are used where the primary consideration is one of high voltage gain. Such tubes usually operate into high impedance loads.

Power amplifier tubes are those which have a relatively low value of amplification factor and fairly low values of plate resistance. They are capable of controlling appreciable currents at reasonably high plate voltages.

Amplifiers also may be classifed according to their frequency range, the method of tube operation, and the method of interstage coupling. For example, amplifiers may be classed as direct coupled, audio frequency, video, or tuned RF amplifiers, if some indication of frequency of operation is desired. Also, the position of the quiescent operating point and alternating grid voltage may determine the class of amplifier. This will specify whether the tube is being operated in class A, class AB, class B, or class C.

A class A amplifier is an amplifier in which the grid bias and alternating grid voltage are such that plate current flows at all times (fig. 7-37A). A class AB amplifier is an amplifier in which the grid bias and alternating grid voltage are such that plate current flows for appreciably more than half but less than the entire electrical cycle (fig. 7-37B).

A class B amplifier is an amplifier in which the grid bias is approximately equal to cutoff value so that the plate current is approximately zero when no alternating grid voltage is applied, and so that plate current flows for approximately one half of each cycle when an alternating grid voltage is applied (fig. 7-37C).

A class C amplifier is an amplifier in which the grid bias is appreciably greater than the cutoff value, so that plate current is zero when no alternating grid voltage is applied, and so that plate current flows for appreciably less than one half of each cycle when an alternating grid voltage is applied (fig. 7-37D).

The suffix 1 may be added to the letter or letters of the class identification to denote that grid current does not flow during any part of the input cycle. The suffix 2 may be used to denote that grid current flows during part of the cycle. This being so, then figure 7-37A is actually class A_1; 7-37B is AB_1; 7-37C is actually class B_2; and 7-37D is class C_2 operation. Amplifiers may also be classified according to circuit configuration. The three basic electron tube circuit configurations are illustrated in figure 7-38.

The grounded or common cathode configuration (fig. 7-38A) is the most extensively used. This configuration has good voltage and power gain and high input and output impedances. The voltage gain of the common cathode configuration utilizing a triode is given by

$$A_v = -\frac{\mu R_L}{r_p + R_L}$$

This will be true for frequencies up to approximately 500 kHz, depending on the tube used.

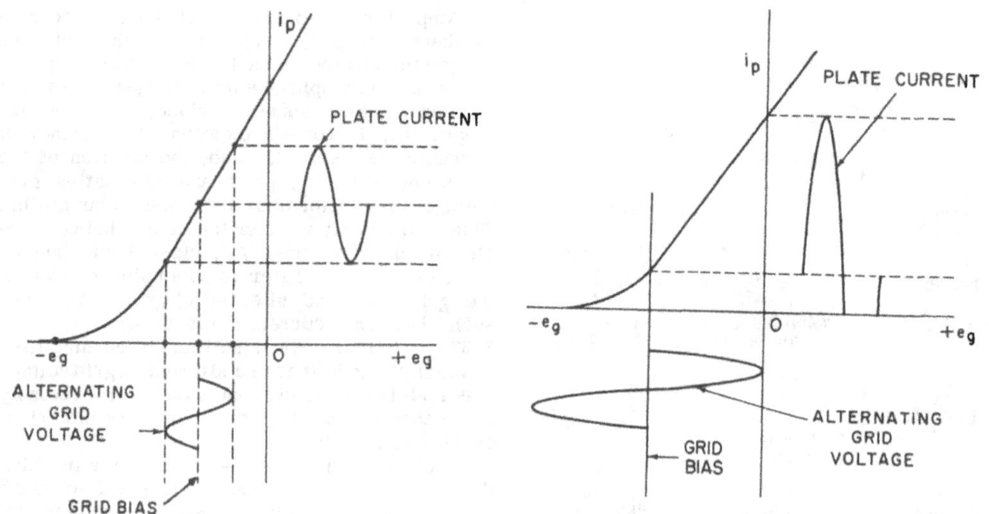

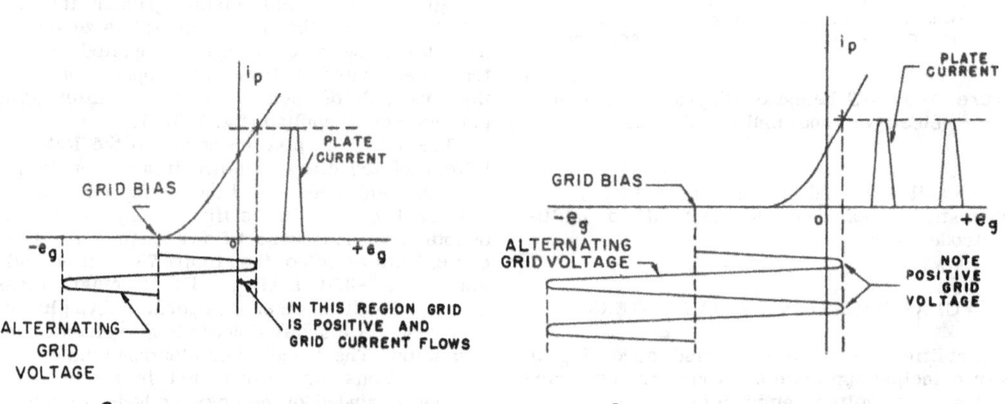

(A) 20.78 (B) 20.80 (C) 20.79 (D) 20.81

Figure 7-37.—Graphic illustration of classes of amplifier operation.

The gain equation shows two things: First, there is a reversal of phase in the amplifier. Second, the voltage amplification is equal to the amplification factor of the tube multiplied by $R_L/(r_p + R_L)$, a factor which is always less than 1. Thus, the amplification is always less than the amplification factor of the tube. The amplification may be increased (although it cannot exceed the amplification factor) by making R_L larger, but a practical limitation exists. Usually

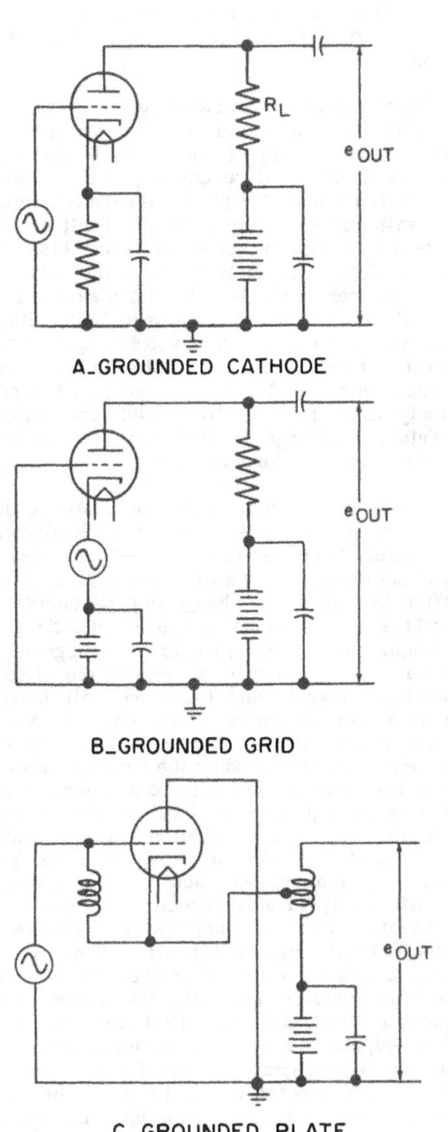

Figure 7-38.—Amplifiers classified according to circuit configurations.

the direct plate current flows through this resistor. If R_L is made too large, the direct plate voltage is reduced, with an attendant reduction of g_m and an increase in r_p. The increase in r_p rapidly becomes large enough to offset the increase in load resistance.

The voltage gain of a pentode common cathode configuration is approximately

$$A_v = - g_m R_L$$

Generally, the pentode amplifier will be capable of much higher gain than the triode amplifier, as was previously explained.

The common or grounded grid configuration (fig. 7-38B) has characteristics substantially different from the common cathode configuration. It has a low input impedance. It has a somewhat lower power gain but a high voltage gain (equal to $\mu + 1$). It is more stable and is used at high frequencies (VHF & UHF) because the grid effectively acts as a shield between the input cathode connection and the output plate connection. This configuration does not produce a phase reversal from input to output.

The common plate or cathode follower (fig. 7-38C) (so named because the cathode potential very closely follows the grid potential) has a high input and low output impedance. For this reason it is often used as an impedance matching or isolating stage. This configuration has a voltage gain of less than one, and a power gain lower than that of a common cathode configuration. This configuration also does not produce a phase reversal.

METHODS OF BIASING

In all amplifier circuits previously discussed, bias has been supplied from a fixed source. This type of bias, known as fixed bias, may be obtained from a power supply. Fixed bias is either shown as a battery (as illustrated previously) or as illustrated in figure 7-39.

The type of bias most commonly used is self bias. With self bias the bias voltage is developed across a resistance by the tube current or input signal.

CATHODE BIAS

One method of obtaining self bias is known as cathode bias, and is illustrated in figure 7-40.

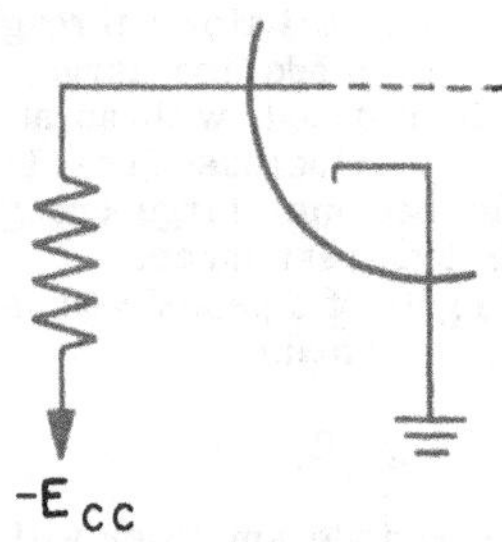

179.165

Figure 7-39.—Schematic representation of fixed bias.

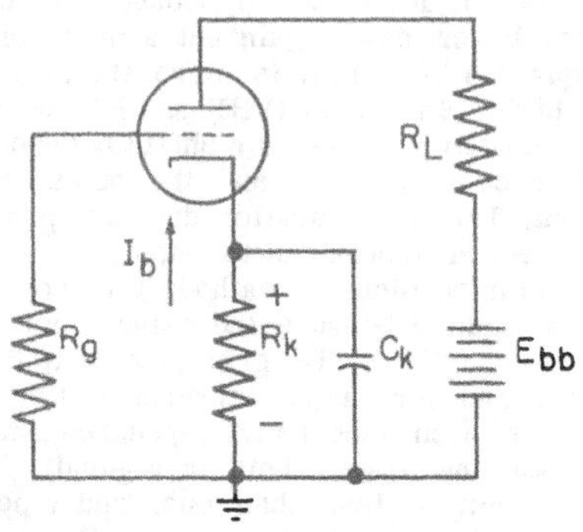

179.166

Figure 7-40.—Obtaining cathode bias.

When the cathode of an electron tube is biased positively with respect to the grid, the electron tube operates exactly as though an equivalent negative bias is applied to the grid. Since current flow within an electron tube is from cathode to plate, cathode resistor R_k can be inserted in the cathode line to produce a voltage drop (cathode bias) as long as the plate current flows continuously. Since cathode current always flows in the same direction, the voltage drop remains more positive at the cathode. Thus, plate current flow within the electron tube itself produces a positive cathode bias.

The value of the resistance for cathode biasing a single tube may be determined from the following formula:

$$R_k \text{ (ohms)} = \frac{\text{desired } E_c}{\text{rated } I_k}$$

For example, the resistance required to produce 5 volts bias for a triode which operates at 10 ma. plate current is $5/10 \times 10^{-3} = .5 \times 10^{3} = 500$ ohms.

This type of bias is restricted in use to amplifiers in which plate current flows for more than half of each cycle of operation, because plate current flow cannot be cut off, or the developed bias will be lost. But it can be used in combination with a fixed minimum bias to limit maximum plate excursions or as a protective bias which limits plate current to a safe value when grid drive is removed. An inherent disadvantage of this circuit is that the developed bias voltage is subtracted from the plate voltage applied to the electron tube. Thus, if 30 volts cathode bias is needed with a 250-volt plate supply, the power supply must be able to furnish 280 volts. Its greatest advantage is that it is simple to use and is the most economical.

When used in audio, video, or radio-frequency circuits, the cathode resistor must always be adequately bypassed (C_k fig. 7-40) to prevent a constant change of bias with signal. If the cathode is not bypassed, any change in plate current will produce a corresponding change in cathode bias voltage. The change in cathode voltage will be in such a direction as to oppose the effects of the input signal, and therefore will have the same effect as degenerative feedback. While a controlled amount of degenerative feedback can be beneficial in extending the over-all frequency response and in reducing distortion, a large amount of degenerative feedback will result in a serious loss of amplification. When the cathode bias resistor is adequately bypassed, the fluctuating a.c. plate current caused by the input signal is effectively shunted around the cathode bias resistor, through a much lower reactance path offered by the bypass capacitor. Thus, only the d.c. component of plate current flows through the bias resistor, and the total cathode current remains constant at the initial static (d.c.) value of no-signal current and is unchanged by the varying input signal. For satisfactory bypassing, the bypass reactance should be about 10% of the d.c. bias resistance at the lowest frequencies used. Cathode bias is usually used in audio-frequency or video-frequency amplifiers, in low power RF amplifiers, and in test equipment. In some applications, such as high fidelity audio amplifiers or video IF amplifiers, a relatively small value of unbypassed cathode resistance may be used. The degenerative action of this resistance increases the fidelity (frequency response) of the stage and reduces the possibility of overloading from strong signals.

GRID-LEAK BIAS

Another type of self bias is grid-leak bias. Grid-leak bias is obtained by allowing grid current flow, produced by the a.c. input signal, to charge an RC network in the grid-cathode circuit. Two basic circuits are used to develop this form of bias—the shunt type and the series type. The methods of developing grid voltage in these circuits are similar, but the physical connections of the network components are different. See figures 7-41 and 7-42.

The grid-cathode circuit, of both types, is used as a diode rectifier to develop a d.c. voltage that is proportional to the positive peak input (driving) signal amplitude. Grid-leak capacitor C_c operates as a coupling capacitor to apply the input (driving) signal to the grid. On the positive input signal excursions the grid is driven positive causing grid current to flow between grid and cathode and through grid-leak resistor R_g. The result is to produce a d.c. voltage across R_g which is polarized negatively at the grid.

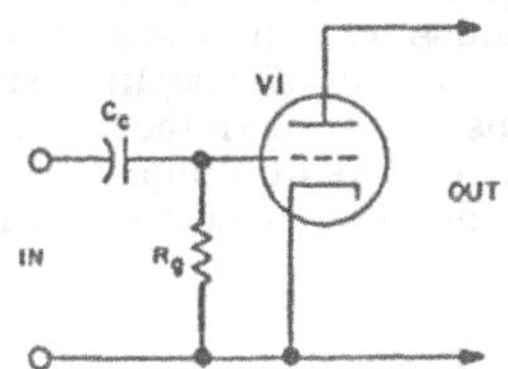

179.167
Figure 7-41.—Shunt grid-leak bias.

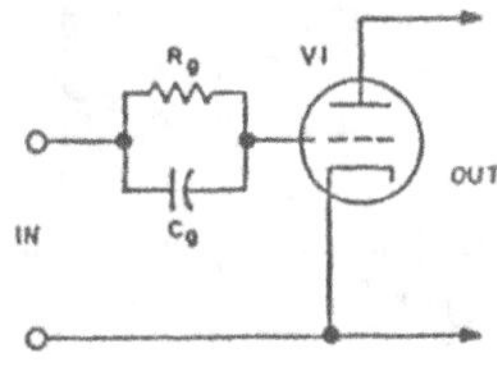

179.168
Figure 7-42.—Series grid-leak bias.

The grid-leak capacitor, C_c, is effectively connected in series with the applied input signal as shown by the equivalent charging circuit of figure 7-43A. As long as the input signal remains positive the coupling capacitor charges. When the input signal becomes negative, grid capacitor C_c begins discharging through the grid-leak resistor (fig. 7-43B); during the discharge period the grid is held negative by the charge remaining in the capacitor plus the negative input signal, and no grid current flows. The grid-leak time constant is made long with respect to the signal frequency (usually seven times the period for one input cycle) so that the discharge is relatively slow. The charge time is much faster than the discharge time because the grid leak is effectively shunted by the flow of grid current and offers a low resistance of about 500 ohms (r_{gk}, figure 7-43A) in comparison with the large resistance of R_g.

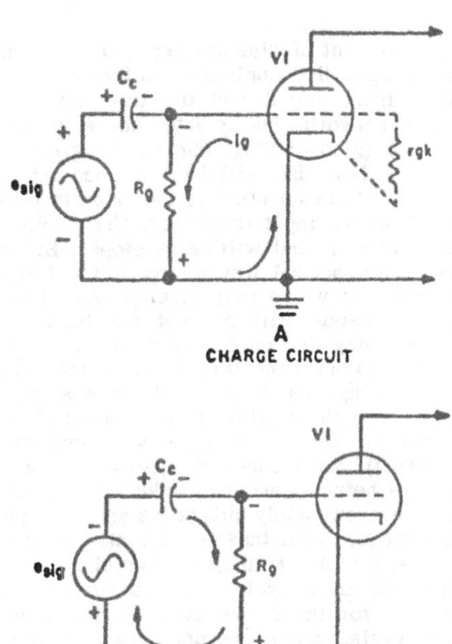

179.169
Figure 7-43.—Simplified charge and discharge circuits.

When the signal begins its next positive excursion, C_c again starts to charge while still retaining a certain amount of residual charge. The cycle resumes, and the action is repeated. After only a few more cycles, the grid bias stabilizes, since only a small portion of each charge leaks off before the application of the next positive half of the input signal. The residual charge on the coupling capacitor will not permit the next positive portion of the input signal to drive the grid as far positive as the positive portion did in the first half cycle. Thus, the additional charge placed on C_c is not as great as it was for the first half cycle, but when added to the residual charge that remained, it causes the total bias to increase. This process of rapid charging and slow discharging of C_c leads to the steady state condition of operation, in which the amount of charge deposited on each positive half cycle exactly equals the amount of charge lost through discharge on each negative half cycle.

The amount of bias developed in this manner depends upon the amplitude and frequency of the applied input signal and the time constant of the grid network. The greater the amplitude of the input (grid-driving) signal, the greater the amount of bias that will be developed. Also, the longer the time constant of the grid circuit with respect to the input frequency, the greater the amount of bias that will be developed. However, there are practical limitations to the length of time constant which may be employed. Too long a time constant will prevent the bias voltage from changing enough during a cycle, or group of cycles, to bring the stage out of cutoff. Therefore, the stage must wait until the bias has been reduced enough to allow it to come out of cutoff; as a result, the stage will operate intermittently (motorboat). When functioning properly, a grid-leak bias network should produce a d.c. voltage that is approximately 90% of the applied positive peak voltage. Under these conditions grid current will flow for only a small portion of the positive peak of the input cycle. Plate current, however, may flow for the entire cycle or for only part of the cycle, depending upon the amplitude of the input signal and the cutoff voltage of the electron tube.

Since the amplitude of the developed grid-leak bias is determined by the driving-signal amplitude, it is evident that class A, B, or C operation may be achieved with grid-leak bias merely by increasing the amount of drive (and by selecting the proper RC value). The limiting factor in this application is the amount of distortion that may be tolerated, because the plate current will be distorted for the portion of the cycle during which grid current flows. It is important to remember that grid-leak bias is developed by the rectification of the input (driving) signal and that loss of the input signal will result in a complete loss of this form of bias. In power amplifier circuits where loss of bias means excessive tube currents and possible damage, a second method of biasing (either fixed or cathode bias) is usually employed as a protective feature. Any of the above described types of bias may be used in combination. Figure 7-44 illustrates one such possibility.

COUPLING METHODS

Usually more than one amplifier, operated in cascade, is needed to increase the amplitude of the feeble input signal to the required output value. Cascaded amplifier stages are connected (coupled) together by resistance-capacitance networks, impedance-capacitance networks, transformers, or direct coupling as described in the following paragraphs. While all coupling networks are frequency responsive, some coupling methods provide better response than others, but their basic principles of operation remain the same regardless of whether or not they are employed singly as input or output coupling devices, or in cascade. Discussion in this section will be limited to the actual types of coupling and important considerations involved in the various functional classes on a relative or comparative basis. Circuit discussions in other sections will fully cover

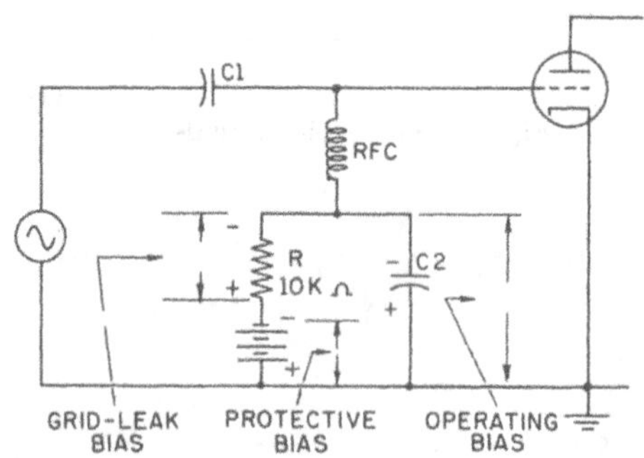

Figure 7-44.—Combination grid-leak and battery bias.

the limiting parameters for the specific circuit arrangement.

RC COUPLING

RC coupling involves the use of two resistors and a capacitor, as shown in figure 7-45. Because of the rather wide frequency response offered by this form of coupling, in addition to small size and economy, it finds almost universal application where voltage amplification is required with little or no power output. Basically, the RC coupler is a pi-type, hi-pass network, with plate resistor R_L and grid resistor R_g forming the legs of the pi and capacitor C_c the body. Since the plate and grid resistors are not frequency responsive, it can be seen that, basically, over-all frequency response is limited by the capacitive reactance of C_c between the plate and grid circuits, plus the effect of shunt wiring and electrode-to-ground capacitances across the network. At d.c. or zero frequency the coupling capacitor separates, or blocks, the plate voltage of the driving stage from the grid bias of the driven stage, so that bias and plate or element voltages are not affected between stages. Note that, with respect to a.c., the tubes are in parallel with R_L.

In the conventional RC amplifier signal voltage variations on the grid produce plate current variations through plate load resistor R_L, and the resulting voltage developed across R_L represents an amplified replica of the input signal but 180 degrees out of phase. The amplified signal is coupled through capacitor C_c and applied to the grid of the next stage across grid load resistor R_g. The same cycle of operation is repeated for each stage of the cascaded amplifier.

Since load resistor R_L is in parallel with the internal tube plate resistance, r_p, the load resistor of a triode is never made less than twice r_p. Usually the maximum undistorted output for a triode is obtained with a value of R_L that is 5 to 10 times the plate resistance of the tube. With pentodes the extremely high r_p makes this impractical; best results are usually obtained with a load resistor of 1/4 to 1/10 the value of r_p, or with a value that is as high as is practical for the plate voltage and current needed without requiring excessive supply voltage.

At low frequencies (below 100 Hz), the reactance of C_c, plus R_g in series, parallels the load resistor. But the reactance of C_c is in series between the plate and grid, and produces a large voltage drop (or loss of gain), while R_g shunts the grid and has the least voltage developed across it. Therefore, the interstage grid resistance has less over-all effect on the gain than the reactance of C_c, and the distributed circuit and electrode capacitance to ground is not effective and may be neglected. Low-frequency response, therefore, is controlled by the size of the coupling capacitor.

A simplified equivalent low-frequency circuit is shown in figure 7-46A. Only those circuit elements which are of significance at low frequencies are shown. When the coupling capacitor is excessively large, distortion is caused by

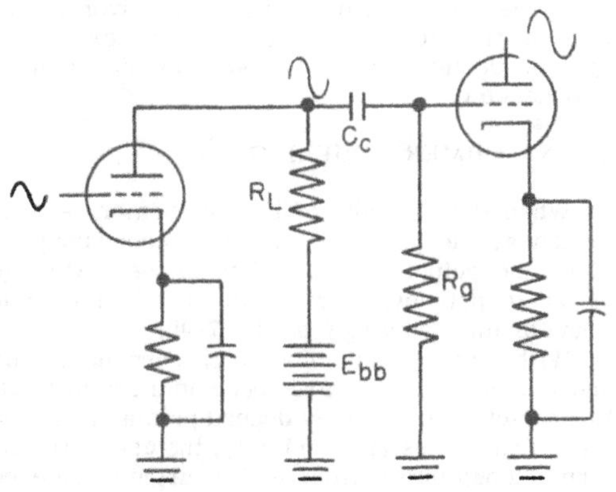

179.170

Figure 7-45.— RC coupling.

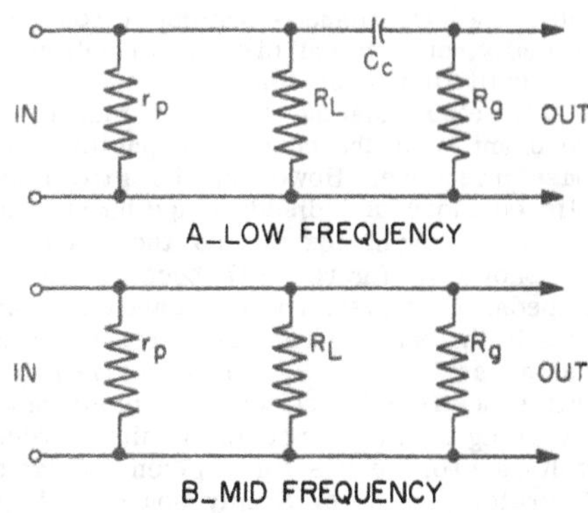

179.171

Figure 7-46.— Equivalent circuits of figure 7-45.

excessive phase shift, and motorboating may occur. Motorboating is caused by regenerative feedback at low frequencies through impedance coupling in multiple cascaded RC stages with a common power supply.

Over the mid-frequency range (100 to 20,000 Hz), amplification is relatively constant, and the coupling capacitor reactance is much lower than the resistance of grid resistor R_g. Therefore, the reactance of coupling capacitor C_c may be neglected as in the simplified equivalent mid-frequency circuit of figure 7-46B. To insure that the plate resistor is not shunted excessively by the grid resistance of the following stage and to maintain a high input resistance, R_g is made two to four times R_L and never less than R_L. Usually in low-level stages the grid resistor is in the megohm range (from 1 to 5 megohms), with its value for grid-leak bias ranging from 5 to 10 megohms; whereas in power amplifier stages, the grid resistor is on the order of 0.5 megohm and usually never more than 1 megohm. High values of grid resistance tend to cause adverse effects due to the possibility of grid current flow caused by imperfect evacuation, grid emission, and leakage effects in the electron tube.

Because of the low amount of gain available with RC coupling at radio frequencies, it is seldom used in RF amplifiers; but it may be encountered in special cases, particularly in test equipment.

IMPEDANCE COUPLING

When an impedance is substituted for the plate load resistor in a resistance-coupled circuit, the impedance-coupling circuit results. Actually, an original circuit derivation involved a substitution of an impedance for the grid resistor also. Making the grid inductor series resonant with the coupling capacitor improved bass response. However, the grid impedance offered more of a disadvantage because of high-frequency shunt losses, so the LCR circuit as shown in figure 7-47 became the standard impedance coupler. The impedance coupler operates in the same manner as the resistance coupler as far as C_c and R_g are concerned; the basic difference is in the effect of the plate impedance. By using an impedance in the plate circuit, less voltage drop occurs for a given voltage supply. Therefore, a lower supply source will provide the same effective plate voltage, there is less I^2R (power) loss, and a better over-all efficiency results. Low-frequency response is dependent

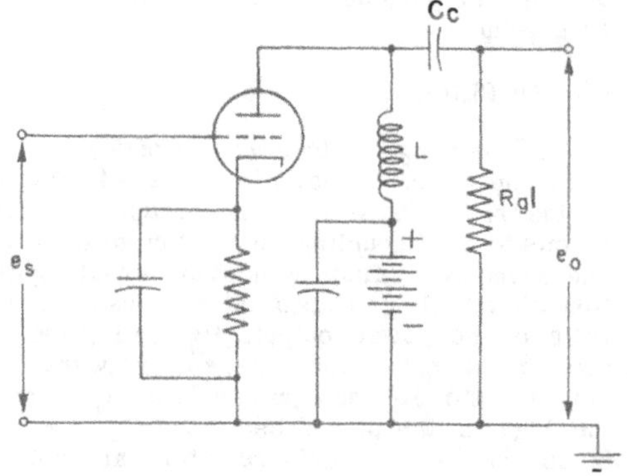

Figure 7-47.—Impedance-coupled amplifier.

upon obtaining a high inductive reactance in the plate circuit and requires a large number of turns for good low-frequency response. The distributed capacitance associated with a winding of many turns produces a large shunting capacitive reactance with a consequent drop in high-frequency response. Since the impedance of the plate reactor varies with frequency, the response is not as uniform as that of the resistance coupler.

The impedance-coupling circuit is used where a limited response over a relatively narrow band of frequencies is required. As a result, impedance coupling is mostly employed in tuned or untuned amplifier applications such as IF or RF stages, and its use in audio applications has generally been discontinued in favor of the RC or transformer-coupled circuit, except for special designs. Figure 7-48 illustrates tuned impedance coupling.

TRANSFORMER COUPLING

When the primary of a transformer is connected as the plate load, and the secondary provides the output signal, either to the next stage or an output device, we have what is known as transformer coupling (see fig. 7-49).

With transformer coupling, response, gain and output considerations become more difficult to predict, because they depend primarily on the transformer design. Basically, the use of transformer coupling provides additional gain, achieved through the use of a step-up turns ratio of primary to secondary, but this gain usually does not exceed 2 or 3 to 1. Since there is no physical d.c. connection between stages, plate and bias voltages are kept separate, and the a.c. signal is coupled from the plate of one stage to the grid of the following stage by mutual inductive coupling between primary and secondary windings.

Low-frequency response is primarily dependent on the inductive reactance of the primary, dropping off approximately 3 db at the frequency where the inductive reactance is equal to the plate resistance plus the primary resistance ($2\pi fL = R_p + R_{pri}$).

Gain at the mid frequency is the highest and is essentially equal to the amplification factor of the electron tube multiplied by the transformer turns ratio. Response is considered to be relatively flat over a range of frequencies above and below the mid frequency, but such results are usually achieved only in the ideal case. In practice, the response curve is a continually changing curve, dropping off on either side of the mid frequency, with possible humps produced by circuit and transformer resonances (fig. 7-50). Since the secondary contains more turns than the primary, a larger shunt capacitance to ground is produced and, together with the primary and secondary leakage reactances, limits the high-frequency response.

Transformer coupling is generally used for interstage applications with electron tubes having plate resistances of 5 to 10 thousand ohms maximum, since higher plate resistances require excessively large transformer primary inductances. For output stages, lower plate resistances are used, and the transformer is carefully designed to handle larger plate currents. Generally speaking, a lower plate current produces less d.c. core saturation effects.

Since the impedance transformation in a transformer varies as the square of the turns ratio between primary and secondary, output and input matching is possible and is extensively employed. For interstage applications, matching is not always used, because power output is not required; in this case, more attention is given to the step-up ratio to provide a higher voltage gain.

The limitations of frequency response generally restrict the use of transformer coupling to audio circuits which do not require an exceptionally wide bandpass or frequency response, but do require voltage or power outputs. Wide use is found for transformer coupling in the radio and intermediate frequency ranges where selective; high-Q bandpass filters and tuned transformers are universally used.

DIRECT COUPLING

Direct coupling is characterized by the direct connection of tube elements; that is, the plate of the driver stage is physically connected to the grid of the driven stage, and the coupling network is eliminated. A basic two-stage d.c. amplifier is shown in figure 7-51.

Since the plate and bias circuits are not isolated by a transformer or coupling capacitor the direct-coupling circuitry is slightly complicated by the arrangement necessary to produce an effective negative bias. Usually a voltage-divider bias arrangement similar to that shown in figure 7-51 is used.

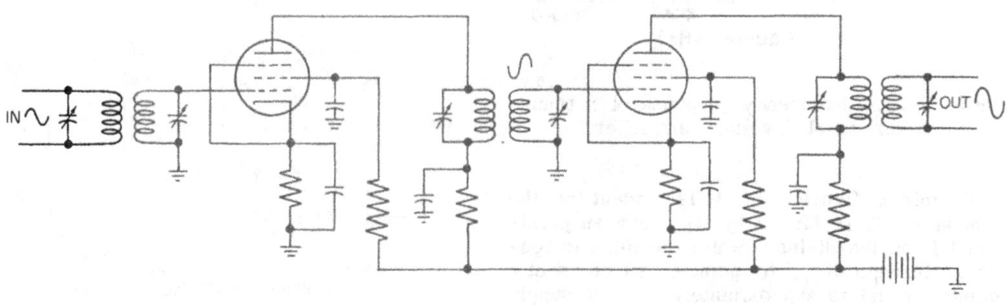

179.172

Figure 7-48.—Tuned impedance coupling.

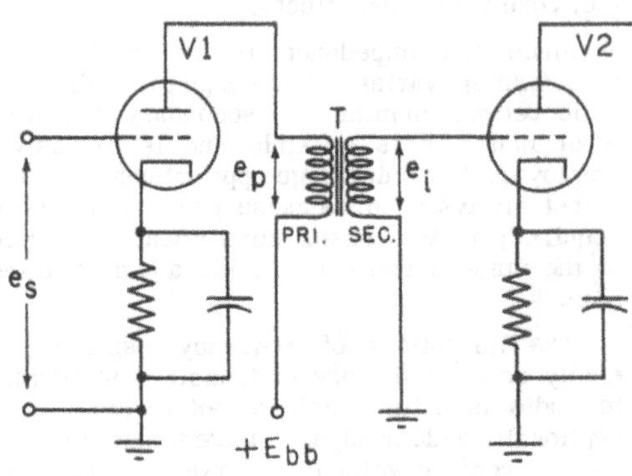

20.97
Figure 7-49.—Transformer-coupled voltage amplifier.

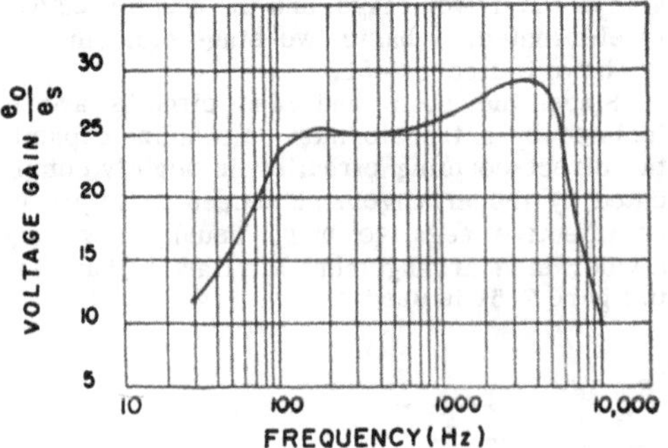

20.98
Figure 7-50.—Frequency response of a transformer-coupled voltage amplifier.

Examining figure 7-51, it is evident that the cathode of V1 is biased by R_{k1} between points A and B on the divider, while the plate is connected through R_{L1} to point C, which is at a potential equal to approximately half the supply voltage. Plate current through load resistor R_{L1} (which is also the grid resistor for V2) provides a negative voltage drop and bias on V2. The cathode of V2 is tapped back at point D on the divider, and produces a voltage in opposition to the drop across R_{L1}. When R_{k2} is correctly adjusted, the voltage on the cathode of V2 is always more positive than the plate of V1 (and the grid of V2), producing an effective negative bias on V2 of the desired value. Voltages must be chosen so that the grid of V2 is not driven positive by the input signal. Since the plate of V2 is at the highest positive point, it is more positive than the cathode and conduction occurs. Thus the plates are maintained positive and the grids negative, to provide the conditions required for operation.

Because there is no coupling network inserted between the output of one tube and the input of the following tube, there is no phase distortion, time delay, or loss of frequency response. Since the plate and grid of the tubes are directly connected, the low-frequency response is extended down to d.c. (zero frequency). The high-frequency response is limited only by the tube interelectrode-to-ground capacitance, plus the circuit distributed wiring capacitance. By appropriate matching (or mismatching) of tubes, high values of amplification and power output may be obtained.

Since the use of more than two stages requires plate voltages two or more times the normal value for one tube, plate supply considerations limit direct coupling to a few stages. Any change in the supply voltage affects the bias of all the tubes and is cumulative; therefore, special voltage supply regulation circuits are necessary. Noise and thermal effects in tubes produce circuit instability and drift that limit the use of this type of coupling in audio or RF amplifiers.

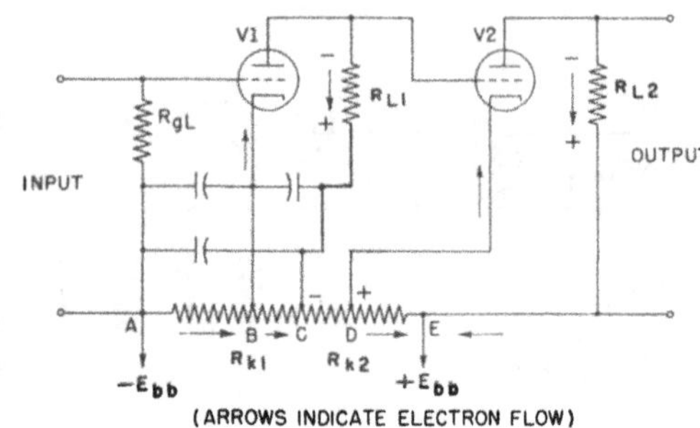

20.102
Figure 7-51.—Direct-coupled amplifier.

Because of its ability to amplify direct current or zero frequency, direct-coupled circuitry is often used in computer circuits, and in the output circuits of video amplifiers. Because response is practically instantaneous and no time delay occurs, it is especially valuable for pulse circuits.

FREQUENCY COMPENSATION

Frequency compensation networks may be used with RC coupled amplifiers to extend their frequency response. These networks are identical to those previously discussed under transistor frequency limitation (chapter 6) and are shown in figure 7-52.

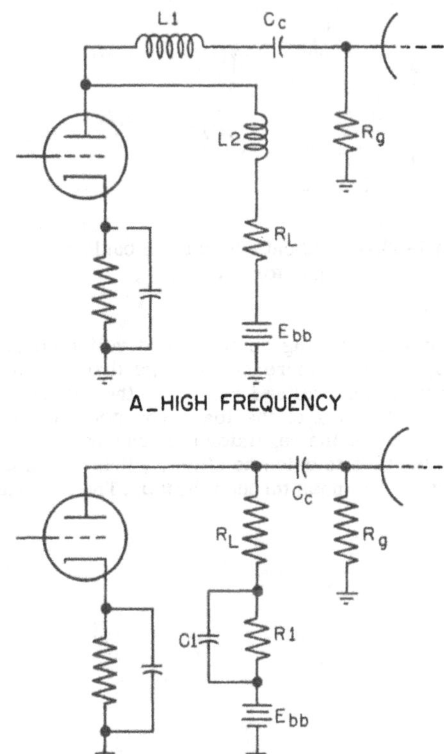

Figure 7-52.—R-C coupled amplifiers with frequency compensation networks.

COLD CATHODE VOLTAGE REGULATOR TUBES

As previously explained, certain "cold cathode" gas filled diodes will maintain, within limits, a constant voltage drop across their elements, though the current through them varies. Such a tube is illustrated in the basic voltage regulator circuit of figure 7-53. This circuit is sim'lar to the zener regulator previously discussed in chapter 5. Cold cathode voltage regulator tubes, also called VR tubes, come in various operating voltages. Typical voltage ratings are 75, 90, 105, and 150 v. d.c. Typical maximum current ratings are 30, 40, and 50 ma.

There are two voltages to be considered when discussing the operation of the basic cold cathode regulator circuit. These are the firing voltage and the operating voltage. The firing voltage is that amount of voltage, applied across the tube, that will cause ionization of the gas and, therefore, conduction. The operating voltage is that voltage dropped across the tube when it is conducting. The firing voltage will be 30 to 40 percent greater than the operating voltage and this difference in the two voltages is dropped across the series resistance, R_s (fig. 7-53), once the tube ionizes.

The ionization of the gas within the tube changes, depending on the applied voltage. The greater the electron flow, the more ionization and the greater the current through the tube. When the applied voltage is decreased, the electron flow decreases and there is less ionization and less current. The gas tube regulator must be ionized at all times for proper regulation to occur. The amount of ionization might be changed due to a change in the unregulated input voltage, or a change in the load current, but in either case, the regulated output voltage will remain relatively constant at all times.

An increase in the input voltage causes an increase in the regulator tube current (a decrease in tube resistance). This increase in tube current, passing through R_s, increases the voltage drop across R_s leaving the voltage across the tube and the load constant. A decrease in the input voltage results in a decrease in tube current (increase in tube resistance) causing less total current to pass through R_s. This lowers the voltage drop across R_s and the output voltage remains constant.

For load changes the results are the same—an increase in load current is an increase in

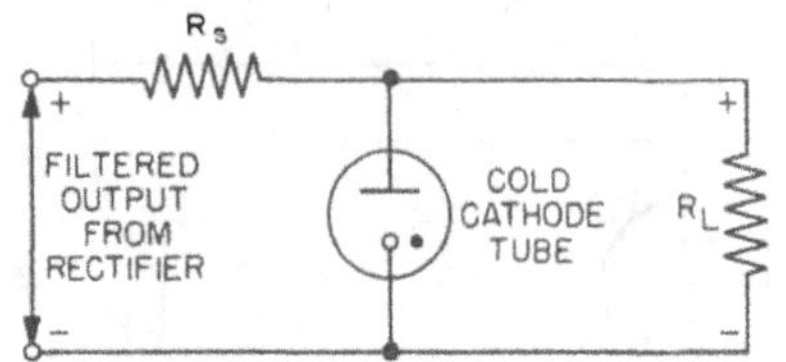

20.43
Figure 7-53.—Cold cathode tube in basic voltage regulator circuit.

total current causing an increased voltage drop across R_s. This increased voltage drop across R_s leaves less voltage applied to the regulator tube, and thereby to the load. The decrease in voltage across the regulator tube causes the gas within the tube to deionize slightly, thus, causing less current to pass through the tube. The current through series resistor R_s decreases by the same amount as the decrease through the tube. This results in a decrease in the voltage drop across R_s and thus, an increase in the voltage across the load to its original value.

If the load current should decrease, the gas within the tube would further ionize causing an increase in current through the tube, and conditions opposite to those described above would occur to maintain the voltage across the load constant.

The limits within which the VR tube will operate are a maximum and minimum tube current. If the maximum current is exceeded, the tube will be destroyed. If the tube current decreases below the minimum value, the tube will de-ionize, and the voltage across the tube must again be raised to the firing potential in order to reionize the tube.

CHAPTER 8

ELECTRONIC VOLTAGE REGULATORS AND METERS

Most electronic equipment performs satisfactorily with some variations in supply voltage. However, operation of many circuits is sensitive to minimal changes in supply voltage. Thus, use of a voltage regulator is required. This chapter discusses the common types of electronic voltage regulators. Also discussed are the electronic multimeter and the digital voltmeter.

ELECTRONIC VOLTAGE REGULATORS

An electronic voltage regulator is connected between the power supply and load impedance to maintain the output voltage at a specific value. The regulator circuit reacts automatically within its design limits to compensate for deviations in output voltage due to changes in line voltage or variations in load impedance.

Regulation of power supply output voltages using zener diodes and gas tubes has previously been discussed. While these devices are suitable for circuit loads requiring small currents, they cannot regulate over a wide current range. Current handling capabilities, as well as percent of regulation, can be improved by using transistors or electron tubes in conjunction with the zener diode or gas tube.

Depending upon their circuit relationship to the load impedance, electronic voltage regulators may be grouped into two general types. These are shunt regulators and series regulators. Series regulators are the most efficient, and therefore, are the most widely used.

SHUNT REGULATOR

Figure 8-1 illustrates the equivalent resistive network of a shunt regulator. The location of the regulating device is shown in relation to the load impedance (R_L). Regulation is accomplished in the shunt regulator by division of current between the regulating device and load impedance. When the load demands more current, less current is diverted through the regulator (R_v), allowing more current flow through R_L.

Efficiency of shunt regulators is low under light load conditions (i.e., when current flow through the load is minimal) because the majority of current is being drawn by the regulator. Under full load conditions regulator efficiency is high because minimum current is drawn by the regulator. A significant advantage of a shunt regulator, as opposed to a series regulator, is that it will not be overloaded under load short circuit conditions.

Solid State Shunt Voltage Regulator

Figure 8-2 depicts a solid state shunt regulator. Q1, in parallel with the load impedance, is an NPN transistor with the collector connected to the positive side of the voltage supply and the emitter connected to the negative side through CR1, a zener diode. CR1, when reverse

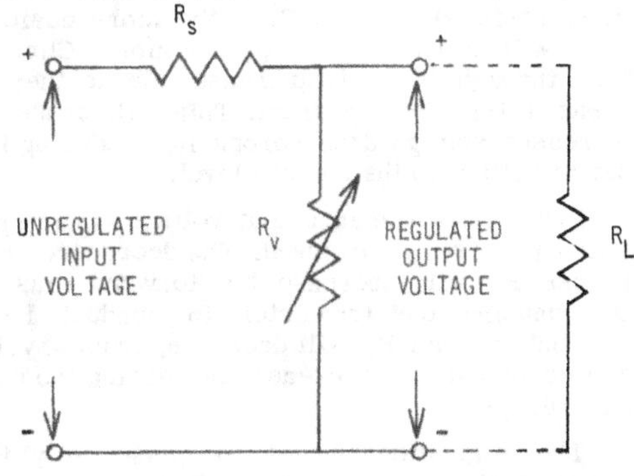

Figure 8-1.—Shunt regulator.

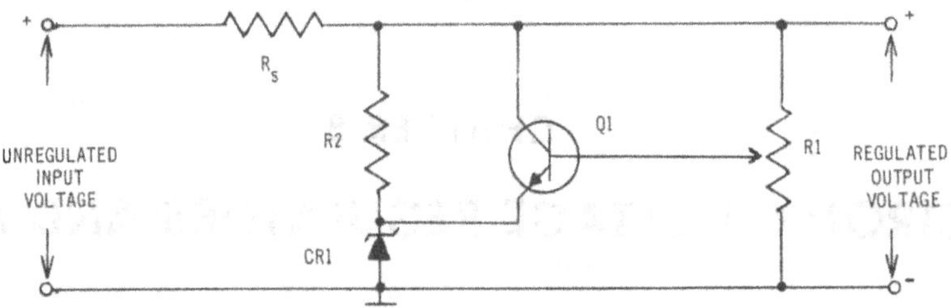

179.175

Figure 8-2.—Solid state shunt voltage regulator.

biased to its breakdown voltage by R2, maintains a constant reference voltage at the emitter of Q1. The base voltage of Q1 is determined by the setting of potentiometer R1. This voltage is adjusted so that the base is positive with respect to the fixed emitter potential, forward biasing Q1, and causing it to conduct. The setting of R1 determines the amount of current through Q1.

The regulated output voltage is equal to the available supply voltage minus the drop across R_s, the series dropping resistor. The voltage drop across R_s is controlled by the amounts of current drawn by Q1. Thus, the setting of R1 will determine the value of the regulated output voltage.

Regulation is accomplished in the following manner. If, for any reason the output voltage increases, the drop across R1 will increase. This will cause an increased positive potential at the base of Q1. Since Q1's emitter is at a fixed potential, due to CR1, the more positive base will cause Q1 to conduct more. Current flow through R_s will increase due to the increased transistor current. This will cause an increased voltage drop across R_s, reducing the output voltage to the desired level.

For a decrease in output voltage, the regulation process is reversed. The decreased drop across R1 will decrease the forward bias of Q1, causing the transistor to conduct less. Current through R_s will decrease, causing voltage across R_s to decrease, increasing the output voltage.

The regulation process is essentially the same as that occurring in a shunt zener diode regulator. However, the current handling capabilities are greatly increased.

Electron Tube Shunt Voltage Regulator

An electron tube shunt voltage regulator is illustrated in figure 8-3. V1's cathode is maintained at a constant reference voltage due to the action of V2 and R2. The difference between the unregulated input voltage and the regulated output voltage is dropped across R_s. The output voltage level is determined by the setting of R1. V1's bias is the difference between the voltage at the wiper arm of R1 with respect to ground and the voltage at the plate of the VR tube with respect to ground. Just as in the solid state version, changes in output are sensed by V1; through R1 varying the conduction of V1 and controlling the voltage drop across R_s in a manner that will maintain a relatively constant output voltage. This type electron tube shunt regulator is used extensively in high voltage power supplies of video receivers.

SERIES REGULATOR

Regulation is achieved in a series regulator (fig. 8-4), by division of voltage between the regulator R_s and the load impedance (R_L), depending on the needs of the load. Efficiency of series regulators is high under light load conditions and low under full load conditions. The series regulator has no inherent overload protection. A short circuit in the load would cause heavy current through the regulator circuit overloading most types.

Transistor and electron tube configurations of series, shunt, and shunt detected series voltage regulators will be presented in the remainder of this chapter.

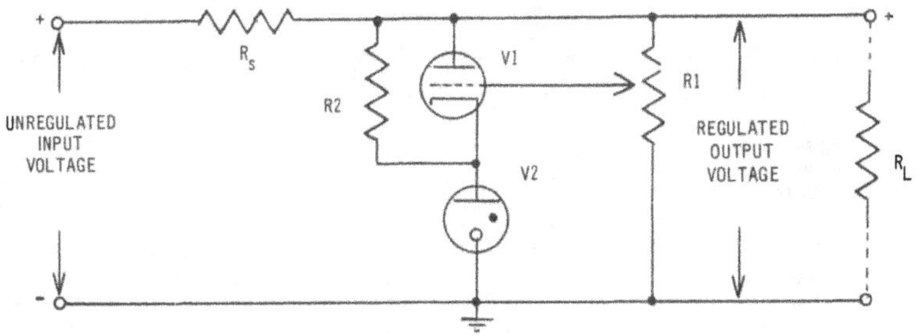

179.176
Figure 8-3.—Electron tube shunt voltage regulator.

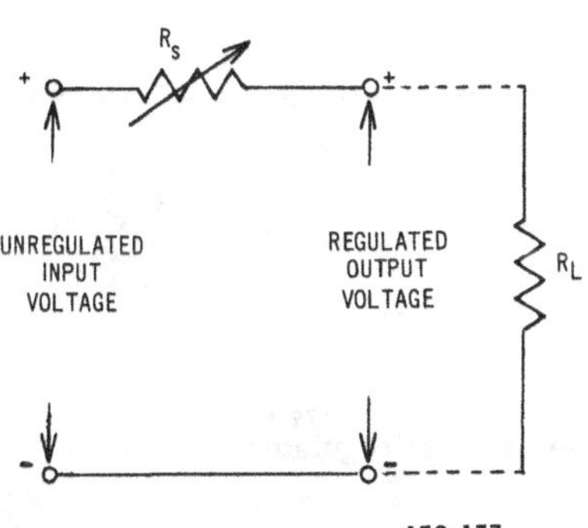

179.177
Figure 8-4.—Series regulator.

Solid State Series Voltage Regulator

A schematic diagram of a solid state series voltage regulator is shown in figure 8-5. Q1 functions as a variable resistance between the source and load impedance. CR1, a zener diode, in conjunction with R1 maintains a constant voltage at the base of Q1.

The unregulated input voltage is applied across the series network of Q1 and R_L. The fixed base voltage is of sufficient value to forward bias Q1. Under normal operating conditions, the base bias is fixed at a value that will produce the desired voltage across the load impedance. The load voltage is equal to the unregulated input voltage minus the drop across Q1.

Regulation occurs in the following manner. Assume that the line voltage increases. This will cause a momentary increase in load voltage which causes the emitter of Q1 to appear more negative with respect to ground, decreasing the forward bias of Q1. This decrease in forward bias causes the internal resistance of the transistor to increase, producing an increased drop across Q1, thereby returning the voltage drop across R_L to its normal level. The action of the regulator is reversed with a decrease in applied line voltage.

The following will explain the operation of the regulator under varying load conditions. Assume an increase in load (more current). Voltage across the load impedance will decrease and the emitter of Q1 will become less negative, increasing the forward bias. This reduces the transistors internal resistance, effectively reducing the voltage drop across Q1. This allows more current flow through R_L, thus returning the output voltage to its normal level. A decrease in load will cause a reverse reaction of the regulator.

Electron Tube Series Voltage Regulator

An electron tube version of a series regulator is illustrated in figure 8-6. The operation is similar to a solid state series regulator. V1 acts as a variable resistance between the source and the load impedance to compensate for changes in line and load voltage. V2 maintains the grid of V1 at a fixed reference voltage. If the output voltage increases, the cathode potential goes more positive with respect to ground, thereby increasing the bias. This increase in bias causes V1 to decrease conduction, decreasing the drop across R_L to the desired level. The

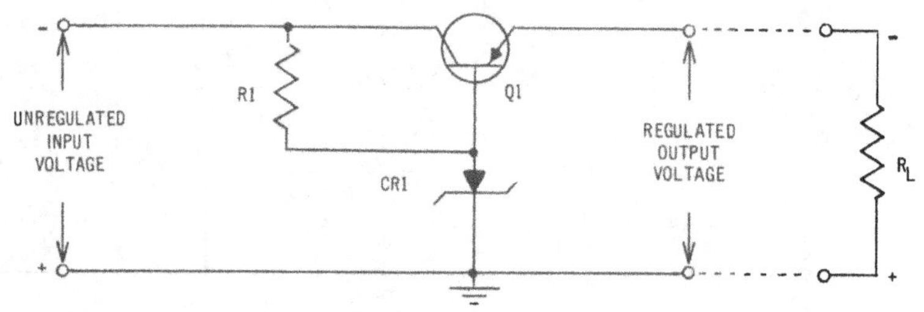

179.178
Figure 8-5.—Solid state series voltage regulator.

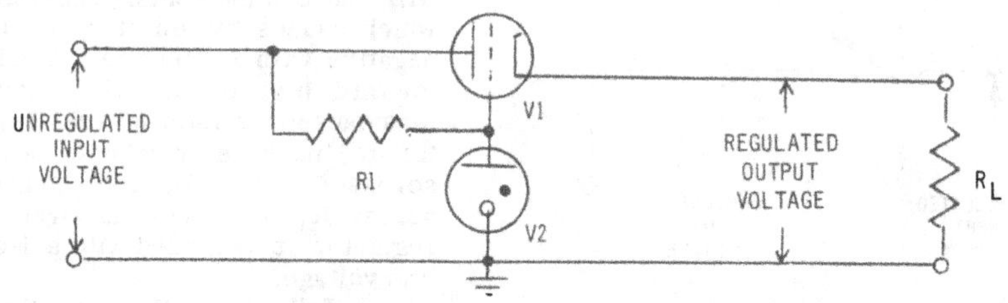

179.179
Figure 8-6.—Electron tube series voltage regulator.

opposite action will occur for a decrease in output voltage.

A disadvantage of simple series regulators is that they do not rapidly respond to small changes in voltage. The effectiveness of a series regulator is improved by the addition of circuitry that detects and amplifies small changes, thus allowing the regulator to respond more rapidly. The shunt detected series regulator is such a circuit.

SHUNT DETECTED SERIES VOLTAGE REGULATOR

Figure 8-7 depicts a block diagram of a shunt detected series voltage regulator. The series regulator acts as a variable resistance in series with the supply voltage and load impedance to compensate for any changes in source or load voltage. A portion of the output voltage is fed back to the amplifier and detector stage, which compares the sampled voltage with a previously set reference voltage and senses any change. This change is then amplified and sent to the series regulator, which changes the conduction level, thereby returning the regulated voltage to the desired level.

Solid State Shunt Detected Series Voltage Regulator

A solid state shunt detected series voltage regulator is illustrated in figure 8-8. The circuit consists of two sections—the regulator circuit and the control circuit. Q1 acts as a series regulator and operates in the same manner as the circuit in figure 8-5. The control circuit is composed of Q2 and its associated circuitry.

The emitter of Q2 is held at a constant negative potential with respect to point A due to the action of R1 and CR1. Resistors R3, R5 and potentiometer R4 act as a voltage divider

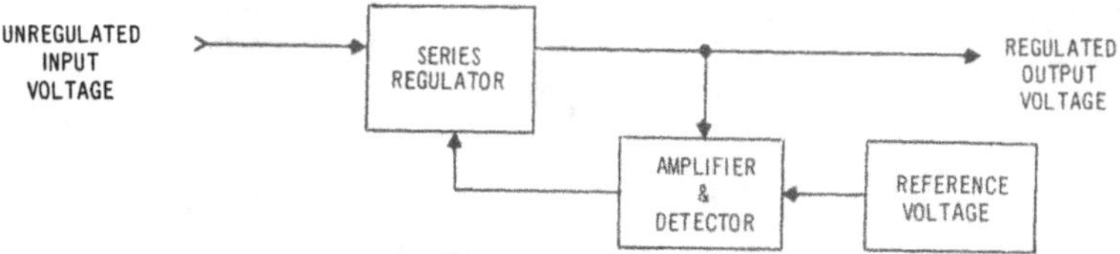

179.180
Figure 8-7.—Shunt detected series voltage regulator, block diagram.

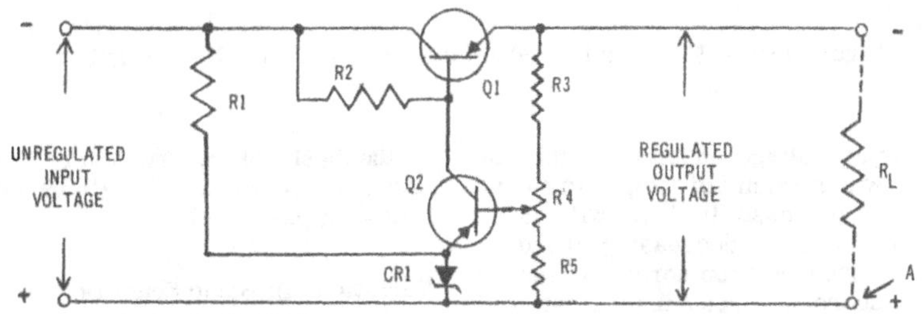

179.181
Figure 8-8.—Solid state shunt detected series voltage regulator.

network in parallel with R_L. A negative potential with respect to point A is tapped from R4 and applied to the base of Q2, controlling its forward bias. R2 establishes the forward bias level for Q1. Thus, the intensity of current flow through Q2 determines the ultimate conduction level of Q1, either manually or automatically.

R4 is initially adjusted to fix a conduction level of Q1 which establishes the desired load voltage. If load voltage increases, due to either an increase in source voltage or a decrease in load current, the negative potential with respect to point A at the wiper arm of R4 will increase, increasing the forward bias of Q2. This increase in forward bias causes Q2's conduction level to increase and collector voltage to decrease, reducing the negative potential at the base of Q1 with respect to point A. This decrease in negative potential causes the forward bias on Q1 to drop, decreasing current flow through the series network of Q1 and R_L, and returning the output voltage to its preselected level. A decrease in load voltage will produce the opposite effects of those outlined above.

Electron Tube Shunt Detected
Series Voltage Regulator

A schematic diagram of an electron tube shunt detected series voltage regulator is shown in figure 8-9. V1 performs as a series regulator and the control circuit consists of V2 and its associated circuitry. The plate of V3, a VR tube, is connected to $+E_{bb}$ through R2. This allows the VR tube to ionize, keeping the cathode of V2 at a fixed potential with respect to ground, when the input voltage is applied. A positive voltage with respect to ground is tapped from R4 and applied to the grid of V2, controlling the tube's bias. R1, plate load resistor for V2, establishes the grid bias level of V1. Since V2's plate is directly connected to the grid of V1, the amount of current flow through V2 determines the conduction level of V1.

Once R4 is initially set, any change occurring in source voltage or load voltage will cause a change in the potential across R4, changing the bias of V2. Assume a decrease in the source voltage. This will cause a momentary

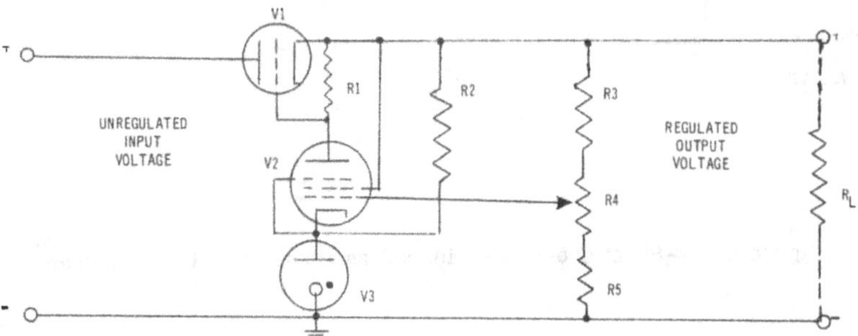

179.182
Figure 8-9.— Electron tube shunt detected series voltage regulator.

decrease in output voltage and reduce the potential at the wiper arm of R4, increasing bias of V2. This increase in bias will cause V2 to reduce conduction, decreasing the drop across R1. The decreased drop across R1 causes bias on V1 to decrease, reducing its internal impedance. With less impedance, the voltage drop across V1 decreases, counteracting the decrease in load voltage. An increase in load voltage would cause the opposite effects throughout the circuit.

A pentode is used in the control circuit to take advantage of its high amplification factor. Even rapid variations in load voltage will be amplified sufficiently by V2 to change the conduction level of V1. A pentode will give a better percentage of regulation and a faster response than a triode. There are many variations and modifications of the basic shunt detected series voltage regulator. A few examples will be discussed in the following paragraphs.

Variations In Basic Shunt Detected Series Voltage Regulators

Since total load current flows through the series regulator, this device must be capable of passing a high value of current. Several transistors or electron tubes may be connected in parallel if the current capability of a single device is not sufficient. This configuration is illustrated in figure 8-10.

Another modification is to insert an additional d.c. amplifier as shown in figure 8-11. The amplifier detector stage performs the same

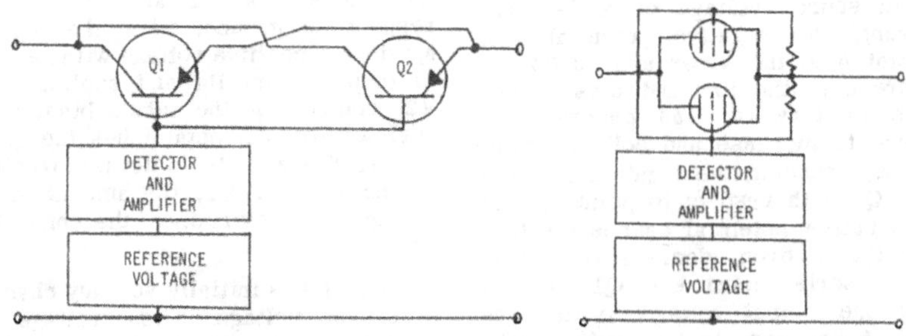

179.183
Figure 8-10.— Parallel shunt detected series voltage regulators.

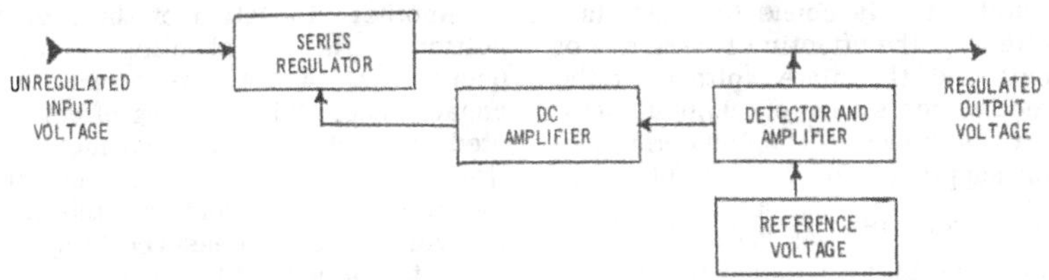

179.517
Figure 8-11.—Additional d.c. amplifier.

function as in the basic configuration, but now the output is fed to the second d.c. amplifier. This increased amplification provides greater sensitivity, thereby improving percent of regulation and speed of response.

If a stable voltage source is available, such as another regulated supply, the zener diode or VR tube can be eliminated. The reference voltage would then be provided by the constant voltage source.

CURRENT REGULATOR

Certain equipments require a stable current supply. Figure 8-12 illustrates a simple solid state series current regulator. This circuit will supply a constant value of current under varying conditions. It may be noted that operation is similar to a simple series voltage regulator.

Regulation occurs in the following manner. If current increases, the voltage across R2 will increase with respect to ground. This increase will reduce the forward bias of Q1, increasing the transistor's internal impedance, thus reducing current flow to the predetermined level.

ELECTRONIC MULTIMETER

The multimeter combines a voltmeter, ammeter, and ohmmeter in one unit. Nonelectronic type multimeters are discussed in chapter 3. The ammeter and ohmmeter sections of the electronic multimeter are similar to the corresponding sections of the nonelectronic multimeter. The voltmeter section of the electronic multimeter however, uses an electron tube. It is much more sensitive than the nonelectronic type, and will give an accurate reading without loading down the unit being tested.

The ordinary voltmeter has several disadvantages that make it practically useless for measuring voltages in high-impedance circuits. For example, suppose that the plate voltage of a pentode amplifier is to be measured. When the meter is connected between the plate of the electron tube and ground, the meter current constitutes an appreciable part of the total current through the plate load resistor. Because of the shunting effect of the meter on the pentode, the plate voltage decreases as the current through the plate load resistor increases. As a result, an incorrect indication of plate voltage is obtained.

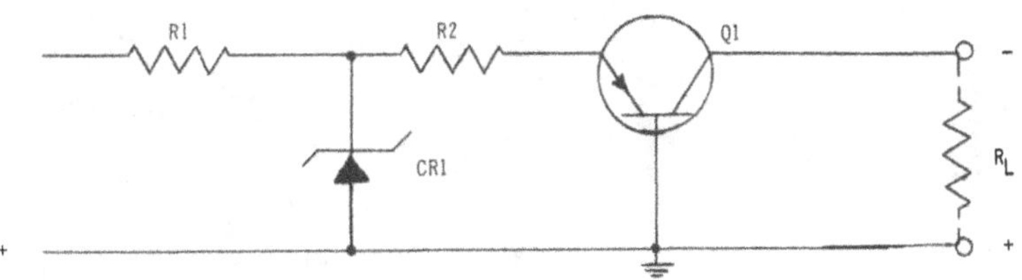

179.518
Figure 8-12.—Series current regulator.

Before the voltmeter is connected, the plate current is limited by the effective resistance of the plate circuit and the plate voltage. If the tube has an effective resistance of 100,000 ohms, the plate load a resistance of 100,000 ohms, and the plate power supply is constant at 200 volts, then the plate current is $\frac{200}{200,000}$, or 0.001 ampere. The plate voltage is 0.001 x 100,000, or 100 volts.

Assume that the voltmeter used to measure the plate voltage of the tube has a sensitivity of 1,000 ohms per volt and that the range is from 0 to 250 volts. The meter will then have a resistance of 250,000 ohms. This resistance in parallel with the tube resistance of 100,000 ohms produces an effective resistance of 71,400 ohms in series with the plate load resistor. The total resistance across the plate supply is therefore 171,400 ohms and the current through the plate load resistor is $\frac{200}{171,400}$, or about 0.00117 ampere. Across the plate load resistor the voltage drop is 0.00117 x 100,000, or 117 volts and the plate-to-ground voltage on the tube is 200-117, or 83 volts when the meter is connected, thus causing an error of 17 percent. The lower the sensitivity of the meter the greater this error will be.

A meter having a sensivitity of 20,000 ohms per volt and a 250-volt maximum scale reading would introduce an error of about 1 percent. However, in circuits where very high impedances are encountered, such as in grid circuits of electron tubes, even a meter of this sensitivity would impose too much of a load on the circuit.

Another limitation of the a.c. rectifier type voltmeter is the shunting effect at high frequencies of the relatively large meter rectifier capacitance. This shunting effect may be eliminated by replacing the usual metallic oxide rectifier with an electron tube amplifier in which the plate circuit contains the d.c. meter, and the voltage to be measured is applied to the grid circuit. Voltages at frequencies up to and greater than 100 megahertz can be measured accurately with this type of meter. THE INPUT IMPEDANCE IS LARGE, and therefore the current drawn from the circuit whose voltage is being measured is small and in most cases, negligible. Simplified diagrams of the a.c. and d.c. electron tube voltmeter sections of a representative electronic multimeter are shown in figure 8-13. The a.c. voltage to be measured is applied to the a.c. probe (fig. 8-13A). It is rectified by V1 and filtered by the RC network in the probe.

The meter circuit is a balanced bridge network. Diode V1 (fig. 8-13A) causes a contact potential to be established across the voltage divider network connected to the grid of V2A. This voltage would unbalance the bridge. Therefore, a similar contact potential is introduced across the grid of V2B from V3 and its associated voltage divider to balance the bridge before the a.c. voltage to be measured is applied to the diode probe.

When the input voltage between the probe and ground is zero, the bridge is balanced and the voltages across the two arms containing the plate load resistors of V2 are equal. Thus, the d.c. meter indicates zero. If a voltage is applied between the probe and ground, the bridge becomes unbalanced, and current flows through

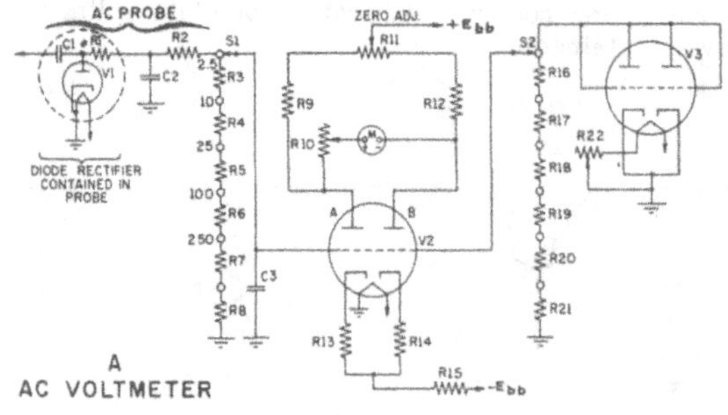

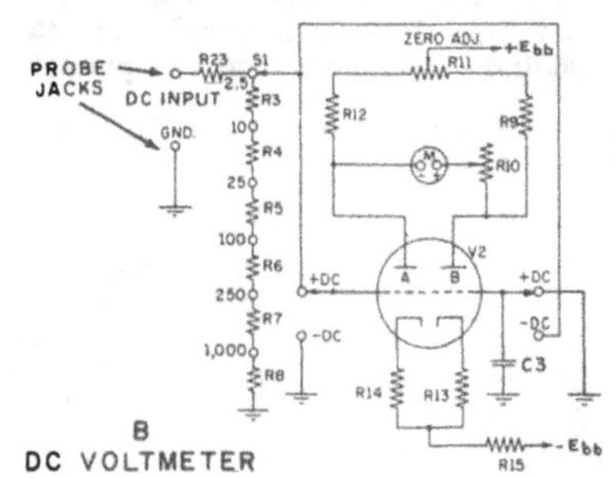

20.349
Figure 8-13.— Electron tube voltmeter.

the meter. The meter is calibrated in rms volts.

Bias is obtained for V2A and B through the voltage drop across R13, R14, and R15. The cathodes are positive with respect to B- by an amount equal to the bias. Thus the grids are correspondingly negative with respect to the cathodes. Adjustable resistor R10 is used for calibration purposes.

The input impedance is very high. At the lower frequencies, the input capacitance is negligible, but as the frequency increases the input capacitance introduces an additional load on the circuit under test and causes an error in the meter reading.

The d.c. electron tube voltmeter circuit is shown in figure 8-13B. The d.c. voltage to be measured is applied between the d.c. input terminal and ground. The d.c. input voltage is therefore applied through R23 to the divider network feeding the grid of V2A. The grid of V2B is grounded. The meter is connected across a normally balanced bridge and causes the meter to deflect. The calibration is in d.c. volts.

In figure 8-13B, no diode is used in the probe, hence, no contact potential is established so that V3 and its associated voltage divider network (fig. 8-13A) are omitted from the circuit. The bias and ZERO ADJ. circuits are the same as for the a.c. voltmeter.

DIGITAL VOLTMETER

The digital voltmeter measures voltages and presents the measured value directly in numerical form on a dial. It is essentially a self-balancing potentiometer.

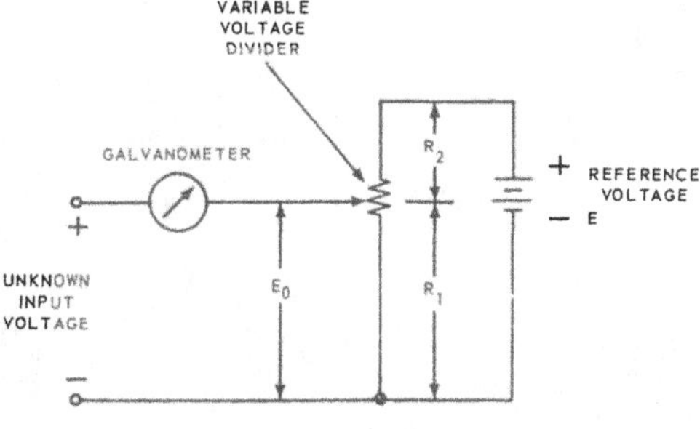

151.105
Figure 8-14.—Basic potentiometer circuit.

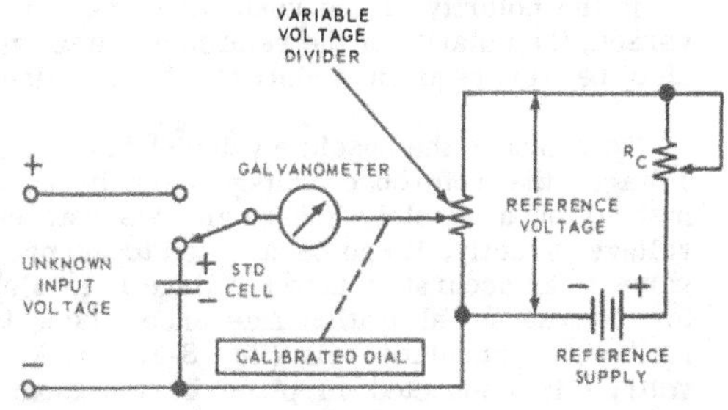

151.106
Figure 8-15.—Potentiometer with calibration rheostat.

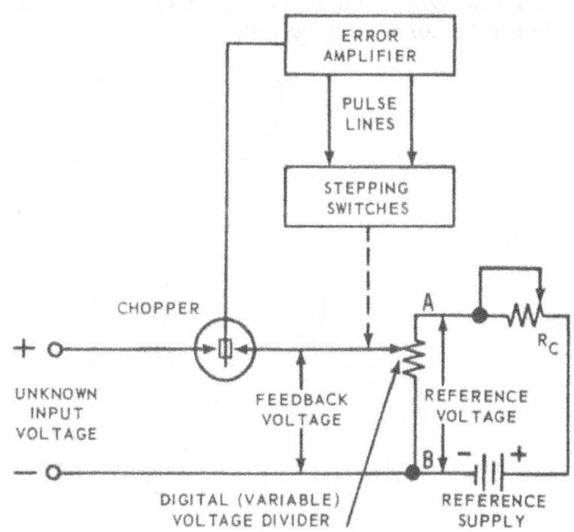

151.107
Figure 8-16.—Single-range digital voltmeter.

To understand the operation of a voltmeter of this type, consider the basic circuit shown schematically in figure 8-14. If an unknown voltage is applied at the input terminals and a variable divider adjusted until the voltage E_0 is equal in magnitude to the unknown voltage, zero current will flow through the galvanometer. This condition is called balance. Note that in the balanced condition no current is drawn from the input (circuit under test). The unknown voltage can be computed from the formula:

$$\text{Unknown voltage} = \frac{R_1}{R_1 + R_2} \times \text{reference voltage.}$$

If the polarity of the unknown voltage is reversed, the polarity of the reference voltage must also be reversed in order to obtain balance.

To measure the absolute value of the unknown voltage, the reference voltage must be known. Instead of accurately measuring the reference voltage directly, it can be adjusted to the proper value if an accurately known voltage is available for use as a calibration reference. Using this method, a standard cell (fig. 8-15) of known voltage is connected in place of the unknown voltage. The variable voltage divider dial reading (not shown) is set to this known voltage, and the voltage across the variable voltage divider is adjusted by means of rheostat, R_C, to obtain balance. The potentiometer is now calibrated, or standardized, and ready for use as an absolute voltage-measuring device.

The digital voltmeter is essentially an extension of the basic principles of a self-balancing digital potentiometer as described above. A block diagram of a single-range digital voltmeter is shown in figure 8-16. A stepping switch operated variable voltage divider divides the reference voltage into a large number of precisely equal parts in the decimal number system. An error amplifier compares this divided voltage, called the feedback voltage, to the input voltage and commands stepping switch motion in a logical sequence until the two voltages are equal within the limits of resolution of the digital voltmeter. When this occurs, stepping switch motion ceases and the meter is balanced. Input to the error amplifier is through an electromechanical chopper. Automatic range and polarity changing features are accomplished by a combination range and polarity switch (not shown).

CHAPTER 9

FUNDAMENTALS OF COMMUNICATION THEORY

Radio communications is the nerve center of today's modern Navy. This fact makes necessary devices for sending and receiving intelligence which are rapid, accurate and dependable. The word "radio" can be defined briefly as the transmission and reception of signals through space by means of electromagnetic waves.

Of the several methods of radio communications available those utilized most commonly by the Navy are radiotelegraphy, radiotelephony, radioteletype, and radiofacsimile. These modes are defined as follows:

1. RADIOTELEGRAPHY (CW): The transmission of intelligence coded radio frequency waves in the form of short transmissions (dots) and long transmissions (dashes).
2. RADIOTELEPHONY (AM): The transmission of sound intelligence (voice, music, or tones) by means of radio frequency waves.
3. RADIOTELETYPE (FSK): The transmission of messages from a teletypewriter or coded tape over a radio frequency channel by means of coded combinations of mark and space impulses.
4. RADIOFACSIMILE (FAX): The transmission of still images (weather maps, photographs, sketches, typewritten pages, and the like) over a radio frequency channel.

Radio equipment can be divided into two broad categories, transmitting equipment and receiving equipment. Both transmitting and receiving equipments consist basically of electronic power supplies, amplifiers, and oscillators.

A basic radio communication system may consist of only a transmitter and a receiver, which are connected by the medium through which the electromagnetic waves travel (fig. 9-1). The transmitter comprises an oscillator (which generates a basic radio frequency), RF amplifiers, and the stages required to place the audio intelligence on the RF signal (modulator).

The electromagnetic variations are propagated through the medium (space) from the transmitting antenna to the receiving antenna. The receiving antenna converts that portion of the transmitted electromagnetic energy received by the antenna into a flow of alternating radio frequency currents. The receiver converts these current variations into the intelligence that is contained in the transmission.

FREQUENCY SPECTRUM

Radio transmitters operate on frequencies ranging from 10,000 hertz to several thousand megahertz. These are divided into eight bands as shown in table 1-1, in chapter 1, and explained in chapter 20.

Because the VLF and LF bands require great power and long antennas for efficient transmission, the Navy normally uses these bands for shore station transmission. (The antenna length varies inversely with frequency.) Only the upper and lower ends of the MF band have naval use because the commercial broadcast band extends from about 550 kHz to 1700 kHz.

Most shipboard radio communications are conducted in the HF band. Consequently a large percentage of shipboard transmitters and receivers are designed to operate in this band.

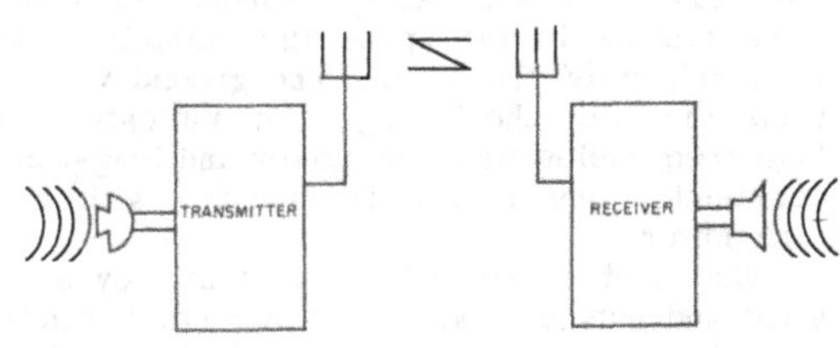

31.6

Figure 9-1.—Basic radio communication system.

The HF band lends itself well to long-range communications.

A large portion of the lower end of the VHF band is assigned to the commercial television industry and is used by the Armed Forces only in special instances. The upper portion of the VHF band (225 to 300 MHz) and the lower portion of the UHF band (300 to 400 MHz) are used extensively by the Navy for its UHF communications. The frequencies above 400 MHz in the UHF band through the SHF and EHF bands are normally used for radar and special equipment.

ANTENNAS AND PROPAGATION

Before we proceed into transmitters and receivers a basic concept of antennas and propagation will be presented.

An antenna is a conductor or system of conductors that radiates or intercepts energy in the form of electromagnetic waves. In its elementary form, an antenna may be simply a length of elevated wire. For communications work, however, other considerations make the design of an antenna system a more complex problem. For instance, the height of the radiator above ground, the conductivity of the earth below the radiator, and the shape and dimensions of an antenna all affect the radiated field pattern in space.

When RF current flows through a transmitting antenna, radio waves are radiated from the antenna. Part of each radio wave moves outward in contact with the ground to form the groundwave, and the rest of the wave moves outward and upward to form the skywave. The ground and sky portions of the radio wave are responsible for two different methods of carrying signals from transmitter to receiver.

Commonly, the groundwave is considered to be made up of two parts, a surface wave and a direct wave. The surface wave travels along the surface of the earth, whereas the direct wave travels in the space immediately above the surface of the earth. The ground wave is used for both short-range communications at high frequencies with low power and long-range communications at low frequencies with very high power.

That part of the radio wave that moves upward and outward, but is not in contact with the ground, is called the skywave. The ionosphere refracts (bends) some of the energy of the skywave back toward the earth. A receiver in the vicinity of the returning skywave receives strong signals even though the receiver is several hundred miles beyond the range of the ground wave. The skywave is used for long-range, high-frequency, daylight communications. It also provides a means for long-range contacts at somewhat lower frequencies. Figure 9-2 shows the components of the propagated wave.

The direct wave is that portion of radiated energy which contains no sky or ground wave components. It attempts to travel in a straight line; however, it is refracted (bent) slightly downward due to atmospheric density. All VHF and UHF communications are conducted via the direct wave.

AM TRANSMITTER

One of the simplest types of radio transmitters is the amplitude modulated transmitter. This AM transmitter is designed to send audio intelligence. An AM transmitter has four essential components: (1) a generator of RF oscillations (2) a means of amplifying, and, if necessary, multiplying the frequencies of these oscillations, (3) a method of modulating the RF carrier with the audio intelligence, and (4) a power supply to provide the operating potential to the various transistors and/or electron tubes.

In order to explain the operation of complex electronic equipment, it is often necessary to graphically separate the major functional units into functional blocks. This type of diagram is often referred to as a functional block diagram. An equipment functional block diagram is an overall symbolic representation of the functional units within an equipment and the signal and power paths between them.

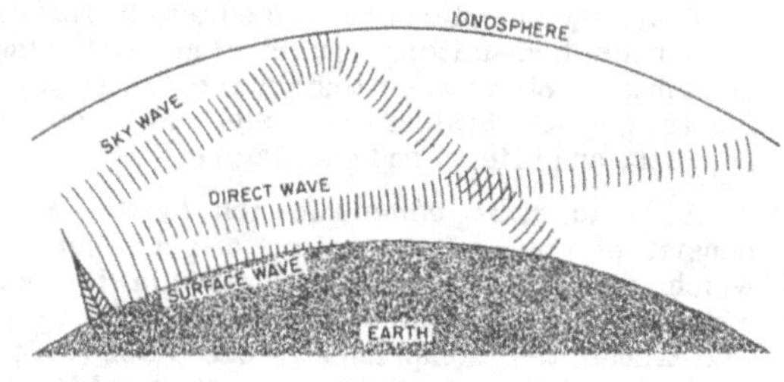

Figure 9-2.— Propagated wave paths.

A block diagram of a basic transmitter used for amplitude modulated transmission is illustrated in figure 9-3. The signal paths are represented by light lines, and the power supply paths to individual functional units are shown by heavy lines. The oscillator is the basic frequency determining element of the transmitter. It is here that the RF signal is generated. If the oscillator fails to function, no RF signals will be produced.

Frequently the transmitter output frequency is so high that it is difficult to maintain a stable oscillator. To overcome this difficulty the oscillator is operated on a submultiple of the transmitters output frequency. A process called frequency multiplication is used to increase the transmitter frequency.

Present-day transmitters may contain several oscillators to perform various functions. In general, only one of these is used to generate the basic transmitter radio frequency. This oscillator usually is called the master oscillator (MO) to distinguish it from any other oscillator in the transmitter.

The power amplifier (PA) is operated in such a manner that it greatly increases the magnitude of the RF current and voltage. The output of the modulator is coupled to the power amplifier to produce the desired AM signal. The output from the PA is fed to the antenna via RF transformers and transmission lines. The modulator receives the intelligence to be transmitted. This normally weak signal is amplified to a level sufficient to modulate the RF in the power amplifier.

Transmitters require d.c. voltages ranging from hundreds of volts negative to thousands of volts positive. Additionally, they need a.c. voltages at smaller values than those available from the ship's normal power source. It is the function of the power supply to furnish these voltages at the necessary current ratings.

SUPERHETERODYNE RECEIVER

The modulated RF carrier wave produced at the transmitter travels through space as an electromagnetic wave. When the wave passes across a receiving antenna, it induces small RF voltages (and associated currents) in the antenna wire at the frequency of the transmitted signal. The signal voltage is coupled to the receiver input via the antenna coil or antenna transformer.

Electromagnetic energy is received from many transmitters simultaneously by the receiving antenna. The receiving circuits must select the desired transmitted signal from those present at the antenna and amplify this signal. The RF stages must isolate the internally generated frequencies of the receiver from the antenna to prevent radiation of these signals particularly while radio silence is being observed. Further, the receiver must extract the audio component from the carrier frequency by a process called demodulation, or detection, and amplify the audio component to the proper magnitude to operate a loudspeaker or earphones.

Figure 9-4 illustrates a block diagram of a superheterodyne receiver (discussed in more detail in chapter 11). The RF amplifiers increase the strength of the signal. The output of the RF amplifier is fed to the mixer stage. Another input into the mixer is a locally generated, unmodulated RF signal of constant amplitude from the local oscillator. The intermediate frequency is produced by a process called heterodyning. This action takes place in the mixer, so called because it receives and combines (mixes) these two frequencies.

The heterodyning action in the mixer produces four frequencies at the mixer output. These frequencies are: (1) the incoming RF signal, (2) the local oscillator signal, (3) the sum of the incoming signal and local oscillator signal, and (4)

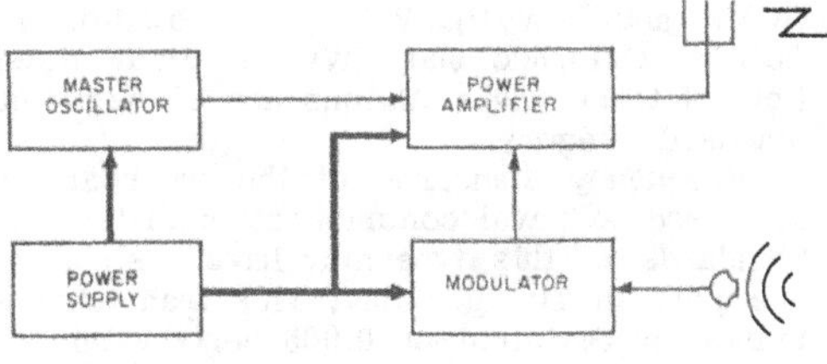

13.53(20C)
Figure 9-3.—Basic AM transmitter.

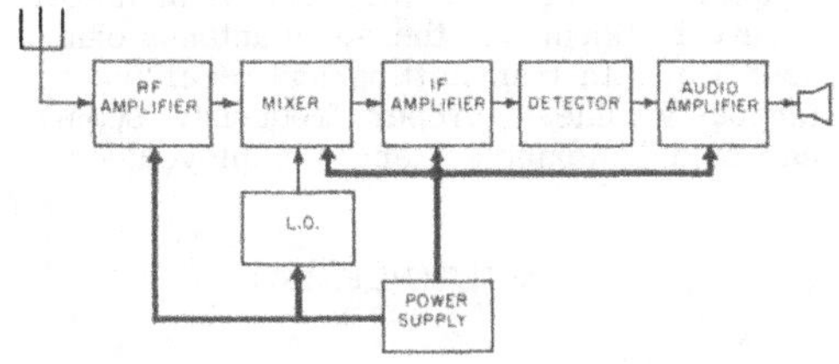

13.66(20C)
Figure 9-4.—Basic superheterodyne receiver.

the difference of these signals. Both the sum and the difference frequencies contain the amplitude modulation. Usually, the difference frequency is used as the intermediate frequency, although the sum frequency can be used.

In the detector stage, the intelligence component of the modulated wave is separated from the RF carrier. The separation process is called detection or demodulation. The function of the audio frequency section of the receiver is to amplify the audio signal from the detector. In most instances, the amount of audio amplification necessary depends on the type of reproducer.

A transmitter-receiver comprises a separate transmitter and receiver mounted in the same rack or cabinet. The same antenna may be used for the transmitter-receiver arrangement. When so used, the capability for simultaneous operation of both the transmitting and receiving equipment does not exist. The equipments may be operated independently, using separate antennas.

A transceiver is a combined transmitter and receiver in one unit which uses switching arrangements in order to utilize parts of the same electronic circuitry for both transmitting and receiving. Hence, a transceiver cannot transmit and receive simultaneously.

ANTENNA MULTICOUPLERS

Because of the large number of transmitters and receivers on board ships, it is not feasible to use a separate antenna for each equipment. One satisfactory approach to the problem is provided by multicouplers.

Antenna multicouplers are devices that permit the simultaneous operation of several transmitters or receivers into (or from) the same antenna. The term "multicoupler" is descriptive of two or more couplers stacked or grouped together to from a single equipment, which then is connected to a broadband antenna. A separate coupler is required for each transmitter or receiver. Normally, the same antenna cannot be used for both transmitting and receiving simultaneously unless proper frequency separation and/or a duplexing system is employed.

MULTIPLEXING

The number of communication networks in operation per unit of time throughout any given area is increasing constantly. Not too long ago, each network was required to operate on a different frequency. As a result, all areas of the radio frequency spectrum had become highly congested.

The maximum permissible number of intelligible transmissions taking place in the radio spectrum per unit of time can be increased through the use of multiplexing. The main purpose of a multiplex system is to increase the message-handling capacity of radio communication, or teletypewriter channels and the transmitters and receivers associated with them. This increase in capacity is accomplished by the simultaneous transmission of several messages over a common channel.

There are basically two ways this is accomplished, frequency division and time division. The frequency division multiplexing telegraph terminal employs a number of tone channels slightly displaced in frequency. The time division multiplexing telegraph terminal employs a number of time sharing channels slightly displaced in time. Each channel carries the signals from a separate teletypewriter circuit and modulates a common carrier frequency. Receiving equipment at a distant station accepts the multiplex signals, converts them to mark-space signals, and distributes them in the proper order to a corresponding number of circuits. Frequency division is the most common and widely used multiplexing.

FREQUENCY STANDARDS

In modern communication networks, the frequency of operation is precise and exacting. To keep within the close tolerances required of the equipment it is necessary to check it against a known frequency standard. Frequency standards belong to two general categories, primary and secondary.

The National Bureau of Standards provides frequency standards from two radio stations (WWV and WWVH). WWV is located in Fort Collins, Colorado and WWVH in Maui, Hawaii. Both of these radio stations are of the primary standard category.

Frequency standards of the secondary type are used at naval communication installations. Standards of this type may have a stability of one part in 10^9 per day. This translates to a maximum deviation of 0.005 hertz within a 24 hour period for a frequency of 5 MHz.

Frequency standards are highly stable and accurate references against which other signal

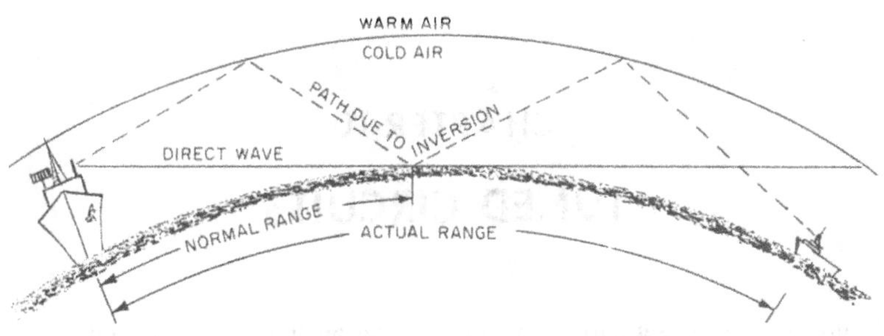

179.185

Figure 9-5.—Tropospheric ducting.

sources may be compared. The actual comparison may be performed by a technician using a frequency deviation meter. The comparison may also be performed automatically by a comparison circuit built into the equipment being serviced. This does not mean that the necessary adjustment is accomplished automatically. The point to remember is that the standard is a reference.

ENVIRONMENTAL EFFECTS ON WAVE PROPAGATION

In modern communication systems, both weather and atmospheric conditions will affect the RF signal between the transmitter and receiver. The effect in some cases may improve reception, and in other cases reception will be hampered. At the receiver the energy may arrive over a number of paths. The signal from each path has a different attenuation, phase and polarization. This multipath effect leads to variations in the net received signal strength and is referred to as FADING.

To obtain a steady signal we use a system called DIVERSITY RECEPTION. Diversity reception is reception in which the effects of fading are minimized by combining two or more sources of signal energy carrying the same modulation. To obtain and combine these sources of signal energy we may use one or both of the following methods.

SPACE DIVERSITY, which takes advantage of the fact that fading does not occur simultaneously for antennas spaced several wavelengths apart, involves the use of two or more antennas feeding individual receivers whose outputs are combined. This gives an essentially constant output despite fading.

FREQUENCY DIVERSITY takes advantage of the fact that signals differing slightly in frequency do not fade simultaneously. It involves the use of carrier frequencies separated 500 Hz or more and having the same modulation. The receiver minimizes the effects of fading by using at each instant the frequency having the higher signal strength.

Another condition which affects the transmitted signal is TEMPERATURE INVERSION. A temperature inversion is caused by a region in the trophosphere where temperature increases rather than decreases with altitude. Temperature inversions cause abnormal refractions that make the beam refract up and down between the two air layers forming a duct. When this condition exists, microwave signals may travel several times the normal limit. The duct formed by temperature inversion is called a TROPHOSPHERIC DUCT. Tropospheric ducting is shown in figure 9-5.

Wave propagation is discussed further in chapter 29.

CHAPTER 10

TUNED CIRCUITS

A tuned circuit consists essentially of a coil and capacitor, connected either in series or in parallel (fig. 10-1A & B). The resistance in the circuit is usually limited to the inherent resistance of the circuit components (particularly the coil).

The action of coils and capacitors in a.c. circuits is explained in Basic Electricity, NavPers 10086 (Revised), and is the basis of the following discussion of their action in tuned circuits.

Before tuned circuits can be analyzed, an elementary understanding of vectors and vector algebra is required. Accordingly, a brief review of vectors as they are expressed both in the rectangular and the polar form follows.

EXPRESSING VECTORS ALGEBRAICALLY

Many common physical quantities such as temperature, the speed of a moving object, or the displacement of a ship can be expressed as a certain number of units. These units define only the magnitude and give no indications of the direction in which the quantity acts. Such quantities are called SCALAR quantities. If both the magnitude and the direction in which the quantity acts are indicated, it is called a VECTOR quantity and may be represented by a vector. For example, a vector representing the speed and heading of a ship having a speed of 10 knots and a heading of 45° (northeast) is a straight line extending upward and to the right. The length of the line is proportional to the speed of 10 knots. The angle that the line makes with the vertical (north at the top) is 45° clockwise from the vertical.

Electrical vectors are commonly used to represent a.c. currents and voltages and their phase relations. The length of the vector represents the magnitude of the quantity involved, and the direction of the vector, with respect to a reference axis, represents the lapse in time between the positive maximum values of current and voltage.

Impedance triangles, the sides of which represent vector quantities, are also used to represent the resistance and reactance components of a.c. circuits. These are right triangles having a base equal to the resistive component, an altitude equal to the reactive component, and a hypotenuse

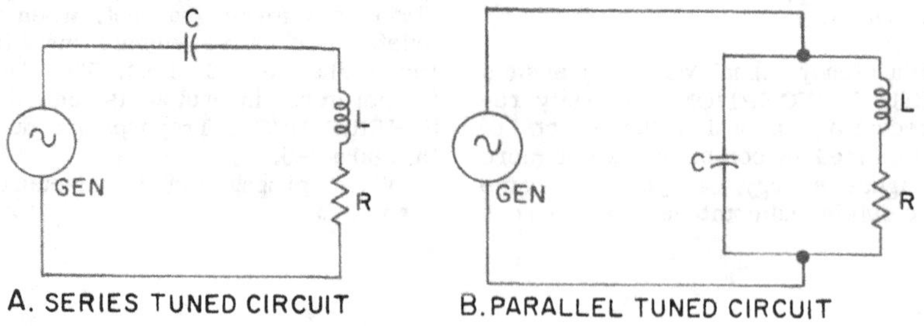

Figure 10-1.—Basic tuned circuits.

equal to the combined impedance. The angle between the combined impedance and the resistive component (hypotenuse and base) is equal to the phase angle between the voltage across the impedance and the current flowing through it.

In this chapter it is necessary to determine circuit impedance by the addition, subtraction, multiplication, and division of vector quantities. When it is inconvenient to express the quantity by simple algebra, a system of complex notation is used.

OPERATOR J

In calculations in electronics it is often necessary to perform operations involving the square root of a negative number — for example, $\sqrt{-9}$, $\sqrt{-5}$, and $\sqrt{-x}$. Because no number when multiplied by itself will produce a negative result, the roots of numbers such as the foregoing cannot be extracted. It therefore becomes necessary to introduce a new type of notation to indicate the square root of a negative number. These numbers are called IMAGINARY NUMBERS to distinguish them from the so-called REAL NUMBERS. Actually, the numbers that we call imaginary in the mathematical sense are real in the physical sense. The term is merely one of convenience, as will be pointed out in the succeeding paragraphs.

In algebra, the foregoing quantities are treated as $\sqrt{-1}\sqrt{9}$, or $\sqrt{-1}(\pm 3)$, $\sqrt{-1}\sqrt{5}$; and $\sqrt{-1}\sqrt{x}$. The term, $\sqrt{-1}$, is expressed as i (for imaginary) in mathematics books, but when working with electrical circuits it is convenient to use the term j (called the J OPERATOR), because i is used to indicate instantaneous current value.

GRAPHICAL REPRESENTATION

In order to present a quantity graphically, some system of coordinates must be employed. Quantities involving the j operator may be conveniently expressed by the use of RECTANGULAR COORDINATES, as shown in figure 10-2. In order to specify a vector in terms of its X

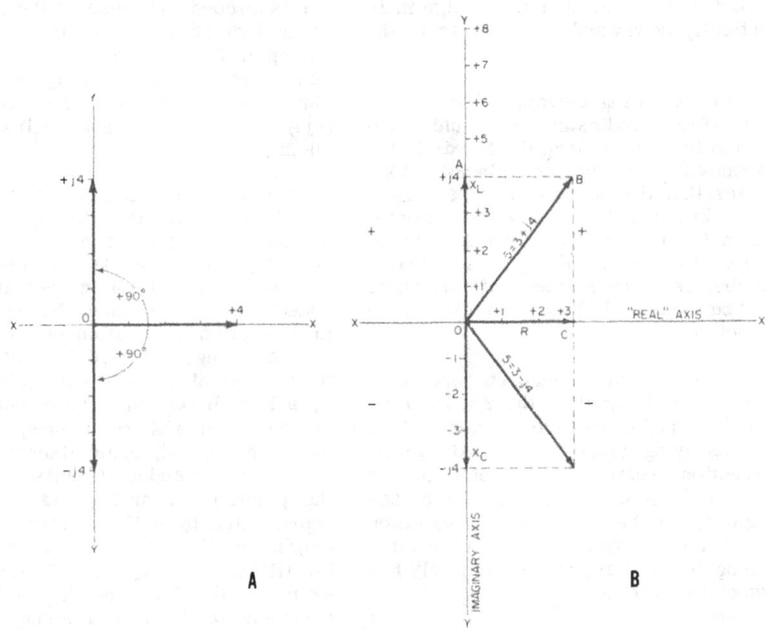

Figure 10-2. — Coordinates showing real and imaginary axes.

and Y components, some means must be employed to distinguish between X axis and Y axis projections. Because the +Y axis projection is +90° from the +X axis projection, a convenient operator is one that will, when applied to a vector, rotate it without altering the magnitude of the vector. Let +j be such an operator that produces 90° COUNTERCLOCKWISE rotation of any vector to which it is applied as a multiplying factor. Also, let -j be such operator that produces 90° CLOCKWISE rotation of any vector to which it is applied as a multiplying factor.

Successive applications of the operator +j to a vector will produce successive 90° steps of rotation of the vector in the counterclockwise direction without affecting the magnitude of the vector. Likewise, successive applications of the operator -j will produce successive 90° steps of rotation in the clockwise direction. This rotation is shown in table 10-1.

In the four quadrants (fig. 10-2) (upper right, upper left, lower left, and lower right) the signs indicate the direction of the vertical (j) component. The + sign indicates a vertically upward direction from the X axis and the - sign indicates a vertically downward direction from the X axis.

Consider the following example: The number +4 in figure 10-2A, indicates that 4 units are measured from the origin along the X axis in the positive direction. A +j operator placed before the 4 indicates that the number is to be rotated 90° counterclockwise and will now be measured along the Y axis in a positive direction. Likewise, a -j operator placed before the 4 indicates that the number is to be rotated 90° clockwise, and will now be measured along the Y axis in the negative direction.

It may be recalled that inductive reactance, X_L, is indicated as lying along the Y axis in the positive direction, and capacitive reactance, X_C, is indicated as lying along the Y axis in the negative direction; resistance in each case is measured along the X axis in the positive direction. Therefore, +j has a direct association with X_L in that both are measured in the same direction along the Y axis, and -j similarly has a direct association with X_C.

The function of the j operator may be shown as follows: The expression, 4 ohms, indicates that pure resistance is involved. In order to

Table 10-1.—Relation of operation j to vector rotation

Operator	Mathematical equivalent	Direction of rotation	Degree of rotation
j	$\sqrt{-1}$	Counter-clockwise	90°
j^2	-1	"	180°
j^3	$-\sqrt{-1}$	"	270°
j^4	1	"	360°
-j	$-\sqrt{-1}$	Clockwise	-90°
$(-j)^2$ . .	-1	"	-180°
$(-j)^3$ . .	$\sqrt{-1}$	"	-270°
$(-j)^4$ . .	1	"	-360°

179.187

indicate that the 4 ohms represent capacitive reactance or inductive reactance a special symbol is needed. The use of the j operator gives a clear indication of the type of reactance. For example, if the j operator is not used, the 4 ohms are resistive. If +j is used (+j4), the 4 ohms are inductive reactance. If -j is used (-j4), the 4 ohms are capacitive reactance (fig. 10-2B).

The so-called COMPLEX NUMBER contains the "real" and the "imaginary" terms connected by a plus or a minus sign. Thus, 3+j4 and 3-j4 are complex numbers. This means that the 3 and the 4 in each instance are to be added vectorially, and the +j and -j indicate the direction of rotation of the vector following it. Following figure 10-2B, the real number in these examples is 3 and could be represented by a line drawn three units out from the origin on the positive X (resistance) axis. The imaginary number, +j4, could likewise be represented by a line extended 4 units from the origin on the positive Y, or X_L, axis; and -j4 could be represented by a line extended 4 units from the origin on the negative Y, or X_C, axis. The IMAGINARY, or QUADRATURE, quantities (for example, the X_L and X_C values) are always assumed to be drawn along the Y axis, and the REAL quantities (for example, the R values) are always assumed to be drawn along the X axis.

ADDITION AND SUBTRACTION OF COMPLEX NUMBERS

Values that are at right angles to each other cannot be added or subtracted in the usual sense of the word. Their sum or difference can only be indicated, as is done in the case of binomials (an expression involving two terms). Thus, assume that it is desired to add 3+j4 to 3-j4.

$$\begin{array}{r} 3+j4 \\ \underline{3-j4} \\ 6-0 \end{array}$$

The imaginary term disappears, and only the real term, 6, remains. If 3+j4 is added to 3+j4, the sum is the complex quantity, 6+j8.

One complex expression may also be subtracted from another complex expression in the same manner that binomials are treated. For example, 3-j2 may be subtracted from 3+j4 as

$$\begin{array}{r} 3+j4 \\ (-)\ \underline{3-j2} \\ 0+j6 \end{array}$$

The real term disappears, and the result is 6 units measured upward from the origin on the Y axis. If 3-j2 is subtracted from 6+j4, the difference is the complex quantity, 3+j6.

MULTIPLICATION AND DIVISION OF COMPLEX NUMBERS

Complex numbers are multiplied the same way that binomials are multiplies—for example, if 3-j2 is multiplied by 6+j3

$$\begin{array}{r} 3-j2 \\ \underline{6+j3} \\ 18-j12 \\ \underline{+j9-j^2 6} \\ 18-j3-j^2 6 \end{array}$$

Because $j^2 = -1$, the product becomes 18-j3-(-1)6, 18-j3+6 or 24-j3.

Complex numbers may be divided in the same way that binominals containing a radical in the denominator are divided. The denominator is rationalized (multiplied by its conjugate—a term that is the same as the denominator except that it has the opposite algebraic sign before the j term), and the quotient is expressed as a term having only a real number as the divisor. For example, if 4+j3 is divided by 2-j2,

$$\frac{4+j3}{2-j2} = \frac{(4+j3)\ (2+j2)}{(2-j2)\ (2+j2)}$$

$$= \frac{8+j6 + j8 + j^2 6}{4-j4 + j4 - j^2 4}$$

$$= \frac{8+j14-6}{4 + 4}$$

$$= \frac{2 + j14}{8}$$

$$= \frac{1 + j7}{4}$$

$$= \frac{1}{4} + \frac{j7}{4}$$

$$= 0.25 + j1.75$$

RECTANGULAR AND POLAR FORMS

Sometimes it is more convenient to use polar coordinates than rectangular coordinates. In RECTANGULAR FORM the vector is described in terms of the two sides of a right triangle, the hypotenuse of which is the vector. Thus, in figure 10-3, vector OB (from fig. 10-2B) is described in rectangular form by the complex number 3+j4. In other words, the end of the vector, OB, is 3 units along the +X axis and 4 units along the +Y axis and its length is 5 units.

The vector, OB, may also be described if its length and the angle, θ, are given. When a vector is described by means of its magnitude and the angle it makes with the reference line it is expressed in the POLAR FORM. In this instance the length is 5 units and the angle, θ, is approximately 53.1°. The vector, OB, may then be expressed in the polar form as $5\angle +53.1°$. If the rectangular form is 3-j4 (fig. 10-2B), the polar form is $5\angle -53.1°$.

The plus sign is shown with positive angles in this chapter in order to emphasize positive angles as contrasted with negative angles. The negative sign preceding the angle indicates clockwise rotation of the vector from the zero position.

CONVERTING FROM ONE FORM TO ANOTHER

Assume that the rectangular form is expressed by the complex number, 3+j4. The

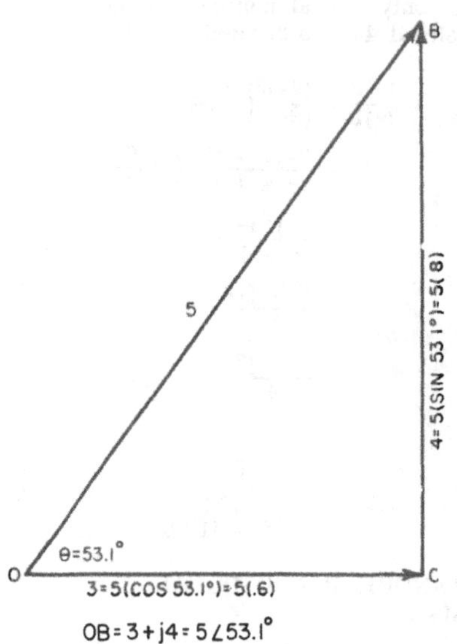

20.62
Figure 10-3.—Rectangular and polar forms.

angle, θ, and the actual length of the vector, OB, are not given. The length, OB, can be determined by use of the Pythagorean theorem ($OB = \sqrt{3^2 + 4^2}$), but it is usually simpler to determine first the angle, θ, by finding the angle whose tangent is the closest to $\frac{4}{3} = 1.33$. The angle is 53.1° from the table of trigonometric functions (appendix). From the same table, sin 53.1° = 0.7997. Since sin (sine) $\theta = \frac{BC}{OB}$ it follows that

$$OB = \frac{BC}{\sin 53.1°} = \frac{4}{0.7997} \approx 5$$

and the vector may be expressed in the polar form as 5∠+53.1°.

If the vector is originally expressed in the polar form as 5∠+53.1°, it may be converted to the rectangular form by the use of cos (cosine) 53.1° and sin 53.1°. In this instance the vector is 5 units in length and makes an angle of approximately 53.1° with the +X axis. Thus,

$$\sin 53.1° = \frac{BC}{5}$$

or

$$BC = 5 \sin 53.1° = 5 \times 0.7997 \approx 4$$

$$\cos 53.1° = \frac{OC}{5}$$

or

$$OC = 5 \cos 53.1° = 5 \times 0.6004 \approx 3$$

Therefore, with BC and OC known, the vector may be expressed as the complex number 3+j4 (fig. 10-2B).

The polar form may be converted to the rectangular form more concisely in the following manner:

$$5\angle +53.1° = 5 \cos 53.1° + j5 \sin 53.1°$$

$$= 5 \times 0.6004 + j5 \times 0.7997$$

$$\approx 3+j4$$

ADDITION AND SUBTRACTION OF POLAR VECTORS

Unless polar vectors are parallel to each other they cannot be added or subtracted algebraically. Therefore, the polar form is converted first to the rectangular form. Then the real components are added algebraically, and likewise, the imaginary components are added algebraically. Finally, the result may be converted back to the polar form. Vector summation is indicated in this text by the symbol $\oplus$.

As an example, find the resultant vector, OR (fig. 10-4) of vectors OA and OB when OA = 10∠30° and OB = 8∠60°. OR = OA $\oplus$ OB. Converting to rectangular form:

$$OA = 10 \cos 30° + j10 \sin 30°$$
$$OB = 8 \cos 60° + j\,8 \sin 60°$$

$$OA = 10 \times .866 + j10 \times .500$$
$$OB = 8 \times .500 + j\,8 \times .866$$

$$OA = 8.66 + j\,5.0$$
$$OB = 4.00 + j\,6.93$$

$$OR = 12.66 + j11.93$$

$$OR = \sqrt{12.66^2 + 11.93^2}$$

$$OR \approx \sqrt{160.28 + 142.32}$$

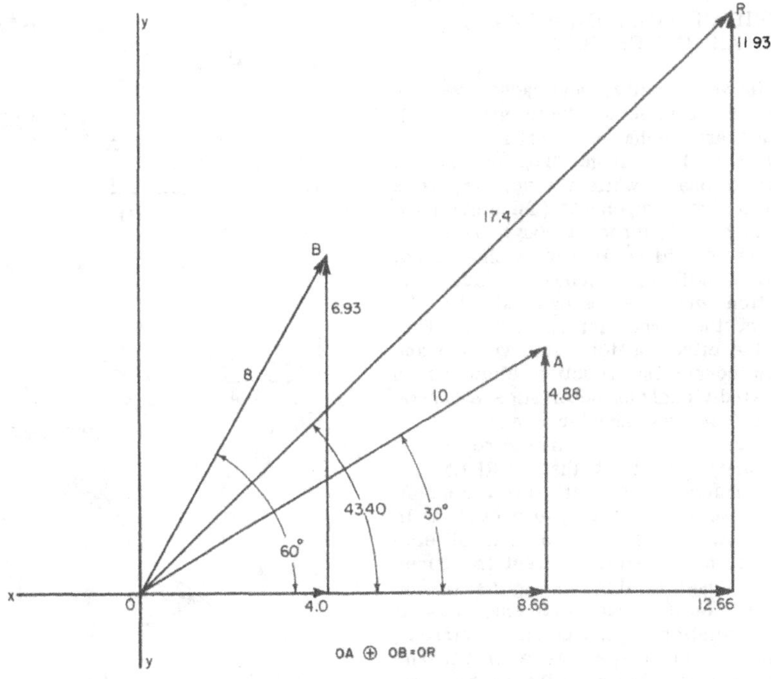

20.63
Figure 10-4.—Addition of vectors.

$$OR \approx \sqrt{302.6}$$

$$OR \approx 17.4$$

and $\tan \theta = \frac{11.93}{12.66} \approx 0.943$ from which $\theta \approx 43.3$. In poar form $OR \approx 17.4\angle 43.3$.

MULTIPLICATION AND DIVISION OF POLAR VECTORS

The method of multiplying and dividing complex numbers by treating them as binomials and rationalizing the denominators may be simplified considerably by first converting the vectors into polar form and then proceeding to combine them in the following manner:

To obtain the product of two vectors, multiply the numbers representing the vectors in polar form and add their corresponding angles algebraically. The resultant vector is in polar form. Thus,

$$(5\angle + 53°)\ (5\angle - 53°) = 25\angle 0°.$$

To obtain the quotient of two vectors, divide the numerator by the denominator as in ordinary division, then subtract algebraically the angle of the denominator from the angle in the numerator. The resultant vector is in polar form. Thus,

$$\frac{10\angle + 25}{5\angle - 20} = 2\angle +45°.$$

RESISTANCE, INDUCTANCE, AND CAPACITANCE IN SERIES

A detailed treatment of circuits containing resistance, R, inductance, L, and capacitance, C, in series is contained in the Basic Electricity, NavPers 10086-B training manual. The discussion below is therefore intended only as a review of some of the most important points presented in that coverage prior to a consideration of series and parallel resonance.

RELATIONSHIP OF VOLTAGE AND CURRENT IN AN RLC CIRCUIT

When resistive, inductive, and capacitive elements are connected in series, their INDIVIDUAL characteristics are unchanged. That is, the current through and the voltage drop across the resistor are in phase while the voltage drops across the reactive components (assuming pure reactances) and the current through them are 90 degrees out of phase. However, this is not true of their COMBINED characteristics; and a new relation must be recognized with the introduction of the three-element circuit. This pertains to the effect on total line voltage and current when connecting reactive elements in series, whose individual characteristics are opposite in nature, such as inductance and capacitance. Such a circuit is shown in figure 10-5A.

In the figure, note first that CURRENT is the common reference for all three element voltages, because there is only one current in a series circuit, and it is common to all elements. The common series current is represented by the dashed line in figure 10-5A. The voltage vector for each element, showing its individual relation to the common current, is drawn above each respective element. The total voltage E_t (fig. 10-5D) is the vector sum of the individual voltages of IR, IX_L, and IX_C.

The three element voltages are arranged for summation in part B of figure 10-5. Since IX_L and IX_C are each 90 degrees away from I, they are, therefore, 180 degrees from each other. Vectors in direct opposition (180 degrees out of phase) may be subtracted directly. The total reactive voltage E_X is the difference of IX_L and IX_C.

or: $E_X = IX_L - IX_C = 45-15 = 30$ volts

The final relationship of line voltage and current, as seen from the source, is shown in part D. Had X_C been larger than X_L, the voltage would lag, rather than lead. When X_C and X_L are of equal value, line voltage and current will be in phase.

IMPEDANCE OF RLC SERIES CIRCUITS

The DIFFERENCE of X_L and X_C must be determined prior to computing the total impedance of an RLC (three element) series circuit. When employing the Pythagorean theorem-based formula for determining series impedance, the net reactance of the circuit is represented by the quantity in parenthesis ($X_L - X_C$). Application of this formula to a series RLC circuit where $X_L = 45$ ohms, $X_C = 15$ ohms and $R = 40$ ohms, yields an impedance of:

$$Z = \sqrt{R^2 + X^2}$$

$$Z = \sqrt{R^2 + (X_L - X_C)^2}$$

$$Z = \sqrt{40^2 + (45-15)^2}$$

$$Z = \sqrt{1600 + 900}$$

$$Z = \sqrt{2500} = 50 \text{ ohms}$$

Series impedance can also be determined by use of the vector triangle method. In an impedance triangle for a series circuit (fig. 10-5C) the base always represents the series resistance, the altitude represents the NET reactance

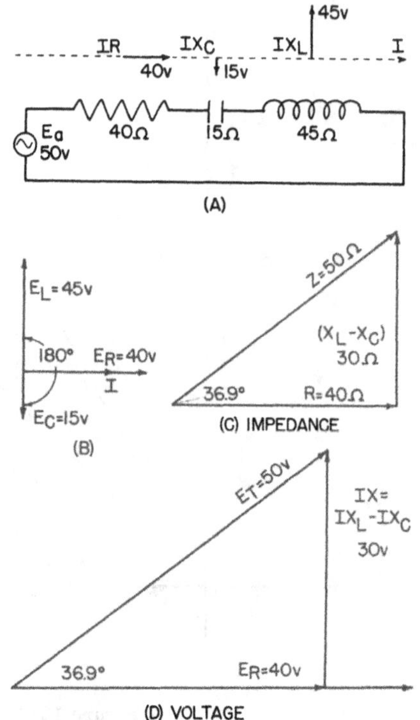

Figure 10-5.—Resistance, inductance, and capacitance in series.

$(X_L - X_C)$, and the hypotenuse represents total impedance.

It should be noted that as the DIFFERENCE of X_L and X_C becomes greater, total impedance also increases. Conversely, when X_L and X_C are equal, their effects cancel each other, and impedance is minimum, equal only to the series resistance. When X_L and X_C are equal their INDIVIDUAL voltages are 90 degrees out of phase with current, but their COLLECTIVE effect is zero because they are equal and opposite in nature. Therefore, when X_L and X_C are equal, line voltage and current are in phase. This condition is the same as if there were only resistance and no reactances in the circuit.

SERIES CIRCUIT RESONANCE

When the reactances in a series circuit cancel, and, as far as the source is concerned, the circuit appears to contain only resistance, the circuit is said to be in a condition of RESONANCE. When resonance is established in a series circuit certain conditions will prevail.

1. The inductive reactance will be equal to the capacitive reactance.
2. The circuit impedance will be minimum.
3. The circuit current will be maximum.

RESONANT FREQUENCY

The reactance of capacitors and inductors is determined by their physical construction and the applied frequency. X_L varies directly with frequency and X_C varies inversely with frequency. Due to this relationship any combination of inductance and capacitance will have a specific frequency at which the reactances will be equal. The relationship of X_L, X_C, frequency, and the resonant frequency point (f_0) is shown in figure 10-6. In a series RLC circuit the largest reactance value determines the appearance and phase angle of the circuit. It can be seen (from fig. 10-6) that below the resonant frequency point X_C is the larger reactance and above f_0 the X_L is the larger reactance. Therefore, the circuit will appear capacitive below f_0 and inductive above f_0.

An equation for determination of the resonant frequency can be developed in the following manner:

At resonance $X_L = X_C$

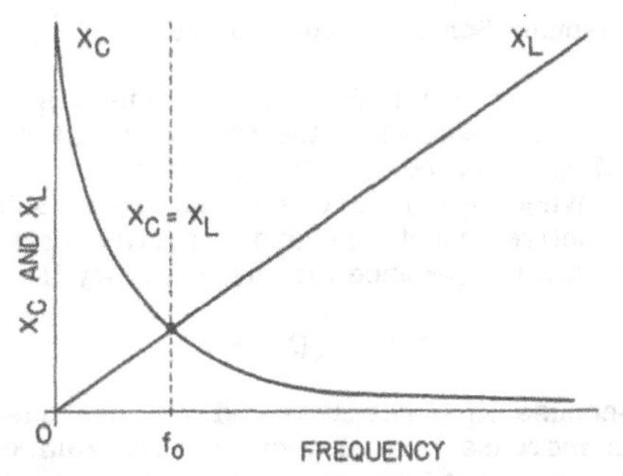

Figure 10-6.—Reactance curves for series RLC circuit.

Substituting the equation for X_L and X_C:

$$2\pi f L = \frac{1}{2\pi f C}$$

Transposing: $$1 = 4\pi^2 f^2 LC$$

$$f^2 = \frac{1}{4\pi^2 LC}$$

Solving for f: $$f_0 = \frac{1}{2\pi\sqrt{LC}}$$

where: f_0 = resonant frequency in hertz.
L = inductance in henrys
C = capacitance in farads

Example. Determine the resonant frequency of a series RLC circuit consisting of an 80 microfarad capacitor, 10 microhenry coil, and a 10 ohm resistor.

Solution: $$f_0 = \frac{1}{2\pi\sqrt{LC}}$$

$$f_0 = \frac{1}{6.28 \times \sqrt{80 \times 10 \times 10^{-12}}}$$

$$f_0 \approx \frac{1 \times 10^6}{177.5}$$

$$f_0 \approx 5.63 \text{ kHz}$$

Thus, when a frequency of 5.63 kHz is applied to the circuit in the example the capacitive and inductive reactances will be equal.

Resonant Series Circuit Analysis

In order for the series circuit (fig. 10-7A) to be in resonance, the frequency of the applied voltage must be such that $X_L = X_C$.

When a series circuit contains resistance, inductive reactance and capacitive reactance, the total impedance for any frequency is:

$$Z = R + j(X_L - X_C)$$

Because X_L increases and X_C decreases with an increase in frequency, at a certain frequency (the resonant frequency) X_L will equal X_C, they will cancel, the j term will drop out, and Z will equal R. Furthermore, because the total impedance is now only the resistance, R, of the circuit, the circuit current is maximum. In other words, at resonance the generator is looking into a pure resistance.

At frequencies below resonance, X_C is greater than X_L and the circuit contains resistance and capacitive reactance; at frequencies above resonance, X_L is greater than X_C and the circuit contains resistance and inductive reactance. At resonance, the current is limited only by the relatively low value of resistance.

Because the circuit is a series circuit (fig. 10-7A), the same current flows in all parts of the circuit (fig. 10-7B), and, therefore, the voltage across the capacitor is equal to the voltage across the inductor, because X_L is equal to X_C. These voltages (fig. 10-7C), however, are 180° out of phase, since the voltage across a capacitor lags the current through it by approximately 90°, and the voltage across the inductor leads the current through it by approximately 90°. The total value of the input voltage, E_t, then appears across R and is shown as E_R in phase with the current, I (fig. 10-7C).

Assume that at a given instant corresponding to angle 0°, the current through the circuit is a maximum as indicated in figure 10-7B. During the first quarter cycle (from 0° to 90° the circuit current falls from maximum to zero. The capacitor is receiving a charge, as is indicated by the rising voltage, e_c, across it. The product of the instantaneous values of e_c and i for this interval indicates a positive power curve (p). The shaded area under this curve represents the energy stored in the capacitor during this time it is receiving a charge.

During the first quarter of a cycle (0° to 90°), when the capacitor is receiving a charge, the magnetic field about the inductor is collapsing because the circuit current is falling, and the inductor acts like a source of power that supplies the charging energy to the capacitor. The voltage, e_L, across the coil, is opposite in phase to the voltage building up across the capacitor and is shown below the line. Therefore, the product of the instantaneous values of the current and voltage across the inductor indicates a negative power curve for the coil between 0° and 90°.

During the second quarter cycle (90° to 180°) the capacitor discharges from maximum to zero, as indicated by the capacitor voltage curve, e_c, and the coil reverses its function and acts like a load on the capacitor. Thus, the capacitor now acts as a source of power. The product of a negative current and a positive voltage (e_c) indicates a negative power curve for the capacitor for this interval. During the same quarter cycle the current is rising through the inductor (in the opposite direction), and energy is being stored in the magnetic field. The product of the negative current and negative voltage, e_L, for the second quarter cycle indicates a positive power curve for the inductor.

A similar interchange of energy between the capacitor and inductor takes place in the third and fourth quarter cycles. Therefore, the average power supplied to the inductor and capacitor by an external source is essentially zero. All circuit losses are assumed to be in the resistor,

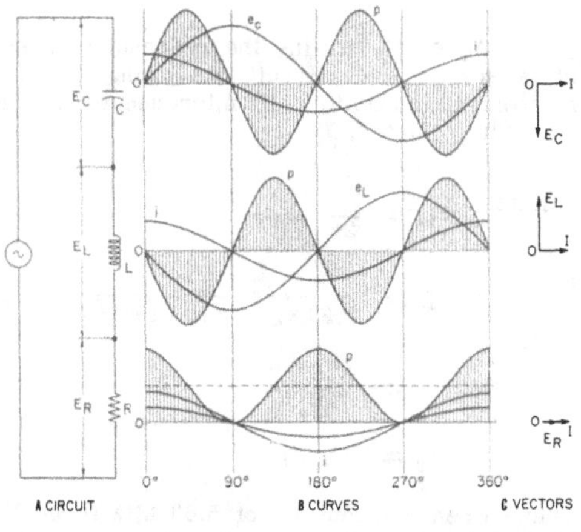

Figure 10-7.—Series resonance.

R. The voltage across the resistor and the current through it are in phase. The product of the voltage and current curves associated with the resistor indicates a power curve that has its axis displaced above the X axis. The displacement is proportional to the true average power which is equal to the product, EI (where E and I are effective values). Whatever power is dissipated in R is supplied by the source.

CIRCUIT Q

The Quality or Q of a circuit is an important consideration in determining the actual merit or efficiency of an inductor, capacitor, or a combination of these components.

The Q of an Inductor

The ratio of the energy stored in an inductor during the time the magnetic field is being established to the losses in the inductor during the same time is called the QUALITY, or Q of the inductor, it is also called the FIGURE OF MERIT of the inductor. This ratio is:

$$Q = \frac{I^2 X_L t}{I^2 Rt}$$

cancellation yields:

$$Q = \frac{X_L}{R}$$

where: Q = a number representing the quality of the inductor.
X_L = inductive reactance of the coil in ohms
R = combined d.c. and a.c. resistances of the coil in ohms

The Q of the inductor is therefore equal to the ratio of the inductive reactance to the effective resistance in series with it, and it approaches a high value as R approaches a low value. Thus, the more efficient the inductor, the lower the losses in it and the higher is the Q.

In terms of the impedance triangle (fig. 10-8):

$$Q = \frac{X_L}{R}$$

$$Q = \tan \theta$$

where θ is the phase angle between the hypotenuse, Z, and the base, R. As θ approaches

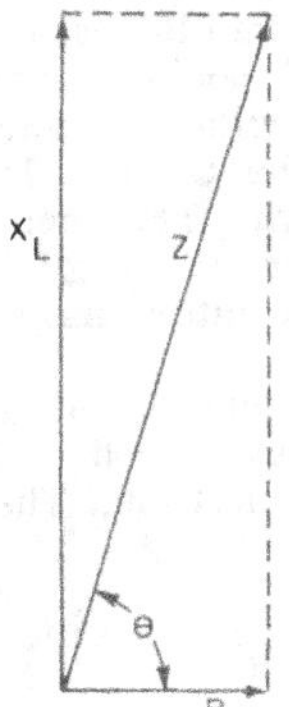

179.191
Figure 10-8.—Impedance triangle.

90°, tan θ approaches infinity, and the coil losses approach zero.

The Q of a coil does not vary extensively within the operating limits of a circuit. It would seem, that since X_L is a direct function of frequency that Q also must be a direct function of frequency. Such is not the case. It is true that as frequency increases the X_L will increase, but as frequency increases the effective resistance of the coil also increases. Since Q is an inverse function of the effective resistance the net effect of a frequency increase is to leave Q relatively unchanged.

The Q of a Capacitor

A capacitor's Q, is a measure of the ratio of the energy stored to the energy dissipated in heat within the capacitor for equal intervals of time. This ratio is reduced by algebraic manipulation to the equation expressed below:

$$Q = \frac{X_C}{R}$$

where: R = the effective resistance of the capacitor dielectric (losses).
Q = a number representing the quality of the capacitor.
X_C = capacitive reactance.

The effective resistance is low with respect to the capacitive reactance, and is such that, when multiplied by the square of the effective capacitor current, it equals the true power dissipated in heat within the capacitor.

Since most of the losses in a solid-dielectric capacitor occur within the dielectric rather than in the plates, the Q of a low-dielectric-loss capacitor is high. The losses of an air-dielectric capacitor are negligible, and thus the Q of such a capacitor may have a very high value.

The Q of a series resonant circuit is the ratio of the energy stored to the energy lost in equal intervals of time. The expression becomes:

$$Q = \frac{X_L}{R} = \frac{X_C}{R}$$

where R represents the total effective series resistance of the entire circuit. Since the capacitor has negligible losses, the circuit Q becomes equivalent to the Q of the coil. The circuit Q may be maintained satisfactorily high by keeping the circuit resistance to a minimum. This may be expressed mathematically in the following manner. The inductive voltage drop is:

$$E_L = IX_L$$

Since $$I = \frac{E}{R}$$

then $$E_L = \frac{EX_L}{R}$$

and $$Q = \frac{X_L}{R}$$

then $$E_L = QE$$

transposed $$Q = \frac{E_L}{E}$$

where: Q = a number representing the quality of the circuit.

E_L = the voltage across the inductor (can be E_C, voltage across the capacitor).

E = the voltage across the effective series resistance.

Therefore, the Q of the circuit is the ratio of the voltage across either the inductor or capacitor to that across the effective series resistance. In other words, the voltage gain (V.G.) of the series resonant circuit depends on the circuit Q. Expressed mathematically:

$$V.G. = \frac{E_L}{E} = \frac{IX_L}{IR} = \frac{X_L}{R} = \frac{IX_C}{IR} = \frac{X_C}{R} = Q$$

SERIES RLC CIRCUIT ANALYSIS

Figure 10-9 shows the relation between the effective current and frequency in the vicinity of resonance for a series circuit containing a 159 μh coil, a 159 pf capacitor, and an effective series resistance of either 10 ohms, or 20 ohms.

The resonant frequency, f_o, is:

$$f_o = \frac{1}{2\pi\sqrt{LC}}$$

$$f_o = \frac{1}{6.28\sqrt{159 \times 10^{-6} \times 159 \times 10^{-12}}}$$

$f_o = 1 \times 10^6$ hertz or 1,000 kHz.

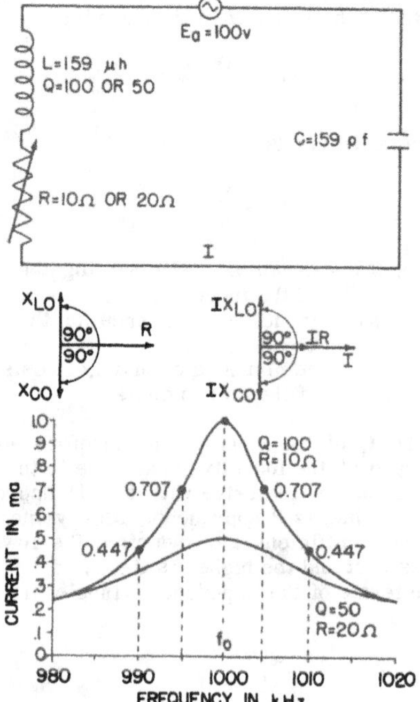

20.66

Figure 10-9.—Resonance curves of a series RLC circuit.

The reactances and impedance at resonance may likewise be determined. Thus:

$$X_{Lo} = 2\pi fL = 6.28 \times 10^6 \times 159 \times 10^{-6}$$

$$X_{Lo} = 1000 \angle +90^\circ$$

where X_{Lo} is the inductive reactance at resonance. The +90° angle indicates that the IX_{Lo} and X_{Lo} vectors are plotted vertically upward because the current vector is horizontal and extends to the right. The current vector thus lags the voltage, IX_{Lo}, across the coil by 90° (counterclockwise rotation is positive, fig. 10-9).

$$X_{Co} = \frac{1}{2\pi fC} = \frac{1}{6.28 \times 10^6 \times 159 \times 10^{-12}}$$

$$X_{Co} = 1000 \angle -90^\circ$$

where X_{Co} is the capacitive reactance at resonance. The -90° angle indicates that the vectors X_{Co} and IX_{Co} are plotted vertically downward because the current vector is the horizontal reference vector extending to the right and leads the voltage drop, IX_{Co}, across the capacitor by 90° (fig. 10-9). Note that current is a common factor to both voltage and impedance vectors. When the effective resistance of the circuit is equal to 10 ohms the impedance at resonance (Z_o) will be:

$$Z_o = R + j(X_L - X_C)$$

$$Z_o = 10 + j(1000 - 1000)$$

$$Z_o = 10 + j0 \text{ ohms}$$

$$Z_o = 10 \angle 0^\circ \text{ ohms}$$

If the applied voltage is assumed to be 10 millivolts (mv.) at a frequency of 1000 kHz, the circuit current is:

$$I = \frac{E}{Z}$$

$$I = \frac{0.01}{10}$$

$$I = 1 \text{ ma.}$$

At the resonant frequency, the voltage across the inductor is:

$$E_L = IX_L$$

$$E_L = 0.001 \times 1000$$

$$E_L = 1 \text{ volt}$$

The voltage across the capacitor is the same, except it is 180° out of phase with the voltage across the coil. The losses in the coil and capacitor are assumed to be lumped in the effective series resistance. The circuit Q is:

$$Q = \frac{X_L}{R}$$

$$Q = \frac{1000}{10}$$

$$Q = 100$$

The voltage gain at resonance is:

$$V.G. = \frac{E_L}{E}$$

$$V.G. = \frac{1.00}{0.01}$$

$$V.G. = 100$$

The Q of a tuned circuit can be found by this equation:

$$Q = \frac{1}{R} \cdot \sqrt{\frac{L}{C}}$$

This can be proven by substituting the values from figure 10-9 into the equation and solving for the results. It is important to note that the higher the resistance in a tuned circuit the lower will be the circuit Q.

The resonance curves of current vs frequency are symmetrical about a vertical line (f_0) extending through the point of maximum current, (fig. 10-9). The shape of the resonance curve may be approximated in the vicinity of resonance by applying the following rules that can be derived from the resonant circuit equations. (The derivation is not given because of its length.)

RULE 1. If the frequency of the applied voltage is decreased by an amount $\frac{1}{2Q}$ times the resonant frequency, f_0, the current in the tuned circuit decreases to 0.707 of its value at the resonant frequency and leads the applied voltage by 45°.

Example. The input frequency of the circuit (fig. 10-9) is decreased from the resonant frequency (f_0) of 1000 kHz by an amount equal to:

$$f_{decrease} = \frac{1}{2Q} \times f_0$$

$$f_{decrease} = \frac{1}{2 \times 100} \times 1000 \text{ kHz}$$

$$f_{decrease} = 5 \text{ kHz}$$

The new applied frequency is then:

$$f_{new} = f_0 - f_{decrease}$$

$$f_{new} = 1000 - 5$$

$$f_{new} = 995 \text{ kHz}$$

The X_L at the new applied frequency is:

$$X_L = 2\pi f L$$

$$X_L = 6.28 \times 0.995 \times 10^6 \times 159 \times 10^{-6}$$

$$X_L \approx 995 \angle +90° \text{ ohms}$$

The X_C at the new applied frequency is:

$$X_C = \frac{1}{2\pi f C}$$

$$X_C = \frac{1}{6.28 \times 9.995 \times 10^6 \times 159 \times 10^{-12}}$$

$$X_C \approx 1005 \angle -90° \text{ ohms}$$

The circuit impedance at 995 kHz is:

$$Z_t = R + j(X_L - X_C)$$

$$Z_t = 10 + j(995 - 1005)$$

$$Z_t = 10 - j10 \text{ ohms}$$

Converting to polar form:

$$Z_t = 14.14 \angle -45°$$

The circuit current at 995 kHz is:

$$I_t = \frac{E_t}{Z_t}$$

$$I_t = \frac{0.01 \angle 0°}{14.14 \angle -45°}$$

$$I_t \approx 0.707 \angle 45° \text{ ma.}$$

At this frequency (which is the cutoff frequency) the voltage across the coil, or the capacitor, is reduced to approximately 70 percent of its value at resonance.

The voltage across the coil is:

$$E_L = IX_L$$

$$E_L = 0.707 \times 995$$

$$E_L \approx 705 \text{ mv.}$$

RULE 2. If the frequency of the applied voltage is decreased by an amount 1/Q times the resonant frequency, the current decreases to 0.447 of its value at resonance and leads the applied voltage by 63.4°.

Example. The input frequency of the circuit (fig. 10-9) is decreased from the resonant frequency at an amount equal to:

$$f_{decrease} = \frac{1}{Q} \times f_0$$

$$f_{decrease} = \frac{1}{100} \times 1000$$

$$f_{decrease} = 10 \text{ kHz}$$

The new applied frequency is then:

$$f_{new} = f_0 - f_{decrease}$$

$$f_{new} = 1000 - 10$$

$$f_{new} = 990 \text{ kHz}$$

The X_L at the new applied frequency of 990 kHz is:

$$X_L = 2\pi f L$$

$$X_L = 6.28 \times 0.990 \times 10^6 \times 159 \times 10^{-6}$$

$$X_L \approx 990 \angle +90° \text{ ohms}$$

The X_C at the new applied frequency of 990 kHz is:

$$X_C = \frac{1}{2\pi f C}$$

$$X_C = \frac{1}{6.28 \times 0.990 \times 10^6 \times 159 \times 10^{-12}}$$

$$X_C \approx 1010 \angle -90° \text{ ohms}$$

The impedance of the series circuit at 990 kHz is:

$$Z_t = R + j(X_L - X_C)$$

$$Z_t = 10 + j(990 - 1010)$$

$$Z_t = 10 - j20$$

Converting to polar form yields:

$Z_t \approx 22.4 \angle -63.4°$ ohms

At this frequency the circuit current is:

$$I_t = \frac{E_t}{Z_t}$$

$$I_t = \frac{0.10 \angle 0°}{22.4 \angle -63.4°}$$

$$I_t \approx 0.447 \angle +63.4° \text{ ma.}$$

The voltage across the coil is:

$$E_L = IX_L$$

$$E_L = 0.447 \times 990$$

$$E_L \approx 444 \text{ mv.}$$

Decreasing the applied frequency is seen (by the positive phase angles of I_t) to cause the total current to lead the applied voltage. Corresponding increases in the frequency of the applied voltage above the resonant frequency will produce the same reductions in circuit current and voltage across the reactive portions of the circuit. In this case, however, the current lags the applied voltage instead of leading it. Thus, the resonance curve is symmetrical about the resonant frequency in the vicinity of resonance.

The series resonant circuit increases the voltage gain at the resonant frequency. If the circuit losses are low the circuit Q will be high and the voltage gain relatively large. For resonant circuits involving iron-core coils the Q may range from 20 to 100. In practice, because nearly all of the resistance of a circuit is in the coil, the ratio of the inductive reactance to the resistance is especially important. The higher the Q of the coil, the better is the coil and the more effective is the series resonant circuit that utilizes it.

POWER IN RLC SERIES CIRCUITS

Total true power in an RLC series circuit is the product of line voltage and current times the cosine of the angle between them. When X_L and X_C are equal, total impedance is at a minimum, and thus current is maximum. When maximum current flows, the series resistor dissipates maximum power. When X_L and X_C are made unequal, total impedance increases, line current decreases, and moves out of phase with line voltage. Both the decrease in current and the creation of a phase difference cause a decrease in true power.

In the example above, decreasing the applied frequency caused a leading current, therefore, a positive phase angle. A positive phase angle causes a leading power factor and vice versa. Thus, the P. F. is leading when the frequency is decreased and lagging when the frequency is increased (with respect to the resonant frequency).

BANDWIDTH

If the circuit Q is low, the gain at resonance is relatively small and the circuit does not discriminate sharply between the resonant frequency and the frequencies on either side of resonance, as is shown by the lower curve in figure 10-9. The range of frequencies included between the two frequencies at which the current drops to 70 percent of its value at resonance is called the BANDWIDTH for 70 percent response.

Frequencies beyond (outside of) the 70% area, referred to as the HALF-POWER points on the curve, are considered to produce no usable output. The series resonant circuit is seen to have two half-power points, one above the resonant frequency point and one below. The two points are designated upper f_{co} and lower f_{co} or simply f_1 and f_2. The range of frequencies between these two points comprises the bandwidth. Figure 10-10 illustrates the bandwidths for high and low Q series resonant circuits. The bandwidth may be determined by the equation:

$$BW = \frac{f_o}{Q}$$

$$BW = f_2 - f_1$$

where: BW = bandwidth of a series resonant circuit in units of frequency.
f_o = resonant frequency.
f_2 = the highest frequency the circuit will pass.
f_1 = the lowest frequency the circuit will pass.
Q = as defined previously.

Example. Determine the bandwidth for the curve shown in figure 10-10B.

Solution: $BW = f_2 - f_1$

$$BW = 483.6 \text{ kHz} - 426.4 \text{ kHz}$$

$$BW = 57.2 \text{ kHz}$$

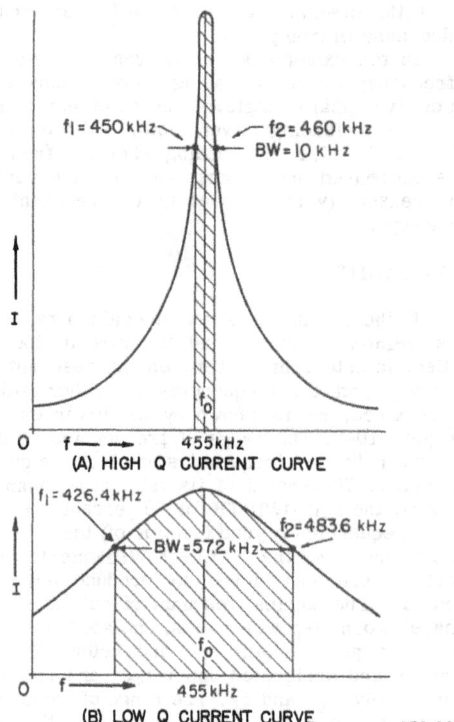

Figure 10-10.—Bandwidth for high and low Q series resonant circuits.

The Q of the curve in part A is given as 45.5: determine the bandwidth.

$$BW = \frac{f_o}{Q}$$

$$BW = \frac{455 \text{ kHz}}{45.5}$$

$$BW = 10 \text{ kHz}$$

APPLICATIONS OF SERIES RESONANT CIRCUITS

Series resonant circuits are used largely as filters (to be treated later) for audio and radio frequencies. With proportionately larger component values the series circuit may be used as a power-supply filter. For example, assume that a d.c. generator has a ripple frequency of 400 Hz. A series resonant circuit tuned to 400 Hz may be connected across the terminals of the generator and thus effectively short-circuit the ripple voltage. The coil and capacitor insulation must be able to withstand the relatively high a.c. voltage caused by the series resonant action.

The series-tuned circuit may also be used to give an indication of frequency if the capacitor is calibrated for the appropriate frequency range. The capacitor and the inductor are connected in series with a current-indicating device across the source of the unknown frequency. At resonance the current as indicated by the device, will be maximum. Therefore, if the value of L and C are known, the value of the unknown frequency may be calculated.

SUMMARY OF RESONANT CONDITIONS FOR SERIES RLC CIRCUITS

The major characteristics of series RLC circuits at resonance are summarized in table 10-2.

PARALLEL RLC CIRCUITS

Parallel RLC circuits may be analyzed in the same manner as parallel RC and RL circuits. (A detailed treatment of these circuits is contained in Basic Electricity, NavPers 10086-B.) They are simply three branch circuits consisting of a resistor, a capacitor, and an inductor connected in parallel. A representative circuit arrangement with ammeters (connected to indicate the current in each branch) is shown in figure 10-11A.

The vector diagram in figure 10-11B has much in common with the vector diagrams for series RLC circuits (treated earlier in this chapter) because the two reactive components are 180° out of phase with each other. The reader is encouraged to solve for branch currents, I_1, I_2, and I_3, and total impedance, Z_t, in order to better understand the relationship of forces active in this circuit before moving on to parallel resonance. Recommended values of frequency, resistance, capacitance, and inductance are as follows:

F = 30 kHz R = 4.8 kohms C = 6000 pf L = 6 mh;

whereupon $I_t = 19.3 \angle 49.6°$ ma.,

$Z_t = 3108 \angle -49.6°$ ohms

and, $P_t = 750$ mw.

Table 10-2.—Major characteristics of series RLC circuits at resonance

QUANTITY	SERIES CIRCUIT
At resonance: Reactance (X_L-X_C)	Zero, because X_L=X_C
Resonant frequency	$\frac{1}{2\pi\sqrt{LC}}$
Impedance	Minimum; Z = R
I_{LINE}	Maximum value
I_L	I_{LINE}
I_C	I_{LINE}
E_L	Q x E_{LINE}
E_C	Q x E_{LINE}
Phase angle between E_{LINE} and I_{LINE}	0°
Angle between E_L & E_C	180°
Angle between I_L & I_C	0°
Desired value of Q	10 or more
Desired value of R	Low
Highest selectivity	High Q, low R, high $\frac{L}{C}$
When f is greater than f_o: Reactance	Inductive
Phase angle between I_{LINE} and E_{LINE}	Lagging current
When f is less than f_o: Reactance	Capacitive
Phase angle between I_{LINE} and E_{LINE}	Leading current

179.193

IDEAL PARALLEL RESONANCE

The ideal parallel resonant circuit is one that contains only inductance and capacitance. Resistance and its effects are not considered in an ideal parallel resonant circuit. One condition for parallel resonance is the application of that frequency which will cause the inductive reactance to equal the capacitive reactance. The formula used to determine the resonant frequency of a parallel LC circuit is the same as the one used for a series circuit.

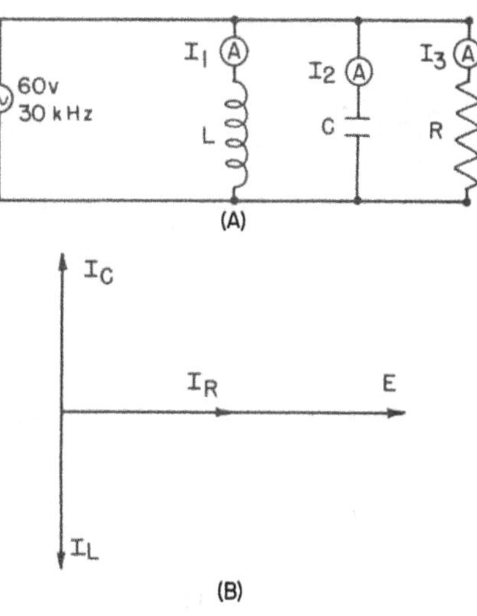

179.194

Figure 10-11.—Parallel RLC circuit and its associated vector diagram.

$$f_o = \frac{1}{2\pi\sqrt{LC}}$$

where: f_o = resonant frequency in Hz
L = inductance in henrys
C = capacitance in farads

If the circuit values are those shown in figure 10-12A, the resonant frequency may be computed as follows:

$$f_o = \frac{1}{2\pi\sqrt{LC}}$$

$$f_o = \frac{1}{6.28\sqrt{(3.5 \times 10^{-3})(2.5 \times 10^{-6})}}$$

$$f_o = 1700 \text{ hertz}$$

At the resonant frequency:

Determine X_L:

$$X_L = 2\pi f L$$

$$X_L = (6.28)(1.7 \times 10^3)(3.5 \times 10^{-3})$$

$$X_L \approx 37.4 \text{ ohms}$$

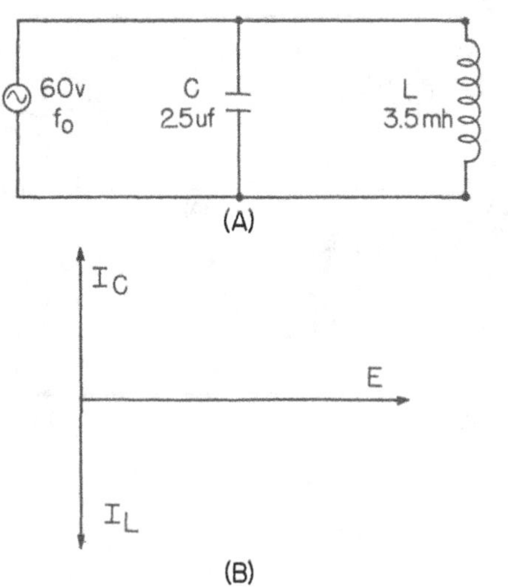

179.195

Figure 10-12.—Parallel LC circuit at resonance.

Determine X_C:

$$X_C = \frac{1}{2\pi f C}$$

$$X_C = \frac{1}{(6.28)(1.7 \times 10^3)(2.5 \times 10^{-6})}$$

$$X_C \approx 37.4 \text{ ohms}$$

Determine I_C:

$$I_C = \frac{E_a}{-jX_C}$$

$$I_C = \frac{60}{-j\,37.4}$$

$$I_C \approx 0 + j1.605 \text{ amps}$$

Determine I_L:

$$I_L = \frac{E_a}{jX_L}$$

$$I_L = \frac{60}{j\,37.4}$$

$$I_L \approx 0 - j1.605 \text{ amps}$$

The total current is determined by addition of the two currents in rectangular form:

$$\begin{aligned} I_C &\approx 0 + j1.605 \\ I_L &\approx 0 - j1.605 \\ \hline I_t &\approx 0 \end{aligned}$$

Therefore, in an ideal resonant parallel circuit the line current (I_t) is zero. If total current is zero then:

$$Z = \frac{E_{app}}{I_t} = \frac{60}{0} = \text{UNDEFINED}$$

or: it may be said that the impedance approaches infinity.

At frequencies other than the natural resonant frequency of the circuit, X_C will not be equal to X_L and some amount of current will be drawn from the source. If the applied frequency is lower than the resonant frequency of the circuit, X_L will be smaller than X_C and a lagging source current will result. When the applied frequency is above the resonant frequency, X_C is smaller than X_L, and the source current leads the source voltage.

To obtain an overall view of the operation of a parallel LC circuit, a graph can be constructed in which impedance and current are plotted as a function of frequency. To obtain the information for the graph, the capacitive and inductive reactances, impedance, and total current are computed for a group of frequencies centered about the resonant frequency. These computations have been performed for a parallel circuit containing a 2.5 microfarad capacitor and a 3.5 millihenry inductor, and have been tabulated in table 10-3.

At 700 hertz, for example, the inductive and capacitive reactances are found to be 15.39 ohms and 90.95 ohms respectively. The impedance of the circuit can be computed by the product over the sum method as follows:

$$Z = \frac{(jX_L)(-jX_C)}{(jX_L) + (-jX_C)}$$

$$Z = \frac{(0 + j15.39)(0 - j90.95)}{0 - j75.56}$$

$$Z = \frac{1399.72\angle 0^\circ}{75.56\angle -90^\circ}$$

$$Z = 18.52\angle 90^\circ \text{ ohms}$$

Table 10-3.—Reactance, impedance, and current as a function of frequency

f (Hz)	X_L (ohms)	X_C (ohms)	Z (ohms)	I_t (amps)
700	15.39	90.95	18.53	3.238
800	17.59	79.58	22.59	2.657
900	19.79	70.74	27.48	2.183
1000	21.99	63.66	33.60	1.786
1100	24.19	57.88	41.56	1.444
1200	26.39	53.05	52.51	1.142
1300	28.59	48.97	68.69	0.874
1400	30.79	45.47	95.34	0.629
1500	32.99	42.44	148.08	0.405
1600	35.19	39.79	314.25	0.197
*1700	37.39	37.45	22221.00	Zero
1800	39.58	35.34	332.18	0.181
1900	41.78	33.51	169.17	0.355
2000	43.98	31.83	115.22	0.521
2100	46.18	30.32	88.23	0.680
2200	48.38	28.94	72.01	0.833
2300	50.58	27.68	61.13	0.982
2400	52.78	26.53	53.33	1.125
2500	54.98	25.47	47.44	1.265
2600	57.18	24.49	42.83	1.401
2700	59.38	23.58	39.11	1.534

*Resonant frequency

179.196

If the applied voltage is 60 volts the total current is:

$$I_t = \frac{E_a}{Z}$$

$$I_t = \frac{60\angle 0^\circ}{18.52\angle 90^\circ}$$

$$I_t \approx 3.24 \angle -90^\circ \text{ amps}$$

Notice, that, at 700 Hz the impedance and current have phase angles of 90 degrees indicating that the circuit appears as a pure reactance. At all frequencies below the resonant frequency of 1700 Hz, the ideal circuit appears as a pure inductance and the source current lags the source voltage by exactly -90 degrees.

As the frequency is increased above 700 Hz the impedance rises and the source current decreases. At 1700 Hz the inductive branch current becomes equal to the capacitive branch current causing the total (line) current to diminish to zero. Since no current is drawn from the source the impedance of the circuit moves toward infinity.

To investigate the operation of the circuit for applied frequencies above the natural resonant frequency of the circuit, the impedance and current will be computed for a frequency of 2100 Hz.

$$Z = \frac{(jX_L)(-jX_C)}{(jX_L) + (-jX_C)}$$

$$Z = \frac{(0 + j46.18)(10 - j30.32)}{0 + j15.86}$$

$$Z = \frac{1400\angle 0^\circ}{15.86\angle 90^\circ}$$

$$Z \approx 88.27\angle -90^\circ$$

The current at 2100 Hz is:

$$I_t = \frac{E_a}{Z}$$

$$I_t = \frac{60\angle 0^\circ}{88.27\angle -90^\circ}$$

$$I_t \approx 0.68\angle 90^\circ \text{ amp}$$

The phase angles of the impedance and current at 2100 Hz show that the LC circuit appears as a pure capacitive reactance to the source. This condition exists for all applied frequencies which are higher than the resonant frequency, since, for these frequencies I_C is greater than I_L.

At this point certain conclusions can be drawn concerning the characteristics of a parallel LC circuit. These are as follows:

1. For every possible parallel combination of inductance and capacitance, a frequency exists which will make X_C equal to X_L. This frequency is called the resonant frequency of the circuit.
2. When the circuit is operated at its resonant frequency I_C is equal to I_L.
3. At resonance the source current is minimum (zero current in the ideal LC circuit).
4. At resonance the impedance of the circuit is maximum (infinite in the ideal circuit).
5. At resonance the circuit appears resistive to the source, the phase angle is zero, and the power factor is unity.
6. When the applied frequency is below the resonant frequency of the circuit, I_L is greater than I_C, and the circuit appears inductive.
7. When the applied frequency is above the resonant frequency of the circuit, I_C is greater than I_L, and the circuit appears capacitive.

PRACTICAL PARALLEL RESONANT CIRCUITS

The primary difference that exists between the ideal parallel resonant circuit and the practical parallel resonant circuit is that the practical parallel LC circuit contains resistance. This resistance exists throughout the circuit; however, most of it is located in the inductive branch of the circuit. For purposes of analysis all of the circuit resistance will be represented by a single resistor placed in parallel with the inductive branch. This resistance will be assumed to account for all of the circuit losses, both a.c. and d.c.

The schematic diagram of a practical parallel LC circuit is shown in figure 10-13A. In this circuit the impedance of the capacitive branch is equal to X_C, while, the impedance of the inductive branch is equal to the vector sum of X_L and R. If the source is adjusted to the frequency at which X_L is equal to X_C, the current through the inductive branch will be smaller than the current through the capacitive branch. The resulting vector cancellation of I_C and I_L yields a low value of I_C (the total amount being determined by the size of the resistance, R). Therefore, the total current will lead the applied voltage by a small angle making the circuit appear slightly capacitive to the source.

If the applied frequency is reduced slightly the current through the capacitive branch decreases and the current through the inductive branch increases. The current through the inductive branch will cancel a greater percentage of the capacitive branch current, and the total current will attain its minimum value.

Due to the energy losses which occur in a practical parallel LC circuit, some current must be drawn from the source. Because of this current the circuit will have a finite impedance at the resonant frequency.

EFFECTS ON Q CAUSED BY LOADING A PARALLEL RESONANT CIRCUIT

As for series RLC circuits, the Q of a parallel resonant circuit is determined by the ratio of X_L/R (or X_C/R). The parallel circuit impedance at resonance varies inversely with the resistance in the coil branch. This (series) resistance may act as the total load on the parallel circuit. Hence, an increase in series resistance inherent in the coil or produced by increasing the physical resistance in the coil branch of the circuit represents an increased load. This lowers the Q and the total impedance of the circuit.

The total or line current (I_t) is equal to the sum of the branch currents I_C and I_L. These currents must be converted to rectangular form before they can be added.

The impedance is found by dividing the source voltage by the total current. At resonance, the phase angle of the impedance is zero and the circuit is purely resistive to the source.

The combination of L, C, and R in figure 10-13A form a parallel resonant circuit in which a parallel resistance has been added. Notice that there is no resistance shown in series with the inductive branch. At resonance, the vector diagram of the branch currents will be as shown in figure 10-13B. As before, if the lengths of the vectors representing the

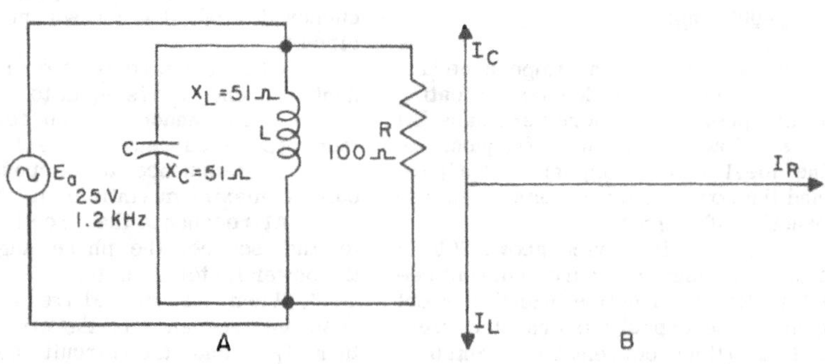

179.197

Figure 10-13.—Parallel resonant circuit with parallel resistance.

reactive currents are equal, the circuit will appear to be a purely resistive circuit having a resistance value that is equal to R.

The circuit shown in figure 10-13A is to be used to solve for current, impedance, circuit Q and other characteristics in the following manner:

Find: I_C = ? I_{tank} = ?
I_L = ? I_{line} = ?
Q = ?

Solution: Determine I_C.

$$I_C = \frac{E_a}{X_C}$$

$$I_C = \frac{25}{51}$$

$$I_C \approx 490 \text{ ma. or } 0 + j\,490 \text{ ma.}$$

Determine I_L:

$$I_L = \frac{E_a}{X_L}$$

$$I_L = \frac{25}{51}$$

$$I_L \approx 490 \text{ ma. or } 0 - j\,490 \text{ ma.}$$

Determine I_R:

$$I_R = \frac{E_a}{R}$$

$$I_R = \frac{25}{100}$$

$$I_R = 250 + j\,0 \text{ ma.}$$

To determine the current drawn from the source (line current) insure all currents are in rectangular form and add:

I_C =	0 +j 490	
I_L =	0 -j 490	
I_R =	250 +j 0	
I_{line} =	250 +j 0	

Since the reactive currents cancel, the line current is equal to the current drawn by the resistive component. At resonance, in a three branch circuit:

$$I_{line} = I_R$$

Since the circulating current of the tank is the same for either reactive component, then tank current may be found by determining current flow through the capacitor or inductor.

$$I_{tank} = \frac{E_a}{X_C}$$

$$I_{tank} = \frac{25}{51}$$

$$I_{tank} \approx 490 \text{ ma.}$$

or at resonance:

$$I_{tank} = I_C = I_L$$

For the purpose of discussion, it can be assumed that the LC combination in figure 10-13A forms an ideal parallel resonant network. The impedance of such a circuit would be higher than any pre-assigned value. Thus, the equivalent resistance (impedance) offered to the source by the parallel arrangement of the extremely high impedance of the tank circuit in parallel with a known value of resistance is equal to the value of the known resistance. The circuit impedance can therefore be determined in the following manner:

$$Z = \frac{E_a}{I_{line}}$$

$$Z = \frac{25}{0.25}$$

$$Z = 100 \text{ ohms}$$

Determining the value of Q for a three branch circuit:

$$I_{line} = \frac{I_{tank}}{Q}$$

transposing:

$$Q = \frac{I_{tank}}{I_{line}}$$

$$Q = \frac{490}{250}$$

$$Q \approx 1.96$$

Q may be determined by:

$$Q = \frac{I_{tank}}{I_{line}}$$

$$I_{tank} = \frac{E_a}{X_L}$$

$$I_{line} = \frac{E_a}{R}$$

Therefore, by substitution:

$$Q = \frac{\frac{E_a}{X_L}}{\frac{E_a}{R}}$$

To divide, invert denominator and multiply:

$$Q = \frac{E_a}{X_L} \times \frac{R}{E_a}$$

Cancelling E_a:

$$Q = \frac{R}{X_L}$$

$$Q = \frac{100}{51}$$

$$Q \approx 1.96$$

The banopass of a three branch RLC circuit such as that shown in figure 10-13A undergoes slight modification, i.e., it will not be the same as for a two branch circuit if expressed as a function of R, L, and C. The shape of the current versus frequency curve is the same as for the series resonant circuit since as the resistance, R, is increased the line current decreases. Since:

$$BW = \frac{f_o}{Q}$$

and: $Q = \frac{R}{X_L}$ in the three branch circuit,

$$BW = \frac{f_o \times X_L}{R}$$

If the parallel resistance is increased, the line current goes down. As the shunting resistance approaches infinity, the line current approaches zero. As resistance is increased, the bandwidth becomes narrower, and the selectivity increases. Therefore, it can be seen how regulation of the bandwidth may be accomplished by variation of the "shunt" resistance. This shunt resistance is sometimes referred to as a "swamping" resistance.

The inverse relationship between resistance and bandwidth may be seen by examination of the BW (bandwidth) formula above or the Impedance and Current versus Frequency curves in figure 10-14, where f_{co} defines the maximum and minimum cutoff frequencies.

APPLICATIONS OF PARALLEL RESONANT CIRCUITS

The parallel resonant circuit is one of the most important circuits used in electronic transmitters, receivers, and frequency-measuring equipment.

The IF transformers of radio and television receivers employ parallel-tuned circuits. These are transformers used at the input and output of each intermediate frequency amplifier stage for coupling purposes and to provide selectivity. They are enclosed in a shield and provided with openings at the top through which screwdriver adjustments may be made when the set is being aligned.

Parallel-tuned circuits are also used in the driver and power stages of transmitters, as

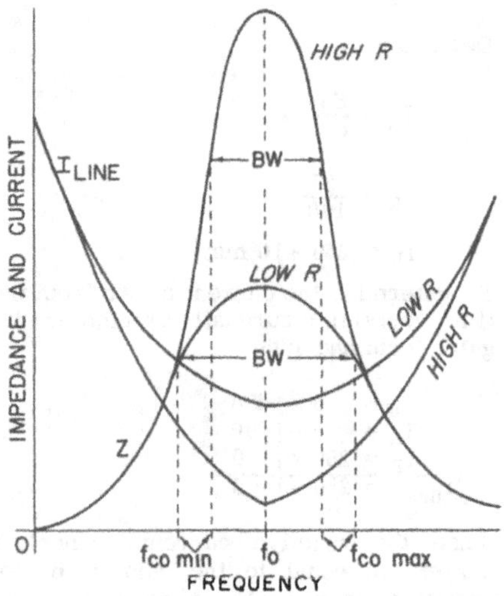

Figure 10-14.—Effect of shunt resistance on BW, I_{line}, and Z.

well as in the oscillator stages of transmitters, receivers, and frequency-measuring equipment.

Various types of filter circuits employ parallel-tuned circuits as well as series-tuned circuits.

SUMMARY OF RESONANT CONDITIONS FOR PARALLEL RLC CIRCUITS

The major characteristics of parallel RLC circuits at resonance are summarized in table 10-4.

Table 10-4.—Major characteristics of parallel RLC circuits at resonance

Quantity	Parallel Circuit
At resonance: Reactance (X_L-X_C)	Zero; because nonenergy currents are equal
Resonant frequency	$f_o = \frac{1}{2\pi\sqrt{LC}}$
Impedance	Maximum; $Z = \frac{L}{CR}$
I_{LINE}	Minimum value
I_L	$Q \times I_{LINE}$
I_C	$Q \times I_{LINE}$
E_L	E_{LINE}
E_C	E_{LINE}
Phase angle between E_{LINE} and I_{LINE}	0°
Angle between E_L and E_C	0°
Angle between I_L and I_C	180°
Desired value of Q	10 or more
Desired value of R	Low
Highest selectivity	High Q, low R, $\frac{L}{C}$
When f is greater than f_o: Reactance	Capacitive
Phase angle between I_{LINE} and E_{LINE}	Leading current
When f is less than f_o: Reactance	Inductive
Phase angle between I_{LINE} and E_{LINE}	Lagging current

179.198

TUNED CIRCUITS AS FILTERS

Tuned circuits are employed as filters for the passage or rejection of specific frequencies. Bandpass filters and band-rejection filters are examples of this type. Tuned circuits have certain characteristics that make them ideal for certain types of filters, especially where high selectivity is desired. A series-tuned circuit offers a low impedance to currents of the particular frequency to which it is tuned and a relatively high impedance to currents of all other frequencies. A parallel-tuned circuit, on the other hand, offers a very high impedance to currents of its natural, or resonant, frequency and a relatively low impedance to others.

BANDPASS FILTERS

A bandpass filter is designed to pass currents of frequencies within a continuous band, limited by an upper and lower cutoff frequency, and substantially to reduce, or attenuate, all frequencies above and below that band. A simple bandpass filter is shown in figure 10-15A.

The curves of current vs. frequency are shown in figure 10-15B. The high Q circuit gives a steeper current curve; the low Q circuit gives a much flatter current curve.

The series- and parallel-tuned circuits are tuned to the center frequency of the band to be passed by the filter. The parallel-tuned circuit offers a high impedance to the frequencies within this band, while the series-tuned circuit offers very little impedance. Thus, the desired frequencies within the band will travel on to the load without being affected, but the currents of unwanted frequencies, that is, frequencies outside the desired band, will meet with a high series impedance and a low shunt impedance so that they are in a greatly attenuated form at the load.

There are many circuit arrangements for both bandpass and band-elimination filters. For the purpose of a brief analysis the bandpass circuit shown in figure 10-15A will be considered. Let it be assumed that a band of frequencies extending from 90 kHz to 100 kHz is to be passed by the filter. For an input and output resistance of 10,000 ohms, the values of

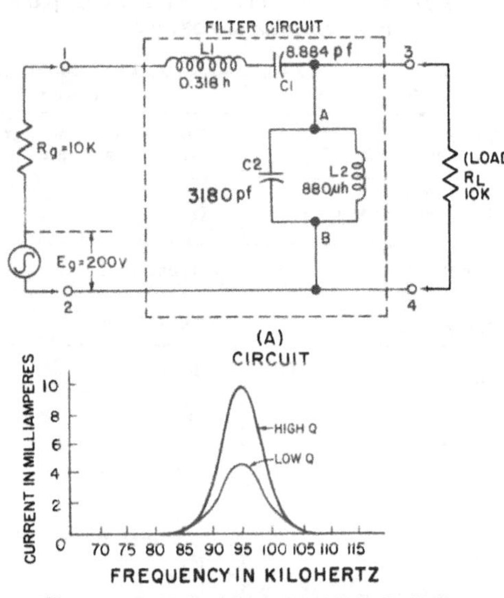

13.25

Figure 10-15.—Bandpass filter.

inductance and capacitance are as indicated in the figure. The formulas by which these values are obtained may be found in handbooks on the subject.

The resonant frequency of the series circuit, L1C1, is:

$$f_o = \frac{1}{2\pi\sqrt{LC}}$$

$$f_o = \frac{1}{6.28\sqrt{0.318 \times 8.884 \times 10^{-12}}} \approx \frac{0.159 \times 10^{+6}}{\sqrt{2.825}}$$

$$\approx \frac{159{,}000}{1.68} \approx 95{,}000 \text{ Hz},$$

and for the parallel circuit, L2C2, is

$$f_0 \approx \frac{0.159}{\sqrt{3{,}180 \times 10^{-12} \times 8.8 \times 10^{-4}}} \approx 95{,}000 \text{ Hz}.$$

Thus, both circuits are resonant at the center frequency of the bandpass filter, the upper limit of which is 100 kHz and the lower limit 90 kHz.

At the resonance the impedances of L1 and C1 cancel and maximum current flows through the load, R_L; also, the parallel circuit, C2L2, offers almost infinite impedance and may be considered an open circuit. The inherent resistances associated with the filter components are neglected. Thus, at resonance, with an assumed source voltage of 200 volts and a total impedance of 20,000 ohms (10,000 ohms at the source and 10,000 ohms at the load) the current through the load is approximately

$$I = \frac{E}{R} = \frac{200}{20{,}000} = 0.01 \text{ a., or } 10 \text{ ma.}$$

Below resonance—for example, at 90 kHz—the impedances of L1C1 and C2L2 are such that with the assumed source voltage of 200 volts only about 4 milliamperes flow through the load. The calculations are as follows:

at 90 kHz:

$$X_{L1} = 2\pi f L1 = 6.28 \times 90 \times 10^3 \times 0.318$$

$$\approx 180{,}000 \text{ ohms}$$

$$X_{C1} = \frac{1}{2\pi f C1} = \frac{10}{6.28 \times 90 \times 10^3 \times 8.884}$$

$$\approx 200{,}000 \text{ ohms}$$

$$X_{C1} - X_{L1} = j\,20{,}000 \text{ ohms}$$

$$X_{L2} = 2\pi f L2 = 6.28 \times 90 \times 10^3 \times 0.88 \times 10^{-3}$$

$$\approx 497 \text{ ohms } \angle 90°$$

$$X_{C2} = \frac{1}{2\pi f C2} = \frac{10^{12}}{6.28 \times 90 \times 10^3 \times 3180}$$

$$\approx 556 \text{ ohms } \angle -90°$$

The impedance offered to the flow of current entering and leaving terminals AB is equal to the product of L2C2 impedances divided by their sum.

$$Z_{AB} = \frac{X_{L2}\,X_{C2}}{+jX_{L2} - jX_{C2}} = \frac{(497 \angle 90°)(556 \angle -90°)}{+j497 - j556}$$

$$= \frac{(497 \times 556) \angle 0°}{59 \angle -90°} \approx 4700 \angle 90° \text{ ohms.}$$

The parallel impedance, Z_{AB34}, of R_{LOAD} and Z_{AB} is equal to their product divided by their sum.

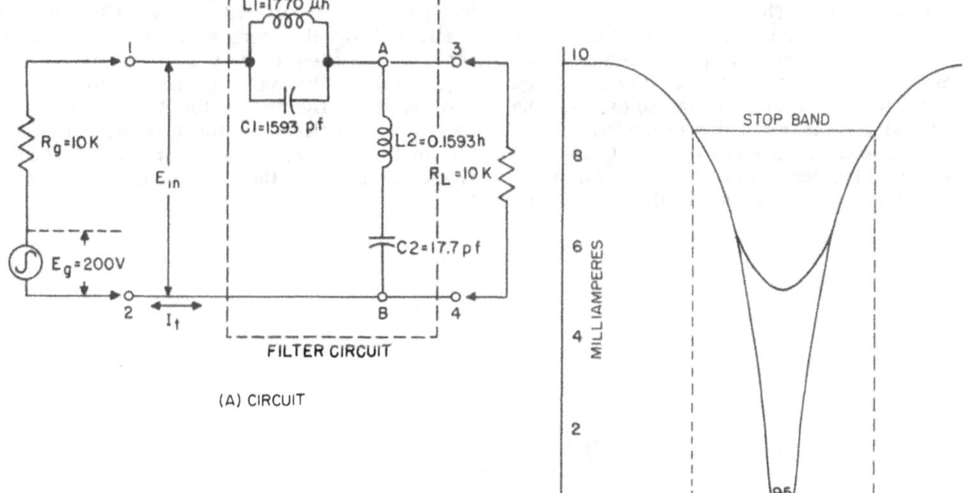

20.69
Figure 10-16.—Band-elimination filter.

$$Z_{AB34} = \frac{Z_{AB}\,R_L}{R_L + Z_{AB}} = \frac{(4700\angle 90°)(10{,}000\angle 0°)}{10{,}000 + j4700}$$

$$= \frac{(4700 \times 10{,}000)\angle 90°}{11080\angle 25.4°} \approx 4240\angle 64.6°$$

$$\approx 1830 + j3830 \text{ ohms.}$$

The total impedance Z_t of the circuit is equal to the vector sum of R_g, X_{L1}, X_{C1}, and Z_{AB34}.

$$Z_t = 10{,}000 + j180{,}000 - j200{,}000 + 1830 + j3830$$

$$= 11{,}830 - j16{,}170$$

$$\approx 20{,}000\angle -53.7° \text{ ohms}$$

$$I_t = \frac{E_g}{Z_t} = \frac{200}{20{,}000} = 0.010 \text{ a.}$$

$$E_{34} = I_t Z_{AB34}$$

$$= 0.010 \times 4240$$

$$= 42.4 \text{ v.}$$

$$I_{LOAD} = \frac{E_{AB34}}{R_{LOAD}}$$

$$= \frac{42.4}{10{,}000}\ 0.00424 \text{ a., or } 4.24 \text{ ma.}$$

Further below resonance the current through the load is even less; and at 75 kHz the load current drops to approximately 0.0024 milliamperes. The same relative decrease in current occurs through the load with a corresponding increase in frequency. Figure 10-15B is a graph of the load current vs frequency characteristic of the filter shown in figure 10-15A.

BAND-ELIMINATION FILTERS

A band-elimination filter (or band-suppression filter) is designed to suppress current of all frequencies within a continuous band, limited by the lower and upper cutoff frequencies, and to pass all frequencies below or above that band. A simple band-suppression filter is shown in figure 10-16A. This type of filter is just the opposite of the bandpass filter; currents of frequencies within the band are greatly attenuated or weakened. The series- and parallel-tuned circuits are tuned to the center of the band to be eliminated. The parallel-tuned circuit in series with the source offers a high impedance to this band of frequencies, and the series-tuned circuit in shunt with the load offers very low impedance; therefore, the signals are blocked and diverted from the load. All other currents, that is, currents at all frequencies outside the stop band (fig. 10-16B) pass through the parallel circuit with very little opposition unaffected by the series-tuned circuit since it acts as an open circuit at these frequencies.

Representative voltage and component values are shown in figure 10-16A. Proof of the selectivity or rejectivity of the circuit, however, is left to the reader. The current versus frequency characteristics of the filter are shown in figure 10-16B.

CHAPTER 11

INTRODUCTION TO RECEIVERS

An AM receiver processes amplitude modulated signals received by its antenna, and delivers as an output a reproduction of the original signal that modulated the RF carrier at the transmitter. The signal can then be applied to some reproducing device such as a loudspeaker or a set of headphones. Actual AM receivers vary widely in complexity. Some are very simple but contain a relatively large number of circuits.

A receiver must perform certain basic functions in order to be useful. These functions, in order of their performance, are: reception, selection, detection, and reproduction.

Reception involves having a transmitted electromagnetic wave pass through the receiver antenna in such a manner as to induce a voltage in the antenna.

Selection involves being able to select a particular station's frequency from all the transmitted signals that happen to be induced in the receiver's antenna at a given time. This is called the selectivity of the receiver. The better a receiver is at differentiating between desired and undesired signals, the better is the receiver's selectivity rating. This function is accomplished in the frequency-selection circuits.

Detection is the action of separating the low frequency intelligence from the high frequency carrier and is accomplished in the detector circuits.

Reproduction is the action of converting the electrical signals to sound waves which can then be interpreted by the ear as speech, music, etc.

BASIC RECEIVER

Figure 11-1 shows the block diagram of a simple receiver which will perform all the functions required of a receiver. The inputs to the receiver are the electromagnetic waves propagated from the antennas of the transmitters. These waves will induce small a.c. voltages in the receiver antenna. These a.c. voltages cause signal current variations in the primary, L1, of the input transformer, creating magnetic fields around L1. The section of the receiver formed by the antenna and L1 perform the function of reception.

The magnetic fields of L1 induce voltages into the secondary, L2. L2 and C1 form a tuned tank circuit with C1 being variable to permit tuning so that only one of the RF signals within its tuning range is selected. When the tank is tuned to the desired frequency, it resonates and develops a relatively large circulating current. The input to the detector is then a relatively large signal voltage at the resonant frequency and minimum signal voltage at other frequencies. Thus, L2 and C1 perform the function of selection.

It should be remembered that the amplitude modulated signal, as it leaves the transmitter with a single audio frequency as the intelligence, is actually composed of energy at three distinct frequencies. These frequencies are the carrier frequency, and two (upper and lower) sideband frequencies. The separation between the maximum limits of the upper and lower sideband frequencies constitutes the bandwidth of the transmitted signal, while the separation between either sideband and the carrier is equal to the intelligence frequency. To perform its function properly, the frequency-selection circuit must pass both sidebands.

Figure 11-2 shows the characteristics of an ideal tuned circuit and an actual circuit. The circuit is tuned to receive an RF carrier at 820 kHz containing a 5-kHz audio signal. In the ideal circuit, all the frequencies contained in the 10-kHz bandwidth receive equal amplification, while any frequency outside the bandwidth would receive zero amplification. In the actual circuit though, all the frequencies contained in the bandwidth (measured from .707 points) do not receive equal amplification and frequencies outside the bandwidth receive some amplification.

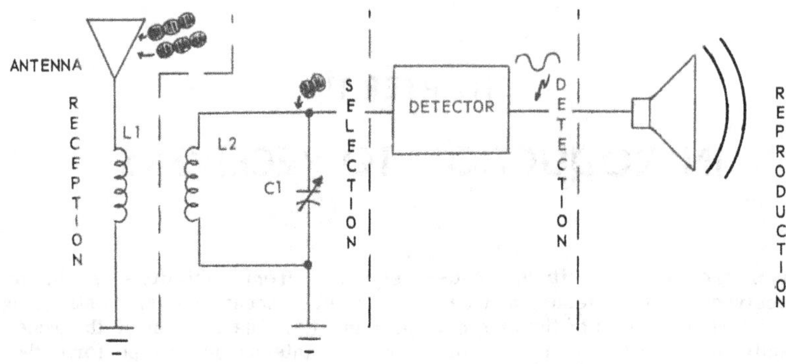

179.200
Figure 11-1.—Basic receiver.

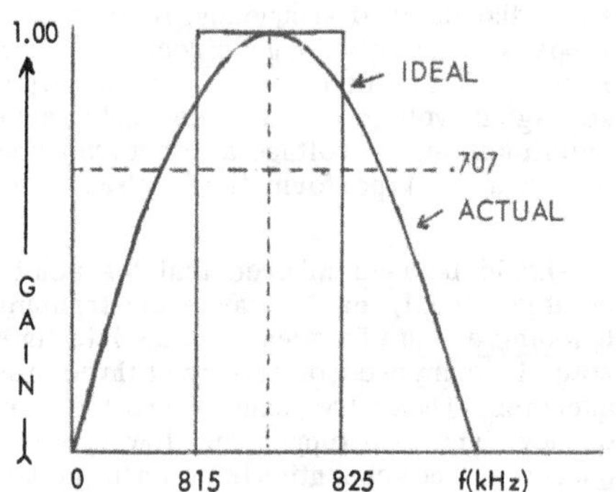

179.201
Figure 11-2.—Tuned circuit characteristic curve.

The ability of a tank circuit to select a band of frequencies is a function of its Q. The higher the Q, the smaller the bandwidth. Normally, the tank circuit represents a compromise between high Q for good selectivity and low Q for fidelity. The Q of the tuned circuit should be high enough to reject signals outside of the desired bandwidth but no so high as to reject the desired sidebands. This is necessary to have linear reproduction of the intelligence contained in the sidebands.

The signal that has been selected by the frequency-selection circuits still consists of the carrier and the two sidebands. The next function of the receiver is to separate the desired intelligence from the carrier. This is accomplished in the detector stage.

When a radio signal is amplitude modulated, the intelligence is contained in the relationship between the carrier and sideband frequencies. This composite signal results in a symmetrical RF envelope, of which the average value of amplitude is zero. As long as the average value of the signal is zero, no useful energy may be extracted. The detector is a non-linear device which will cause the RF signal to be distorted. The distortion introduced by the detector will cause the average value of the waveform to be other than zero and vary in accordance with the original modulating signal. This distortion, called heterodyning, takes place within the detector. The output will be a series of pulses containing many frequencies. The predominant components will be the original carrier and sideband frequencies plus their sums and differences. A filter circuit will then remove all of the unwanted frequencies while preserving the desired intelligence.

The output from the detector circuit consists of a weak signal voltage like the original signal used to modulate the transmitter. This signal is fed to a reproducer which then converts the signal to sound waves. The reproducer may be a pair of headphones or a loudspeaker.

RECEIVER CHARACTERISTICS

Receiver characteristic measurements are useful in determining operational conditions and as an aid for comparison to other units. Important receiver characteristics are sensitivity, selectivity, fidelity, and noise.

The ability of a receiver to reproduce the signal of a very weak station is a function of the receiver's sensitivity. In other words, the weaker a signal that can be applied to a receiver, and still achieve the same value of signal output, the better is that receiver's sensitivity rating. The sensitivity of a receiver is expressed in terms of the signal voltage, usually in microvolts, that must be applied to the antenna input terminals to give a standard output. The output may be an a.c. or d.c. voltage measured at the detector output or a power measurement at the loudspeaker or headphone terminals.

This sensitivity measurement, however, could be made only on an ideal receiver. All receivers generate a certain amount of noise which must be taken into account. This noise is a limiting factor on the minimum usable signal that the receiver can process and still deliver a usable output. An indication of this sensitivity, taking the receiver noise into account, is the amplitude of the signal at the receiver input required to give a signal-plus-noise output some stated ratio above the noise output of the receiver.

The ability of a receiver to reject unwanted signals is determined by its selectivity. This is controlled by the frequency-selection circuits which are tuned tank circuits. Two characteristics of a tank circuit are important in determining the selectivity. They are the slope and the passband, which in turn are governed by the Q of the tank. Figure 11-3 shows the characteristic curves of two tank circuits, one with a high Q, and the other with a low Q.

The passband of the low Q circuit (BW1) is much wider than that of the high Q (BW2) circuit. Due to the slope of the sides of the response curves, frequencies falling outside the passband of the high Q circuit will receive greater attenuation with respect to f_o than will frequencies falling outside the passband of the low Q circuit. Thus, a receiver employing high Q circuits will have good selectivity.

The fidelity of a receiver determines the highest modulating frequency, contained in the sidebands, that can be reproduced and is a function of bandwidth. A low Q circuit has a wide passband and will pass sidebands containing high frequency signals. As a result, receiver design represents a compromise between high selectivity and good fidelity.

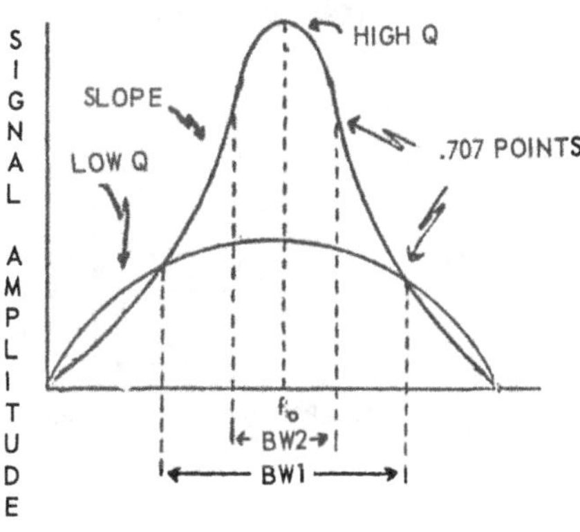

179.202

Figure 11-3.—Characteristic curves of two tank circuits.

CRYSTAL RECEIVER

In the early days of radio, receivers were rather simple devices. They performed the minimum basic requirements of a receiver, which are reception, selection, detection, and reproduction. Figure 11-4 shows a block diagram of this simple receiver. This was a crystal set, so called because the detector was a diode made from a Galena crystal.

The antenna performs the function of reception; selection of the desired signal is accomplished by tuning the frequency-selection circuit. The detector separates the audio component from the RF component. The audio component is then passed on to the earphones, which perform the function of reproduction by converting the electrical signal into sound waves.

The basic crystal set has several disadvantages that limit its usefulness. Among these are poor sensitivity and selectivity. Also the earphones can be used by only one person at a time. This last problem led to another function being added to the receiver—audio amplification. The audio amplifier produced enough power to operate a speaker so that everybody could hear the reproduced sound waves at the same time. Figure 11-5 shows the block diagram of the receiver with a stage of audio amplification added.

By adding stages of audio amplification, weaker stations that could not be heard when headphones were used alone could now be heard. This was due to the amplifier responding to a smaller

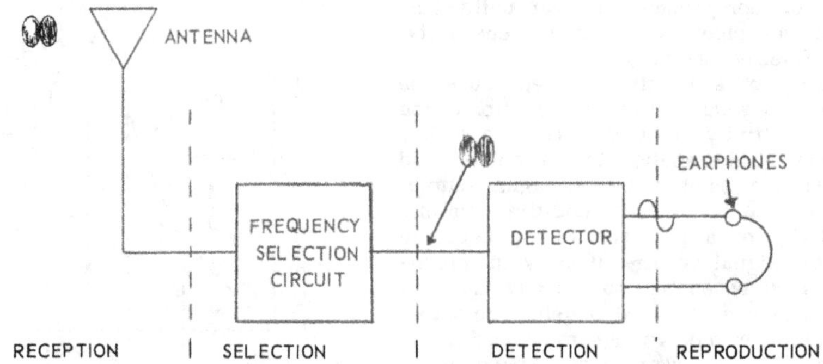

Figure 11-4.—Block diagram of a simple receiver.

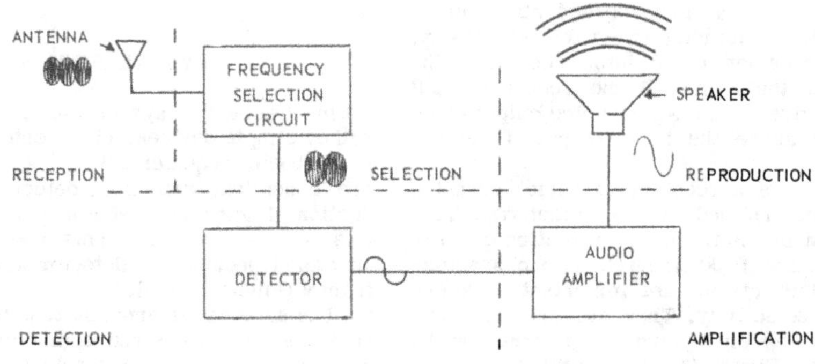

Figure 11-5.—Audio amplified receiver.

audio output component than the detector, but this did not improve receiver sensitivity. Of course, any noise present would be amplified along with the signal, and if the noise was larger in amplitude than the signal, it would drown out the signal. These facts limited the number of stages of audio amplification.

TRF RECEIVER

The sensitivity of the basic receiver is a function of the detector. The detector requires a minimum level to perform its function. If signal level is below this minimum, detection will not take place.

It was discovered that the addition of tuned RF amplifier stages increased not only the sensitivity of the set, but also its selectivity. Figure 11-6 shows the block diagram of a tuned radio frequency (TRF) receiver.

The amplitude of the AM signal at the input of the receiver is relatively small. The RF amplifier stage amplifies the waveform, but does not change its basic shape. The detector separates the audio component from the RF component, and passes the audio component to the audio frequency amplifier. The AF amplifier increases the amplitude of the signal to a value sufficient to operate the loudspeaker.

The sensitivity and selectivity of a TRF receiver are improved by increasing the number

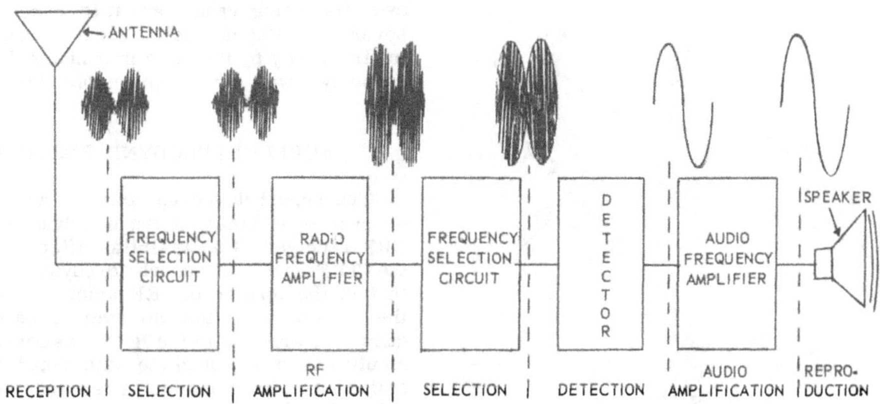

Figure 11-6.—Block diagram of a TRF receiver.

of RF stages. Figure 11-7 shows the block diagram of a TRF receiver having two stages of RF amplification. The tuned tanks of each RF stage must be tuned to the same frequency. Therefore, when changing frequency, you must change all of the tanks to the new frequency. When two or more stages are used, the tuning capacitors are ganged (gang tuning) on the same shaft. This is indicated by a dashed line connecting the tanks in figure 11-7.

The improved selectivity of the TRF receiver is due to the presence of several tuned stages. Figure 11-8 shows the overall response curve, stage by stage, for figure 11-7. Curve 1

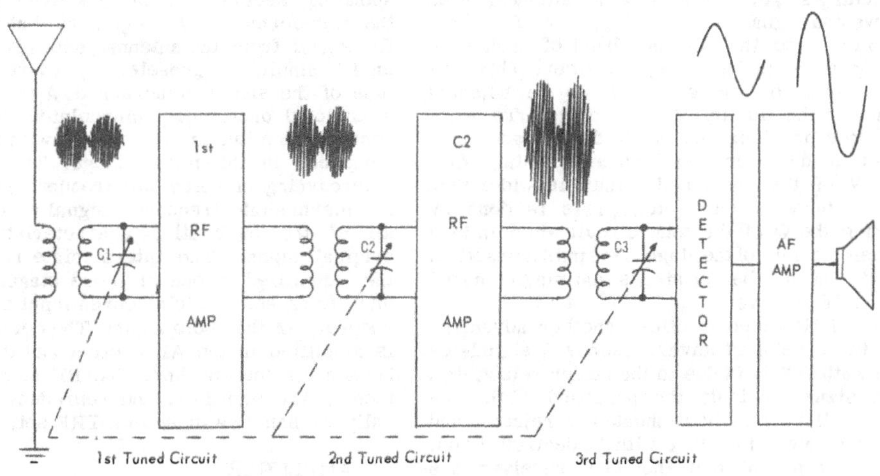

Figure 11-7.—TRF receiver with two stages of RF amplification.

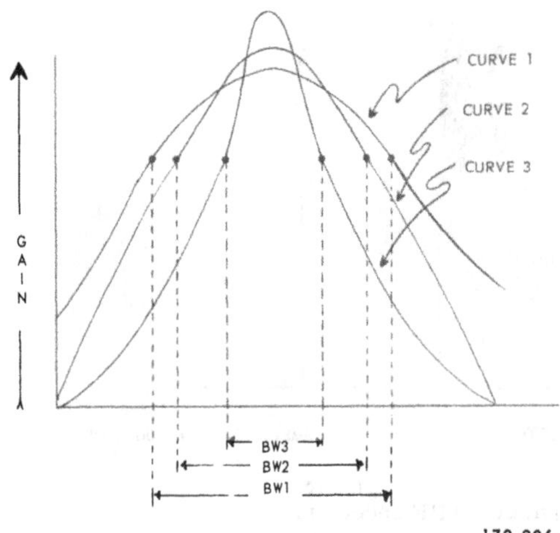

Figure 11-8.—Response curve for figure 11-7.

represents the response cruve of the 1st tuned circuit. Curve 2 is the overall response curve of the 1st RF amplifier stage (including the 1st and 2nd tuned circuits). Curve 3 is the overall response curve of the two RF amplifiers in cascade (including the three tuned circuits). Notice that slope of the sides becomes steeper for each succeeding stage. This results in better rejection of unwanted signals.

Notice also that the passband of each succeeding stage is becoming narrower. This will cause some of the sidebands to be rejected and lower the fidelity of the receiver. To overcome this problem, the bandwidth of each tuned stage is made wider than the previous stage (i.e., the BW of the 2nd tuned circuit is wider than the 1st tuned circuit, etc.). This is done by lowering the Q of the tank circuit which in turn reduces the gain of the stage. This puts a practical limit to the number of stages that may be added to the TRF receiver.

The TRF receiver offers another advantage over the crystal receiver, improved signal-to-noise ratio. This is due to the better selectivity. Noise signals outside the passband of the receiver will be greatly attenuated or rejected and therefore cannot interfere with the desired signal.

It is apparent that the TRF receiver has several advantages over the crystal receiver. These are: better sensitivity, selectivity, and signal-to-noise ratio. There are disadvantages, though. Selectivity does not remain constant over its tuning range, and it is difficult to make several RF stages track (all tuned circuits changing frequency by the same amount simultaneously) properly over any appreciable tuning range.

SUPERHETERODYNE RECEIVER

The superheterodyne receiver was developed to overcome many of the disadvantages of the TRF receiver. The essential difference between the TRF and the superheterodyne receiver is that in the former the RF amplifiers preceding the detector are tunable over a band of frequencies, whereas in the latter the corresponding amplifiers are tuned to one fixed frequency called the Intermediate Frequency (IF). The principle of frequency conversion by heterodyne action is employed to convert any desired station frequency within the receiver range to this intermediate frequency. Thus, an incoming signal is converted to the fixed intermediate frequency before detecting the audio signal component, and the IF amplifier operates under uniformly optimum conditions throughout the receiver range. The IF circuits thus may be made uniformly selective, high in voltage gain, and of satisfactory bandwidth to contain all of the desired sideband components associated with the AM carrier.

The block diagram of a typical superheterodyne receiver is shown in figure 11-9. Below corresponding sections of the receiver are shown the waveforms of the signal at that point. The RF signal from the antenna passes first through an RF amplifier (preselector) where the amplitude of the signal is increased. A locally generated (local oscillator) unmodulated RF signal of constant amplitude is then mixed with the carrier frequency in the mixer stage. The mixing or heterodyning of these two frequencies produces an intermediate frequency signal which contains all of the modulation characteristics of the original signal. The intermediate frequency is then amplified in one or more stages called IF amplifiers and fed to a conventional detector for recovery of the audio signal. The detected signal is amplified in the AF section and then fed to a headset or loudspeaker. The RF amplifier, detector, AM amplifier, and reproducer are basically the same as those in a TRF set.

RF AMPLIFIER

The RF stage amplifies the small a.c. voltages induced in the antenna by the electromagnetic

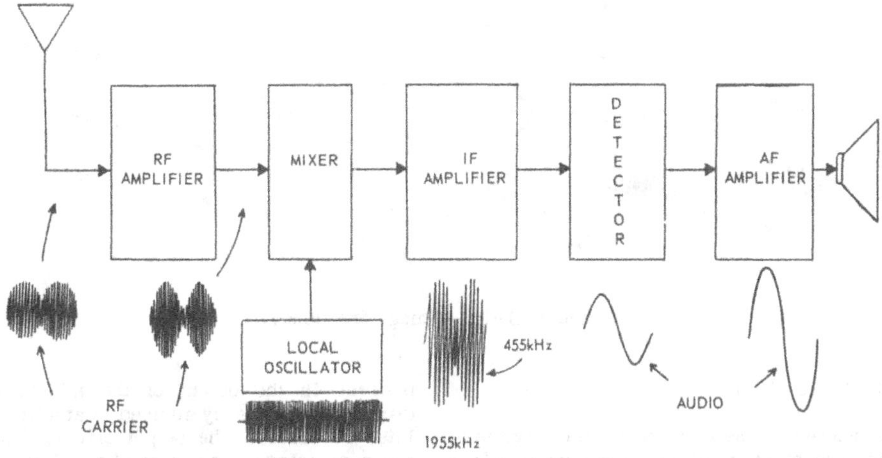

Figure 11-9.—Typical superheterodyne receiver block diagram.

wave from the station transmitter. Utilizing a tuned circuit between the antenna and the input of the RF amplifier permits selection of the desired station frequency from among the many present in the antenna.

Besides amplifying the RF signal, the RF amplifier isolates the local oscillator from the antenna. If the antenna were connected directly to the mixer stage, a part of the local oscillator signal might be radiated into space.

If the mixer stage were connected directly to the antenna, unwanted signals, called images, might be received, because the mixer stage produces the intermediate frequency by heterodyning two signals whose frequency difference equals the intermediate frequency.

The image frequency always differs from the desired station frequency by twice the intermediate frequency. Image frequency = station frequency ± (2 x intermediate frequency). The image frequency is higher than the station frequency if the local oscillator frequency tracks above the station frequency, as generally used for lower frequencies. The image frequency is lower than the station frequency if the local oscillator tracks below the station frequency, as generally used for the higher frequencies. This is shown in figure 11-10. For example, if a superheterodyne receiver having an intermediate frequency of 455 kHz is tuned to receive a station frequency of 1500 kHz (fig. 11-10), and the local oscillator has a frequency of 1045 kHz, the output of the IF amplifier may contain two signals—one from the 1500-kHz station and the other from an image station of 590 kHz: 1500 - (2 x 455) = 590 kHz. The same receiver tuned near the low end to a 900-kHz station has a local oscillator frequency of 1355 kHz. The output of the IF amplifier contains the station signal and may contain an image signal: 1810 kHz image signal = 900 kHz station signal + (2 x 455 kHz IF signal). Thus, the 1810-kHz signal is an image which may be heard simultaneously with the 900-kHz station signal.

It may also be possible for any two signals having sufficient strength and separated by the intermediate frequency to produce unwanted signals in the reproducer. The selectivity of the preselector (RF stage preceding the mixer stage) tends to reduce the strength of these unwanted signals. However, there is a practical limit to the degree of selectivity obtainable in the preselector due to the fact that the RF stage must have a wider bandwidth than the bandwidth of the desired signals. The ratio of the amplitude of the desired station signal to that of the image is called the image rejection ratio and is an important characteristic of a superheterodyne receiver.

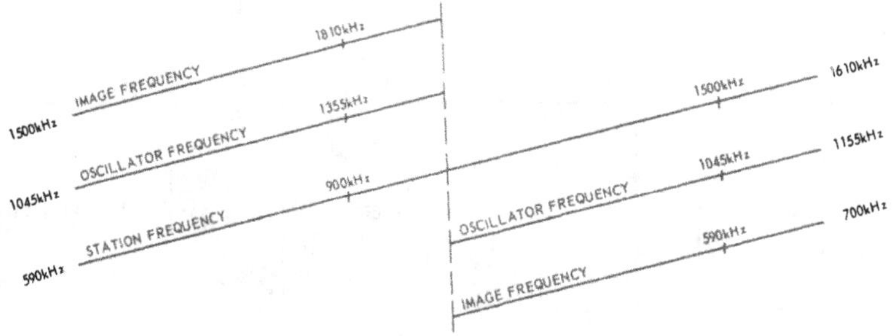

Figure 11-10.—Image frequency. 179.207

LOCAL OSCILLATOR

The function of the local oscillator stage is to produce a constant amplitude sine wave of a frequency which differs from the desired station frequency by an amount equal to the intermediate frequency of the receiver. Although the oscillator may be operated either above or below the station frequency, in most broadcast band receivers the oscillator is operated above the station frequency. In order to allow selection of any frequency within the range of the receiver, the tuned circuits of the RF stage and the local oscillator are variable. By using a common shaft, or ganged tuning for the variable component of the tuned circuits both circuits may be tuned in such manner as to maintain the difference between the local oscillator frequency and the incoming station frequency equal to the receiver IF.

MIXER

The function of the mixer stage is frequency conversion by heterodyne action. The input to the mixer consists of two signals: the modulated RF signal and the unmodulated local oscillator signal. The mixer then combines, or mixes, these two signals. As a result of this mixing action, the output of the mixer will contain four major frequencies plus many minor frequencies. The four major frequencies are: (1) the original signal frequency, (2) the local oscillator frequency, (3) the sum of the signal and oscillator frequencies, and (4) the difference of the signal and oscillator frequencies. The additional frequencies present are produced by combinations of the fundamentals and harmonics of the signal and oscillator frequencies. Of the frequencies present in the output of the mixer, only the difference frequency is used in amplitude modulated receivers. The output circuit of the mixer stage contains a tuned circuit which is resonated at the difference frequency. The output of the mixer is fed to the IF amplifier and consists of a modulated intermediate frequency signal.

IF AMPLIFIER

Superheterodyne receivers employ one or more IF amplifiers depending on design and quality of the receiver. Transformers are usually used for interstage coupling in the IF section. The IF circuits are permanently tuned to the difference frequency between the incoming RF signal and the local oscillator. As previously stated, all incoming signals are converted to the same frequency by the mixer stage, and the IF amplifier operates at only one frequency. The tuned circuits, therefore, are permanently adjusted for maximum signal gain consistent with the desired bandpass and frequency response. These stages operate as class A voltage amplifiers, and practically all of the selectivity (adjacent channel and interference frequencies, not image frequency) of the superheterodyne receiver is determined by them.

The values of intermediate frequencies for broadcast receivers range from 130 kHz to 485 kHz, with the most popular value being around 455 kHz. The output of the IF amplifier is fed to the detector.

DETECTOR, AF AMPLIFIER, AND REPRODUCER

The receiver functions of detection, AF amplification, and reproduction are performed by the detector, AF amplifier, and speaker in the same manner previously described for the TRF receiver.

CHARACTERISTICS OF THE SUPERHETERODYNE RECEIVER

Since the IF stages operate at a single frequency, the superheterodyne receiver may be designed to have better selectivity and sensitivity across the entire broadcast band and better gain per stage than the TRF receiver. Although a superheterodyne receiver contains more tuned stages than a TRF receiver, the majority of the stages are fixed tuned at one frequency. This reduces the tracking problem and makes alignment of the tuned stages much easier. The major disadvantage of the "superhet" is the reception of image frequencies.

THE DECIBEL

Because of the need for some unit which expressed the comparison of output power to input power, a unit known as the bel was originated for the use in power measurements. If the larger value of power to be compared is designated as P_2 and the smaller value as P_1, then bel = $\log \frac{P_2}{P_1}$. Since this unit is too large for general use, it is more convenient to use a smaller one, the decibel, which is equal to one-tenth of a bel, or db = $10 \log \frac{P_2}{P_1}$. Both of these units are relative measurements and do not specify any definite amount of power. They are the logarithmic ratio between any two values of power.

Decibels indicate a ratio change of power, and are useful as a measure of how an electronic device affects the transmission of energy through itself. Since the decibel is a logarithmic unit, the gains (+db) and losses (-db) in a circuit can be added algebraically to determine the overall net effect of the circuit.

Decibels express power ratios, not voltage ratios or current ratios. Power ratios, in decibels, may be computed when appropriate voltages, currents, and impedances are known. When $Z_{in} = Z_{out}$, one of the following equations may be used: db = $20 \log \frac{E_2}{E_1}$; db = $20 \log \frac{I_2}{I_1}$. To properly apply decibels, a review of logarithms is helpful.

LOGARITHMS

Until the 17th century, people who used mathematics in their work were constantly faced with the necessity of laboriously carrying out their computations. In the 17th century, through the contributions of men such as Napier, Briggs, and others, a system called LOGARITHMS based on the use of exponential laws was developed. With this system, the number of mathematical computations was reduced. For example, in astronomy, the use of logarithms reduced computation time from months to a few days.

The word "logarithm" is derived from two Greek words, LOGOS and ARITHMOS, which mean, respectively, "proportion" and "number." This combination of words was selected because of the way through which the first table of logarithms came into being. By the use of logarithms, the process of multiplication is reduced to addition and the process of division is reduced to subtraction; raising a number to a power is reduced to simple multiplication; and extracting a root is reduced to simple division.

This section on logarithms is to serve as a review for those who need it and as a source of material for all. A more detailed discussion is contained in Mathematics, Volume I, NP 10069( ).

A Logarithm Is an Exponent

In the example, $10^2 = 100$, it will be recalled that 10 is the base, 100 is the number, and 2 is the exponent or power. The exponent, 2, in this application may also be called a logarithm of the number 100 in the base 10 system. Specifically, the logarithm of a number to a given base is the exponent by which the base must be raised to yield that number. In the previous example, the exponent of the base 10 is the number 2. The logarithm of the number 100 must, therefore, be the number 2 when the base of the system is assumed to be 10.

Written as a logarithm:

$$10^2 = 100 \text{ is } \log_{10} 100 = 2$$

Thus, by definition, the logarithm and the exponent are the same number. This name "logarithm" for an exponent was developed to denote a special application of the laws of exponents.

Any number may be used as base for a system of logarithms. The selection of a base is a matter of convenience. Briggs, the originator of the common logarithm system presently used, found in 1617, that the base ten possessed many advantages not obtainable in ordinary calculations with other bases. The selection of 10 as a base proved as satisfactory that it is used almost exclusively in ordinary calculations. Logarithms with a base 10 are called common logarithms.

Common Logarithms

When the number 10 is used as a base, it is not necessary to so indicate. It is understood to be 10 when no other base is given. As an example, instead of writing $\log_{10} 100 = 2$, the expression, $\log 100 = 2$ is satisfactory.

Table 11-1 illustrates the relationship between the powers of 10 and the logarithms of certain numbers. Notice that only numbers which can be represented as a power of ten have whole numbers for logarithms. Also notice that the log of any number between 100 and 1000 would be between the numbers 2 and 3. That is, the logarithm of a number between 100 and 1000 would be the number 2 plus some decimal quantity.

Example: The log of the number 67 would be 1 plus some decimal value.

Because:

Exponential Form	Logarithmic Form
$10^1 = 10$	$\log 10 = 1$
$10^{1+x} = 67$	$\log 67 = 1 + x$
$10^2 = 100$	$\log 100 = 2$

Table 11-1.—Typical conversions from exponential form to logarithmic form

EXPONENTIAL FORM	LOGARITHMIC FORM
$10^3 = 1000$	$\log 1000 = 3.00000$
$10^2 = 100$	$\log 100 = 2.00000$
$10^1 = 10$	$\log 10 = 1.00000$
$10^0 = 1$	$\log 1 = 0.00000$
$10^{-1} = 0.1$	$\log 0.1 = -1.00000$
$10^{-2} = 0.01$	$\log 0.01 = -2.00000$
$10^{-3} = 0.001$	$\log 0.001 = -3.00000$

179.208

Log 10 equals 1 and the log of 100 equals 2. Therefore, the log 67 (a number between 10 and 100) would be between the numbers 1 and 2, or the number 1 plus a decimal.

To present the logarithm of any number, it is necessary to utilize decimal powers.

EXAMPLE:

$$10^{2.5563} = 360$$
$$\log 360 = 2.5563$$

Likewise:
$$10^{2.5224} = 333$$
$$\log 333 = 2.5224$$

Using the same logic, it follows that the logarithm of a number between 0.1 and 0.01 would be between -1 and -2.

EXAMPLE:

log 0.03 =

Exponential Form	Logarithmic Form
$10^{-1} = 0.1$	$\log 0.1 = -1$
$10^{-2+x} = 0.03$	$\log 0.03 = -2 + x$
$10^{-2} = 0.01$	$\log 0.01 = -2$

Every logarithm is composed of two parts. The whole part is called a CHARACTERISTIC. The characteristic will always be positive for any number greater than one, and negative for any number less than one. The decimal portion of the logarithm is called the MANTISSA and by convention is always kept positive. The characteristic of a number being converted to a logarithm may be determined by inspection as shown in table 11-1, or it may be determined in the following manner:

Place the number to be expressed as a logarithm in standard form (exponent notation—chapter 3). The characteristic of the number will be equal to the exponent to which the number 10 is raised.

EXAMPLE: What is the characteristic of the number 684? Expressing the number in standard form:

$$684 = 6.84 \times 10^2$$

The exponent, 2, is the characteristic of the number 684.

EXAMPLE: What is the characteristic of the number 0.0684? Expressing the number in standard form:

$$0.0684 = 6.84 \times 10^{-2}$$

Therefore, the characteristic of the number 0.0684 is -2.

If the characteristic is negative, it is customary to place the negative sign above the number which is the characteristic. In the previous example, the characteristic would be written $\overline{2}$.

EXAMPLE:

$$\log 0.0461 = \log 4.61 \times 10^{-2} = \overline{2}.6637$$

The $\overline{2}.6637$ means -2 + 0.6637. If the minus sign precedes both the characteristic and the mantissa, it would tend to indicate that both the characteristic and the mantissa are negative. This, of course, would be incorrect because the mantissa must be positive. Therefore, confusion regarding the sign is eliminated by placing the sign above the characteristic.

Another method used to indicate a negative characteristic is to add a positive number (usually ten) to the characteristic and subtract the same number from the mantissa. If the absolute value of the characteristic is greater than ten, the number chosen to be added to the characteristic and subtracted from the mantissa is the next higher multiple of ten.

EXAMPLE: log 0.0461 = $\overline{2}$ plus 0.6637

May be written:

$$10 - 2 + 0.6637 - 10$$

Simplifying:

$$8.6637 - 10$$

Since the characteristic of the logarithm of a number may be found by inspection, it is necessary to calculate only the mantissas of the logarithms. Mantissas can be derived by use of advanced mathematics. However, for convenience, the decimal part of the logarithm has been computed and placed in tables. A list of numbers with their associated mantissas can be found in the Table of Logarithms in the appendix. To find the logarithm of a number:

1. Place the number in standard form.
2. Write the characteristic before locating the mantissa in the table.
3. Find the mantissa which corresponds to the significant figures of the number in the log table.
4. Add the characteristic and mantissa to produce the logarithm of the number.

EXAMPLE: Find the log of the number 232.

$$232 = 2.3 \times 10^{2}$$

The exponent 2 is the characteristic. The significant figures are 232. The first two significant figures are found under the column marked N in the log table. The third significant figure, 2, is found under the column marked 2. The mantissa of the significant figures 232 is .3655. Therefore, the log 232 equals the characteristic, 2, plus the mantissa, .3655; or 2 + .3655.

EXAMPLE: Find the log of 0.00248

$$0.00248 = 2.48 \times 10^{-3}$$

Therefore, $\overline{3}$ is the characteristic. The significant figures are 248. The first two significant figures are found in the column marked N in the log tables. The third significant figure, 8, is found under the column marked 8. The mantissa is the decimal number in the row containing 24, and in the column under 8. The mantissa of the significant figures 248 is .3945. Therefore, the log of 0.00248 is -3 + .3945 or $\overline{3}.3945$, or 7.3745 - 10.

Finding a Number Corresponding to a Given Logarithm

In most problems involving logarithms, it is necessary to find not only the logarithm of numbers, but also to use the inverse process to find the number corresponding to a logarithm—the ANTILOGARITHM. The word "antilogarithm" is abbreviated antilog.

When a mantissa of a logarithm is given exactly in the table, finding the antilog is relatively simple.

EXAMPLE: What number has 3.3874 for its logarithm? The process of determining the antilog is as follows:

1. Find the mantissa in the tables. Remember, the characteristic is not part of the mantissa.
2. Find the significant figures that correspond to the mantissa, 3874. Write the significant figures as a number between 1 and 10 (244 is the significant figure). It should be written as 2.44.
3. Since the characteristic is 3, the significant figures will be multiplied by 10 raised to the third power.
4. Therefore, antilog 3.3874 = 2.44×10^{3}, or equal to 2,440.

EXAMPLE: Find the number which has 2.3522 for its logarithm.

The mantissa corresponds to a number equal to 2.25 as found in the log tables. Since the characteristic is the number 2, the number will be written as:

$$2.25 \times 10^2 = 225$$

Therefore, the logarithm of the number 225 is 2.3522.

When the logarithm tables are used to find the antilogarithm, you may occasionally find that the mantissa falls between two numbers. If you encounter a situation of this type and the number is exactly one half or more of the difference between the two numbers, then you should use the next higher number. If the number is less than half, use the lower number.

EXAMPLE: Find the numbers which have 2.4560 and 2.4539 for their logarithms.

The mantissa for 2.4560 corresponds to a number that falls between 2.85 and 2.86 as found in the log tables. The mantissa for 2.85 is 4548 and the mantissa for 2.86 is 4564. Since the mantissa 4560 lies more than one half the difference between the two numbers, you would use the higher number, 2.86. As shown previously, since the characteristic is the number 2, the number will be written as:

$$2.86 \times 10^2 = 286$$

The mantissa for 2.4539 corresponds to a number that falls between 2.84 and 2.85 as found in the log tables. The mantissa for 2.84 is 4533 and the mantissa for 2.85 is 4548. Since the mantissa 4539 lies less than one half the difference between the two numbers, you would use the lower number, 2.84. The characteristic is again the number 2, and the number is written as:

$$2.84 \times 10^2 = 284$$

Addition and Subtraction of Logarithms

All logarithms are added and subtracted arithmetically, but, the mantissa must be kept positive. Therefore, every negative characteristic is added to a negative number.

EXAMPLE: Add the logarithm of 4.3010 and $\overline{6}.8513$.

The negative characteristic must be changed to a positive characteristic before the addition can be effected.

Therefore:

$$\overline{6}.8513 = 4.8513 - 10$$

Completing the additions arithmetically:

$$\begin{array}{r} 4.3010 \\ \underline{4.8513 - 10} \\ 9.1523 - 10 \end{array}$$

or:

$$9.1523 - 10 = \overline{1}.1523$$

EXAMPLE:

Add the logarithm 3.7076 and 6.6893 - 10.

$$\begin{array}{r} 3.7076 \\ \underline{6.6893 - 10} \\ 10.3969 - 10 \end{array}$$

or:

$$10.3969 - 10 = 0.3969$$

EXAMPLE:

Subtract the logarithm 4.6721 from the logarithm 3.7076.

To subtract a larger logarithm from a smaller, add 10 to the smaller logarithm and subtract the same number from the logarithm by writing the number added with a minus sign to the right of the logarithm.

$$\begin{array}{rr} 3.7076 = & 13.7076 - 10 \\ & \underline{4.6721 } \\ & 9.0355 - 10 \end{array}$$

or:

$$9.0355 - 10 = \overline{1}.0355$$

EXAMPLE:

Subtract the logarithm 3.4298 from the logarithm 1.5224.

$$\begin{array}{rr} 1.5224 = & 11.5224 - 10 \\ & \underline{3.4298 } \\ & 8.0926 - 10 \end{array}$$

or:

$$8.0926 - 10 = \overline{2}.0926$$

Multiplication Using Logarithms

LAW: The logarithm of a product is equal to the sum of the logarithms of the factors.

Consider the two factors 10 and 100. The common logarithms of two factors are 1 and 2 respectively.

Therefore:

$1 = \log 10$ (1)
and: $2 = \log 100$ (2)

Writing equation (1) in exponential form:

$10^1 = 10$

Writing equation (2) in exponential form:

$10^2 = 100$

Therefore:

$10 \times 100 = 10^1 \times 10^2$

When multiplying two factors with exponents, add their exponents. Keep in mind that an exponent is a logarithm.

Therefore:

$10^1 \times 10^2 = 10^{1+2} = 10^3$
$\log (10 \times 100) = 1 + 2 = \log 10 + \log 100$

The latter equation stated as a law is:

Multiplication factors may be accomplished by adding their logarithms, or

$\log A \times B = \log A + \log B.$

EXAMPLE:

Find the product of 2.1 x 336.
Let X equal the desired product.

Then:

$X = 2.1 \times 336$
Taking the logarithm of both sides:
$\log X = \log (2.1 \times 336)$

Applying the law of logarithms
$\log AB = \log A + \log B$:

$\text{Log } X = \log 2.1 + \log 336$

Determining the logs from the tables and adding:

$$\begin{aligned} \log 2.1 &= 0.3222 \\ \log 336 &= \underline{2.5263} \\ & \quad 2.8485 \end{aligned}$$

Therefore:

$\log X = 2.8485$

To solve the equation for X, take the antilog of both numbers.

$\log X = 2.8485$
$\text{antilog} \log X = \text{antilog } 2.8485$
The antilog of log X equals X.

Therefore:

$X = \text{antilog } 2.8485$
$X = 7.06 \times 10^2$
$X = 706$

EXAMPLE:

Given E = IR. Find the value of E when I equals 0.000326 and R equals 621,000.

$\log E = \log IR$
$\log E = \log I + \log R$
$\log E = \log 0.000326 + \log 621{,}000$
$\log E = \bar{4}.5132 + 5.7931$

$$\begin{array}{r} 6.5132 - 10 \\ \underline{5.7931} \\ 12.3063 - 10 \end{array}$$

$\log E = 2.3063$
$\text{antilog } E = \text{antilog } 2.3063$
$E = 2.02 \times 10^2$
$E = 202$

Division Using Logarithms

The logarithm of the quotient of two numbers is the logarithm of the dividend minus the logarithm of the divisor. As with multiplication, this rule is simply an application of the laws of exponents.

Consider the two factors 10 and 100. The common logarithms of the numbers are 1 and 2 respectively.

Therefore:

$1 = \log 10$ (1)
$2 = \log 100$ (2)

Writing equation (1) in exponential form:

$10^1 = 10$

Writing equation (2) in exponential form:

$$10^2 = 100$$

Therefore:

$$\frac{100}{10} = \frac{10^2}{10^1} = 10^{2-1} = 10^1$$

Since the exponents are logarithms:

$$\log \frac{100}{10} = 2 - 1 = \log 100 - \log 10$$

Stating the latter equation as a general law:

Division of factors may be accomplished by subtracting their logarithms, or

$$\text{LOG} \frac{A}{B} = \text{LOG } A - \text{LOG } B.$$

EXAMPLE:

Find the quotient of 37.4 ÷ 1.7 by the use of logarithms. Let X equal the desired quotient.

Therefore:

$$X = 37.4 \div 1.7$$

Stated in the terms of logarithms:

$$\log X = \log 37.4 - \log 1.7$$

Finding the logarithms of the numbers and subtracting:

$$\begin{array}{rl} \log 37.4 = & 1.5729 \\ \text{(minus)} \log 1.7 = & \underline{0.2304} \\ & 1.3425 \end{array}$$

Therefore:

$$\log X = 1.3425$$

Solving the equation by taking the antilog of both members of the equation:

$$\begin{aligned} \text{antilog} \log X &= \text{antilog } 1.3425 \\ X &= \text{antilog } 1.3425 \\ X &= 2.20 \times 10^1 \\ X &= 22 \end{aligned}$$

or:

$$\frac{37.4}{1.7} = 22$$

EXAMPLE:

Find the quotient of $\frac{16.3}{0.008}$. Let X equal the desired quotient:

Therefore:

$$X = \frac{16.3}{0.008}$$

Taking the log of both members of the equation:

$$\log X = \log \frac{16.3}{0.008}$$

Applying the law of logarithm:

$$\log X = \log 16.3 - \log 0.008$$

Finding and subtracting:

$$\begin{array}{rl} \log 16.3 = & 11.2122 - 10 \\ \log 0.008 = & \underline{(-)7.9031 - 10} \\ & 3.3091 \end{array}$$

Notice that the characteristic in the subtrahend 7 was changed from the characteristic -3.

Therefore:

$$\log X = 3.3091$$

Taking the antilog of both members:

$$\begin{aligned} \text{antilog} \log X &= \text{antilog } 3.3091 \\ X &= \text{antilog } 3.3091 \\ X &= 2.04 \times 10^3 \\ X &= 2040 \end{aligned}$$

or:

$$\frac{16.3}{0.008} = 2040$$

Raising a Number to a Power
by the Use of Logarithms

LAW: The logarithm of a power of a number equals the logarithm of the number multiplied by the exponent of the number.

$$\text{LOG } B^A = A \text{ LOG } B$$

EXAMPLE:

Find the value of $(18.5)^5$. Let X equal the desired result.

Then:

$X = (18.5)^5$

Taking the logarithm of both members:

$\log X = \log (18.5)^5$

Applying the law of logarithms:

$$
\begin{aligned}
\log X &= 5 \log (18.5)\\
\log 18.5 &= 1.2672\\
\log X &= 6.3360\\
\text{antilog} \log X &= \text{antilog } 6.3360\\
X &= \text{antilog } 6.3360\\
\text{antilog } 6.3360 &= 2.17 \times 10^6\\
X &= 2{,}170{,}000
\end{aligned}
$$

or:

$(18.5)^5 = 2{,}170{,}000$

Extracting a Root by the Use of Logarithms

LAW: The logarithm of the root of a number is equal to the logarithm of the number divided by the root.

EXAMPLE:

Evaluate $\sqrt[5]{327}$. Let X equal the desired result.

Therefore:

$$
\begin{aligned}
X &= \sqrt[5]{327}\\
\log X &= \log \sqrt[5]{327}\\
\log X &= \log (327)^{1/5}\\
\log X &= \frac{1}{5} \log 327\\
\log 327 &= 2.5145
\end{aligned}
$$

Therefore:

$$
\begin{aligned}
\log X &= \frac{1}{5}(2.5245)\\
\log X &= 0.5029\\
\text{antilog} \log X &= \text{antilog } 0.5029\\
X &= \text{antilog } 0.5029\\
\text{antilog} \log 0.5029 &= 3.18\\
X &= 3.18
\end{aligned}
$$

or:

$\sqrt[5]{327} = 3.18$

Summary

Any number can be used as the base of a system of logarithms. A logarithm whose base is 10 is called common logarithm, abbreviated log.

LAW I: The logarithm of the product of two numbers is the sum of the logarithms of the factors. For example,

$$
\begin{aligned}
\log 2 &= 0.3010\\
\log 3 &= \underline{0.4771}\\
\log 6 &= 0.7781
\end{aligned}
$$

Referring to the log table, it is found that the number whose log is 0.7781 is six. What has been done is this:

$$
\begin{array}{ccccc}
10^{.3010} & \times & 10^{.4771} & = & 10^{.7781}\\
'' & & '' & & ''\\
2 & & 3 & & 6
\end{array}
$$

LAW II: The logarithm of a quotient is the logarithm of the dividend minus the logarithm of the divisor. For example,

$$
\begin{aligned}
\log 6 &= 0.7781\\
\log 3 &= \underline{0.7441}\\
\text{(subtracting)} \quad & 0.3010
\end{aligned}
$$

Referring to the log table, it is found that the number whose log is 0.3010 is two.

LAW III: The logarithm of the nth root of a number is the logarithm of the number divided by n. For example,

$\log 4 = 0.6020$

dividing by two gives 0.3010

Referring to the log table, it is found that the number whose log is 0.3010 is two.

The actual log of a number consists of two parts: An integer called the characteristic, and a decimal part called the mantissa. The mantissa is the part which is found in the log table. The characteristic is found by applying the rules:

(1) If a number N is greater than or equal to one, the characteristic of its log is positive and is numerically equal to the number of digits between the exponent notation and N.

(2) If a number N is less than one, the characteristic of its log is negative and numerically equal to the number of digits between the exponent notation and N. For example,

N	Characteristic
1214	3
62	1
8	0
0.01	-2
0.0036	-3

The mantissa is found in a table of logarithms. The mantissa of the common log of a positive number is independent of the position of the decimal point. The value of the mantissa depends only upon the arrangement of the digits in the number. This can be seen by the following examples:

$5670 = 5.67 \times 10^3 = 10^{.7536} \times 10^3 = 10^{3.7536}$;

therefore, log 5670 = 3.7536

$567 = 5.67 \times 10^2 = 10^{.7536} \times 10^2 = 10^{2.7536}$;

therefore, log 567 = 2.7536

$56.7 = 5.67 \times 10^1 = 10^{.7536} \times 10^1 = 10^{1.7536}$;

therefore, log 56.7 = 1.7536

$5.67 = 5.67 \times 10^0 = 10^{.7536} \times 10^0 = 10^{0.7536}$;

therefore, log 5.67 = 0.7536

When the number is positive but less than one, it might be expected that the log would be written as a negative number. For example,

$0.567 = 5.67 \times 10^{-1} = 10^{.7536} \times 10^{-1} = 10^{-0.2464}$

or log 0.567 = -0.2464

However, in practice, such logs are normally written as a negative characteristic plus a positive mantissa, as log 0.567 = $\bar{1}.7536$, or more commonly 9.7536 - 10. It can be seen that 9.7536 - 10 is the same as minus 0.2464. The form 9.7526 - 10 is more easily manipulated in problems.

USING THE DECIBEL

Decibels represent the logarithmic ratio between any two values of power. If an amplifier with a 10 watt input delivers a 100 watt output, the gain in db would be: db = 10 log P_2/P_1 = 10 log 100/10 = 10 log 10 = 10X1 = 10 db gain.

A second amplifier with a 10 watt input delivers a 1,000 watt output. The db gain would be; db = 10 log 1000/10 = 10 log 100 = 10 x 2 = 20 db gain.

A third amplifier with a 10 watt input delivers a 10,000 watt output. The db gain would be; db = 10 log 10000/10 = 10 log 1000 = 10 x 3 = 30 db gain.

Every time the input power was increased by a factor of 10 by the amplifier (e.g. 10 watts to 100 watts) the decimal point moved one place, and the gain changed by 10 db. Thus, whenever the decimal point shifts one place, a 10 db change takes place. Since the decimal point can be moved in either direction, an increase in power indicates a gain (+db), while a decrease in power indicates a loss (-db). For example, a change in power from 100 watts to 1,000 watts would be indicated by a +10 db while a change in power from 1,000 watts to 100 watts would be indicated by a -10 db. Whether there is a gain (amplification) or loss (attenuation) can usually be determined by inspection, and the appropriate sign (plus or minus) applied.

By always placing the larger number in the numerator, the calculations will be less involved. This is shown in figure 11-11.

An amplifier with a 10 watt input delivers a 20 watt output. The gain would be; db = 10 log 20/10 = 10 log 2. From the table of logarithms in the appendix, the mantissa of 2 is .3010, and the characteristic by inspection is 0. Therefore, log 2 = 0.3010, db = 10 x 0.3010 = 3.010 db gain. For most calculations, this can be rounded off to 3 db. A doubling of power, then, is represented by a 3 db change.

A device with a 20 watt input attenuates the power to a 10 watt level. The db of attenuation is: -db = 10 log 20/10 = 10 log 2 = 10 x 0.3010 = -3.010 db. Reducing the power by one-half is a -3 db change.

Doubling the power or cutting it in half is a 3 db change. Increasing the power from 1 to 2 watts is a 3 db change; from 100 watts to 200 watts is a 3 db change; from 1,000,000 to 2,000,000 watts is a 3 db change. No matter how great the power is, if it is doubled, only a 3 db change occurs.

By using logs and decibels, the computations to find gain or loss are greatly simplified. For example, the six stage amplifier shown in figure 11-12 has a gain in watts equal to 1 x 10 x 2 x 100 x 2 x 10 x 10 = 400,000 watts. If the amplifier gains were expressed in db, they would be added for a gain of 10 + 3 + 20 + 3 + 10 + 10 = 56 db. To find the output power

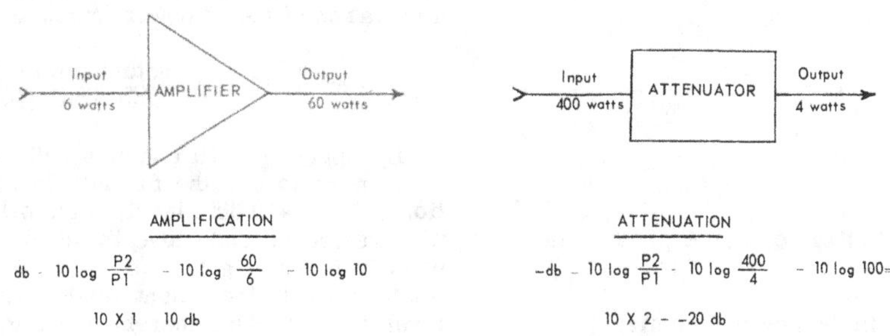

179.209
Figure 11-11.—Calculating db.

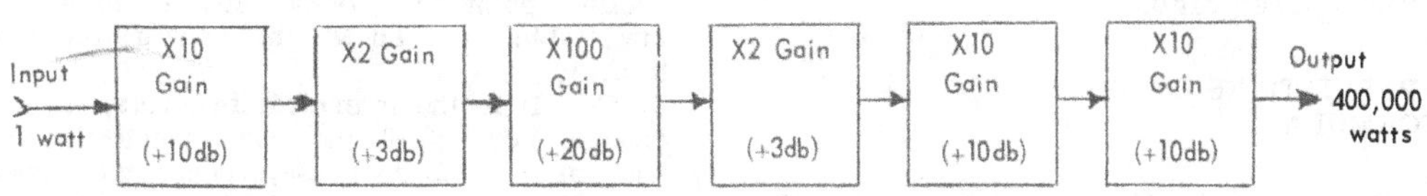

179.210
Figure 11-12.—Six stage amplifier.

in watts, convert 56 db in 10 db and 3 db steps. For every 10 db, move the decimal point one place. For a 50 db change, move the decimal point 5 places. 1 watt increased by 50 db = 100,000 watts. Power is doubled for every 3 db increase, so 53 db = 200,000 watts, and 56 db = 400,000 watts.

Gain and attenuation are handled by algebraic addition. For example, the four stage amplifier in figure 11-13 has an overall gain of 13 - 3 + 10 - 3 = 17 db. To convert to power, first make a 20 db change. 1 watt increased by 20 db = 100 watts. Now, reduce this level by 3 db. 20 db - 3 db = 17 db = 100/2 = 50 watts.

Voltage or current changes are handled in the same manner using the appropriate formulas. For voltage: db = 20 log E_2/E_1. This formula applies only when $Z_{in} = Z_{out}$. For example, the amplifier in figure 11-14 has a 10 volt input and a 100 volt output. The db gain is: db = 20 log 100/10 = 20 log 10 = 20 x 1 = 20 db. Note that for the decimal point to move one place, 10 volts to 100 volts, there was a 20 db change.

Earlier it was stated that decibels expressed power ratio only. Referring to figure 11-14, to find the input powers (assuming Z = 10 ohms) use the formula $P = E^2/Z$.

$$\text{Input } P = (10)^2/10 = 10 \text{ watts}$$
$$\text{Output } P = (100)^2/10 = 1{,}000 \text{ watts}$$

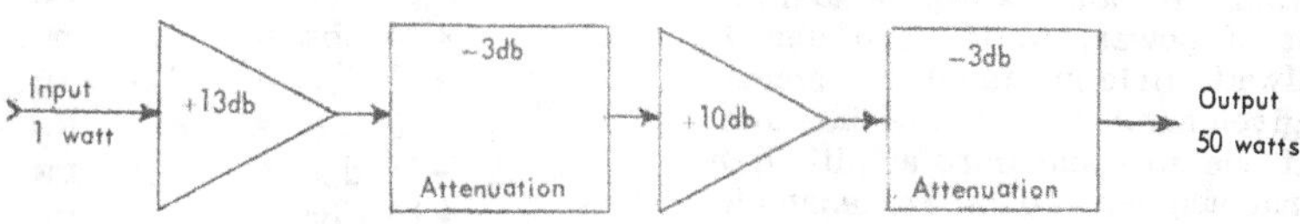

179.211
Figure 11-13.—Four stage amplifier.

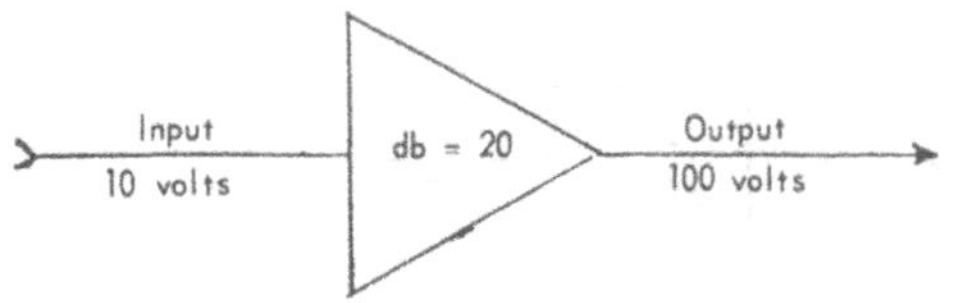

179.212

Figure 11-14.— Voltage change expressed in db.

Substituting in the power formula:

$$db = 10 \log 1000/10 = 10 \log 100 = 10 \times 2 = 20 \text{ db}$$

No matter which formula is used, decibels express a power ratio.

APPLICATIONS OF DB FORMULA

There are two general methods of applying the db formula: the first method compares the input to the output and expresses the result in db's, as shown above, and the second relates either the input or the output power to a stated reference level.

In the first method the decibel cannot be used to represent actual power, but only to represent the ratio of one power to another. To say that an amplifier has a 3 db gain means that the output power is twice the input power, which gives no indication of the actual power represented. Thus, to be meaningful the input power must be stated. However, in many applications it is desirable to have a logarithmic expression which will represent the actual power, not a power ratio, as in the second method.

One standard reference is the dbm. The dbm is a "coined" term used to represent power levels above or below one milliwatt. Minus dbm (-dbm) represents power levels below one milliwatt and plus dbm (+dbm) represents power levels above one milliwatt. In other words, a dbm is a specific amount of power, while zero dbm is equal to 1 milliwatt. Briefly stated, the amount of power in a given number of dbm is the power which results if one milliwatt were amplified or attenuated by that number of db's. For example, 40 dbm represents an actual power level (watts or milliwatts) 40 db above 1 milliwatt, while -10 dbm represents a power level 10 db below one milliwatt. The formula for finding dbm is a variation of the db power formula:

$$\text{dbm} = 10 \log \frac{\text{actual power } (P_2)}{.001 \text{ watt } (P_1)}$$

By applying a 10 db, or a 3 db change, there is no need to use the formula in most applications. For a +10 dbm level, start with the 1 milliwatt reference and move the decimal point one place to the right, i.e., +10 dbm = 10 mw. Another 10 db increment would bring the power level to +20 dbm, thereby moving the decimal point another place to the right, i.e., +20 dbm = 100 mw. For a -10 dbm level, again start with 1 milliwatt, but this time move the decimal point one place to the left, or -10 dbm = .1 mw. An additional 10 db decrease would result in another decimal point shift to the left, or -20 dbm = .01 mw (table 11-2 shows the conversion of dbm to mw).

To fill in the intermediate levels, use a 3 db change. For a 3 db increase, double the power, and for a 3 db decrease, reduce the power by one-half, i.e., +3 dbm = 2 mw. and -3 dbm = .5 mw. A +6 dbm is a 3 db change from +3 dbm, so double the power level of the +3 dbm, i.e., +6 dbm = 4 mw.

The db change can be made in either direction. For example, +7 dbm is a decrease from +10 dbm. By reducing the +10 dbm power by one-half, we have +7 dbm = 5 mw. A +4 dbm power level is a 3 db decrease from +7 dbm, so +4 dbm = 2.5 mw. By using this method, any power level corresponding to a given dbm can be quickly found.

The dbm is handled in circuits in the same manner as the db (fig. 11-15). To find the output level, algebraically add the db's starting with

Table 11-2.—Conversion of dbm to mw

dbm		mw	
+20 dbm	=	100	mw.
+10 dbm	=	10	mw.
+ 7 dbm	=	5	mw.
+ 6 dbm	=	4	mw.
+ 4 dbm	=	2.5	mw.
+ 3 dbm	=	2	mw.
0 dbm	=	1	mw.
- 3 dbm	=	.5	mw.
-10 dbm	=	.1	mw.
-20 dbm	=	.01	mw.

179.213

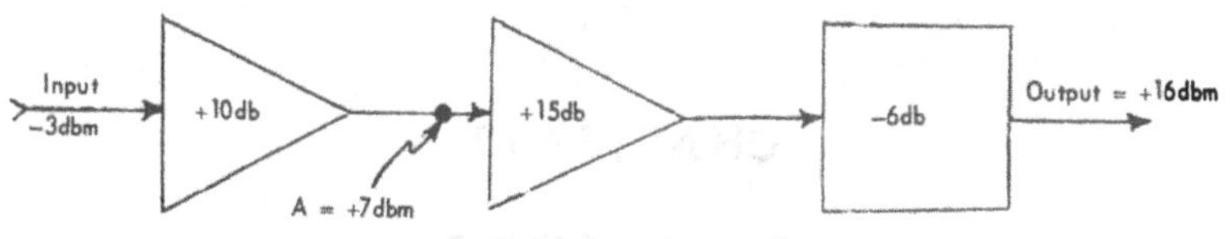

179.214
Figure 11-15.—Three stage amplifier.

the -3 dbm input, i.e., -3 dbm + 10 db + 15 db - 6 db = +16 dbm. To convert this to power, simply use table 11-2 and interpolate, i.e., + 10 dbm = 10 mw., +13 dbm = 20 mw., and +16 dbm = 40 mw.

A -3 dbm input indicates that the power level at that point is .5 mw. Note that dbm gives the actual power level, while db merely indicates how much gain or attenuation will take place. Again referring to figure 11-15, at point A the input level has been increased by 10 db. The power level at this point in dbm would be -3 + 10 = +7 dbm. One milliwatt represents just one reference value. There are several others in use and when working with decibels, it is necessary to check what value of power is being used as the reference.

CHAPTER 12

DETECTORS

DETECTION, also called DEMODULATION, is the process of re-creating original modulating frequencies (intelligence) from radio frequencies which are present in the composite IF signal. In general, a signal present at the output of the IF amplifier consists of an IF carrier having modulation sidebands symmetrically located above and below the carrier frequency by an amount equal to the original modulating frequencies. Applying the composite IF signal to a circuit containing a non-linear conducting device (or non-linear impedance) allows re-creation of the original modulating signal. The circuit, in which this re-creation is achieved, is called the DETECTOR or DEMODULATOR.

The term demodulator is used because the process of detection is considered to be the opposite of modulation. The non-linear conducting device used in solid state detector circuits may be either a PN junction diode or the input junction of a transistor. In electron tube circuits either a diode or the grid or plate circuits of a triode electron tube may be used as the non-linear device.

The IF signal consists of many radio frequencies which shift continuously in their phase with respect to each other. Graphing the instantaneous sums of these frequencies with respect to time will result in a symmetrical radio frequency envelope. The amplitude variations in the positive half of this envelope are equal and opposite to amplitude variations in the negative half of the envelope. Thus, the average value of the composite IF signal is zero. As long as the average value of the signal is zero, no useful energy may be extracted.

The non-linear characteristics of the detector circuit will cause the IF signal to be distorted. The distortion introduced by the detector circuit will cause the average value of the waveform to be other than zero and vary in accordance with the original modulating signal. The output of the detector (before filtering) will contain, in addition to the carrier and sideband frequencies, the original modulation frequencies.

This chapter will discuss the theory and operation of series and shunt detectors, utilizing both PN junction diodes and electron tube diodes as the non-linear device, first. Next, a detector circuit utilizing a transistor as the non-linear device will be discussed. Then, the use of electron tube triodes as detectors will be covered and, finally, methods of output coupling will be analyzed.

ICO CONCEPT

In order to show how a signal is developed and altered within a block, the Input-Conversion-Output (ICO) concept is used. This is a method of illustrating the comparison between input and output signals within a particular segment of an equipment. This concept is used to indicate the conversion which occurs within that specific block. Frequently, the only input to a block is the potential(s) from a power supply.

The conversion that takes place within a block would be one or more of the following. Output phase could lead or lag input phase by a given amount. Amplitude and/or frequency of the output signal could be increased or decreased. The output waveform could be altered due to limiting, differentiating, integrating or by clamping it to a different reference level. All of the conversions will be individually presented in subsequent chapters.

Figure 12-1 depicts a block diagram of a transmitter with the input and output waveforms of each block included. The power supply provides the necessary a.c. and/or d.c. voltages for proper operation of the equipment. An analysis of the equipment illustrated in figure 12-1 using the ICO concept, is presented in the following discussion.

The microphone converts sound energy into electrical impulses which vary at an audio rate.

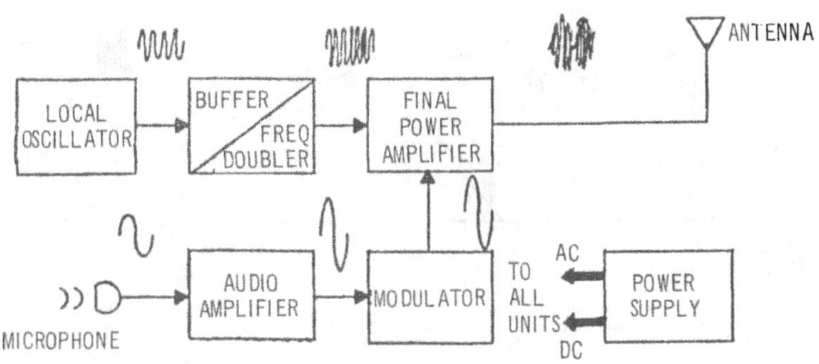

179.215
Figure 12-1.—Transmitter block diagram.

This small audio signal is coupled to the audio amplifier where the audio signal is increased in amplitude. The conversion that takes place within the amplifier is one of amplification. The amplified audio signal is applied to the modulator and again the conversion of the signal is amplification. This large audio signal is then coupled to the final power amplifier.

The local oscillator generates an RF signal of constant frequency and amplitude. This signal is applied to the buffer which isolates the oscillator and provides some amplification of the signal. The RF signal is doubled in frequency and increased in amplitude in the frequency doubler. This RF signal is then applied to the final power amplifier.

The final power amplifier has two signal inputs, the audio signal from the modulator and the RF signal from the frequency doubler. Those two signals are combined in a process called modulation to produce a modulated carrier. The modulated carrier is coupled to the antenna which causes the signal to be propagated.

The ICO concept can be used effectively in troubleshooting electronic equipment. Given a basic equipment block diagram showing ideal input and output waveforms, the technician can use an oscilloscope to observe waveforms for each functional block of the equipment and compare them with the ideal waveforms shown on the block diagram. Any component failure within one of the functional units of the equipment would result in an abnormal or altered output waveform for that particular functional unit. This same concept can be used to then isolate the trouble to a particular circuit within that functional unit. Once a trouble is isolated to a particular circuit, voltage and/or resistance measurements will normally isolate the defective component.

To aid the technician in troubleshooting, more information is required than what is available on a functional block diagram. Information concerning voltages, amplitudes, frequencies, and waveshapes are contained on a SERVICING BLOCK DIAGRAM. Pertinent controls, testpoints and lead or pin connections, for transistors or electron tubes, are also illustrated. Figure 12-2 illustrates a typical servicing block diagram.

DETECTOR

Figure 12-3 shows the ICO block diagram of a representative detector circuit. The input to the detector circuit is obtained from the output of the IF amplifier. This composite IF signal contains the IF carrier and the sidebands which are located above and below the IF carrier frequency. The sideband frequencies deviate from the carrier frequency by an amount equal to the original modulating frequencies. As an example, if the IF carrier frequency is 455 kHz and the original modulation consisted of a 5-kHz frequency signal, then the input to the detector would consist of frequencies of 450, 455, and 460 kHz. Had the original modulating signal been 1 kHz then the input to the detector would consist of frequencies of 454, 455, and 456 kHz.

The purpose of the bandpass filter is to increase receiver selectivity, to pass the IF carrier and its principal sidebands which fall

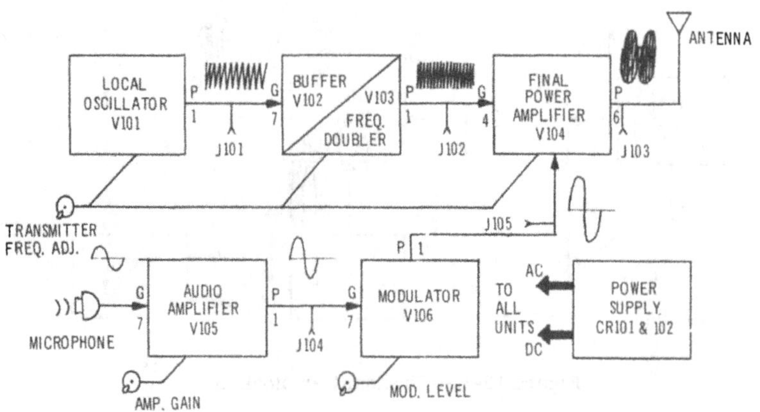

179.216
Figure 12-2.—Servicing block diagram.

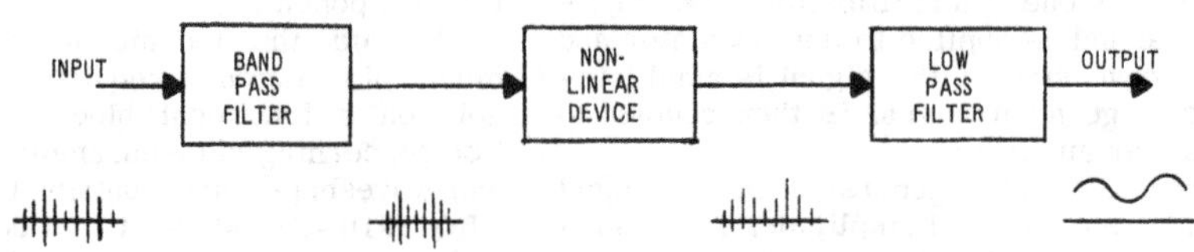

179.217
Figure 12-3.—ICO block diagram of a detector circuit.

within the bandpass of the filter, and to reject other frequencies which occur above or below the filter's bandpass. The output of the bandpass filter is essentially the same as the input with respect to amplitude, but unwanted frequencies will have been attenuated. The nonlinear device output (before filtering) will contain, in addition to the carrier and sideband frequencies, the original modulation frequencies. The purpose of the low pass filter is to eliminate any remaining RF pulses and to retain only the original audio frequencies. The output of the low pass filter will be an a.c. variation of a d.c. level. The a.c. variations will occur in accordance with the original modulation signal.

COMPOSITION OF INPUT WAVEFORM

As previously stated, the input to the detector stage consists of the IF carrier and the various sideband frequencies which contain the intelligence. As the original modulating signal may have been complex (as in the case of a human voice), then the composition of the detector input could also be complex.

Figure 12-4 shows the frequency spectrum of several signals. Figure 12-4A shows the spectrum of the IF carrier alone. It occurs at a single frequency and is therefore shown as a single line. In figure 12-4B the frequency spectrum is shown for an IF carrier which contains 1-kHz modulation. In this diagram it is seen that a 1-kHz modulation produces an upper sideband which occurs at 456 kHz and a lower sideband which occurs at 454 kHz. Figure 12-4C shows the frequency spectrum of a complex audio signal which simultaneously contains frequencies of 500 Hz, 1 kHz, and 2 kHz. The heights of the spectral lines indicate the relative amplitudes of the various frequency components. Figure

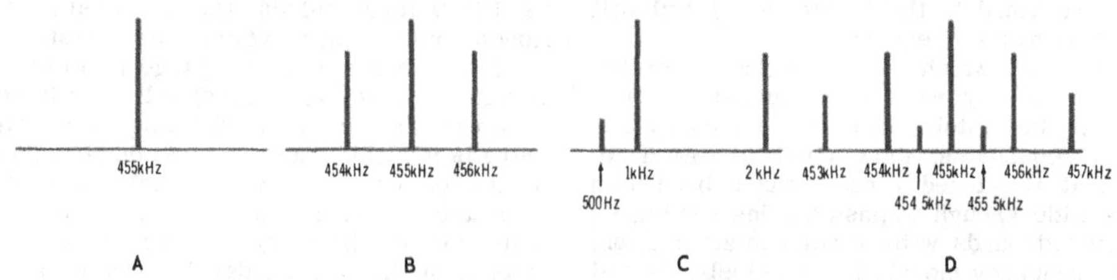

93.33(20C)
Figure 12-4.—Frequency spectrums.

12-4D shows an IF carrier and the sidebands which the complex modulation of figure 12-4C would produce riding on a 455-kHz carrier. Note that for each frequency component of the modulating signal both an upper and a lower sideband have been produced.

As the sidebands depend on the original modulation, it should be obvious that the frequency spectrum is continuously shifting as the frequency and amplitude of the modulating signal changes.

It should be noted here that the input for the detector circuit has been amplified in one or more preceding tuned amplifiers. Recalling that such devices have a tuned circuit in either their input or output circuits (and in some cases both), it becomes apparent that changes in the frequencies and relative amplitudes of the modulating signals will affect the input signal to the detector stage.

BANDPASS FILTER

Figure 12-5 shows the schematic diagram of a series diode detector with a resistive load. The tuned circuit will be recognized as the bandpass filter, and the solid state diode is the nonlinear device. The input signal is obtained from the preceding IF amplifiers (or in the case of TRF receivers, the RF amplifiers).

As the input is coupled by means of tuned circuits it is obvious that frequency changes will affect the input to the detector. The tuned

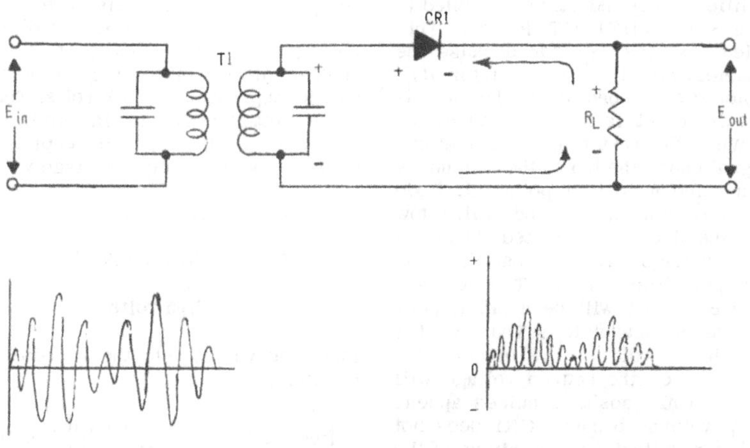

179.218
Figure 12-5.—Series diode detector with resistive load.

circuits are tuned to the IF frequency and will respond best to that frequency.

Frequencies which are different from the resonant frequency will be attenuated to some degree. As the intelligence is carried by the sideband frequencies, care must be taken to assure that the tuned circuit has a bandwidth which is wide enough to pass the desired higher frequency sidebands without serious attenuation. For low frequency modulation the sidebands will be near the resonant frequency, and little attenuation will occur. However, if the modulation signal is, as an example, 10 kHz, then the sidebands will be 10-kHz above and below the IF carrier frequency. This would require the tank circuit to have at least a 20-kHz bandwidth if serious attenuation is to be prevented. Also, the amplitude of the modulating frequency will affect the response of a tuned circuit. A very large amplitude sideband can withstand attenuation and still be detectable even though it may be far from the resonant frequency of the tank. On the other hand, if a weak sideband occurs at or near the half power point of the tank circuit, it may be attenuated to such a degree as to not be detectable. For a complete discussion of tuned circuit response in a receiver refer to the chapter on IF amplifiers.

SERIES DETECTOR OPERATION

The circuit in figure 12-5 shows the output of the IF amplifier being inductively coupled to the input of the SERIES DETECTOR stage. This circuit is called a series detector because the diode is in series with the load resistor R_L. The voltage which is developed across the secondary tuned circuit of T1 is now the source for the detector. When the resulting RF variations of the input signal cause the top of the secondary tuned tank to assume a positive potential, diode CR1 will be forward biased. The diode will allow current flow in the direction indicated. The current will cause voltage drops across R_L and CR1 with the polarities shown. The forward biased resistance of CR1 will be a small percentage of the resistance of R_L. Thus, during the conducting half cycle of the circuit, the greatest percentage of the source voltage will appear at E_{out}. Only positive pulses appear in the output waveform because CR1 does not conduct when the resultant RF variations of the input signal are going through their negative alternations. All of these voltage pulses are represented in the output waveform (fig. 12-5) as RF voltage pulses. Due to reverse current flow a small negative voltage may exist.

The average output voltage (average of the resultant RF voltage pulses) will vary in accordance with the original modulating signal. Although there is a usable audio voltage component present in the output, it is very small in amplitude.

Detector circuit efficiency (not to be confused with power efficiency) is determined by the amount of d.c. voltage developed across the load resistor for a given peak value of a.c. voltage input. This is expressed in the form of an equation as:

$$\% \text{ Rect. eff.} = \frac{\text{d.c. output volt.}}{\text{a.c. input volt.}} \times 100$$

Where:

% Rect. eff. = the percentage of circuit rectification efficiency.

d.c. output volt. = the d.c. voltage developed across the diode load resistor.

a.c. input volt. = the peak input amplitude of the applied a.c. voltage.

This equation can be used to determine the efficiency of the circuit in figure 12-5. For simplicity it will be assumed that the input to the detector is an UNMODULATED (constant amplitude) carrier of 10 volts peak amplitude. Since the forward resistance of CR1 is very small in comparison to the resistance of R_L, virtually all of the rectified voltage will appear across R_L. For an input voltage of 10 volts peak, the peak amplitude of the voltage pulses across R_L will be approximately 9.9 volts. The action of the diode detector circuit is similar to that of a half-wave rectifier power supply. Therefore, the average, or d.c., output voltage will be:

$$E_{AV} = .318 \times E_{PEAK}$$

$$E_{AV} = .318 \times 9.9$$

$$E_{AV} = 3.15 \text{ volts}$$

Inserting values into the rectification efficiency equation:

$$\% \text{ Rect. eff.} = \frac{\text{d.c. output volt.}}{\text{a.c. input volt.}} \times 100 =$$

$$\frac{3.15}{10} \times 100 = 31.5\%$$

It can be seen that the rectification efficiency of the diode detector with only a resistive load is very poor.

The efficiency of the circuit can be improved considerably by connecting a capacitor, of the proper value, in parallel with the load resistor. An arrangement such as this is shown in figure 12-6. This capacitor in parallel with the load resistor R_L will be recognized as the low pass filter of figure 12-3. Before examining the circuit efficiency with the capacitor added, a brief description of circuit action will be given.

When the input signal causes the top of T1's secondary tank to become positive, CR1 will be forward biased. Current, indicated by the solid arrows in figure 12-6, will flow from the bottom of the tank through the parallel combination of R_L and C1, through CR1 and back to the top of the tuned circuit. Since the only resistance in the charge path of C1 is the low forward resistance of the diode, C1 will charge rapidly to almost the peak value of the drop across R_L.

As the input signal decreases and polarity reverses, the voltage developed across the tuned circuit decreases. This causes CR1 to cease conducting and C1 will begin discharging in an attempt to maintain the value of voltage established across R_L. However, the only path available for this discharge current is through the load resistor, R_L. The charge time of C1 is short due to the low forward resistance of CR1. The pulses which control the conduction of CR1 occur at the intermediate frequency. Therefore, C1 will lose only a small percentage of its charge between pulses. This will cause the output voltage to follow the peak value of the rectified voltage as seen in figure 12-6.

It must be realized that the pulses in the waveform are expanded tremendously for the purpose of illustration. The sawtooth appearance, which depicts the charge and discharge of the capacitor, would not be visible in the actual output waveform. Comparison of the output waveforms for a diode detector with and without the capacitor shows that the voltage in both cases varies in accordance with the original modulating signal.

An example problem will be used to show how the addition of a capacitor improves the efficiency of the diode detector circuit. For simplicity it will again be assumed that the input to the detector is an UNMODULATED carrier of 10 volts peak amplitude. The voltage divider action of R_L and CR1 will again cause the peak amplitude of the voltage across R_L to be approximately 9.9 volts. C1 will charge very close to this value. Due to the action of C1 and R_L in maintaining the value of voltage between charging pulses the average d.c. output voltage will remain at

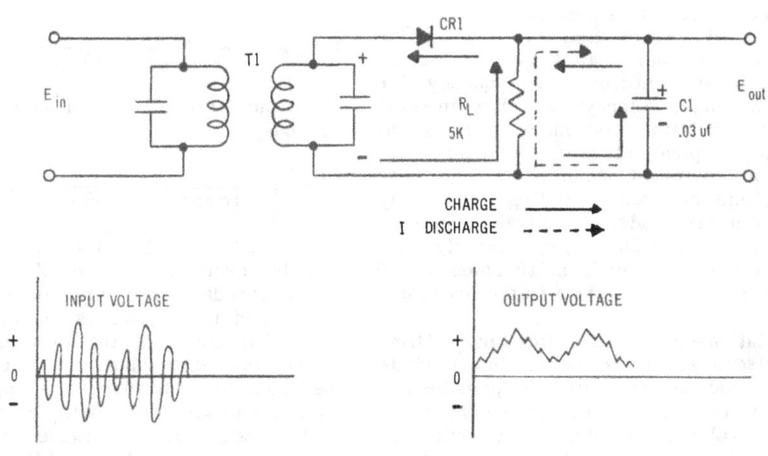

Figure 12-6.—Series diode detector with low pass filter.

approximately 7.8 volts. Inserting this information into the efficiency equation yields:

$$\% \text{ Rect. eff.} = \frac{\text{d.c. output volt.}}{\text{a.c. input volt.}} \times 100$$

$$\% \text{ Rect. eff.} = \frac{7.8}{10} \times 100 = 78\%$$

The rectification efficiency of the diode detector has been increased from 31.5% to 78% by the addition of C1. C1 also acts as a filter for the output. Even higher efficiency ratings could be obtained by adding additional filtering networks to the output.

CHOICE OF R AND C VALUES FOR LOW PASS FILTER

In all electronic circuits the choice of component values is usually a compromise between values for an ideal situation and a practical situation. Such is the case in choosing the values for R and C to be used in the output of the detector circuit. The rectification efficiency is highest when the value of R_L is as large as possible in relation to CR1. Filtering action is best when the value of C1 is large in comparison to the value of the diode capacitance. It would appear that the value of both R and C should be large. On the other hand, one of the prime qualifications for the detector circuit, for faithful reproduction, is that the charge and discharge of C1 be able to follow the modulation envelope as closely as possible.

Thus, the discharge time constant of the circuit should be such that it appears long in comparison to the intermediate frequency (for good filtering and efficiency) and short in relation to the modulation frequencies (for faithful reproduction). Typical values of components used, in transistor broadcast superhet receivers, to achieve a time constant exhibiting this quality are: C = .03 microfarads, R = 5 kohms, and CR1 forward resistance = 50 ohms. Actually, the diode may range in value from 10 ohms to 100 ohms or more. Thus, 50 ohms is a representative value.

The relationship between the charge time constant, discharge time constant, intermediate frequencies, and the modulating frequencies can be shown by using typical component values in an example problem. The detector circuit used in the example will contain a 5 kohm load resistor, a .03 microfarad filter capacitor, and a diode with 50 ohms of dynamic forward resistance (fig. 12-6). The input signal will be a composite waveform containing a 455-kHz IF carrier and sideband frequencies of 454.6 kHz and 455.4 kHz (representing the original signal of 400 Hz). For the circuit to operate properly, the time constant for capacitor discharge should appear relatively long at the carrier frequency of 455 kHz and relatively short at the modulating frequency of 400 Hz. The charge time constant of the circuit must be very short. The charge time constant of the circuit is determined by the product of the diode's forward resistance and the value of capacitance of C1. Inserting values:

$$TC = RC = 50 \times 3 \times 10^{-8} = 1.5 \times 10^{-6}$$

Thus, the time constant of the circuit during charge is 1.5 microseconds. The discharge time constant of the circuit is determined by the load resistance R_L and the value of capacitance C1. Inserting values:

$$TC = RC = 5 \times 10^{3} \times 3 \times 10^{-8} = 150 \times 10^{-6}$$

Thus, the time constant of the circuit during discharge is 150 microseconds.

In order to examine the relationship between the discharge time constant and the IF and modulating frequency, it is necessary to determine the time for one cycle of each of these frequencies. The period for one cycle is equal to the reciprocal of the frequency. Therefore, the time for one cycle of the IF will be:

$$\text{Period} = \frac{1}{\text{frequency}} = \frac{1}{455 \times 10^{3}} = 2.19 \text{ microseconds}$$

The time for one cycle of modulating frequency will be:

$$\text{Period} = \frac{1}{\text{frequency}} = \frac{1}{400} = 2.5 \text{ milliseconds}$$

From the results of the example problem it can be seen that the discharge time of 150 microseconds is long in respect to the 2.19 microseconds of the IF, but is short in respect to the 2.5 milliseconds of the modulating frequency.

The correct choice of values for R and C in the diode detector circuit is important if maximum sensitivity and fidelity are to be obtained.

If either R or C are made too large, the time constant will be too long and the discharge of the capacitor will not follow the modulating frequencies. This will result in a situation such as illustrated in figure 12-7. It can be seen that too long a discharge time constant will cause the negative peaks of the desired waveform to be clipped. This type of distortion is referred to as DIAGONAL CLIPPING.

Distortion will also be introduced if the discharge time constant is too short. Figure 12-8 illustrates the effect of too short a time constant. The dotted line in the figure indicates the desired waveform. The solid line (actual waveform) shows that the short time constant allows the capacitor to discharge considerably between pulses. The result is a loss of rectification efficiency and the presence of a high frequency component in the output. As previously stated, this problem can be corrected either by an increase in the capacitance of C1 or by the addition of more filter sections.

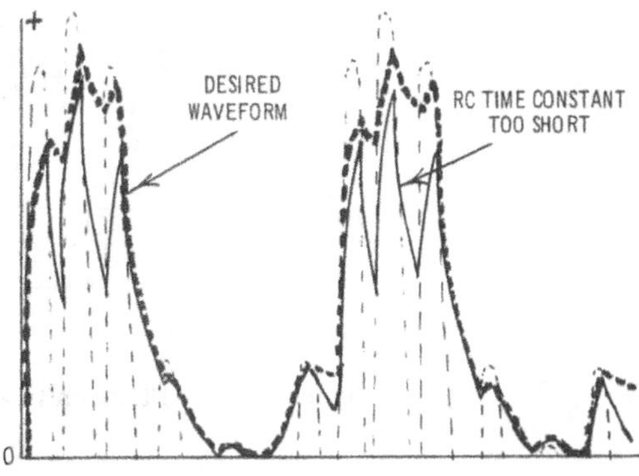

179.221

Figure 12-8.—Waveforms showing effect of too short a time constant.

ELECTRON TUBE DIODE OPERATION

Figure 12-9 shows the schematic diagram of a series detector using an electron tube diode as the non-linear detector element. When the incoming signal causes the top of the tuned circuit to become positive with respect to the bottom, V1 will conduct and allow C1 to charge to almost the peak value of the a.c. input signal. There will be a small voltage drop across the diode, but it is usually insignificant. C1 will charge quickly as the internal conduction resistance of V1 is the only resistance in C1's charge path. Current will flow from the bottom of the tank circuit (which is acting as the detector source) through C1 and V1, and back to the top of the tuned circuit. As the input signal decreases and reverses polarity V1 quits conducting, and C1 begins to discharge through R_L. C1 will maintain the majority of its charge due to the relatively large discharge time constant of R_L and C1. If C1 has been properly chosen with regard to the size of R_L, the output voltage will closely follow the peak value of the input signal for all audio frequencies. The output voltage therefore varies in accordance with the original modulating signal.

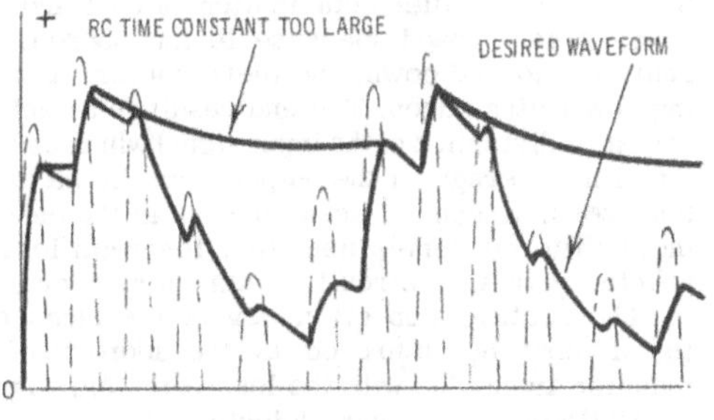

179.220

Figure 12-7.— Effect of too large a time constant.

Figure 12-10 shows the spectrum analysis of a detector output with and without the low pass filter. In both cases it is assumed that the IF frequency is 225 kHz and that the modulating signal is a single 5-kHz frequency.

Note that the low pass filter provides a very strong 5-kHz signal (the original modulation) while effectively shunting the 10-kHz harmonic and the RF frequencies. In the unfiltered output, changes in carrier and sideband amplitudes would be felt, but in the output of the filtered detector these changes would be almost unnoticeable. Perhaps the most noticeable change in the output of the filtered circuit would occur if one of the primary sidebands were lost. There is a 5-kHz component due to the difference between the 225-kHz IF carrier and the 220-kHz sideband. There is an additional 5 kHz component due to the 230-kHz sideband. The loss of one of these primary sidebands would eliminate a portion of the 5-kHz signal strength and would reduce its amplitude in the output of the detector. The amplitudes shown in figure 12-10 are measured in dbs and are purposely distorted for the sake of clarity.

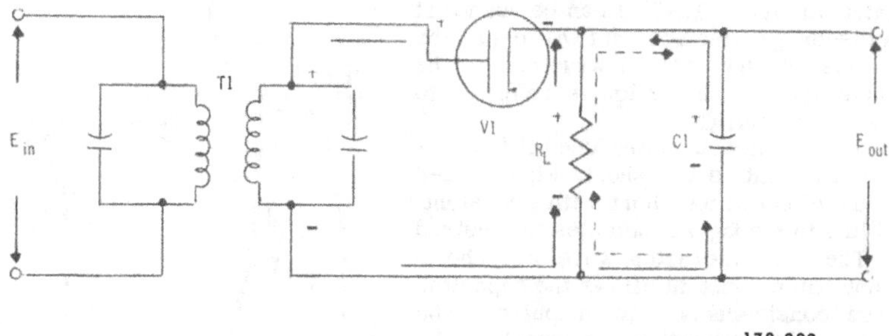

179.222

Figure 12-9.—Electron tube series detector with low pass filter.

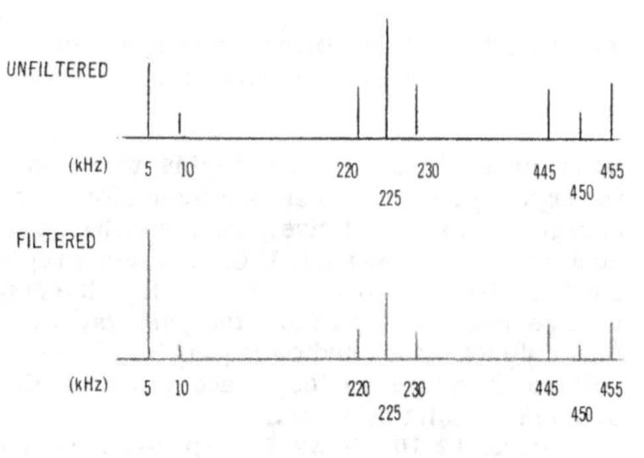

179.223

Figure 12-10.—Spectrum analysis.

SOLID STATE SHUNT DETECTOR

The SHUNT DETECTOR also produces a voltage and current output which is proportional to the modulation signal. The shunt detector is characterized by the diode being in shunt (parallel) with the load resistor rather than in series with it. Figure 12-11 illustrates a simple shunt detector with only a resistive load.

When the resultant RF variations of the input signal cause point A to become positive with respect to point B (solid polarity signal), diode CR1 will be reverse biased and will appear as a very high-resistance in parallel with the load resistance R_L. Thus, the current flow in the circuit will follow the path depicted by the solid arrow (fig. 12-11). This current flow will cause a voltage drop across R_L with the polarity (solid signs) shown. There will be a very slight current flow through the parallel path of CR1; however, for all practical purposes, its magnitude is so small that it is neglected. Therefore, the peak of the resultant voltage pulse developed across R_L is almost equal to the peak of the input voltage.

When the resultant RF variation of the input signal causes point A to become negative with respect to point B (dotted polarity signs), diode CR1 will be forward biased and will appear as a very low resistance in parallel with the load resistor. As a result the majority of the current will flow (dotted arrow) through the diode. A very small percentage of the current will flow through the parallel path of R_L causing a small voltage drop with the polarity shown by the dotted signs. This small reverse voltage drop is evident in the output waveform (fig. 12-11) by the extension of the voltage pulses slightly below the reference line. This action may be summarized as follows:

When the diode is reverse biased the parallel combination of CR1 and R_L appear as a high impedance to the source, with a corresponding high voltage drop. When the diode is forward biased, the parallel combination of CR1 and R_L appears as a low impedance to the source, the source is loaded down, and there is a corresponding low voltage drop. The end result is a rectification or distortion of the input signal which causes the time average of the output signal to be other than zero. It can be seen that, as in the case of the unfiltered series detector, the rectification efficiency of this circuit is also very poor.

The rectification efficiency of the shunt detector may be improved by the addition of an inductor in series with the load resistor. Figure 12-12 illustrates a shunt detector with an inductor added to the circuit. When point A becomes positive with respect to point B the current path is

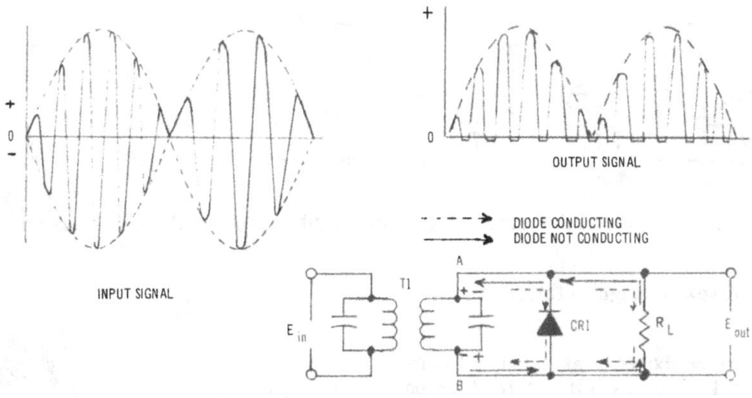

179.224
Figure 12-11.—Solid state shunt detector.

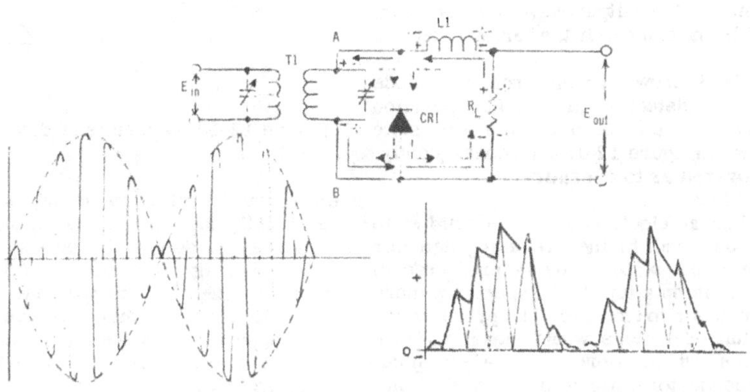

179.225
Figure 12-12.—Shunt detector with inductor added.

shown by the solid arrows and the voltage drops by the solid polarity signs. During this period of time the diode is in an non-conducting state and the majority of the current flow is building a magnetic field around L1. When point A becomes negative with respect to point B, the source current (dashed arrows) flows through the diode and back to point B. By virtue of the collapsing magnetic field which attempts to maintain current flow, L1 now appears as the source for current flow through R_L. The path for current flow caused by the collapsing magnetic field of L1 is shown by the dotted arrows in figure 12-12. Notice that this current maintains a voltage drop across R_L which is the same polarity as the original half cycle.

The values of L1 and R_L are chosen with the same regard to the time constant as was shown in the series detector. In other words, the magnetic field of L1 must not be allowed to collapse too much between RF current pulses, and, on the other

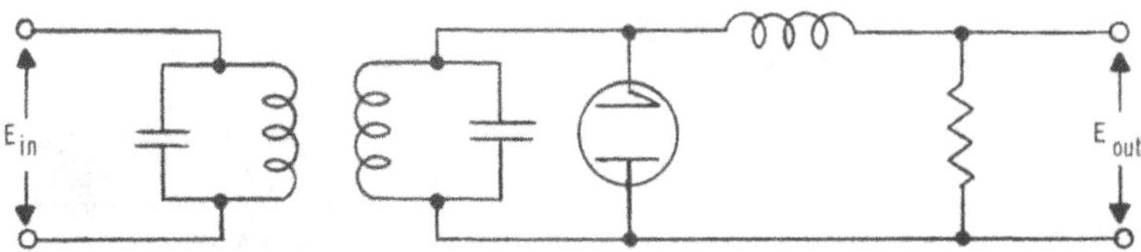

179.226

Figure 12-13.—Shunt detector using electron tube diode as the non-linear device.

hand, it must collapse enough to follow the modulation envelope.

It must be remembered that the equation for the time constant of an RL circuit is different from that of an RC circuit. In an RL circuit the time constant is a direct function of L and an inverse function of R. This is shown by the equation $TC = \frac{L}{R}$. The effect of too long or too short a time constant on the output waveform of shunt detector will be the same as in the series detector.

Figure 12-13 shows an electron tube diode used in a shunt detector circuit. Its operation is essentially the same as that of the solid state detector shown in figure 12-12 and, consequently, will not be covered as to operation.

While either an electron tube or a solid state device may be used in the preceding detector circuits, the solid state diode is preferred in most cases. It is smaller, less expensive, more efficient, and has a longer life. The primary use of electron tube detectors would therefore be in older systems which were built when small signal semiconductors were not too reliable.

NON-LINEARITY IN DIODES

Now that a basic understanding of detector circuits has been achieved, the property of non-linearity in diodes will be more thoroughly described. The process of HETERODYNING can be achieved only in a non-linear device. It is acceptable if the device in question is linear over a portion of its parameters, but it must not be totally linear. If a modulated IF carrier (such as shown in fig. 12-14A) were to be applied to a linear device such as a resistor, both the positive and negative alternations of the signal would be developed. The time average of the signal in this case would still be zero and the output would be essentially the same as the input.

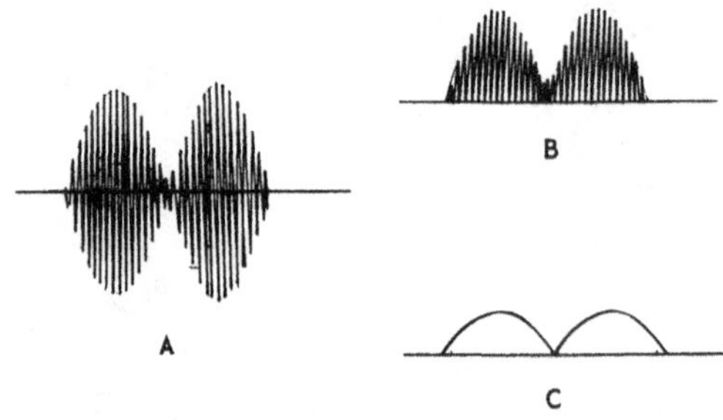

179.227

Figure 12-14.—Results of diode non-linearity.

However, due to the one-way conduction of a diode, the signal is distored (clipped), as shown in figure 12-14B, and will have a time average which is a value other than zero (fig. 12-14C). It is apparent that the time average of the distorted signal closely follows the original modulation envelope.

Figure 12-15 shows the characteristic curve of a rectifier current (a) that is completely linear in the positive voltage region. The non-linearity occurs due to the non-conduction of the diode in the reverse direction. A rectifier current that is presented on a graph according to a law known as the SQUARE LAW (b) is also shown.

In the first, or linear, case (a), each time the voltage is doubled, the current is doubled. This graph would represent the idealized case for a solid state diode or electron tube. The graph is totally linear in the region of positive voltage and non-existent in the negative voltage region. If a linear diode could be produced with such characteristics, the detector would introduce no distortion other than the clipping effect of the normal

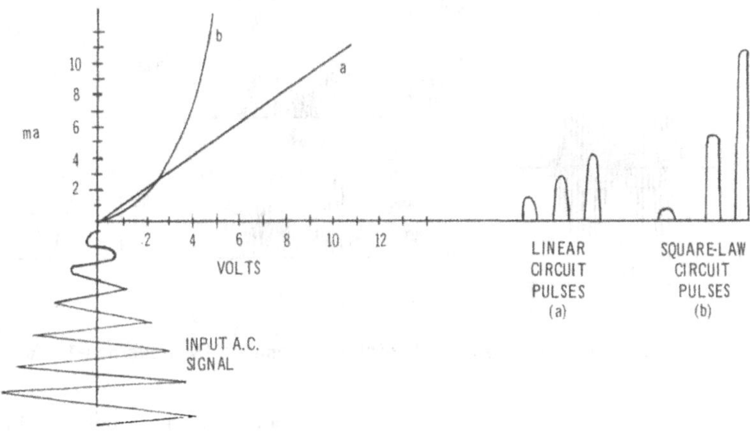

179.228

Figure 12-15.—Currents in a linear rectifying circuit and a square-law rectifying circuit.

rectification process. The second curve (b) shows that each time the voltage is doubled the current increases FOUR times, or as the square of the voltage increases.

A linear rectifier circuit will produce a current waveshape which is essentially the same as the input voltage waveshape. Little distortion will be produced in this case (fig. 12-15A). With a square-law detector the waveshape of the output current will differ from the input voltage waveform, and distortion will be produced (fig. 12-15B).

A diode is a nearly linear device, except when operated close to the zero-voltage-zero-current point, where it rounds off in a square-law manner. Also, in the case of a solid state diode, the characteristic curve extends slightly into the negative voltage region.

As long as diodes are operated well up on their curves, they will produce reasonably undistorted outputs. However, when they are operated near the zero-current value, they become square-law devices and produce a distortion of the signal in addition to the normal rectification (fig. 12-16). This is an amplitude distortion which will affect the waveshape and produce harmonics in the output which were not present in the original modulating signal. In figure 12-16 this form of amplitude distortion has been exaggerated in order to show it more clearly. The diode can operate in either the linear or square law regions depending upon the design of the circuitry (fig. 12-17).

FAILURE ANALYSIS FOR DIODE DETECTORS

Troubleshooting and repairing malfunctioning equipment are the most important jobs of a technician. For this reason it would be well for the student to be aquainted with the three major areas of diode detector failure analysis.

No Output

A no-output condition is usually limited to an open or short-circuited component or a defective diode. With an open input transformer, stray capacitive coupling may feed enough signal through to produce an output. With an open detector load resistor or a short-circuited bypass capacitor, however, no output will be obtained. Usually, a resistance analysis will quickly locate the defective component.

Low Output

Lack of sufficient input signal will cause a low output, and could be due to poorly soldered (high-resistance) joints or to a defective input transformer. In the case of electron tube diodes, low tube emission may also cause a weak output, although this usually shows up as a fading signal on a strong local station. An open load bypass

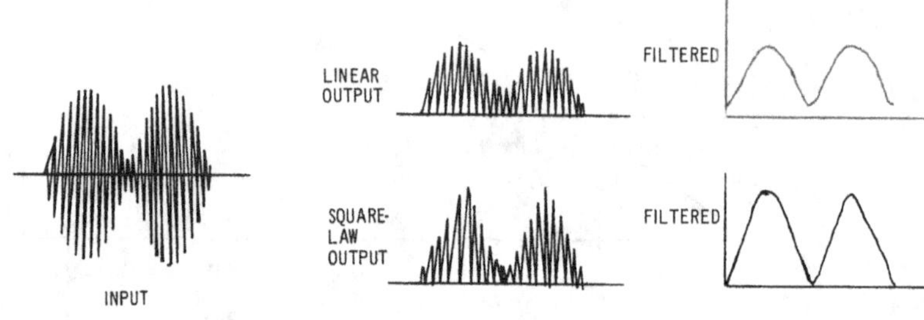

179.229

Figure 12-16.—Amplitude distortion caused by diode non-linearity.

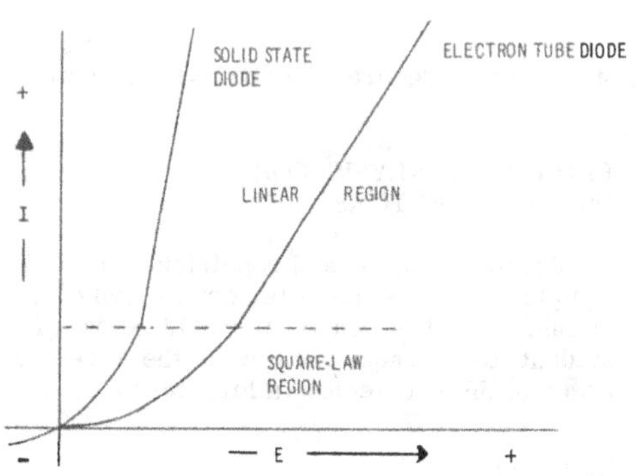

179.230

Figure 12-17.—Representative diode characteristic curves.

capacitor will reduce the output, as only the average current flow through the load resistor will now produce an output. A lack of RF or IF amplification preceding the detector can also cause low output; therefore, it is necessary to isolate the detector by checking with a VTVM for adequate input and output. Since no amplification is produced in the diode detector, but the detection efficiency is relatively high, a d.c. output indication lower than 10 percent of the input indication would be indicative of possible detector trouble.

Distorted Output

Since diode detection is relatively linear, a distorted output signal usually indicates component changes. Amplitude distortion is a definite indication of non-linearity in the detector, provided that the input signal to the detector is linear. Lack of high audio frequency response would be directly traceable to excessive selectivity, caused by regeneration in the preceding IF or RF stages or to excessive capacitive shunting of the detector, particularly in high frequency applications. Distortion at high volume levels with a strong, heavily modulated local signal would indicate normal peak-clipping effects. Poorly soldered (high-resistance) joints can be suspected when component values and diode conduction are normal. Fringe howl or a tendency toward oscillation would indicate a lack of RF or IF filtering. Because of the simplicity of the circuit, an oscilloscope waveform check should quickly locate the trouble.

ADDITIONAL DETECTOR CIRCUITS

Other detection methods exist. Transistor detectors, grid-leak detectors, plate detectors and full-wave diode detectors are examples.

TRANSISTOR DETECTOR

It has been shown that the output voltage for a diode detector is always slightly less than the input voltage (rectification efficiency will never equal 100 percent). This slight loss of signal, due to detection, is of no concern in the average broadcast receiver at the output of the IF stage. However, there are many cases in electronics where it is necessary to detect a signal of very small amplitude. Two problems arise in the detection of small signals. First, since the signal is very small to begin with, the slight loss due to detection is not desirable, and in some cases is intolerable. Second, it is necessary to use a device which is much more sensitive to small signal variations than the diode. In transistor equipment this problem is solved by the use of a TRANSISTOR DETECTOR CIRCUIT. The transistor detector circuit combines the actions of detection and amplification within the same device.

A transistor detector connected in the common-emitter configuration is illustrated in figure 12-18. Before discussing circuit operation a brief explanation of component functions will be given.

R1 and R2 constitute a voltage divider network to establish the no signal forward bias for Q1. C1, C2, and C3 are RF bypass capacitors and act to prevent the development of an RF voltage across bias resistor R1, emitter resistor R3, and the power supply, respectively. Transformer T1 couples the signal from the IF stage to the detector. Load resistor R4 and the demodulator filter capacitor C4 perform the function of modulation signal development and RF filtering in the same manner as the load resistor and filter capacitor in the series diode detector. Transistor Q1 is the device which performs the actions of partial rectification and amplification.

Application of a modulated IF signal to the primary of T1 will cause voltage variations at the IF rate to be developed across T1's secondary. The positive alternations of this induced signal will oppose forward bias and reduce collector current. Thus, the collector current will be caused to vary at an IF rate. Due to the heterodyning action of the various radio frequencies (present in the input signal) within the non-linear transistor, the collector current variations will also contain a modulation component. As the collector current variations are applied to the parallel network R4-C4, an action similar to that of the load resistor and filter capacitor network of the series diode detector takes place. The high frequency components are filtered out, and a d.c. voltage, which varies in accordance with the original modulation, is developed between collector and ground.

The dynamic operation of the transistor detector circuit can best be demonstrated by use of the transfer characteristic curve. A typical curve (graph of collector current I_C versus base-emitter voltage V_{BE}) with the resultant waveforms is shown in figure 12-19. The graph depicts the extreme non-linearity of collector current with respect to base-emitter voltage. The symmetrical IF signal with 75 mv peaks is used as the input signal to the detector circuit. The operating point of the circuit is established with a base-emitter voltage of 175 mv and a collector current of 1 ma.

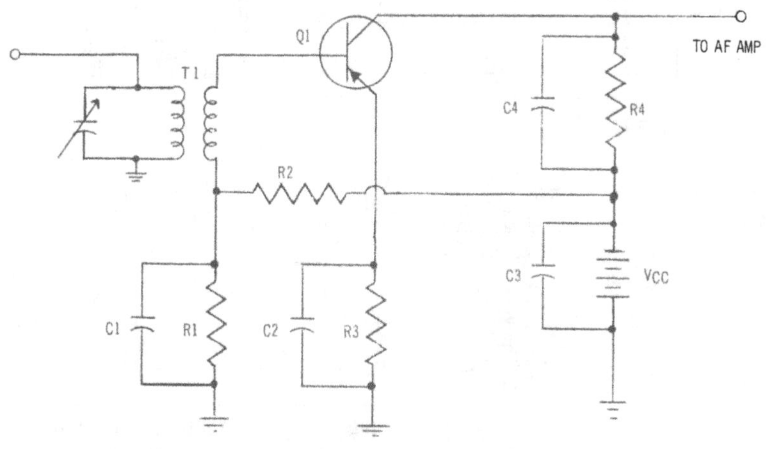

Figure 12-18.—Transistor detector.

Figure 12-19 shows that negative excursions (increase in V_{BE}) of the input signal produce much larger variation in I_C than do the positive excursions (decrease of V_{BE}). Notice that the output waveform of figure 12-19 is the collector current variations BEFORE application to the R4-C4 filter network. The long time constant, at IF frequencies, and short time constant, at audio frequencies, of the network will convert the output to a voltage which very closely follows the peaks of the I_C variations. The output developed across the R4-C4 network is shown in figure 12-19B.

The main advantage of the transistor detector over the diode detector is its ability to provide a usable output signal with an extremely small input signal. The main disadvantage is that the transistor detector cannot handle as large a signal input as the diode detector.

GRID-LEAK DETECTOR

An electron tube detector which has some similarities to the transistor detector is the GRID-LEAK DETECTOR. The grid-leak detector functions like a diode detector combined with a triode amplifier. It is convenient to consider detection and amplification as two separate functions. In figure 12-20 the grid functions as the diode plate. The values of C_d and R_d must be chosen so that C_d charges during the positive peaks of the incoming signal and discharges during the negative peaks. The time constant of R_d-C_d should be long with respect to the RF cycle and short with respect to the AF cycle.

An approximate analysis of the waveform existing in the diode (grid circuit) is shown in figure 12-20B. Part 1 shows the input waveform which is also the waveform in the input tuned circuit. Because RF current i_g flows in only one direction in the grid circuit, part 2 shows a rectified current waveform in this circuit. Part 3 shows the waveform developed across C_d. This audio waveform is produced in the same way as the audio waveform in the diode detector. However, the waveform shown in part 3 is not the output voltage. In the grid-leak detector the

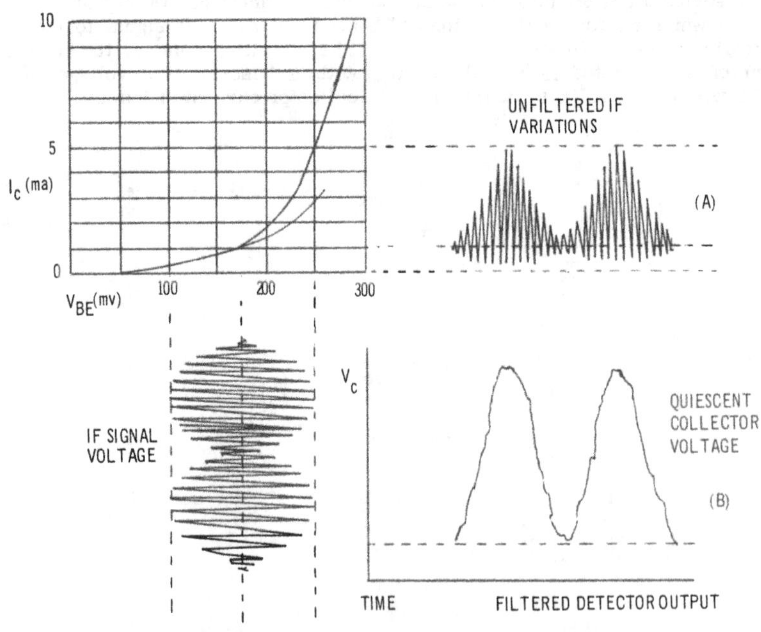

Figure 12-19.—I_C-V_{BE} curve with resulting waveforms.

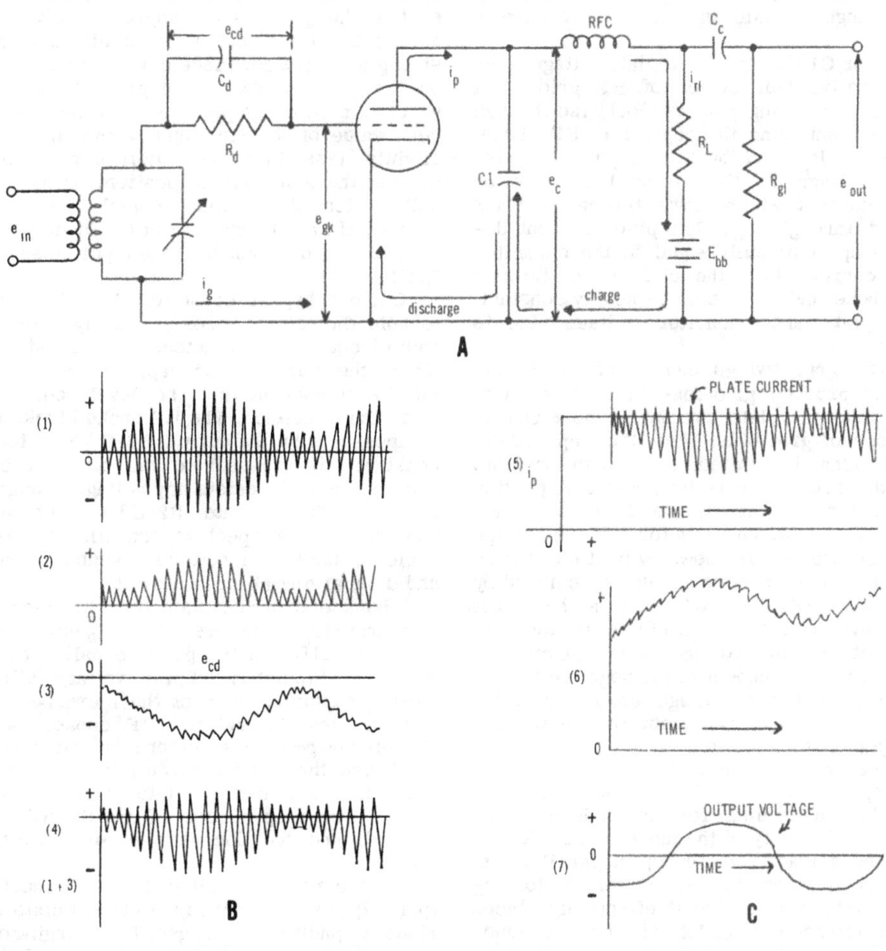

179.233

Figure 12-20.—Grid-leak detector and waveforms.

waveform produced across C_d is combined in series with the RF waveform in the tuned circuit to produce the grid-to-cathode waveform shown in part 4.

An approximate analysis of the waveforms existing in the triode plate circuit is shown in figure 12-20C. Part 5 is the plate current waveform and part 6 is the plate voltage waveform. Capacitor C1 discharges on the positive half cycles of grid input voltage. The discharge path is clockwise through the circuit including the tube and capacitor. The time constant of the discharge path is the product of the effective tube resistance and the capacitance of capacitor C1, and this time constant is short because the effective resistance of the tube is low. The increase in plate current is supplied by the capacitor rather than the E_{bb} supply, thus preventing any further increase in current through the RF choke and plate load resistor R_L. Therefore, any further change in plate and capacitor voltage is limited.

Capacitor C1 charges up as plate voltage rises on the negative half cycles of RF grid input voltage. The charging path is clockwise through the circuit containing the capacitor, RF choke, load resistor R_L, and the E_{bb} supply. The rise in plate voltage is limited by the capacitor charging current which flows through the RF choke and through R_L. The plate current decrease is approximately equal to the capacitor charging current; thus, the total current through the RF choke and R_L remains nearly constant, and the plate and capacitor voltage rise is checked.

Positive grid swings cause sufficient grid current to produce grid-leak bias. Low plate voltage limits the plate current on no signal in the absence of grid bias. Thus, the amplitude of the input signal is limited, since with low plate voltage the cutoff bias is low, and that portion of the input signal that drives the grid voltage below cutoff is lost. The waveform of the voltage across capacitor C1 is shown by part 6 of figure 12-20C. The plate voltage ripple is removed by the RF choke (RFC). Part 7 shows the output voltage waveform. This waveform is the difference between the voltage at the junction of R_L and RFC with respect to the negative terminal of E_{bb} and the voltage across coupling capacitor C_c, which for most practical purposes is a pure d.c. voltage.

Because the operation of the grid-leak detector depends on a certain amount of grid current flow, a loading effect is produced which lowers the selectivity of the input circuit. Recall that placing a load (resistance) in parallel with a tuned circuit widens the bandwidth. The loading which occurs in this circuit effectively places such a resistance in parallel with the input tuned circuit. However, the sensitivity of the grid-leak detector is moderately high for low-amplitude signals, and this partially offsets the disadvantage of lowered selectivity.

PLATE DETECTOR

In a grid-leak detector the incoming RF signal is partially rectified in the grid circuit and the resulting AF signal is amplified in the plate circuit. In a PLATE DETECTOR, the RF signal is first amplified in the plate circuit and then it is detected IN THE SAME CIRCUIT.

A plate detector circuit is shown in figure 12-21A. The cathode bias resistor, R1, is chosen so that the grid bias is approximately at cutoff during the time that an input signal of proper strength is applied. Place current then flows only on the positive swings of grid voltage, during which time average plate current increases. The peak value of the a.c. input signal is limited to slightly less than the cutoff bias to prevent driving the grid voltage positive on the positive half cycles of the input signal. Thus, no grid current flows at any time in the input cycle, and the detector does not load the input tuned circuit, L1C1.

Cathode bypass capacitor C2 is large enough to hold the voltage across R steady at the lowest audio frequency to be detected in the plate circuit. C3 is the demodulation capacitor across which the AF component is to be developed. R2 is the plate load resistor. The RF choke blocks the RF component from the output. R2-C3 has a long time constant with respect to the time for one RF cycle so that C3 resists any voltage change which occurs at the RF rate. R2-C3 has a short time constant with respect to the time for one AF cycle so that the capacitor is capable of charging and discharging at the audio rate.

The action of the plate detector may be demonstrated by the use of the i_p-e_g curve in figure 12-21B. On the positive half cycle of RF input signal (point 1) the plate voltage falls below the E_{bb} supply because of the increased voltage drop across R2 and the RF choke. Capacitor C3 discharges. The discharge current flows clockwise through the circuit including the tube and C3. The change in plate current is supplied by C3. The drop across R2 and the RF choke is limited, and the decrease in plate voltage is slight.

On the negative half cycle of RF input signal (point 2) plate current is cut off and plate voltage rises. Capacitor C3 charges. The charging current flows clockwise around the circuit including the RF choke, R2, and the E_{bb} supply. The voltage drop across R2 and the RF choke, caused by the charging current of C3, checks the rise in plate voltage. Thus, C3 resists voltage changes at the RF frequency. Because C3-R2 has a short time constant with respect to the lowest AF signal, the voltage across C3 varies at the AF rate.

The plate detector has excellent selectivity. Its sensitivity (ratio of AF output to RF input) is also greater than that of the diode detector. However, it is inferior to the diode detector in that it is unable to handle strong signals without overloading. Another disadvantage is that the

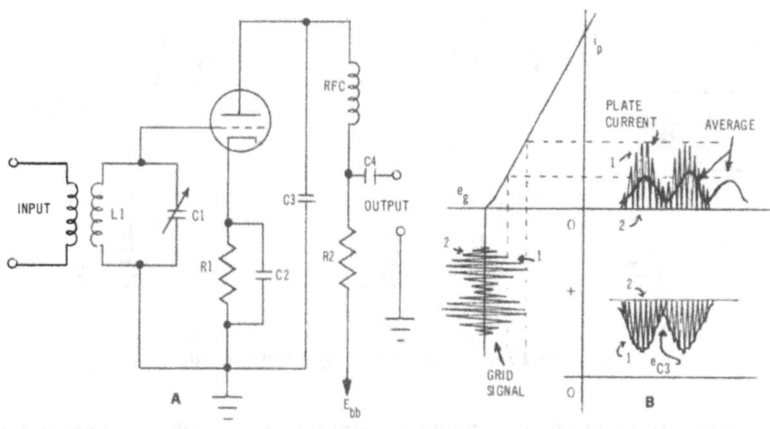

179.234
Figure 12-21.—Plate detector and waveforms.

operating bias will vary with the strength of the incoming signal and thus cause distortion unless a means is provided to maintain the signal input at a constant level. Thus, automatic volume control or manual RF gain control circuits usually precede the detector.

FULL WAVE DIODE DETECTORS

Figure 12-22 shows two forms of full wave diode detectors. While this type of circuit is seldom if ever used in standard broadcast receivers, it finds many applications in higher frequency circuits such as television and radar. It is particularly useful in circuits designed for pulse reception.

Figure 12-22A shows the standard full wave rectifier configuration while figure 12-22B shows the bridge circuit configuration. The main advantage of the bridge detector is use of the entire secondary induced voltage, whereas a centertap is required in the circuit of figure 12-22A, and only half of the input is utilized. Therefore, for the same input to each circuit, the output of the bridge detector would be approximately twice as large as the output of the full wave detector.

Both of these circuits provide a ripple voltage of twice the IF. This permits easier filtering of the ripple voltage in the detector load circuit without attenuation of the higher video frequencies present in the received signals. These detectors also provide a more accurate reproduction of pulses because of the full wave rectification. This

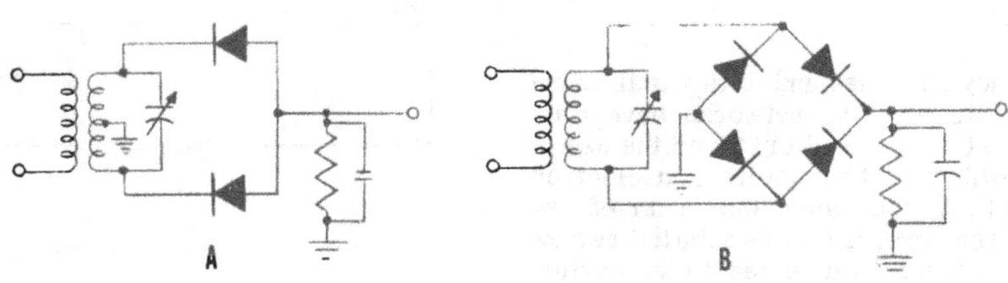

179.235
Figure 12-22.—Full wave diode detectors.

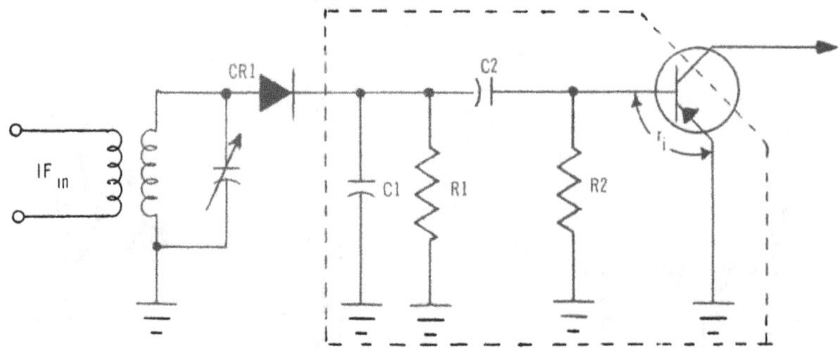

179.236
Figure 12-23.—RC coupling components.

may be an important factor in certain applications, such as moving target indicator radar systems.

OUTPUT COUPLING

There are a number of methods of coupling the audio signal from the detector stage to the audio stage, such as RC coupling, and impedance coupling. Each type of coupling has certain inherent advantages and disadvantages. Thus, the choice of type of coupling to be used for a particular application depends on the specific requirements of the circuit. In other words, some circuits may require a good frequency response while voltage gain is not important. In other circuits high voltage gain may be the prime consideration while exceptional frequency response is not too important, and in yet another circuit a compromise between these factors may be required. RC and transformer coupling are the two basic methods and are the only ones to be considered here.

RC COUPLING

The frequency response and voltage gain characteristics of RC coupling networks have been covered in great detail earlier. Since the use of RC coupling with both transistors and electron tubes is essentially the same, only a brief review will be given here. If a more detailed review is desired the reader is directed to the earlier text material on RC coupling.

Figure 12-23 is a partial schematic illustrating the components involved in the RC coupling of a signal from the detector stage to the audio stage. The coupling components are included within the dotted area of the figure. C1 and R1 have been discussed previously in connection with the series diode detector.

These components are analogous to the plate load resistor and electron tube output capacitance and affect the characteristics of the coupling network in approximately the same manner. C2 is the coupling capacitor in other transistor and electron tube circuits. The parallel combination of R2 and r_i (input junction resistance of the transistor) are analogous to the grid resistor. It must be realized that in comparing the action of coupling networks for semiconductor and electron tube devices, that while the characteristics and general operation may be the same, the component values will differ greatly.

Figure 12-24 illustrates a typical response curve of an RC coupling network. Depending on component values and design accuracy the low

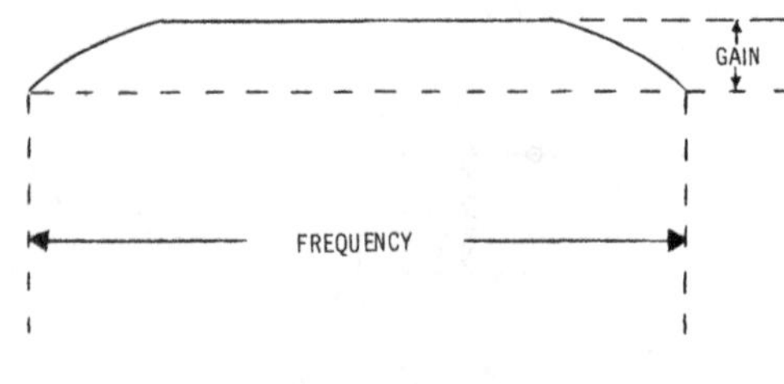

179.237
Figure 12-24.—RC coupling frequency response curve.

frequency limit may range from approximately 30 to 60 Hz and the high frequency limit from approximately 100 to 200 kHz.

The response is measured in terms of voltage gain over a range of frequencies. The gain falls off at very low frequencies because of the increase in the capacitive reactance of the coupling capacitor. This capacitor acts in series between the source and the load and has developed across it an increasing percentage of the signal voltage as the frequency is decreased.

At midband frequencies the coupling capacitor is considered to be a short for all practical purposes. The reduction in gain at the high frequencies is caused by the shunting effect of the output capacitance of one stage, the input capacitance of the next stage, and the distributed capacitance of the coupling network. The combined effect of these capacitances is to increase the percentage of the total signal voltage that is developed across the internal resistance of the input stage and decrease the percentage of signal voltage which appears as the output signal.

TRANSFORMER COUPLING

Diode detector circuits use RC coupling almost exclusively. On the other hand, a transistor or electron tube amplifier detector may use EITHER RC or transformer coupling. The common-emitter transistor detector shown previously in figure 12-18, would, by necessity, use RC coupling, since connecting a low resistance transformer winding across the output would shunt the load resistance and V_{CC}. It is sometimes more desirable, in a transistor detector, (or electron tube detector) to use transformer coupling rather than RC coupling. In the case of transistor detectors, this would allow higher collector voltages.

Figure 12-25 illustrates a common-base transistor detector using transformer coupling. The non-linear and unilateral properties of the input junction provide the rectification (distortion) and heterodyning actions necessary for detection. Filtering of the high frequency component is provided by the C2R1 network, with additional high frequency filtering provided by C3. The base-emitter voltage will vary in accordance with the audio modulation signal. This audio variation will be amplified and applied to the primary of T2 and, thus, transformer coupled to the audio amplifier.

A transformer coupled stage has certain advantages over other types of coupling. The voltage amplification of the stage may exceed that of the transistor if the transformer has a step-up turns ratio. Direct current isolation of the base, or emitter, of the next transistor is provided without the need for a blocking capacitor, and the d.c. voltage drop across the coupling resistor (which is necessary when RC coupling is used) is avoided. This type of coupling lends itself well to coupling high impedance sources to low impedance loads, or vice versa by choosing a suitable turns ratio.

Transformer coupling has the disadvantages of greater cost, greater space requirements and the possibility of poorer frequency response at the higher and lower frequencies. The voltage

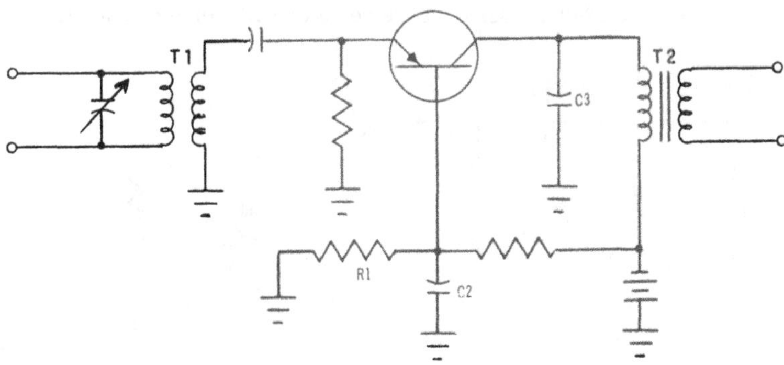

179.238

Figure 12-25.—Transformer coupled CB transistor detector.

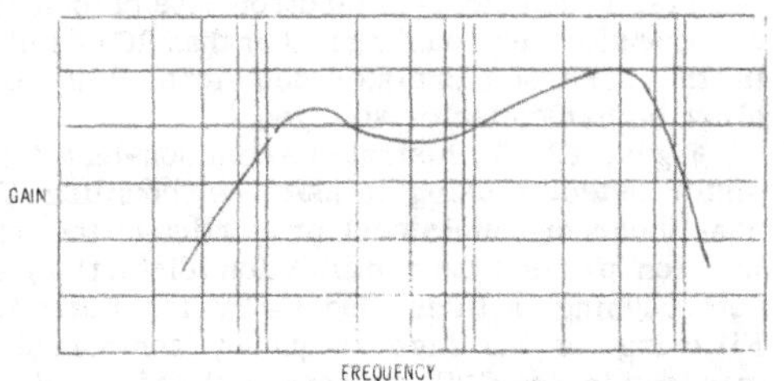

179.239

Figure 12-26.—Percent of voltage gain versus frequency for transformer coupling.

gain as a function of frequency throughout most of the audio range is shown in figure 12-26. The curve shows that the transformer coupled amplifier has a relatively high gain and uniform frequency response over the middle range of audio frequencies, but poor response for both low and high audio frequencies.

The decrease in response at the low frequencies is due to the decrease in inductive reactance of the transformer primary winding. The decrease in response at the high frequencies is due, mainly, to the shunting effect of transistor interelement capacitance, stray capacitances, and distributed capacitance of the transformer windings. If proper design precautions are observed, a peak in the response curve, such as observed in figure 12-26 (around 4 kHz), will occur due to a series resonant condition being established between the leakage inductance and the stray capacitances. This series resonant condition usually occurs near the high frequency limit.

VOLUME CONTROL

There are many different methods of achieving volume control in a broadcast receiver. One of these many methods is illustrated in figure 12-27. The circuit shown is a standard series diode detector circuit in which the load resistor has been replaced with a potentiometer. The output signal is RC coupled to the audio amplifiers. The

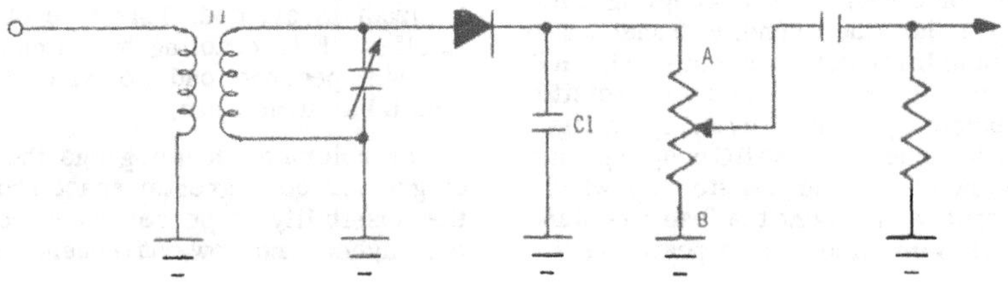

179.240

Figure 12-27.—Series diode detector with volume control.

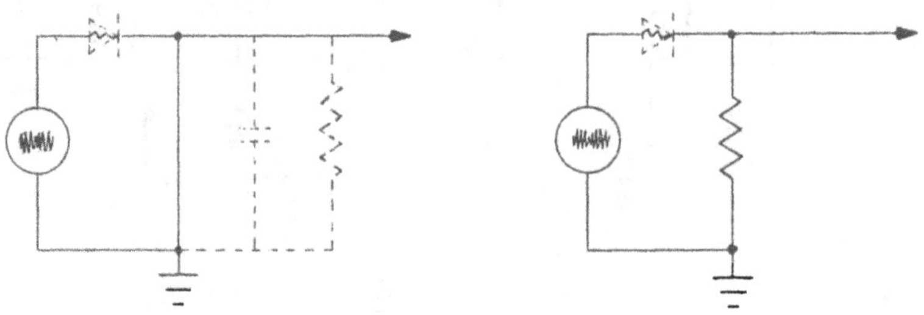

179.241

Figure 12-28.—Equivalent circuits for series diode detector.

detector operation is the same as before, with the IF signal being developed in the secondary of T1, the series diode acting as the non-linear device, and capacitor C1 and the potentiometer acting as the low pass filter. The total audio signal is developed across the filter circuit and the potentiometer is used to allow the operator to determine the amount of this output audio signal which will be fed to the audio amplifiers. If the wiper arm of the potentiometer is set near point A, a large signal is fed to the audio amps and the volume will be loud. As the wiper arm is moved toward point B, the volume decreases. If the wiper arm is set all the way down to point B there will be no signal to the audio amps, and no audio will be heard.

EQUIVALENT CIRCUITS

By treating the input tank circuit as the source of the various frequencies which are applied to the detector, simplified equivalent circuits may be constructed. Two such equivalent circuits are shown in figure 12-28.

Figure 12-28A shows the equivalent circuit of a series diode detector at high frequencies. At the IF and sideband frequencies the filter capacitor has a negligible capacitive reactance and acts as an effective short circuit. Under these conditions, the source sees only the low forward impedance of the diode and the high frequencies are effectively shorted out of the output.

Figure 12-28B is the equivalent circuit of the detector at audio frequencies. In this case the impedance of the filter network is high in comparison to the forward impedance of the diode. As a result, those frequency components which occur at the modulation rate will be developed across the filter network and applied to the audio amplifiers.

CHAPTER 13

AUDIO AMPLIFIERS

Audio amplifier circuits are designed to amplify signals which lie within the audio frequency spectrum (20 to 20,000 Hz). In practice, to insure that these signals are amplified to sufficient power levels, audio amplifiers are designed to respond to frequencies above and below the audio frequency, often from 10 to 100,000 Hz. Both transistors and electron tubes are used in audio frequency amplifiers for voltage or power amplification.

AUDIO VOLTAGE AMPLIFIERS

Figure 13-1 shows the location of the AF voltage amplifier in a complete superheterodyne receiver. Note that the AF amplifier receives the demodulated output of the detector. The purpose of the audio voltage amplifier is to provide a sufficient input to the audio power amplifier. The input to the audio voltage amplifier is a low voltage signal while the output is the amplified version of this signal.

SINGLE STAGE TRANSISTOR AUDIO AMPLIFIER

The circuit of figure 13-2 shows a basic single stage class A audio amplifier using a transistor as the amplifying device. The input will be a low amplitude audio voltage (normally in the millivolt range) which will be amplified by Q1 and coupled either to another voltage amplifier or, if sufficient amplification has been obtained in this stage, to a power amplifier. Q1 is a PNP transistor in the common-emitter configuration. R_L serves as the collector load resistor, R_E and C_E are the emitter bias stabilization network, and R_B limits the bias current and thereby establishes the operating point. Coupling capacitor C2's value is such so as to assure good frequency response. When $-V_{CC}$ is applied to the circuit, the current flow through the base-emitter junction develops a voltage which forward biases the input circuit.

During the positive alternation of the input signal, forward bias is being decreased. This causes a resulting decrease in collector current and the collector voltage becomes more negative (with less of a voltage drop across R_L, the collector voltage will increase towards the negative value of V_{CC}). The output signal at the collector is an amplified, negative-going signal.

During the negative alternation of the input signal the forward bias is increased. This increases the collector current through R_L, with the result that the collector voltage becomes less negative (more positive). The output signal during this half cycle is an amplified positive half cycle. The paths for base and collector currents (d.c.) are shown in figure 13-2. The dotted path is for collector current and the solid line shows the base biasing current.

IMPEDANCE CONSIDERATIONS

When an audio amplifier is used to "build up" very weak audio signals, it is generally referred to as a preamplifier. Preamplifiers find their greatest application with weak audio signal sources such as microphones or magnetic tape pickup. The internal impedance of these devices may be either high or low depending on their type. In order to attain a maximum transfer of energy the input impedance of the transistor should be as close as possible to the internal impedance of the device being used as the voltage source.

The most desirable method of matching source impedance to the transistor input impedance is by transformer coupling; however, this is not always practical. When a transistor amplifier must be fed from a low impedance source, (20 to 1500 ohms), without the benefit of transformer

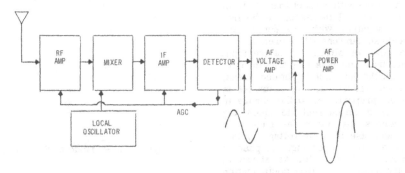

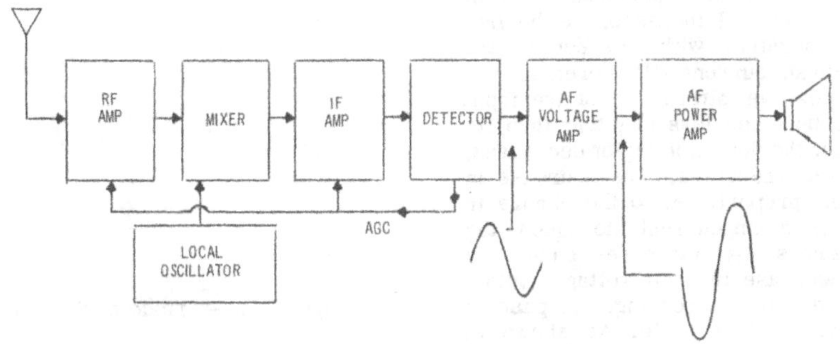

Figure 13-1.—Audio signal amplifier in a superheterodyne receiver.

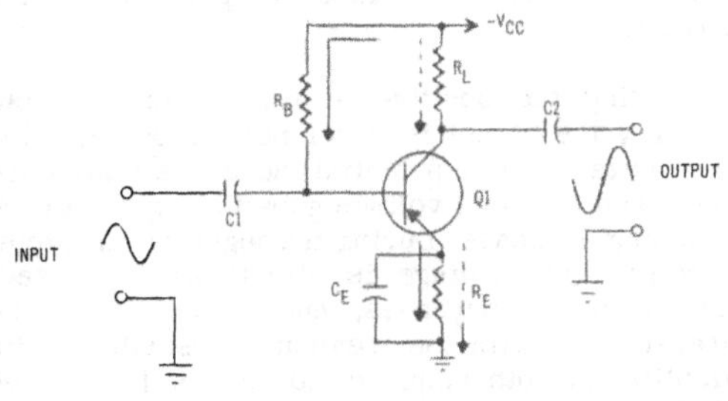

Figure 13-2.—Transistor audio amplifier.

coupling, either the CB or the CE transistor configuration can be used.

If the signal source has a high internal impedance, an amplifier having a high input impedance is required. This requirement may be met by using the common-collector configuration, by the addition of a series resistance to the base lead of a CE configuration, or by leaving the emitter bias resistor unbypassed in the CE configuration. One of the two CE configurations is normally used as it provides a greater gain than the common-collector configuration. Also, the frequency response of the CE circuit arrangement is satisfactory for audio work. Finally, the CE circuit's input impedance is less dependent on the value of load impedance than either the CB or the CC configuration. Thus, the CE circuit is by far the most common one for general purpose audio amplification.

SINGLE STAGE FET AMPLIFIER

The circuit of figure 13-3 shows a basic FET (field effect transistor) audio amplifier. This circuit utilizes an N-channel FET as the amplifying device. It should be recalled (chapter 6) that the FET is a high gain device which is quite similar in characteristics to the electron tube pentode. The circuit shown is a class A audio amplifier utilizing the common-source configuration. Self bias is developed by the source-bias circuit consisting of R1 and C2. C1 is the input coupling capacitor, R_G is the input gate resistance used to develop the input signal, C3 is the output coupling capacitor, and R_D is the drain load resistor. In figure 13-3, as the gate

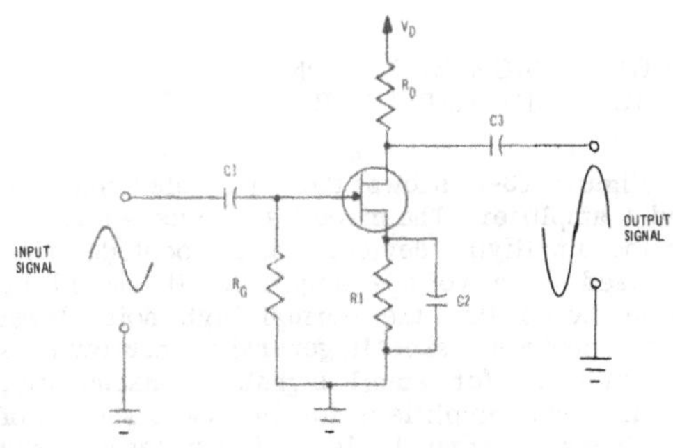

Figure 13-3.—FET audio amplifier.

becomes less negative with respect to the source, source-drain current will increase. As the gate becomes more negative with respect to the source, source-drain current will decrease.

During the positive alternation of the input signal the gate becomes less negative than its quiescent voltage, the depletion region decreases, and drain current increases. The increase in drain current is proportional to the change in junction bias. As drain current increases, the voltage drop across R_D increases causing a corresponding decrease in drain voltage. Output voltage is consequently decreasing, or, passing through its negative half cycle. As shown in figure 13-3, the common source configuration provides a 180 degree phase inversion.

During the negative alternation of the input signal, the gate voltage becomes more negative than its quiescent value, the depletion region increases causing drain current to decrease. The decreased drain current will drop less voltage across R_D resulting in an increased drain voltage. Therefore, the output voltage is going through its positive alternation.

As mentioned previously, the FET is a high gain device, with gain values of 400 to 1000 being common for practical amplifier circuits. FETs have a high input impedance which immediately makes them acceptable for use with high impedance input devices. Their output impedances are rated as medium to high and, therefore, are applicable for driving high-impedance earphones or high power amplifiers. FET circuits, as a rule, are less complex than standard transistor circuits as they require fewer circuit components. They have low noise ratings for the audio frequency range and produce minimal leakage currents.

SINGLE STAGE ELECTRON TUBE AUDIO AMPLIFIER

Figure 13-4 shows the basic electron tube audio amplifier. The circuit shown uses a triode as the amplifying device. While a pentode could be used as a voltage amplifier, it should be remembered that the normal high noise level of the pentode makes it generally unsuitable as an amplifier for small signals. Consequently, small signal amplifiers for the audio section of receivers are normally limited to triode electron tubes.

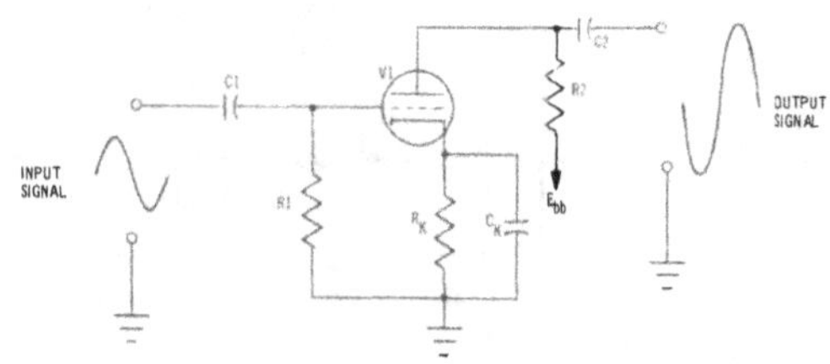

179.244
Figure 13-4. — Triode audio amplifier.

C1 and C2, figure 13-4, are the input and output capacitors, R1 is the grid-input resistor for signal development and returning the grid to ground, R2 is the plate load resistor, and R_k-C_k is the self-bias network used to develop the necessary bias for class A operation of the circuit.

During the positive alternation of the input voltage, the signal is developed across R1, the less negative grid potential increases plate current, and the plate voltage goes through a corresponding decrease. During the negative half cycle, negative grid voltage is effectively increased, tube current decreases, and plate voltage increases. As with the previously described audio amplifiers, both amplification and a 180 degree phase inversion have taken place. The gain of triode amplifiers is normally in the range 10 to 50. Also, the internally generated noise of the triode is low enough to be acceptable.

The grounded cathode amplifier exhibits a relatively high input impedance which makes it suitable for use with high-impedance voltage sources. Its output impedance, while not as high as that of the tetrode or pentode, is sufficiently high to allow output coupling to high-impedance earphones. When low impedance sources or loads are used, coupling is normally accomplished through impedance matching transformer coupling.

AUDIO FREQUENCY COUPLING NETWORKS

There are four types of coupling networks which may be used within the audio range. They are: RC coupling, direct coupling, impedance coupling, and transformer coupling. As each of these methods has previously been covered in detail, only a brief review will be given here.

Bias Network Frequency Response

In each of the three preceding audio amplifiers, self bias was utilized to develop class A operating bias. The amount of bias developed is determined by the quiescent (no-signal) current flow through the amplifying device. The bypass capacitor is in the circuit to prevent fluctuations in the bias voltage which would otherwise occur as the current varied. These unwanted fluctuations, if present, would cause degeneration and a loss of amplifier gain.

At the higher audio frequencies, it is relatively easy for a capacitor to bypass the signal variations to ground. However, at the lower frequencies, if the capacitor is improperly chosen, bias voltage variations will be present and degeneration will occur.

In order to prevent unwanted bias variations, degeneration, and a loss of gain at the lower audio frequencies, the capacitor must bypass even the variations of the lowest audio frequency to be encountered in the circuit. A rule of thumb is: The capacitive reactance of the capacitor must be 1/10 or less than the self bias resistance value, AT THE LOWEST AUDIO FREQUENCY (20 Hz).

RC Coupling

By far the most common method of coupling two stages of audio amplification is RC coupling. RC coupling in both transistor and electron tube circuits is shown in figure 13-5. The RC network (shown within the dashed line in fig. 13-5) used between the stages consists of a load resistor (R1) for the first stage, a d.c. blocking capacitor (C1), and a d.c. return resistor (R2) for the input element of the second stage.

Because of the dissipation of d.c. power in the load resistors, the efficiency (ratio of a.c. power out to d.c. power delivered to the stage) of the RC coupled amplifier is low.

The d.c. blocking capacitor prevents the d.c. voltage component of the output of the first stage from appearing on the input terminal of the second stage. To prevent a large signal voltage drop across the d.c. blocking capacitor, the reactance of the capacitor must be small compared to the input resistance of the following stage with which it is in series. Since the reactance must be low, the value of capacitance must be high. However, because of the low voltages used in transistor circuits, the physical size of the capacitor may be kept small. Physically larger capacitors are required in electron tube circuits due to the higher voltages. The resistance of the d.c. return resistor is usually much larger than the input resistance of the second stage. The upper limit of the value of this resistor is dictated by the d.c. bias considerations in the case of the transistor circuit, and by the shunting capacitance in the case of the electron tube circuit.

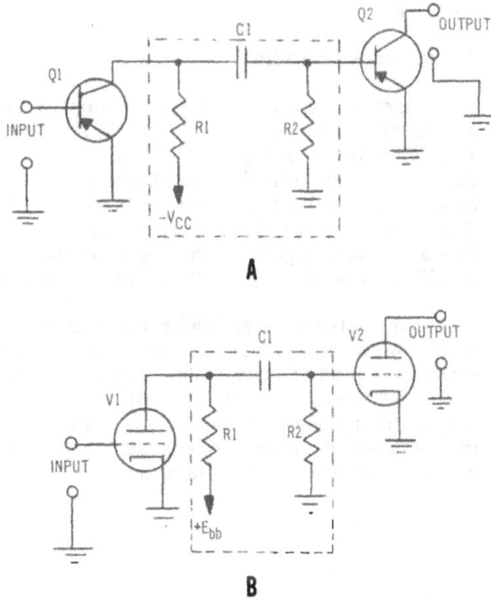

Figure 13-5.—RC coupling.

The frequency response of the RC coupled audio amplifier is limited by the same factors that limited frequency response in other RC coupling circuits. In other words, the very low frequencies are attenuated by the coupling capacitor whose reactance increases with low frequencies. The high frequency response of the amplifier is limited by the shunting effect of the output capacitance of the first stage and the input capacitance of the second stage. Circuit techniques (such as peaking) that can be used to extend the low frequency and high frequency response of RC coupled transistors and electron tube amplifiers are covered elsewhere in this text.

RC coupling is used extensively in audio amplifiers because of good frequency response, economy of circuit parts, and the small physical size which can be achieved with this method of coupling.

Transformer Coupling

Interstage coupling of audio amplifiers by means of transformers is shown in figure 13-6. The primary winding of transformer T1 (including the a.c. reflected load from the secondary winding) is the load impedance of the first stage. The secondary winding of transformer T1 introduces the a.c. signal to the input of the second amplifier and also acts as the input d.c. return path.

Because there is no load resistance to dissipate power, the power efficiency of the transformer coupled amplifier approaches the theoretical maximum of 50%. For this reason, the transformer coupled amplifier is used extensively in portable equipment where battery power is used. Transformers facilitate the matching of the load to the output of the amplifier and the source to the amplifier to bring about maximum available power gain in a given stage.

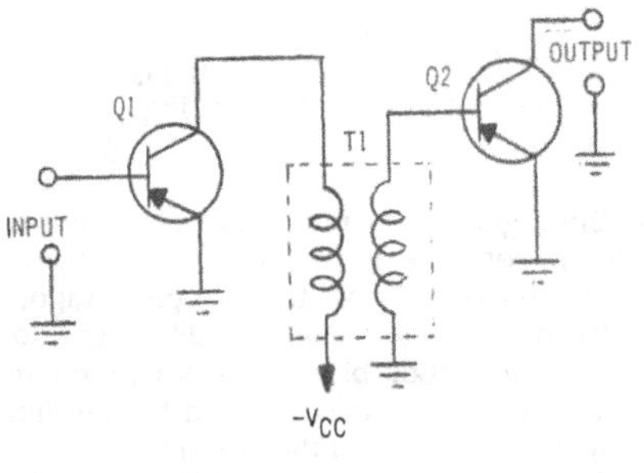

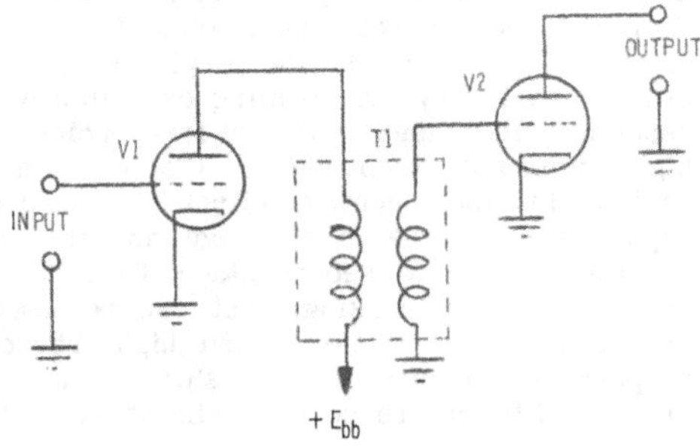

179.246

Figure 13-6.—Transformer coupling in audio amplifiers.

The frequency response of a transformer coupled stage is not as good as that of the RC coupled stage. The shunt reactance of the primary winding at low frequencies causes the low frequency response to fall off. At high frequencies the response is reduced by the input capacitance and the leakage reactance between primary and secondary windings.

In addition, transformers are more expensive, heavier, and larger in size compared to resistors and capacitors required for coupling. While transformer coupling may be used for audio amplifiers, its use is normally limited to the output power stage of the receiver. Power amplifiers are discussed later in this chapter.

Direct Coupling

Figure 13-7 shows direct coupling between two stages of the audio amplification. In the transistor circuit, figure 13-7A, the first stage is biased so as to provide a quiescent value of 7 volts on the collector of Q1. R1 is the base biasing resistor. R2 serves both as a collector load for Q1 and also as the base input resistor for Q2, R3 is the collector load for Q2. R4, R5 and R6 form a voltage divider network to provide the necessary circuit voltage requirements. The 7 volt collector voltage of Q1 is also the base voltage of Q2. The emitter of Q2 is at a potential of 6 volts which is developed by the voltage divider network. This forward biases Q2 and determines its quiescent operating point.

Essentially the same configuration is used in the case of the electron tube direct-coupled audio amplifier (fig. 13-7B). R1C1 is the cathode bias network for the first stage. This network develops (in this example) 10 volts of bias. R2 serves a dual purpose: (1) it is the plate load resistor for the first stage and (2) it is the grid input resistor for the second stage. R3 is the plate load resistor for the second stage. R4C2 and R5C3 are filtering and decoupling networks which are required to form a voltage dividing network which provides the required voltage of 150 volts from the 400 volt source. The input signal is applied to the grid of the first stage and is taken off the plate of the second stage.

In the first stage, of the tube type, the bias (10 volts), and the effective E_{bb} for the first stage (150 volts), set the operating point of the tube. Assume the quiescent plate voltage to be 135 volts. With an input signal, this first stage

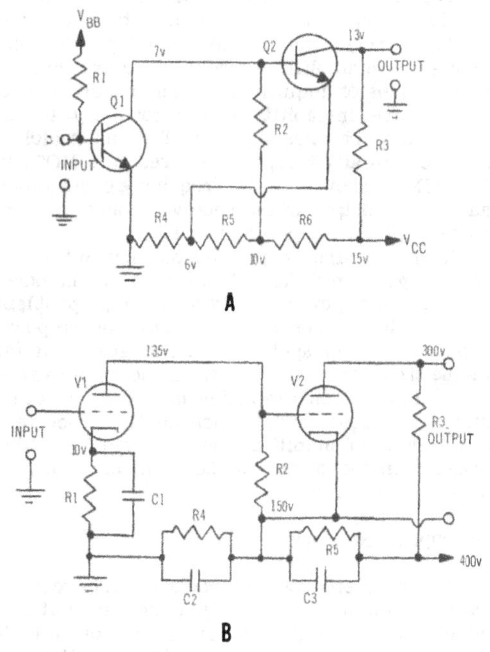

179.247
Figure 13-7.—Direct coupling in audio amplifiers.

of amplification would act as any other electron tube amplifier.

Notice that the plate voltage of the first stage is applied directly to the grid of the second stage (hence the name direct coupling). This places a positive 135 volt potential on the grid. In most cases this would be highly undesirable. However, as the diagram shows, the cathode of the second stage has been connected to the 150 volt supply of the first stage. This results in a 15 volt bias on the second stage. Therefore, grid current will not be drawn and the circuit can operate as a normal audio amplifier. Note that where both amplifiers could have previously been operated with a single 150 volt power supply, it is now necessary to provide a much higher supply voltage.

The major advantage of direct coupling is that reactive coupling elements (capacitors and inductors) are not required. Transformer coupling quite often gives a poor high frequency response and, likewise, capacitors tend to give a poorer low frequency response than desirable. Direct coupled audio amplifiers provide a very flat frequency response for the range of frequencies that they are designed to cover. The major problem of direct coupling is the larger than normal requirements for power supplies. Each succeeding stage requires a higher source voltage than the preceding one. Usually direct coupling is a very inefficient method due to the high power losses in the load resistors, coupled with the intricate biasing requirements. These losses become even greater as the number of stages are increased. It is also often difficult to match the output impedance of one stage to the input impedance of a following stage utilizing only a single resistor. This adds to the inefficiency of the circuit operation.

Impedance Coupling

Figure 13-8 shows an electron tube, impedance coupled, audio amplifier. This method of coupling is seldom used for audio amplifiers and is covered here only so that all four coupling methods for audio amplifiers may be reviewed at this time. As can be seen from figure 13-8, impedance coupling differs from resistance coupling only in that the load is an inductor rather than a resistor.

Since the d.c. resistance of L1 is low, this method of coupling is more efficient than resistive coupling. However, the frequency response is not as flat as is resistive coupling. The frequency response of the impedance coupled amplifier stage is very similar to that of a transformer coupled stage for the same reasons.

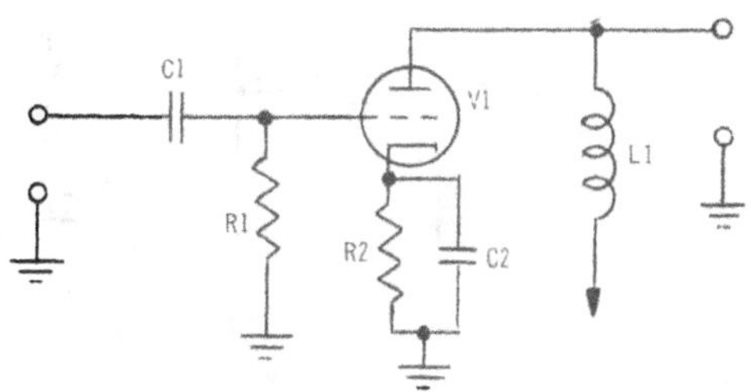

179.248
Figure 13-8.—Impedance coupling in an electron tube amplifier.

FEEDBACK IN AUDIO AMPLIFIERS

Figure 13-9 is a partial schematic used to illustrate the problem of regenerative feedback which sometimes occurs in high gain audio amplifiers. To the left of the diagram is shown a power supply filter consisting of C1, C2, and R1. R2, R3, and R4 are the plate load resistors for V1, V2, and V3 respectively. C3 and C4 are standard interstage coupling capacitors. A positive signal is shown being applied to the grid of V1. V1 amplifies the signal and sends it to the grid of V2 as a negative signal. V2 in turn amplifies it and sends it on as a positive going signal to the grid of V3. After two stages of previous amplification the signal is quite large. The signal now causes a large increase in current through V3. As V3's current is increasing, it is obvious that there has been a tremendous reduction in the impedance of the third stage of amplification. The increased current of V3 causes a very heavy load on the power supply with the result that the output voltage of the filter circuit is reduced.

Note the effect that a reduced E_{bb} will have on the first stage of amplification. The V1 plate signal was negative going as shown, and a reduction in E_{bb} will further reduce the plate voltage on V1. This change will be fed through the circuit as a signal, causing a negative going signal on the grid of V2, a further positive going signal on the grid of V3, and the loading effect on the power supply is further increased. The entire process is regenerative with the possible effects of V2 cutoff and V3 saturation. The same regenerative effects will be noted if the signal applied to the grid of V1 were negative instead of positive. In this case, however, there would be a reduction in the loading effect on the power supply output and E_{bb} would tend to rise.

These regenerative effects tend to occur only in high gain amplifiers where there are three or more stages of amplification. The results of such effects are instability and a tendency toward oscillation. The resulting oscillations (which do not occur in all cases) are referred to as MOTORBOATING. They are low frequency oscillations, and at the output of the receiver sound, as their name implies, like a motorboat.

It should also be pointed out that in the case of high gain amplifier chains there is an excellent chance of power supply hum being a problem. As any ripple present in the output of the power supply filter is applied to all stages, it is felt in the first stage. In this stage, the ripple variations may be on the order of amplitude of the signal voltages and, consequently, receive a great deal of amplification. The ripple is then present in the output in the form of a low frequency "hum".

Decoupling Networks

The problems of motorboating and excessive hum are often eliminated by using a circuit such as is shown in figure 13-10. In this circuit, the full voltage of the power supply is applied only to the third stage. The earlier stages receive less plate voltages and have additional filtering networks (called decoupling networks) which help isolate the amplifiers from power supply variations.

Stage 3, figure 13-10, receives the largest input signal and therefore requires the highest plate voltage. The full 400 volt E_{bb} is applied to this stage. State 2 receives only a 300 volt plate voltage. This is accomplished through the

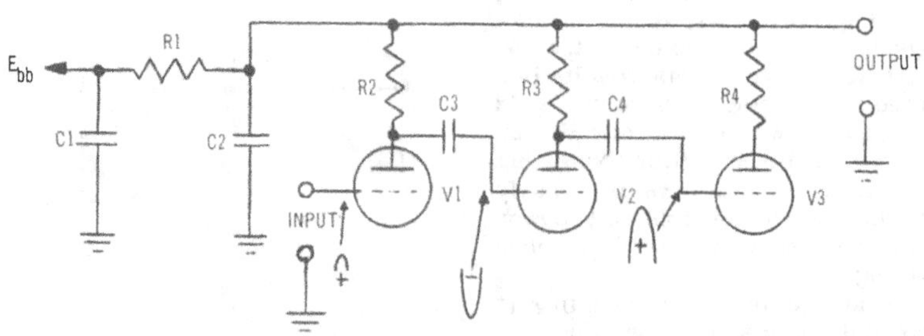

179.249

Figure 13-9.—Illustration of regenerative feedback.

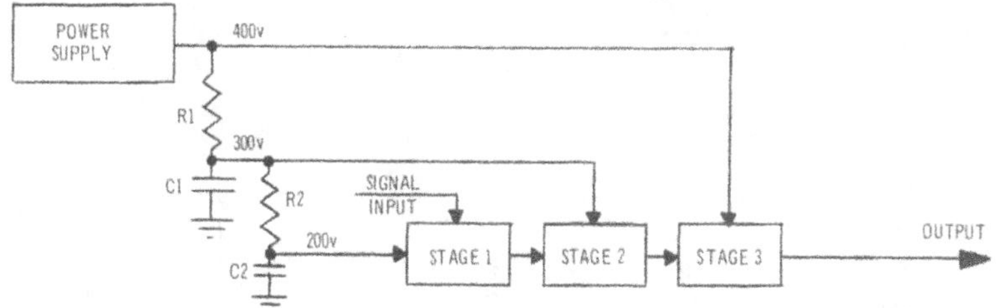

179.250
Figure 13-10.—Decoupling networks in audio amplifiers.

filtering action of R1C1. R1 drops the voltage down to 300 volts and C1 provides additional filtering action so that power supply variations will not reach stage 2. Stage 1, which has the lowest input signal, and therefore the smallest plate voltage requirements, receives only a 200 volt d.c. source. Here, R2 is used to drop the 300 volt source down to the 200 volt level, and, as before, additional filtering is provided by C2. The additional filtering action thus reduces the likelihood that power supply variations will be regeneratively fed back to the first stage. Therefore, motorboating is very unlikely when decoupling networks are utilized.

The additional filtering action of the decoupling networks also reduces the effects of power supply ripple (usually 60 or 120 Hz.). Serious "hum" problems normally occur only when the first audio amplifier is not adequately isolated from power supply variations. With the decoupling circuits of figure 13-10, however, adequate isolation is obtained for practically all receiver applications.

Negative Feedback

As just discussed, high gain audio amplifiers are subject to regenerative (POSITIVE) feedback when several stages of amplification are used. In addition to the required decoupling, it may be necessary to provide some form of degenerative feedback (NEGATIVE FEEDBACK). Negative feedback can then be used to cancel the effects of the regenerative feedback and provide increased amplifier stability and frequency response. This is because at the ends of the frequency response curve of an amplifier the gain decreases. This decrease causes a decrease in negative feedback and tends to increase the effective gain, thus extending the frequency response.

The most common method of achieving negative feedback is to leave the emitter/source/cathode resistor unbypassed. An example circuit using this method, in addition to decoupling, is shown in figure 13-11. As illustrated, the emitter resistors are left unbypassed. This causes degeneration (negative feedback) by limiting the gain of the various amplifier stages. In addition to the prevention of regenerative feedback being aided, the fidelity of the amplifiers is improved.

High gain audio amplifier chains are always subject to signal distortion. That is, large gains may often result in peak clipping, where one or more amplifiers are inadvertently allowed to be driven either to cutoff or a near saturation condition. By leaving the emitter resistors unbypassed (no emitter capacitors), the gain of the amplifiers is reduced, and the chances of a loss of fidelity have been reduced. The purpose of the audio section of any receiver is to reproduce the audio signal without any appreciable distortion. Clipping must therefore be eliminated in the circuitry, and this requires some means of reducing the overall amplifier gain. The method illustrated in figure 13-11 adequately meets these requirements.

An alternate method of obtaining negative feedback is illustrated in figure 13-12. In this method of obtaining negative feedback, a portion of the output signal is fed back to the input circuitry in such a way as to degenerate the input signal. In this way, the gain of the various amplifiers remains the same but it is the signal which is attenuated.

The majority of the circuitry has been left out of figure 13-12 for the purpose of simplification. A positive input signal is shown applied

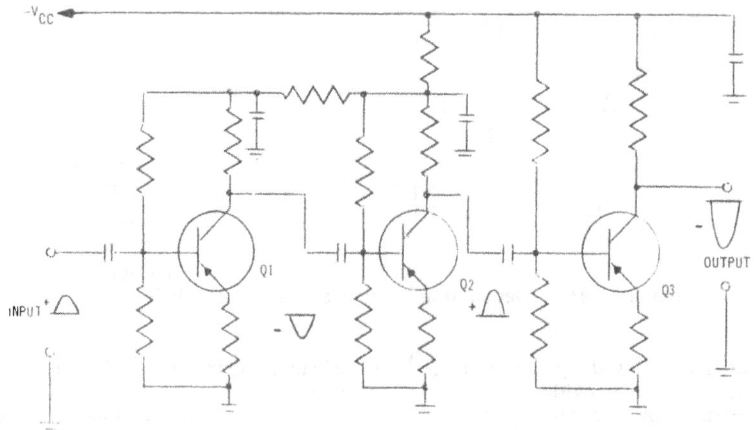

179.251
Figure 13-11.—Three-stage audio amplifier utilizing degeneration.

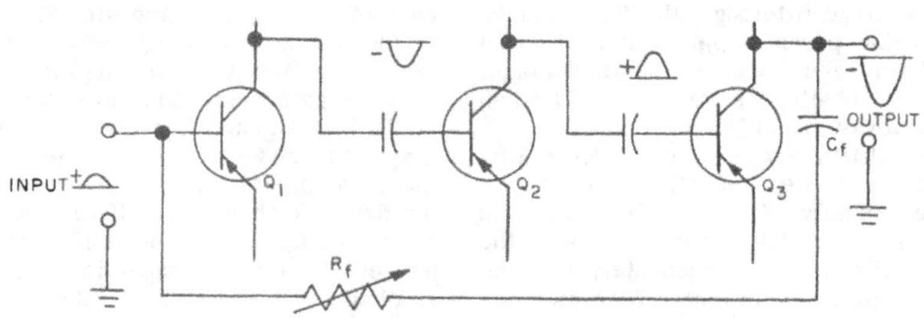

179.252
Figure 13-12.—Negative feedback utilizing resistive feedback network.

to the base of Q1. The input to Q2 is negative and the input to Q3 is positive. The final output signal is negative going. C_f is the feedback capacitor and R_f is an attenuating resistor which is used to reduce the amplitude of the feedback voltage to reasonable values. As the original input signal on the base of Q1 is positive, it will be partially cancelled by the negative feedback voltage coming from the collector of Q3. R_f is shown as a variable resistor so that the amplitude of the degenerative feedback may be controlled.

It should be pointed out that the application of the feedback voltage is not limited to the base of Q1. The feedback voltage may be applied to any point in the preceding circuit where it will produce the desired effects. As an example, it could be applied in the collector circuit of Q2 to decrease the positive collector signal of that stage or, it could be applied to the emitter of Q2 if that emitter had been left unbypassed.

A final method of obtaining feedback is through transformer action. This method is seldom used as most audio amplifiers are RC coupled. In transformer obtained feedback an extra winding is placed on the interstage coupling transformers. A portion of the output energy is then fed back to one or more of the preceding stages as the required feedback. A typical method is shown in

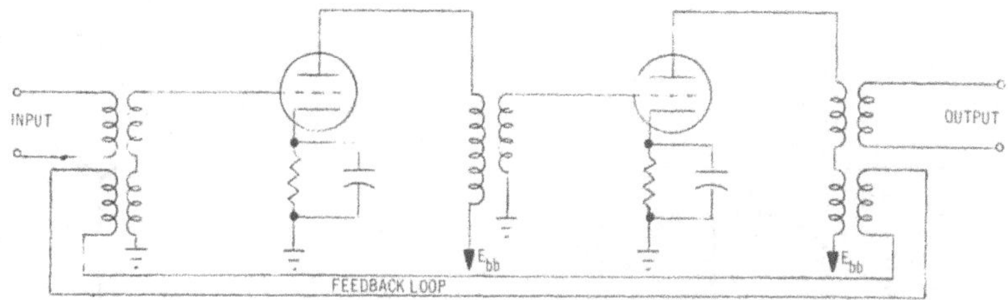

179.253
Figure 13-13.—Feedback through transformer action.

figure 13-13. The correct amplitude of feedback voltage is controlled by the turns ratios of the feedback windings and the correct polarity is obtained by the wiring connections of the feedback loop. As opposed to the predecing forms of feedback (voltage feedback), transformer obtained feedback is essentially a current feedback.

IMPEDANCE MATCHING IN AUDIO AMPLIFIERS

Audio amplifiers must sufficiently amplify an input signal to the point where it may be applied to a power amplifier. Due to their high input impedance, electron tubes cause few impedance matching problems. While the input impedance for transistors may be high in some configurations, transistor circuits are nevertheless more troublesome in this respect. For this reason, a more detailed discussion will now be given on the manner in which the effects of the input impedance determine the choice of transistor configurations.

The most desirable method of matching source impedance to input impedance is by transformer coupling; however, as mentioned previously, this is not always practical. When the preamplifier must be fed from a low resistance source (20 to 1500 ohms), without the benefit of transformer coupling, either the CB or the CE configuration may be used. The CB configuration has an input impedance which is normally between 30 and 150 ohms; the CE configuration has an input impedance which is normally between 500 and 1500 ohms.

If the signal source has a high internal impedance, a high input impedance can then be obtained by using one of the three following circuit arrangements.

It would appear that the easiest configuration to use would be the common collector. The input resistance of the CC configuration is high because of the large negative voltage feedback in the base-emitter circuit. As the input voltage rises, the opposing voltage developed across the load resistance (fig. 13-14) substantially reduces the net voltage across the BE junction. By this action, the current drawn from the signal source remains low. From Ohm's law it is known a low current drawn by a relatively high voltage represents a high impedance. If a load impedance of 500 ohms is used, the input resistance of a typical CC configuration will be over 30,000 ohms. The disadvantage of the CC configuration, however, is that small variations in the current drawn by the following stage cause large changes in the input impedance value.

The variation of input impedance, as a function of load impedance, for the CE, CB and CC configurations is shown in figure 13-15. Notice that the CC configuration with an R_L of 500 ohms has an input impedance of approximately 15,000 ohms. Within the operating range of 1,000 to 100,000 ohms of load resistance the input impedance of the CC circuit will increase, r_i of the CE circuit will decrease with an R_L

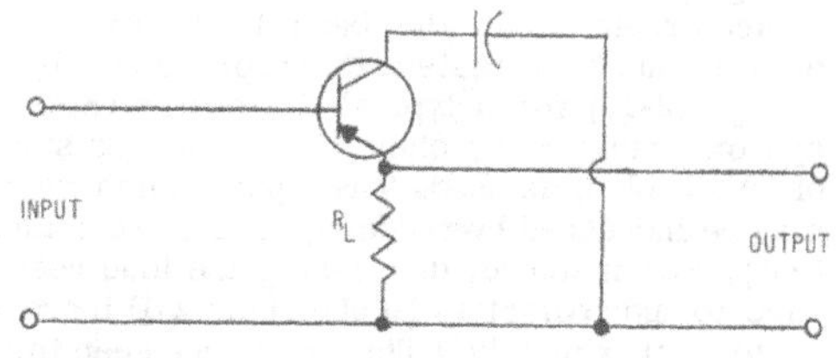

179.254
Figure 13-14.—Simplified schematic of CC preamplifier.

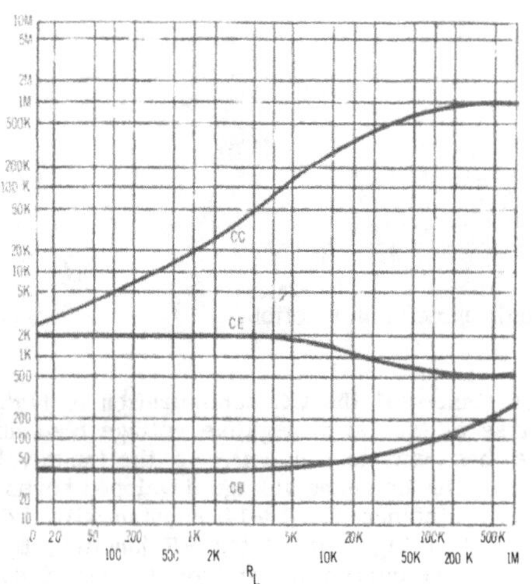

179.255
Figure 13-15.—Variations of r_i with R_L for each configuration.

increase, and r_i of the CB circuit will increase with an R_L increase.

The highest power gain is realized using a CE configuration. Its frequency response is adequate for use in most audio stages. The relationship of its input and output impedances are the most convenient of the three configurations, and when used with transformer coupling it permits easier impedance matching. Finally, its input impedance is less dependent on the value of load resistance than either the CB or CC configurations. Thus, the CE circuit is most often used for general purpose audio amplification.

The CE configuration may be used to match a high source resistance by the addition of a series resistor in the base lead. The base-emitter junction resistance (represented by r_i in fig. 13-16) for a typical CE configuration is approximately 1,000 ohms, if a load resistance of 30,000 ohms is used. The input resistance, r_i, may be increased by reducing the load resistance (R_L). For instance, decreasing the load resistance to approximately 10,000 ohms will increase r_i to approximately 1,500 ohms, as seen in the curve of figure 13-15.

Assume a source impedance of approximately 20,000 ohms. In order to use a CE configuration having a 10,000 ohm R_L, it would be necessary

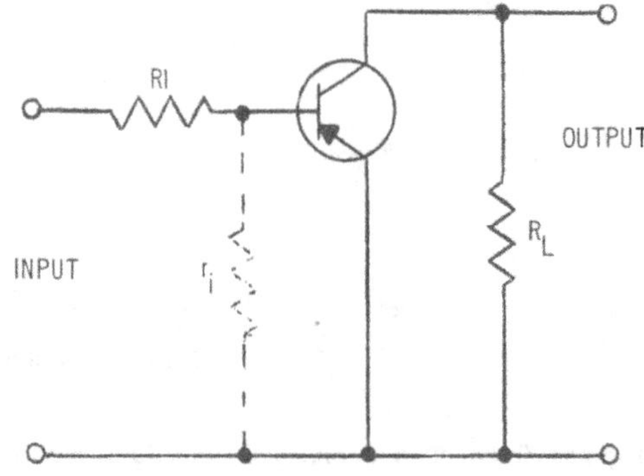

179.256
Figure 13-16.—Simplified schematic of CE preamplifier with series resistor.

to insert a series resistance (R1 in figure 13-16) of approximately 18,500 ohms in the base lead. Thus, the input resistance (r_i) of the amplifier will appear equal to the source. Notice that before the series resistor was added the input resistance r_i varies considerably with change in R_L. When R1 is added to the circuit, r_i will still vary with a change in R_L, however, r_i is now only a small portion of the total input resistance so that r_i remains essentially constant with load resistance changes. The fact that the total input resistance remains relatively constant, even with large variations in transistor parameters or current drain by the following stage, is the main advantage of this circuit arrangement. The disadvantage of this circuit, in addition to a small loss in current gain, is the large resistance in the base lead. This resistance leads to poor bias stability if the bias voltage is fed to the transistor through this resistor.

Another method of increasing the input resistance r_i of a common-emitter configuration is shown in figure 13-17. This type of circuit is the previously mentioned DEGENERATED CE configuration. If an unbypassed resistor (R_E) is inserted in the emitter lead, the signal voltage developed across this resistor opposes the input signal voltage.

As in the case of the common collector configuration, this negative-feedback voltage or degenerative voltage, causes an increase in the input resistance. With a bypassed resistor in the emitter lead the input resistance of the CE configuration would be 2,000 ohms, if a load resistor

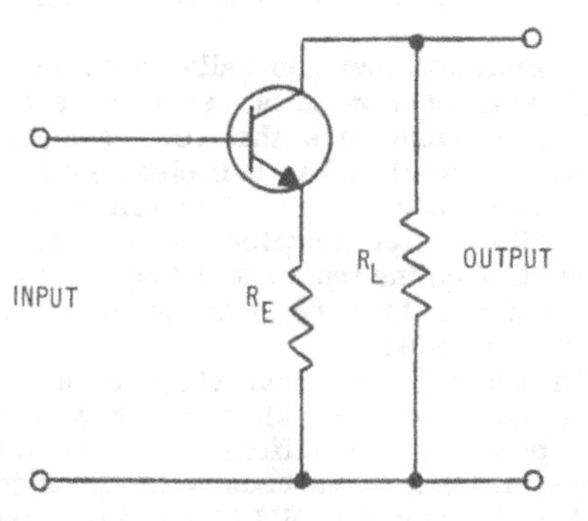

179.257
Figure 13-17.–Degenerated C E configuration.

of 500 ohms were used (fig. 13-15). With an unbypassed resistor (R_E) of 500 ohms, the input resistance (r_i) will appear as approximately 20,000 ohms. The input resistance may be made to appear as any desired value (within practical limits) by the proper choice of R_L and R_E. Like the CE circuit with the series resistor the total input resistance of the degenerated CE circuit will remain relatively constant with a varying load. However, the advantage of the degenerated CE configuration is that the unbypassed resistor (R_E) also acts as an emitter stabilizer and aids in stabilizing the transistor bias.

Figure 13-18 shows the normal frequency response for an audio amplifier chain (dotted lines). With proper feedback circuits this response curve can be improved to that shown in the solid line curve of figure 13-18.

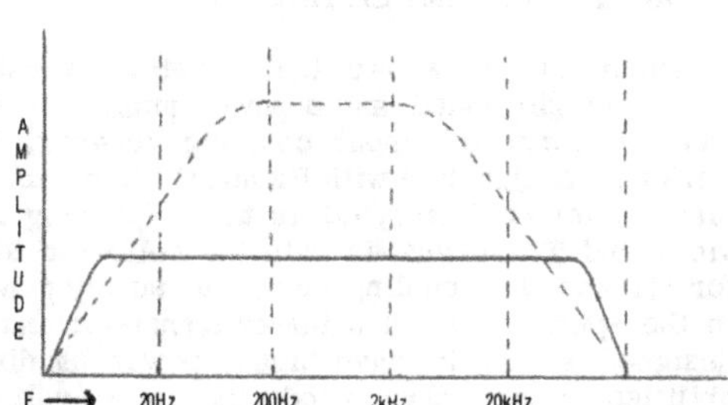

179.258
Figure 13-18.—Audio amplifier response curves.

In each of the feedback networks considered previously, it was considered that each feedback network was both proportional to signal amplitude and non-frequency selective. It is, however, possible to develop a feedback network which will provide very little feedback at the low and high audio frequencies (allowing gain to remain high), while more feedback is developed for the middle audio frequencies (thus reducing mid-band amplifier gain). While such a system will reduce the overall gain of the system, it, nevertheless, provides a more uniform frequency response over the entire range of desired amplification.

A simplified system for developing frequency sensitive feedback voltages is shown in figure 13-19. At low frequencies R1 and R2 act as a voltage divider which selects the initial amount of feedback. The inductance (L) is chosen so that it is essentially a short at these low frequencies. Capacitor (C) is chosen so as to offer a very high reactance at the lower audio frequencies. As the frequencies go into the mid-band range, the gain of the amplifier chain increases. This will increase the voltage applied to the feedback network. However, the capacitor at these frequencies begins to partially shunt the signal around R2. This has the effect of holding the output at roughly the same amplitude rather than allowing it to rise, as the input signal does. As the frequency continues to increase, the shunting action of the capacitor increases. However, the reactance of the inductor now begins to become appreciable. The sum of the remaining voltage across R2, plus the drop across L manages to keep the output amplitude roughly the same as before. In effect, the increasing reactance of L counteracts the increasing shunting effect of the capacitor.

Finally, at high frequencies, (normally above the regular audio range) the interwinding capacitance shorts (effectively) the inductance. A capacitor may be added in parallel with the inductor to insure that this happens.

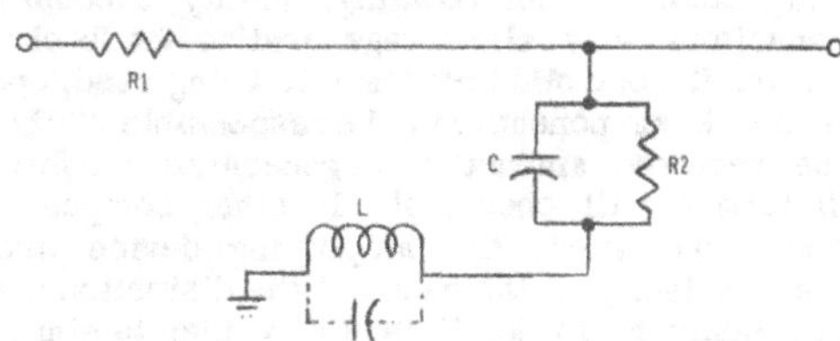

179.259
Figure 13-19.—Frequency sensitive feedback.

By using a circuit similar to the one shown in figure 13-19 it is therefore possible to obtain a much flatter response curve and a much wider bandpass. While the gain of the normal amplifier chain is much higher, note the bandwidths as shown in figure 13-18. As can be seen, the bandwidth is much wider when a frequency sensitive feedback network is utilized in conjunction with the amplifiers.

FAILURE ANALYSIS

If there is no output, check for a loss of source voltage, open load resistors, or open biasing components which would cause the amplifying device to cut off. The transistor or tube could also be faulty. A check of the circuits d.c. voltages will normally indicate the exact trouble. If the d.c. voltages are normal, check for faulty soldering connections of the coupling capacitors.

If the output is reduced, the amplifying device is probably defective. This is especially true if the reduction has taken place over a period of time. A second cause could be a change in value of the biasing components which in turn changes the operational characteristics of the amplifier. A reduced load impedance (which lowers amplifier gain) could also be responsible. Turning off the power and making a resistance check should show any change in resistance which has taken place. Faulty soldering connections could cause a bypass capacitor to be effectively open and thereby introduce unwanted degenerations. If degenerative feedback is being used, a quick check of feedback attenuators is in order. Excessive degenerative feedback will also cause a reduction in output.

If the output is distorted, check for a loss of bias (open biasing resistors), or excessive regenerative feedback. Regenerative feedback can result in motorboating. Faulty decoupling capacitors may allow regenerative feedback to occur. If controlled feedback is being used, open feedback components may be responsible. Without the required amount of degenerative feedback, distortion will occur. If all other components check out good, the amplifying device itself may be faulty. If the cause of the distortion cannot easily be found, it is a good idea to start at the output and work backwards through the circuitry with an oscilloscope until the stage in which the trouble originates is located.

AUDIO POWER AMPLIFIERS

A situation that generally exists in the audio amplifying section of a receiver is to have a series of amplifiers that receive a signal and amplify it until it has sufficient power to perform some useful work. Regardless of whether the audio section contains few, or many, amplifiers between the input and output, the function of each stage is to increase the level until the final stage is reached.

In general, the final stage of a series of audio amplifiers is called the POWER STAGE. The power stage differs from the preceding stages in that it is usually designed to obtain MAXIMUM POWER OUTPUT rather than maximum voltage gain. The discussion of audio power amplifiers will begin with the single-ended power amplifiers.

The purpose of a power amplifier is to efficiently deliver distortion free power to the load. In the case of the single-ended power amplifier the load will be a speaker which is transformer coupled to the amplifier. Through the impedance matching properties of the transformer a reasonably good efficiency is obtained. The circuit must, if possible, deliver the electrical power to the speaker in a distortion free manner. Some distortion of the audio waveforms always takes place, but in a properly designed circuit this distortion will be held to a minimum. Figure 13-20 shows the location of the power amplifier in a standard superheterodyne receiver. The input to the power amplifier is high amplitude voltage as developed by the preceding voltage amplifiers. The voltage output of the power amplifier is lower in amplitude but very high in power level. This output is sent to the speaker in order to develop audio sound power.

TRANSISTOR SINGLE ENDED AUDIO POWER AMPLIFIER

Figure 13-21 shows the transistor version of the single-ended audio power amplifier. C1 and R1 form the input coupling network, R3 works in conjunction with R2 and C2 to establish class A forward bias, Q1 is the amplifying device, and T1 serves as both the collector load for Q1 and the coupling device to supply power to the speaker. Q1 is a power transistor and is designed so as to have higher power handling abilities rather than producing large voltage gains. The transistor is operated class A so as to introduce as little distortion of the audio waveforms as possible.

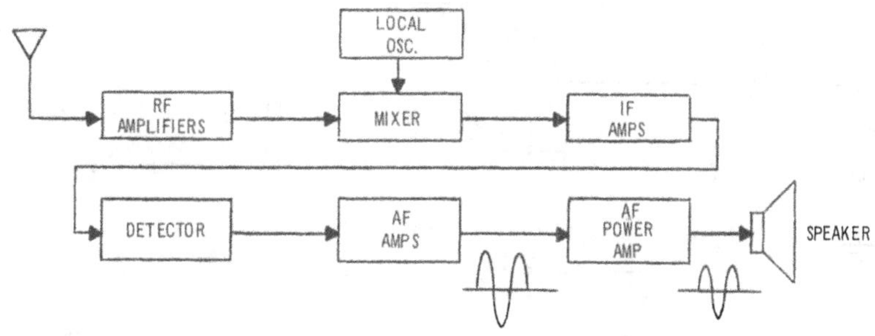

179.260
Figure 13-20.—Power amplifier location in a superheterodyne receiver.

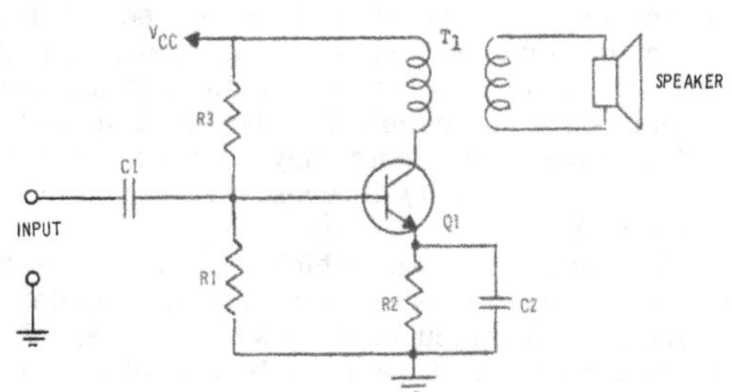

179.261
Figure 13-21.—Transistor power amplifier.

As the positive half-cycle of the input signal is applied the forward bias on the transistor is increased and collector current through the primary of T1 expands the magnetic field of the primary and induces a voltage into the secondary. During the negative half-cycle of the input, the forward bias is reduced, collector current reduces, and the magnetic field about the primary of T1 begins to collapse. The collapsing field induces the opposite polarity voltage (as compared to the previous half-cycle) into the secondary of T1. Thus, a.c. audio signal power is supplied to the speaker, and the electrical audio signals are reproduced in the form of sound power.

ELECTRON TUBE SINGLE ENDED AUDIO POWER AMPLIFIER

In audio power amplifiers, a large power output is of more importance than the voltage gain; therefore, voltage gain is sacrificed in order to obtain higher-power handling capabilities. In general, triodes designed for this application have low amplification factors, low plate resistances, and the ability to handle high plate currents. In order to obtain the low plate resistance, the spacing between the plate and cathode is smaller in power amplifier tubes than in a voltage amplifier tube. Also, in a power tube, the area of the plate is increased and the cathode is designed to handle higher currents. The grid mesh is widely separated so as not to interfere with the current flow, and this, in turn, reduces the amplification factor. The electrodes of a power tube are physically larger than those found in most voltage tubes because higher currents will be encountered, and higher power handling capabilities are required.

Figure 13-22 shows a schematic diagram of a basic single-ended power amplifier using a triode driving a speaker. In this circuit C1 and R1 form the input coupling network. C1 blocks the d.c. component from the preceding stage while passing the a.c. component to be developed across R1. R2 and C2 are used to develop self bias for the stage. T1 is the output transformer acting as both the plate load impedance for the tube and providing impedance matching and power transfer between the plate circuit and the speaker. This circuit differs from that of a voltage amplifier in the type of triode used and the function of the plate transformer.

As to circuit operation, on the positive half cycle of the input, current increases, and the magnetic field about the primary of T1 expands. On the negative half cycle of the input voltage, plate current decreases, and the field about the primary winding decreases. The expanding and

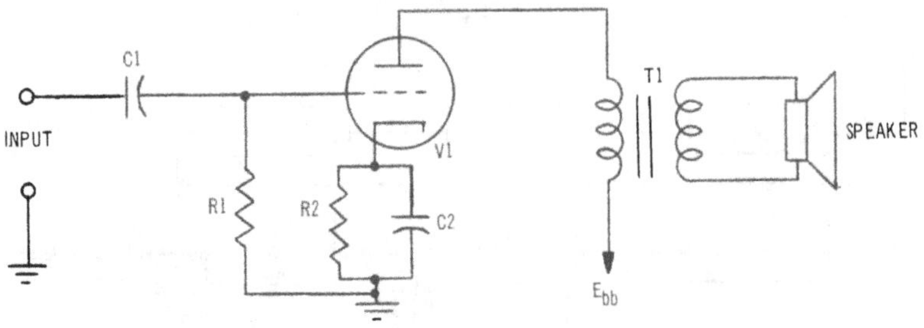

179.262
Figure 13-22.—Basic single-ended electron tube power amplifier.

collapsing field about the primary of T1 induces an a.c. voltage into the secondary which provides the required power to operate the speaker.

TRANSFORMER COUPLING
IN POWER AMPLIFIERS

In order to transfer maximum power to a load, as explained previously, the load impedance must equal the source impedance. In audio power stages which are used to drive a speaker, the output impedance of the transistor or tube is considered to be the source impedance. The load impedance is that of the speaker, which is usually from 4 to 16 ohms. If this load were connected directly to the collector or plate of a power amplifier having a relatively large a.c. output impedance, the transfer of power to the load would be small. Therefore, a method of improving the power transfer to the load must be used. This is the purpose of the IMPEDANCE MATCHING TRANSFORMER. The transformer will match the impedance of the load to the impedance of the source. While perfect impedance matching will not be achieved and the theoretical 50% efficiency cannot be obtained, it is, nevertheless, far superior to connecting the low impedance speaker directly to the amplifying device.

One of the disadvantages of using a single ended power amplifier with transformer coupling is the possibility of core saturation. A direct current is flowing through the primary of T1 (fig. 13-22) at all times and tends to permanently magnetize the core material of T1. If this is allowed to occur, the inductive reactance of the transformer will be reduced which may cause distortion in the form of a loss of gain at low frequencies. The variations in collector/plate current are required to expand and contract the magnetic field of the primary. However, if the core material is already highly magnetized in one direction, an increasing current will not cause the field expansion that is required, and peak clipping will result. In order to eliminate this problem, special core materials may be used, quiescent collector/plate currents may be held to a minimum. or an entirely different circuit configuration may be used.

A second problem which may occur is the previously mentioned motorboating. In order to provide large amounts of power to the speaker, wide current variations must be provided through the primary of T1. This causes a heavy loading effect on the power supply and in the cases of multiple audio stages, regenerative feedback occurs rather easily. This problem may be corrected through better power supply regulation or through the correct use of degenerative feedback networks.

FAILURE ANALYSIS

NO OUTPUT—This could be caused by a defective amplifying device (tube or transistor), by an open in the primary of T1, an open in the emitter/cathode resistor, or faulty soldering connections for the input coupling capacitor. An open in the secondary of T1 or a faulty speaker could also produce no output. A short across the secondary of T1 while unlikely, would also cause no output. It must be remembered that in this case, shorting the secondary will reflect an effective short to the primary of T1 and possibly destroy the amplifying device.

REDUCED OUTPUT—A reduced output could be the result of a reduced input which indicates a problem in other stages. This, of course, also applies to the no output condition. Changes in the values of biasing resistances can cause a reduced output as can shorted turns in the primary of T1. Improper soldering connections to the emitter/cathode bypass capacitor may allow degeneration to occur and thereby reduce the output.

DISTORTED OUTPUT—A distortion in the form of motorboating may occur if the power supply lacks sufficient regulation, or the preamplifiers are inadequately decoupled. Poor power supply filtering may also allow a 60 or 120 Hz hum to appear in the output. Distortion of the signal may also occur in the form of clipping. This may result from excessive gain in the preamplifier stages. It may also result from improper biasing in the power amplifier produced by emitter/cathode resistors which change in value or open; or changed in value resistors in the base circuit of transistor power amplifiers. A leaky input capacitor can also change the operating point of the amplifier and cause distortion. Also, if a number of turns in either the primary or secondary of T1 short, the turns ratio will be affected. This, in turn, changes the reflected impedance causing an impedance mismatch and possibly causing signal distortion.

PHASE SPLITTERS AND INVERTERS

In order to overcome the disadvantages of the single-ended power amplifier, push-pull amplifiers are often used. In the cases where the power amplifier is a push-pull circuit it is necessary to use a PHASE SPLITTER or PHASE INVERTER as a driving stage. DRIVER STAGE is the term used to describe the amplifier which is used to supply the "driving", or input, signal to the final power amplifier. The purpose of phase splitters and inverters, as drivers, is to supply TWO equal amplitude output signals, which differ in phase by 180 degrees with respect to each other. These signals must be developed from a single source. The various methods used to accomplish this are discussed in this section.

Through the use of a center tapped secondary transformer, a single-ended audio stage can produce the signal requirements for push-pull operation. Figure 13-23 shows both the transistor and electron tube version of an audio driver stage with a transformer phase splitter.

The audio signal is coupled from the volume control (R1) to the input of the driver stage by C1. In the transistor version, R2 and R3 provide fixed bias in conjunction with emitter network R4C2. In the tube version, R2 is the standard input resistor and the bias is developed only by C2R4. T1 is used for signal coupling, phase splitting, and impedance matching (it should be noted that while a maximum power transfer is not desired at this point in the circuit, T1 must be designed to reflect the required load impedance to Q1 and V1).

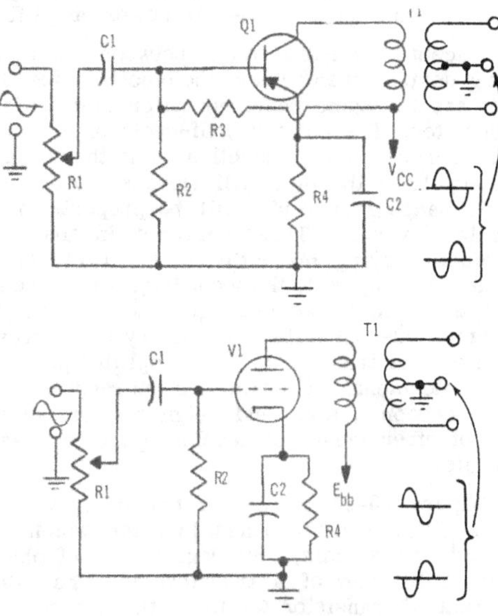

Figure 13-23.—Driver stage with transformer phase splitter.

As the input signal goes through its positive alternation the forward bias on Q1 will be decreased with a resultant decrease in transistor conduction. Collector current, from V_{CC} through T1 primary decreases, and the collector potential (top of T1 primary) becomes more negative. Through transformer action, the bottom of the secondary goes negative (with respect to the secondary center tap) while the top of the secondary goes positive.

When the input signal goes through its negative alternation, the forward bias of C1 is increased. Transistor conduction increases, and the top of T1 primary goes in a positive direction (less negative). Through transformer action the bottom of T1 secondary produces a positive output whereas the top of the secondary produces a negative output. It can be seen that the secondary signals are of equal amplitude, but 180 degrees out of phase, satisfying the input signal requirements of a push-pull power amplifier.

Except for the differences between the amplifying devices themselves, the electron tube circuit has the same basic operation as that of the transistor. The positive half-cycle of the input will increase tube conduction while the negative half-cycle of the input will decrease conduction. The changing currents will be proportional to the input voltage. These currents, in turn, will produce varying magnetic fields in T1. These fields are coupled to the secondary. In the secondary, there exists two a.c. signals which are equal in amplitude but differ in phase by 180 degrees. Although the transformer phase splitter provides a simple means of developing the required signals, economy, size, and weight may dictate the use of other means of developing the necessary signals.

Figure 13-24 shows the transistor version of a SPLIT-LOAD phase inverter which develops two signals, 180 degrees out of phase, without the use of a transformer. The output current of transistor Q1 flows through both the collector load resistor R3 and the emitter load resistor R2. Resistors R2 and R3 are equal in value. Resistor R1 establishes the base bias voltage. C1 and R4 are the standard input coupling components. C2 and C3 are the output coupling capacitors for the two signals which are to be developed.

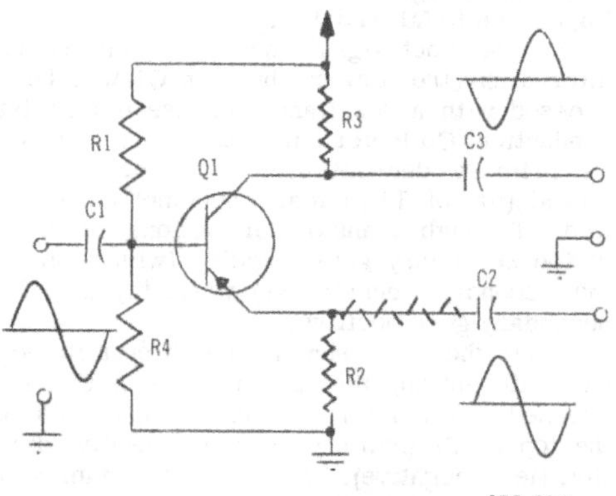

Figure 13-24.—Split load phase splitter.

When the input signal aids the forward bias (base becomes more negative), the output current increases. The increased output current causes the collector side of resistor R3 to become more positive with respect to ground, and the emitter side of resistor R2 to become more negative with respect to ground. When the input signal opposes the forward bias, the output current decreases and causes voltage polarities across resistors R3 and R2 opposite to those just indicated. This action produces two output signals that are reversed 180 degrees with respect to each other.

In this circuit, equal voltages are obtained by making resistor R2 equal in value to resistor R3. However, an unbalanced impedance results because the collector output impedance of transistor Q1 is higher than its emitter output impedance. This disadvantage is overcome by the addition of a series resistor between C2 and the top of R2. The values of R2 and the series resistor (indicated by dotted lines) are chosen so that the output impedance of the collector and emitter are balanced. This eliminates distortion at strong signal currents. The signal voltage loss across the series resistor is compensated for by making R2 higher in value than R3.

Notice that emitter resistor R2 is unbypassed in order to develop one of the output signals. Because of the large negative feedback voltage developed across R2 (degeneration), a large signal input is required to drive the one-stage phase inverter. This disadvantage can be overcome by using two-stage phase inverters. In addition, a two-stage phase inverter provides more power output than a one-stage phase inverter. This advantage is important if the driver stage must feed a large amount of power to a high-level push-pull power output stage.

It should be noted that the single stage phase inverter is also often referred to as a PHASE SPLITTER while the two stage circuitry (covered next) is called only a phase inverter. Finally, it should be noted that due to the degeneration which occurs in the emitter circuit, a voltage gain of less than unity is all that is available from the one-stage phase inverter.

The electron tube version of the one-stage phase inverter is shown in figure 13-25. With the exception of the circuit arrangements for biasing, its configuration is identical with the transistor version. C1R1 is the input coupling network, C2R2 is a biasing network used to develop self bias, R3 is the plate load resistor, R4 is the cathode load resistor, and C3 and C4 are output coupling capacitors. This circuit provides, as

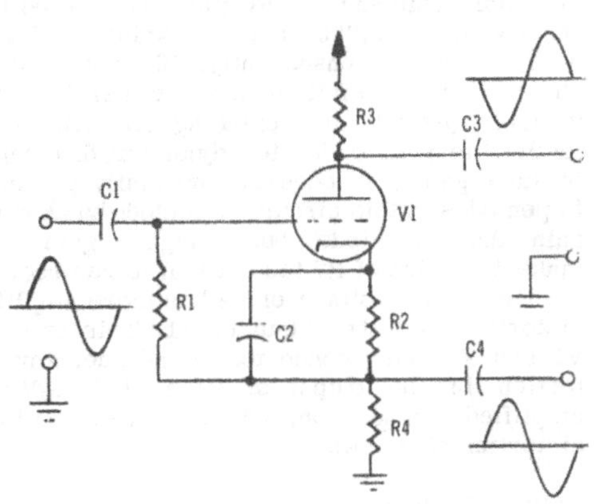

179.265

Figure 13-25.—Electron tube phase splitter.

does the transistor version, two signals which are approximately equal in amplitude and 180 degrees out of phase with respect to each other. Again, because of the degenerative effect of R4 being unbypassed, there is a voltage gain of less than unity.

TWO STAGE PHASE INVERTER

The two-stage inverter, also called the PARAPHASE AMPLIFIER, provides, as the previous circuit, two signals of equal amplitude which are 180 degrees out of phase with each other. The transistor version of this circuit is shown in figure 13-26. The circuit consists of two identical CE configurations.

R1 is utilized to develop forward bias for Q1, C1 and R2 aid in the selection of the operating point, Q1 is the first stage amplifying device and R3 is its collector load resistor. C2 is an interstage coupling capacitor, R4 is an attenuator resistor which reduces signal amplitude to Q2, R5 establishes forward bias for Q2, C3R7 is the self bias network for Q2, R6 is the collector load for Q2, and C4 and C5 are the coupling capacitors for the two output signals.

Assume that each transistor configuration produces a voltage gain of 20 and that R4 produces a 20 to 1 voltage reduction. Assume that a negative half cycle is applied to the input at a .1 volt amplitude. Because of the 180 degree phase reversal in the CE configuration, the signal at the collector of Q1 will be positive going. The signal will have been amplified and will appear in the collector circuit as a 2 volt signal. This 2 volt positive going signal is coupled through C4 as one of the required outputs. It is also coupled through C2 to R4 and Q2. The action of R4 will be such that it will drop 1.9 volts and allow only a positive going .1 volt signal to be felt on the base of Q2. Q2 will amplify the signal and a negative going 2 volt signal will appear on its collector. This signal is coupled out through C5 as the second of the two required signals.

Since two identical CE configurations are used, the source impedances are equal for the two input circuits of the push-pull output stage. In addition, the amount of power that can be

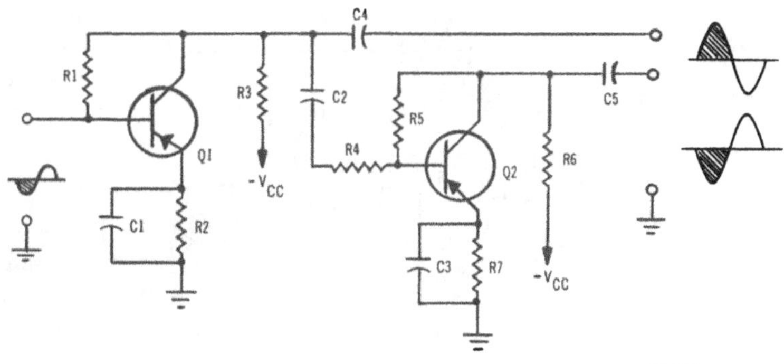

179.266

Figure 13-26.—Two stage transistor paraphase amplifier.

delivered by the two-stage paraphase amplifier is much greater than that of the split-load phase inverter. Also, rather than having a voltage gain of less than unity, the paraphase amplifier is capable of producing voltage amplification at the same time that the phase-splitting action is taking place. In the case where the gains of Q1 and Q2 are not quite equal, R4 may be made adjustable so that it can provide additional, or less, attenuation as required.

Figure 13-27 shows the electron tube version of the two tube paraphase amplifier. Again, circuit purposes and operational characteristics are essentially the same as those of the transistor version. C1R1 is the input coupling network for V1. R2C2 is the self bias network for V1 while R3C3 serve the same function for V2. R4 and R5 are the plate load resistors for the two stages, C4 and C5 are interstage coupling capacitors, and R6R7 forms a voltage divider network which develops the correct input voltage for V2. R3 is a simple output voltage developing resistor.

For a numerical example of the operation of this circuit, assume that V1 and V2 each have a gain of 20, R8 has a resistance of 40,000 ohms, and the resistance of the combination of R6 plus R7 is also 40,000 ohms. The output voltage of V1 would be 20 volts if there is an input voltage of one volt. In order to produce equal output voltages, the input voltages to the two grids must be identical. The input to V1 is assumed to be 1 volt so it is therefore necessary to have a 1 volt input for V2. As there is a 20 volt output of V1, R6R7 must form a voltage divider network which will provide 1 volt to the grid of V2 with a 20 volt signal applied to the voltage divider. Therefore, R7 must be 1/20 of the total resistance of R6 plus R7. As the total resistance is 40,000 ohms, the value of R7 must be 2,000 ohms. Consequently, R6 must be 38,000 ohms. It is possible to make either R7 or R6 variable (usually by replacing R7 with a potentiometer) in order to adjust for differences in tube gains. Triodes are normally preferred to pentodes in this circuit. A pentode has a higher gain than a triode, but a higher gain would make the value of R7 too critical to be practical.

The output voltage of V2 has more amplitude distortion than the output of V1. This is due to V1 and V2 being connected in cascade. Any distortion in the output of V1 is fed to V2 and amplified. In addition, V2 introduces additional distortion of its own.

Failure Analysis

NO OUTPUT—In the single-stage phase-splitter, a no output condition would often be the result of an open collector/plate resistor, an open emitter/cathode resistor, or a defective amplifying device. Open coupling capacitors (poorly soldered leads) in either the input or output would also cause a no signal condition. It should be noted that as there are TWO outputs from the phase splitter, the loss of only one output would immediately indicate the problem area (collector/plate or emitter/cathode).

In the two-stage phase inverter, the lack of one of the two outputs would immediately indicate a trouble in the second stage. If a no signal condition existed in the first stage, there would be no input (and consequently no output) to the second stage. Once it is determined which stage

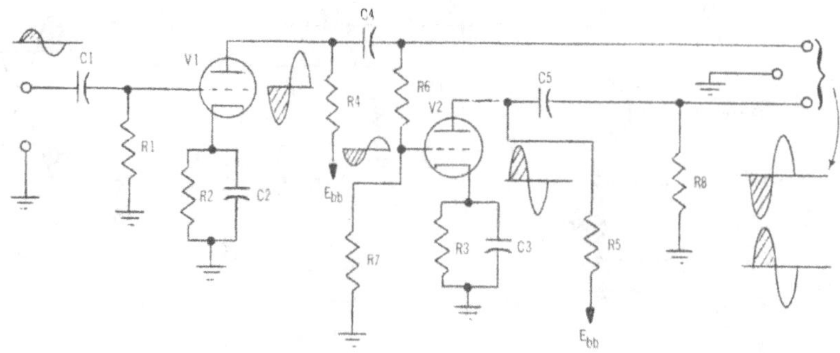

Figure 13-27. — Two-stage electron tube paraphase amplifier.

is at fault, the possible troubles are the same as for the standard voltage amplifier.

REDUCED OUTPUT—In the single stage phase splitter, a reduced output could be caused by an amplifying device which has weakened with age. A change in value for the input resistance could possibly reduce the level of input signal. Finally, the following stage could load the circuitry so heavily that its output could be reduced. This, however, would be indicative of a problem in the following stages rather than in the phase splitter. Also, poorly soldered connections on the coupling capacitors could cause attenuation which would reduce the output signal.

In the two stage phase inverter, as before, if both outputs are reduced, the problem is either in the original input or in the first stage. If a problem exists in the second stage, only its output will be reduced in amplitude. The possible troubles which could reduce the output amplitude of the two stage phase inverter are the same as those for a standard voltage amplifier.

DISTORTED OUTPUT—In the single stage phase splitter, improper bias, low supply voltages, or a defective amplifying device can cause a distorted output. If the bias is either too high or too low, peak clipping is likely to result. Low supply voltage can also result in peak clipping. Finally, too large an input signal will cause peak clipping by instantaneously causing cutoff of the amplifying device. An oscilloscope can be used to trace the signal through the circuitry and the origin of the distortion located.

The two stage phase inverter, as previously covered, is two cascaded amplifiers. As such, they are subject to the same forms of distortions as other voltage amplifiers. If only one stage shows distortion, it will be a trouble in the second stage. If a series distortion is present in both stages, the trouble is to be located either in the first stage or the preceding circuitry. It should be noted that about 1 or 2% higher distortion is expected from the second stage. Distortion beyond this amount, however, indicates a malfunction which should be corrected.

CLASS A, PUSH-PULL POWER AMPLIFIERS

The push-pull amplifier is normally the last stage of a receiver as shown in figure 13-28. Its purpose is to develop the necessary driving power to operate the loudspeaker. It produces a higher distortion-free output power than does the single ended amplifier. It is a two stage device consisting of a transformer coupled to the speaker.

The input to the push-pull amplifier comes from a one or two stage phase inverter. It requires two input signals which are equal in amplitude and 180 degrees out of phase.

Transistor Push-Pull Amplifier

A transistor, class A, push-pull amplifier is shown in figure 13-29. It consists essentially of two transistors connected back to back. Both transistors are biased for class A operation. In this circuit, a transformer phase splitter is used to develop two signals which are equal in amplitude and 180 degrees out of phase with each other. These signals are applied as the two inputs for the push-pull amplifier. The transformer also provides the path used by R1 to develop class A bias for both Q1 and Q2.

The output of the amplifier is transformer coupled to the speaker. One half of the primary of T2 represents the collector load impedance for Q1 while the second half represents the collector load for Q2. T2 also provides impedance matching between the relatively high output impedance of the transistors and the low impedance of the speaker voice coil.

On the positive alternation of the input signal, the potential on the base of Q1 will increase in a positive direction, while the potential on the

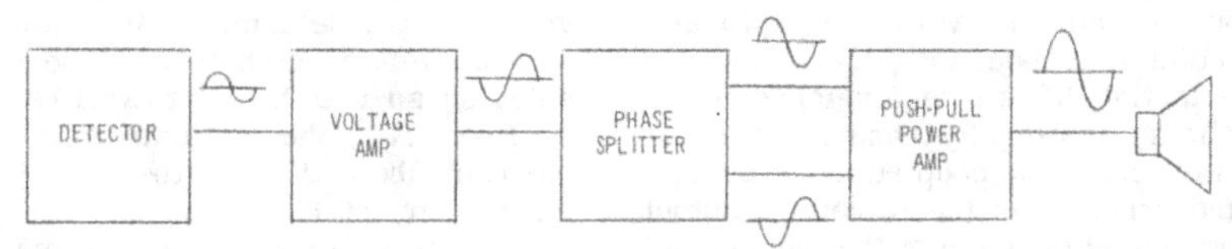

Figure 13-28.—Push-pull amplifier location in a receiver.

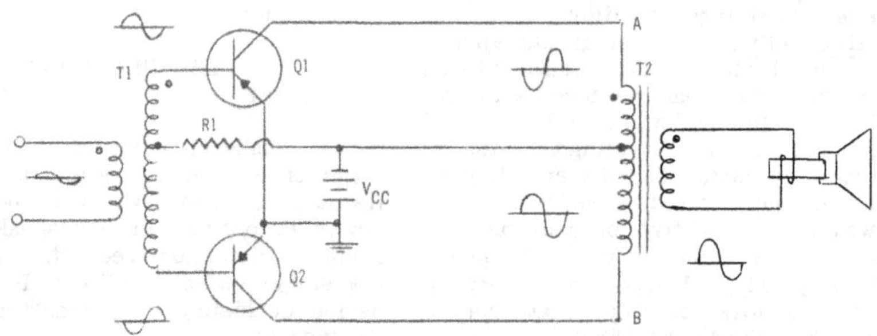

Figure 13-29.—Transistor class A push-pull amplifier.

base of Q2 will increase in a negative direction. Since both transistors are of the PNP type, the voltage applied to their respective base elements will cause the conduction of Q1 to decrease and the conduction of Q2 to increase.

Currents flow from the center tap of T2 TOWARD points A and B. Under quiescent operating conditions these currents are equal and the magnitude and polarity of the voltage drops they produce are such that there is no difference of potential between points A and B. The center tap of T2 is fixed at the maximum negative value in the circuit by virtue of being connected to V_{CC}. Thus, the potentials at points A and B are less negative by some amount than the center tap. As stated, the positive alternation of the input signal causes the conduction of Q1 to decrease. In other words, point A becomes more negative because it is approaching the value of the potential at the center tap. The collector voltage waveform of Q1, which is actually a graph of the potential at point A, is seen to be increasing in a negative direction at this time.

The positive alternation of the input signal also causes the conduction of Q2 to increase. This causes the potential difference between point B and the center tap to increase. Thus, the potential at point B is increasing in a positive direction (as shown by the collector voltage waveform of Q2). Since point A is becoming more positive, there is a potential difference developed across the entire primary winding. By transformer action this potential difference is coupled to the secondary of T2 and appears as the waveform output.

On the negative alternation of the input signal, the reverse of the above actions occurs, as can be seen by the collector voltage waveforms of Q1 and Q2, and as the polarity of the voltage developed across the primary of T2 is reversed.

The power output from this class A push-pull circuit is more than twice that obtainable from a single ended class A power amplifier. An added advantage of this circuit is that, due to the push-pull action of the output transformer, all even order harmonics are eliminated in the output if the transistor circuits are balanced. Balancing will be covered later. Since the distortion is caused mainly by second order harmonics, elimination of these harmonics will result in a relatively distortion-free output signal.

Electron Tube Push-Pull Amplifier

Figure 13-30 shows the schematic diagram of a class A electron tube push-pull amplifier. This circuit is also shown in the transformer-coupled configuration. T1 and T2 are the coupling transformers and R_k provides a common cathode bias for both V1 and V2.

D.C. CURRENT PATHS.—The path of the quiescent plate currents will be analyzed beginning at the ground connection of T1. Quiescent current for both tubes flows through R_k to the cathodes. This total current, times the resistive value of R_k, determines the cathode bias which will be felt by both tubes. The value of R_k is selected so that the tubes will be operated class A. From R_k, the currents split equally through the two tubes, through the respective halves of the primary of T2, to E_{bb}.

The two tubes are the same type and are tested to insure that their conduction characteristics are as near as possible the same. When two

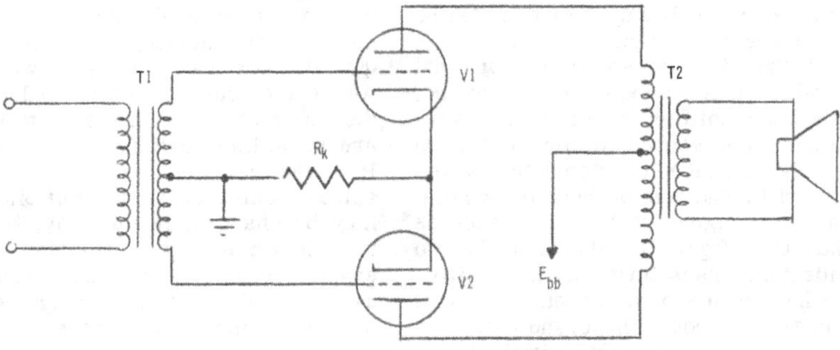

Figure 13-30.—Electron tube push-pull amplifier.

tubes of the same type possess the same conduction characteristics, they are said to be MATCHED. The same is true of transistors. The two tubes (fig. 13-30) that are matched, have equal plate loads connected to a common power supply, and have the same bias voltage. Therefore, the quiescent plate currents in both V1 and V2 are equal.

Since equal plate currents flow through each half of the primary in opposite directions, the resulting magnetic fields are equal in intensity but are opposing each other. Thus, the magnetizing effect of the direct currents on the iron core is cancelled and there can be no d.c. core saturation of the output transformer. This is the same condition as that which existed in the transistor version of this circuit.

A.C. OPERATION.—As previously stated, the circuit shown in figure 13-30 is biased for class A operation. This means that current will flow through each tube for 360 degrees of the input signal. The signals applied to the tubes will have opposite polarities. That is, the grid of V1 will be positive with respect to ground, while the grid of V2 will simultaneously be negative with respect to ground. Consequently, plate current through V1 will be increasing while the plate current of V2 will be decreasing. The proportion of increase and decrease through each tube is equal. If the current of V1 increases by 10 ma., the plate current of V2 will decrease by 10 ma.

Assume that the plate current of V1, in flowing through the top half of the primary of T2, produces a given direction magnetic field. Then the plate current of V2, which flows through the bottom half of T2, would produce the opposite direction magnetic field. Thus, if an expanding upper primary field induces a positive going voltage in the secondary (caused by an increase in i_1), the collapsing lower primary field caused by the corresponding decrease in i_2 will also induce a positive voltage into the secondary of T2.

The above conditions will occur for one half cycle of the input. The input signal then reverses polarity, and the opposite effects occur. This makes the outputs of the two tubes additive at all times, as was the case with the transistor version of this circuit.

Because the magnetic fields have opposite polarities, it should be noted that if they were to both expand or contract at the same time, there would be no voltage induced into the secondary. This explains the necessity of two input signals of opposite polarity applied to the tube grids.

CANCELLATION OF HARMONICS.—Second order harmonics are eliminated in the push-pull output, as previously stated. Also, power supply hum will not appear in the output of the push-pull amplifier. If supply voltage is increased, then the plate currents of both tubes will increase. Consequently, both halves of the magnetic field expand, their effects cancel, and the change is not felt in the output.

When it is stated that second order harmonics are eliminated by using push-pull amplifiers, the harmonics referred to are those which are generated within the amplifiers themselves. If the input signal contains distortion, the push-pull amplifiers will also amplify the distorted signal. The reason for the elimination of the even harmonics in the push-pull amplifier can best be understood by reference to figure 13-31.

Part A of the diagram shows the original input signal which is assumed to be undistorted. Assuming that the amplifier is sufficiently overloaded so that some second order distortion occurs, the individual tubes develop output waves as shown in part B. The sum of these two waves, which would be the signal felt by the speaker, is shown in part C of figure 13-31. When the two tubes have identical characteristics, the positive and negative half cycles of this resultant output wave differ in sign but not in shape, and therefore contain only odd harmonics. This similarity in the shapes of the positive and negative half cycles in the combined output results from the fact that at time intervals that are exactly one-half cycle apart the tubes have merely interchanged function. That is, at the later time, tube 2 is operating under exactly the same conditions as was tube 1 a half-cycle earlier, and vice versa.

Balancing Networks

As mentioned previously, matched amplifying devices (transistors or tubes) are used in push-pull amplifier circuits to insure equal values of current and voltage in the output. No matter how well matched the two devices may be, there will be some differences in their characteristics. Therefore, unequal outputs will cause excessive distortion. The center tap on the transformer is critical to the output. If the tap is not physically located at the exact center of the winding there will be unequal outputs from the amplifiers. A means must therefore be developed to insure that the outputs of the amplifying devices are the same. This is provided by the use of BALANCING NETWORKS.

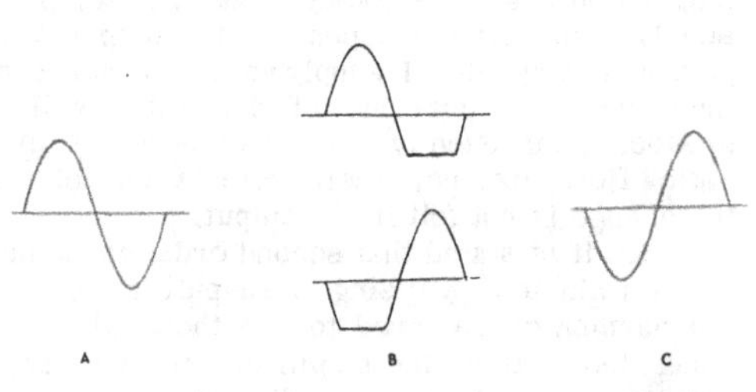

Figure 13-31.—Elimination of harmonics.

When the d.c. component of the collector/plate currents through the devices are not equal, an emitter (or cathode) balancing network is required. Such a circuit is shown in figure 13-32. As these circuits are used to balance the quiescent d.c. currents through the matched devices, they are sometimes called static balancing networks. By adjusting the potentiometer arm of R3, the value of emitter/cathode self bias to each device may be changed. In this way, the transistor or tube currents may be adjusted so that they are exactly equal to each other. This adjustment is usually made while observing accurate ammeter measurements of the currents through the devices.

A second form of balancing is shown in figure 13-33. The network is called BASE BALANCING in the transistor version and GRID BALANCING in the electron tube version of the push-pull amplifier. The purpose of this balancing network is to insure that the outputs of both amplifying

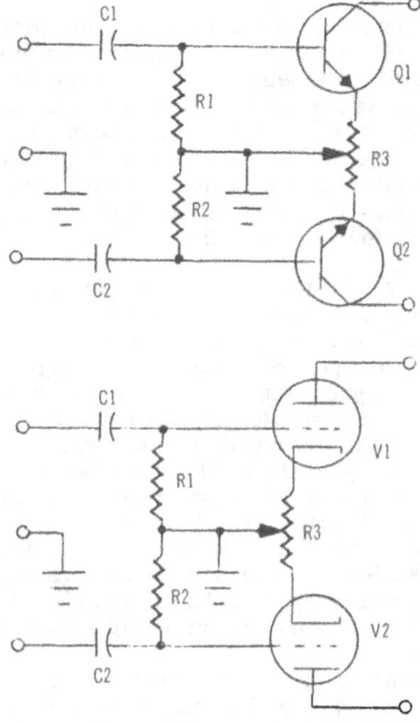

Figure 13-32.—Balancing networks.

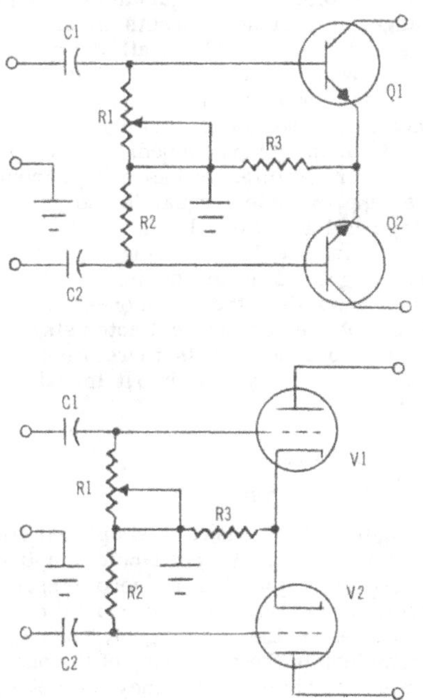

179.273

Figure 13-33.—Dynamic signal balancing.

devices are equal. This network is necessary because no matter how well two amplifier devices are matched, they will not possess exactly the same parameters.

By changing the position of the wiper arm of potentiometer R1, the amplitude of the input signals to the devices may be varied to cause equal outputs. For example, if Q1/V1 has a slightly greater gain than Q2/V2, the wiper arm should be moved toward the top of the potentiometer to decrease the magnitude of the input to Q1/V1. This adjustment is made by observing the voltage waveform across the emitter/cathode resistor, R3, on an oscilloscope. When the circuits are balanced, the current increase through one device will equal the current decrease through the other device and vice versa. Thus, the current through the emitter/cathode resistor is constant, developing only a d.c. voltage equal to that at quiescence. Therefore, the circuits are balanced when a minimum signal appears across resistor R3.

Comparison of Power Amplifiers

Some of the advantages of using the push-pull amplifiers instead of the single-ended amplifier have been mentioned. There are no adverse effects due to core saturation in the push-pull amplifier because of the balanced direct currents flowing in the primary of the output transformer. It was also shown that the push-pull amplifier eliminated even-order harmonic distortion by the process of cancellation. One other advantage of the push-pull amplifier is that there is no hum in the output caused by the power supply ripple. In a single-ended power amplifier driving a loudspeaker the power supply ripple will cause the plate current to increase and decrease at the ripple frequency, causing a hum to be heard in the output speaker. This will not occur in a push-pull stage, since an increase in power supply voltage will cause an increased plate current through both amplifying devices. Equal and opposite fields will result about the transformer primary, and the effects will cancel. The same will occur when the power supply decreases its voltage. Providing the push-pull stage is balanced, no voltage will be induced in the secondary due to the power supply ripple.

Another advantage of a push-pull stage is that a decoupling network is not necessary, providing the circuit is properly balanced. The purpose of the decoupling network (as covered previously) is to prevent plate current variations from causing the output of the power supply to vary at an audio a.c. rate. The reason this network is not necessary in a push-pull stage is because the input signals are of opposite polarity causing the current in one of the amplifying devices to increase at the same time that the current in the other device is decreasing. Thus, the current through the power supply is constant and a varying voltage is not developed across it. For this same reason, an emitter/cathode bypass capacitor is not required when using a common emitter/cathode resistor.

Failure Analysis

NO OUTPUT—The no output condition is seldom encountered in the push-pull amplifier. The most common cause of trouble in any amplifier circuit is the failure of the amplifying device. It is highly unlikely that both amplifying devices of the push-pull amplifier should fail at the same time. Therefore, a no output condition will indicate the failure of a component common to both amplifying devices. The emitter/cathode resistor could be open, the power supply may have failed, or there may be no input signal. The latter trouble would indicate that the trouble is located in one of the preceding amplifiers and not in the push-pull stage itself.

REDUCED OUTPUT—If one of the two amplifying devices fails, or components which would disable one half of the push-pull stage, a reduced output will result. An oscilloscope should be used to determine which of the two amplifiers is not operating. The individual device is then subject to the same failure analysis as a single stage amplifier. If both devices are found to be operating satisfactory, check the input signal. A preceding stage may be responsible for the problem. A reduced output could also be caused by a change in value of the balancing resistors. This would allow a mismatch in amplifier operation and possibly cause degeneration through the presence of a degenerative signal across the emitter/cathode.

DISTORTED OUTPUT—A distorted output could be caused through the improper biasing of one of the amplifiers, improper balancing or changes in component values in one of the amplifier circuits. An oscilloscope should be used to determine the source of the distortion and then analyze the trouble the same as for a single stage amplifier.

CLASS B, PUSH-PULL POWER AMPLIFIERS

The electron tube version of the push-pull power amplifier may be operated class A, AB, or B. The class of operation previously described for electron tube circuits has been class A. Class A operation is used where plate efficiency is not essential. Class AB operation exists when the tube conducts more than 180 degrees but less than 360 degrees of the input cycle. It is used when class B operation causes excessive distortion due to the non-linearity of the characteristic curves near plate current cutoff. Class B operation is used when high efficiency from an audio amplifier is desired. When Class B operation is used, each tube only conducts for 180 degrees of the input cycle. Class C operation is never used with an audio amplifier of any type because of the high degree of distortion it produces.

The common-emitter configuration is the most used for transistor circuits. However, on some occasions either the common-base or the common-collector configurations may prove advantageous. In many respects all three configurations are similar. They all deliver about the same power output, have the same maximum efficiency, and require about the same load impedance. They differ mainly, as might be expected, in their input impedance and power gain. The common-collector has a high input impedance, approximately equal to the peak collector voltage swing divided by the peak base current. The common-collector stage generally has a greater gain than the common-base stage, although less than the common-emitter. Another feature of the common-collector stage is its low output impedance. It is therefore attractive in cases where very low output impedance is desirable.

Transistor Push-Pull Amplifier

Figure 13-34 shows a simplified circuit of a transistor class B push-pull amplifier. The emitter-base junction is zero biased. In this circuit each transistor conducts on alternate half cycles of the input signal. The output signal is combined in the secondary of the output transformer. Maximum efficiency is obtained even during idling (no input signal) periods, because neither transistor conducts during this period.

An indication of the output current waveform for a given signal current input can be obtained by considering the dynamic transfer characteristics for the amplifier. It is assumed that the two transistors have identical dynamic transfer characteristics. This characteristic, for one of the transistors, is shown in figure 13-35A. The variation in output (collector) current is plotted against input (base) current under load conditions. Since two transistors are used, the overall dynamic transfer characteristics for the push-pull

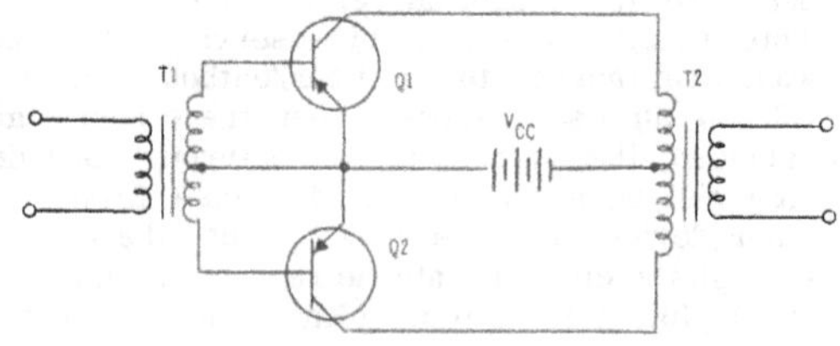

179.274

Figure 13-34.—Transistor class B push-pull amplifier.

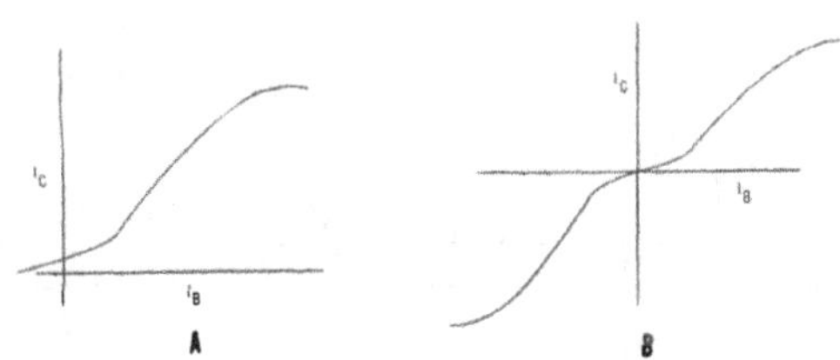

179.275
Figure 13-35.—Dynamic transfer characteristic curves.

amplifier is obtained by placing two of the curves (fig. 13-35B) back to back.

Note that the zero line of each is lined up vertically to reflect the zero bias current. In figure 13-36 points on the input base current (a sine wave) are projected onto the dynamic transfer characteristic curve. The corresponding points are determined and projected as indicated to form the output collector current waveform. Note that severe distortion occurs at the crossover points. That is, the distortion occurs where the signal passes through zero value. This type of distortion becomes more severe with low input currents. Crossover distortion can be eliminated by using a small forward bias on both transistors of the push-pull amplifier.

A class B push-pull amplifier with a small forward bias applied to the base-emitter junction is shown in figure 13-37. A voltage divider is formed by resistors R1 and R2. The voltage

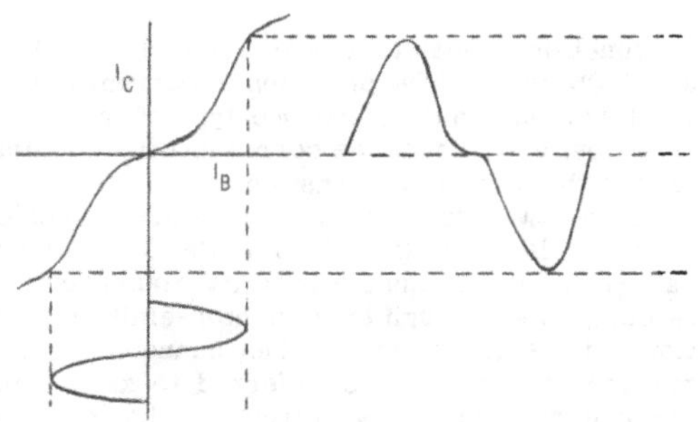

179.276
Figure 13-36.—Dynamic transfer curves of class B push-pull amplifiers.

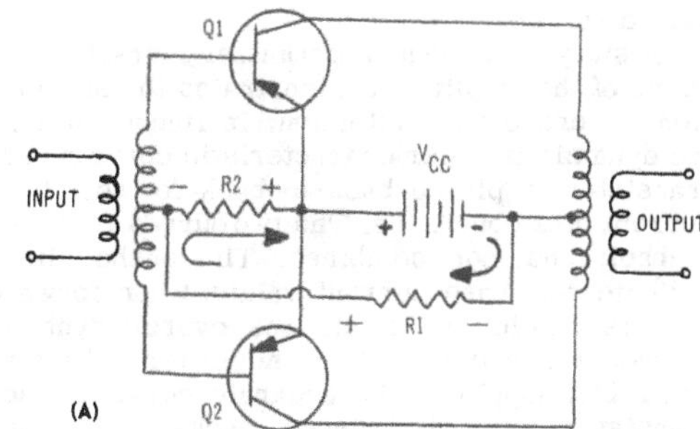

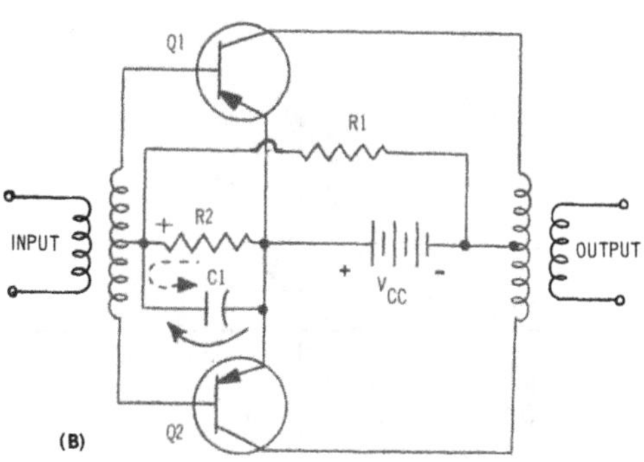

179.277
Figure 13-37.—Class B, push-pull amplifier with small bias voltage applied.

across R2 supplies the base-emitter bias for both transistors. This small forward bias eliminates crossover distortion.

In the voltage divider (fig. 13-37A), current flow from the battery is in the direction of the arrow. This current established the indicated polarity across resistor R2 to furnish the required small forward bias. Note that no bypass capacitor is across resistor R2. If a bypass capacitor were used (fig. 13-37B), the capacitor would charge (solid line arrow) through the base-emitter junction of the conducting transistor (during the presence of a signal), and discharge (dashed-line arrow) through resistor R2. The discharge current through resistor R2 would develop a d.c. voltage with the polarity indicated. This is a reverse bias polarity that could drive the amplifier into class C operation with the resultant distortion even more severe than crossover distortion.

A study of the dynamic transfer characteristic curve of the amplifier demonstrates the elimination of crossover distortion. In figure 13-38A, the dynamic transfer characteristic curve of each transistor is placed back-to-back for zero base current bias conditions. The two curves are back-to-back and not combined. The dashed lines indicate the base current values when forward bias is applied to obtain the overall dynamic characteristic curve of the amplifier. With forward bias applied, the separate curve of each transistor must be placed back-to-back and aligned at the base bias current line (dashed line). The zero base current lines (solid lines) are offset (fig. 13-38B).

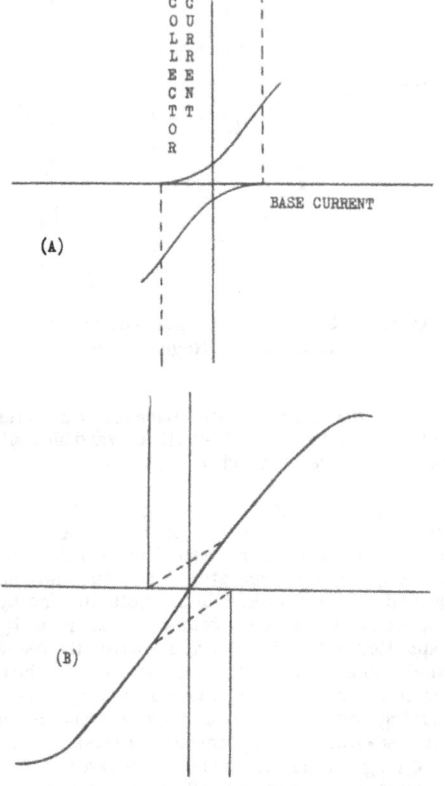

179.278

Figure 13-38.—Dynamic transfer curves, low bias, class B.

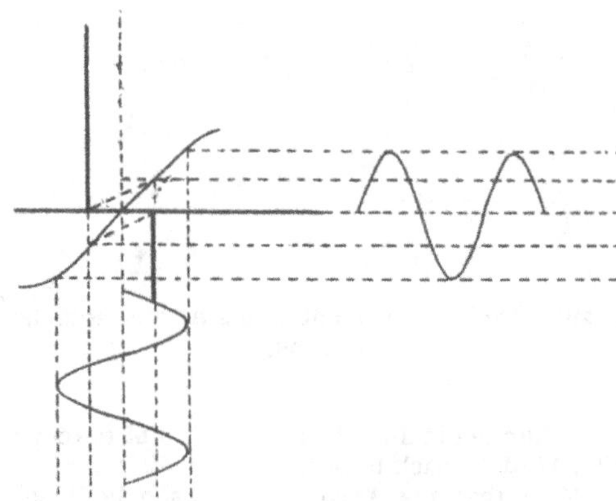

179.279

Figure 13-39.—Dynamic transfer characteristic curves of class B, push-pull amplifier with small forward bias, showing input and output waveforms.

In figure 13-39 points on the input base current (a sine wave) are projected onto the dynamic transfer characteristic curve. The corresponding points are determined and projected as indicated to form the output collector current waveform. Compare this output current waveform with that shown in figure 13-36. Note that the crossover distortion does not occur when a small forward bias is applied.

Complementary Symmetry

Junction transistors are available as PNP and NPN types. The direction of current flow in the terminal leads of the one type of transistor is opposite to that in the corresponding terminal lead of the other type transistor.

If the two types of transistors are connected in a single stage (fig. 13-40), the d.c. current path (indicated by the arrows) in the output circuit is completed through the collector-emitter junctions of the transistors. When connected in this manner, the circuit is referred to as a complementary symmetry circuit. The complementary symmetry circuit provides all the advantages of conventional push-pull amplifiers without the need for a phase-inverter driver stage, or for a center-tapped input transformer.

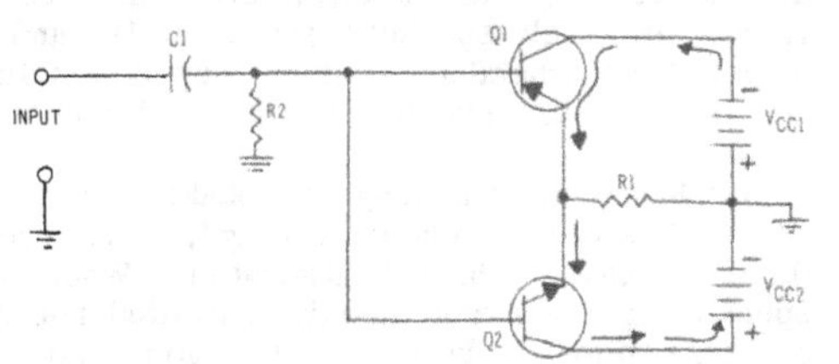

179.280

Figure 13-40.—Zero bias complementary symmetry circuit.

Transistor Q1, figure 13-40, is a PNP transistor and transistor Q2 is an NPN type. A negative going input signal forward biases transistor Q1 and causes it to conduct. As one transistor conducts, the other is nonconducting, because the signal that forward biases one transistor, reverse biases the other transistor.

The resultant action in the output circuit can be understood by considering the circuit of figure 13-41. This is a simplified version of the output circut. The internal emitter-collector circuit of transistor Q1 is represented by a variable resistor R1 and that of transistor Q2 by variable resistor R2. With no input signal and class B operation (zero emitter-base bias), the arms of the variable resistors can be considered to be in the OFF positions. No current flows through the transistors or through load resistor R_L. As the incoming signal goes positive, transistor Q2 conducts, and transistor Q1 remains nonconducting. Variable resistor R1 remains in the OFF position. The variable arm of resistor R2 moves toward point 3 and current passes through the series circuit consisting of battery V_{CC2}, resistor R_L and variable resistor R2. The amount of current flow depends upon the amplitude of the incoming signal, the variable arm moving toward point 3 for increasing forward bias and toward point 4 for decreasing forward bias. The current flows in the direction of the dashed arrow, producing a voltage with the indicated polarity.

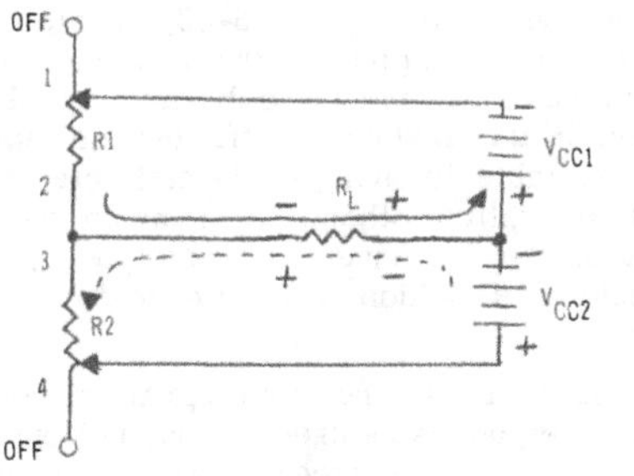

179.281

Figure 13-41.—Simplified version of output circuit of complementary symmetry circuit.

When the input signal goes negative, transistor Q1 conducts and transistor Q2 becomes nonconducting. The same action is repeated with variable resistor R1. Current flows through battery V_{CC1}, variable resistor R1, and load resistor R_L with the polarity indicated.

For class A operation of a complementary symmetry circuit (such as fig. 13-40), a voltage divider network is used to apply forward bias to the two transistors so that collector current is not cut off at any time. In the simplified circuit (fig. 13-41), the variable resistors will not be in the OFF position at any time. Bias current in the output circuit flows out of the negative terminal of battery V_{CC2} into the positive terminal of battery V_{CC1}, through variable resistor R1, variable resistor R2, and into the positive terminal of battery V_{CC2}. No resultant current flows through resistor R_L.

Under the above conditions, the output circuit can be considered a balanced bridge, the arms of the bridge consisting of resistors R1 and R2 and batteries V_{CC1} and V_{CC2}. When the input signal goes positive, transistor Q2 conducts more and transistor Q1 conducts less. In the simplified circuit, the variable arm of resistor R1 moves toward point 1 and that of resistor R2 moves toward point 3.

This action results in an unbalanced bridge and resultant current flows through resistor R_L in the direction of the dashed-line arrow, producing a voltage with the indicated polarity. When the input signal goes negative, transistor Q1 conducts more and transistor Q2 conducts less. In the simplified circuit, the variable arm of resistor R1 moves toward point 2 and that of resistor R2 moves toward point 4. Again the bridge is unbalanced, and current flows through resistor R_L in the direction of the solid-line arrow, producing a voltage with the indicated polarity.

In either class B or class A operation no resultant d.c. current flows through the load resistor or the primary of the output transformer, whichever is used as the load for the amplifier. Thus the voice coil of a loudspeaker may be connected directly in place of resistor R_L. The voice coil will not be offset by d.c. current flow through it, and thus distortion will not occur.

AUDIO FREQUENCY REPRODUCTION DEVICES

The purpose of audio reproduction devices such as loudspeakers and headphones is to convert electrical audio signals to sound power. Figure 13-42 shows a diagram of a loudspeaker called the PERMANENT MAGNET SPEAKER. This speaker consists of a permanent magnet which is mounted on soft iron pole pieces, a voice coil which will act as an electromagnet, and a loudspeaker cone which is connected to the voice coil. The audio signal has been previously amplified (in terms of both voltage and power) and is applied to the voice coil. The voice coil is mounted on the center portion of the soft iron pole pieces in an air gap so that it is mechanically free to move. It is also connected to the loudspeaker cone, and as it moves, the cone will also move. When audio currents flow through the voice coil, the coil is moved back and forth proportionally to the applied a.c. current. As the cone (diaphragm) is attached to the voice coil, it also moves in accordance with the signal currents and, in so doing, periodically compresses and rarifies the air and thus produces sound waves.

Most speakers of the above type receive their input by means of transformer coupling.

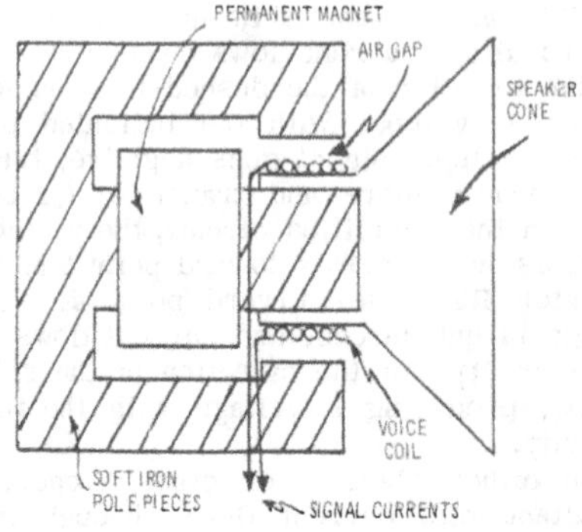

179.282

Figure 13-42.— Permanent magnet speaker.

This is necessary because of the normally low impedance of the voice coil. Standard impedance values for such speakers are 4, 8, 16, and 32 ohms. Other impedance values may be obtained but those listed are by far the most common.

While permanent magnet speakers perform reasonably well in the audio range, they, nevertheless, have inherent limitation. When the speaker is constructed, only a limited number of turns may be built into the voice coil. It, therefore, has a fixed inductance. At low frequencies, the inductive reactance of the voice coil will be relatively low, and large audio currents will flow. This provides a strong magnetic field around the voice coil and a strong interaction with the field of the permanent magnet. Low frequency response is therefore excellent. At mid-band frequencies, the inductive reactance has increased and less current can flow in the voice coil. This produces less magnetic field and less interaction. Mid-band response, however, is still acceptable in a properly designed speaker. At high audio frequencies, however, inductive reactance is quite high and very little current flows in the voice coil. This results in a greatly reduced voice coil field and very little interaction with the permanent magnet field. Also, at high frequencies the inter-winding capacities of the voice coil tend to shunt some of the high audio frequencies and further reduce the response.

In summary, it may be said that the frequency response of most permanent magnet speakers will fall off at the higher audio frequencies. This problem is normally overcome either by the procurement of an expensive, specially designed speaker, or through the use of two speakers, one of which is designed to operate well at the higher audio frequencies.

As shown in figure 13-43, an electromagnet may be used in place of the permanent magnet to form an electromagnetic dynamic speaker. However, in this instance, sufficient d.c. power must be available to energize the field electromagnet. The operation otherwise is much the same as that of the permanent magnet type. This type of speaker is seldom encountered in Navy equipment.

Figure 13-44 shows a diagram of representative headphones as used in conjunction with Navy equipment. The device consists of a permanent magnet and two small electromagnets through which the signal currents pass. A soft iron diaphragm is used to convert the electrical

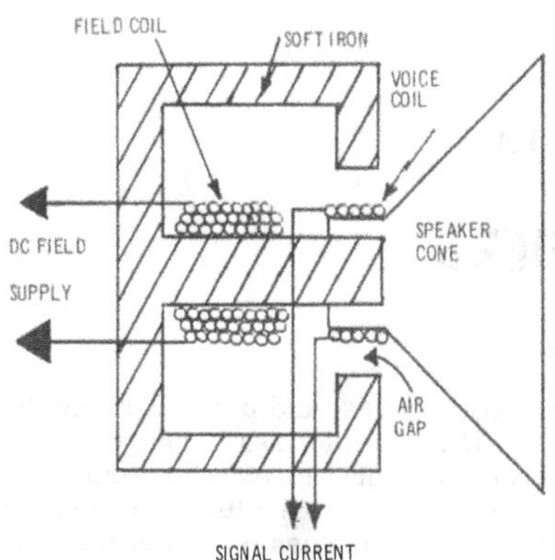

179.283
Figure 13-43.—Electromagnetic speaker.

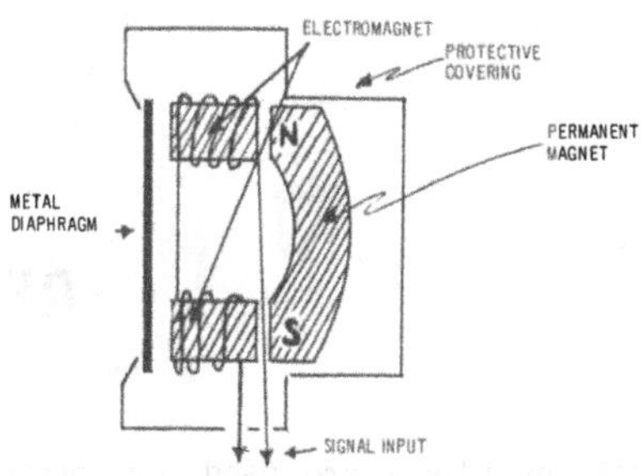

179.283
Figure 13-44.—Headphone.

effects of the device into sound power. When no signal currents are present, the permanent magnet exerts a steady pull on the soft-iron diaphragm. Signal current flowing through the coils mounted on the soft iron pole pieces develops a magnetomotive force that either adds to or subtracts from the field of the permanent magnet. The diaphragm thus moves in or out according to the resultant field. Sound waves are then reproduced which have an amplitude and frequency (within the mechanical capability of the reproducer) similar to the amplitude and frequency of the signal currents.

As compared to the permanent magnet speakers, standard headphones are considered to be high-impedance devices. Headphone electromagnets are normally wound of small wire with many turns providing the larger impedance. Because of the physically small size and inflexibility of the metal diaphragm the headphones often give poor response to the lower audio frequencies. In the voice range of audio, however, most headphones are adequate reproducers.

CHAPTER 14

RF AMPLIFIERS

RF amplifiers are used for both receiving and transmitting. There are two general classes of RF amplifiers, the untuned amplifier and the tuned amplifier. In the untuned amplifier, response is desired over a broad frequency range, and its sole function is amplification. In the tuned RF amplifier, very high amplification is desired over a small range of frequencies, or at a single frequency. Thus, in addition to amplification, selectivity is also desired to separate the wanted from the unwanted signals. The use of the tuned RF amplifier is generally universal, while that of the untuned RF amplifier is relegated to a few special cases. Consequently, when RF amplifiers are mentioned, they are ordinarily assumed to be tuned unless specified otherwise.

In receiving equipment, the RF amplifier circuit serves to both amplify the signal and select the proper frequency; in addition, it serves to fix the signal-to-noise ratio. A poor RF amplifier will make the equipment able to respond only to large input signals, whereas a good RF amplifier will bring in the weak signals above the minimum noise level (determined by the noise generated in the receiver itself only) and thus permit reception which would otherwise be impossible.

RF amplifiers and IF amplifiers are almost identical. Both are actually RF amplifiers, but the IF amplifier operates at a fixed frequency, which is usually lower than the frequency of the RF amplifier. The RF amplifier normally consists of only one stage, whereas the IF amplifier uses a number of cascaded stages to obtain high gain with the desired selectivity at the fixed IF. In some cases more than one stage of RF amplification is used to obtain greater sensitivity.

As might be suspected, since the RF amplifier is employed over the entire RF spectrum, careful attention to design parameters is necessary to obtain proper operation. In the medium-, low-, and high-frequency ranges, conventional transistors, tubes, and parts are used. In the VHF, UHF, SHF, and microwave ranges, specially designed transistors, tubes, and parts are required to obtain optimum results. This chapter will be devoted to the discussion of tuned RF amplifiers used in receivers in the low-, medium-, and high-frequency ranges. Microwave circuits and components will be discussed in a later chapter.

Figure 14-1 shows the position of the RF amplifier in relation to the other stages of the superheterodyne receiver. The RF amplifier stage receives its input from the antenna and sends an amplified output to the mixer stage for conversion to the receiver's intermediate frequency (IF).

The RF stage, or preselector, is optional in a superheterodyne receiver but practically all Navy receivers have at least one RF stage to amplify the RF signal and to improve their signal-to-noise ratio.

Noise can originate externally and/or be generated within the receiver. Although most of the receiver's gain is in the IF section, the RF gain is of primary importance in determining the ratio of received signal voltage to noise voltage generated in the receiver. The amplitude of receiver noise produced by the RF amplifier is in the order of microvolts. The mixer stage generates the majority of receiver noise voltage. Enough RF amplification is needed, therefore, with a low noise level in the RF amplifier, to supply an adequate signal with a high signal-to-noise ratio for the input to the mixer stage. As a result, the signal level at the mixer input is the limiting factor in the ability of the receiver to reproduce an acceptable output with a weak signal input from the antenna.

Most of the receiver's selectivity (ability to reject adjacent channel frequencies) is in the IF section, but RF amplification assists in rejecting interferring RF signals that can produce beat frequencies within the IF bandpass of the receiver. This is especially important in rejecting image frequencies. Filters and wavetraps are

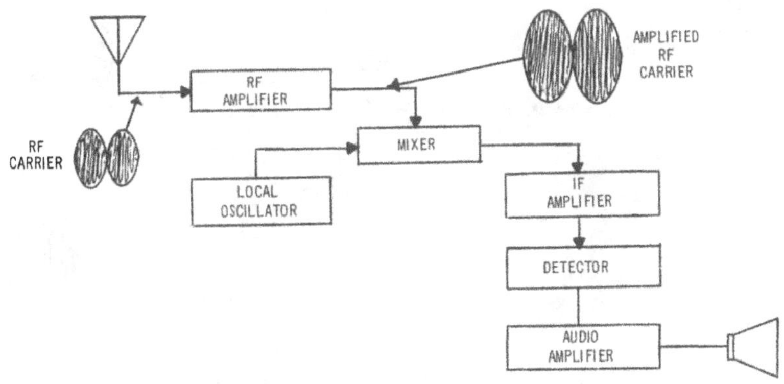

179.284
Figure 14-1.—Superheterodyne receiver.

sometimes used in the RF input circuit to reject unwanted image signals. The advantage of additional RF stages lies in the improved selectivity which is possible against image and other undesired frequencies.

In the transistor RF amplifier, the effect of the base-collector capacitance and the development of negative resistance through a change in internal parameters can cause oscillation. Neutralization circuits are used to prevent this oscillation and to obtain maximum gain.

Triode or pentode electron tubes can be used for the RF stage. The pentode RF amplifier tubes allow more gain and their lower plate-to-grid interelectrode capacitance results in less coupling of the local oscillator output in the mixer grid circuit back to the antenna input circuit. Triode RF amplifier tubes produce less tube noise than pentodes, but their greater interelectrode capacitance provides less isolation between the local oscillator and antenna and may provide feedback for oscillations. Thus, neutralization of the triode RF amplifier may be necessary to prevent excessive feedback of the RF signal from plate-to-grid which could cause the RF amplifier to oscillate.

Figure 14-2 shows the block diagram of an RF amplifier and its associated tank circuits. The reasons why an RF amplifier, or preselector, is used in a receiver are to provide: (1) amplification, (2) isolation, (3) increased selectivity, (4) image rejection, and (5) improved signal-to-noise ratio. Amplification and isolation are provided by the action of the transistor or tube used in the preselector. On the other hand, selectivity and image rejection (and to a certain extent, signal-to-noise ratio) are mainly a function of the bandwidth of the stage which is determined primarily by the tuned circuits and coupling networks of the stage, as discussed in chapter 10.

In figure 14-2, the input tank selects the desired signal from among the many signals received by the antenna. From the input tank, the desired signal is applied to the RF amplifier. The RF amplifier increases the amplitude of the desired signal but does not change its frequency. From the amplifier, the signal is coupled by the output tank to the mixer stage. The output tank improves the selectivity and the signal-to-noise ratio.

COMMON-EMITTER RF AMPLIFIER

Figure 14-3 shows the schematic diagram of a common-emitter RF amplifier using a PNP transistor. The common-emitter configuration is used in order to take advantage of the greater gain and the simpler biasing arrangement. At the start of the cycle of operation, the transistor is in a quiescent condition, with the collector current determined by the voltage divider resistors R1, R2, and emitter resistor R3. Current flow is from the negative terminal of the source through R1 and R2 to ground. The values of R1, R2 and R3 are chosen so that the difference in potential between the base

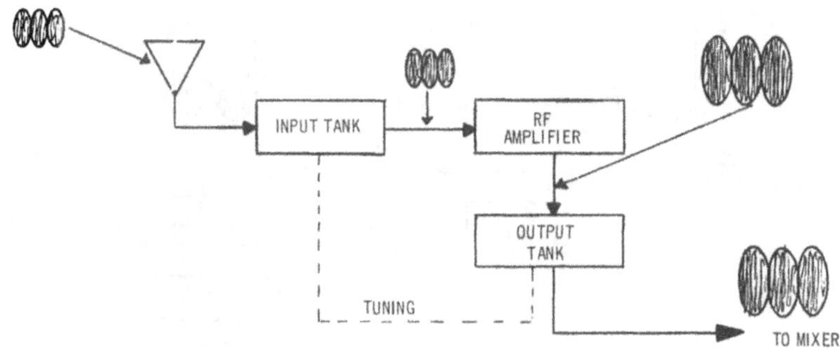

179.285
Figure 14-2.—RF amplifier.

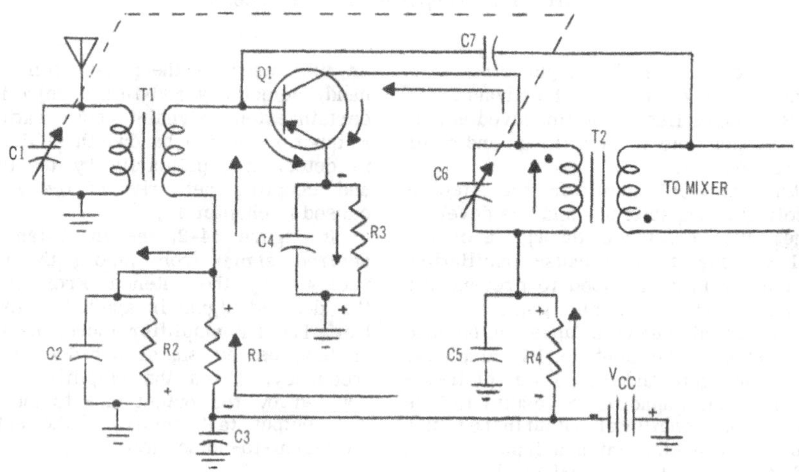

179.286
Figure 14-3.—Common-emitter RF amplifier using a PNP transistor.

and emitter (forward bias) is only a few tenths of a volt, and is set at the center of the forward transfer characteristic cruve (discussed in chapter 6) for class A operation.

Resistor R4 determines the collector voltage for Q1 while emitter bias and temperature stabilization are provided by R3. Capacitor C4 prevents degeneration across the emitter resistor by placing the emitter at a.c. ground. C7 is the neutralizing capacitor while capacitors C2, C3 and C5 bypass the RF currents around the biasing network and power supply. The input tank consists of C1 and the primary of T1 while C6 and the primary of T2 make up the output tank. C1 and C6 are variable so that the tanks can be tuned to resonate at the desired frequency and are ganged together so that they may be tuned (tracked) simultaneously.

When the antenna receives an RF signal at the resonant frequency of the input tank circuit, the tank appears as a high impedance and develops the maximum signal voltage across it.

For any other RF signals present, the tank appears as a low impedance and very little signal voltage is developed.

The signal is coupled from the primary to the secondary of T1 and applied to the base of Q1. Transformer coupling is used to match the high impedance of the tank circuit to the low impedance of the transistor, Q1. As the input signal swings negative, it adds to the forward bias, increasing the conduction of the transistor, which causes the collector current to increase. As the input signal swings positive, forward bias is reduced decreasing collector current. The collector current follows the changes in the input signal but is larger in amplitude since it is approximately equal to beta times the input signal.

The output tank circuit is the load impedance for the collector of transistor Q1. Since it is tuned to the same frequency as the input tank circuit, it will present a high impedance to the desired signal. When the input signal increases the conduction level of Q1, the increasing collector current causes the voltage drop across the tank to increase, reducing the collector voltage (fig. 14-4). When the collector current decreases, the voltage drop across the tank decreases, and the collector voltage increases.

The output signal is coupled to the mixer stage or another RF amplifier by T2. Transformer coupling is used to match the high output impedance of the stage to the input impedance of the succeeding stage. Part of the output signal is coupled back from the secondary of T2 by C7 to the base of Q1 for neutralization. Neutralization is provided to prevent oscillation and therefore insure maximum stability. In the transistor RF amplifier, the effect of the base-collector capacitance causes a portion of the output signal to be fed back to the input thereby causing the stage to oscillate. The neutralizing capacitor feeds back part of the output signal out of phase with the input to prevent this oscillation.

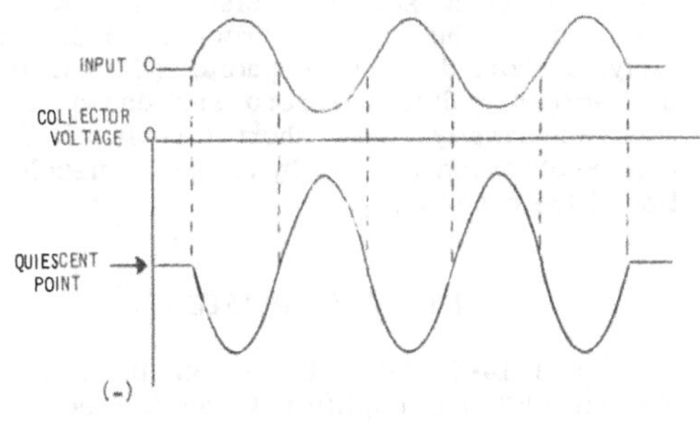

Figure 14-4.—Voltage waveforms for common emitter RF amplifier.

Lack of an input signal, improper bias, lack of supply voltage, or a defective transistor can cause loss of output. Check the bias and supply voltages with a voltmeter. If R1 is open (fig. 14-3), the base bias will be near zero, which corresponds to class B operation, and there is a possibility that strong signals may produce a partial output. If the secondary of T1 is open, the base circuit of Q1 is also open, and no output can occur. If R3 is open, the emitter circuit will be incomplete and no output will be obtained. If a voltage exists at the base of Q1 but not at the collector, either the primary of T2 or R4 is open, or C5 is shorted. If neutralizing capacitor C7 is shorted, the base of the following stage will be directly connected to the base of Q1 and will shunt transistor Q1. There will be a loss of amplification, but not necessarily a complete loss of output.

If C4 is open, the emitter current flowing through R3 will produce degeneration and oppose the bias. The output will be reduced in proportion to the amount of degeneration. With a collector output equal to or less than the input at the base of Q1, either T2 or Q1 is defective.

PENTODE RF AMPLIFIER

Figure 14-5 is the schematic of a pentode RF amplifier. T1 is an RF transformer which matches the impedance of the antenna to the input impedance of the pentode control grid. The secondary of T1 is tuned to select a specific frequency from the incoming signals by adjusting the value of variable capacitor C1 for resonance at the desired frequency. Resistor R1 and capacitor C2 form the cathode bias resistor and bypass capacitor. Resistor R2 is the screen voltage dropping resistor, and capacitor C3 is the screen bypass capacitor, which stabilizes the screen voltage and prevents it from being affected by the signal. The suppressor element of V1 is grounded directly. In some circuits it is connected externally to the cathode. RF transformer T2 acts as the plate load and couples the output to the next stage. The output winding is tuned by C4 to the desired RF output frequency.

When a signal appears on the antenna, it is coupled through the primary of T1 to the tuned

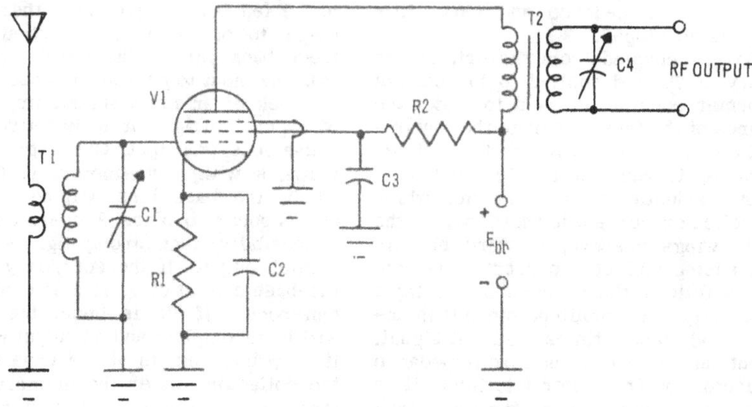

179.288
Figure 14-5.— Pentode RF amplifier.

secondary (grid input) circuit. With capacitor C1 tuned to the frequency of the incoming signal, a relatively large RF voltage is developed across the tuned circuit and applied to the grid of V1. The positive half-cycle decreases grid bias instantaneously, causing a plate current increase. The negative half-cycle increases grid bias instantaneously, causing a plate current decrease. This changing plate current flowing through the primary of output transformer T2 induces an output in the tuned secondary winding.

The RF amplifier operating as the first stage in the receiver is usually a small-signal amplifier. That is, the input voltage is usually on the order of microvolts, except in strong signal areas. Therefore, a small signal voltage change causes only a very small grid voltage change, and it is necessary to employ high-transconductance electron tubes to produce effective amplification. The pentode tube is suited for this purpose, since it has both a high amplification factor and a high transconductance. By using a large value of inductance and as small a tuning capacitance as practicable, the tuning circuit exhibits a high Q. Thus, its effective impedance is much larger than that presented by a tuning tank of low Q. Hence, a large input voltage is developed between grid and ground across the tuned circuit. With a step-up turns ratio from transformer primary to secondary, if closely coupled, a still larger input voltage is produced. The step-up of voltage in the transformer and the high Q tuned grid tank increase the small input voltage before it is applied to the tube for further amplification. Normally, class A bias is used to produce linear operation and to minimize distortion. In addition, the low grid-to-plate capacitance of the pentode reduces the tendency toward plate-to-grid feedback and self-oscillation. This eliminates the necessity for neutralization.

Loss of plate, screen, or heater voltage, or a defective tube can cause loss of output. The voltages can be checked with a voltmeter, and an open heater can sometimes be observed by noting that the tube is not illuminated (glass envelope) and/or feels cold to the touch. If the plate and screen voltages are normal and the heater is good, substitute a tube known to be good. An open screen resistor (R2) will be indicated by the lack of screen voltage. Similarly, a shorted screen capacitor (C3) will drop the screen voltage to zero and cause R2 to heat abnormally. The short circuit condition may be observed visually by smoke or discoloration of the resistor.

FET RF AMPLIFIER

Figure 14-6 shows the schematic of an N-channel FET RF amplifier. C1 and the secondary of T1 make up the input tank circuit while C3 and the primary of T2 make up the output tank circuit. C2 is the bypass capacitor for source resistor R1, while C4 bypasses RF currents

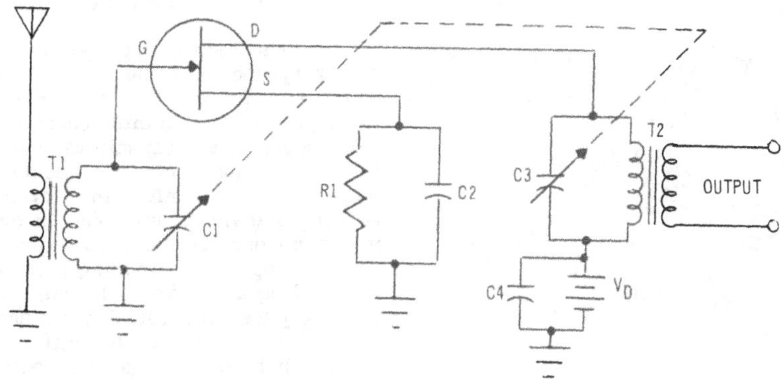

179.289
Figure 14-6.—N-channel FET RF amplifier.

around the power supply. Source resistor R1 establishes the quiescent gate-to-gate source voltage in the linear operating region and is a form of self-bias.

When a signal appears on the antenna, it is coupled from the primary to the secondary of T1. The signal voltage developed across the input tank circuit is fed to the gate, changing the conduction level of the FET. As the input signal swings positive, the difference of potential between the gate and source is reduced causing the drain current to increase. When the input signal swings negative, the difference of potential between the gate and source is increased, reducing the drain current. The drain current change corresponds to the input signal voltage change but is much larger in amplitude.

The drain current changes appearing in the output tank are coupled to the secondary and fed to the next stage. Transformer T1 is used to match the low impedance of the antenna to the high input impedance of the FET while T2 is used to match the high output impedance of the FET to the input impedance of the next stage.

If the primary or secondary of T1 opened, no signal could be coupled to the amplifier resulting in a loss of output. If C2 opened, degeneration would occur and the output amplitude would be reduced. Should the primary of T2 open, the drain voltage would be removed with a resulting loss of output.

TUNING

One of the functions of the tuned tanks in the RF amplifier is that of selecting the desired signal frequency from among the many frequencies available. The process of adjusting the inductance (L) or the capacitance (C) of a tuned circuit in order to select a specific station frequency is called tuning. A receiver may be tuned by varying either the inductive or the capacitive component of the tuned circuit.

As shown previously, a receiver has more than one tuned circuit. If all the tuned circuits are tuned by varying the capacitance, then an air dielectric variable capacitor having two or more sections is used. The sections are mechanically connected (ganged) to the same shaft, but electrically isolated from each other. Each section is used for a separate tuned circuit. When the shaft is moved, the capacitance will vary by approximately the same amount. Figure 14-7 illustrates a ganged air dielectric variable capacitor having two sections.

In circuits that are tuned inductively, the cores of the inductors are mechanically connected together to permit changing the inductance of all tuned circuits simultaneously. Inductive tuning is called "permeability" or "slug" tuning.

Figure 14-8A shows the schematic of an inductively tuned RF amplifier while figure 14-8B shows a capacitively tuned RF amplifier. The dotted line indicates the mechanical linkage.

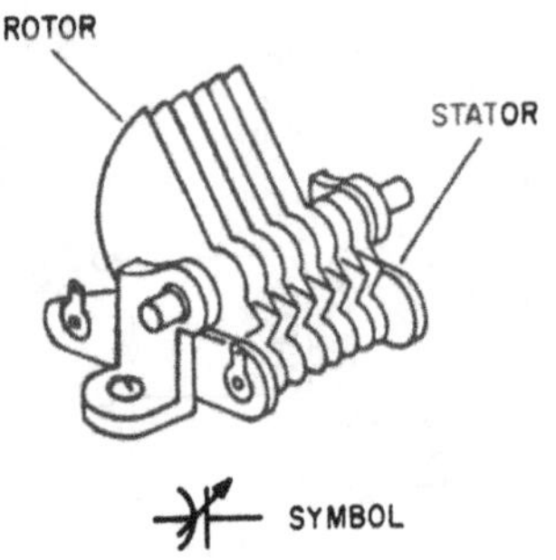

179.290
Figure 14-7.—Ganged air dielectric variable tuning capacitor.

TRACKING

It was stated that a ganged variable capacitor will vary the capacitance of all the tuned circuits by approximately the same amount. Due to slight mechanical differences (manufacturing, tolerances) in the capacitors, coils, and variations in wiring from one receiver to another, there will be variations in electrical adjustment between individual sections. In order to obtain maximum performance, all tuning circuits must "track" together. Tracking means that the resonant frequency for each tuning circuit will be relatively the same for all positions of the tuning shaft. To compensate for differences in capacitance between the ganged sections, trimmer capacitors are used.

The trimmer capacitor is usually of the compression mica type, consisting of two plates

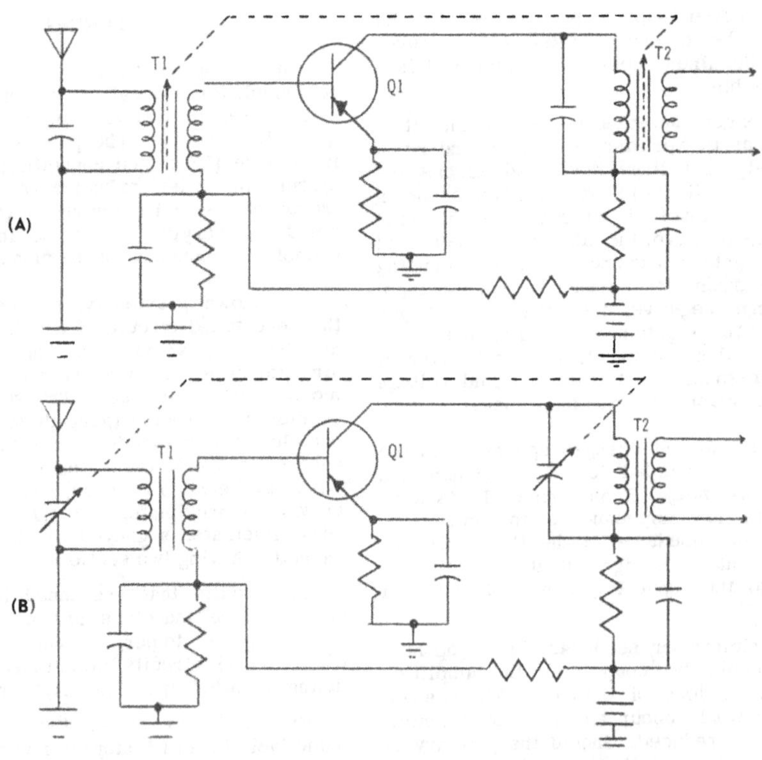

179.291
Figure 14-8.—RF amplifiers.

separated by a sheet of mica with a small adjustable screw through the center to vary the capacitance. The capacitance is varied by turning the screw, which either moves the plates closer together or further apart. The trimmers are usually mounted on the main tuning capacitor, as shown in figure 14-9A, and electrically are in parallel with each section as shown in figure 14-9B. The value of the trimmer capacitor is smaller than its associated capacitor.

The trimmer is used to compensate for tracking variations at the high end of the tuning range as demonstrated by the following example (fig. 14-9B). To determine the effect of trimmer variation on total tank capacitance at the high frequency end of the tuning range, set both capacitors at minimum (C2 = 10.6 pf and C1 = 2 pf). Thus, the minimum capacitance of the parallel combination will be:

$$2 + 10.6 = 12.6 \text{ pf}$$

Maintaining the tuning capacitor at minimum and setting the trimmer to maximum (C1 = 17 pf) will cause the minimum capacitance of the combination to increase to:

$$17 + 10.6 = 27.6 \text{ pf}$$

Thus at the high frequency end of the range the trimmer capacitor can cause a change in total capacitance of:

$$27.6 - 12.6 = 15 \text{ pf}$$

The percentage of change in capacitance by the trimmer is:

$$\frac{15}{27.6} \times 100 = 54.3\%$$

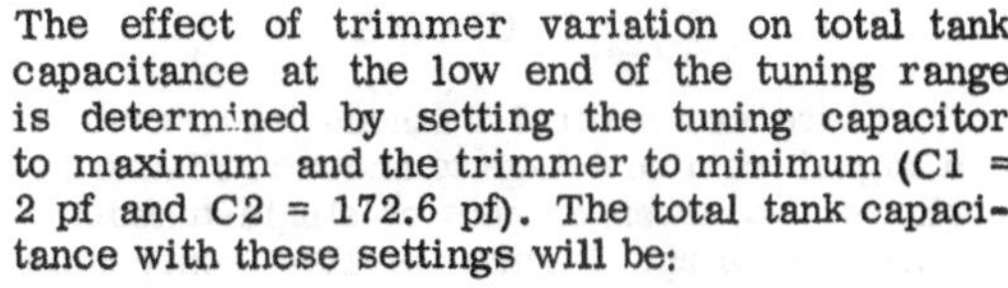

The effect of trimmer variation on total tank capacitance at the low end of the tuning range is determined by setting the tuning capacitor to maximum and the trimmer to minimum (C1 = 2 pf and C2 = 172.6 pf). The total tank capacitance with these settings will be:

$$2 + 172.6 = 174.6 \text{ pf}$$

Holding the tuning capacitor at maximum and setting the trimmer at maximum will increase the tank capacitance to:

$$17 + 172.6 = 189.6 \text{ pf}$$

The change in tank capacitance at the low frequency end caused by varying the trimmer will be:

$$189.6 - 174.6 = 15 \text{ pf}$$

The change in capacitance of 15 pf is the same as occurred at the high end, however, the percentage of change is vastly different:

$$\frac{15}{189.6} \times 100 = 7.06\%$$

Therefore, it can be seen that varying the trimmer capacitor affects the percentage of tank capacitance to a much greater extent at the high end than at the low end. For this reason, the trimmer is adjusted at the high frequency end of the receiver's tuning range.

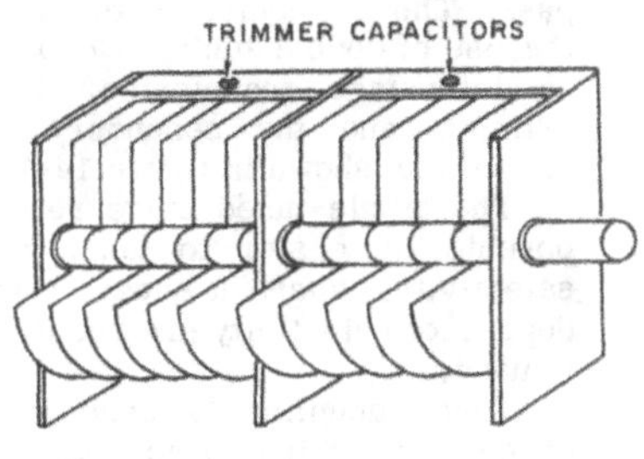

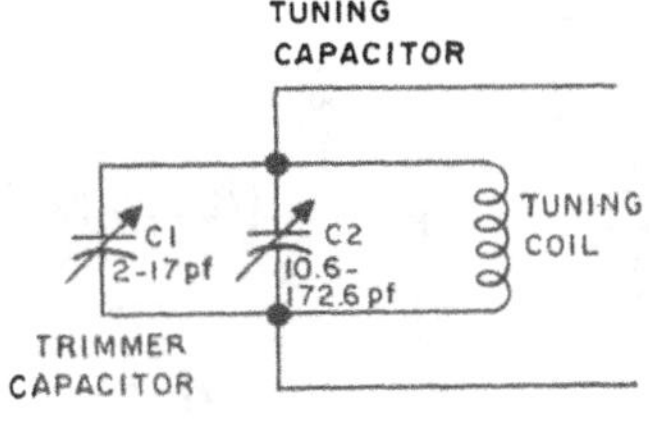

179.292
Figure 14-9.—Trimmer capacitor.

COUPLING

There are several methods for transferring a signal from one stage to another. Figure 14-10 shows two arrangements of single-tuned interstage coupling. Figure 14-10A represents a series fed arrangement, so called because the source voltage is fed in series with the tuned tank, while figure 14-10B represents a shunt fed arrangement. Use of the single-tuned stage yields very satisfactory results in many applications. With a high Q circuit, the selectivity is good, but the bandwidth is somewhat narrow. By reducing the Q, the bandwidth can be increased. The intended application governs the choice of Q and bandwidth.

Figure 14-11 shows two arrangements of single-tuned interstage transformer coupling. Figure 14-11A shows a tuned primary-untuned secondary arrangement. Figure 14-11B has an untuned primary and a tuned secondary. The turns ratio can be arranged to provide a step-up or step-down ratio.

In some applications the bandwidth requirements are such that a single-tuned stage will not be able to meet them. Another application may require that all sideband frequencies within the bandpass receive relatively equal amplification. Still another application may require a wide bandpass but also a high degree of selectivity for frequencies immediately outside the bandpass. These requirements can be satisfied by the use of double-tuning. Double-tuning refers to an interstage transformer in which both the primary and the secondary contain resonant circuits as shown in figure 14-12.

The single-tuned stage requires fewer components, is easier to tune, and provides good selectivity. Where a wider bandwidth and a high degree of selectivity are required, double-tuning is used.

Tuned coupling is used in circuits for impedance matching to achieve maximum gain as well as for selectivity. Although matching impedances will provide maximum power gain, this may not be practical or desirable in some cases. It is sometimes necessary to sacrifice gain to obtain the desired selectivity. In some instances, by mismatching stages with individual high gains

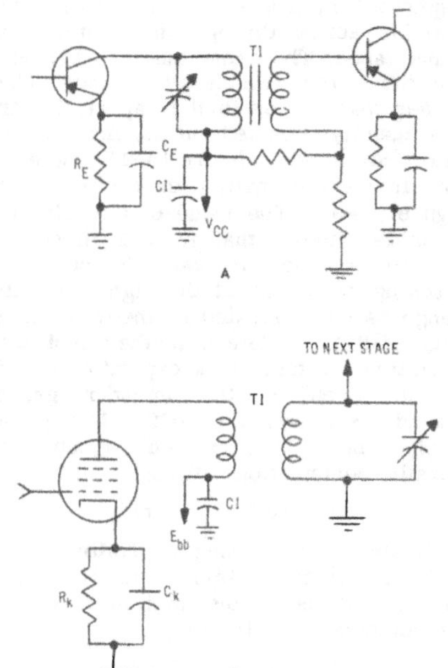

179.293
Figure 14-11.—Single-tuned interstage transformer coupling.

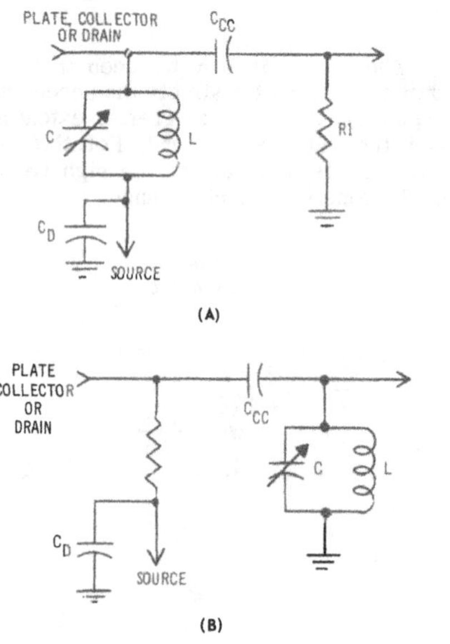

179.294
Figure 14-10.—Single-tuned interstage coupling.

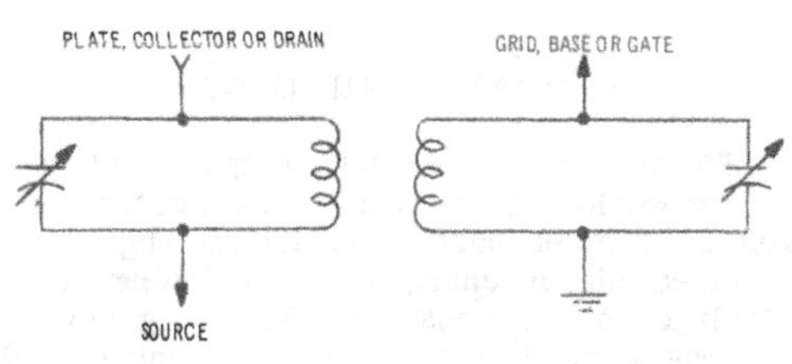

179.295
Figure 14-12.—Interstage transformer with double-tuning.

it becomes unnecessary to neutralize the stages. It is also difficult to obtain a high Q circuit if the input or output impedance of a stage which shunts the coupling circuit is low. Therefore, a variety of circuit arrangements are used to obtain the desired impedance while maintaining a high Q for optimum selectivity. Figure 14-13 shows typical forms of coupling circuits used.

In circuit A, a single-tuned, transformer coupled circuit is shown; the impedance relationships vary as the square of the turns ratio. In B, a tapped auto-transformer is used; L1 plus L2 are the primary, while L2 is the secondary producing a step-down impedance ratio. In C, a capacitance divider is used to reduce the impedance between the input and the output. In D, both the input and output impedances are

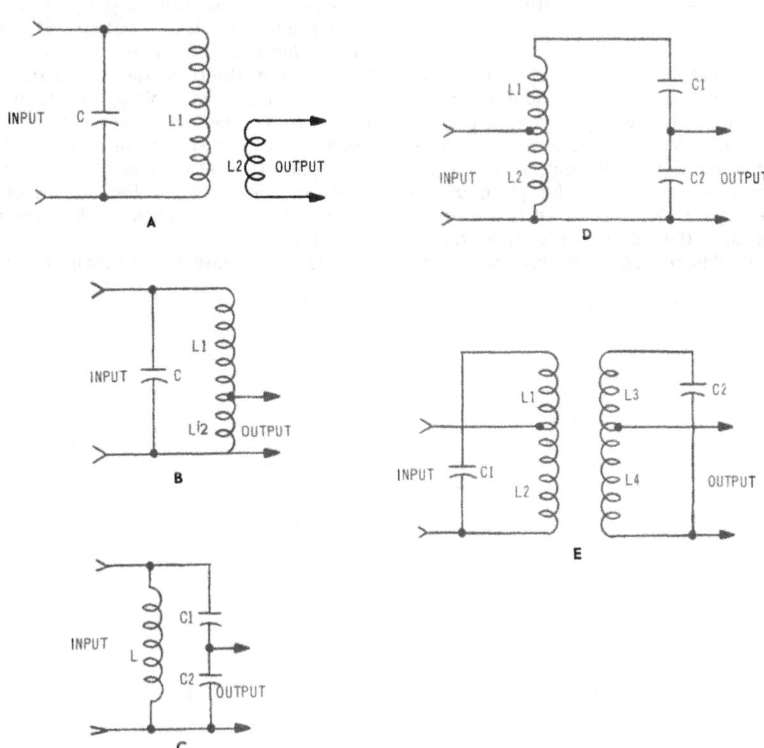

179.296
Figure 14-13.—Coupling circuits.

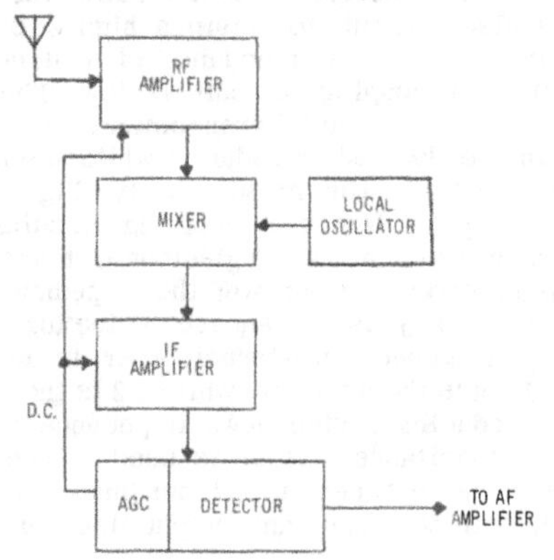

Figure 14-14.—AGC distribution.

different, and L1 helps to improve the over-all Q of the tank circuit. In E, a double-tapped transformer is used to supply high-Q primary and secondary tanks with the proper impedance matching for the input and the output. Since the Q of the circuit depends upon the ratio of reactance to resistance (X_L) in a coil, the lower the resistance and the higher the inductance, the larger the Q. Therefore, to match the values of input and output impedance of a circuit, the inductance is tapped at the appropriate point.

AUTOMATIC GAIN CONTROL

To maintain a constant output level, a receiver would require readjustment of the manual volume control each time the strength of the received signal changed due to fading or as a result of station changes. Also, signals from stations operating on different frequencies but at the same power may not reach the receiver antenna with the same power because of differences in transmission distances. The function of Automatic Gain Control (AGC) is to reduce variations in the amplitude of the audio/intelligence output of the receiver caused by variation in strength of the received signal.

AGC is accomplished by filtering a portion of the detector circuit output with an RC network to produce a d.c. voltage that is proportional to the amplitude of the audio output of the detector. This d.c. voltage is then used to vary the gain of the preceding RF and IF amplifier stages (as shown in fig. 14-14). The effect on the controlled stages is to reduce the stage gain with an increase in the d.c. voltage or to increase the stage gain with a decrease in d.c. voltage.

AGC is discussed in more detail in chapter 18.

CHAPTER 15

OSCILLATORS

The primary function of an oscillator is to generate a given waveform at a constant amplitude and specific frequency, and maintain this waveform within certain limits. The oscillators described in this chapter will produce a sinusoidal waveform. A mechanical generator or alternator will also a produce a sinusoidal waveform at a given frequency (such as 60 Hz for an a.c. power line), and it is possible to maintain the frequency within very strict limits. However, generators and alternators are NOT oscillators. It is true that before the advent of the transistor and electron tube some alternators were made which could produce a radio frequency (the Alexanderson alternator produced 100 kHz), but even their top frequency limit was still very low in comparison to RF. Along with the disadvantage of their frequency limitations they were very expensive, and could be used at only one frequency. The development of the transistor and the electron tube, however, solved the problem of generating high frequency a.c.

By definition an oscillator is "a non-rotating device for producing alternating current, the output frequency of which is determined by the characteristics of the device." The definition includes the word non-rotating in order to eliminate the possibility of including an alternator as an oscillator. An alternator is a rotating device and would NOT be classified as an oscillator.

The basic definition could be enlarged upon in the following manner: "An oscillator is a passive electronic component, or group of passive electronic components, which when supplied with a source of energy in the proper phase and of sufficient amplitude to overcome circuit losses, will furnish an electrical periodic function repetitively." Since transistors and electron tubes are amplifiers they may be used in oscillator circuits to provide the energy required.

The mere fact that a circuit is capable of oscillating does not make it a useful oscillator. In order to qualify as an oscillator, a circuit must be able to generate sustained oscillations in a desired and controllable manner. Most people are familiar with the undesired oscillation in a public address system (loud whistle) that occurs when the microphone is moved too close to the speaker.

A transistor or electron tube oscillator is essentially an amplifier that is designed to supply its own input through a feedback system. Two conditions must be met by this feedback network if sustained oscillations are to be produced. First, the voltage feedback from the output circuit must be in phase with the excitation voltage at the input circuit of the amplifier; in other words, the feedback must aid the action taking place at the input. This is called POSITIVE or REGENERATIVE feedback. Second, the amount of energy fed back must be sufficient to compensate for circuit losses.

Feedback may be accomplished by inductive, capacitive, or resistive coupling between the output and input circuits. Various circuits have been developed to produce feedback with the proper phase and amplitude. Each of these circuits has certain characteristics that make its use advantageous under given circumstances.

Without the electronic oscillator very few advanced electronic circuit applications would be possible. Because of this fact, the importance of a thorough understanding of oscillator theory and operation is stressed. The list of types of equipment employing oscillators (in one form or another) is a long one, including equipment such as radar, sonar, guided missiles, communications gear, test equipment, home entertainment products, etc.

DESIRED OSCILLATOR CHARACTERISTICS

Virtually every piece of equipment that utilizes an oscillator necessitates two main requirements for the oscillator — AMPLITUDE STABILITY and FREQUENCY STABILITY. The rigidity with which these requirements must be met depends on the accuracy demanded of the equipment.

Amplitude stability refers to the ability of the oscillator to maintain a constant amplitude output waveform—the less the deviation from a predetermined amplitude, the better the amplitude stability.

Frequency stability refers to the ability of the oscillator to maintain the desired operating frequency—the less the oscillator deviation from the operating frequency, the better the frequency stability.

REVIEW OF PARALLEL RESONANCE

Figure 15-1 shows a resonant LC circuit and its operation. It is assumed that a battery has previously been used to charge the capacitor and the inductor is then connected across the capacitor. Part A of the diagram shows the tank circuit immediately after the inductor has been connected. The capacitor is acting as a voltage source and produces a current in the direction shown. The inductor produces a counter EMF which opposes the voltage across the capacitor. As the capacitor discharges through the inductor, the inductor is storing energy in the form of a magnetic field. At the time when the capacitor has completed its discharge there is zero voltage across the circuit and all circuit energy is stored in the inductor's magnetic field.

Figure 15-1B shows the second quarter cycle of operation. The magnetic field of the inductor is collapsing in order to maintain inductor current. It is now the inductor that is acting as a voltage source. The collapsing magnetic field produces a current in the same direction as that previously established. This current now charges the capacitor with a polarity opposite to that it originally had. When the magnetic field of the inductor is exhausted the capacitor is fully charged and all circuit energy is stored in the form of an electrostatic field across the capacitor dielectric.

Figure 15-1C shows the third quarter cycle of operation. The capacitor is now discharging and once again the inductor is building up its magnetic field. The direction of current flow is in the direction shown in the diagram.

In figure 15-1D the capacitor is seen to be fully discharged with the inductor acting as a voltage source once again. The collapsing magnetic field of the inductor will continue the circuit current and recharge the capacitor back to the original condition shown in figure 15-1A. Future cycles of operation will be identical to the one just described.

In any electrical circuit there exists some inherent resistance. This resistance will constitute a power loss which will eventually kill the oscillations in the tank circuit. The output of a tank circuit in which the amplitude of oscillations decreases to zero is called a DAMPED WAVE. Damping can be overcome only by resupplying energy to the tuned circuit at a rate comparable to that at which it is being used. This re-supplying of energy is usually accomplished through the use of a transistor or tube acting as a switching device.

If more detailed analysis is required refer back to the previous chapter on LC tank circuits. The inductance and capacitance of the tank circuit determine the natural frequency of resonance for the tank circuit. The resonant frequency may be found using the equation:

$$f_o = \frac{1}{2\pi\sqrt{LC}}$$

LC OSCILLATIONS

The LC type of oscillator uses a tuned circuit consisting of lumped and distributed inductances

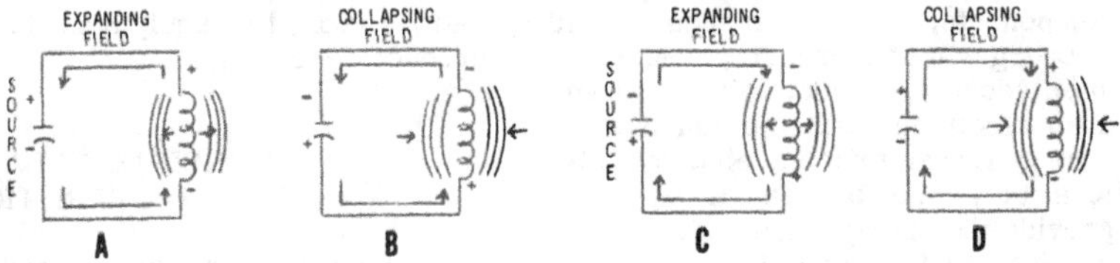

179.298

Figure 15-1.—Operation of resonant LC circuit.

and capacitances. These inductances and capacitances are connected as resonant circuits and are used to determine the frequency of operation. Operation is normally in the radio frequency range (operation in the audio range may occasionally be used). Oscillation is achieved by the application of positive (regenerative) feedback from the output circuit to the input circuit. The feedback is provided by either external or internal components, which may be either capacitive or inductive in nature.

Class A operation is used for test equipment oscillators where waveform linearity is important. A class similar to class C is used for power oscillators. Class B operation may occasionally be used but is not normally encountered.

Efficient circuit operation and maximum frequency stability are achieved with a tank circuit having a high loaded Q (this is achieved by a high C to L ratio). This is produced for parallel tuned tanks by using a large tuning capacitance with a small inductance. Normally, the inductance of the tank circuit remains fixed, and the capacitance is the variable tuning element. Inductance tuning will normally be encountered only in the low and very-low RF ranges.

There are a number of types of LC oscillators, and each type has a particular advantage or feature claimed for it. All LC oscillators may be designed to operate at any frequency within the entire frequency spectrum. However, for practical reasons, each type is normally designed to operate within a specific frequency range, and depending on the type these frequency ranges may or may not overlap. In cases where their frequency ranges do overlap, the type of oscillator to be chosen for a particular application is generally left up to the designer.

BASIC OSCILLATORS

A basic oscillator can be broken down into three main sections. The frequency determining device is usually an LC tank circuit. While the tank circuit is NORMALLY found in the input circuit of the oscillator (both transistor and electron tube), it should not be considered out of the ordinary if it appears in the output circuit of a transistor oscillator. The differences in magnitude of collector and plate currents and shunting impedances are partly responsible for this. In both types of circuits (solid state and tube) oscillations take place in the tuned circuit. Both the transistor and tube function primarily as an electric valve that amplifies and allows the feedback network to automatically deliver the proper amount of energy to the input circuit to sustain oscillations. In both transistor and tube oscillators the feedback circuit couples energy of the proper amount and phase from the output to the input circuit in order to sustain oscillations. A basic block diagram is shown in figure 15-2. The circuit is essentially a closed loop utilizing d.c. power to maintain a.c. oscillations. All oscillators that will be studied in this chapter, use this principle of operation and differ mainly in the type and method of feedback used.

SERIES FED TRANSISTOR HARTLEY OSCILLATOR

The circuit shown in figure 15-3 is a series fed Hartley oscillator using an NPN transistor. It is designated as a series fed oscillator because the feedback component is connected in series with the power supply. This transistor oscillator can be operated class A or B. Where stability and pureness of waveform (low harmonic content) are the major considerations, the oscillator would be operated class A. If efficiency is the

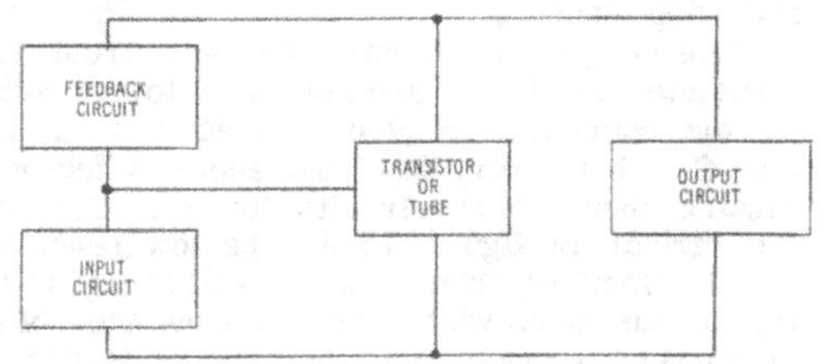

179.299

Figure 15-2.—Basic oscillator circuit.

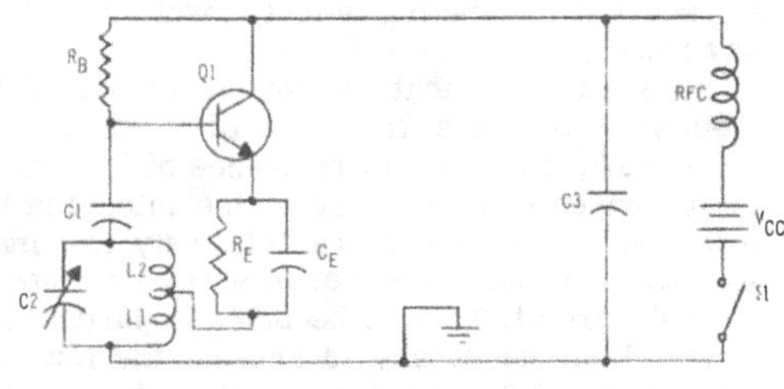

179.300

Figure 15-3.—Transistor Hartley oscillator.

primary concern, a class of operation similar to class C operation is chosen.

The purpose of the component parts of a transistor oscillator should become apparent upon examination of the schematic diagram in figure 15-3. The sine wave is generated in the tank circuit consisting of C2, L1, and L2. C2 is made variable so that the oscillator can be tuned to the desired frequency. The RFC and capacitor C3 are used for decoupling. Resistor R_E and capacitor C_E form a bias stabilization network and are connected in series with the emitter. Degeneration, which would otherwise occur, is minimized by this network.

In order for oscillations to begin, a small amount of initial forward bias must be applied to the emitter-base junction. This bias establishes a flow of current through the transistor when the circuit is first energized, and insures that the oscillator will be self starting. R_B provides the path for this bias current.

The magnitude of this bias current is determined by the values of R_B, emitter-base resistance R_E, V_{CC}, and to some extent the small resistance contained in L1 and RFC. The choice of values used for R_B and R_E represents a compromise. This compromise arises in the following manner.

The components shunting the tank circuit in a transistor oscillator are relatively low in value. At the operating frequency capacitors C1, C3, and C_E have very low reactances, effectively making them short circuits for a.c. Consider the circuit in figure 15-3. The low reactance of C1 effectively connects one end of R_B to the top of the tank while the low reactance of C2 effectively connects the other end of R_B to the bottom of the tank (with respect to a.c.). Therefore, if too small a value is chosen for R_B it will swamp the tank circuit, lowering the Q to the point where oscillations cannot be sustained. If, on the other hand, R_B is made very large to reduce its shunting effect, insufficient bias may result.

Consider also that the low reactance of C1 effectively connects the base of Q1 to the top of the tank, and the low reactance of C_E effectively connects the emitter to the middle of the tank coil. This causes the relatively low input impedance of the transistor to shunt a portion of the tank circuit. Thus it becomes apparent that, while C1 is necessary to prevent the low d.c. resistance of L2 from placing a d.c. short across the series combination of the emitter-base junction and the emitter resistor, C1 must also have sufficient reactance to provide a degree of a.c. isolation between the tank circuit and the shunting effects of R_B and the low input impedance of Q1. Due to the large variation in component values, from one circuit to another, no attempt will be made to assign typical values to the components in the transistor oscillators.

A complete cycle of operation for the series fed Hartley transistor oscillator will now be discussed. Refer to figure 15-4 for this discussion. At the instant the switch is closed all the capacitors in the circuit, having no charge, are effectively short circuits. Thus, a surge of displacement current attempts to flow. This attempted surge of current is minimized by the inductive action of the radio frequency choke (RFC). Due to the different values of resistance in the capacitor charge paths, each capacitor will have a different value time constant. However, after a few cycles of operation all of the capacitors will have acquired an average, or d.c., value of charge. Also, the resistive components of the circuit will exhibit voltage drops caused by d.c. biasing currents. The distribution of and polarities of these d.c. voltage drops are shown in figure 15-4. Representative values of voltage have been assigned merely to give the reader an approximation of relative magnitudes.

For this explanation the resistance of the coils L1, L2, and RFC will be considered too small to give a measurable d.c. voltage drop, although, in actuality a very small voltage drop will exist across each one.

Notice that C3 charges to the value of the supply voltage. While actually very complex, the circuit can be simplified by being thought of as essentially two main branches in parallel with the 6 volts across C3. The first branch consisting basically of transistor Q1's collector-emitter resistance (r_e) and the stabilizing resistor R_E. The feedback coil L1 is in series with this branch but since it is considered a d.c. short (a direct connection with very low resistance) it effectively connects the bottom of the stabilization resistor to d.c. ground potential. Notice that the sum of the voltage drops V_{CE} and the stabilization network V_{RE} equals 6 volts.

The second branch consists primarily of the base biasing resistor R_B and the isolation capacitor C1. Coils L1 and L2 are considered to be d.c. short circuits; they effectively connect one end of C1 to the d.c. ground potential. Capacitor C1 is in parallel with the stabilizing network and the base-emitter junction; it will charge to the sum of the voltage drops V_{BE} and V_{RE}, or 3 volts. An important point to note is that the tank capacitor (C2) accumulates no

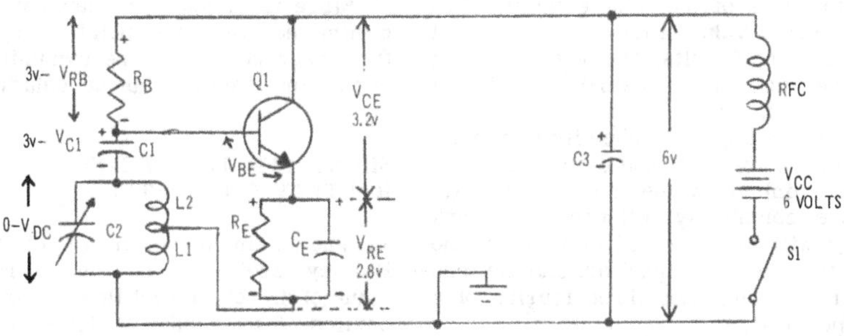

179.301
Figure 15-4.—Operation of transistor Hartley oscillator.

measurable average charge because it is shunted by the low resistance of the tank coil. The voltage drop V_{BE} is of such a polarity as to apply forward bias to the transistor.

Detailed Operation

Previous chapters have explained the production of an alternating voltage by a tank circuit and the internal transfer of energy between the tank coil and the tank capacitor. The action of the tank circuit in the various oscillators remains the same. Therefore, the greatest emphasis will be placed on the method of replacing dissipated energy in order to sustain oscillations. Refer to figure 15-4.

1. Assume that at a given instant the voltage across the tank circuit is zero. All the energy is contained within the magnetic field of the tank inductor. The magnetic field begins to collapse and produces a current flow. The current flow is in such a direction as to produce a positive voltage at the top of the tank with respect to the bottom of the resonant circuit (ground potential).

2. As the top of the tank becomes positive with respect to ground this potential is coupled to the base of Q1 by capacitor C1. The positive potential on the base results in an increase in the forward bias on Q1 and causes collector current to increase. An increase in collector current causes an increase in emitter current, which flows through the feedback winding L1 of the tank coil. An increased current flow through L1 will result in energy being supplied to the tank circuit, which, in turn, will increase the positive potential at the top of the tank and, hence, the forward bias.

3. This regenerative action will continue until such time as the combined actions of transistor non-linearity, the effect of R_E, and the d.c. resistance of L1 and the power supply cause the rate of change of current to decrease. The voltage developed in the tank circuit is now at its positive peak, and is ready to begin a decrease which is caused by normal tank action.

4. As the rate of change of current through L1 decreases, less is supplied to the tank circuit, and through normal tank action, the positive potential begins to decrease. This causes a decrease in the forward bias, which in turn, causes the collector and emitter currents to decrease.

5. At the instant the potential of the tank circuit has decreased to zero the energy of the tank is once again contained in the magnetic field of the coil. The d.c. voltage conditions will again be as depicted in figure 15-4. The tank circuit will have completed a half cycle of operation, tank circuit conditions have been reversed, and the magnetic field is ready to collapse and begin the negative half cycle of its operations.

6. As the magnetic field of the inductor collapses, a current is produced which will charge the capacitor with a negative potential on the top of the tank with respect to ground. This negative alternation is fed to the base of Q1 by coupling capacitor C1 and will reduce the forward bias. This will reduce the collector and emitter current, and in addition, the feedback current which flows through L1. It should be noted that most transistor oscillators are operated class A in order to limit the effect of varying parameters on the stability of the oscillators. Therefore, the value of tank potential will rarely be great enough to cause reverse biasing. For

instance, in the case of the value established in figure 15-4 the peak value of the tank signal will not be greater than .2 volts. The negative going signal does decrease the forward bias, but it will never exceed it.

7. When the tank signal reaches its maximum negative value, the collector and emitter current will be at their minimum values and the magnetic field will have completely collapsed. The tank circuit energy will be contained entirely in the electrostatic field of the capacitor, and the tank circuit will have completed three fourths of a cycle of its operation.

8. Through normal tank action the capacitor will now begin to discharge. As it discharges the negative potential across the tank circuit decreases. This allows an increase in forward bias. As forward bias begins to increase, collector and emitter currents increase, and energy is re-supplied to the tank circuit through L1. If the energy replaced is equal to, or greater than that dissipated, oscillations will be sustained. The capacitor continues to discharge and eventually reaches zero potential. At this time tank circuit voltage is zero, current has been reversed, all tank circuit energy is stored in the inductor, and a complete cycle of oscillation has been completed.

Biasing

In the transistor circuit, bias depends on the amount of current flowing through the base-emitter junction. The value of this bias current is determined by the supply voltage V_{CC}, and the resistance of the bias current path. The path of the bias current, in the circuit just presented, is from the negative terminal of V_{CC}, through L1, R_E, the base-emitter junction, R_B, RFC, and back to the positive terminal of V_{CC}. The RFC and L1 cause no measurable d.c. voltage drop. Knowing the value of voltage drop across R_B and R_E (caused by the bias current) simplifies the determination of the bias voltage. The bias voltage (V_{BE}) equals the supply voltage (V_{CC}) minus the sum of the other voltage drops. Mathematically:

$$V_{BE} = V_{CC} - (V_{RB} + V_{RE})$$

It is usually desirable to operate the transistor circuit class A. It should be recalled from previous material that parameters vary greatly when the transistor is operated on the non-linear portion of its characteristic curve. This would cause oscillator instability.

Since the amount of parameter variations is minimized when the point of operations is on the linear portion of the dynamic curve, better oscillator stability can be achieved by class A operation.

SHUNT FED TRANSISTOR HARTLEY OSCILLATOR

The main disadvantage of the series fed Hartley oscillator is that a relatively large value of direct current flows through the feedback portion of the tank coil. This disadvantage may be overcome by a variation of the connection of the feedback coil, isolating it from the d.c. component of emitter current by capacitors C1 and C3 as shown in figure 15-5. The oscillator derives its name from the fact that the feedback component (L1) is connected in shunt rather than in series with the power supply. By preventing the emitter current from flowing in the feedback winding, the stability of the oscillator is improved. Also, if emitter current were flowing through the feedback section of the tank coil, there might be a requirement for L1 to be designed differently than L2. These problems are not encountered with the shunt fed Hartley oscillator.

The functions of the components shown in figure 15-5 are essentially the same as those of figure 15-3. In addition to providing decoupling between the power supply and the transistor, the radio frequency choke (RFC) also acts as the collector load for Q1. The RFC develops the variations in collector voltage. These variations are coupled to the feedback coil L1 by the isolation capacitor C3. The value of C3 partially

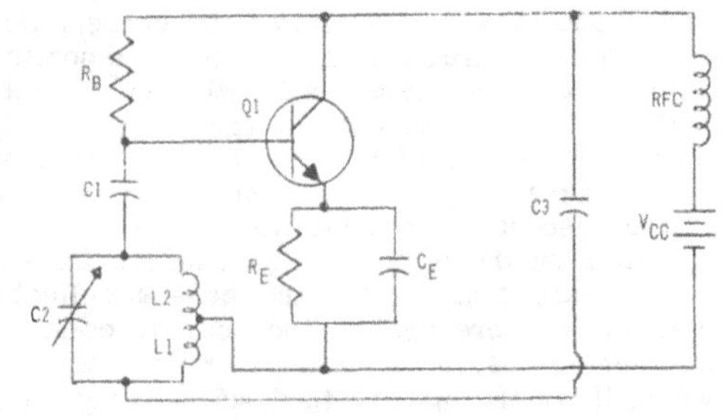

179.302

Figure 15-5.—Shunt fed transistor Hartley oscillator.

determines the amplitude of the feedback signal. If the value of C3 is relatively large its X_C will be low. A low X_C will cause very little of the feedback signal to be dropped across C3 and, therefore, will provide more feedback to the resonant circuit. A relatively small capacitive value for C3 will cause a larger value of reactance and will reduce the amplitude of the feedback signal to the tank. C3 has a second purpose in that it also acts as an isolation capacitor to prevent the d.c. component from flowing through the feedback coil.

There is a continuous path for d.c. bias current through R_E, the base-emitter junction, R_B and RFC. This allows an operating point to be established for class A operation. It also insures the starting of the oscillator.

Operation of the Shunt Fed Hartley Oscillator

When the oscillator is first turned on there will be relatively high currents in the circuit due to the uncharged condition of the capacitors. While these surge currents are large, they will nevertheless be limited to reasonable values by the inductance of RFC. The initial charge path for C3 is of special interest. The path begins at the negative terminal of V_{CC}, continues through L1, C3 and the RFC, and ends at the positive terminal of V_{CC}. The important factor to note here, is that the charging current for C3 passed through L1. In doing so, this current has provided energy to the resonant circuit. By normal flywheel action the tank circuit will now begin to oscillate.

POSITIVE ALTERNATION.—As the top of the tank circuit goes more positive with respect to the bottom, the positive potential is coupled through C1 to the base of Q1. This increases forward bias, and transistor conduction increases. As collector voltage decreases, C3 must discharge. Its discharge path will be through L1, R_E and Q1 and, as such, will aid the potential already present in the tank circuit. The energy which is introduced into L1 by the discharge of C3 will further the positive potential at the top of the tank circuit. This further increases forward bias and the transistor conduction. This, in turn, causes a further discharge of C3. This regenerative action will continue until such time as the combined actions of transistor non-linearity, the effect of R_E and the d.c. resistance of L1 cause the rate of change of current to decrease. As the rate of change of current through L1 decreases, less energy will be supplied to the resonant circuit and, through normal tank action, the positive potential at the top of the tank circuit will begin to decrease. This causes a decrease in forward bias which, in turn, causes the collector current to decrease, collector voltage begins to rise, and C3 begins to recharge.

NEGATIVE ALTERNATION.—As the tank circuit goes through its negative alternation the negative potential is coupled to the base of Q1. This has the effect of reducing forward bias. As forward bias decreases, transistor conduction decreases and collector voltage increases (goes more positive). C3 couples the positive going signal from the collector of Q1 to the bottom of the tank circuit. This increases the potential across the resonant circuit, and forward bias is further reduced. This is regenerative action. Oscillations will now continue as long as sufficient energy is provided to the tank circuit to replace internal power losses and energy lost due to external power drains (energy coupled to other circuits).

FAILURE ANALYSIS FOR TRANSISTOR HARTLEY OSCILLATOR

The following is a discussion of some of the most common failures of transistor Hartley oscillators.

No Output

A lack of oscillation may be due to a shorted or an open-circuited transistor. Deterioration with age causing a lack of transistor gain may result under high-temperature conditions. Shorting of the collector and base blocking capacitors will short-circuit the biasing arrangement through the tank coil and prevent operation. Open-circuit conditions of the biasing resistor (R_B) may stop oscillation, though it is more likely that reduced output will result. An open R_E would definitely stop oscillation. Failure of the RFC would either prevent the transistor from obtaining its required operating voltages or, in the case of a short, would shunt the RF to ground through the power supply capacitors. A shorted condition of the tuning capacitor will stop oscillations, and it cannot be detected by a continuity check without first disconnecting the tank coil. When trouble-shooting with test equipment containing line filter capacitors, care should be exercised to prevent application of excessive voltage to the transistor by using a common ground on both the transistor and test chassis. Use high-impedance meters to avoid placing a d.c. shunt or return path in the circuit and causing improper current flows or voltage distributions.

Reduced or Unstable Outputs

Instability should be resolved into one of two types—frequency or amplitude. Frequency instability will most likely result from poor tank circuit connections, poor insulation between turns of the tank coil, or changes in L or C values. Also, changes in the supply voltage will produce frequency changes because of changes in the operating point of the transistor and changes in the internal capacitance of the transistor with different applied voltages. Excessive bias will probably cause a reduction in output and will most likely be produced by an increase in the value of the bias resistor or opening of the emitter bypass capacitor, with the consequent production of degeneration. Temperature changes are usually evidenced by increased current in the collector circuit, and can be caused by a shorted emitter resistor.

Incorrect Output Frequency

Normally, a small change in output frequency can be compensated for by realigning the variable component of the LC resonant tank circuit, assuming that all parts of the circuit are known to be operating satisfactorily. A change in transistor parameters will also affect frequency. Increased collector voltages will reduce the collector-base capacitance, and increased emitter current will increase the collector-base capacitance. However, these effects are not equal and therefore do not compensate each other. Reflected load reactances can cause a change in frequency depending on the type and degree of coupling between the oscillator and its load. Changes in distributed circuit capacitance or in the tank inductor will cause frequency changes. A comparison of operational indications against operational standard values will generally indicate the area of fault. For example, frequency changes that vary with power supply voltage fluctuations indicate that the power supply must be regulated and that the trouble is not in the oscillator. Major frequency changes will involve the transistor elements and components associated with the tank circuit, since they constitute the major frequency-determining portion of the circuit.

GRID LEAK BIAS

Grid leak bias is used almost universally in triode oscillators rather than fixed bias. The use of grid leak bias makes the oscillator self starting and provides more stable operation than fixed bias. This type of bias is very suitable for an application where it is desired to have zero bias under no signal conditions and a negative bias when an input signal is applied. This method of bias commonly appears in one of two forms: SERIES or SHUNT grid leak bias. Before entering into a discussion of the operation of oscillator circuits employing grid leak bias, a brief review of RC time constants will prove advantageous.

Charge Time Constant

If a potential is applied across a capacitor, the capacitor will charge almost instantaneously to the value of the applied potential. However, if a circuit contains resistance in series with the capacitor, the charge time will be altered. The time involved for a series resistor and capacitor combination to reach 63.2% of its maximum charge is called the TIME CONSTANT (TC) of the circuit. The mathematical relationship of the resistance, capacitance, and time constant is expressed in the equation: $TC = R \times C$, where TC is the time constant in seconds, R is the circuit resistance measured in ohms, and C is the circuit capacitance measured in farads.

For most practical applications a capacitor can be assumed to have reached full charge (99.3%) after five time constants have elapsed (5TC). The significance of the RC charge time constant, as regards grid leak bias and oscillators, may be examined by use of the partial circuit in figure 15-6.

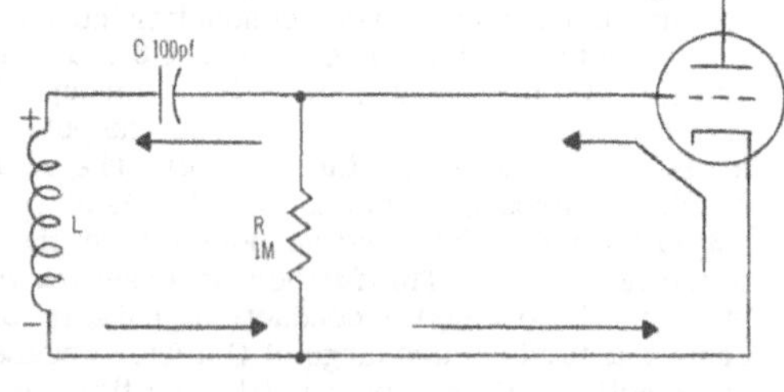

179.303

Figure 15-6.—RC charge time.

Assume that a potential is applied to L having the polarity as shown in figure 15-6. Current will be caused to flow in the direction indicated. Since the capacitor appears as a short at the first instant the full potential of the source (L) will appear across the parallel combination of resistor R and the grid to cathode impedance of the tube. This voltage drop will cause the grid to assume a positive potential and draw grid current. The magnitude of the grid to cathode impedance (while grid current is being drawn) may be on the order of 1 kohm. Since R is 1 megohm, the total impedance of the parallel combination is, for all practical purposes, 1 kohm. The time constant of this circuit during charge would then be:

$$TC = R \times C = (1 \times 10^{3}) \times (100 \times 10^{-12}) = 1 \times 10^{-7} \text{ sec.}$$

The capacitor will assume a full charge in 5 time constants. Therefore, the full charge time will be: 5TC or 5×10^{-7} seconds.

Thus, the capacitor will assume a full charge in approximately one half a microsecond. From the preceding it can be seen that if a sine wave (up to 500 kHz) were applied to L, the rate of the charge on the capacitor would very nearly follow the rise of voltage across L. Figure 15-7 shows the circuit with a sine wave input signal. From zero degrees to 90 degrees of the input cycle the capacitor is charging. The current path and polarities are as shown. As stated, the charge time is so small as to allow the capacitor charge to follow the voltage rise.

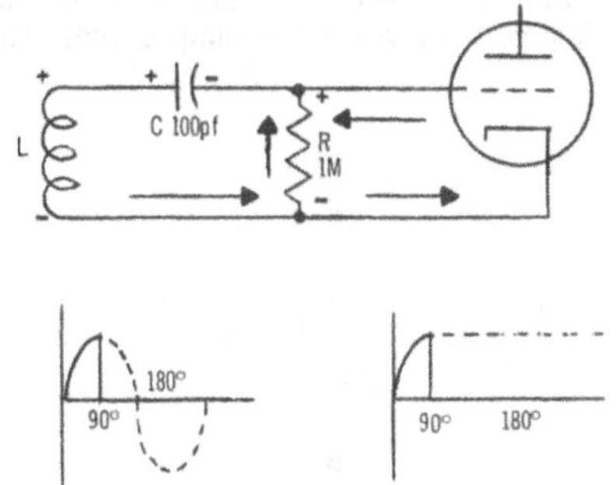

179.304

Figure 15-7.—Sine wave input (0° to 90°) capacitor charge.

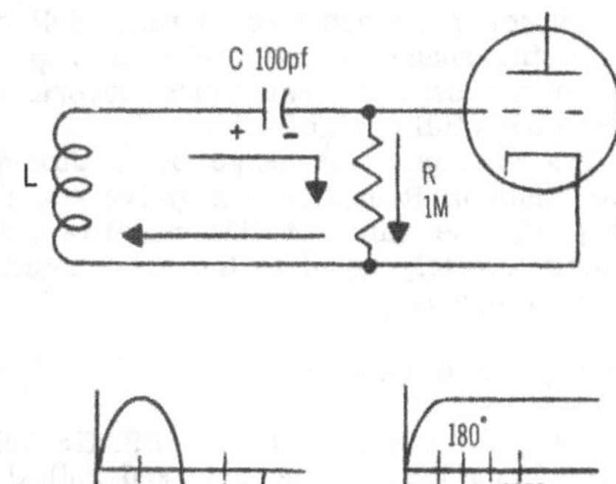

179.305

Figure 15-8.—Sine wave input (90° to 360°) capacitor discharge.

Discharge Time Constant

Figure 15-8 shows the conditions existing in the circuit between 90 and 360 degrees of the input signal. It is assumed that the capacitor charge reached maximum at the same time the input signal reached its positive peak (90 degrees). As the input signal passes 90 degrees and becomes less positive, the potential on L becomes less than the charge on the capacitor. The capacitor will begin to discharge. However, since the tube does not allow current flow from grid to cathode, the low resistance path between grid and cathode does not exist for the discharge current. Instead, the path for the capacitor discharge current is through the 1 megohm resistor, transformer winding L, and back to the capacitor. The time constant of the circuit during discharge is:

$$TC = R \times C = (1 \times 10^{6}) \times (100 \times 10^{-12}) = 1 \times 10^{-4} \text{ sec.}$$

As 5 time constants are required for a complete discharge, the total discharge time is equal to 5×10^{-4} seconds.

Thus, the charge time of the circuit is a half microsecond while the discharge time is 500 microseconds. If the frequency of the input signal were 500 kHz, one cycle would take approximately 2 microseconds. The capacitor will be unable to discharge a significant amount of its charge prior to the next positive peak. The graph of voltage across the capacitor (fig. 15-8) shows that the capacitor will rapidly charge to the peak value of the input signal and then maintain this value. NOTE: At this point in the explanation the capacitor is assumed to reach full charge within the space of one cycle. In a practical circuit the capacitor will take several cycles to acquire a full charge.

The attempted discharge of C through the grid resistor R applies a negative potential to the grid. The value of this negative potential is approximately equal to the peak amplitude of the input signal.

Development of Grid Leak Bias

Bias is defined as the AVERAGE value of d.c. voltage between the grid and cathode of a tube. If the average potential between grid and cathode of a tube is zero, the bias will be zero. Application of a sine wave will cause the instantaneous value of grid voltage to vary but the average grid voltage, or bias, will remain at zero. This is shown by use of a partial schematic in figure 15-9.

Part A of the figure shows the input circuit of a tube with a coil connected between the grid and cathode. The coil is the secondary of a transformer, thereby permitting an external signal to be coupled into the circuit. A 20 volt peak signal is applied across the coil. Part B of figure 15-9 illustrates a plot of the voltage between grid and cathode. Notice that the average voltage varies between plus 20 and minus 20 volts.

The addition of a resistor and capacitor in series with the coil will alter the bias. It was previously shown that the capacitor will charge rapidly to the magnitude of the applied voltage, and (for all practical purposes) maintain this charge as long as the magnitude of the applied voltage is not changed. This application of a 20 volt peak sine wave to the coil in figure 15-8 will cause the capacitor to charge to 20 volts with the polarity as shown. The capacitor will attempt to discharge through the large value grid resistor, and thereby produce a negative potential, -20 volts, on the grid of the tube.

In order to better understand the operation of the bias circuit when a signal is applied, figure 15-8 is redrawn as an equivalent circuit in figure 15-10. Since the voltage between grid and cathode is the same as that across R, the output of the equivalent circuit is taken across R. The 20 volts across the capacitor is assumed to stay relatively constant while the voltage across the coil varies from 20 to -20 volts. The output voltage, e_c, is graphed in part B of figure 15-10.

At the instant of time zero, the output voltage, e_c, is the algebraic sum of the capacitor voltage and the coil voltage. Thus, at t_0 the instantaneous grid voltage (e_c in fig. 15-10) is -20 volts. As the input signal increases in the positive direction a voltage is developed across the coil, with a polarity of positive at the top (capacitor end) with respect to the bottom of the coil. The polarity of the coil voltage (during the positive half cycle of the input signal) is such as to subtract from the capacitor voltage.

At t_1 the capacitor voltage is still -20 volts, but the input signal has now developed 20 volts of the opposite polarity across the coil. Thus, at t_1, the voltage e_c across the grid resistor, is zero volts.

At t_2 the input signal is again passing through zero. There is no voltage present across L, and the output voltage is again equal to the capacitor voltage (-20 volts).

As the input signal goes through its negative half cycle the polarity of the induced voltage across the coil is such as to aid the capacitor voltage. Therefore, when the input signal attains its

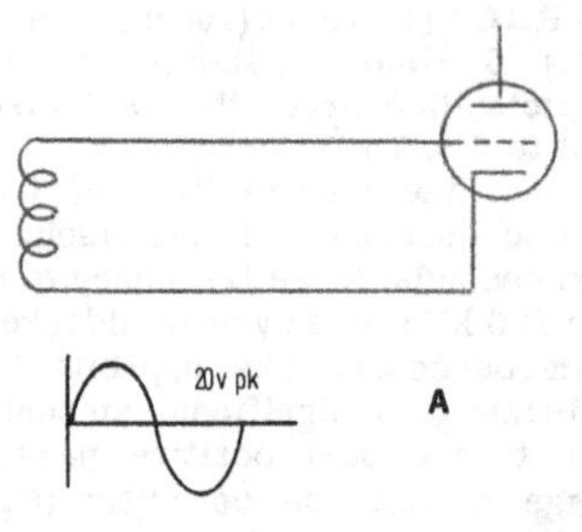

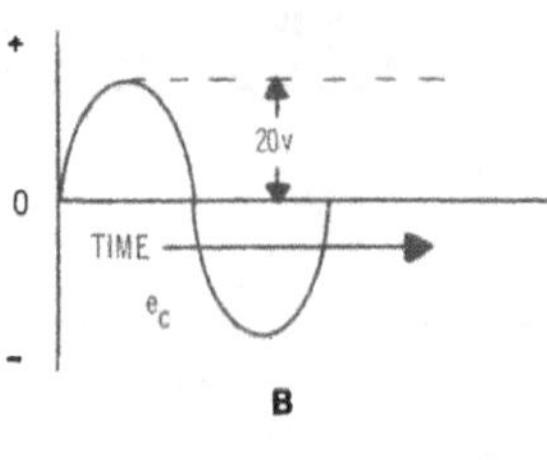

179.306
Figure 15-9.—Zero bias.

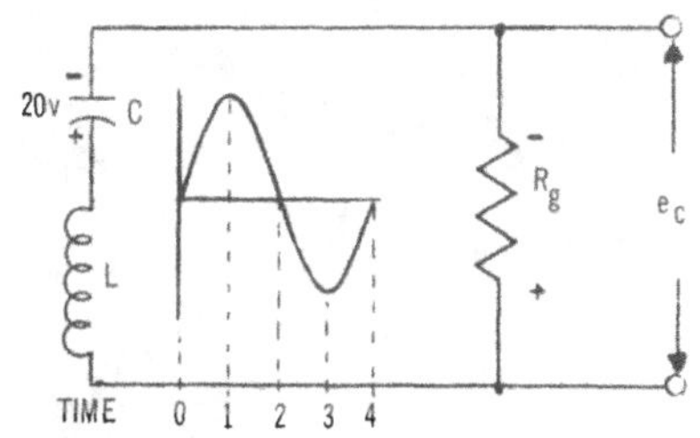

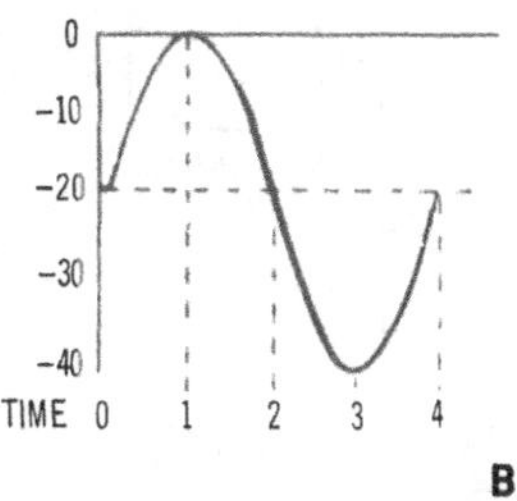

179.307

Figure 15-10.—Equivalent circuit for grid leak bias.

maximum negative peak of -20 volts at t_3, the e_c will be -40 volts. At t_4 e_c will again be equal to -20 volts.

It can be seen from the above that the addition of the resistor and capacitor combination has caused the average voltage between grid and cathode (bias) to become -20 volts.

It has been assumed, during this explanation, that the capacitor was capable of maintaining a constant voltage. In actuality the capacitor will lose a small percentage of its charge between the positive peaks of the input signal. This lost charge is replaced by driving the grid SLIGHTLY positive for a brief interval during each positive peak of the input signal. For instance, assume that between t_1 and t_4 (fig. 15-10) the capacitor voltage decreases to -19 volts. The positive peak of the next cycle of the input signal will cause L to develop 20 volts. The instantaneous grid voltage will then be +1 volt. During this interval the grid is driven 1 volt positive; grid current will be drawn, and the capacitor will again be charged to 20 volts (as explained previously).

Therefore, the bias voltage will continue to be approximately -20 volts as long as the magnitude of the input signal does not vary. If the input signal were to increase in amplitude, the capacitor (in a few cycles) would charge to the new value and then maintain this new voltage as the bias voltage. On the other hand, if the input signal were to decrease in magnitude the capacitor charge would decrease (in a few cycles of operation) to the new value and maintain this as the new bias. Under proper operation grid leak bias will maintain bias relatively near the peak value of the input signal.

SHUNT GRID LEAK BIAS.—Figure 15-11 shows the grid circuit of an operating oscillator. It is assumed that the necessary excitation and feedback circuits are available, but are not shown for the sake of simplicity. This circuit utilizes SHUNT grid leak bias.

When the tank voltage becomes more positive than the charge on C_{g1}, grid current will be drawn and C_{g1} will charge. When tank voltage is less positive than the charge on C_{g1} the capacitor will discharge through resistor R_{g1}, causing a negative potential to be applied to the grid. Shunt grid leak bias derives its name from the fact that the grid leak resistor is in parallel, or shunt, with the tank circuit.

SERIES GRID LEAK BIAS.—Figure 15-12 illustrates a simplified oscillator grid circuit

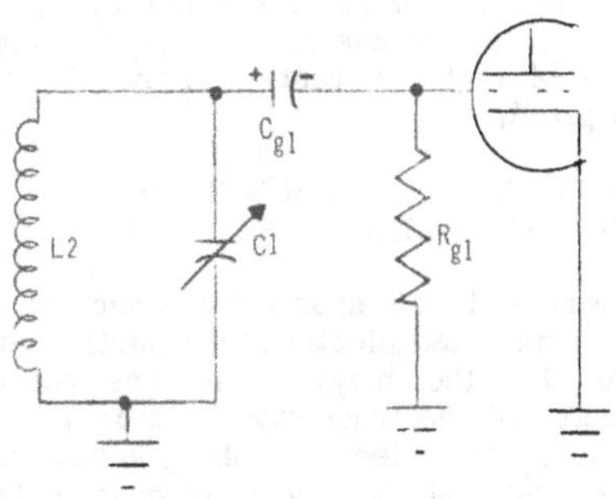

179.308

Figure 15-11.—Simplified circuit of an oscillator using shunt grid leak bias.

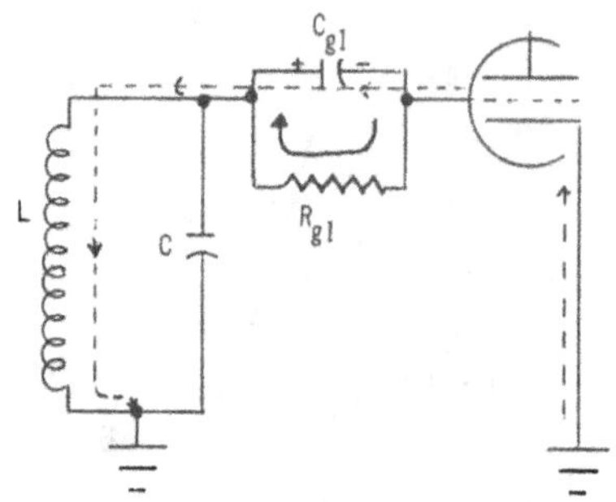

179.309

Figure 15-12.—Series grid leak bias.

which utilizes SERIES grid leak bias. The operation is similar to the shunt type, differing only in discharge path of C_{g1}. The charge and discharge paths are shown in figure 15-12. The charge path is dotted and the discharge path is the solid arrow. Series grid leak bias is so named because the grid leak resistor is in series with the tank circuit.

One of the advantages of grid leak bias for oscillators is that the circuit is self starting. This is possible because when the circuit is first energized C_{g1} has no charge. Therefore, bias is zero and plate current can flow. This will initiate feedback and start oscillations.

Another advantage is that the circuit is self adjusting. This is due to the bias level being dependent on the input signal amplitude. One disadvantage of grid leak bias is due to the zero bias with no applied signal. Should a malfunction occur in the tank circuit or bias network, the bias will become zero, causing a steady large value of plate current until the circuit is deenergized.

SERIES FED ELECTRON TUBE HARTLEY OSCILLATOR

Figure 15-13 shows the schematic diagram of a series fed electron tube Hartley oscillator. Note that the major difference between this circuit and the transistor version is the biasing method. The electron tube Hartley oscillator uses grid leak bias and, of course, has much higher operating voltages than the transistor circuit. However, component purposes and circuit operation are quite similar to the transistor Hartley oscillator.

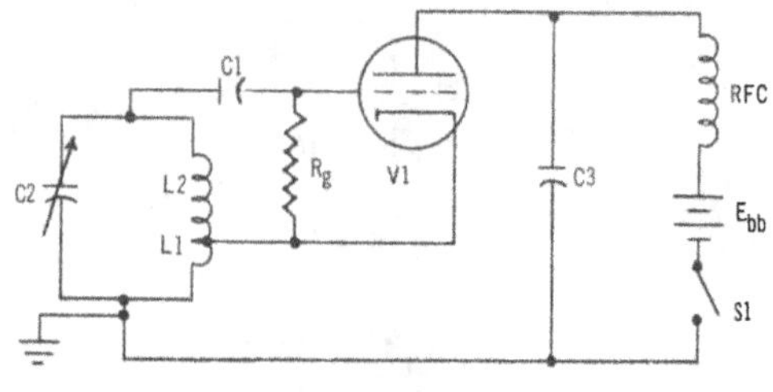

179.310

Figure 15-13.—Electron tube series fed Hartley oscillator.

Capacitor C3 and the RFC provide decoupling between the power supply and the oscillator. Although not required for operation of the oscillator itself, C3 and RFC improve the stability of the oscillator by minimizing variations in plate voltage and current. In this respect they act much in the same manner as the "L" type filter discussed in the chapter on power supplies. Any tendency for the plate voltage to change as a result of a change in plate current will be suppressed. The choke offers a high resistance to changes in current, and the capacitor presents a low reactance to changes in plate voltage. When properly designed, the decoupling network greatly minimizes the effects of plate voltage and current changes on the operation of the oscillator.

A close examination of figure 15-13 shows that both the direct and alternating components of plate current flow through L1. Since the d.c. plate current flows through a portion of the tank circuit the oscillator is series fed. Capacitor C1 in conjunction with grid resistance R_g provides grid leak bias, the value of which determines the operating point of the tube. The tank components are L1, L2, and C2.

Consider that the filaments of V1 are at operating temperature but switch S1 is open. When S1 is closed the following events occur:

1. The tube has no initial bias, therefore, the tube will begin to conduct causing a current flow from ground, through L1, V1, RFC, and back to the power supply. At the same time capacitor C3 is assuming a charge with a negative potential on the bottom plate and positive on the top plate.

2. The current flow through the feedback component, L1, induces a voltage in L2. The

induced voltage in L2 causes a current to flow in the tank which begins to charge tank capacitor C2 with a negative potential on the bottom plate and a positive potential on the top plate.

3. The voltage across the tank is essentially the charge across C2. Therefore, grid current will be drawn, to charge capacitor C1, as the top of the tank assumes a positive potential.

4. The regenerative cycle of increasing grid voltage causing increased plate current will continue until limited by the non-linear characteristics of the tube.

5. As the rate of change of current through L1 decreases, the induced voltage of L2 decreases. When the value of induced voltage falls below the value of charge on capacitor C2, the tank capacitor will begin discharging.

6. The reduction of magnitude of the positive potential at the top of the tank will cause capacitor C1 to attempt to discharge through grid resistor R_g, thus applying a negative potential to the grid of V1.

7. At the time when C2 is completely discharged, circulating tank current will be maximum, and the tank voltage will have completed a half cycle of operation. The magnetic fields of L1 and L2 will now collapse, maintaining current flow in the same direction and subsequently charging C2 with the opposite polarity (negative on the top plate and positive on the bottom plate).

8. At the instant the magnetic fields are completely collapsed the tank current will be zero and the voltage across C2 will be maximum. The tank circuit has now completed three-fourths of a cycle.

9. Tank capacitor C2 will begin discharging, causing tank current to reverse direction. The discharge of C2 will establish a magnetic field around L1 and L2. When C2 is completely discharged, tank voltage will be zero, an entire cycle of oscillator operation has been completed, and the circuit is ready to begin the next cycle.

In order to maintain the oscillations as explained in the preceding section, it is necessary that the tube conduct at the correct time so that the energy can be resupplied to the tank circuit via L1. A careful look at the schematic diagram will show that during the positive half cycle of the tank's operation, the positive potential which occurs at the top of the tank will oppose and reduce the effect of grid leak bias. If the bias is already sufficiently small, the positive signal from the tank circuit will be sufficient to bring the tube out of cutoff and into conduction. (It is assumed that the oscillator appears to be operating class C.) It can be easily seen that tube conduction on the positive alternation of the tank circuit would be regenerative and tend to sustain oscillations.

SHUNT FED ELECTRON TUBE HARTLEY OSCILLATOR

The main disadvantage of the series fed Hartley oscillator is the fact that a relatively large value of direct current flows through the feedback portion of the tank coil. This disadvantage can be overcome by a variation of the connection of the feedback coil as shown in figure 15-14. When connected in this manner the feedback coil is isolated from the d.c. component of plate current by capacitor C3. The oscillator derives its name from the fact that the feedback component (L1) is connected in shunt rather than series with the power supply. The operation of the electron tube shunt fed Hartley is very similar to that of the transistor shunt fed Hartley which was previously covered.

It will be assumed that the oscillator has been in operation for a period of time and that all capacitors have assumed their normal operating charges. It will further be assumed that at the instant the explanation begins, the oscillator tube is cut off by the action of grid leak bias, the tank voltage is at the zero point of its cycle, the energy of the tank is stored in the magnetic field of the tank coil, and the positive alternation is about to begin.

During the positive alternation of the tank voltage the grid-to-cathode potential becomes less negative. This is due to algebraic addition of the tank voltage and the charge on C1. Plate current will flow when this negative potential becomes less than that value required for tube cutoff. The increase in plate current causes a decrease in plate voltage.

Since the value of plate voltage is now less than E_{bb}, C3 will start discharging through L1, increasing the tank voltage, causing the grid voltage to become even less negative. This action is cumulative and the plate current soon reaches a maximum value as determined by the non-linear characteristics of the tube. At this time the tank voltage will be maximum positive and the plate voltage will be minimum. C3 no longer discharges; thus, there is no change in current flow through L1 and no voltage induced in L2. C2 now starts to discharge causing the tank voltage to decrease. This allows the grid to become more negative, plate current will

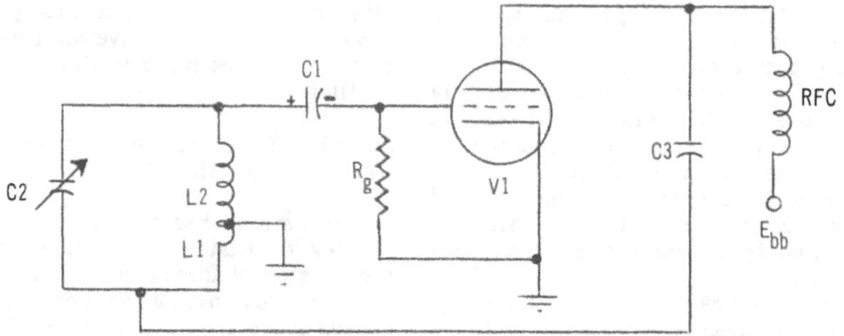

179.311

Figure 15-14.—Shunt fed electron tube Hartley oscillator.

decrease, and plate voltage will increase. C3 now recharges to the increased value of plate voltage. The charge current of C3 flows through L1 causing a voltage drop which further reduces the tank voltage.

When the positive tank voltage decreases to a point where the instantaneous grid-to-cathode potential becomes negative enough to cut the tube off, plate current stops, and the plate voltage rises to E_{bb} potential. C3 charges to E_{bb} and the feedback cycle stops. The tank continues to oscillate due to flywheel action.

In addition to providing decoupling, the RFC acts as a plate load device for V1. It develops the variations of plate voltage which C3 couples back to the tank circuit. The size of C3 will play a part in determining the amplitude of the feedback. If the size of C3 is increased, X_C will be reduced, less voltage drop (a.c.) will occur across it, and feedback amplitude will increase. If the size of C3 is decreased, its X_C will be larger and the feedback amplitude will decrease due to an increased a.c. drop across it.

It can be seen from figure 15-14 that the path for plate current is from the cathode to the plate, through RFC, through the power supply, and back to the cathode of V1. Thus, the plate current does not flow through any part of the tank circuit.

FAILURE ANALYSIS FOR ELECTRON TUBE HARTLEY OSCILLATORS

The following is a discussion of some of the most common failures of electron tube Hartley oscillators.

General

When the circuit is in a non-oscillating condition, negative grid bias cannot be developed; excessive plate current will flow; and plate voltage will be reduced to a below normal value. Also, unless the power supply is well regulated the additional current drain will cause a reduction in its output voltage.

Excessive circuit losses in the resonant tank circuit will prevent sustained oscillation. Changes in value of the grid leak components will directly affect the operating bias, and hence the class of operation and overall gain of the tube. Such changes, if substantial, may cause a loss of oscillation. Too high a value of grid leak resistance may cause intermittent operation or "motorboating." Shorted turns of the oscillator coil(s), in addition to affecting output amplitude and frequency, may cause loss of oscillation because of loading effects. A leaky C2 may also load the oscillator sufficiently to stop operation. In the shunt fed circuit, a defective RFC or C3 may stop oscillation since the oscillator is dependent on these components for development and application of feedback voltages.

Reduced Output

A relative indication of oscillator output is provided by the amount of bias voltage developed across the grid resistor. Variation from the normal value is indicative of abnormal operation. A reduction in applied plate voltage will cause a reduced output. Therefore, an unregulated voltage source will produce output amplitude variations and probably frequency changes or instability. Losses in feedback, due to shorted turns of poorly soldered connections may cause reduced output or unstable operation. A leaky C3 (plate capacitor) may cause reduced or unstable output by loading the oscillator or by reducing plate voltage by adding to the normal current flow through a series resistance. Care should be used in selecting a replacement for a defective RF choke, since an improper replacement choke may cause unwanted oscillations by resonating with distributed circuit capacitances or with the distributed capacitance of its own windings. Similar care should be exercised in replacing a defective tube in those oscillators operating in the higher frequency ranges where interelectrode capacitances may constitute a considerable portion of the tuned circuits.

Variations of this physical capacitance from one tube to another may, in addition to affecting output frequency, cause the oscillator to shift from one type of operation to another as the oscillator is tuned through its frequency range. This shifting may cause frequency jumps and/or dead spots in the tuning range of the oscillator. At the higher frequencies it is good practice to try more than one replacement tube if the first substitution does not achieve the desired results in frequency of operation and stability. Realignment of circuit components to compensate for a tube substitution should be avoided wherever possible.

Incorrect Output Frequency

Normally, a small change in output frequency can be compensated for by realigning or adjusting the variable component of the LC resonant circuit, assuming that all component parts of the circuit are known to be satisfactory. Changes in distributed circuit capacitance or reflected load reactance will affect the frequency to some extent. Thus, an increase in capacitance will lower the frequency, and a decrease in capacitance will increase the frequency. Therefore, care must be used in the removal and replacement of parts in order not to disturb the distributed capacitance of the circuit which is inherent in the placement of physical parts and the wiring of the circuits. Extreme changes in room temperature may affect the operating frequency of the oscillator.

COLPITTS OSCILLATORS

The Colpitts oscillator is used to produce a sine-wave output of constant amplitude and fairly constant frequency within the RF range and occassionally within the audio range. The circuit is generally used as a local oscillator in receivers, as a signal source in signal generators, and as a variable frequency oscillator for general use over the low, medium, and high frequency ranges.

The Colpitts oscillator uses an LC circuit to establish its frequency of operation. Inductive rather than capacitive tuning is employed, and feedback is obtained through a capacitive voltage divider circuit. It is operated similar to class C, where waveform purity is not important, and class A, where good waveform linearity is required. Its frequency stability is good (considered better than that of the Hartley at the lower and medium frequency). The Colpitts oscillator oscillates easily at the higher frequencies (inductive feedback oscillators, such as the Hartley often have difficulty obtaining sufficient feedback at higher frequencies).

TRANSISTOR COLPITTS OSCILLATOR

Figure 15-15 shows the schematic diagram for the transistor Colpitts oscillator. Notice that the tank circuit is composed of a variable tank inductive (L1) and tank capacitances C1 and C2. The voltage developed across C2 is the input signal to the transistor and the amount of feedback depends on the ratio between C1 and C2. For a specific amount of signal, the ratio of C2 and C1 should remain the same. Since it is easier to use fixed capacitors in this application, the Colpitts oscillator lends itself

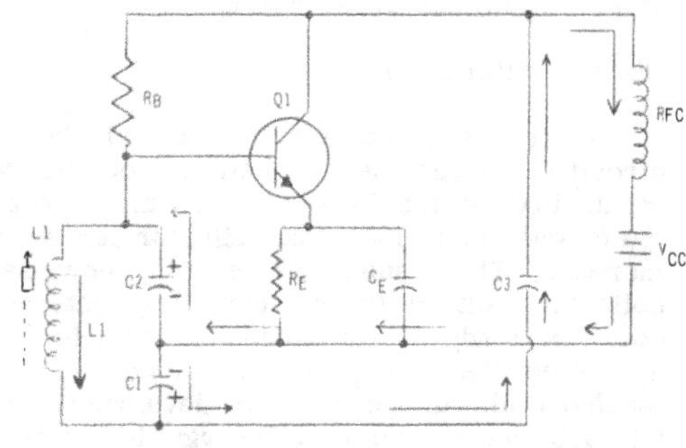

179.312
Figure 15-15.—Transistor Colpitts oscillator.

well to the use of permeability tuning (indicated by the arrow and dotted core of L1).

The transistor Colpitts oscillator is operated class A and the value of bias current necessary to establish the desired operating point is determined by the standard voltage divider, formed by R_E, the base-emitter junction, and R_B. It will be noted that no coupling capacitor is used between the base and tank circuit in figure 15-15. In the transistor Colpitts oscillator, d.c. isolation of the tank coil is provided by capacitors C1 and C3; therefore, no base coupling capacitor is required. The RFC in figure 15-15 acts as a collector load for Q1. Capacitor C3, as well as providing isolation, acts as the coupling agent for the feedback voltage. Notice also, that it would be impossible for a Colpitts oscillator to be series fed. As capacitive coupling is used in the feedback path, d.c. emitter current does not flow in any portion of the tank circuit. Therefore, all Colpitts oscillators will be SHUNT FED.

When the circuit is first turned on, collector voltage is applied to Q1 through the RFC, and bias current is developed through a path consisting of R_E, the base-emitter junction, R_B, and RFC. The forward bias for Q1 is found, as in the case of the Hartley oscillator by: $V_{BE} = V_{CC} - (V_{RB} + V_{RE})$.

At the same time C1, C2 and C3 will begin to charge. The charge paths are shown in figure 15-15. Note that a portion of this charging current flows through L1. As this charging current flows through L1, energy will be stored in the form of a magnetic field. The charging currents soon cease and the magnetic field of L1 will begin collapsing, setting up an oscillating current in the resonant circuit. This oscillating current will cause the polarity shown on C1 to reverse as the circuit starts oscillating.

Positive Alternation

During the positive half cycle of the tank circuit's oscillations, a positive potential will be applied to the base of Q1. This increases the forward bias of Q1, and collector current will increase. This causes a resultant decrease in collector voltage (recall that RFC acts as a collector load). The decrease in collector voltage is coupled through C3 to C1. The portion of the feedback voltage which is developed across C1 acts as a source of energy to re-supply energy to the tank circuit. It should be noted that the amount of feedback is controlled by the relative sizes of C1 and C3. It should also be noted that the feedback is regenerative. That is, the negative going feedback signal which is coupled to C1 will have the effect of increasing the potential across the tank, therefore, increasing the amplitude of the positive going signal on the base of Q1. This increases the forward bias and the collector current, decreases collector voltage, and continues the regenerative process.

This regenerative action will continue until such time as the combined actions of transistor non-linearity, the effect of R_E, etc., cause the rate of change of the feedback voltage to decrease. Through normal tank circuit operation the potential across the resonant circuit now begins to decrease. This causes a decrease in forward bias which, in turn, will cause collector current to decrease and collector voltage to increase.

Negative Alternation

As the oscillator is being operated class A, the decreasing potential at the top of the tank circuit will not be sufficient to reverse bias the base-emitter junction of Q1. But the negative going tank signal will decrease the forward bias on the base of Q1, and the collector current will be reduced. As the collector current decreases, the collector voltage will rise. As before, C3 couples this changing collector voltage to the bottom of the tank circuit. As the top of the tank circuit is going more negative, the positive going increase felt at the bottom of the tank circuit again increases the effective voltage across the tank. This, again, is a regenerative action and tends to sustain oscillations in the tank circuit.

Failure Analysis

The following is a discussion of the most common failures of transistor Colpitts oscillators.

NO OUTPUT.—Although oscillation may fail because of a shorted or open-circuited transistor, deterioration with age similar to the decrease of conduction in electron tubes does not normally occur. High-temperature operation may cause premature failure or drop off in performance of the transistor; however, this can normally be avoided by proper ventilation of the equipment. Reflected reactance from heavy loading due to tight coupling may occur and prevent oscillation, but it should not occur with properly designed oscillators. Changes in the value of R_B will probably reduce the output rather than prevent oscillation completely. Shorting of either of the tank capacitors would prevent oscillation as would a malfunctioning tank inductor.

REDUCED OR UNSTABLE OUTPUT.—As in the oscillators discussed previously, it is important to determine whether the instability is associated with frequency or amplitude. When amplitude instability is caused by temperature effects, the biasing network is probably at fault. Reduced output will most likely be due to excessive biasing caused by changes in the value of the bias resistors (R_B and R_E) or to excessive degeneration caused by an open C_E in the emitter bypass circuit. A leaky feedback coupling capacitor (C3) will cause a change in bias by reflecting, through the tank inductance, a parallel resistance between the base and collector elements of the transistor.

Frequency instability will indicate that the tank circuit or supply voltage is at fault. Although this circuit minimizes capacitance changes and supply voltage changes, it should be noted that a varying supply voltage also changes the transistor's operating point and may therefore affect the frequency to some extent. As in the other oscillators, poor mechanical connections and shorted turns or deteriorated insulation in the tank circuit may cause unstable operation.

INCORRECT OUTPUT FREQUENCY.—As in the other oscillators discussed previously, small changes in frequency can be corrected by adjusting the tuning device, assuming that all components are in good condition. Variations of frequency with supply voltage changes indicate trouble in the tank circuit since it is the primary frequency determining circuit.

Selection of Configurations

It is possible to use any of the three basic transistor configurations (CC, CB, CE) in the construction of an oscillator. However, certain considerations in the application of the circuit (high operation frequency, large output power required, etc.) usually make the selection of one configuration more desirable than another.

COMMON-COLLECTOR CONFIGURATION.—Since there is no phase reversal between the input and the output circuits of a common-collector configuration, it is not necessary for the feedback network to provide a phase shift. However, since the voltage gain is less than unity and the power gain is low, the common-collector configuration is very seldom used in oscillator circuits.

COMMON-BASE CONFIGURATION.—The power gain and voltage gain of the common-base configuration are high enough to give satisfactory operation in an oscillator circuit. A transistor used in the common-base configuration exhibits better high frequency response than when used in the common-emitter configuration. Therefore, the oscillator will usually be of the CB type when the operating frequency is equal to or greater than the alpha cutoff frequency of the transistor.

It has been stated that a transistor produces no phase shift when used in the CC and CB configurations and 180 degree phase shift when used in the CE configuration. This is true ONLY when the transistor is operated far below its alpha cutoff. The transistor introduces an additional phase shift when operated above alpha cutoff. The additional phase shift can be as much as 58 degrees. Since CB oscillators are used when the operating frequency is at, or beyond, the alpha cutoff frequency, this phase shift must be compensated for by the feedback network.

COMMON-EMITTER CONFIGURATION.—Due to the high power gain of the CE configuration and the relatively low frequency of operation (in relation to the 2-10 MHz alpha cutoff of transistors used at broadcast band frequencies), the CE configuration is most often used for receiver oscillators.

In order for energy to be fed back from the output, and be in phase with the energy at the input, the feedback of a CE oscillator must provide a phase shift of approximately 180 degrees. Usually a CE transistor oscillator is operated at a frequency sufficiently below the alpha cutoff. Therefore, the additional phase shift is negligible. However, if an additional phase shift, due to the transistor, is present, it must be compensated for by the feedback network. An additional advantage of the CE configuration is the moderate range between the input and output impedances which simplifies the job of impedance matching.

ELECTRON TUBE COLPITTS OSCILLATOR

The Colpitts electron tube oscillator, which is illustrated in figure 15-16, is similar to the electron tube Hartley oscillator. The primary

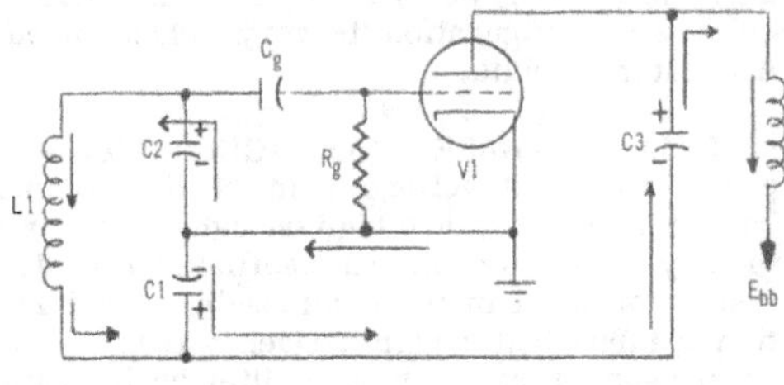

179.313
Figure 15-16.—Electron tube Colpitts oscillator.

difference is in the type of feedback used. The Colpitts oscillator uses a split tank capacitor rather than a tapped inductor. Capacitors C1 and C2 act as a capacitive voltage divider across the tank circuit. Since the voltage across C2 is the input signal to the tube, the amount of feedback depends on the ratio of C1 and C2. Capacitor C3, which is the feedback coupling capacitor, also determines the amplitude of the feedback voltage to some degree.

The tank circuit is composed of L1, C1 and C2. The feedback path is through C3 and C1. The RFC provides decoupling and develops the changes in plate voltage necessary for feedback. C_g and R_g form a shunt grid leak bias network. Shunt grid leak bias must be used in order to provide a d.c. path from the grid to cathode. In the case of the Hartley circuit, the feedback was applied to a part of the tank inductor. In the Colpitts, the feedback will increase the charge on the tank capacitor. In the Hartley, the first cycle of oscillation started with the initial surge of tube current. In the Colpitts, the first cycle starts with the collapse of the magnetic field of the tank inductor L1. The operation of the Colpitts is shunt fed as there is no d.c. current flow through the tank circuit.

In order to examine a cycle of operation, assume that V1's filament is warm and that plate voltage has not been applied (fig. 15-16).

1. Plate voltage is supplied to V1. Since bias is zero, plate current will start to flow. C1, C2, and C3 will begin to charge. The path of the charging currents and the resulting polarities are shown in figure 15-16.

2. The charging current produces a magnetic field around L1. When the capacitors complete their charge, the charging current ceases.

3. The magnetic field of L1 will start collapsing in an attempt to maintain current flow. The current caused by the collapsing field of L1 is now the circulating current of the tank circuit. The collapsing magnetic field and the discharge of C1 increase the positive potential on the grid side of C_g. Tank voltage is now starting its positive alternation.

4. The positive tank voltage causes plate current to increase and plate voltage to decrease. The decrease of plate voltage causes C3 to discharge into C1 which further increases the difference of potential across the tank circuit.

5. This action will continue until the field around L1 has completely collapsed. At this point tank voltage and plate current will stop increasing, and plate voltage will stop decreasing. The tank voltage is now at the positive peak of its first cycle.

6. Capacitors C1 and C2 start to discharge. Tank current reverses direction. The positive tank voltage begins to decrease causing plate current to decrease and plate voltage to increase. C3 starts to charge, aiding tank action.

7. This action will continue until the tube is cut off by the action of grid leak bias. At this time plate voltage is equal to E_{bb} (C3 is no longer charging) and the feedback cycle has stopped. Tank oscillation is now continued by the flywheel effect.

Note the similarity between this circuit, and the shunt fed Hartley, in the method of feedback. In both circuits, feedback is produced when a change in plate voltage is coupled through a capacitor to a tank circuit. The major difference lies in the tank component to which the feedback is applied.

Failure Analysis

The following is a discussion of some of the most common failures of electron tube Colpitts oscillators.

NO OUTPUT.—If the circuit is in a non-oscillating condition, the negative grid bias will be much less than it is in the oscillating condition, because it will consist of contact potential bias only, which will allow the tube to operate at approximately zero grid volts rather than at cutoff or higher. Thus the plate current will be much higher than normal. Excessive losses present in the resonant tank circuit will prevent sustained oscillations. Reduced tube gain, if sufficient, will also affect oscillation, but will not be as noticeable as in the inductive feedback circuits. Changes in value of grid leak components R_g and C_g will directly affect the operating bias, and change the amplitude of the oscillation; if the change is large, oscillation may be prevented. A defective radio frequency choke, or coupling capacitor C3, may cause loss of oscillation since the circuit depends upon these components for the development and application of the feedback voltage.

REDUCED OR UNSTABLE OUTPUT.—A relative indication of oscillation output is provided by the amount of bias voltage being developed across R_g. Variation from the standard operating value is an indication of abnormal operation. A reduction of applied plate voltage will cause a reduced output. Therefore, an unregulated voltage source will produce output amplitude variations and probably some frequency instability. Losses in feedback due to shorted turns, poorly soldered connections, or changing values of feedback capacitance can also cause reduced output or unstable operation. Care should be used in selecting a replacement for a defective RFC, as in improper replacement may result in unwanted oscillations, due to its distributed capacitances.

INCORRECT OUTPUT FREQUENCY.—Normally, a small change in output frequency can be corrected by adjusting the tunable component of the tank circuit. Changes in distributed capacitance of the circuit or reflected load reactances will affect the circuit operation to some extent, most noticeably in higher-frequency circuits. Care should be used in replacing a defective tube. It is a good practice to try more than one replacement tube if the first substitution does not give satisfactory results. Tank circuit realignment is to be avoided whenever possible. Large changes in ambient temperature may also affect the operating frequency by changing component values or circuit operating parameters.

CLAPP OSCILLATORS

Clapp oscillators are used to produce a sine wave of constant amplitude and frequency within the RF range. The circuit is normally used as a signal source in signal generators and as a variable frequency oscillator for general use over the high and very-high frequency ranges.

The Clapp oscillator uses a SERIES-RESONANT LC circuit to determine the frequency of operation. This is in contrast to other oscillator types which use PARALLEL resonant circuits for frequency determination. Feedback is obtained through a capacitive type voltage divider. Its frequency of operation is relatively independent of variations in transistor parameters.

The Clapp oscillator operates similar to class C, where waveform linearity is not a prime requirement, and class A where waveform linearity is important. Its frequency stability is better than that of the Colpitts oscillator.

TRANSISTOR CLAPP OSCILLATOR

The Clapp oscillator circuit is considered to be a variation of the Colpitts circuit discussed previously. While either of the three transistor configurations may be used, the common-emitter configuration is the most common and is shown in figure 15-17.

The circuit utilizes the stabilizing effect of a series-resonant tuned tank circuit, loosely coupled to the feedback loop, to provide good stability. It also offers capacitive tuning using only one capacitor, which will not affect the feedback ratio.

In the resonant tank circuit, tuning is accomplished by a capacitor (C3) which is essentially the sum of series capacitors C1, C2 and C3. Since the total value of series capacitors is always less than the value of the smallest capacitor, it can be seen that when C1 and C2 are much larger than C3, the effective tuning capacitance is essentially the capacitance of C3 alone. Thus, the series resonant circuit consisting of L and C3 is the basic frequency determining circuit.

Since the tank circuit is controlled by L and C3, it should be clear that the feedback voltage divider (which consists of C1 and C2) determines only the feedback amplitude and, therefore, changes of component values in the tank circuit will have a negligible effect on the amplitude of oscillations. The tuned circuit is designed to have a high Q under loaded conditions and, therefore, has a greater stability than the basic Colpitts oscillator.

Operation

The shunt fed Clapp transistor oscillator uses a single power supply, with fixed bias being supplied by R1 and R2. Emitter components R_E and C_E are used to prevent instability and degeneration. Collector current flows from the negative terminal of V_{CC}, through the RFC,

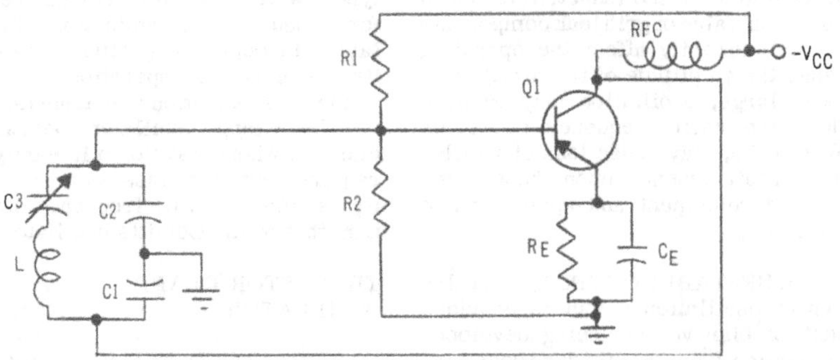

179.314
Figure 15-17.—Shunt fed transistor Clapp oscillator.

collector-emitter, and R_E, to ground. The RFC provides power supply isolation and acts as a collector load for Q1. A quick check of figure 15-17 will show that, as capacitive feedback is used, collector and base blocking capacitors are not required. If desired, operation similar to class C may be achieved in this circuit by providing sufficient feedback so that the amplitude of oscillations in the tank circuit will exceed the level of fixed bias provided. If these conditions are met, the emitter components (R_E and C_E) will operate in the same manner as the grid leak components of an electron tube circuit and provide a bias sufficiently high to maintain the transistor in cutoff for the greater part of each cycle.

When voltage is first applied to the circuit, the various capacitors will begin to charge. The most important of these charge paths begins at the negative terminal of V_{CC}, goes through the RFC, L, C3, and C2 to ground. This charging current will cease as soon as C3 and C2 are charged. Notice however, that this charging current produces a magnetic field around the tank inductor. As the charging current diminishes, the magnetic field about L collapses, and oscillations are started by normal tank action.

As the top of the tank circuit goes more positive with respect to ground, the forward bias of Q1 is reduced. This causes a resulting decrease in collector voltage. (Notice that V_{CC} is negative in this circuit, and, therefore, an increase in collector voltage would be a negative going signal.) The increasingly negative collector voltage is coupled to the bottom of the resonant circuit and has the effect of increasing the total voltage drop across the tank circuit. This makes the top of the tank circuit appear more positive, which will further decrease the forward bias and collector current. This regenerative action will continue until such time as the combination of transistor non-linearity, biasing voltages, and normal tank action reverses the process.

During the next half cycle a negative going voltage is applied to the base of Q1. This will increase the forward bias, increase collector current, and cause a corresponding decrease (positive going) in collector voltage. Notice that coupling a positive going signal to the bottom of the tank circuit again produces a regenerative effect. As mentioned before, a proper choice of R1, R2, and the emitter components R_E and C_E, will allow this circuit to be operated either class A or similar to C.

Failure Analysis

The following discussion treats some of the most common failures of transistor Clapp oscillators.

NO OUTPUT.—Since no d.c. blocking capacitors are required, and the tank circuit is essentially unaffected by changes in transistor parameters (such as junction capacitances), a lack of output is usually limited to a lack of feedback or to improper biasing voltages. An open or shorted transistor or an open or shorted feedback capacitor will stop oscillation. A shorted tuning capacitor will also stop oscillation because it will allow the d.c. collector voltage across inductor L to be shorted to the base of Q1. Poorly soldered connections may produce circuit losses which are sufficient to prevent oscillation, while changes in value or open bias resistors will probably reduce rather than stop the oscillations completely.

REDUCED OR UNSTABLE OUTPUT.—Reduced output would most likely indicate a change in bias resistor value or a defective emitter bypass capacitor. An open or partially open emitter bypass capacitor would produce excessive degeneration and perhaps completely cut off the transistor, although the latter is very unlikely. Frequency instability is directly traceable to the tank circuit components and connections. An unstable output amplitude could be caused by an intermittent open or short in the bias circuitry, or by a poor connection to the transistor or supply voltage. Lack of supply voltage regulation would normally be indicated by amplitude changes rather than by frequency changes.

INCORRECT OUTPUT FREQUENCY.—Changes in distributed circuit capacitance or reflected load reactances will affect the frequency of operation to some extent, but can normally be corrected by resetting or tuning the variable capacitor. Large frequency changes will be produced by variation of tank circuit inductance that may be caused by shorted turns or poorly soldered connections. At first glance it might appear that a shorted tuning capacitor would cause the frequency of oscillation to be determined by the inductance of the tank circuit alone (plus some distributed capacitance). However, for this condition oscillation could not occur because the shorted capacitor would short the collector voltage through the tank inductor to the base of Q1.

ELECTRON TUBE CLAPP OSCILLATOR

Figure 15-18 shows an electron tube version of the Clapp oscillator. This schematic shows a few of the variations which may be encountered in the Clapp oscillator. For example, an RFC may be inserted between cathode and ground to keep the cathode at a high impedance above ground, and yet offer a d.c. return path to the cathode. Usually this choke is chosen to resonate with its distributed capacitance at the operating frequency of the oscillator. Occasionally, a trimmer is added to this circuit to adjust the parallel resonant choke circuit for the correct value of impedance. While these and other variations may often be encountered, it should be noted that they are not essential to the operation of the Clapp oscillator.

The tuned resonant circuit consists of L, C1, C2, and C3. Capacitors C3 and C4 act as the feedback path. The amplitude of the feedback is determined by the capacitive voltage divider composed of C2 and C3. C_g and R_g form the grid leak bias network. The RFC in the plate circuit acts to isolate the power supply from the high frequency variations and also acts as a plate load device for V1. Operation of the electron tube Clapp oscillator is similar to the

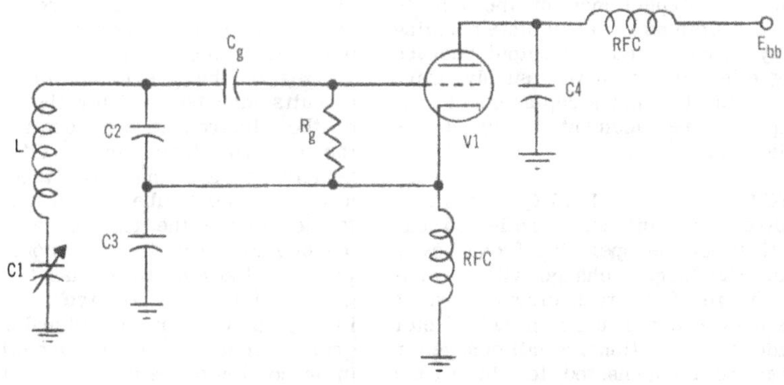

Figure 15-18.—Electron tube shunt fed Clapp oscillator.

oscillators previously covered and consequently will not be discussed here.

Failure Analysis

Some of the most common failures of electron tube Clapp oscillators will be covered in this section.

NO OUTPUT.—High plate current will indicate non-oscillation and a consequent loss of bias. Reduction in tube gain, if sufficient, will also affect oscillation. Large changes in the grid leak bias network may also prevent oscillation. A shorted or open RFC will prevent plate voltage variations from being developed with a resulting loss of feedback voltage and oscillations. Shorted or open components in the tuned circuit may also prevent oscillations.

REDUCED OR UNSTABLE OUTPUT.—A relative indication of oscillator output is provided by the amount of bias being developed across R_g. Variation from the standard operating value is an indication of incorrect operation. A reduction of plate voltage will decrease the output amplitude. Therefore, unregulated power supplies will produce possible amplitude and frequency variations, even though the Clapp oscillator is considered to be more stable than the Colpitts oscillator. A faulty plate coupling capacitor may affect the amplitude of the feedback voltage and, therefore, cause instability.

At the higher radio frequencies it may be possible for the plate coupling capacitor to be open and yet have sufficient feedback coupled through the inter-electrode capacitance of the tube to sustain weak oscillations. A leaky plate coupling capacitor may cause a reduced output because of its loading effect on the power supply. Care should be taken in choosing a replacement RFC since an improper replacement can cause unwanted oscillations.

INCORRECT OUTPUT FREQUENCY.—Changes in circuit capacitance or reflected load reactance will affect the operating frequency to some degree. The largest change will be produced by a change of the tank circuit Q which may be caused by shorted turns in the inductor or poorly soldered connections. Small changes in frequency may be compensated for by circuit adjustments while larger frequency changes may require the replacement of one or more components.

ELECTRON COUPLED OSCILLATORS (ECO)

The electron coupled oscillator is used to produce an RF output of constant amplitude and an extremely constant frequency, usually within the RF range. This circuit is normally used in any type of electronic equipment where good stability is required, and where the output waveshape is not critical, as some distortion of the waveform is normal for this oscillator.

The electron coupled oscillator uses the shielding effect between the plate and screen grid in a tetrode or pentode to isolate the plate load device. While this circuit will operate with other configurations, the LC type oscillator is used most frequently. The frequency stability of the electron coupled oscillator is better than that of previously discussed oscillators.

General Analysis

In the triode type oscillator, variation of the plate supply voltage and reflected load reactance causes slight frequency variations because the plate is involved in the feedback loop. By using the cathode, grid, and screen grid of a tetrode or pentode as the oscillator, the plate is eliminated from the feedback loop. Since the plate current in the tetrode or pentode is relatively independent of plate voltage changes and remains nearly constant for a specific value of screen voltage, variations in the plate load circuit have practically no effect on the other elements. Thus, connecting the screen grid as the plate of the triode oscillator electrostatically shields the plate of the pentode or tetrode, and couples the output of the oscillator to the load through the plate electron stream.

Any of the previously discussed oscillator circuits may be used for the oscillating portion of the electron coupled oscillator circuit, and the electron tube can be either a tetrode, a pentode, or a beam-power tube. The tetrode is a four-element tube, which uses a screen grid to accelerate the flow of electrons from the cathode to the plate under control of the control grid. As the electrons pass through the screen grid to the more positive plate, some of them impinge on the screen grid and produce a screen-grid current, which will normally be quite small in proportion to the plate current.

These electrons, upon striking either the plate or screen grid generally have sufficient force to cause secondary electrons to be emitted by these elements. If the plate voltage is maintained at a more positive value than the screen grid these secondary electrons do not materially affect operation since they will be reabsorbed by the nearest positive element (either plate or screen grid). However, should the plate voltage become equal to or less than the screen voltage, a space charge (cloud of electrons) will form between the screen grid and plate. This space charge which is also known as a "virtual cathode" then acts as a source of electrons for the plate and there is no coupling from the oscillator to the load via the electron stream since the control grid and screen grid now have little or no control over the plate current. Therefore, to reduce or eliminate the effects of secondary emission the screen grid is usually supplied from the plate source through a series voltage-dropping resistor, and it is bypassed to ground with a capacitor to provide an electrostatic shield.

The beam-power tetrode uses special construction and beam forming plates connected to its cathode to provide the same effect. The pentode utilizes a fifth element (the suppressor grid), usually connected to the cathode internally and located between the screen grid and plate, to eliminate the effects of secondary emission. In some tubes the suppressor grid connection is brought out externally, and may be connected to ground, screen grid, or plate as desired. In the ECO the plate load may be another tuned tank circuit, an RFC, or a resistor. The tuned plate tank may be operated at the same frequency as the grid tank, or it may be utilized as a frequency multiplier by operating it at a harmonic of the grid tank frequency. In test equipment applications, a resistor or an RFC (instead of a tank) is usually utilized in order to assure minimum loading effects and maximum frequency stability.

When a resistor or RFC is used in the output circuit, the waveform is not precisely sinusoidal because the plate current does not vary linearly with the screen grid current. Although the control grid varies the screen grid current linearly, not all the current controlled by the control grid reaches the plate. Some of the current in the screen-plate region is attracted back and is absorbed by the screen grid, while other portions of the current are shunted to the cathode by the suppressor grid.

Therefore, at any particular instant, when the screen grid current is being slightly increased, the plate current may or may not be increased by the same amount—in fact, it may actually decrease. Hence, the output waveform, which is the result of plate current flow through the plate load device, will not be exactly like the screen grid waveform; it will be similar, but distorted. If a plate tank or an LC filter is employed, the distortion will be corrected and a nearly sinsoidal waveform will result.

Operation

Figure 15-19 shows the schematic of an ECO using a pentode tube. The oscillator section utilizes the basic Hartley shunt fed circuit with series grid leak bias. The screen of the pentode acts as the triode plate.

RFC2 acts as the plate loading device and also isolates the power supply from plate voltage variations. The signal output to the following stage is capacitively coupled. To insure that the screen grid of the pentode remains at a lower potential than the plate, a dropping resistor, R1, is normally used. Since the screen grid is above ground at RF potential, RFC1, and C3 are necessary to isolate the power supply from screen grid voltage variations. Note that the suppressor grid is connected to the cathode, which is grounded. If the cathode were above-ground, the suppressor could not be connected to the cathode or it would capacitively couple the oscillator into the plate circuit and destroy the shielding effect. Since in this instance the cathode is grounded, the suppressor performs its normal function of eliminating secondary emission effects. It also isolates the plate and screen grid from each other.

The oscillator normally operates similar to class C, and for each pulse or oscillation there is a corresponding pulse produced in the plate circuit through control grid modulation of the plate current. Thus, electron coupling exists between the screen grid and plate. A change in pentode plate current has little or no effect on the screen voltage and current; thus, oscillator stability is relatively unaffected by plate load variations. Note, while the pentode circuit is considered to be independent of power supply variations, only a properly designed tetrode circuit can be made completely independent. The advantage of the pentode circuit is that the suppressor forms an effective electrostatic screen between the screen grid and plate, and secondary emission effects need not be considered.

When power is first applied to the circuit of figure 15-19 various charging currents will flow

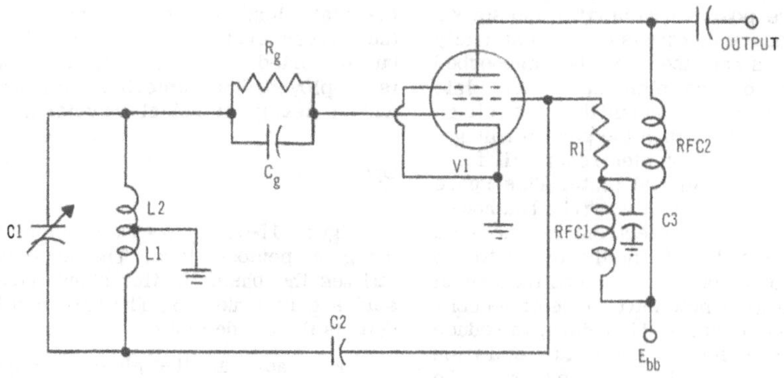

179.316

Figure 15-19.— ECO oscillator.

due to the uncharged condition of the capacitors. Of prime importance is the charging current which flows from ground, through L1, coupling capacitor C2, R1, and RFC1 to E_{bb}. This charging current builds up a magnetic field around L1. As the charging current ceases, the magnetic field of L1 collapses and begins oscillations in the resonant tank circuit, which is composed of L1, L2, and C1. The oscillations in the tank circuit will allow the bias network (R_g and C_g) to build up grid leak bias after a few cycles, and the circuit is now fully operational.

It is assumed that the circuit is being operated similar to class C; thus, the tube can conduct only during the positive peaks of the tank circuit's oscillations. When a positive peak is present the tube will conduct, screen grid current will flow, and the screen current flow through R1 and RFC1 causes a resulting reduction in screen grid voltage. A reduction (negative going) in screen voltage will be coupled through C2 to the bottom of the tank circuit. This has the effect of making the bottom of the tank go more negative at the same time the top of the tank is in its positive half cycle. The effect is to increase the total voltage across the tank, placing a more positive signal on the control grid and increasing the conduction of the pentode. This is obviously a regenerative action. The amount of feedback which reaches the tank circuit is determined by the L1-L2 ratio, as L1 is the actual feedback element. The regenerative feedback cycle will take place during the positive peaks of the tank circuit's alternations. Normal tank action keeps the oscillations going during those portions of the cycle when the pentode is kept cutoff.

Up to this point oscillations have been sustained without giving any consideration to the plate circuit. It must be remembered that screen grid current is only a small fraction of the total current flow in a pentode, most of the current goes to the plate. In passing through the area of the control grid this plate current is also affected by the variations in voltage on the control grid and even a slight variation in plate current will be converted to a relatively large voltage change in the output. In other words, any variations in control grid voltage are amplified in the plate circuit and applied to the output.

Failure Analysis

Common ECO oscillator failures are discussed in the next paragraphs.

NO OUTPUT.—Failure of oscillation would cause no output, and could result from the same effects described in the failure analysis section of the Hartley oscillator. In addition, lack of supply voltage caused by open components or a short from plate to ground would effectively stop output. Actually, in this type of circuit, tube emission could drop to the point where there is sufficient current for the oscillator section, but practically none for the plate. Therefore, the circuit should be considered as basically a two-section tube, and checked for oscillator output first and then for continuity through the plate circuit. If the plate output circuit is known to be complete with all components in good operating condition, and the oscillator is operating, a lack of plate output will be indicative of a faulty tube.

REDUCED OR UNSTABLE OUTPUT.—In addition to those causes indicated for the basic Hartley oscillator, the following should be considered: If the basic oscillator is unstable or has a reduced output, then with all other conditions equal, the plate output will also be reduced or unstable. Thus, the oscillator should first be investigated for proper operation. Since the screen voltage determines the amount of plate current that will be drawn by the tube, a reduction of screen voltage would have a more noticeable effect on the output before it dropped below the point where oscillation could be sustained. An increase in resistance of the screen dropping resistor, if used, is a primary cause of reduced output. Where the plate voltage is supplied through a load resistor, it is common for a high resistance to develop, reducing both the plate voltage and the output. This condition can result in instability if the plate voltage becomes lower than the screen voltage. If the output circuit of the plate includes a tank circuit, a normal increase in output is to be expected when the tank is tuned to the same frequency as the oscillator or to a low-order harmonic. Therefore, any effects causing tank de-tuning will immediately result in a reduced output.

When instability is produced by load variations, it can immediately be assumed that the energy coupling is involved; in this case, the circuit should be checked for capacitive coupling between screen and plate or for a change in voltage ratios due to an excessive voltage drop or a short circuit in either the screen or plate circuits.

INCORRECT OUTPUT FREQUENCY.—Since the output frequency is determined basically by the oscillator portion of the tube, any change in frequency with normal output amplitude indications would indicate that a change of oscillator components or voltages is the most probable cause. If the oscillator frequency change is not due to components in the oscillator itself, the electrostatic shielding between the plate and screen sections has probably deteriorated because of open or shorted bypass capacitors, changes in tube voltages, or the presence of stray capacitances from other causes.

TUNED-PLATE TUNED-GRID OSCILLATOR

The tuned-plate tuned-grid (TPTG) oscillator is used to produce a sine-wave output of constant amplitude and fairly constant frequency within the RF range. This circuit is generally used as a variable-frequency oscillator over the high and very-high RF ranges. It is not often encountered in Navy equipment but its similarity to an oscillating RF amplifier makes it well worth covering.

The TPTG oscillator contains two parallel tuned LC circuits, one in the grid circuit and the other in the plate circuit of the electron tube, and uses the internal grid-plate capacitance of the tube to accomplish feedback.

It operates similar to class C for ordinary or power operation and is seldom used for linear waveform equipment. The frequency stability of the TPTG oscillator is good when it is properly adjusted.

General Analysis

When a tuned LC tank circuit is connected between the grid and cathode of an electron tube and a similar tank circuit is connected between the plate and cathode of the tube with coupling existing between the two tuned circuits, sustained oscillation will occur when these tuned circuits are resonated at the same frequency. The amplitude of the oscillations is determined by the amount of feedback from plate to grid, and is also affected by the tuning of the two circuits. In order to provide proper operation, both the grid tank and the plate tank are tuned to a slightly higher frequency than the resonant frequency at which operation is desired. At the operating frequency both tanks then appear slightly inductive.

To sustain oscillation it is necessary only to supply the losses in the grid circuit, which because of the high Q grid tank, are relatively small. Consequently, only a small capacitance is needed to supply sufficient feedback. The grid-plate interelectrode capacitance of the triode tube is adequate for this purpose in most cases. When a pentode is employed in the circuit, it will be necessary to supply an external capacitor to provide the feedback. This is because of the low interelectrode capacitance inherent in the pentode.

Figure 15-20 shows the schematic diagram of a tuned-plate tuned-grid oscillator using a triode. The grid tank consists of C1 and L1, while the plate tank consists of C2 and L2, with

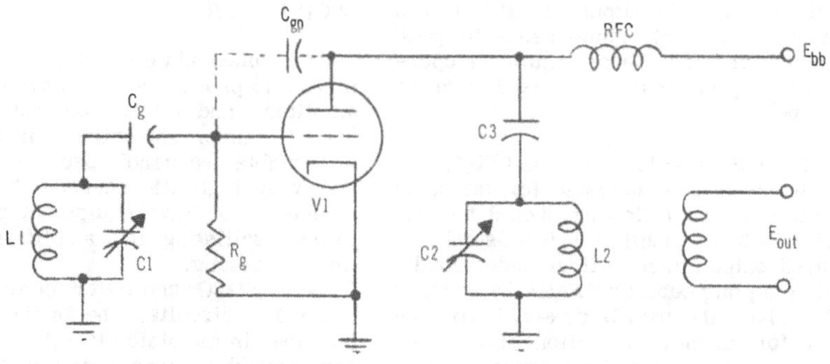

Figure 15-20.—Tuned-plate tuned-grid oscillator.

no mutual inductance existing between L1 and L2. Feedback occurs through the plate-grid capacitance, C_{gp} (shown in dotted lines), when both tanks are tuned to the same frequency. The feedback amplitude can be controlled by detuning the grid tank with C1. Thus, differences in interelectrode capacitance between different tubes can be accommodated with no adverse effects.

Upon examination of the circuit, it is evident that shunt grid leak bias (R_g and C_g) and shunt plate feed are used, with C3 serving as the plate coupling and blocking capacitor. The RFC isolates the power supply from RF variations. The grounded-cathode connection is shown since it permits the grounding of the tuning capacitor rotors in order to eliminate any body capacitance effects during the tuning procedure. The use of shunt grid leak bias isolates the grid tank as far as d.c. is concerned, and reduces circulating grid current, although series grid leak bias could be used as well. Sometimes a series resistor is included between cathode and ground to provide a protective self-bias in the event of non-oscillation. Both the plate and grid capacitors are sufficiently large so that the tank circuits are effectively coupled to the grid and plate for radio frequencies, yet isolated as far as d.c. is concerned. Grid leak bias action and automatic amplitude regulation are the same as described previously. The feedback action takes place, making use of the coupling between the plate and grid tanks provided by the grid-plate tube capacitance. At radio frequencies which require the use of relatively large inductors, and when appreciable power is generated, it is important for the tanks to be shielded or turned at right angles to each other. This is to prevent possible inductive coupling effects which would deteriorate circuit performance. The output is obtained by inductive coupling to the plate tank, although a capacitive connection may be made if desired.

Detailed Analysis

When power is first applied, a charging current will flow from ground, through C2, C3, and the RFC to E_{bb}. The charging of C2 provides energy for the plate tank circuit and oscillations begin. As shown in the simplified diagram in figure 15-21, the plate tank is capacitively coupled to the grid tank by the interelectrode capacitance of V1. Both C3 and C_g are relatively large capacitors by comparison and C_{gp} will therefore determine the initial amount of feedback that reaches the grid tank. At the beginning of oscillations in the plate tank, C2 is discharging in the direction shown in figure 15-21, the L2 voltage is of the polarity shown and as a result, the top of the grid tank will feel a positive potential. The positive potential on the grid will cause grid current flow, which establishes grid bias and provides additional energy for the grid tank. Oscillations are now established in the grid tank.

The oscillations present in the grid tank, in conjunction with grid leak bias, now allow the grid tank to control the operation of V1. As the potential at the top of the grid tank (with respect to ground) reaches its positive peak, the tube will conduct, C_g will recharge any lost energy through grid current, and the variation in plate

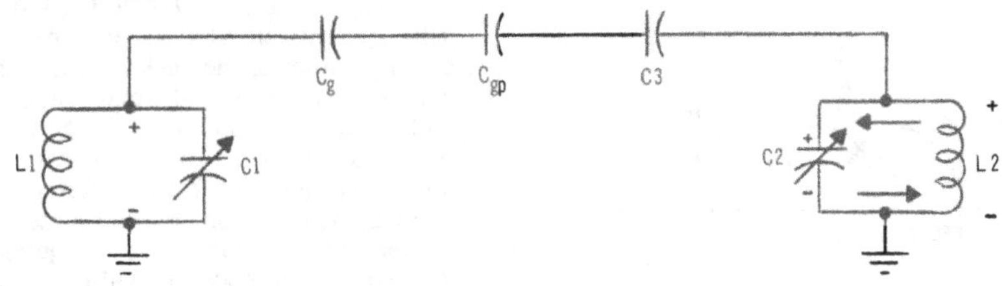

179.318
Figure 15-21.—Simplified feedback network for TPTG oscillator.

voltage of V1 will be coupled through C3 to supply the necessary energy to keep the plate tank in operation. The plate tank, in turn, will maintain oscillations in the grid tank through interelectrode capacitance coupling.

As was mentioned previously, both the grid and plate tank circuits were tuned to a frequency somewhat higher than the actual operating frequency of the circuit. For this reason, both circuits appear slightly inductive. The explanation of this, while complicated, may be simplified through the use of figure 15-22. It should be noted that figure 15-22 shows the grid-to-cathode capacitance, and the plate-to-cathode capacitance, of V1. As C_g is a relatively large capacitance, having a negligible reactance at radio frequencies, C_{gk} is effectively in parallel with the grid tank and must, therefore, be considered as part of it. The total capacitance in the grid tank is, therefore, $C1 + C_{gk}$. It would be incorrect to calculate the resonant frequency only on the basis of L1 and C1.

The tuning procedure, therefore, requires that the external (L1 and C1) portion of the grid circuit be tuned to a frequency slightly higher than the desired circuit frequency. The additional effect of C_{gk} and distributed grid circuit capacitance will lower the frequency to that actually desired.

The same argument holds for the plate circuit where C_{pk} and distributed plate circuit capacitances will form the extra capacity to reduce the frequency to that desired. Due to the different Q's, loading effects, etc., it is found that while both tuned circuits are adjusted to frequencies higher than that at which the circuit will operate, they will not be adjusted to the same frequency. See figure 15-23.

Tuning of the TPTG oscillator is accomplished by first adjusting the grid tank to the APPROXIMATE desired operating frequency. This approximation is obtained through calibration marks on the tuning dial, or operator experience. The plate tank is then adjusted until the circuit begins to oscillate. Once oscillations are obtained, the frequency of operation is compared with a frequency standard, and the grid tank READJUSTED slightly to bring the circuit to the desired frequency of operation. The plate tank is now READJUSTED for maximum output. Although it is possible to use the TPTG at relatively low frequencies, its use is normally confined

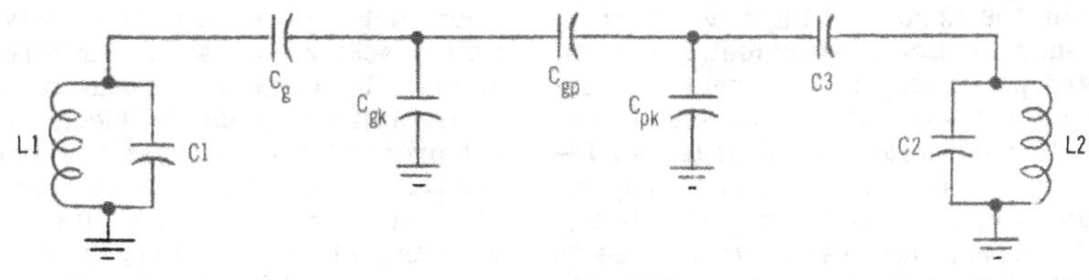

179.319
Figure 15-22.— Equivalent circuit of TPTG oscillator.

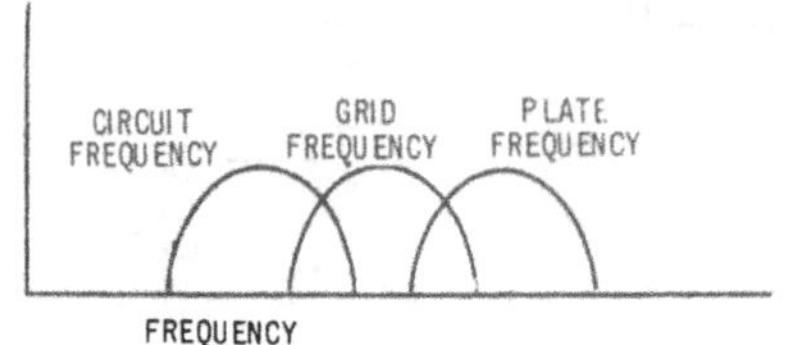

179.320

Figure 15-23.—Tank circuit tuning in TPTG oscillator.

to the high frequency ranges by the value of the tube's interelectrode capacitance.

Failure Analysis

The most common failures of the TPTG oscillator are discussed below.

NO OUTPUT.—If something happens to either of the tank circuits so as to cause a large enough shift in that tank's resonant frequency, the circuit will not oscillate. Plate current, therefore, will be higher than normal, and the standard operating value of grid bias will not be obtained. Excessive losses, if present in the tank circuit, will prevent oscillation. Changes in R_g and C_g will affect operating bias and change the amplitude of oscillation. Too tight coupling of the output circuit will produce the same effect as losses in the tank circuit, and reflected reactance may cause sufficient detuning to prevent oscillation. A malfunctioning RFC may not isolate the power supply adequately, resulting in a low impedance path for the RF, and a consequent loss of oscillation.

REDUCED OR UNSTABLE OUTPUT.—Any variation from the normal voltage and current readings is an indication of abnormal operation. An unregulated power supply may cause amplitude variations and some frequency instability. A leaky coupling capacitor would cause an increased current drain on the power supply, lowering plate voltage, and causing instability. Interelectrode capacitance varies from tube to tube, and instability may result when replacements are made. Care should also be taken in replacing the RFC, as unwanted oscillations may result.

INCORRECT OUTPUT FREQUENCY.—Changes in output frequency can be compensated for by retuning the tank circuits. Detuning effects caused by reflected load reactances are often noticeable on the very-high frequency ranges. Shorted turns in the grid tank inductance will cause frequency changes (assuming oscillations continue). Changes in ambient temperature sometimes affect operating frequency due to the change in component values. Care should also be taken in replacing circuit components, as the inherent distributed capacitance may be changed and cause resulting frequency changes.

Series Fed Tuned-Plate Tuned-Grid Oscillator

A series fed TPTG oscillator is easily created by placing the plate tank in series with the RFC and then bypassing it to ground with capacitor C3. The series fed oscillator will eliminate the possible problem of RFC oscillations but the effects of d.c. in the tank circuit (insulation requirements) usually pose bigger disadvantages than the advantages presented.

CRYSTAL OSCILLATORS

In order to obtain a higher degree of frequency stability in UHF transceivers a CRYSTAL OSCILLATOR is used. A properly cut crystal possesses the characteristics of a resonant circuit, and may therefore be used in place of a tuned circuit as the frequency controlling element. Before discussing the use and operation of a crystal in a circuit, it will prove advantageous to discuss some of the properties and preparation of various types of crystals.

The control of frequency by means of crystals is based upon the PIEZOELECTRIC EFFECT. When certain crystals are compressed or stretched in specific directions, electric charges appear on the surface of the crystal. Conversely, when such crystals are placed between two metallic surfaces across which a difference of potential exists, the crystals expand or contract. This interrelation between the electrical and mechanical properties of a crystal is termed the piezoelectric effect. If a slice of crystal is stretched along its length, so that its width contracts, opposing electrical charges appear across its faces; thus, a difference of potential is generated. If the slice of crystal is now compressed along its length, so that its width expands, the charges across its faces will reverse polarity. Thus if alternately stretched and compressed, the slice of crystal becomes a source of alternating voltage.

Also, if an alternating voltage is applied across the faces of a crystal, it will vibrate mechanically. The amplitude of these vibrations will be very vigorous when the frequency of the a.c. voltage is equal to the natural mechanical frequency of the crystal. If all mechanical losses are overcome, the vibrations at this natural frequency will sustain themselves and generate electrical oscillations of a constant frequency. Accordingly, a crystal can be substituted for the tuned tank circuit in transistor and tube circuits.

Practically all crystals exhibit the piezoelectric effect, but only a few are suitable as the equivalent of tuned circuits for frequency control purposes. Among these are quartz, Rochelle salt, and tourmaline. Of these, Rochelle salt is the most active piezoelectric substance; that is, it generates the greatest amount of voltage for a given mechanical strain. However, the operation of a Rochelle salt crystal is affected to a large extent by heat, aging, mechanical shock, and moisture. Although these factors render this substance unsuitable for use in frequency control applications, it does find use in some microphones and phonograph pickups.

Tourmaline is almost as good as quartz over a considerable frequency range, and is somewhat better than quartz in the 3 to 90 MHz range, but it has the disadvantage of being a semiprecious stone, and its cost is therefore prohibitive.

Quartz, although less active than Rochelle salt, is used universally for oscillator frequency control because it is inexpensive, mechanically rugged, and expands very little with heat. Crystals used in oscillator circuits are cut from natural or artificially grown quartz crystals which have the general form of a hexagonal prism as shown in figure 15-24A. However, they are rarely as symmetrical as that shown. Assuming a symmetrical crystal, the cross section is hexagonal, as in part B of figure 15-24.

The axis joining the points at each end, or apex of the crystal, is known as the optical or Z axis. Stresses along this axis produce no piezoelectric effect. The X axis passes through the hexagonal axis and cross sectional area at right angles to the Z axis and these are known as the electrical axes. These are the directions of the greatest piezoelectric activity.

The Y axes, which are perpendicular to the faces of the crystal, as well as to the Z axis, are called the MECHANICAL axes. A mechanical stress in the direction of any Y axis produces an electrostatic stress, or charge, in the direction of that X axis which is perpendicular to the Y axis involved. The polarity of the charge depends on whether the mechanical strain is a compression or a tension. Conversely, an electrostatic stress, or voltage applied in the direction of any electrical axis, produces a mechanical strain, either an expansion or a contraction, along that mechanical axis which is at right angles to the electrical axis.

For example, if a crystal is compressed along the Y axis, a voltage will appear on the faces of the crystal along the X axis. If a voltage is applied along the X axis of a crystal, it will expand or contract in the direction of the Y axis. This interconnection between mechanical and

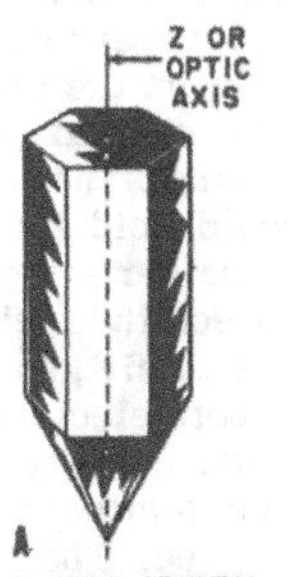

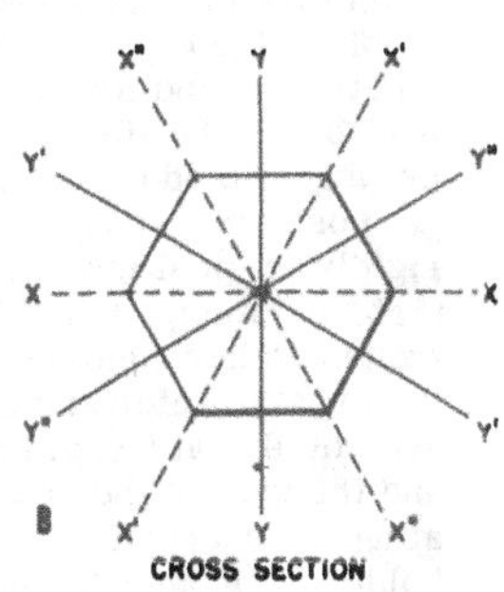

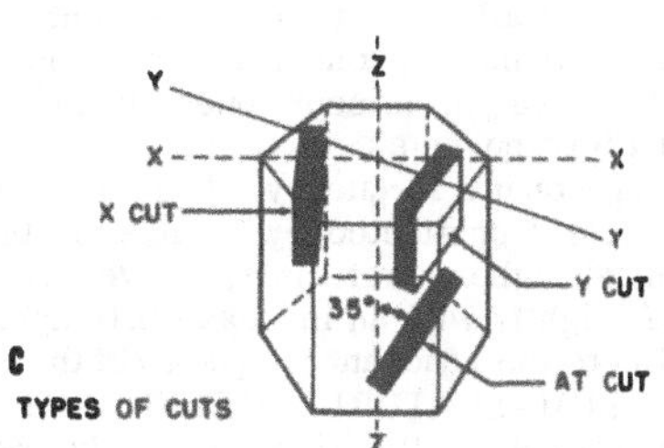

20.148

Figure 15-24.—Quartz crystal axes and cuts.

electrical properties is exhibited by practically all sections cut from a piezoelectric crystal.

Crystal wafers can be cut from the mother crystal in a variety of directions along the axes. They are known as CUTS and are identified by designations such as X, Y, AT, etc. (fig. 15-24C). Each has certain advantages, but, in general, one or more of the following properties are desirable: ease of oscillation at the intended frequency, a single operating frequency, and minimum frequency variation due to temperature changes.

Both the X and Y cuts have unfavorable temperature characteristics, as will be explained later. Better characteristics can be obtained by cutting wafers at different angles of rotation about X axes. The Y cut serves as the zero-degree reference, since it is lined up with both the X and Z axes; that is, it lies in a plane formed by the X and Z axes. By rotating this slice from its starting point, different cuts may be formed. Many other cuts exist, but they will not be discussed here.

Crystals used in oscillator circuits must be cut and ground to accurate dimensions. For instance, a typical quartz which is resonant at 1 MHz would measure 2.54 cm by 2.54 cm by .286 cm. Crystals may be cut in various shapes, with crystals in the lower frequency range being square or rectangular and some of the crystals in the higher frequency range being disc shaped, similar to a coin. For precision test work, crystals may be cut in the form of a flat ring. The type of cut also determines how active the crystal will be. Also, for any given crystal cut, the thinner the crystal, the higher the resonant frequency. The more care that is taken in cutting and grinding the crystal, the more accurate will be the final result.

Crystals which are not ground to a uniform thickness may have two or more resonant frequencies. Usually one of these resonant frequencies will be more pronounced than the others, with the less prominent ones being termed SPURIOUS frequencies.

The resonant frequency of quartz crystals is practically unaffected by changes in the load. Like most other materials, however, quartz expands slightly with an increase in temperature. This affects the resonant frequency of the crystal.

The TEMPERATURE COEFFICIENT of the crystal refers to the increase or decrease in the resonant frequency as a function of temperature. The temperature coefficients vary widely from one crystal cut to another. A positive temperature coefficient is assigned to those cuts which produce an increase in frequency with an increase in temperature. A negative coefficient is assigned to those crystals which decrease their natural resonant frequency when the temperature increases. One cut, the GT, has a practically zero temperature coefficient over a wide range of temperature variations. The temperature coefficient also depends on the surrounding temperature at which it is measured. Heating of the crystal can be caused by external conditions such as the high temperature of transmitter tubes. Heating can also be caused by excessive RF currents flowing through the crystal. The slow shift of the resonant frequency resulting from crystal heating is called FREQUENCY DRIFT. This is avoided by use of crystals with nearly zero temperature coefficient, and also by maintaining the crystal at a constant temperature.

To maintain the extremely close frequency tolerances required, the general practice is to construct the entire oscillator assembly in such a manner as to provide for nearly constant temperature. This helps avoid frequency drifts. Power supply voltages are kept as constant as possible by suitable voltage regulator circuits. In addition, the quartz crystal is operated within a constant temperature oven. The oven is electrically heated and temperature controlled by special thermostats. The entire assembly usually is constructed of an aluminum shell enclosed by thick layers of insulating material to insulate the assembly. For extreme stability, the entire compartment may be placed inside still another temperature controlled box. In this way, frequency stabilities as high as 1 part in 10,000,000 or better can be attained.

Crystals become practical circuit elements when they are associated with a crystal holder. In a holder the crystal is placed between two metallic electrodes and forms a capacitor, the crystal itself being the dielectric. The crystal holder is arranged to add as little damping of the vibrations as possible, and yet it should hold the crystal rigidly in position. In some holders, the crystal plate is clamped firmly between the metal electrodes; other holders permit an air gap between the crystal wafer and one or both electrodes. The size of the air gap, the pressure on the crystal, and the size of the contact plates affect the operating frequency to some degree. The use of a holder with an adjustable air gap permits slight adjustments of frequency to be made. For the control of appreciable amounts of power, however, a holder which clamps the plate firmly in place is preferred.

At its resonant frequency, a crystal behaves like a tuned circuit, as far as the electrical circuits associated with it are concerned. The crystal and its holder (fig. 15-25A) can be replaced by an equivalent circuit (fig. 15-25B). C_m represents the capacitance of the mounting, with the crystal in place between the electrodes but not vibrating. C_g is the effective series capacitance introduced by the air gap when the contact plates do not touch the crystal. The series combination, L, R, and C, represents the electrical equivalent of the vibrational characteristics of the quartz plate. The inductance, L, is the electrical equivalent of the crystal mass effective in the vibration. C is the electrical equivalent of the mechanical elasticity, and R represents the equivalent of the mechanical friction during vibration. The capacitance of the holder, C_m, is about 100 times as great as the vibration capacitance, C, of the crystal itself.

The presence of both series and parallel resonant frequencies is revealed by the crystal resonant curve, in part C of figure 15-25. This curve is very sharp, and extremely high Q's are easily attainable. It is found in practice that the L/C ratio of the equivalent circuit is extremely large compared with that of a conventional tank circuit.

SOLID STATE CRYSTAL OSCILLATORS

Although the quartz crystal and semiconductor are both solid state devices, there is no direct relationship so far as frequency stability or piezoelectric effects are concerned. The frequency stability of the quartz crystal is derived from its basic mechanical vibration, and due to the piezoelectric effect, we have a means for controlling the vibration, and hence for controlling an electronic circuit.

The quartz crystal is cut to enhance the mechanical vibrating properties at a specific frequency. The semiconductor material is cut to a convenient, economical size, and has electrodes attached in such a manner that it cannot easily oscillate mechanically. Because the semiconductor is a partial conductor, it exhibits little, if any, piezoelectric effect. Thus, it has no inherent frequency stability even though it is a solid state device. Actually, the semiconductor is less frequency stable than an electron tube, mostly because of its inherent ability to act as a variable capacitor with a change of collector voltage.

In an LC oscillator the tank circuit is the frequency determining and stabilizing element, but in the crystal oscillator, it is the crystal itself, because the crystal operates essentially as a parallel tuned tank or series tank depending on the method of operation. Therefore, whether semiconductors or electron tubes are used, the basic principles of operation of the resulting electromechanical oscillators are identical. The chief differences are the effects on feedback and the circuitry used for the effective feedback control.

Since there are numerous circuit variations now in use and the state of the art is such that changes occur constantly, the circuits have been classified into three arbitrary groups. The first

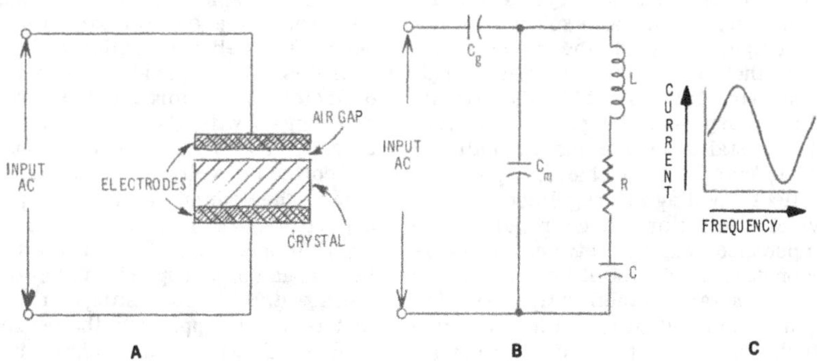

Figure 15-25.— Equivalent circuit of crystal and mounting.

group, known as the TRANSFORMER COUPLED GROUP, includes those oscillators using mutually coupled separate coils to provide feedback. The oscillators in this group have the advantage of extreme flexibility. They can be used in either CB, CE, or CC configuration, since polarities and impedances can be completely controlled by the number of turns and directions of windings.

The second group uses a capacitive voltage divider and is called the COLPITTS TYPE CRYSTAL OSCILLATOR. The third group is the OVERTONE GROUP, which generally uses a tuned tank with a tapped coil in the Hartley type feedback arrangement operating at a desired harmonic or overtone. In any of these circuits, series or parallel resonance of the crystal may be employed. However, the transistor is a low-impedance device, operating with relatively low voltage and high current; thus, it conveniently lends itself to the use of series resonant crystal operation, using only one stage instead of two stages as is often necessary in tube circuitry.

For extreme stability, temperature compensation is necessary, particularly since the transistor itself is somewhat temperature sensitive. At normal room temperatures (or lower) and with average low-temperature or zero-temperature co-efficient crystals, the transistor crystal oscillator is stable enough for average use without compensation. At higher temperatures, compensation is necessary in most cases.

In considering the output requirements, we find that the load is generally in shunt with the crystal or the tuned tank circuit; as a result, increased power output usually affects the frequency stability by reducing the effective Q of the resonator or by increasing the effective coupling impedance between the crystal and the feedback circuit. Or, as in self excited oscillators, the less output required, the more stable the frequency; that is, with a smaller load, changes in load have less affect on the circuit. For all ordinary purposes, the good mechanical stability of the crystal resonator makes it almost free from the effects of load changes, particularly if the collector voltage is regulated.

Good crystal oscillator design requires the use of low-impedance coupling between the oscillator and resonator, the use of either resistance stabilization or reactance stabilization, regulation of supply and bias voltages, and temperature control of both the transistor and the crystal.

As there are many solid state crystal oscillators and several variations of each type, only the Colpitts crystal oscillator will be discussed in the text. It is representative of most other types and only minor modifications are required in order to understand further circuit configurations.

COLPITTS CRYSTAL OSCILLATOR

The Colpitts crystal oscillator is used primarily at the higher radio frequencies as an extremely stable oscillator. However, it may also be used at low and medium radio frequencies. This circuit uses the piezoelectric effect of a quartz crystal to control the oscillator frequency. Feedback is provided through a capacitive voltage divider arrangement, which is usually external, but it may be provided through the transistor element capacitances. This circuit does not normally utilize a tuned tank circuit, but a tank circuit may be required in special applications. The Colpitts crystal oscillator operates similar to class C in those circuits where waveform linearity is not important, and class A when good waveform linearity is required.

Circuit Operation

A typical common-emitter Colpitts circuit using the external capacitive divider feedback method is shown in figure 15-26. Voltage divider bias is used for easy starting and is supplied by R1 and R_B. C3 is a base bypass capacitor and effectively isolates the base from RF variations. With the base bypassed by C3, the capacitive voltage divider consisting of C1 and C2 is connected between the collector and emitter for feedback purposes. Both feedback capacitors are variable to permit adjustment and control of feedback. Capacitor C1 also serves to bypass RF around the emitter resistor. The crystal is connected between collector and ground and operates as a parallel resonant circuit. The collector is fed through the RFC which acts in conjunction with C4 as a power supply isolation circuit. The RFC also develops the variations in collector voltage by acting as a collector load. The output (not shown) may be taken capacitively from the collector.

The operation of the Colpitts crystal oscillator basically depends upon the action of the voltage divider consisting of C1 and C2. Assume that power is applied to the oscillator. With bias applied through R_B, collector current flows, and a voltage appears across the C1-C2 divider. The voltage appearing across C1 is in parallel with emitter resistor R_E. Now assume that a noise pulse occurring within the transistor causes

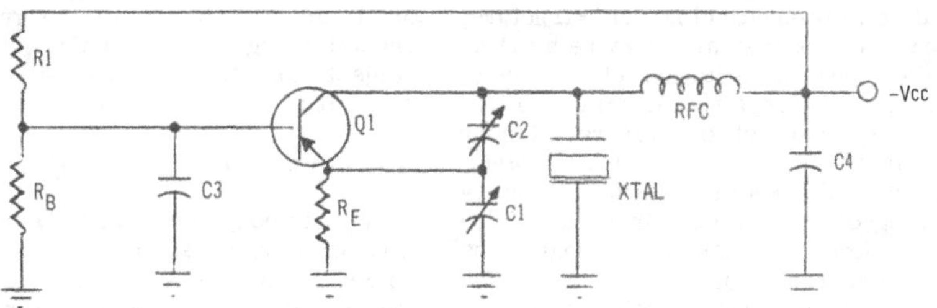

179.322

Figure 15-26.—CE configuration Colpitts oscillator.

collector current to increase. This will cause collector voltage to decrease (less negative or more positive), and capacitor C2 couples this voltage change to the emitter. The decreasing negative signal (positive going) applied to the emitter is regenerative. This will cause a further increase in collector current. The collector voltage continues to decrease (a negative voltage which is changing in a positive direction) and capacitor C2 continues to couple this changing voltage to the emitter. At the same time, the change in collector voltage appears across the crystal. Thus, the crystal is slightly strained (mechanically) by piezoelectric action. When collector current reaches the saturation level, no further change occurs, and the regenerative action ceases.

At this time the electrostatic strain across the crystal begins to decrease, C1 begins a slight discharge through R_E, emitter to ground voltage is reduced, forward bias is reduced, and the collector current starts to fall. This action is also regenerative, and the transistor quickly reaches cutoff. As the collector current decreases, collector voltage increases (becomes more negative), and the crystal now stresses in the opposite direction. As a result, as each cycle of this action continues, the crystal oscillates at its parallel resonant frequency. Since oscillation of the crystal produces a voltage across it, once started into vibration the crystal will continue to oscillate. Since the crystal is connected in shunt from collector to ground, it effectively functions as a parallel resonant circuit and smooths out the pulses of oscillations into approximate sine waveforms. On the conducting portion of the transistor operating cycle, oscillation is effectively reinforced by the following pulses of collector current. During the nonconducting portion of the cycle, the crystal supplies what would otherwise be the missing half cycle of oscillation by flexing in the reverse direction. Emitter resistor R_E and capacitor C2 form an amplitude limiting device similar to that of grid leak bias in electron tube circuits.

The output is taken across the crystal (collector to ground). It is evident that the crystal must be capable of handling the power developed, since if driven too hard it will fracture. However, at the normally low milliwatt output obtained with standard transistors, this is no problem. It does indicate, however, that the transistor crystal oscillator will require the same RF driving power as electron tube crystal oscillators.

Failure Analysis

Some of the Colpitts crystal oscillator more common failures are analyzed as follows.

NO OUTPUT.—A defective crystal will prevent oscillation. A suspected crystal can easily be checked by the substitution of a crystal known to be good. A simple resistance check will indicate continuity and general parts condition. A defective bias resistor can prevent starting, and will be made evident by a resistance or voltage check. An open RFC will also stop operation. A defective transistor should be suspected only if all other parts check satisfactory, and bias and supply voltages are normal.

REDUCED OUTPUT.—A low supply voltage can cause a reduction of output. A reduction of bias from class B or C to class A will also result in reduced output. Both of these conditions can be determined by a single voltage check, preferably with a vacuum-tube voltmeter. Reduced crystal activity due to aging, semi-fracture, or a dirty crystal container also may be a prime cause. In this case substitution of a crystal known to be good will restore normal operation. A change in the value of emitter resistor or bypass capacitor may cause excessive bias, and consequent motorboating due to the intermittent blocking of oscillations. In circuits designed for the higher frequencies, changing the parts wiring during a repair may change the distributed capacitance sufficiently to detune the circuit and cause a reduced output.

INCORRECT FREQUENCY.—A defective or dirty crystal can cause an abrupt change from one frequency to another; this can sometimes be corrected by adjusting the tank tuning, if such a circuit is being used; otherwise, the crystal should be replaced. Aging of the crystal can also cause a slow change over a long period of time. This change can sometimes be compensated for by means of a small trimmer capacitor, if the frequency is higher than normal. Failure of the collector supply regulation can cause the effective element capacitance of the transistor to change and cause slight off-frequency operation. In this case check the supply voltage with a high-resistance voltmeter and observe whether there are occasional fluctuations of the voltage when the crystal is suddenly removed from its socket and reinserted into its socket, or when the load is suddenly removed or replaced. If all components check normal and the trouble is still present, the crystal is probably defective.

ELECTRON TUBE CRYSTAL OSCILLATORS

As with the transistor crystal oscillators, electron tube crystal oscillators are produced in a variety of circuit configurations. Only two of the possible circuit arrangements will be covered here, the MILLER and the PIERCE crystal oscillators.

MILLER CRYSTAL OSCILLATOR

In the Miller crystal oscillator (fig. 15-27), the crystal is connected between the grid and cathode. This is the most widely used electron tube type crystal oscillator configuration since for a given amount of crystal excitation it will provide the greatest power output as the basic feedback occurs between the grid and plate of the tube and not through the crystal. Also, oscillator reliability will be greater, since the crystal being located in the grid circuit is less subject to stresses and strains which would cause it to crack and fail.

Circuit Operation

R_g in conjunction with the parallel combination of crystal shunt capacitance and crystal holder capacitance provides grid leak bias for the circuit. The tuned tank circuit (L & C2) is located in the plate circuit. As shown, the RFC and C1 form a conventional decoupling circuit. The arrangement shown series feeds the plate through the tank but a shunt feed is easily obtained if required.

Circuit operation in the Miller oscillator is relatively easy to understand. When power is first applied to the circuit, the tube will conduct, plate current will surge through the plate tank, and oscillations will be set up. Assume that the feedback causes the grid to go more negative, thus deforming the crystal. The deforming of the crystal produces a piezoelectric action, increasing the charge on the grid in the same direction, and drives it further negative. When plate current cutoff is reached, the feedback becomes zero, and the accumulated grid voltage discharges through grid leak resistor R_g. As the grid voltage is reduced, the deformation of the crystal is reduced, and the negative piezoelectric charge on the grid is also reduced by a correspondingly positive induced piezoelectric voltage. This continues until plate current again flows. The cumulative action causes the crystal to vibrate mechanically near its parallel resonant frequency. Once started, the vibrations continue and induce, in the grid circuit, an a.c. voltage of a frequency almost equal to the vibration frequency of the crystal. As long as the plate tank is properly tuned, (slightly above crystal frequency) the proper phase for feedback is maintained. When the tank is tuned slightly lower than crystal frequency, the feedback will oppose crystal action and oscillation will be reduced and eventually stopped. Tuning the tank circuit allows the feedback to be controlled from minimum to maximum with a corresponding increase or decrease in output. To produce the proper feedback phase, tuning is approached from the high capacitance side of resonance. Plate current varies in a similar manner from a high value to a low value at the optimum point, then suddenly increases and abruptly reaches its normal static (non-oscillating) value as the series resonant frequency is approached. The action described, although slightly exaggerated for ease of understanding, happens very quickly. That is, the tube reaches its operating condition in one or two cycles of operation.

Since the crystal is essentially the equivalent of a high Q circuit, it resonates only over a vary narrow range of frequencies (tuning is very sharp). Slight changes in tube parameters and supply voltages have a minimal effect, about 100 to 200 times less than in the conventional LC oscillator.

BIASING CONSIDERATIONS.—In the basic Miller oscillator, the bias is supplied by means of grid leak action, as in the conventional LC feedback oscillator. Since the crystal represents a very high Q circuit (100 times or more than that of the conventional LC tank) it is evident that the grid leak resistor effectively acts as a shunt for the voltage generated by the vibrating crystal. Therefore, an RF choke is sometimes placed in series with the grid leak resistor to reduce the load on the crystal and help it start oscillating more easily.

As the bias is increased more excitation is required to overcome the bias, and crystal current will increase. The increased excitation increases the mechanical distortion of the crystal (causes it to vibrate harder). If driven excessively, the crystal will shatter.

OTHER CONSIDERATIONS.—The plate voltage of a crystal oscillator is generally limited to a lower value than that for a similar type LC oscillator in order to reduce the amount of feedback voltage and the resultant current flow through the crystal. This is necessary for two reasons: (1) excessive crystal current flow may cause the crystal to shatter; and (2) unless the crystal is cut to have a zero temperature coefficient, its frequency will be affected by its temperature which, in turn, is affected by crystal current flow. That is, an increase in current flow will cause an increase in temperature, which, in turn, will cause either an increase or decrease in crystal frequency depending on the temperature coefficient.

Temperature changes which are external to the crystal will change the interelectrode capacitance of the tube. Also, changes in supply voltages will change the plate impedance and affect the frequency of operation. This frequency change occurs in parallel-resonant crystal operation because either a change in the feedback capacitance or the plate load impedance will affect the phase in the feedback loop and cause the crystal to operate slightly off its normal resonant frequency.

PIERCE CRYSTAL OSCILLATOR

A Pierce crystal oscillator is usually designed to operate in the lower and medium radio frequency ranges. It produces an output, which is an approximation of a sine wave, having a relatively constant frequency. It is used in applications which require an output of moderate power and good stability at a specific frequency. It is usually interchangeable with the Miller oscillator in low and medium frequency applications, but it is not often used in high frequency applications, primarily due to its low output power.

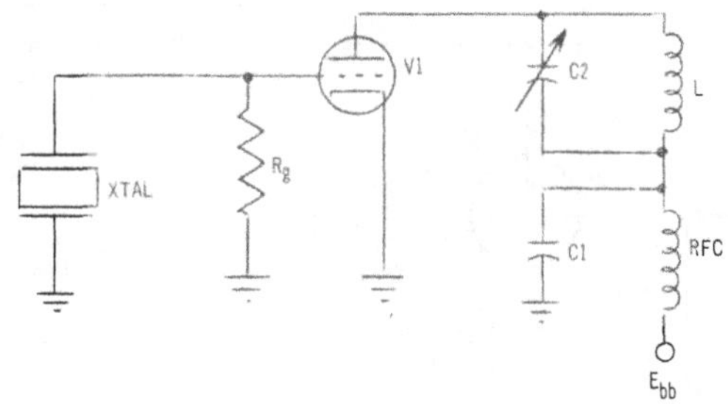

179.323
Figure 15-27.—Miller crystal oscillator.

The Pierce oscillator utilizes the piezoelectric effect of a crystal to control its frequency of operation. In this circuit the crystal is connected between the grid and plate of the tube. It does not require an LC tank circuit for operation (although a tank may be used if desired). The RF feedback occurs only through the crystal, and it normally operates with class B or C self-bias, although it may be operated class A or with a combination fixed and self-bias. The frequency stability of the Pierce oscillator is excellent, with or without temperature compensation, and its output amplitude is relatively constant.

The generalities applicable to the basic crystal oscillator in the discussion of the Miller circuit are also applicable to the Pierce circuit. The simplicity of the Pierce oscillator, with its lack of tuned plate tank circuit and its ability to oscillate easily over a broad range of frequencies with the use of different crystals, make it popular for use in crystal calibrators, receivers and test equipment, and in transmitters which do not require much driving power. The Pierce circuit is sometimes considered as the inverse of the Miller circuit since it exhibits the opposite effects. Thus, when a tank circuit is used, it is always tuned for a lower frequency than that of actual circuit operation. The crystal excitation voltage for the Pierce circuit is approximately half that permissible with the Miller circuit. The plate load of the Pierce oscillator is resistive and is usually large enough in value that minor fluctuations in the tube plate resistance have much less effect on the frequency of operation than in the Miller circuit. This is used to best advantage when a pentode is employed, since its inherently high plate resistance and low grid-plate capacitance permit a greater range of plate load with the use of an external capacitor to fix the amount of excitation.

Circuit Operation

The basic Pierce oscillator is shown in figure 15-28. Conventional grid leak bias is obtained through C_g and R_g, which operate as described previously. The crystal, which is connected between grid and plate, offers a high Q. The plate load is resistor R1. C2 is the conventional plate bypass capacitor used in series plate feed arrangements. The use of R1 in the plate circuit provides a relatively flat response over a wide range of frequencies. Various crystals may be substituted for operation on other frequencies without any tuning being required. However, when single frequency operation is anticipated, R1 is replaced with an RFC. The choke eliminates the d.c. power lost in the resistor and provides a high RF impedance for proper circuit operation. Since this would increase the plate voltage, the output power is correspondingly increased.

Now consider one cycle of operation. At rest, the crystal is unstressed and there is no charge on either plate. When plate voltage is applied, the capacitive effect of the crystal holder provides a positive grid voltage, and high grid and plate current will flow. Simultaneously, the crystal is stressed by this plate potential, and a piezoelectric charge appears across the crystal. The sudden shock of applied voltage causes the crystal to start oscillating at its parallel-resonant frequency. The plate current quickly reaches saturation at some low plate voltage, which is caused by the drop through

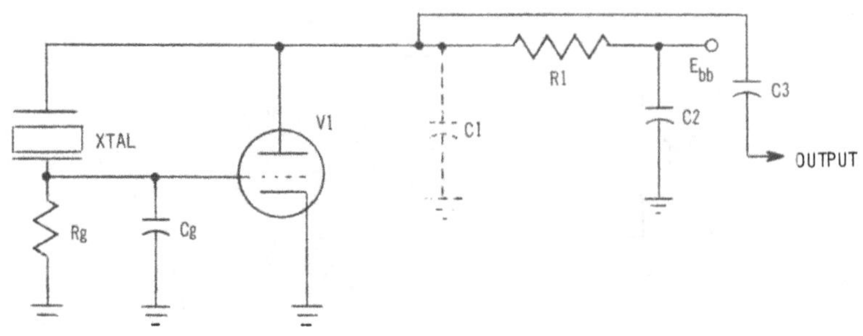

179.324
Figure 15-28.—Basic Pierce oscillator circuit.

plate resistor R1. Grid current has been drawn during this time and C_g has been charged. As the initial surge of charging current is now over, C_g begins to discharge, tube conduction decreases, and plate voltage rises. The crystal now flexes in the opposite direction, and oscillations are begun. Feedback is accomplished through the crystal itself, and the crystal controls grid voltage. This, in turn, controls tube operation.

For oscillations to be sustained, the crystal must be synchronized in its vibration, so that the piezoelectric effect does not oppose oscillation by reducing the feedback between plate and grid. The proper phase relationship is accomplished by connecting the crystal between grid and plate. The inherently high Q of the crystal makes it act as a large inductance to shift the phase of the feedback voltage in the proper direction to cause oscillation. Capacitor C1 may be added to the circuit to provide any additional phase shift required.

Although the Pierce oscillator is normally used without a plate tank circuit, this is not always so. Where the output waveform is important, use of a selective tuned circuit in the plate circuit minimizes the distortion in the output, thus providing a waveform of greater purity than is produced by a resistive plate load. Selection of the circuit with a resistor plate load for use in a crystal calibrator to supply harmonics of 200 to 300 times that of the fundamental proves particularly advantageous. On the other hand, when harmonic (frequency multiplication) operation is desired, the Pierce circuit MUST use a tuned tank circuit to select the desired harmonic if a useful and practical output is to be obtained.

Failure Analysis for Miller and Pierce Oscillators

If the crystal is removed or if poor or defective holder connections cause an open circuit, the crystal will not oscillate, and there will be no output. If the plate load resistor is open or if a short in the plate circuit sufficiently lowers the plate voltage, the crystal will not oscillate. Over-excitation may fracture or burn out the crystal causing non-oscillation. In a poorly sealed container, dust accumulation or moisture from condensation may make crystal replacement necessary. An open series blocking capacitor will disconnect the crystal and stop operation. Insufficient feedback will prevent oscillation, but this will not occur with a tube which has previously oscillated unless the tube becomes defective. An open component in the grid circuit usually stops operation, but a reduced output will sometimes occur. In most cases, only open-circuited or short-circuited conditions will prevent oscillations from taking place. Substitution will quickly determine whether the crystal or the circuit is at fault.

REDUCED OUTPUT.—Since the crystal oscillators operate vigorously when excited, any reduction in output will result from a lack of excitation or low plate voltage. Changes in the grid circuit components may cause a reduced output, as may contaminated crystals. NOTE: Do not attempt to clean or otherwise repair crystals yourself. If such action is required, they should be returned to the authorized repair center.

UNSTABLE OUTPUT.—Instability may be due to an intermittent or poor connection in the feedback circuit, but is likely due to a defective crystal. Crystals improperly cut, and having spurious frequencies, will cause instability, as evidenced by changes in frequency, due to erratic jumping from one frequency to another.

INCORRECT FREQUENCY.—As the frequency is determined primarily by the crystal, an incorrect frequency is probably the result of a change in the crystal characteristics. Normally, these changes are very small. Large frequency changes indicate that the crystal should be replaced. Small changes which occur over a period of time indicate aging in either the crystal or tube. It should be noted that crystal frequencies should not be expected to be absolutely accurate; however, they should be as accurate as their rated tolerance and this far exceeds the standard LC oscillators.

OVERTONE OSCILLATORS

An overtone type oscillator may be used in receivers, converters, test equipment and low powered transmitters. Its use is generally restricted to the high frequency range (in excess of 20 MHz) where crystal operation at the fundamental frequency is impractical. Overtone oscillators are crystal controlled oscillators which make use of crystals cut to have maximum activity at the overtone (normally an odd harmonic of the crystals fundamental frequency) to maintain stable high frequency operation. Overtone oscillators are never untuned. For proper operation, they will make use of one or more tuned circuits which operate at the overtone frequency. Also, in order to produce an output rich in harmonics they must be operated similar to class C.

At the present time, the Butler cathode-coupled two-stage oscillator is probably the most widely used oscillator of this type. This is because of its simplicity, versatility, frequency stability, and comparatively great reliability. This circuit seems to be the least critical as to design and adjustments for operation. Its schematic is shown in figure 15-29. The balanced circuit, plus the fact that twin triodes within a single envelope can be used, contributes to a saving in space and cost and provides for short leads. The problems which exist are that the power output is less than that of the Miller circuit for the same crystal excitation and the broad bandwidth of operation without tuning (as provided by the Pierce circuit) is not possible.

Operation

V2 is a grounded-grid amplifier whose output is fed back through V1 and the series connected crystal. Cathode bias is provided for the tubes by R1 and R2. Although both tubes are identical, circuit design is such that V1 normally conducts harder than V2. The feedback voltage is coupled capacitively through C4 to the grid of V1. The tuned tank circuit (L1 and C3) in the plate circuit of V2 offers maximum impedance at the frequency to which it is tuned. The maximum output voltage (and feedback) occurs at this point, neglecting crystal operation. Resistor R4 and capacitor C2 are a conventional plate dropping resistor and decoupling network for V2. Resistor R3 and capacitor C1 perform a similar function for V1. The output of the circuit is normally taken from the plate of V2 but it could be taken from V1, or the cathodes, without necessarily changing circuit operation.

When power is first applied to the circuit, heavy currents will flow in the circuit. This is due to capacitor charging, and a lack of bias. The difference in conduction levels of V1 and V2 will produce a potential across the crystal, crystal current will flow causing stress, and the resulting stresses will shock it into oscillation.

Assume that the crystal is oscillating in such a manner as to place a positive going signal on the cathode of V2 (this would be equivalent to a negative going signal on the grid). Plate current in V2 will decrease, and plate voltage will increase. The positive going signal on the plate of V2 is coupled through C4 to the grid of V1. V1 will conduct harder, E_{R1} will become more positive than before, and the crystal will couple this positive going signal to the cathode of V2. Note that this is a regenerative effect. Observation should make it clear that regeneration will occur on the negative half cycle of oscillation. Thus, with the crystal providing voltages which will control the conduction of V2, the variations in the plate voltage of V2 can then supply the necessary pulsating energy to the tuned tank circuit to maintain oscillations. It should be noted that this circuit will oscillate without the crystal, if the crystal is replaced with a resistor. Accurate frequency control, however, is possible only with the crystal in place. Failure analysis of this circuit is similar

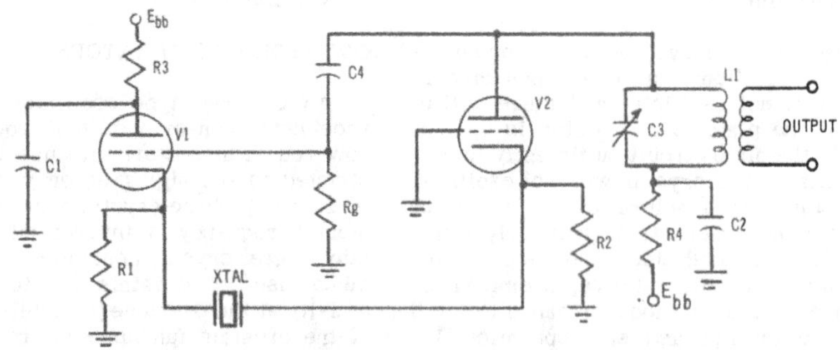

179.325
Figure 15-29.—Butler cathode-coupled crystal oscillator.

to that of other crystal oscillators and will not be covered in this text.

It should be pointed out that there are many variations of the overtone oscillator circuitry, and that overtone oscillators can be developed utilizing solid state devices as well as electron tubes. Due to the wide selection of circuitries available, space allotments preclude their coverage here, and the reader is advised to check locally available references for further information.

TUNNEL DIODE OSCILLATORS

The tunnel diode (also called the Esaki diode, after its discoverer) is one of the most significant solid state devices to emerge from the research laboratory, since the transistor. It is smaller and faster in operation than either the transistor or the electron tube. It also offers a host of additional outstanding features. The high switching speed of tunnel diodes coupled with their simplicity and stability, makes them particularly suitable for high speed operation. They operate effectively as amplifiers, oscillators, and converters at microwave frequencies. In addition, tunnel diodes have extremely low power consumption and are relatively unaffected by radiation, surface effects, or temperature variations.

The two terminal nature of the tunnel diode is something which should be considered. On one hand, this feature allows the construction of circuits which are very simple, and consequently provide a savings in size and weight. This also provides a significant improvement in reliability. On the other hand, the lack of isolation between the input and the output can be a serious problem in some applications. Below the microwave frequency region, transistors are usually more practical and economical. However, at microwave frequencies, tunnel diodes have several advantages and are highly competitive with other high-frequency devices.

GENERAL DESCRIPTION

A tunnel diode is a small two-lead device having a single PN junction which is formed from very heavily doped semiconductor materials. It differs from other PN junction diodes in that the doping levels are from one hundred to several thousand times higher in the tunnel diode. The high impurity levels in both the N and P type materials result in an extremely thin barrier region at the junction. The effects which occur at this junction produce the unusual current-voltage characteristics and high-frequency capabilities of the tunnel diode.

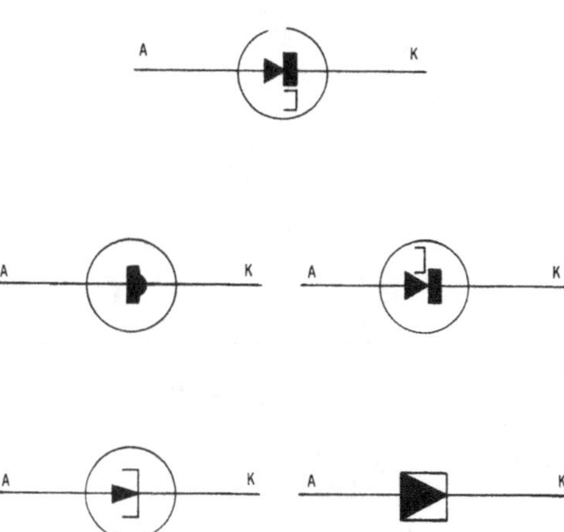

179.326
Figure 15-30.—Schematic symbols for the tunnel diode.

There are various schematic symbols used to represent the tunnel diode as shown in figure 15-30. The MIL-STD symbol is shown at the top of the figure. Figure 15-31 compares the current voltage characteristic curve of the tunnel diode with that of a conventional PN diode. The broken line shows the curve for the conventional diode and the solid line indicates the characteristic curve of the tunnel diode.

In the standard PN diode, the forward current does not begin to flow freely until the forward voltage reaches a value on the order of half a volt. This corresponds to point Y in figure 15-31. This forward voltage is sometimes referred to as the "offset" voltage. In the reverse direction, the conventional diode has a high resistance to current flow until the "breakdown" region is reached. The breakdown region is shown at point X in figure 15-31.

The tunnel diode is much more conductive near zero voltage. Appreciable current flows when a small bias is applied in either the forward or the reverse direction. Because the active region of the tunnel diode is at a much lower voltage than the standard semiconductor devices, it is an extremely low power device.

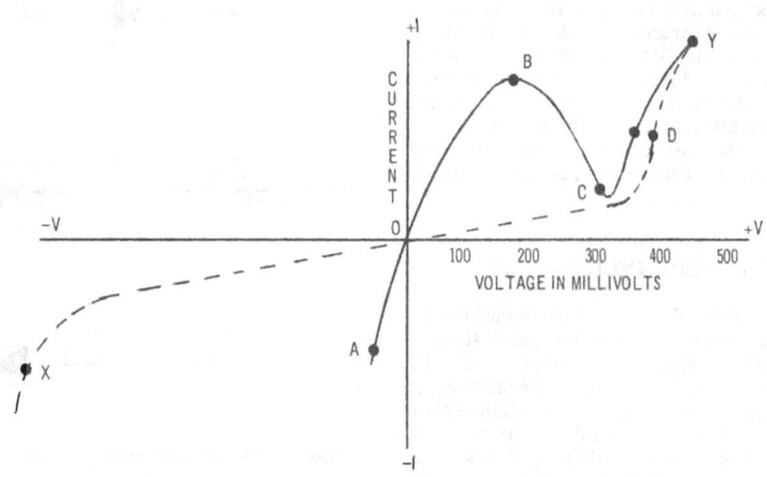

179.327

Figure 15-31.—Characteristic of a tunnel diode compared to a standard PN junction.

As the forward bias on the tunnel diode is increased, the current reaches a sharp maximum as shown at point B in figure 15-31. This point is referred to as peak voltage and peak current point. The curve then drops to a deep minimum at point C. This is called ''valley point voltage'' and ''valley point current'' respectively. The curve now increases exponentially with the applied voltage and finally coincides with the characteristic curve of a conventional rectifier. This is shown at point Y on the characteristic curve. The drop in current with an increasingly positive voltage (area B to C) gives the tunnel diode the property of negative resistance in this region. This negative resistance enables the tunnel diode to convert d.c. power supply current into a.c. circuit current, and thus permits its use as an oscillator.

Operation of the Tunnel
Diode Junction

An mentioned previously, tunnel diodes use much higher doping levels than conventional diodes. Typically, the transistor region in tunnel diodes (potential barrier) is 100 times less than that of a standard PN junction. This difference in width and doping level accounts for the major difference in the tunnel diode and other types of junctions. Because of the narrowness of the potential barrier, charge differences may easily make themselves felt across the junction when the junction voltages are applied.

Figure 15-32 shows a graph which indicates the rate of electron orbital distortion which occurs in P and N type materials. The graphs have been greatly expanded and slightly distorted for the sake of clarity. Notice, that initially, the N material electrons undergo the greatest orbital distortion. This is the area on the graph from the origin to point A. During this portion of the graph the electrons of the N material have a greater orbital distortion than those in the P material. Consequently, energy is transferred across the relatively narrow junction. This corresponds to the area from O to B in figure 15-31. In this area, relatively large currents are produced with the application of very small voltages.

At point A in figure 15-32, the rate of orbital distortion of the electrons in the P material has increased and has caught up with the rate of distortion in the N material. The electron orbits in the P material are now being distorted faster than the N material electron orbits can distort and intrude across the junction. The total transfer of energy must therefore decrease. The area from A to B in figure 15-32 corresponds to the area

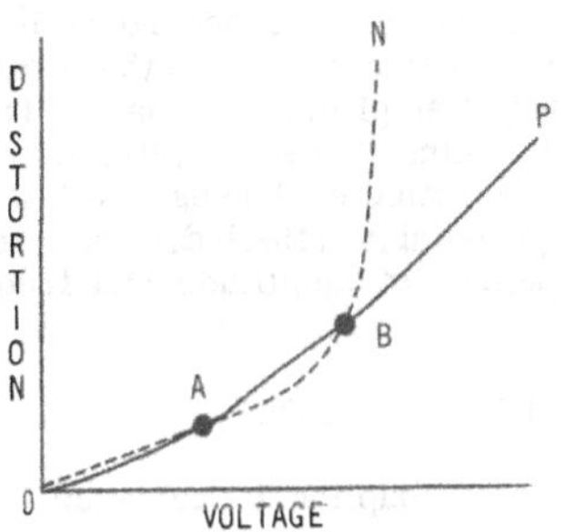

179.328
Figure 15-32.—Orbital distortion.

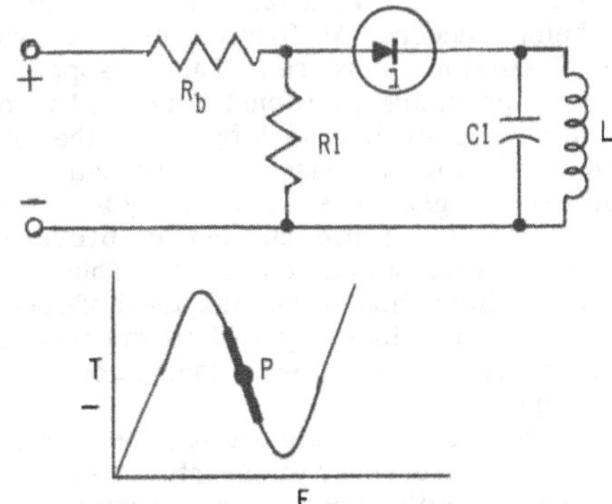

179.329
Figure 15-33.—Basic tunnel diode oscillator and characteristic curve.

B to C in figure 15-31. This is the negative resistance region of the tunnel diode.

At point B of figure 15-32 the orbital distortion of the N material electrons has once again caught up with the distortion rate in the P material. Beyond point B the N material electron orbits are distorting faster than those in the P material. There is therefore an increase in the effective intrusion of the N material electron orbits and a corresponding increase in current. From this point on the operation of the tunnel diode is the same as that of a standard PN junction. A logical question at this point would be: Why isn't negative resistance exhibited by a standard diode? It must be remembered that the junction barrier in a standard diode is much wider than in the tunnel diode. The wider barrier prevents any appreciable current flow until sufficient voltage is applied and the diode is operating well above point B of figure 15-32.

TUNNEL DIODE OSCILLATOR CIRCUITS

Figure 15-33 shows the basic circuit for a tunnel diode sinusoidal oscillator, as well as the diode's characteristic curve. R_b and R1 in the schematic are used to adjust the diodes operating point to the negative-resistance portion of its characteristic curve. This is shown as point P on the curve. This point is selected to be as near as possible to the center of the negative-resistance region of the curve. This portion is relatively linear and operation over this linear region is required if the ideal output waveform is to be achieved.

Circuit Operation

When power is first applied there is a brief surge of current through C1, through the junction capacitance of the tunnel diode, and through R_b to the positive terminal of the source. This surge supplies sufficient power to C1 (of the tank circuit) to begin the operation of the resonant circuit. The level of surge current is held to an acceptable value by the limiting action of dropping resistor R_b. The voltage divider action of R_b and R1 now set the d.c. bias for the tunnel diode at point P of the characteristic curve so that the diode will operate on the linear portion of the negative-resistance slope.

Note that the voltage applied across the tunnel diode is determined by not only the potential developed across R1, but also by the varying voltage in the resonant tank circuit. The d.c. bias developed across R1 would (by itself) operate the diode at point P of the characteristic curve. Voltage variations (positive and negative half cycles) of the tank circuit would then shift the operation of the diode back and forth along the negative-resistance portion of the curve. Recall that there is some resistance in the resonant circuit. This positive resistance of the tank circuit absorbs power and would eventually damp the tank circuit oscillations. However, if the negative resistance of the tunnel diode is equal to or greater than the resistance of the tank circuit, the diode will act as a switch which will resupply the lost energy to the tank circuit, and oscillation will be sustained.

During the positive half cycle of the tank's operation, the forward bias on the tunnel diode is being reduced. As forward bias is reduced, diode current increases. This resupplies the tank by providing additional current. During the negative half cycle of tank operation the forward bias is being increased but the current will correspondingly decrease as Valley Point Voltage is approached. Little current is provided and the tank is essentially on its own. Thus, on each positive alternation of the tank, the diode conducts harder, energy is resupplied to the tank, and if this energy is sufficient, oscillations will be sustained.

In this circuit configuration, the tank circuit controls the frequency of operation and the tunnel diode essentially functions as a device to supply the energy required to maintain oscillations. Extremely high frequency operation can be obtained by utilizing the tunnel diode. The narrowness of the potential barrier at the junction allows a very fast reaction time for the device. While tunnel diodes are still a relatively new device, a great deal of research is being done, and it is felt that the full potential of this device is still to be realized.

OSCILLATOR COUPLING METHODS

A circuit which will produce stable, self-sustained oscillations is useless unless there is some way to couple a useful output from it. There are two methods of coupling in general use. They are INDUCTIVE COUPLING and CAPACITIVE COUPLING.

INDUCTIVE COUPLING

Inductive coupling is accomplished when the inductance of a tank circuit is used as the primary of a transformer. The output is then taken from the secondary winding. In some cases the secondary winding is shunted by a capacitor to form a tank circuit which is resonant to the frequency of the oscillator. This is then known as a tuned secondary.

In either case, the output amplitude is determined by the degree of coupling between the tuned primary and the secondary winding. If the coupling is tight (primary and secondary close together), the output amplitude is high, and a considerable amount of power is transferred from the oscillator tank to the secondary. This has the disadvantage of reflecting a high I^2R loss back to the tank circuit, effectively loading it and lowering the tank Q. If the distance between the tank inductance and the secondary is increased, less power is transferred to the oscillator tank, and the output amplitude decreases. This is known as loose coupling. Loose coupling is usually used because the reflected losses are low and tank Q is not appreciably affected. This results in a greater degree of amplitude and frequency stability.

CAPACITIVE COUPLING

Capacitive coupling is achieved by selecting coupling capacitors which present a high capacitive reactance at the resonant frequency. In this manner the signal may be routed from one location to another, without excessively loading the tank circuit and lowering the Q. Slight adjustments may be necessary as capacitive coupling will often have slight effects on the frequency of operation.

STABILITY

AMPLITUDE STABILITY may be affected by changes in bias, transistor or tube gain, supply voltage, and reflected impedance. An increase in bias would cause a decrease in feedback amplitude in most circuits. The decreased feedback would lower the amplitude of oscillations. A possible decrease in bias would produce the opposite effects.

A decrease in the gain of the amplifier device would reduce the gain of the circuit and produce the same effect as increased bias. A change in supply voltage would change the operating point of the circuit and, consequently, change the amplitude of oscillations.

If the impedance of the load decreases it will reflect an increased power drain back to the oscillator tank circuit in most cases. Without a change in feedback amplitude this loss would not be compensated for and the amplitude of the oscillations would decrease. Amplitude stability can be improved by utilizing base leak bias in transistor circuits and grid leak bias in electron tube circuits. This improvement is due to the bias level being dependent on the amplitude of the tank signal.

FREQUENCY STABILITY can be affected by changes in temperature, tank Q, reflected losses, supply voltage and vibration. Changes in temperature will cause the inter-element capacitances to vary in both the transistor and tube. They will also cause slight variations in the values of tanks L and C. These changes will shift the output frequency. Since these changes occur slowly, the resultant shift in frequency is called DRIFT. Frequency drift can be minimized by adequate ventilation, circuit components which can easily dissipate heat, or, in some cases, a temperature control system.

Mechanical vibration can cause the same variations in component values as temperature changes did. In this case, however, the resultant frequency shift is more rapid in nature. The oscillator chassis is often shock mounted to overcome this problem. Variations in supply voltage will change the operating point of the circuit, and cause circuit instability. This can be overcome by use of a regulated power supply.

The load impedance reflected into the tank circuit, has a great effect upon the frequency stability of the oscillator. The reflected impedance may contain both resistive and reactive components. The resistive component will lower the Q of the tank and the reactive component will alter the resonant frequency. When the Q of the tank is very high and the reflected impedance is small, the effect on the resonant frequency and tank Q is negligible. The lower the effective Q of the tank the greater will be the effect of a change in load impedance on the resonant frequency. High Q tanks and loose coupling reduce the frequency instability caused by these factors.

TUNING AND TRACKING

In a superheterodyne receiver, the local oscillator frequency will always differ from the desired station frequency by an amount equal to the intermediate frequency. The local oscillator develops the intermediate frequency. It is usually tuned above the station frequency, although, in some cases, it may be tuned below. When a receiver, having an IF of 455 kHz, is tuned to a station at the low end of the broadcast band (540 kHz for example), the local oscillator frequency will be: 540 kHz + 455 kHz = 995 kHz. When the same receiver is tuned to a station at the high end of the band (1600 kHz for example), the local oscillator will be: 1600 kHz + 455 kHz = 2055 kHz.

Thus, the frequency range of a local oscillator in a superhet receiver, with a 455 kHz IF, will vary from approximately 995 kHz to 2055 kHz. The tuning component, usually the capacitor, of the tank circuit is mechanically ganged with the tuning component of the input circuit. This allows the local oscillator frequency to be changed as the station frequency is changed, thereby maintaining the LO frequency, at all times, separated from the station frequency by an amount equal to the IF.

Since the LO frequency follows the selected frequency as tuning occurs, the LO is said to "track" the capacitors in the RF stage; there will be slight electrical differences (due to manufacturing tolerances) in the oscillator tuning capacitors. For this reason, a trimmer capacitor is usually connected in parallel with the oscillator tuning capacitor. Adjustment of the trimmer at the high end of the band insures better tracking. Tracking adjustments at the low end of the band are usually accomplished either through the use of padder capacitors, slug tuned inductors, or slotted rotor end plates on the tuning capacitor.

Usually, Hartley type oscillators are tuned by means of a variable tank capacitor. This lends itself easily to the ganged tuning mentioned earlier, since the input frequency selector circuit is usually tuned by means of a variable capacitor. The rotors of these variable capacitors can be mechanically connected to a common shaft. It should also be recalled that the varactor is sometimes used to accomplish oscillator tuning.

Tracking is more difficult in a Colpitts oscillator when the tank capacitors are made variable. The two capacitors provide impedance matching between the output and input circuits. The ratio of impedances must be maintained throughout the range of frequencies or an impedance mismatch will occur. This results in a loss of feedback power. The difficulty encountered in varying Colpitts oscillator frequencies may be reduced by making the tank coil variable, or by making the split capacitors of the tank circuit fixed and inserting a variable tuning capacitor in parallel with the tank. Operation of the Colpitts in this manner will allow the tank frequency to be varied, but will maintain the impedance ratio between the two fixed capacitors.

CHAPTER 16

MIXERS AND CONVERTERS

The essential difference between the TRF receiver and the superheterodyne receiver is that in the former the RF amplifiers preceding the detector are tunable over a band of frequencies; whereas in the latter the corresponding amplifiers are tuned to one fixed frequency called the INTERMEDIATE FREQUENCY (IF). The principle of frequency conversion by heterodyne action is here employed to convert any desired station frequency within the receiver range to this intermediate frequency. Thus an incoming signal is converted to the fixed intermediate frequency before detecting the audio signal component, and the IF amplifier operates under uniformly optimum conditions throughout the receiver range. The IF circuits thus may be made uniformly selective, uniformly high in voltage gain, and uniformly of satisfactory bandwidth to contain all of the desired sideband components associated with the amplitude-modulated carrier. This change of frequencies is achieved within the frequency conversion stage of the receiver. There are two basic types of frequency conversion stages used in the "superhet" receiver, one type being the Mixer and the other type being the Converter. The frequency conversion process is the heart of the superheterodyne principle.

Figure 16-1 is the block diagram of a superheterodyne receiver showing the location of the mixer stage. The modulated RF signal from the antenna passes first through an RF amplifier where the amplitude of the signal is increased and passed on to the mixer stage. A locally generated, unmodulated RF signal (local oscillator signal) of constant amplitude is then mixed with the RF carrier in the mixer stage. The local oscillator frequency is usually higher than the incoming signal frequency. The mixing or heterodyning of these two frequencies produces an intermediate frequency (IF) signal which contains all of the modulation characteristics of the original signal. The intermediate frequency is equal to the difference between the station frequency and the oscillator frequency associated with the heterodyne mixer.

The process of combining two or more frequencies in a nonlinear device and producing new frequencies is called mixing, modulating, heterodyning, beating, or frequency conversion. The stage that this process is taking place in may be called a mixer, converter, translator, or first detector.

HETERODYNING

The principle of mixing is not related to electronics alone. The basic principles are discussed in physics courses.

The production of an audible beat note is a phenomenon which is easily demonstrated. For example, if two adjoining piano keys are struck simultaneously, a note will be produced that rises and falls in intensity at regular intervals. This action results from the fact that the rarefactions and compressions, produced by the vibrating strings, will gradually approach a condition in which they reinforce each other. This occurs at regular intervals of time with an accompanying increase in the intensity of the sound. Likewise, at equal intervals of time, the compressions and rarefactions gradually approach a condition in which they counteract each other and the intensity is periodically reduced. Figure 16-2 illustrates graphically how the resultant difference vibration appears.

If wave A were 98 hertz and wave B were 100 hertz, both of the same amplitude, the resultant vibration would rise and fall at a difference frequency of 2 hertz. It is important to note that the mere existence of this amplitude variation does not directly indicate the presence of a difference frequency component. In the diagram shown, the difference frequency is observed, but is not necessarily detected until the ear itself acts as a nonlinear device or mixer.

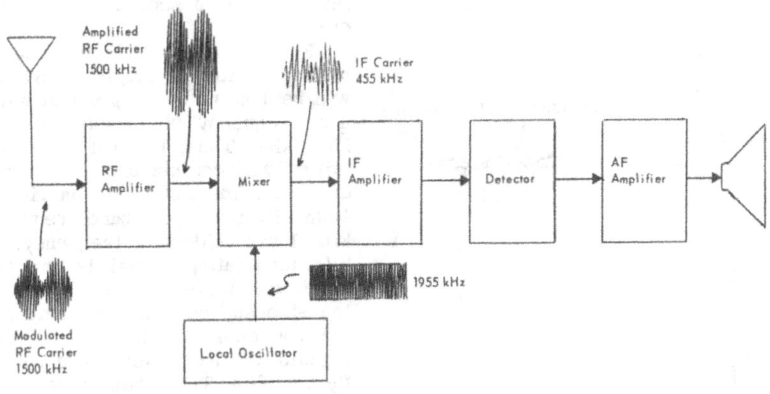

13.66(20C)
Figure 16-1.—Superheterodyne receiver.

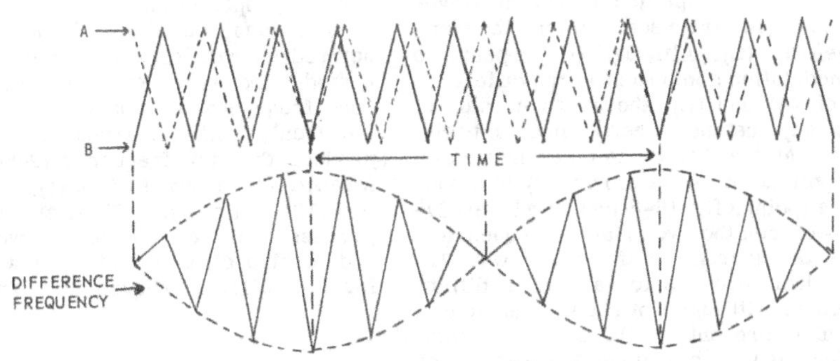

179.330
Figure 16-2.—Resultant difference vibration.

To obtain a difference frequency component, it is necessary to apply the original frequencies to a nonlinear device.

Since the transistor and the tube, like the ear, display nonlinear characteristics, the simultaneous application of various frequencies will result in the reproduction of the original frequencies plus the production of various new frequencies.

The necessity for the use of a nonlinear device to produce the heterodyning process can best be demonstrated by the use of a response curve. The response curve shown in figure 16-3 is a graph of current versus voltage. This curve, a dynamic transfer curve, can represent V_{BE} -I_C of a transistor or E_g -I_p of an electron tube. It should be recalled that the transfer curve was constructed earlier for the purpose of showing the nonlinear characteristics of transistor and tube amplifiers. It was also shown previously that simultaneous application of two different frequencies to a linear device produced an output containing only the original frequencies while simultaneous application of the same two frequencies to a nonlinear device produced not only the original frequencies but also the sum and the difference of two original frequencies. Figure

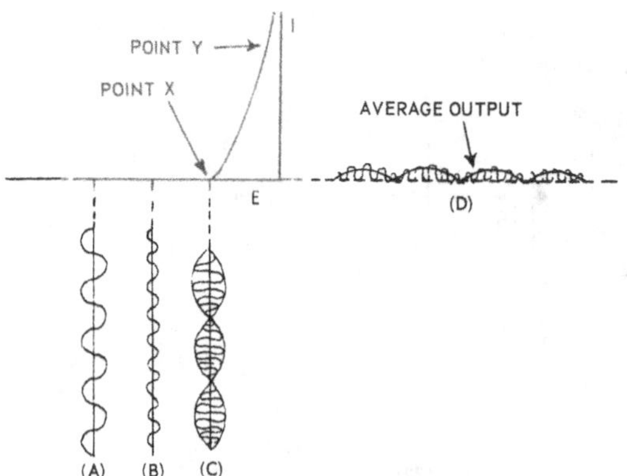

179.331
Figure 16-3.—Current versus voltage.

16-3 will be used to explain this action. Waveforms (A) and (B) are used to form the composite waveform (C). Although (C) appears to contain a modulation component at the difference frequency of (A) and (B), the average value of wave (C) at any instant is zero, thus, no useful energy exists at the difference frequency. However, the application of waveform (C) to a nonlinear device (point X fig. 16-3) causes a heterodyning action between the two original frequencies. This causes the average of the output wave (D) to be other than zero and to vary at the difference frequency. Although not shown graphically, energy is also present at the sum frequency.

If waveform (C) were applied at point Y rather than point X, and the output waveform graphed, it would be noticed that (due to the linear characteristics of the curve in this region) the variations of I would be nearly symmetrical. Thus, very little heterodyning action would take place and the average output would be very nearly zero.

It can be concluded that as the device becomes more linear the average amplitude variations in the output approach zero. If the device used were perfectly linear the average output would be exactly zero and only the original frequencies would be present.

The production of beat frequencies in a "superhet" receiver is similar to the examples previously discussed. In the mixer stage, the process is entirely electrical and the frequencies are much higher. Figure 16-4 illustrates the block diagram of the mixer stage and the actual frequencies involved (for a specific situation) in the process of mixing.

For simplicity, during the explanation of mixing action, a single 5 kHz audio frequency will be used as the original modulating frequency. The frequency of the station carrier signal is 1000 kHz. Thus, the AM signal from the transmitter will contain energy at three distinct frequencies; the 1000 kHz carrier frequency, the 1005 kHz upper sideband frequency, and the 995 kHz lower sideband frequency. The original 5 kHz modulating signal is now contained in the relationship between the carrier and either of the sideband components. This AM signal from the RF amplifier is used as one of the input signals to the mixer stage and is indicated in figure 16-4. The other input to the mixer stage is a constant amplitude RF signal from the local oscillator. A prerequisite of the device used as the mixer is that it be nonlinear. When using a transistor or tube as a mixer, the curved portion of the dynamic transfer curve is utilized to achieve proper mixing action.

The local oscillator frequency will beat or heterodyne with all the individual components of a modulated wave. For example, the upper sideband frequency of 1005 kHz will heterodyne with the local oscillator frequency of 1455 kHz and produce the sum frequency (2460 kHz), the difference frequency (450 kHz), and the original two (1005 kHz and 1455 kHz) frequencies. This process will be repeated between the carrier and local oscillator and the lower sideband and local oscillator frequencies. The net result of

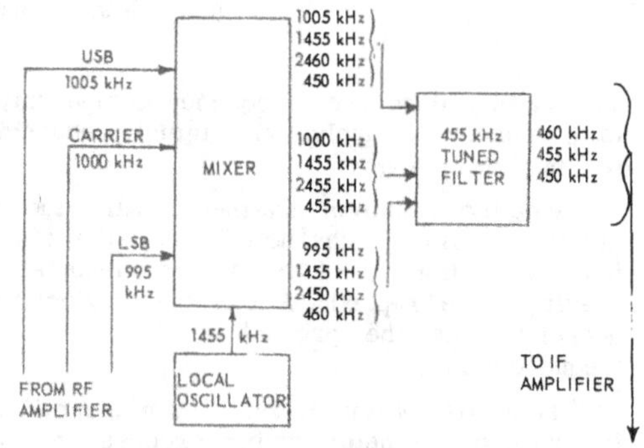

179.332
Figure 16-4.—Mixer stage.

all these actions is an output which contains ten significant frequency components. It should be noted that the output will also contain harmonics of all the sum, difference, and original frequencies. However, for simplicity of explanation they will be neglected. In a practical situation, the AM signal would contain many sideband frequencies, which would in turn produce a multitude of significant output frequencies.

Since the output of the mixer contains many frequencies, a tuned resonant tank circuit acting as a filter is employed to select the desired difference frequencies. This filter consists of components arranged to pass a band of frequencies centered around 455 kHz. The response curve of a typical tuned circuit is shown in figure 16-5. The characteristics of this tank circuit are a relatively uniform gain within the 0.707 (half power) points and increasingly less gain for frequencies further from (above or below) the center frequency. Therefore, only frequencies closely associated with the resonant frequency of the tuned circuit will develop sufficient voltage to be passed on to the IF amplifier. Figure 16-5 indicates that the other frequency components present in the mixer output will fall outside the bandpass of the tank circuit and will not develop sufficient voltage across the tuned circuit to be passed on to the IF amplifier. Figure 16-5 represents the individual bandwidth of the mixer tank circuit, and not the overall bandwidth of a complete receiver. The half power points have been chosen arbitrarily, and are dependent on receiver circuit design and characteristics.

In the mixer block diagram (fig. 16-4), the output of the tuned circuit contains the frequency components 450 kHz, 455 kHz and 460 kHz. The output of the tuned circuit constitutes the modulated intermediate frequency signal. It should be noted that, although the sideband and carrier have been converted to lower frequencies, the same relationship between the sideband and carrier frequencies has been maintained in the difference between the IF and its side bands.

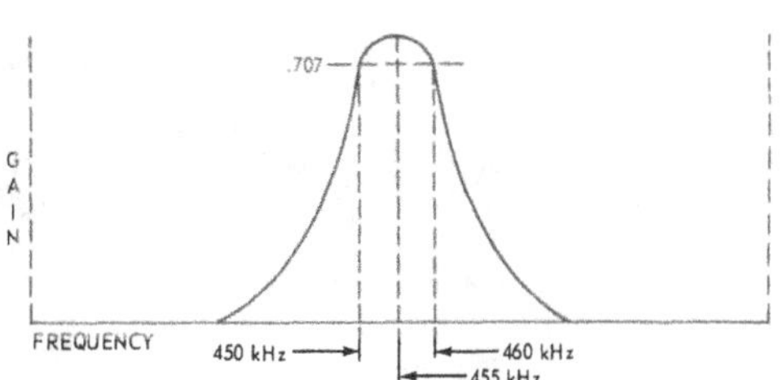

179.333

Figure 16-5.—Bandwidth of mixer tank circuit.

MIXER AND CONVERTER CHARACTERISTICS

Certain characteristics are referred to when describing the performance of mixers and converters in frequency conversion systems. One of these is the conversion transconductance which is defined as the ratio between the IF current in the output of the frequency converter and the RF signal voltage. The value is given in micromhos. The conversion transconductance is an important quantity in that it governs the gain of the stage; the higher this value, the greater the gain.

The amplification achieved in the frequency converter is called the conversion gain and is defined as the ratio of the IF output voltage to the RF input voltage. A high signal-to-noise ratio is another important desired characteristic. All frequency converters introduce a certain amount of noise, reducing the overall signal-to-noise ratio of the receiver.

A highly desirable characteristic in frequency converters is that there be a minimum of interaction between the local oscillator and the RF signal circuit. This interaction results in a change in oscillator frequency, called pulling, under certain conditions. In order that the local oscillator have maximum frequency stability, it is necessary to isolate it from the RF signal circuit.

TRANSFORMER COUPLED MIXER

It has been stated that the mixer has an RF signal input and an input from the local oscillator. These signals may be applied to the mixer in many different ways, depending on circuit configurations, design considerations, etc. Figure 16-6 is a partial schematic of a CE transistor mixer illustrating the injection of the oscillator signal on each of the transistor elements.

Transformer T1 couples the RF signal from the RF amplifier, or in some cases the antenna, to the base of the mixer Q1. The primary of T2 and capacitor C2 form the tuned output circuit of the mixer stage and are resonated at the intermediate frequency. Transformer T3 couples the oscillator frequency into the mixer and is shown in three possible connections. T3A acts in series with T1 and couples the oscillator signal to

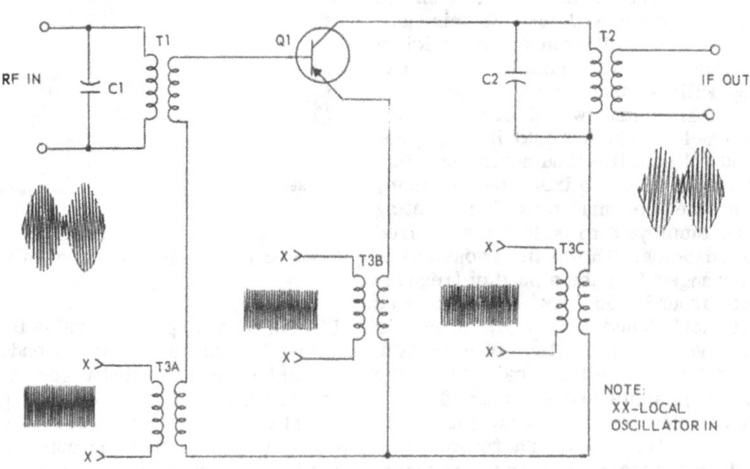

179.334
Figure 16-6.—CE PNP transistor mixer.

the base of Q1. T3B couples the oscillator signal to the emitter and controls the conduction of Q1 by varying emitter bias. T3C shows the oscillator signal coupled in the collector circuit. Regardless of where the oscillator signal is applied, the mixing action described previously remains the same.

Figure 16-7 is a partial schematic of a triode electron tube mixer illustrating the injection of the oscillator signal on each of the tube elements. Transformer T1 couples the signal from the RF amplifier to the grid of mixer V1. The primary of T2 and capacitor C2 form the tuned output of the mixer stage and are resonated at the intermediate frequency. Transformer T3 couples the oscillator frequency into the mixer and is shown in three possible connections. T3A acts in series with T1 and couples the oscillator signal to the grid of V1. T3B couples the oscillator signal to

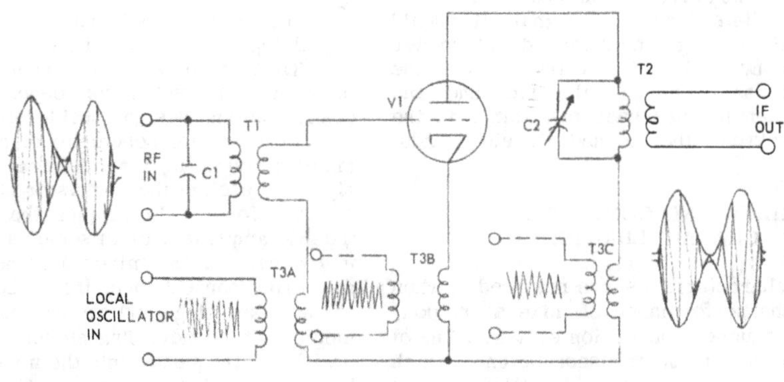

179.335
Figure 16-7.—Triode electron tube mixer.

the cathode while T3C couples the oscillator signal to the plate.

Local oscillator signal injection into the base or grid circuit is generally undesirable because of the interaction between the local oscillator and RF signal. Also, if the input to the mixer is connected directly to the receiver antenna, the radiation of the local oscillator signal by the antenna can cause interference in other receivers.

The use of emitter-cathode or collector-plate injection of the local oscillator signal minimizes the local oscillator voltage appearing across the RF signal source and thus minimizes interaction between these two circuits. In addition, it reduces radiation of the local oscillator signal if the output to the mixer stage is not isolated from the antenna.

RC COUPLED MIXERS

A mixer stage may also employ resistance-capacitance coupling of the local oscillator signal into the stage. Two RC coupled transistor mixers are illustrated in figure 16-8. In figure 16-8A, the oscillator signal is coupled by capacitor C3 to the emitter of the mixer while in figure 16-8B the oscillator signal is coupled by C2 to the base.

Figure 16-9 illustrates two methods of capacitive coupling the oscillator signal into an electron tube triode mixer stage.

DYNAMIC OPERATION

A complete schematic diagram of a mixer stage employing emitter injection of the oscillator signal is shown in figure 16-10.

Capacitor C1 and the primary of T1 make up the RF input tank and are tuned to the radio frequency signal input. T1 couples this signal to the base of Q1. Resistors R1 and R4 form a fixed voltage divider network. Their values are chosen to correctly set the magnitude of forward bias for operation on the nonlinear portion of the transistor's dynamic transfer curve. Resistor R2 provides bias stabilization and capacitors C3 and C4 prevent degeneration. Capacitor C5 and the primary of T3 form a parallel resonant circuit for the intermediate (difference) frequency.

The oscillator output is usually much larger (10 or more times in amplitude) than the incoming RF. For this reason, the collector current of the mixer is controlled primarily by the oscillator signal and the transconductance of the transistor or tube will be a function of the applied local oscillator signal. The signal developed across the

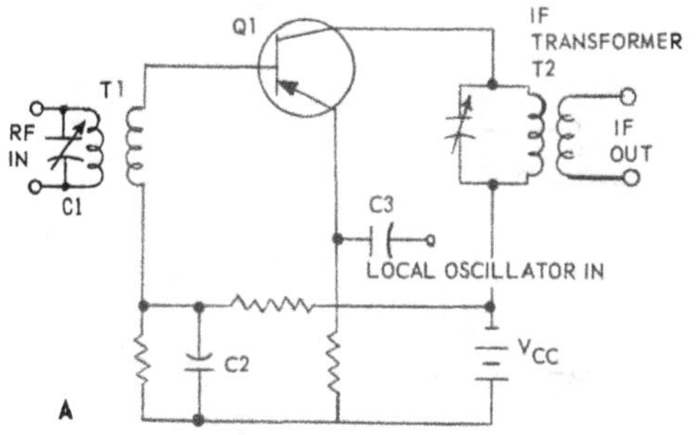

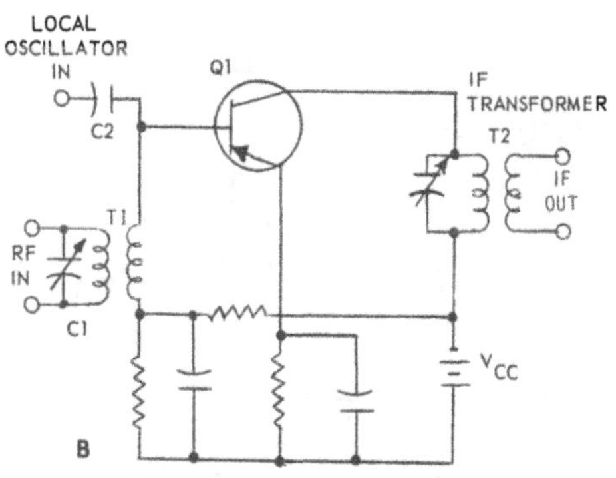

179.336

Figure 16-8.—RC coupled transistor mixers.

emitter-base junction is the instantaneous sum of the oscillator signal and the radio frequency signal. This emitter-base signal will be a complex waveform due to the heterodyning of the oscillator signal and the radio frequency signal in the nonlinear junction of the mixer. The resulting collector current will be a complex waveform consisting of many frequencies including the sum, difference, and the original two frequencies. At resonance the tank consisting of C5 and the primary of T3 will oscillate at, and offer a high impedance to, the difference frequency and its associated sidebands. Because of the high circulating current, the difference frequency will be coupled into the first IF amplifier for further amplification.

Since the useful difference frequency current is just a portion or component of the total collector current signal, the conversion gain is

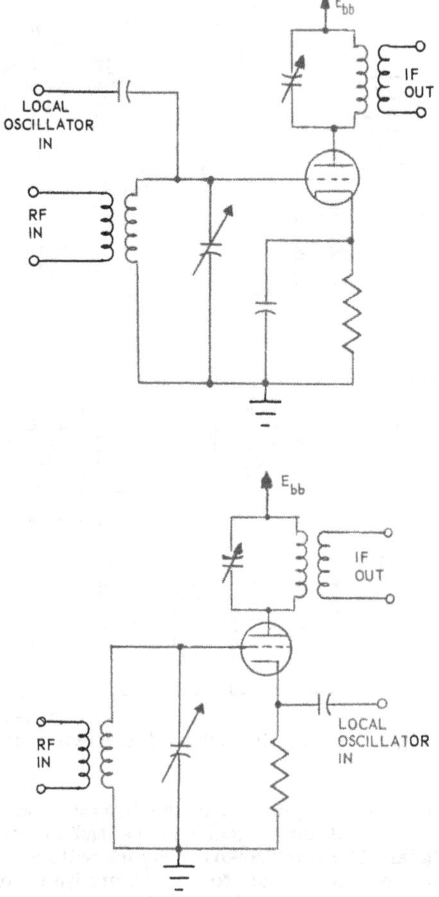

Figure 16-9.—Triode mixer stages.

much smaller when comparing it to the gain of the same stage used as an amplifier. Although the mixer or converter stage does not contribute much to the receivers total gain, the function of frequency conversion has been accomplished. The additional gain required is achieved by the use of IF amplifiers.

Figure 16-11 shows the schematic of a triode electron tube mixer. T1 secondary and C1 make up the RF input tank. R1 is the cathode bias resistor bypassed by C2 to prevent degeneration. C3 and the primary of T2 make up the plate tank and are tuned to the intermediate frequency. C4 bypasses the signal frequencies around the power supply. T3 couples the local oscillator signal to the cathode of the mixer.

Both the oscillator signal coupled into the cathode circuit and the RF signal injected into the grid circuit cause variations in the mixer plate current. The two different frequency components of the plate current beat together and generate the difference frequency (IF). The plate tank begins to oscillate at, and offers a high impedance to, the difference frequency and its associated sidebands. Because of the high circulating tank current, the difference frequency will be coupled into the first IF amplifier for further amplification.

CONVERTERS

Sometimes the function of the oscillator and mixer are incorporated into one stage containing two sections. When this is done, the stage is called a converter. The only external input to the converter stage is the modulated RF signal. The local oscillator signal is generated within the stage. The principal advantage of a converter is the reduction of the number of components and transistors used to accomplish frequency conversion. The disadvantage of a converter is that stable oscillation, which should be a characteristic of the local oscillator, is dependent upon a minimum of nonlinearity, whereas the frequency conversion process requires a high degree of nonlinearity.

A converter circuit using an NPN transistor is shown in figure 16-12. A converter circuit is characterized by only one input signal. The modulated RF signal from the RF amplifier is transformer coupled to the base of Q1. The local oscillator signal is generated internally by the action of L2, C4, and C5. R1 and R2 form a voltage divider network to establish the transistor operating point. R3 provides emitter stabilization. The primary of T3 (L1) provides regenerative feedback to the local oscillator tank circuit. C2 is a bypass capacitor across R1 to prevent degeneration.

OPERATION

The modulated RF signal is transformer coupled to the base of the converter. The constant

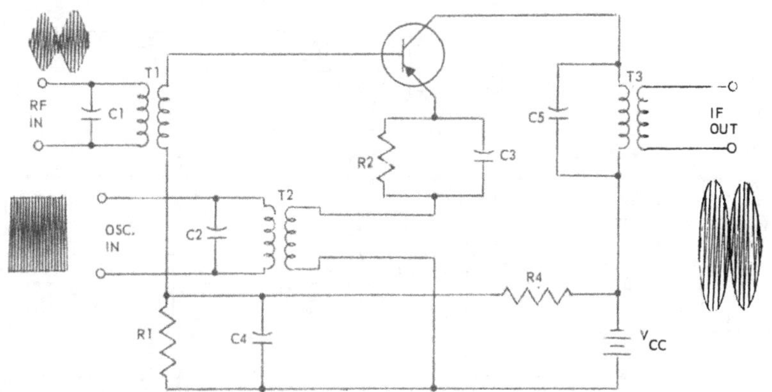

179.338
Figure 16-10.—Mixer stage employing emitter injection of the oscillator signal.

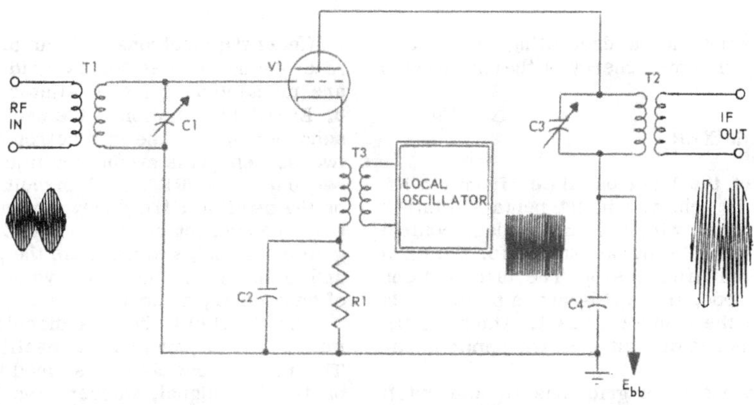

179.339
Figure 16-11.—Triode electron tube mixer.

amplitude local oscillator signal is coupled through C3 to the emitter of the transistor. Transistor Q1 is biased by R1 and R2 on the nonlinear portion of the dynamic characteristic curve for proper heterodyning action. The transistor's collector current is controlled simultaneously by the oscillator signal and the incoming signal. The collector current will be a complex waveform consisting of many frequencies including: the local oscillator signal frequency, the received incoming signal frequency, the sum of these two frequencies, and the difference between these two frequencies.

The primary of the output transformer T2 and C6 form a resonant circuit, which is tuned to the difference frequency. Thus, the IF will develop a relatively large voltage across the tuned circuit while the remaining signals (LO, RF, and the sum frequency) will develop relatively small voltages across the tuned circuit.

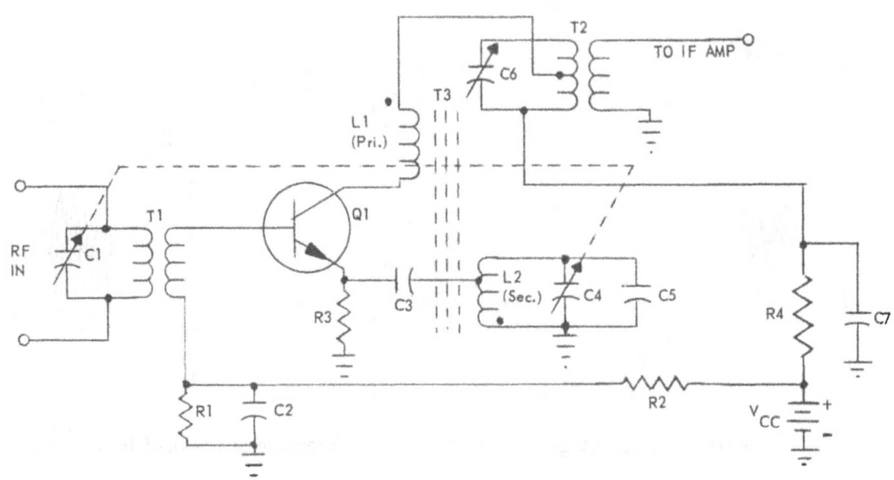

179.340

Figure 16-12.—Converter circuit using an NPN transistor.

R4 and C7 serve as a decoupling network to filter out the remaining energy of the undesirable signals.

PENTAGRID MIXER

Isolation of the local oscillator from the RF input circuit is achieved in the pentagrid mixer which is provided with two independent control grids, one for the RF signal and one for the local oscillator signal (fig. 16-13). The tube contains a heater, cathode, five grids and a plate. Grids 1 and 3 are the control grids to which the RF signal and oscillator voltages are applied, respectively.

Grid 1, the control grid, has remote cutoff (variable-mu) characteristics. Grid 3 is an injection grid used for modulating the plate current in the tube. It has a sharp cutoff characteristic and produces a comparatively large effect on the plate current for a small amount of oscillator voltage. Grids 2 and 4 are screen grids which are connected internally. Their function is to accelerate the electrons and shield grid 3 (oscillator signal grid) from the other electrodes. Grid 5 is a suppressor grid connected to the cathode, just as in an ordinary pentode.

The plate current of the pentagrid mixer is varied by the combined effect of the RF and local oscillator signals. The RF signal on grid 1 affects the electrons as in an ordinary pentode. After being accelerated by grid 2, the electrons are modulated by the oscillator voltage on grid 3. Excellent isolation of the oscillator section is achieved by the use of electron coupling and the two screen grids around the injection grid. Consequently, the RF signal circuit has little effect on the oscillator frequency, and pulling is negligible. The addition of screen grid 4 and suppressor grid 5 also helps to increase the plate resistance and gain of the tube to a value similar to that of an ordinary pentode.

Figure 16-14 shows a circuit of a pentagrid mixer with separate local oscillator excitation. The mixer input circuit is tuned to the frequency of the RF signal, whereas the IF transformer in the plate circuit is tuned to the difference frequency between the RF signal and local oscillator frequencies. The local oscillator tuning capacitor is ganged with that of the signal input circuit so that the frequency difference between them always remains the same. The output of the oscillator is taken from the control grid circuit and is applied to injection grid 3 of the pentagrid mixer through a coupling capacitor.

PENTAGRID CONVERTER

The first detector, or frequency converter, section of a superheterodyne receiver is com-

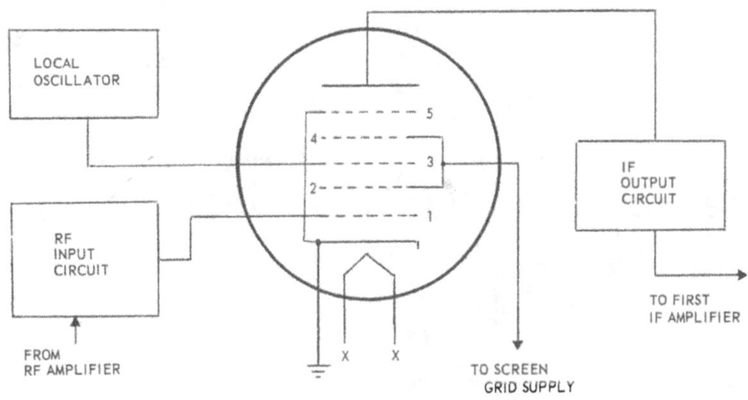

179.341
Figure 16-13. — Pentagrid mixer.

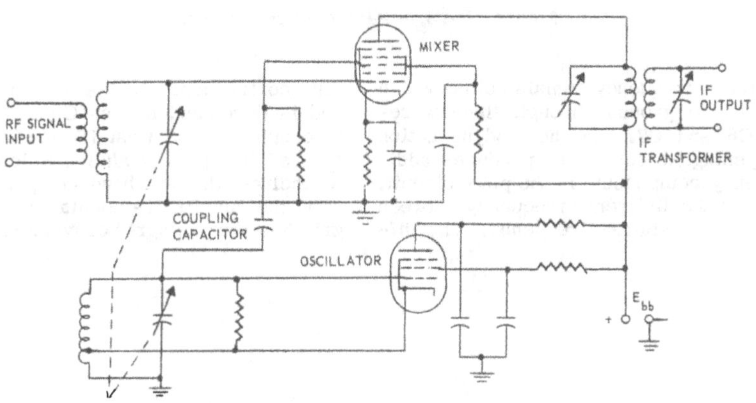

179.342
Figure 16-14. — Pentagrid mixer with separate local oscillator excitation.

posed of two parts — the oscillator and the mixer. In some receivers the same electron tube serves both functions, as in the pentagrid converter shown in figure 16-15.

The oscillator stage employs a Hartley circuit in which C5 and the oscillator coil make up the tuned circuit. C4 is the trimmer capacitor which is used for alignment (tracking) purposes. C3 and R1 provide grid-leak bias for the oscillator section of the tube. Grid 1 is the oscillator grid, and grids 2 and 4 serve as the oscillator plate. Grids 2 and 4 are connected together and also serve as a shield for the signal input grid, 3.

Grid 3 has a variable-mu characteristic, and serves as both an amplifier and a mixer grid. The tuned input is made up of L1 and C1, with the parallel trimmer C2. The dotted lines drawn through C1 and C5 indicate that both of these capacitors are ganged on the same shaft. The plate circuit contains the station frequency

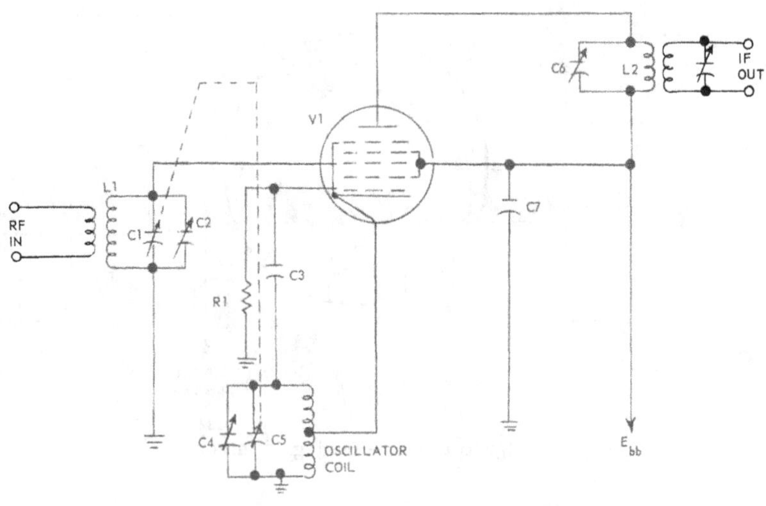

179.343
Figure 16-15. — Pentagrid converter.

and the oscillator frequency signals both of which are bypassed to ground through the low reactance of C6 and C7. The heterodyne action within the pentagrid converter produces additional frequency components in the plate circuit, one of which is the difference frequency between the oscillator and station frequency. The difference frequency is the intermediate frequency and is developed across C6 and L2. This signal is coupled to the first IF amplifier through the desired bandpass coupling which is wide enough to include the sideband components associated with the amplitude-modulated signal applied to grid 3 of the pentagrid converter.

CHAPTER 17

INTERMEDIATE FREQUENCY AMPLIFIERS

The receiver circuits studied thus far have included the RF amplifier, the oscillator and the mixer. The amplitude of the signal from the mixer (or converter) is still comparatively weak. Due to its small amplitude, it is not considered practical to feed this signal directly to a detector stage for demodulation. For this reason the superhet receiver includes one or more stages of INTERMEDIATE FREQUENCY (IF) amplification between the mixer and the detector stages.

Figure 17-1 shows the block diagram of the superheterodyne receiver. It can be seen that the IF amplifier receives its input signal from the mixer stage. This signal is an amplitude modulated wave at a frequency which is lower than the received station signal. The IF amplifier stage, or stages, produces an amplified reproduction of the mixer output signal (a band of frequencies centered around the IF) and applies it to the detector stage.

In many ways the operation of the IF amplifier is similar to that of the RF amplifier. The signals being amplified, however, are at a lower frequency than those in the RF amplifier. Unlike the RF tuned circuits (whose frequency is variable over a wide range), the tuned circuits used in IF amplifiers are fixed at a definite resonant frequency. The tanks are made adjustable only for the purpose of alignment. Since they operate at a fixed band of frequencies, the IF amplifiers can be designed to provide optimum gain and bandwidth characteristics.

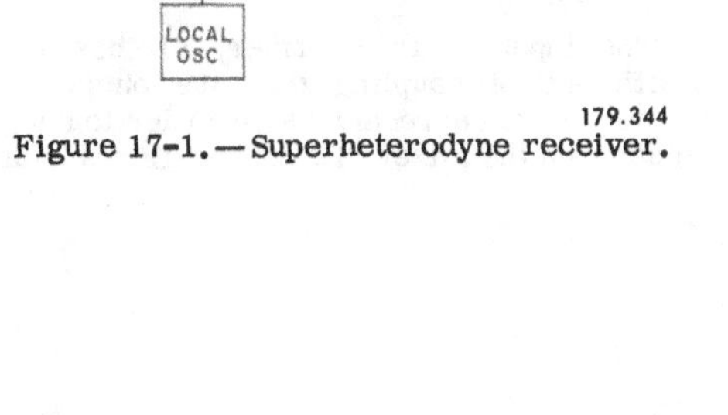

This chapter will consider the operational theory of transistor and electron tube IF amplifiers, including factors affecting the choice of an intermediate frequency and circuit configuration. Transformers used for interstage coupling at the intermediate frequencies will be explained in detail. Finally, the principles of single tuning, double tuning, and stagger tuning will be presented.

179.344

Figure 17-1.—Superheterodyne receiver.

CHOICE OF IF

Many factors are involved in the choice of a receiver IF. In most cases the choice is a compromise and may vary from one brand of receiver to another. The value of IF for a specific receiver is chosen by the designer, or manufacturer, from a wide range of values. In fact since the discovery of the heterodyne principle, intermediate frequencies ranging in value from 130 kHz to 485 kHz have been used in broadcast band "superhet" receivers. Values at the low end of this range are used only occasionally. The use of a low value IF results in slightly better gain and stability characteristics. A low IF also results in a 2 to 3 percent improvement in selectivity. However, the advantages gained are overshadowed by the increased susceptibility to image-frequency reception.

The majority of modern broadcast band receivers use one of three values for the IF: 455 kHz, 456 kHz, or 465 kHz. These values, through experience, have been found to give sufficient gain and stability characteristics while retaining acceptable image-frequency rejection.

TRANSISTOR IF AMPLIFIERS

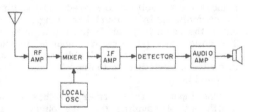

Before discussing the components used in the IF amplifier, a brief, overall description of a typical transistor IF amplifier stage will be presented. This is shown in figure 17-2. The tuned input circuit, which is formed by the primary of T1 and capacitor C1, is actually the output tank circuit of the mixer stage. As previously stated, this circuit is tuned to the receiver IF (455 kHz in this case). The primary of T1 is tuned by the adjustment of a powdered iron core as indicated by the arrow and slug above the primary coil. (This type of tuning is called permeability tuning.) The secondary is untuned. Tapping the primary winding permits matching the impedance of the mixer stage. The turns ratio and the coefficient of coupling are chosen so that the impedance of the secondary winding will match the relatively low input impedance of the transistor Q1.

The operating point for class A operation of the CE configuration is established by a standard voltage divider and bias stabilization network, consisting of R1 and R3. Bypass capacitors C3, C4, and C6 perform their usual functions. Emitter bias is provided by R2. Regeneration is prevented by neutralizing capacitor C2.

The transistor Q1 is a small signal, germanium PNP type. In accordance with standard practice, Q1 is operated in the CE configuration in order to take advantage of the higher power gain. Transformer T2 has its primary tuned to 455 kHz and an untuned secondary. The primary winding is tapped to prevent the moderate output impedance of Q1 from shunting the tuned circuit and lowering its Q (explained later). The secondary of T2 may feed either the detector stage or another IF amplifier stage.

In normal operation, a single stage of IF amplification is impractical due to poor selectivity and poor detector efficiency at low signal levels. However, where cost is a factor and peak performance is not of prime importance it is possible that a single stage of IF amplification may be used.

In most cases the gain, bandwidth, and selectivity requirements of a broadcast band receiver can be satisfied by cascading (connecting in series) two IF stages. When additional gain is needed, it is usually acquired through the addition of audio amplifiers because of the instability resulting when tuned stages are cascaded.

ELECTRON TUBE IF AMPLIFIERS

Figure 17-3 shows an electron tube version of an IF amplifier stage. This is a high gain circuit employing a pentode. As with the transistor version, this amplifier is tuned to the frequency difference between the local oscillator and the incoming RF signal. Pentode tubes are generally employed, with one, two, or three stages, depending on the amount of gain needed. As previously stated, all incoming station signals are converted to the same IF by the mixer stage, and the IF amplifier operates at only one frequency. The tuned circuits, although tunable, are rarely adjusted once set for maximum gain consistent with the desired bandpass and frequency response. These stages operate as class A voltage amplifiers and practically all of the selectivity of the superheterodyne receiver is developed by them.

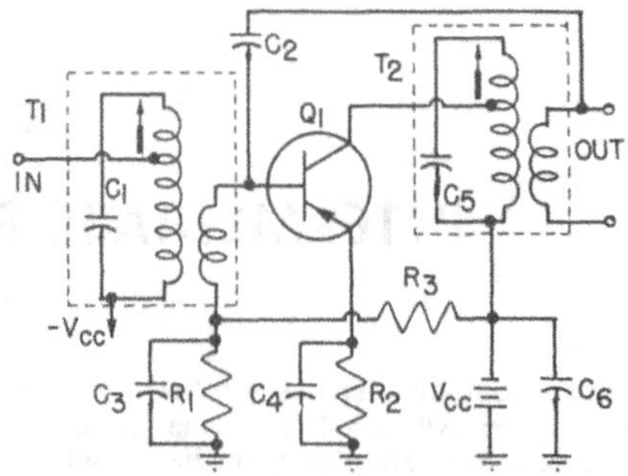

Figure 17-2.—Transistor IF amplifier.

The minimum bias (fig. 17-3) is established by means of R1 and C1. Automatic volume control may be applied to the grid through the secondary of the preceding coupling transformer T1.

The output IF transformer, T2, which couples the plate circuit of this stage to the grid circuit of the second IF stage, is double tuned by means of trimmer capacitors C2 and C3. Mica or air trimmer capacitors may be used. In some instances the capacitors are fixed and the tuning is accomplished by permeability tuning. In special cases the secondary only is tuned (single tuned). The coils and capacitors are mounted in small metal cans which serve as shields, and provision is made for adjusting the tuning without removing the shield.

The input IF transformer, T1, has a lower coefficient of coupling than the output transformer in some receivers in order to suppress noise from the mixer. The output IF transformer

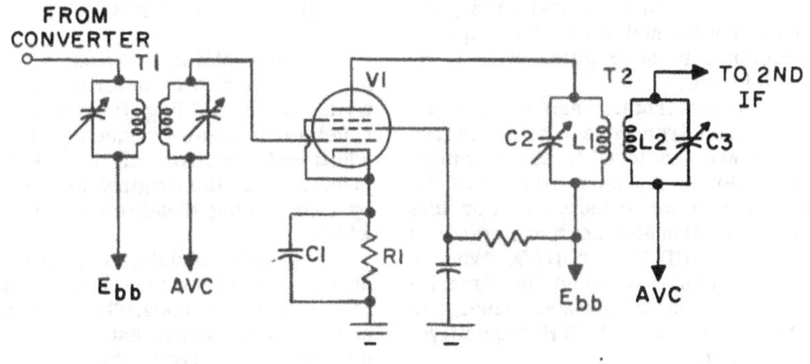

179.346
Figure 17-3.—Electron tube IF amplifier.

is slightly overcoupled with double humps appearing at the upper and lower sideband frequencies. The overall response of the stage is essentially flat, and in a typical broadcast receiver the stage has a voltage gain of about 200 with a bandpass of 7 to 10 kHz when operating with an IF of about 455 kHz.

The chief characteristic of double tuned bandpass coupling is that at frequencies slightly above and slightly below the intermediate frequency the impedance coupled into the primary by the presence of the secondary is reactive. This cancels some of the reactance existing in the primary, and the primary current increases. Thus the output voltage of the secondary does not fall off and the response is uniform within the passband.

INTERMEDIATE FREQUENCY TRANSFORMERS

The characteristics and physical construction of transformers used as coupling devices at high frequencies are quite different from those used at audio frequencies. The purpose of this section is to discuss the characteristics and use of interstage transformers, at intermediate frequencies, in transistor and electron tube receivers.

The primary, or the secondary, (and in some cases both) circuits using IF transformers are tuned for a specific resonant frequency. For any specific resonant frequency the LC product of a tuned circuit is a constant. Thus, the product of the inductance (L) and the capacitance (C), which form the tuned primary of a 455 kHz IF transformer, will be identical whether the transformer is used in a transistor or an electron tube circuit. However, due to the low magnitude of currents and voltages used in transistor receivers, a transistor IF transformer is considerably smaller in physical size than its electron tube counterpart. Figure 17-4 illustrates the relative size of the IF transformers used in the two types of receivers.

Since the selectivity of a superhet receiver is primarily determined by the IF stages, it is desirable that the tuned circuit IF transformers have a relatively high Q. The core material is one of the factors having a pronounced effect on the inductance of a coil and thus affecting Q. Increasing the permeability of the core material increases the inductance of the coil. Use of a

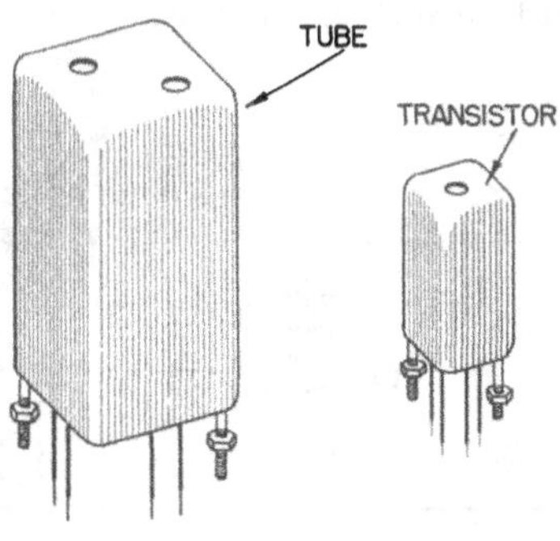

179.347
Figure 17-4.—IF transformers.

powdered iron core (indicated by the dotted lines shown in fig. 17-5) in an IF transformer, increases the permeability and allows the required amount of inductance to be obtained with relatively few turns of wire.

Miniature IF transformers are constructed so that the powdered iron core may be moved in and out of the area enclosed by the windings. This effectively varies the permeability of the core, thereby varying the inductance. For this reason, miniature IF transformers are referred to as being PERMEABILITY TUNED. When a transformer is permeability tuned the capacitance of the tank is usually a fixed value. It is possible, however, to have BOTH capacitive and permeability tuning.

One function of the IF transformer is to provide impedance matching between the output of one stage and the input of the following stage. Another function of the IF transformer and its associated tuned circuit, is to provide the selectivity for the receiver. It can be shown how the relatively high Q, necessary for good selectivity, is obtained by increasing the inductance of the primary winding, while, at the same time, maintaining the impedance match through tapping the primary.

An additional benefit gained is the reduction in the size of the capacitance used to resonate the transformer at the intermediate frequency. The reduction in value allows the physical size of the capacitor to be reduced to the point where it can be included within the metal shield in which the IF transformer is enclosed. To reduce undesirable feedback, the magnetic flux produced by the transformer windings must be prevented from inducing currents and voltages in nearby wires and components. This is accomplished by surrounding the transformer with a grounded metal shield or "can."

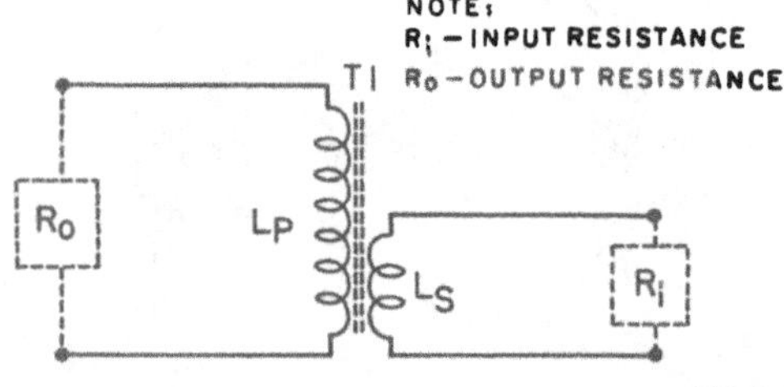

Figure 17-5.—IF transformer impedance matching.

ACHIEVING DESIRED BANDPASS CHARACTERISTICS IN IF AMPLIFIERS

As was stated previously, the IF amplifier has the function of determining receiver selectivity and providing the major portion of the receiver's radio frequency amplification. An additional function of the IF amplifier is to preserve all the original modulating intelligence by maintaining a sufficiently wide overall bandwidth.

Amplitude modulation of a carrier produces an upper and lower sideband frequency for every modulating frequency. These sideband frequencies are situated above and below the carrier, and are separated from the carrier by an amount equal to the original modulating frequency which produced them. Thus, if the highest audio frequency in the original modulating signal were 8 kHz, then an upper sideband frequency of the carrier PLUS 8 kHz, and a lower sideband frequency of the carrier frequency MINUS 8 kHz would be present in the IF stage signal. Therefore, the overall bandwidth of the IF stage must be wide enough to pass these two sidebands, and all sideband frequencies between them, with equal amplification. For this particular application, the receiver would have to have an overall bandwidth of at least 16 kHz.

The overall bandwidth that a receiver is required to have is determined primarily by the job it is required to perform. For instance, a high fidelity receiver, which is required to reproduce the entire aduio frequency range, would by necessity have a wider bandwidth than a military communications receiver which is required to reproduce only voice frequencies.

The purpose of the following sections is to illustrate how the overall bandwidth and response characteristics of the receiver may be altered by variations in the construction and tuning of the IF transformers. The various tuning methods to be discussed are SINGLE TUNING, DOUBLE TUNING, and STAGGER TUNING.

RESONANT CIRCUIT RECOGNITION

Before a discussion of IF tuning, a clarification of resonant circuit configurations will be of benefit. In many cases a circuit that appears to be a parallel resonant circuit, will, in actuality, be a series resonant circuit.

Specifically, a transformer coupled IF stage is just such a case. Consider the simplified circuit of a transformer coupled IF amplifier in figure 17-6. Recognition of series or parallel configurations of tuned circuits is simplified when an obviously defined source of voltage exists. The primary circuit of the transformer in figure 17-6 has a source of applied voltage (E_a). The voltage generator (E_a) represents the voltage produced by an amplifier (transistor or tube) of which the tuned circuit is the load. The tank circuit components (C_P) and (L_P) are in parallel with E_a. C_P and L_P provide a parallel path for the flow of primary current I_P. Therefore, the primary circuit of the transformer is a parallel resonant circuit because the voltage source is in parallel with the tank components.

When the voltage source of the circuit is not clearly defined, the recognition of circuit configuration is more difficult. Consider the secondary circuit of the transformer in figure 17-6. At first glance it appears that the output is being taken across the combination of L_s and C_s. However, when it is realized that the voltage source for the circuit is the induced voltage (represented by voltage generator E_{ind}) of the secondary winding, it can be seen that there is only one path for current flow (I_s) through C_s and L_s. The source voltage (E_{ind}) is effectively applied in series with the tank components. Therefore, the secondary circuit of the transformer is effectively a series resonant circuit with the output being taken across the capacitor.

To simplify the discussion of transformer coupled tuned circuits, it is advantageous to have both the primary and secondary appear as the same configuration. Since any parallel circuit may be converted to an equivalent series circuit, the primary circuit is usually the one converted to an equivalent series circuit during the discussion of overall circuit response.

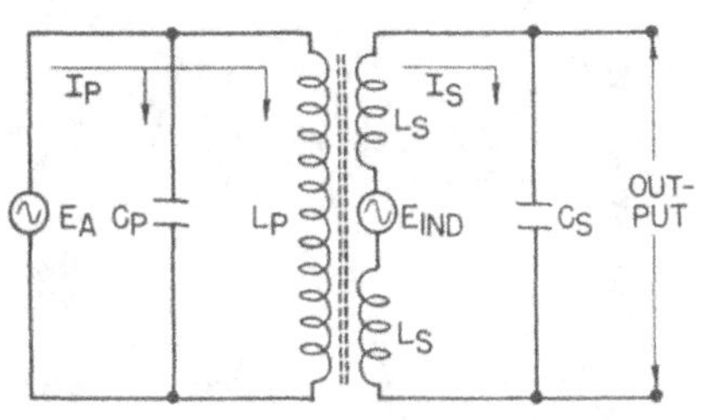

179.349

Figure 17-6.—Simplified transformer coupling of resonant circuits.

SINGLE TUNED TRANSFORMER COUPLED IF STAGE

A SINGLE TUNED, transformer coupled, IF stage is one which contains only a single tuned resonant circuit. A single tuned IF amplifier stage was shown in figure 17-2. Notice that the amplifier stage of Q1 contains only one resonant circuit (C5 and the primary of T2).

It has been shown that when a capacitance and an inductance are connected to form a tuned resonant circuit, this circuit will exhibit frequency discriminatory properties. In other words, the circuit will favor the development of voltage or current at frequencies within its passband and discriminate against frequencies that fall outside its passband.

The behavior of a circuit, over a range of frequencies, is illustrated by a graph of voltage or current versus frequency, called a response curve. The response curve of a single tuned stage, illustrating the effect of circuit Q on voltage amplitude and bandwidth, is shown in figure 17-7. Notice that the bandwidth BW1 of the low Q circuit is much wider than the bandwidth BW2 of the high Q circuit. Due to the slope of the sides of the response curves, frequencies falling outside the passband of the high Q circuit will receive greater attenuation with respect to f_o than will frequencies falling outside the passband of the low Q circuit. Thus a receiver employing high Q circuits will have good selectivity. The method for achieving high Q tuned circuits, and thereby good selectivity was explained previously.

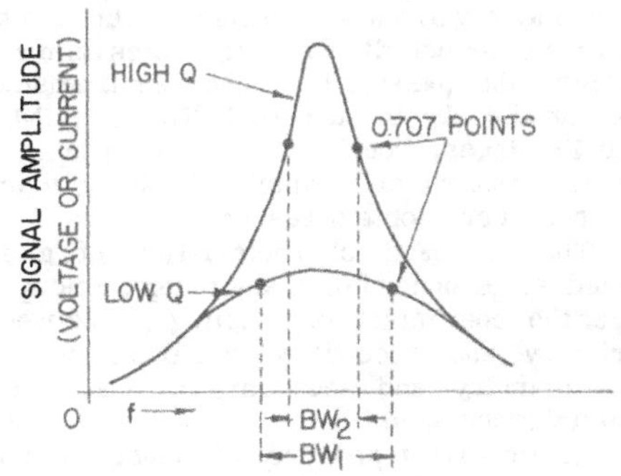

179.350

Figure 17-7.—Passband of single tuned high Q and low Q circuits.

The high Q curve in figure 17-7 represents the response of only a single IF stage containing a single tuned circuit. It was stated previously that when two or more amplifier stages or tuned circuits, all tuned to the same IF are connected in cascade, the overall bandwidth will be less than the bandwidth of any individual tuned circuit. Therefore, while the Q of the IF stages tuned circuits should be high, so as to provide good selectivity, it must not be so high as to reduce the overall bandwidth to the point where the sideband frequencies representing the highest modulating frequencies are attenuated.

Use of the single tuned stage yields very satisfactory results in many applications. High gain is easily obtained. It is evident from the high peak at resonance and the steep slope of the curve for a high Q circuit (fig. 17-7) that selectivity is good. However, it is also evident that all frequencies within the passband will not receive equal amplification. The sideband frequencies, representing the low modulating frequencies, will be amplified to a greater extent than the sidebands representing the higher modulating frequencies.

DOUBLE TUNING OF TRANSFORMER COUPLED IF STAGE

In some applications (such as high fidelity receivers) the bandwidth requirements are such that a single tuned stage will not be able to meet them. A specific application may require a wider bandwidth than is possible with single tuning. Another application may require that all sideband frequencies within the passband receive relatively equal amplification. Still another application may require a wide passband but also a high degree of selectivity for frequencies immediately outside the passband. These requirements can be satisfied by the use of DOUBLE TUNING in the IF stages. Double tuning refers to an interstage transformer in which both the primary and the secondary contain resonant circuits.

The passband characteristics of a double tuned stage depend on many things, among which are the coefficient of coupling (k) between the primary and secondary windings, the Q's of the primary and secondary circuits, and the mutual inductance.

An important property of inductively coupled circuits (transformer) is mutual inductance (M). Mutual inductance is the common property of inductively coupled circuits that determines, for a particular frequency and current in one circuit, the amount of mutually induced voltage in the other circuit.

Increasing the coefficient of coupling will increase the value of mutual inductance and, thereby, increase the value of coupled impedance in the primary, and, UP TO A POINT (explained shortly), the induced secondary voltage will be increased.

It has been stated that the bandwidth characteristics of a double tuned, transformer coupled stage depend in part on the amount of coupling. This can be shown by use of the equivalent circuit and response curves in figure 17-8.

Notice that the coupled impedance (Z_c) is represented as being in series with the primary circuit. Applied voltage (E_a) and the internal resistance (r) represent the amplifier (transistor or tube) voltage and resistance. $R_{P_{eq}}$ is the series equivalent of the original parallel circuit resistance. $R_{S_{eq}}$ is the equivalent resistance of the secondary circuit. It will be assumed, for purposes of explanation, that the primary and secondary circuits have identical Q's, both circuits are tuned to the same resonant frequency, and the amount of coupling between the circuits is variable (by physically moving the coils). It will also be assumed that the Q of both circuits is held constant for all values of coupling.

To demonstrate the effect of coupling variations, the circuits are first moved far enough

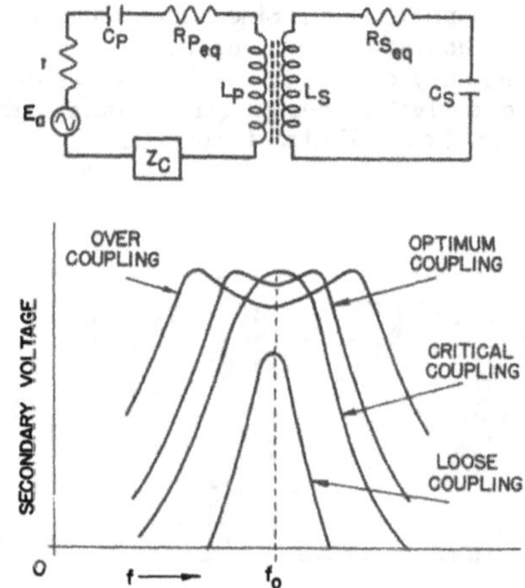

179.351

Figure 17-8.—Equivalent double tuned circuit and response curves.

apart so that very little coupling takes place (low value of k). This condition is called loose coupling. With loose coupling there is very little transfer of energy between the primary and secondary circuits. The value of induced secondary voltage (E_s) is small because the value of mutual inductance M is small. The small value of M also causes the value of Z_C to be small. The small value of Z_C has very little effect on the operation of the primary circuit. Due to the small amount of interaction between the primary and secondary, the two circuits behave essentially as if they were separate tuned circuits. Except for the slope of the sides being slightly steeper, the response curve for loose coupling in figure 17-8 is the same as for a single tuned circuit.

As the coefficient of coupling is increased (the circuits are continually moved closer together), the value of mutual inductance increases. This continually increases the amount of voltage induced in the secondary winding. The value of the coupled impedance is also increasing, and due to its effect on the primary current the response curve becomes wider (bandwidth increases).

As the coefficient of coupling is continually increased, reflected impedance Z_C, which is resistive at the resonant frequency, continues to increase until eventually a value of coupling is reached where Z_C equals the equivalent resistance of the primary ($R_{P_{eq}}$). This condition is called CRITICAL COUPLING (k_c), and since at this point the reflected resistance matches the primary resistance there is a maximum transfer of energy between the circuits. Thus, the induced secondary voltage is at its maximum value, as shown by the response curve for critical coupling in figure 17-8. Notice that the critical coupling curve is flattened slightly on top. At resonance both primary and secondary circuits appear resistive. However, at frequencies above resonance, a series circuit appears inductive, and at frequencies below resonance a series circuit appears capacitive. (It was stated that if the secondary circuit appeared inductive the impedance coupled into the primary would have the same phase angle but would appear capacitive in nature.)

In a single tuned circuit at frequencies slightly off resonance, the primary current begins to decrease; however, a different situation exists in double tuned circuits. At frequencies above resonance, the primary circuit (considered here as an equivalent series circuit) takes on inductive characteristics; however, the capacitive portion of the complex coupled impedance (Z_C) will cancel a portion of the primary inductance. This will, in effect, extend the resonant condition of the circuit slightly beyond f_o. A similar action takes place below resonance, where the primary circuit begins to appear capacitive, the coupled impedance appears inductive in nature and cancels some of the primary capacitance. It could also be said that this action extends the condition of resonance slightly below f_o. In any case, the result is to cause the primary current to remain high within most of the bandwidth (as shown by the flat top of the critical coupling response curve).

Critical coupling (k_C) has been defined as that value of coupling (k) at which maximum transfer of energy between primary and secondary is achieved. When the coefficient of coupling is increased beyond the value of k_C, the primary current begins to decrease due to the continued increase of the coupled impedance. This is indicated by the dip of f_o on the optimum coupling curve. A continued increase in the coefficient of coupling will result in further reduction of the gain at resonance and a further increase in the distance between the two resonant peaks (above and below f_o). The circuits are not said to be overcoupled and the overall response of the stage will take on the appearance of the double humped curve in figure 17-8.

In transistor IF stages (where the output and input resistances are relative low) double tuned stages will yield less voltage gain per stage than single tuned stages (per stage) PROVIDING THE PASSBAND OF THE TWO STAGES ARE THE SAME. In electron tube IF stages, where plate and grid resistances are very high, the voltage gain of one double tuned stage for the condition of critical coupling (assuming that primary and secondary Qs are equal) is approximately half of that for one single tuned stage.

Comparing two cascaded single tuned stages to two double tuned stages, the single tuned stages will have only slightly more gain per stage than the double tuned stages. As additional stages are added, the double tuned stages will have more voltage gain PER STAGE than the single tuned stages. The reason for this is seen in the qualification that the overall passband of the two types of circuits MUST BE THE SAME. The only way the passband of the single tuned stages can be increased, in order to match that of the double tuning, is to decrease the Q of the single tuned circuits, which will, of course, decrease the gain per stage of the single tuned circuits.

For double tuned circuits, the bandwidth is determined primarily by the value of coupling, while the depth of the valley between the resonant peaks (and hence the uniformity of response) is determined by the relation of Q to k. Thus, if the circuit is operated with a value of coupling greater than critical, and k is held constant, a large Q means a deeper valley, and small Q means less of a valley. From the foregoing it can be seen that the desired response curve is a compromise and is achieved by the proper manipulation of BOTH k and Q because they are interrelated.

STAGGER TUNED IF STAGES

In the previous sections, two methods of tuning the IF stages were discussed. It was shown that single tuning of the IF stages provided large voltage gain per stage. However, it was also stated that the large voltage gain of single tuning is possible only with a relatively narrow bandwidth and relatively high Q's. A wide passband can only be obtained, using single tuning, with a sacrifice of selectivity and gain. It was also pointed out that with single tuning the frequency response within the passband is not uniform.

It was then shown that it is possible to obtain a wide passband, while maintaining good selectivity, and an improvement in uniformity of response within the passband, by use of double tuning. However, double tuning does not provide the voltage gain that the narrow band single tuning does.

A third type of tuning, called STAGGER TUNING, is available. This combines the gain of the single tuned stage with the wide, uniform, passband of the double tuned stage. The resonant circuits in stagger tuned, transformer coupled, IF stages are isolated from each other so that no mutual coupling takes place between them. The isolation is necessary since each resonant circuit must have the characteristics of a single tuned circuit.

In previous types of tuning the resonant circuits were all tuned to the same frequency. In stagger tuning two transformer coupled IF stages, the two tuned circuits (one in each stage) are each tuned to a different resonant frequency, one slightly above the receiver IF and the other slightly below the receiver IF.

Figure 17-9 shows a simplified diagram of a two stage, transformer coupled, stagger tuned IF section of a receiver. When tuned in this manner the stages are often called a STAGGERED PAIR. The power supplies, biasing neutralizing capacitors, etc., have been omitted for simplicity.

Notice that, although the receiver IF is 450 kHz, tuned circuit number one is tuned to a resonant frequency of 443 kHz, while tuned circuit

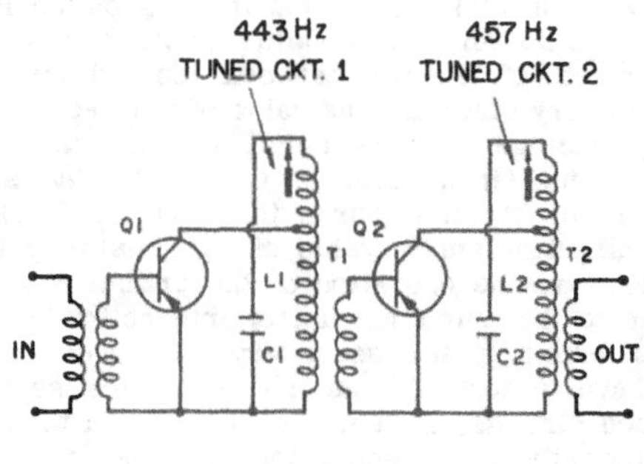

179.352

Figure 17-9.—Staggered pair IF amplifiers (simplified schematic).

number 2 is tuned to a resonant frequency of 457 kHz. In the following discussion it is assumed that the frequency response and amplification of transistors Q1 and Q2 are identical, although in actuality there will always be some differences due to manufacturing tolerances. It will also be assumed that the Q and frequency response of the two tuned circuits are identical. Since there is no inductive coupling between tuned circuit number one and number two, each circuit will have the frequency response and characteristics of a single tuned circuit.

Figure 17-10A illustrates the individual response curves of two tuned circuits. Circuit number one is tuned below the IF of the receiver by 7 kHz and has its resonant peak at 443 kHz. The circuit is so designed to have a bandwidth of 14 kHz. This is evident by the fact that the .707 (half power) points occur at 436 kHz and 450 kHz respectively. It is significant that the UPPER half-power point of tuned circuit number 1 occurs at the receiver IF. If we assume that the relative gain of circuit #1 at its resonant peak ($f_o 1$) of 443 kHz is approximately 140 (139.69), then the relative gain at the half-power points is very close to 100 (98.69).

Figure 17-10A also shows tuned circuit #2 as having its resonant peak ($f_o 2$) at 457 kHz. The bandwidth is also 14 kHz. It might appear, at first glance, that the overall bandwidth will be 28 kHz, however, this is not the case (as will be shown shortly). The amount of amplification, or gain, that a signal receives in passing through a number of stages is the product of the gain of

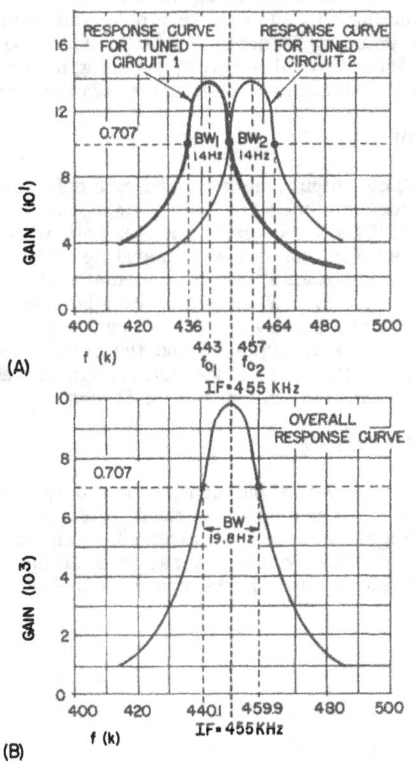

179.353

Figure 17-10.—Response curves of staggered pair IF amplifiers.

the individual stages. One method of obtaining the overall response curve of two or more amplifier stages in series is to apply a number of individual frequencies to the input of the cascaded stages and plot the relative amplitude of the output for each frequency. When all these plotted points are connected together the result will be the overall response curve of gain versus frequency. If the relative gain of the individual stages, over a band of frequencies, are known, the overall response curve may be plotted directly from this data. Thus, the overall response curve of the staggered pair of IF amplifiers (fig. 17-10B) may be plotted from the individual response curves of figure 17-10A. For example, the upper half-power point of circuit #1 and the lower half-power point of circuit #2 coincide at the IF of 450 kHz. Since the gain of the circuit at the .707 point is slightly less than 100, the overall gain of the staggered pair at the receiver IF of 450 kHz will be slightly less than 100 x 100 = 10,000.

This point is shown as the resonant peak of the overall response curve in figure 17-10B. Going above the receiver IF to the resonant frequency of circuit #2 (457 kHz) it is seen that the gain of circuit #1 has decreased to 60 while the gain of circuit #2 is at its maximum point. The overall gain at 457 kHz will be approximately 140 x 60 = 8,400. As the frequency is increased past this point it can be seen that the gain of both circuits is decreasing. When the upper half-power point of circuit #2 (464 kHz) is reached, the overall gain of the staggered pair will be approximately 100 x 42.5 = 4,250, which is well below the half-power point for the overall response curve.

Points for the overall response curve below the receiver IF are plotted in the same manner. Notice, that for a short distance above and below 450 kHz the increasing gain of one tuned circuit and the decreasing gain of the other tuned circuit cause the overall response curve to maintain a relatively flat top. The rapidly decreasing gain of BOTH circuits after the individual resonant peaks have been passed cause the bandwidth of the overall response curve to be less than the sum of the individual bandwidths.

When the upper half-power point of one tuned circuit coincides with the lower half-power point of the other tuned circuit, in a staggered pair, the overall response curve will be similar to the critical coupling response curve of a double tuned stage, and the overall bandwidth will be 1.414 TIMES THE BANDWIDTH OF ONE OF THE TUNED CIRCUITS. Therefore, as seen in figure 17-10B, the bandwidth of the staggered pair is 1.414 x 14,000 = approximately 19.8 kHz. Varying the resonant frequencies of the two tuned circuits, in respect to the receiver IF, will vary the overall response curve and bandwidth.

OUTPUT COUPLING

The output of a transistor or electron tube IF amplifier may be transformer coupled to either another IF amplifier stage or to a detector stage. Although the method of coupling is the same, the characteristics of an IF transformer feeding a detector stage are slightly different than those of a transformer feeding another IF stage, the main difference being the slight variation in turns ratio, coefficient of coupling, etc., to facilitate impedance matching. This is caused by the fact that the input impedance of the detector stage is usually lower than that of an IF amplifier.

FAILURE ANALYSIS

It would be well to keep in mind the general troubleshooting procedures previously covered in the section on RF amplifiers, since the troubles which occur in IF amplifiers are similar.

NO OUTPUT

A defective IF transformer, an open bias resistor, loss of source voltage, or a defective transistor or tube can cause a loss of output. Check the various power supplies with a voltmeter. A lack of the proper operating voltages can result from a defective IF transformer. If the transistor or tube is defective, the bias, collector, and plate voltages will be improper as well. It must be remembered that interelement capacitances in the amplifying devices can have drastic effects on circuit operation, and care should be taken when replacing the transistors or tubes. Detuning will also cause a no output condition if the detuning is serious enough. This trouble is evidenced by all d.c. voltages being near normal, but the signal fails to pass through the amplifier.

LOW OUTPUT

Low output can be caused by a defective transistor or tube, low supply voltages, or too high a bias. Check the amplifier elements with a voltmeter for the correct operating potentials. If all d.c. potentials appear normal and the output is low, either a transistor or tube may be weak or the alignment may be incorrect. If the amplifying device is replaced and there is no increase in gain, check the alignment. If alignment appears to be impossible, replace the IF can.

DISTORTION

Distorted output can be caused by improper bias, or incorrect operating voltages. A d.c. voltage check is recommended. Once the trouble is localized to the general area of the circuit, individual components can be checked further.

CHAPTER 18

RECEIVER CONTROL CIRCUITS

This chapter deals with circuits which control receiver functions. Methods of performing basic receiver functions will be described, as well as manual and automatic controls. The use of control circuits is not limited to radio receivers. Other forms of electronic equipments, such as radar, sonar, direction finders, and navigational aids use these circuits to obtain optimum performance.

BAND SPREADING AND BAND SWITCHING

As previously discussed in chapter 10, an LCR circuit may be made to resonate at any given frequency merely by selecting the proper values of capacitance and inductance. Thus, to make a given LCR circuit resonate at any frequency, within a given range of frequencies, all that would be necessary is to make the capacitance or the inductance, or both, variable. Normally, in a tunable LCR circuit either L or C, but not both, will be made variable. Whether L or C will be selected as the variable is dependent on the end results desired, since each method of tuning has certain inherent characteristics which may make its choice advantageous.

Regardless of the method of tuning chosen, circuit Q and thus bandwidth is dependent primarily on two factors which are: (1) the ratio of L to C; and (2) circuit loading (effective resistance within the circuit). Thus, if a single tuned circuit is required to cover a very wide range of frequencies, tuning such a circuit becomes a major problem. This is true because the ratio of L to C varies widely as the circuit is tuned from one end of its range to the other. That is, at the low end of its tuning range, the ratio of L to C is quite low, while at the high end of its tuning range, the ratio of L to C will be quite high. Assuming circuit loading remains essentially constant over the entire range, then at the low end circuit Q will be low, bandwidth or bandpass will be quite wide, and circuit response at all frequencies including the resonant frequency will be poor. At the other end of the tuning range circuit Q will be quite high, bandwidth or bandpass will be very narrow, and circuit response at the resonant frequency will be very high falling off sharply on either side. Thus, at the low end of the tuning range weak stations will not be received and relatively strong stations, which are close together in frequency, may be received simultaneously, causing garbling. At the high end of the tuning range, the conditions are essentially reversed; that is, circuit response is such that even relatively weak stations may be sufficiently strong enough to drive the following amplifier stages into saturation, causing distortion. In addition, bandpass may become so narrow that the circuit will respond only to the resonant frequency. Thus, the sidebands and any information that they contained will be lost.

Thus, if a given receiver is required to cover a relatively wide range of frequencies then it may become necessary to incorporate bandspreading and band switching in order to compensate for these effects and overcome the mechanical limitations described below.

The tuning range of a particular variable capacitor and inductor combination will depend on the minimum to maximum values of the tuning capacitor and the inductance of the coil. Band spreading provides accurate fine tuning while retaining a wide tuning range. Mechanical bandspreading reduces the tuning rate by mechanical means. A common arrangement is to use a two speed planetary drive. Electrical bandspreading uses suitable circuit networks, as shown in figure 18-1. The main tuning capacitor (C2) is usually large enough to cover a 2 to 1 frequency range. The bandspreading capacitor (C1) is much smaller than C2 and allows only a small change in frequency for a given rotation of its shaft as compared to C2.

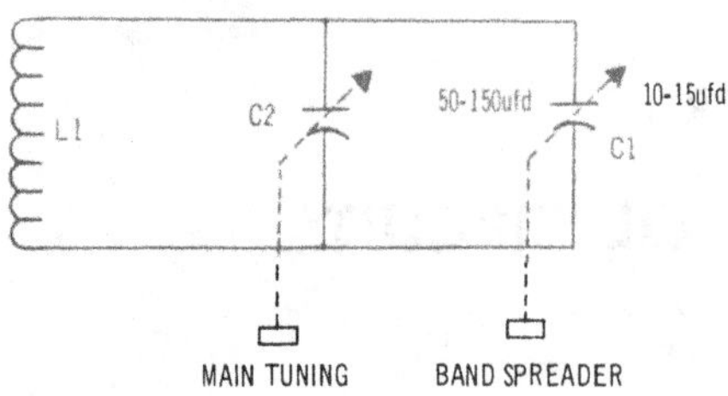

Figure 18-1.— Electrical bandspreading.

The same tuning capacitor and coil cannot be used over a wide frequency range (for example: 2 MHz to 30 MHz) because it is impractical to obtain the minimum to maximum capacitance ratio required in capacitive tuning or the minimum to maximum inductance ratio when using tuning coils. Therefore, the circuit components are changed for different frequency bands. A common method is to retain the same tuning capacitors and insert different coils on each band. An example of switching different tuned transformers is shown in figure 18-2 in which the inductance and capacitance are switched.

MANUAL GAIN CONTROL

Previously covered material indicated that high sensitivity is one of the parameters of a good receiver. In some cases high sensitivity may be a liability. For example, the signal received from a nearby station can be strong enough to overload the RF sections of the receiver. This may cause the audio output to become distorted to the point of complete loss of intelligibility. To overcome this problem, manual gain control of the RF section is utilized. By using a manual gain control, maximum sensitivity is realized and weak input signals are provided with maximum amplification, yet when a strong input signal is received, the RF gain may be reduced to prevent overloading. Typical manual gain control circuits for a receiver are illustrated in figure 18-3.

C1 is an emitter/cathode bypass capacitor. R1 and R2 develop emitter/cathode bias for the amplifier. C2 provides d.c. isolation between the tank and the base of Q1 in the transistor version. (The characteristics of transistors and variable-mu tubes have been previously discussed, and it was illustrated that amplifier gain could be varied by changing the bias.) Gain control (R2) is nothing more than a manual bias adjustment. When the wiper arm of R2 is set at point B, minimum forward bias is applied to the transistor, and maximum bias is developed in the tube circuit. This causes both amplifiers to operate closer to cutoff and thereby reduces their gain. When the control is moved toward point A, the opposite effects occur. R1 limits the maximum conduction of the devices when R2 is short circuited. In transistor circuits, an alternate biasing method may be encountered where the transistor is operated near saturation. In this case, a large change in gain is again a function of bias.

MANUAL VOLUME CONTROL

Figure 18-4 shows the schematics for the standard method of controlling volume in a superheterodyne receiver. C1 and R1 form the input coupling circuit and also the means of controlling the signal level applied to the audio amplifier. R1, R2, and R3 develop forward bias and set the operating point for the transistor amplifier. R4 is the collector/plate load resistor for Q1/V1 and C3 is the output coupling capacitor. By using potentiometer R1 in the configuration shown, the input impedance of the stage remains more constant. The signal from the preceding stage is felt across R1. By varying R1, the input level to Q1/V1 is varied, and the output amplitude is varied.

AUTOMATIC GAIN/VOLUME CONTROL

Variations on the output volume of a receiver may result from variations in the input signal strength. Changes in input signal strength occur as a result of changing stations and from fading which is caused by changing atmospheric conditions. The function of an AUTOMATIC GAIN CONTROL (AGC), also referred to as an AUTOMATIC VOLUME CONTROL (AVC), is to limit unwanted variations in the output of the receiver due to variations in strength of the received signal input. In order to maintain a constant output level, a receiver without AGC would require continuous manual readjustment to compensate for received signal changes.

Signals from stations operating at the same power level may not reach the receiver antenna with the same power because of differences in

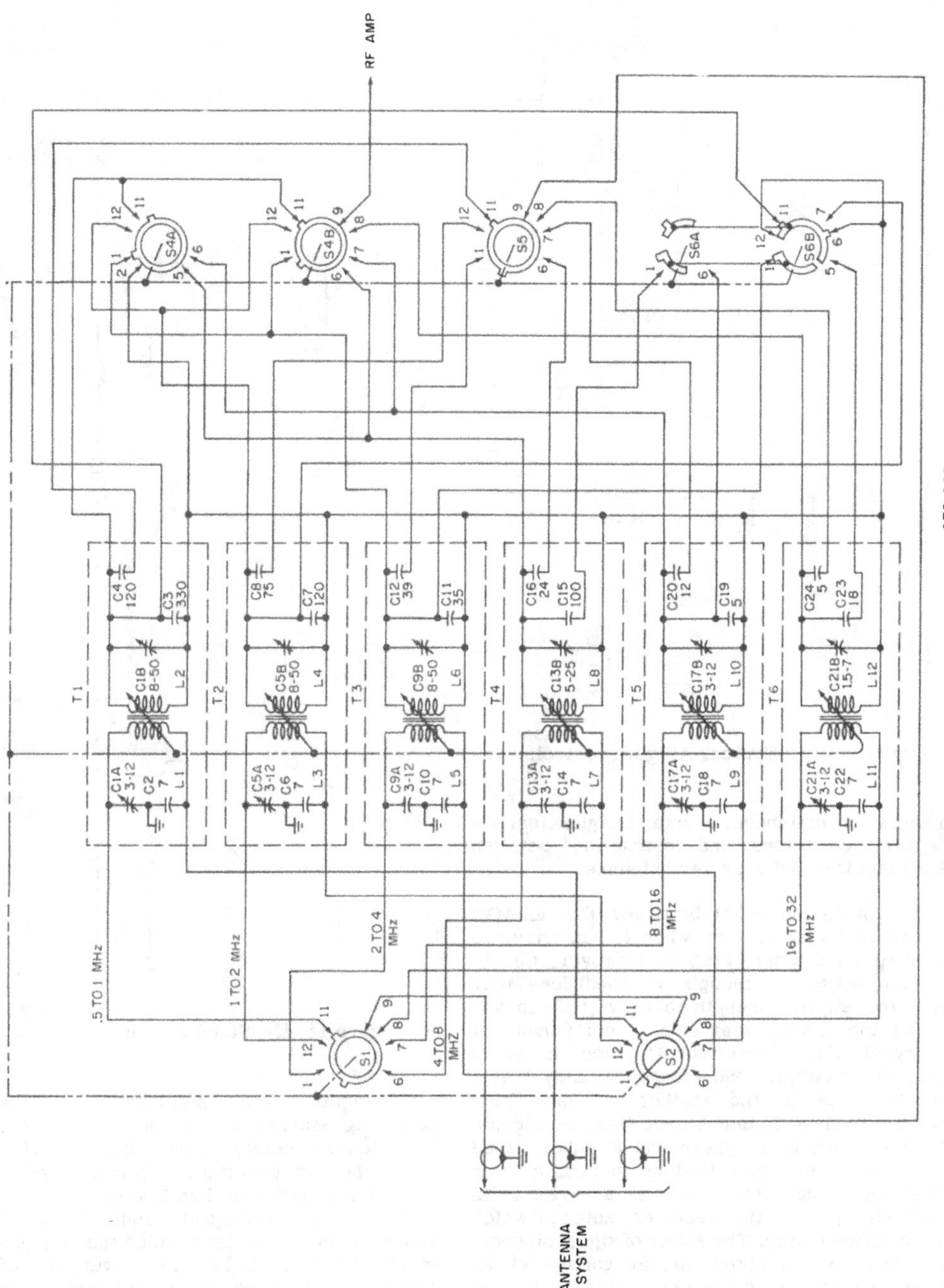

179.355
Figure 18-2.—Example of bandswitching.

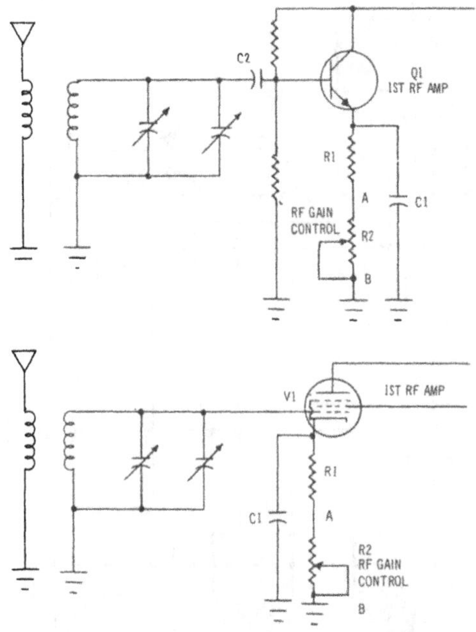

179.356
Figure 18-3.—Typical RF gain controls.

transmission distances, carrier frequencies, atmospheric conditions, and obstructions between the transmitter and receiver antennas.

The conclusion might be drawn that an AGC network is not necessary when the receiver is operating on a single station. However, this is not true because atmospheric conditions may cuase the signal strength to vary (fade in and out), or the antenna may receive components of the signal which have traveled along different paths. For example, one component may travel from the antenna, and another may have been reflected from a distant object. The two signals will sometimes be in phase and at other times be out of phase, thus tending to reinforce or cancel each other. The result is a variation in signal strength at the receiver antenna which is also called fading. The effect of signal strength variations in the output signal voltage of an RF stage may best be demonstrated by an example problem.

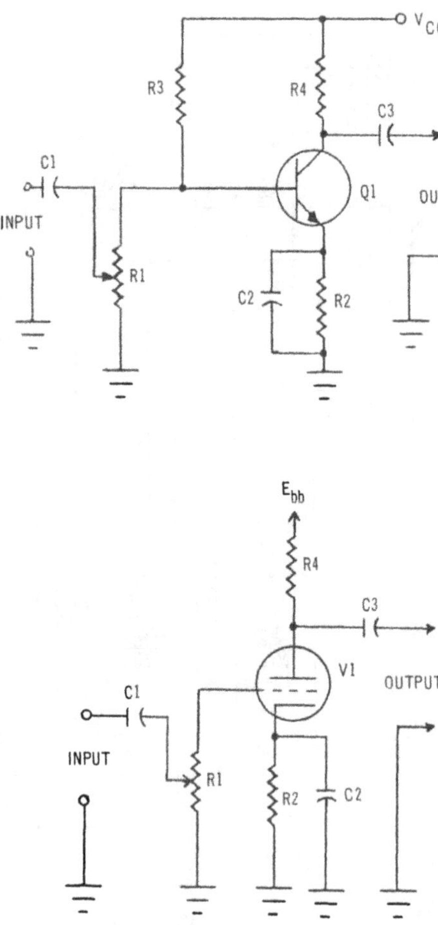

179.357
Figure 18-4.—Standard volume controls.

Example: An RF amplifier, connected to a receiving antenna has a voltage gain of 100. If the antenna receives an input signal of 10 microvolts, the output voltage = 100 x 10 microvolts = 1000 microvolts or 1 millivolt.

The output voltage is equal to 1 millivolt and if fading is to be avoided the output voltage must remain at this level. However, if a reflected signal is received of approximately half the strength (5 microvolts) of the original, and is in phase with the direct signal, the total input signal to the receiving antenna will increase to 15 microvolts. To maintain the desired 1 millivolt of output signal the gain of the RF amplifier must somehow be reduced. With an input of 15 microvolts and a desired output of 1 millivolt (1000 microvolts) the gain of the amplifier must be reduced to:

$$\frac{1000 \times 10^{-6}}{15 \times 10^{-6}} = 66.7$$

When the 10 microvolt signal and the 5 microvolt reflected signal are out of phase with each other, the signal strength at the receiving antenna will decrease to 5 microvolts. To maintain the original 1000 microvolt output signal, the voltage gain of the amplifier must be increased to:

$$\frac{1000 \times 10^{-6}}{5 \times 10^{-6}} = 200$$

The variation of amplifier gain, in the example, is necessary to compensate for the variation of the input signal strength. The required amplifier gain variations can be accomplished automatically by the addition of an AGC circuit to the receiver. The purpose of the following sections is to show the methods and circuits used to produce AGC and the manner in which the AGC (AVC) controls the gain of the receiver.

BASIC AGC CIRCUITRY

The output of the detector circuit contains a d.c. component which is directly proportional to average modulated carrier amplitude. The AGC circuitry utilizes this d.c. component by filtering the detector output to remove the audio and IF components and applying a portion of the d.c. component to the preceding stages. Figure 18-5 is a block diagram representing the AGC portion of the detector output being applied to the preceding stages. The AGC voltage may be used to control the amplification of any or all of the stages preceding the detector stage. A transistor receiver may utilize either positive or negative voltage for AGC, depending on the type of transistors used and the elements to which the control voltage is applied; however, in an electron tube receiver the AGC voltage is usually negative, as electron tubes use negative voltages for bias.

The circuit shown in figure 18-6 produces a positive AGC voltage. The components T1, CR1, C1, and R1 comprise a standard series diode detector as previously described in chapter 5. The AGC network is composed of R2 and C2. In normal operation of the detector circuit, with the potential shown at the input, CR1 conducts. Conduction of the diode will cause a charging current (shown by the dashed line) to flow through the AGC capacitor (C2) and AGC resistor (R2). This charging current will develop a voltage across C2 with the polarity shown in figure 18-6. When the potential across T1 reverses, the diode will be reverse biased and will not conduct. At this time the charging current ceases and C2 begins to discharge. The discharge path for C2 is shown by the solid arrows in figure 18-6. The discharge time constant of C2, R2 and R1 is chosen to be longer than the period (1/f) of the lowest audio frequency present in the output of the detector. Consequently, C2 will not discharge appreciably between peaks of the modulating signal, and the voltage across C2 will be essentially a d.c. voltage. This voltage will be

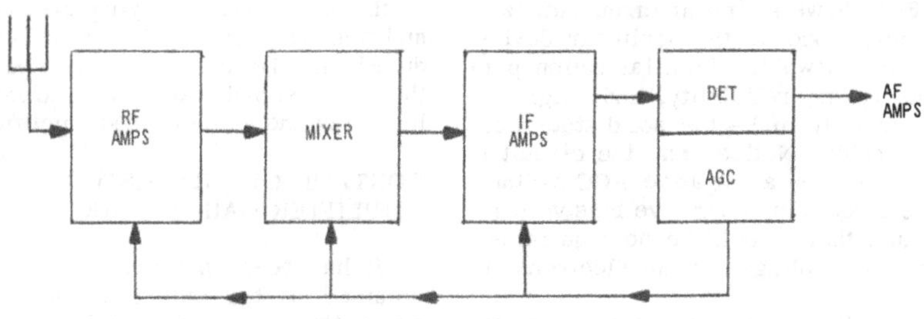

179.358
Figure 18-5.—Block diagram showing AGC application.

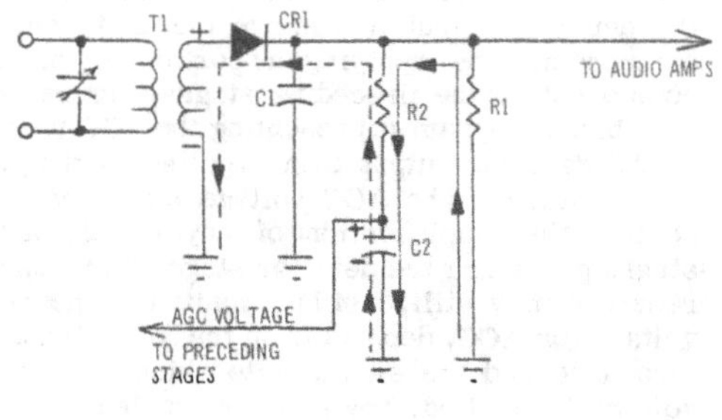

179.359

Figure 18-6.—Series diode detector and simple AGC circuit.

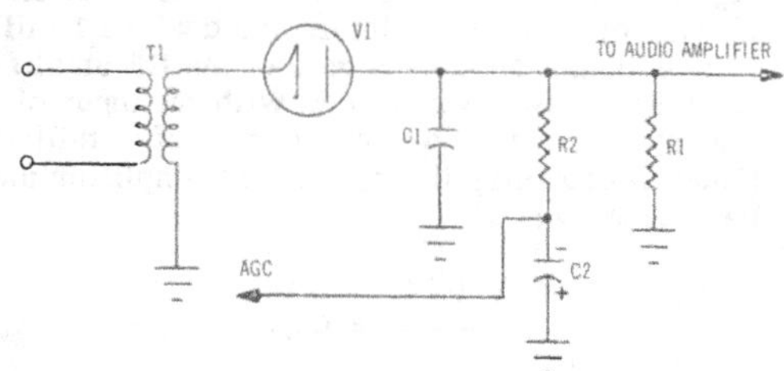

179.360

Figure 18-7.—Electron tube detector with simple AGC circuit.

proportional to the average signal amplitide. Thus, if the signal strength should vary, C2 will either increase or decrease its charge, depending on whether the signal increased or decreased. Since the charge on the AGC capacitor responds only to changes in the average signal level, instantaneous variations in the signal will not affect the AGC voltage.

As mentioned earlier, depending on transistor types, the transistor receiver may require either a positive or a negative AGC voltage. Clearly, a negative AGC voltage could be easily obtained by reversing CR1. Once the values for R2 and C2 have been selected, the voltage divider action of the components is fixed, and the circuit operates automatically without further adjustment. If the average amplitude of the signal increases, the charge on C2 will also increase. If the signal amplitude decreases so does the charge of C2.

Figure 18-7 shows a similar circuit utilizing an electron tube diode as the nonlinear device. This circuit is shown for familiarization purposes only as space, reliability, and weight requirements normally make the solid state diode detector preferable. Notice that the circuit of figure 18-7 produces a negative AGC voltage. Electron tube circuits use negative bias voltages exclusively, and there would be no requirement for positive AGC voltages in an electron tube receiver.

The AGC voltages, in both the transistor and electron tube receivers, act as controlled degenerative feedback. By adjusting the operating point of an amplifier, the gain may be controlled. Under no-signal conditions, the bias of the RF and IF amplifiers is developed by standard means, such as self bias. With an applied signal, an AGC voltage is developed, which in conjunction with the normal biasing methods, develops the operating bias for the amplifiers.

In a transistor receiver, the polarity of the AGC voltage is determined by the type of transistors that are being used. If AGC is applied to the base, a positive AGC voltage is required for PNP transistors and a negative AGC voltage is required when the RF and IF amplifiers are of the NPN type.

Assuming the circuit under discussion is an electron tube receiver; with a signal applied at the input, an AGC voltage is developed. The AGC voltage will have a negative polarity (with respect to ground). The combination of the AGC voltage and the self bias determines the operating point of the amplifiers. If the input signal suddenly increases its strength, the amplitude of the detector output will increase. An increase in the detector output will cause an increased AGC which causes an additional amount of bias on the amplifiers, changing their operating point, and the amplifier gain is correspondingly reduced. In the case of a reduced input signal, the AGC signal would be reduced resulting in less bias and an increased amplifier gain.

CONTROL OF TRANSISTOR AMPLIFIER GAIN BY AGC

It has been shown how a d.c. voltage that is obtained at the output of the AGC network, is proportional to, and will reflect the average variations of the average signal level. All that remains is to use this AGC voltage to control

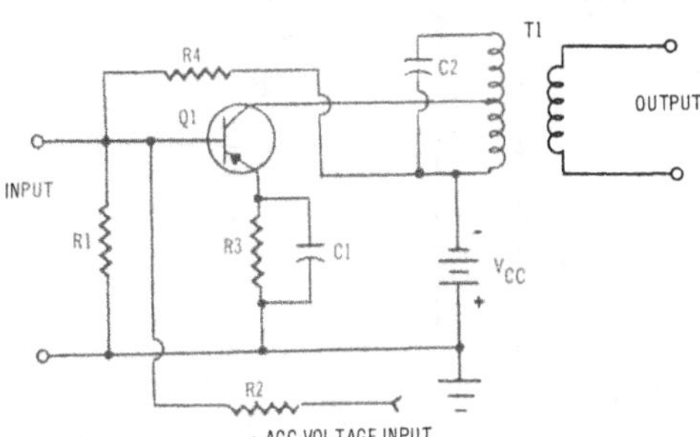

179.361
Figure 18-8.—Common-emitter amplifier with AGC.

the amplification of one or more of the preceding amplifiers in the desired manner. Figure 18-8 illustrates a common-emitter amplifier with AGC applied to the base element. A change in the AGC voltage will change the operating point of the transistor. This will change the d.c. emitter current. In the circuit shown, R1 and R4 form a voltage divider network and establish no-signal (forward) bias on the base. Since a PNP transistor is used the base has a negative potential. The AGC voltage from the detector is positive with respect to ground and is fed to the base through dropping resistor R2. When the d.c. output of the detector increases (due to an increase in the average signal level) the AGC voltage will become more positive. This increased positive potential is applied to the base of Q1, decreasing the forward bias of Q1, thereby decreasing the gain of the amplifier. AGC behaves, in this application, as a controlled degenerative feedback, and the use of an NPN transistor would have required the AGC voltage to possess a negative potential.

FORWARD AND REVERSE AGC

When the AGC voltage is utilized in such a way as to cause degeneration by driving the amplifiers toward cutoff, it is referred to as REVERSE AGC. This is the type of AGC which is normally utilized and has been shown in figure 18-8. A second method of utilizing AGC voltages is possible. This method is called FORWARD AGC. In the case of forward AGC, the amplifier is driven toward the saturation region of its operational characteristics.

As an example, assume that the AGC voltage in figure 18-8 is negative. Q1 would be self biased so as to be operating well up on its characteristic curve under no-signal conditions. When a signal is applied, a negative AGC voltage is developed in the detector circuit, fed back to Q1, and further increases the forward bias. An increase in signal level would cause an increased AGC voltage. This increases conduction which in turn puts the transistor at or near saturation. As the transistor operates nearer saturation its gain is correspondingly reduced.

On the other hand, if the input signal level were to decrease, the negative AGC voltage would decrease, forward bias would be reduced, and the transistor would be allowed to operate on a lower portion of its characteristic curve where its gain would be higher.

While forward AGC provides better signal handling capabilities reverse AGC is simpler to use, causes less loading of the tuned circuits, and produces a smaller variation in input and output capacitances. While there are special applications wherein forward AGC would be preferred, the remainder of this text will deal exclusively with reverse AGC.

DETAILED OPERATIONAL ANALYSIS

Figure 18-9 shows an AGC circuit used with a duo-diode triode in a conventional detector circuit. The two plates of the diode are connected together to form a half-wave rectifier in the RF portion of the circuit. The output of the diode detector is fed to the grid of the triode section which acts as a class A voltage amplifier.

When power is applied to the circuit, there is immediately a path for current from ground to cathode, through V1, through R4 (the plate load resistor) to E_{bb}. Note that the IF signal is not applied to the grid but rather to the diode plates. The diode section of the tube will be capable of conducting when the top of the tank circuit goes positive with respect to the cathode. When this happens, the tank circuit acts as a source of energy and causes a current to flow downward through C3 to ground and back through the diode section of the tube. This places a voltage charge on C3 as shown in figure 18-9. The combination of R2 and C3 form a filter network which will eliminate the IF signal but will retain the audio variations. Figure 18-10 shows the signal voltage present in the tank circuit (part A) and also the voltage present across the C3R2 filter (part B). The filtered signal shown

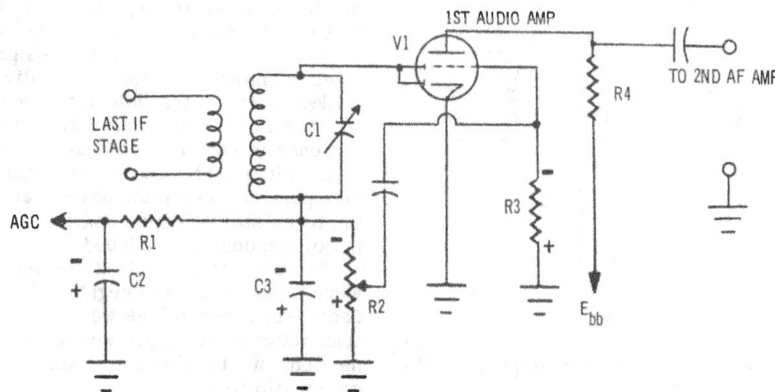

20.268
Figure 18-9.—Detector, AGC, first audio amplifier, and manual volume control circuits.

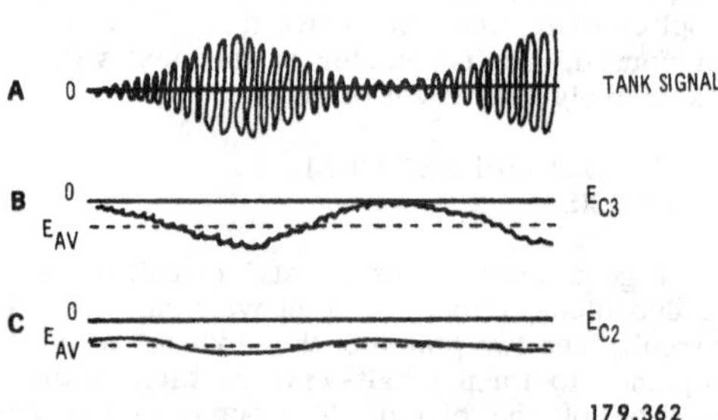

179.362
Figure 18-10.—Waveforms for figure 18-9.

in part B is fed to the grid of the triode section as the audio signal.

Since this voltage is negative, the bias for the triode is automatically determined by the average signal level. If it is undesirable to vary the triode bias with the signal level, the signal from R2 may be capacitively coupled to the grid. If this is done, bias will be achieved by means of contact bias which the grid will automatically develop. It should be noted that the output from R2 to the grid is variable. This is one method of controlling the volume of the receiver output.

The voltage developed across C3 is further filtered by the R1C2 combination and utilized as a negative AGC voltage to control the gain of preceding amplifier stages. The output of C2 is a d.c. voltage which is proportional to the average signal level present across C3 (fig. 18-10C). The filtering is such that the audio variations are filtered out, but slow fading of received signal produces a changing AGC voltage which is sent to preceding stages in the form of a changing AGC voltage.

A disadvantage of ordinary AGC is that even the weakest signals produce some AGC bias which could result in no useable receiver output or a reduction in overall receiver sensitivity. This reduction occurs because a slight increase in the AGC produced by weak signals will cause a decrease in the RF amplifier gain, thereby attenuating very weak signals even more.

DELAYED AUTOMATIC GAIN CONTROL

The disadvantage of automatic gain control, that of attenuating even the very weak signal, is overcome by the use of delayed automatic gain control, as shown in figure 18-11. This type of system develops no AGC feedback until a certain received signal strength is attained. For signals weaker than this value, there is no AGC developed. For sufficiently strong signals, the delayed AGC circuit operates essentially the same as the ordinary AGC circuit.

The delayed AGC circuit is possible using either two diodes and a transistor or using an electron tube (as shown in fig. 18-11). With the exception of different component values, the theory of circuit operation is essentially the same.

The delayed AGC circuit utilizes (fig. 18-11) a duo-diode triode. Diode plate 1 acts as the

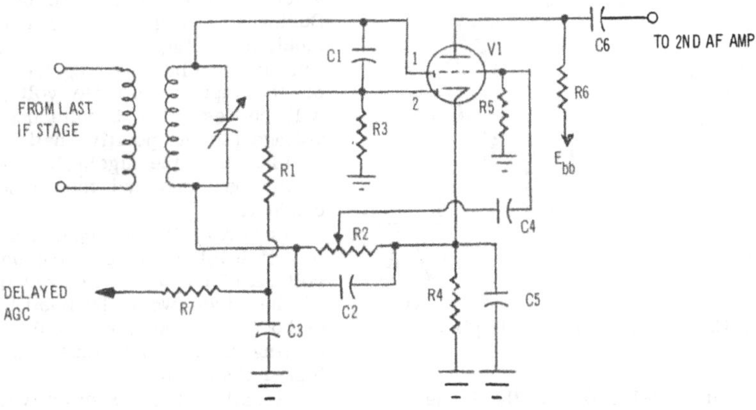

20.269
Figure 18-11.—Delayed AGC circuit with volume control.

detector diode, while plate 2 acts as an AGC diode. The two plates are separated by capacitor C1. The R2C2 network is the standard detector filter, C4 couples the audio signals to the grid of V1 for amplfication, R5 is a standard grid input resistor, R4 and C5 serve as a cathode bias network, R6 is the plate load resistor for V1, C3 is the AGC capacitor, and R7 provides additional filtering as well as isolation.

When power is first applied to the circuit, a charging current flows and a 5 volt bias is developed across R4C5 (fig. 18-12). A second charging current flows from ground, through R3, through C1, the tank, through R2, through the tube, power supply, and back to ground. This charging current establishes a 5 volt bias across C1. This equivalent circuit shows that R3, C1, the tank, and R2, are effectively in parallel with the cathode bias network. Hence, as the cathode bias voltage is 5 volts, C1 must also charge to 5 volts. Compare this equivalent circuit to the complete schematic shown in figure 18-11.

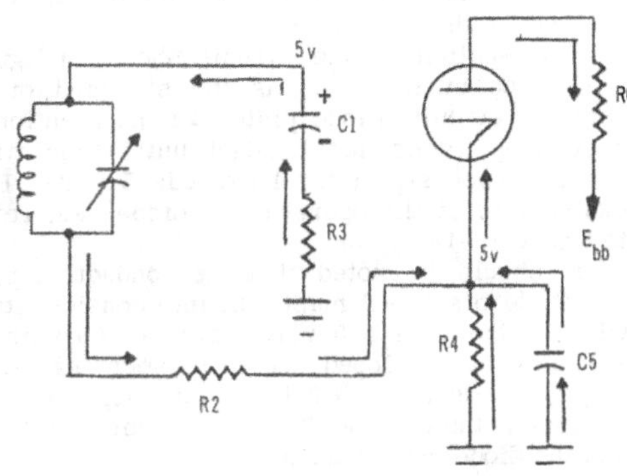

179.363
Figure 18-12.—Charging current.

It should be noted that no quiescent charge develops in the tank capacitor because it is shunted by the tank coil. By the same token, C2 is shunted by R2 and, therefore, develops no quiescent charge.

Under normal operating conditions (to be explained presently) diode plate #1 will conduct and allow detector action to occur. The charge on C1, however will act as a bias to keep diode plate #2 in a cutoff condition. This will prevent the development of AGC until such time as the input signal exceeds 5 volts.

Figure 18-13 shows the operation of the circuit with a signal applied which has an amplitude of less than 5 volts. It is impossible for diode plate #1 to conduct on the negative alternations of the input, but on the positive half cycle, current will flow from the tank (which is acting as the source), charging C2, from the cathode to plate #1, and back to the tank. This action develops a charge across C2 of the polarity

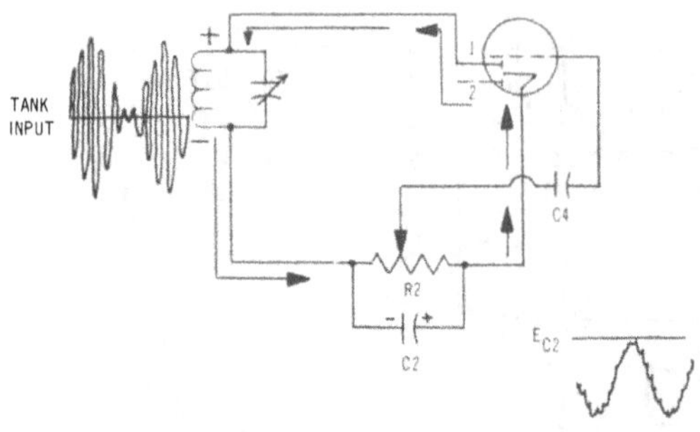

179.364
Figure 18-13.—Operation of diode plate #1.

shown in figure 18-13. Due to the large value of R2, this capacitor does not have time to adequately discharge during the negative half cycles of the input. Therefore, the value of charge across C2 follows the average value of input modulation. The wiper arm allows this filtered audio signal to be fed (in any amplitude desired) to the grid of V1.

The section of the circuit shown in figure 18-13 continues to act as the standard diode detector circuit. Diode plate #2, as mentioned previously, does not conduct until such time as the input signal level exceeds 5 volts. For explanation of the operation of diode #2, refer to figure 18-14.

It should be noted that the conduction path for diode plate #2 must take into consideration C1. As there is a 5 volt charge on C1 with no signal input developed due to cathode bias (fig. 18-12), diode plate #2 is 5 volts negative with respect to the cathode. This is a reverse voltage, and the diode cannot conduct.

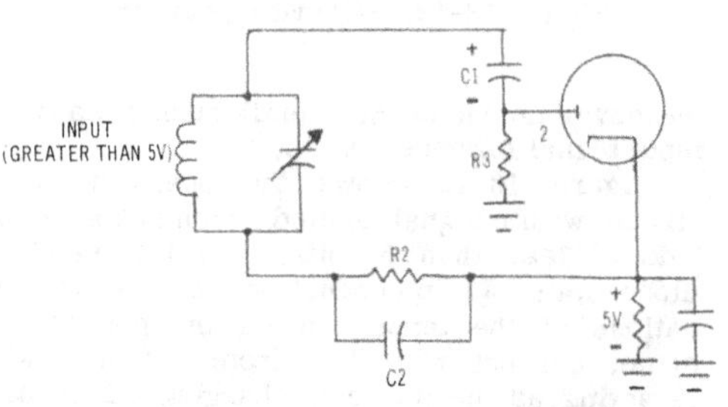

179.365
Figure 18-14.—Operation of diode plate #2.

The voltage applied to the diode will be the algebraic sum of the voltages in the circuit. During the negative half cycle of the signal input, the voltage will be more than 5 volts negative on the plate. During the positive half cycle of the input, the plate voltage for diode #2 will be less than 5 volts negative. If the input voltage (on the positive half cycle) were exactly 5 volts, then the algebraic sum of the voltages would be zero and the diode would still not conduct.

However, if the input voltage exceeded the 5 volt level on its positive half cycle, the bias voltage of C1 would be overcome, the plate (#2) of the diode would be positive with respect to cathode, and the diode would conduct. The conducting diode would thus charge C1 to a value higher than 5 volts.

Recall that C1 is effectively in parallel with the cathode bias circuitry (fig. 18-12). As the cathode circuit is maintaining a 5 volt level, C1 will attempt to discharge any excess charge above 5 volts. The discharge path is down through R3 to ground, up through the cathode bias network, R2, the tank, to the capacitor (fig. 18-15).

This discharging current develops a negative voltage across R3 which is then filtered by R7, C3, and R1, and is utilized as the AGC voltage. C3, of course, is the actual AGC capacitor.

The voltage across C3 constitutes the delayed AGC voltage. It is developed only when the input signal in the tank exceeds the 5 volt level. It is supplied to the grids of the various stages ahead of the second detector in series with the cathode bias developed by the individual tubes.

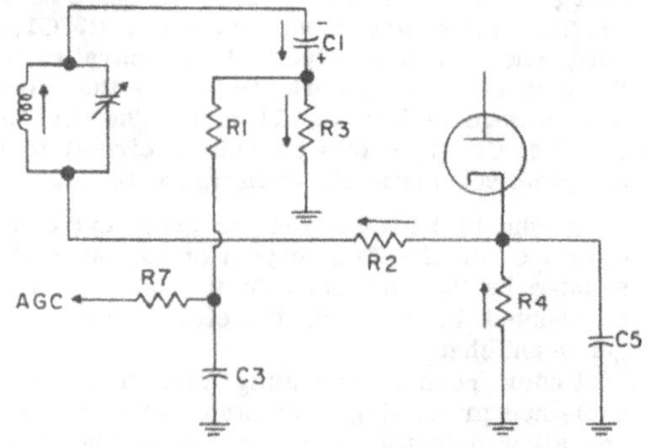

179.366
Figure 18-15.—Delayed AGC action.

Finally, it should be restated that two diodes are required in order to obtain delayed AGC. If only one diode were used, then the AGC would be developed from the detector diode and there would be no delayed action. Or, if a single diode were biased to provide the delaying action desired, NO signal would pass to the audio amplifier until the bias was exceeded by the input signal.

AMPLIFIED AGC

To keep the receiver output constant for large variations in signal strength, a larger AGC voltage variation than is possible with the AGC circuits previously discussed, must be developed. This can be accomplished by using an AMPLIFIED AGC circuit. The amplified AGC also provides better control of low signal level variations.

A typical amplified AGC circuit is shown in figure 18-16. It uses a separate amplifier stage to increase the level of the detected signal. This AGC amplifier receives the rectified signal from the detector through capacitor C1. Its amplified output is rectified by the AGC diode and filtered by resistor R5 and capacitor C4. The resulting amplified AGC voltage is applied to the controlled stages along the AGC line. Variable emitter resistor R3 (cathode resistor in the electron tube version) allows adjustment of the gain of the AGC amplifier. In this way, the AGC voltage can be set at any desired level. If an AGC delay network similar to that described previously is added at the output of the AGC rectifier, the circuit will provide a delayed AGC. Variable resistor R3 then acts to set the level of gain which will set the AGC circuit into action. When used in this form R3 is the "threshold control."

Detailed Circuit Analysis

In both circuits, the tuned tanks couple the modulated IF signal to the detector circuits. The detectors rectify the input signal and C2 and R1 filters out the IF component leaving only the audio signal as the output to the audio amplifiers. The audio signal is also coupled to the AGC amplifiers by the action of C1. The amplified signal appears in the collector or plate circuit where the AGC diode rectifies the signal. The rectified audio signal is then filtered by R5 and C4 and utilized as a d.c. AGC voltage. Additional filtering may be added to the circuit

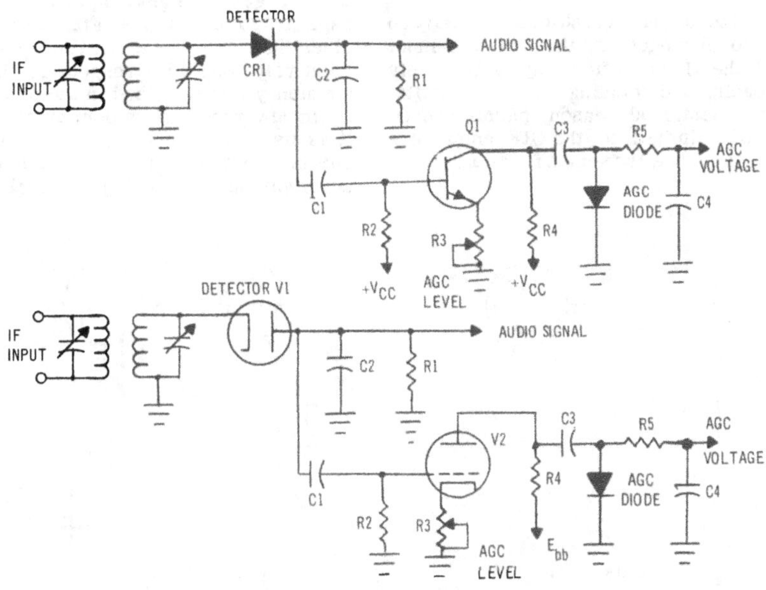

Figure 18-16.—Transistor and electron tube amplified AGC circuit.

if required. In both circuits R3 is used to set the operating point for the AGC amplifier and thus the level for the AGC voltage. As the output (collector/plate) of both circuits is an a.c. variation at a d.c. level it is possible to simply filter the output rather than use the AGC diode. This is often done in the case of the transistor circuit shown in figure 18-16.

METHODS OF FEED

Figure 18-17 shows the two methods of delivering the AGC voltage to the preceding amplifier stages. Part A shows the series feed arrangement while part B shows the parallel feed configuration. Either method is applicable to both transistors and tubes and the choice is a matter of design application.

It should be noted that both circuit configurations require d.c. blocking capacitors to prevent the AGC voltage from being shorted to ground either directly, or through the tank coil. The size of the blocking capacitors must be chosen so as not to affect the tuning range of tank circuits. Also, a close check of the parallel feed circuit will show that the input resistor parallels the tank circuit.

Since the size of the resistor is normally in the range .1 to .5 megohms, the dynamic input impedance of the IF amplifier may be seriously affected by loading and detuning the tank circuits. For the above mentioned reason, parallel feeding is normally limited to the RF amplifiers while IF amplifiers use the series feed configuration.

CONSIDERATIONS OF THE AGC FILTER CIRCUIT

There are several reasons why AGC filter circuits are required. One important reason is to prevent RF and AF voltage variations which appear across the AGC diode load from being applied back to the bases or grids. If these are fed back, then instability troubles are certain to arise. If AF variations are returned to the preceding amplifiers out of phase a reduction in the percentage of modulation of the carrier might result. Or an in phase ripple feedback could lead to an undesirable whistle or motorboating.

From the above discussion, it can be seen that the prime function of the filter circuits is to allow only the d.c. voltage developed in the AGC circuit to be applied, as additional bias, to the controlled stages. The components which make up the filter also serve the necessary function of providing low impedance paths to ground for RF, IF, and AF variations.

It is not permissible to choose values of resistance and capacitance at random if satisfactory circuit operation is to be obtained. The time constant of the filter network must be low enough so that the AGC bias voltage can follow the changes in signal input level with sufficient rapidity to offset the effects of fading. On the other hand, the circuit time constant must be relatively long with respect to the lowest audio frequency to be utilized in the circuit.

In any good quality communications receiver, it is usual to provide a means for selecting any one of a number of AGC time constants. This is accomplished by using several resistance or

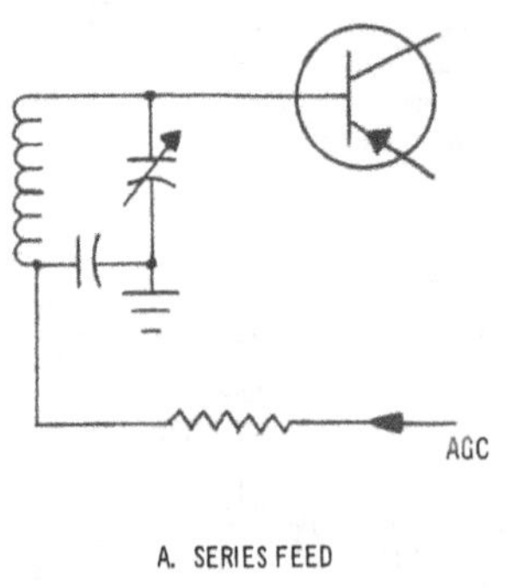

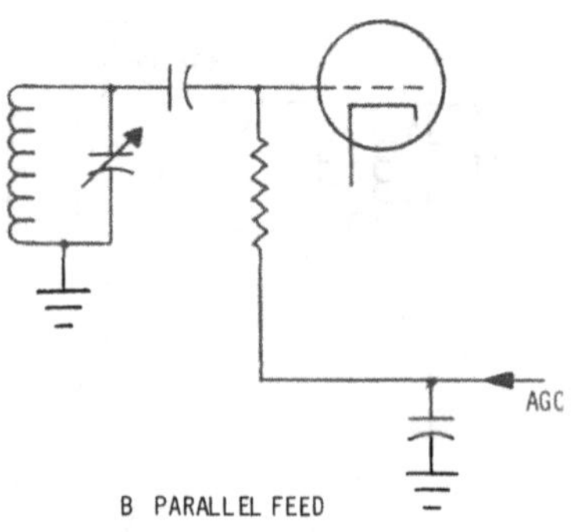

179.368
Figure 18-17.—Methods of feed.

capacitance values which are connected into the AGC filter network by means of a switch. This enables the operator to select the most suitable condition to offset the particular type of fading being encountered. Very short time constants are seldom selected for high quality broadcast reception as rapid fading would cause bass frequency antimodulation, resulting in loss of audio bass (frequencies below approximately 250 Hz) response.

In standard broadcast receivers the charge and discharge time constants are often very nearly equal. However, for some types of reception it may be preferable to have a rapid charge time to prevent the beginning of a signal from being unduly loud, and a slow discharge time to prevent a rapid rise in noise output during intervals between signals.

AUDIO DERIVED AGC

Conventional AGC systems used in AM receivers do not provide satisfactory results when used in single sideband reception. The problem is that the usual AGC system derives its control voltage from the carrier which is transmitted along with the information signal. Since in SSB transmission, no carrier is present to establish an average signal strength level, the AGC system must establish its control voltage from the information contained in pulses. (SSB will be studied in detail in chapter 25.) This information varies greatly in amplitude and it is intermittent with normal voice operation. The time constant of the AGC system is very important. It must be rapid enough to prevent strong signals from coming through too loud at first and yet be slow enough not to follow the syllabic variations of normal speech. Circuits having a charge time of 50 milliseconds and a discharge of 5 seconds have proved successful. Figure 18-18 shows a typical audio derived AGC circuit. The fast charge time (rise time) results from the low impedance charge path for capacitors C4 and C5. A very slow decay time results from the high discharge impedance through R7 and the amplifier circuit Q3/Q4.

OTHER METHODS OF VARYING TRANSISTOR AMPLIFIER GAIN

The use of bias to control the gain of an amplifier has been presented previously in this chapter. However, intermodulation distortion problems occur when the bias is changed to control amplifier gain. Intermodulation distortion is the creation of distortion products based on the sum and difference frequency of harmonics of the desired signals. If signals of 1000 kHz and 1001 kHz were presented to a nonlinear element having second and third order curvature, the difference frequency of 1 kHz could remodulate each frequency to develop distortion products at 999 kHz and 1002 kHz. Since all of

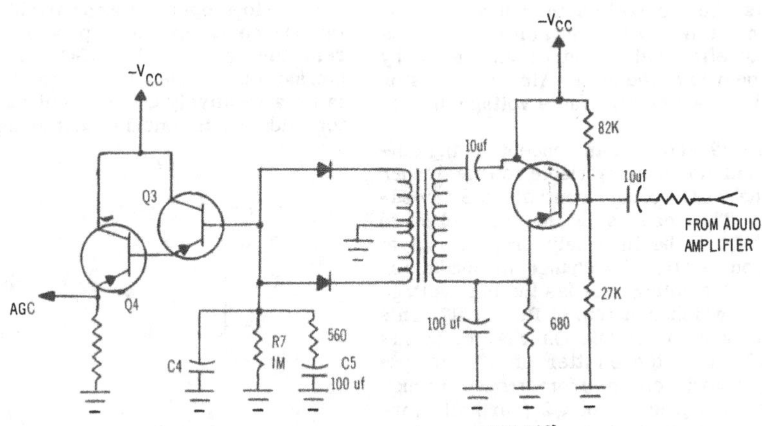

Figure 18-18.—Audio derived AGC.

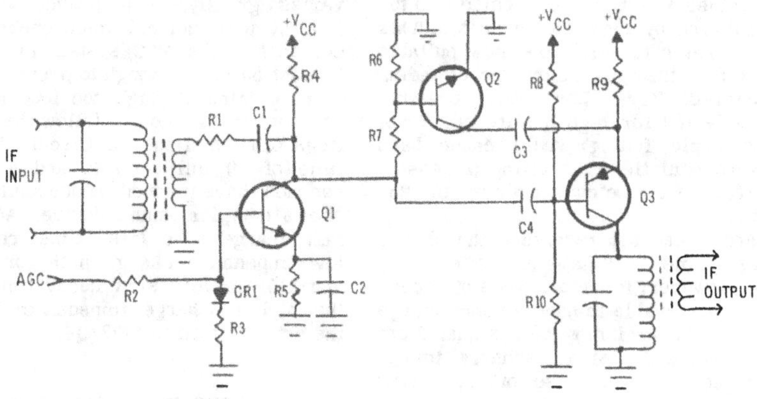

179.370
Figure 18-19.—Gain controlled amplifier.

these signals are within the passband, selectivity is of no help.

As signal input amplitude is increased, the gain of the transistor is reduced, and distortion tends to rise. The best solution is to preserve the linearity of the stage and find other means, besides varying the bias, to control the gain. Removal of the bypass capacitor from across the emitter resistor of a transistor amplifier results in decreased gain due to negative current feedback. This results in increased linearity and thereby reduces distortion as the signal is increased. A transistor can be used as a variable resistance which is a function of a d.c. current change. This d.c. controlled impedance can be connected in series with the emitter bypass capacitor to effectively control the gain by varying the degree of feedback. Also, a transistor can be used as part of a variable voltage divider network.

Figure 18-19 shows both concepts being employed. Q1 and R4 form a variable voltage divider network. The AGC voltage controls the impedance of Q1. This causes the signal applied to the base of Q3 to be inversely proportional to the conduction of Q1. The change in conduction of Q1 changes the voltage across the bias voltage divider of Q2 which consists of R6 and R7. This varies the conduction of Q2. On charge, that is when the voltage at the emitter of Q3 attempts to increase, C3 will charge from ground through the emitter-base junction of Q2 (normally forward biased), through the base-collector junction of Q2 (reverse biased at this time), through C3 to the emitter of Q3. On discharge, that is when the voltage at the emitter of Q3 attempts to decrease, capacitor C3 will discharge through the base-collector junction of Q2 (now forward biased), through R6 to ground and from ground through the primary of the IF output transformer and the base-collector and base-emitter junctions of Q3. Therefore, by varying the conduction of Q2, the charge/discharge rates of C3 and the amount of degenerative feedback applied to Q3 can be varied.

Another method of controlling the signal level is to use a variable H pad consisting of diodes and resistors as shown in figure 18-20. The diode bridge operates as a variable resistor whose impedance is inversely proportional to the current through it. The AGC voltage changes the conduction of the diodes and, therefore, maintains a relatively constant output from the H pad for wide variations in input signal level. In very

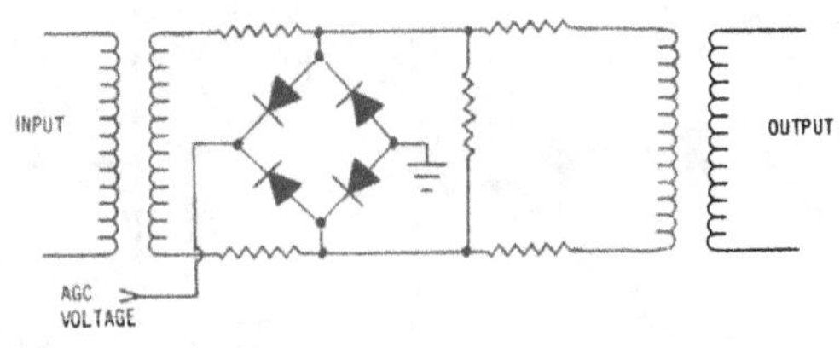

179.371
Figure 18-20.—Variable H pad.

sophisticated receivers, all the methods shown here may be used to provide optimum results.

MULTIPLE CONVERSION

In chapter 14, a higher intermediate frequency was used to improve image frequency rejection, but this reduced the receiver's selectivity which is primarily determined by the IF stages. The selectivity was reduced because the bandwidth of tuned circuits for a given Q increases as the circuit's resonant frequency is increased. To obtain a good image ratio at higher frequencies, and good selectivity, a system known as double conversion is employed. In this system an incoming signal is converted to a rather high intermediate frequency and then amplified. The signal is again converted, this time to a much lower frequency. The first intermediate frequency provides the necessary separation between the image and the desired signal, while the second supplies the majority of the receiver selectivity. Figure 18-21 shows an application of double conversion for the purpose of attaining extremely good stability in a communications receiver, through the use of crystal control of the second oscillator. In this receiver, the overall stability determined by the VFO is equal to the stability of the oscillator which feeds the first mixer, while the selectivity is primarily determined by the bandwidth of the second IF amplifier. Receivers will be encountered which use more than two mixer stages.

FREQUENCY SYNTHESIS

In present day communications systems, long term accuracy of 1 part in 10^6 is required from many of the frequency generators (local oscillators) used in communications equipment. Methods such as using variable frequency oscillators cannot practically achieve this high degree of stability. Therefore, a system known as frequency synthesis has been developed to meet these stringent demands for stability. This system may be defined as a circuit in which a signal frequency is produced by heterodyning and selection of frequencies not necessarily harmonically related to each other or the frequency produced.

In figure 18-22, a multiple crystal frequency synthesizer produces the desired output frequencies by mixing the frequencies from the several crystal oscillators. Each oscillator has 10 or more crystals which may control its operating frequency so that a large number of output frequencies are possible. It is technically and economically unfeasible to maintain all the crystals to the required accuracy. Figure 18-23 shows a practical frequency synthesizer in which the harmonics and subharmonics of a single standard oscillator are combined to provide a multiplicity of output signals, each of which is harmonically related to a subharmonic of the standard oscillator. The primary difficulty encountered in the design of the frequency synthesizer is the presence of spurious signals generated in the "combining mixers." Extensive

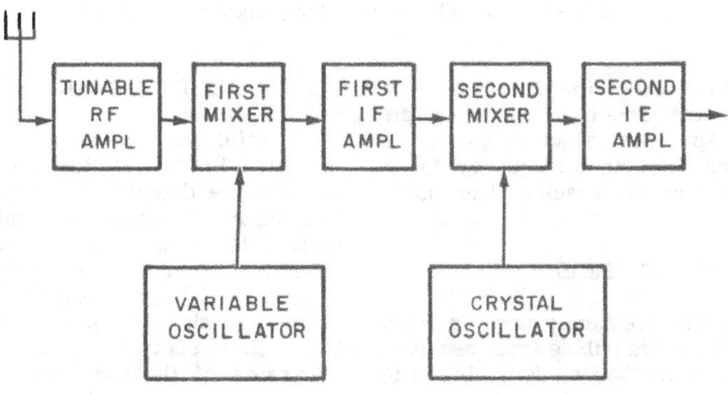

179.372
Figure 18-21.—Double conversion.

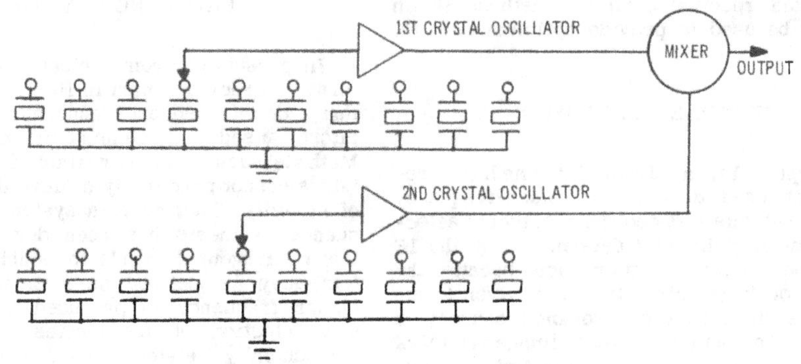

179.373
Figure 18-22.—Multiple crystal frequency synthesizer.

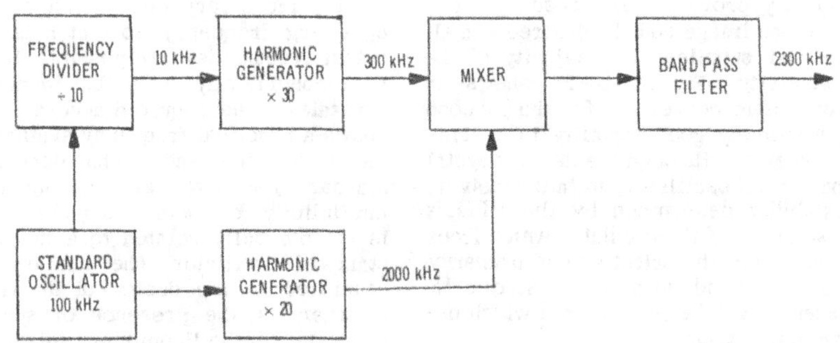

179.374
Figure 18-23.—Single crystal frequency synthesizer.

filtering and extremely careful selection of operating frequencies are required for even the simplest circuits. Spurious frequency problems increase rapidly as the output frequency range increases and the channel spacing decreases.

NOISE LIMITERS

Noise limiters are frequently used in Navy receivers to remove noise pulses from received signals. One common method of doing this is to limit the amplitude of the detector output before it is applied to the audio amplifiers. The noise pulses, which are of relatively large amplitude, are thus eliminated.

SERIES NOISE LIMITER

A solid state series diode limiter is shown in figure 18-24. Limiter diode CR1 is connected between the detector output and the audio amplifier input. The level of limiting is set by resistor R4 by making the cathode of the diode sufficiently negative with respect to the anode. For normal signal levels, CR1 conducts and allows the signal to pass to the audio amplifier. If the limit control is properly set, the negative excursion of the audio does not go sufficiently negative to cut off CR1. However, when a large negative diode pulse is developed across the detector load resistor R1, it drives the anode of the CR1 negative with respect to the cathode.

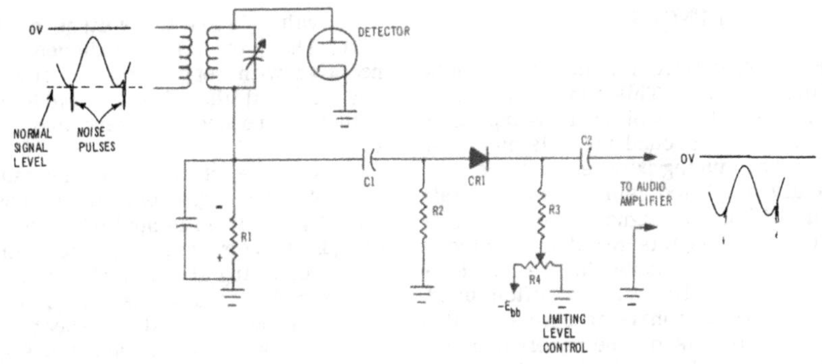

179.375

Figure 18-24.—Series noise limiter.

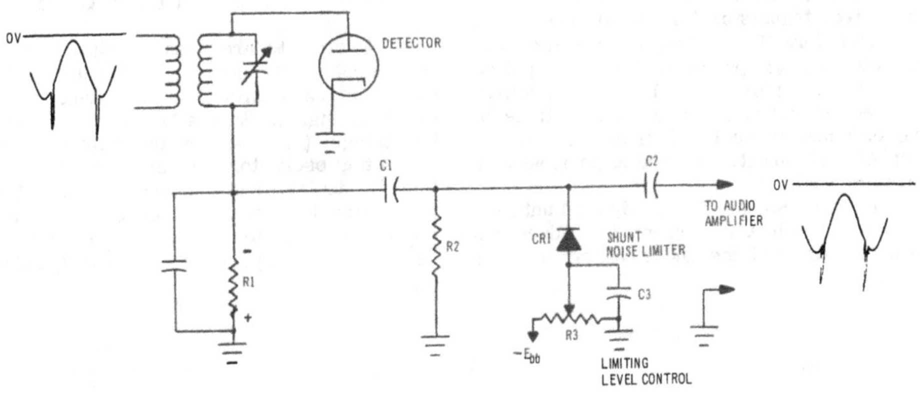

179 376

Figure 18-25.—Shunt noise limiter.

Consequently CR1 cuts off and prevents the noise pulse from reaching the audio amplifier. Positive going noise pulses are not a great problem in the circuit shown because the detector output cannot produce an output above zero reference, therefore, positive noise amplitudes are limited by the detector.

SHUNT NOISE LIMITER

The series diode limiter was so called because the limiter is connected in series with the load. A parallel or shunt limiter as shown in figure 18-25 is placed in parallel to accomplish limiting. The basic difference from the series limiter is that the diode CR1 shunts the noises pulses around the audio input. Therefore, under normal signal conditions the diode is cut off and the detected signal is passed to the audio amplifier. When a large noise pulse is developed across the detector load, it drives the cathode of the limiter diode more negative than the anode. Consequently, the diode conducts and C3 shunts the noise pulse to ground preventing it from reaching the audio amplifier. The diodes can be reversed in both circuits if positive limiting is desired. Also, the solid state diodes can be replaced by tubes.

SILENCER

A receiver's sensitivity is maximum when no signal is being received. This condition occurs, for example, when the receiver is being tuned between stations. Background noise is picked up by the antenna when tuning between stations. The noise is greatly amplified, since receiver gain is maximum without a signal, and is highly annoying. To overcome this problem, a circuit called "noise silencer," "squelch," "noise suppressor," or "noise limiter" is often used. Figure 18-26 shows a transistor circuit of this type. Such a circuit cuts off the receiver output when no input signal is being received. It accomplishes this by blocking either the detector or audio amplifier when no signal is present.

The silencer diode CR1 connects the output of the first AF stage to the input of the second audio amplifier. Silencer amplifier Q1 serves as the control transistor for the silencer. The anode voltage for CR1 is supplied via R2 from the collector of Q1 (which in turn is supplied from the V_{CC} supply via R11) and is positive with respect to ground. The cathode voltage of CR1 is obtained through R3 from the divider made up of R12 and R4 and is also positive with respect to ground.

With no input signal, R9 is adjusted until Q1 draws enough collector current to reduce its collector voltage and the anode voltage of CR1 to a value below the voltage on the cathode of CR1. Thus, the silencer diodes anode voltage is negative with respect to its cathode. Conduction ceases, and the silencer diode cuts off. The output is reduced to zero and the receiver is mute.

The base of Q1 is connected to the AVC line. When a signal enters the receiver, a negative AVC voltage is applied to the base circuitry of Q1, thereby reducing the collector current and increasing the collector voltage, which increases the anode voltage of CR1 until the anode becomes positive with respect to its cathode, the diode will conduct and the signal is passed to the second AF amplifier. Diode CR1 is effectively a switch which is controlled by the AVC voltage.

AUTOMATIC FREQUENCY CONTROL

AFC circuits are used in situations where it is necessary to control, accurately, the frequency of an oscillator in accordance with some external signal. Basically, such a circuit does two things: (1) it senses the difference between the actual oscillator frequency and the frequency that is desired and produces a control voltage proportional to the difference (2) it uses the control voltage to change the oscillator to the desired frequency. There are two types of AFC

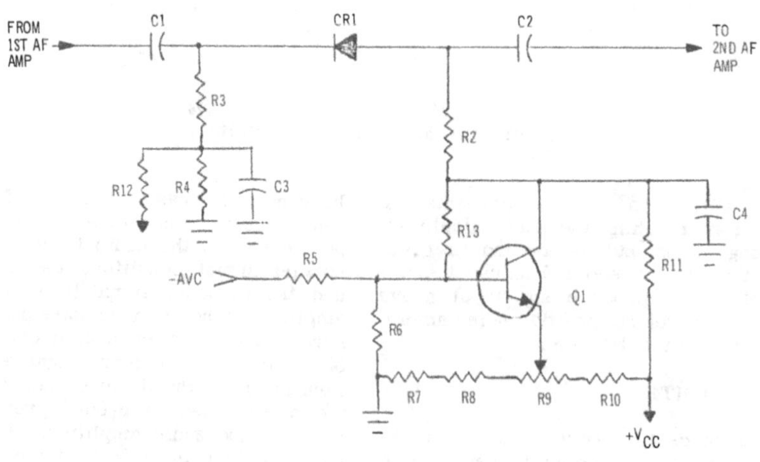

Figure 18-26.—Silencer circuit.

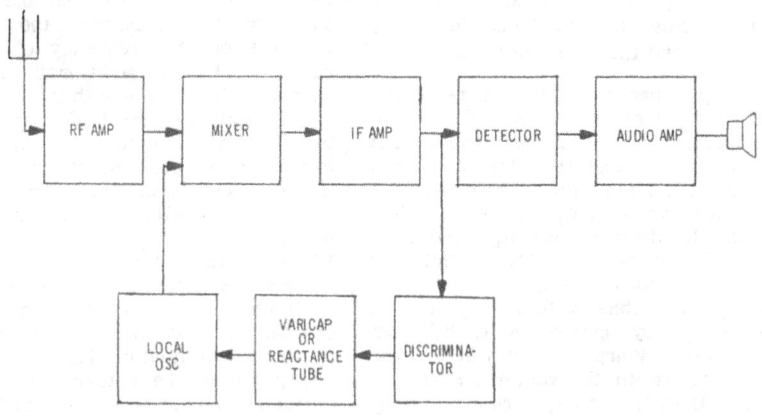

179.378
Figure 18-27.—AFC controlled receiver.

circuits: (1) those that control the frequency of sinusoidal oscillators, and (2) those that are used with nonsinusoidal oscillators. Only the sinusoidal AFC circuit will be presented here. It is used in radio receivers, FM transmitters, and frequency synthesizers to maintain frequency stability. Figure 18-27 illustrates the AFC operation in a receiver.

The frequency discriminator controls the varicap in solid state circuitry and its counterpart, the reactance tube, in electron tube circuitry. The varicap and the reactance tube exhibit an apparent reactance which is included in the oscillator frequency control circuitry. For example, if the IF should be 455 kHz and the local oscillator (LO) is tracking below the incoming station, when the LO output decreases slightly in frequency, the IF will be high. This causes the output of the discriminator to increase the capacitive reactance of the varicap or reactance tube which increases the oscillator frequency to the desired value. The varicap operation will be explained in chapter 25. Its effects on the circuit are the same as those of the reactance tube.

Theory of operation of a reactance tube circuit may be explained with the aid of figure 18-28. The reactance tube, V1 (fig. 18-28A), is effectively in shunt with the oscillator tank, LC, and the phase shift circuit, R_g and C1. The grid current, i_g, in this circuit leads the voltage, e_p, across the circuit by approximately 90 degrees. The voltage, e_p is the alternating component of

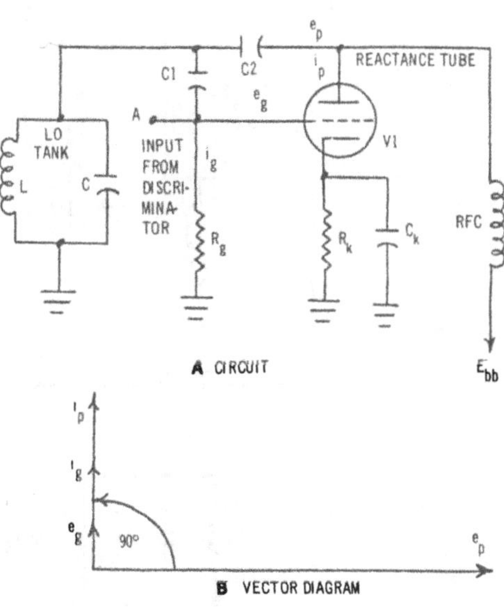

179.379
Figure 18-28.—Reactance tube AFC.

the plate to ground voltage appearing simultaneously across the reactance tube, the phase shift circuit, and the oscillator tank.

The coupling capacitor, C2, has relatively low capacitive reactance to the a.c. component of current through it, and at the same time it blocks the d.c. plate voltage from the phase shift circuit and the tank. The reactance tube receives its a.c. grid input voltage, e_g, across R_g. This voltage is the IR drop across R_g and is in phase with plate current i_p. This relation is characteristic of amplifier tubes.

Because e_g is in phase with both i_g and i_p, and e_g leads e_p by approximately 90°, both i_g and i_p lead e_p by approximately 90°. These reactions are shown in the vector diagram in figure 18-28B. Both i_g and i_p are supplied by the oscillator tank circuit, and because both are leading currents with respect to the tank voltage, e_p, they act like the current in tank capacitor C. Therefore, the effect of these currents on the frequency of the tank is the same as though additional capacitance were connected in parallel with it.

Consider the effect of applying a d.c. voltage to point A in figure 18-28A. When the local oscillator is operating at the correct frequency, the d.c. output of the discriminator is zero, and, consequently, no d.c. voltage is applied to the grid (point A) of V1. The plate current of V1 is varying at the rate of the tank and it offers some value of capacitance to the circuit.

If the local oscillator frequency were to increase, a positive d.c. voltage would be produced by the discriminator. The amplitude of the d.c. voltage would be determined by the amount of frequency deviation. This positive d.c. voltage is applied to the grid of V1, causing an increase in the amplitude of V1 plate current by increasing

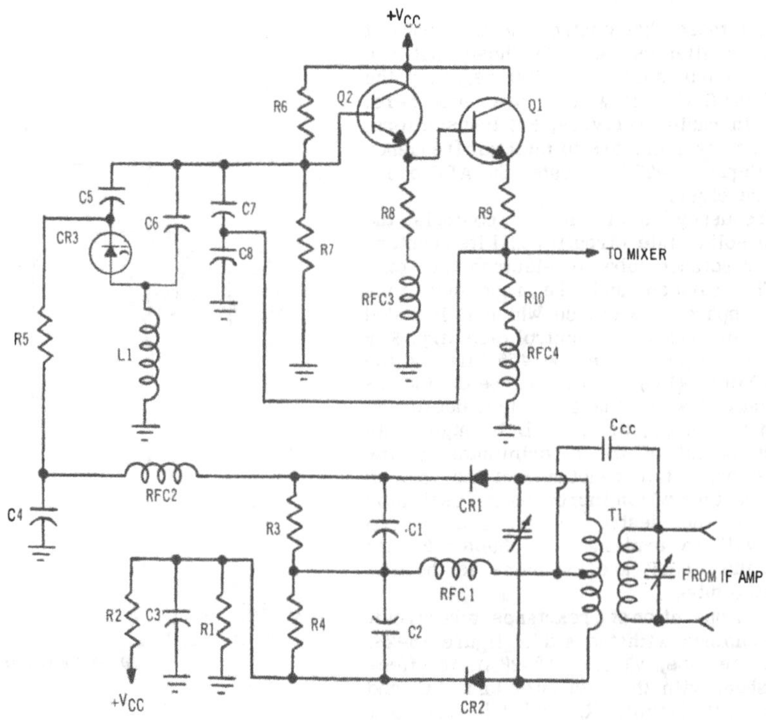

179.380
Figure 18-29.—AFC circuit employing a varicap to control a Clapp oscillator.

the gain. Since the reactance tube circuit appears as a capacitor in shunt with the LO tank, an increase in V1 a.c. plate current causes this apparent capacitance to seem larger. This increase in tank capacitance would decrease the LO frequency until the input voltage to V1, or discriminator output, was again close to zero.

If a decrease in LO frequency occurred, a negative voltage would be produced by the discriminator. The amplitude of the d.c. would once again be determined by the degree of frequency deviation. This negative d.c. voltage, which is applied to the grid of V1, would cause a decrease in plate current. Therefore, the LO frequency increases because the effective capacitance in shunt with the LO tank decreases.

Figure 18-29 shows an AFC circuit employing a varicap (CR3) to control a Clapp oscillator. A voltage divider network consisting of R1 and R2 puts the capacitance of CR3 in the center of its range. The output from the discriminator changes the reactance of CR3 by adding to or subtracting from the level set by R1 and R2. CR3 is part of the frequency control of the CLAPP oscillator.

CHAPTER 19

RECEIVER ALIGNMENT

Alignment may be defined as the process of adjusting the tuned circuits of an electronic equipment to produce resonance at frequencies required or specified for such adjustments. The alignment procedure of a superheterodyne receiver will be discussed in this chapter. Superheterodyne receivers were discussed in chapter 11 of this manual. It is important to remember that the following discussion is a generalization, and the exact alignment procedures for a particular receiver must be obtained from the manufacturer's technical manual for the receiver concerned.

Alignment is necessary to keep the receiver operating at peak performance. Indications of a need for alignment include poor sensitivity and a difference in the frequency received from that of the dial setting. Alignment should also be performed whenever circuit components are replaced.

Before attempting alignment, it must be insured that the receiver is otherwise trouble free, and all available technical literature pertaining to the equipment should be consulted. In addition, a thorough knowledge of the test equipment to be used is of paramount importance.

Prior to the actual alignment, all auxiliary functions, provided in the receiver, which may interfere with proper output indication or circuit resonance, should be disconnected. This includes AGC, silencer or squelch action, noise and output limiters, etc. As this requirement may vary from receiver to receiver, the appropriate technical manuals should be consulted. Also, the antenna should be disconnected and replaced with a dummy antenna.

Alignment is accomplished by injecting a modulated signal, from a signal generator, into the receiver at appropriate points and adjusting the various tuned circuits for a maximum detector rectified output voltage. This voltage may be measured with the use of a high impedance electronic d.c. voltmeter, so as to minimize disturbance to the detector circuit. Alternately, the amplitude of the intelligence at the output of the receiver may be used as an indication of proper alignment. The following discussion will utilize figure 19-1 as a representative receiver block diagram.

ALIGNMENT OF IF SECTION

In general, alignment is best begun in the circuit farthest from the antenna. In the case of a superheterodyne receiver, this is the last IF transformer preceding the detector. In some receivers, it is best to disable the local oscillator prior to attempting IF alignment (S1-fig. 19-1). Again, this requirement varies from receiver to receiver. A modulated signal, usually at the nominal band-center frequency of the particular IF system, is injected at the input electrode of the last IF amplifier stage, and the last or output IF transformer is adjusted for maximum output indication, as previously described (input #1 and adjustment #1 of fig. 19-1). The amplitude of the input from the signal generator should be adjusted to provide a signal output level which is well above the noise level at the output indicator, and also well below the level which would cause overdriving or saturation. If either of these conditions are not met, it will be difficult to get an accurate indication of proper alignment.

After the last or output IF transformer is aligned, the point of signal injection is moved to the input electrode of the preceding IF stage (input #2), and the next to last IF transformer is adjusted (adjustment #2). If there are more than two IF stages, this procedure is continued until the entire IF strip is aligned. Naturally, the amplitude of the injected signal must be decreased as the point of injection moves closer to the antenna and more circuits come into alignment.

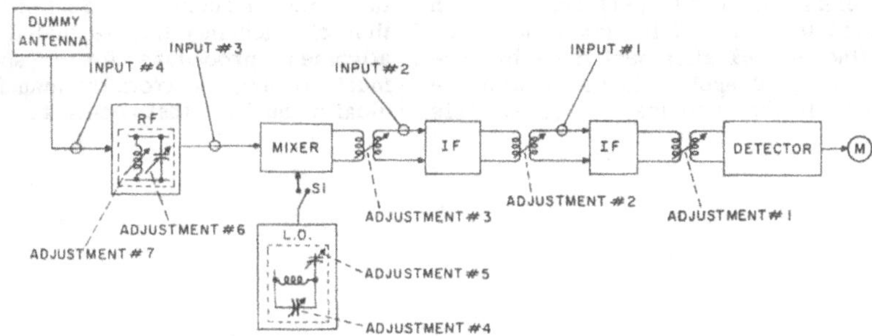

179.381

Figure 19-1.—Receiver block diagram showing signal input and adjustment points for alignment.

Alignment of the first or input IF transformer is achieved with the injected signal placed on the input electrode of the mixer (input and adjustment #3). It should be noted that it is necessary to recheck the overall alignment of the IF strip after the individual stages have been aligned.

It has been previously stated that the injected frequency, used for IF alignment, is usually equal to the nominal band-center of the particular IF system. In some cases, stagger tuning for example, the injected signal to alternate IF transformers must be adjusted above or below the nominal band-center to achieve maximum response of the IF transformers. The end result of such alignment is a wide IF bandpass.

ALIGNMENT OF LOCAL OSCILLATOR

Following the IF alignment, the local oscillator is adjusted. This is done by injecting a modulated signal of the upper alignment frequency, specified for the particular receiver, on to the input electrode of the mixer (input #3 fig. 19-1) and adjusting the shunt trimmer capacitor of the oscillator tank circuit for optimum output indication (adjustment #4). It is now necessary to check oscillator alignment near the low frequency end of the tuning band. The signal generator is reset to the lower alignment frequency specified for the receiver, still inserted at input #3, and the series padder capacitor of the oscillator tank is then adjusted to produce maximum output (adjustment #5). When this process is completed, the shunt trimmer capacitance should be rechecked for maximum output at the high frequency alignment point. Finally, the low end padder adjustment should again be checked.

ALIGNMENT OF RF STAGES

The RF stage(s) or preselector is aligned last. To accomplish this, a dummy antenna is connected to the receiver and a signal of the upper alignment frequency is injected into the antenna input terminal (input #4 fig. 19-1). The RF stage is then adjusted for maximum output indication (adjustment #6). The signal generator is next set at the lower alignment frequency and the RF tuned circuits adjusted for maximum output (adjustment #7). The above steps are repeated until no further improvement is obtained.

ALIGNMENT OF MODULAR RECEIVERS

Special problems are encountered in aligning modular constructed receivers. The impedance of each module must be matched to that of the next. If an impedance mismatch exists, the tuned circuits cannot be properly aligned. In addition, a greater amount of signal amplitude is required to get an output indication from improperly matched modules. The connection of test equipment to a module may cause a mismatch to occur. Therefore, care must be taken to ensure that the test equipment used forms a proper impedance match with the modules being aligned.

Coarse alignment may be performed on each individual module of a modular constructed receiver. A fine overall alignment must be performed with a signal applied to the input of the receiver and all the modules connected. This takes into account the effect of the interconnection of each module. As stated previously, the alignment procedure for a specific receiver must be obtained from the manufacturer's technical manual for that receiver.

CHAPTER 20

INTRODUCTION TO TRANSMITTERS

Equipment used for generating and amplifying a radiofrequency (RF) carrier for transmission through space from an antenna, is called a radio transmitter.

The radio transmitter and its antenna form one link of a communications system. All that is needed to complete the system is a radio receiver and its antenna.

A transmitter may be a simple, low power (milliwatts) unit for sending voice messages a short distance or it may be a highly sophisticated unit utilizing thousands of watts of power for sending many channels of data (e.g., voice, teletype, TV, telemetry, etc.) simultaneously over long distances.

This chapter describes a basic communications system, and presents introductory material on the basic types of communication transmitters. Additional information relating to transmitters is included in later chapters.

BASIC COMMUNICATIONS SYSTEM

Figure 20-1 is a block diagram of a basic communications system. The transmitter generates the RF carrier and amplifies it to some power level. The intelligence then modulates this carrier. The RF signal, which is the output of the transmitter, is applied to the antenna which converts it to electromagnetic energy and radiates it into space.

The receiving antenna receives a small portion of the electromagnetic energy, converts it back to an RF signal and feeds it to the receiver. The receiver then separates the desired intelligence from the carrier.

FREQUENCY BANDS

Transmitters operate from the very-low-frequency (VLF) band to the ultra-high-frequency (UHF) band.

In the VLF band (10 to 30 kHz) signals can be transmitted over long distances and even through magnetic storms that blank out higher radiofrequency channels. This band is used for radio telegraph transmissions, time standard transmissions, and for radio navigation.

The LF band (30 to 300 kHz) is used for long range direction finding, medium and long range communications and aeronautical radio navigation.

In the MF band (300 to 3000 kHz) relatively long distances can be covered. The international distress frequency, 500 kHz, is in this band as well as commercial broadcasting, amateur radio, loran, maritime, and aircraft communication.

The HF band includes the frequencies from 3 to 30 MHz. This band is used for aeronautical, mobile communications, amateur radio, police communications, international broadcasting and maritime communications.

The VHF band extends from 30 to 300 MHz. It is used for aeronautical radio, navigation and communications, early warning radar, television, amateur radio, and FM broadcasting stations.

The UHF band extends from 300 to 3000 MHz and is used for short range communications, such as between ships, amateur radio, television, and radar as well as other services.

This is only a partial list of uses of each of the bands.

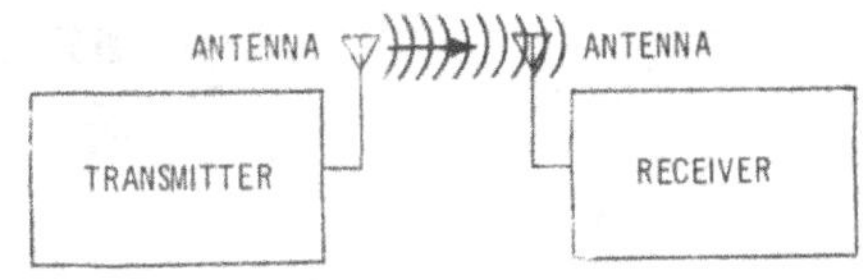

179.382

Figure 20-1.—Block diagram of basic communications system.

TRANSMITTER TYPES

Basic communication transmitters include the continuous wave (CW), amplitude modulated (AM), frequency modulated (FM), and single sideband (SSB) types. A brief description of each type follows.

Continuous Wave (CW) Transmitter

The continuous wave is used principally for radiotelegraphy—that is, for the transmission of short or long pulses of RF energy to form the dots and dashes of the Morse Code characters. This type of transmission is sometimes referred to as interrupted continuous wave (ICW). CW transmission was the first type of radio communication used, and it is still used extensively for long range communications. Some of the advantages of CW transmission are narrow bandwidth, high degree of intelligibility, even under severe noise condition, and long range operation.

The four essential components of a CW transmitter are: (1) a generator of RF oscillations; (2) a means of amplifying these oscillations; (3) a method of turning the RF output on and off (keying) in accordance with the intelligence to be transmitted; and (4) an antenna to radiate the keyed output of the transmitter.

A block diagram of a CW transmitter together with the power supply is shown in figure 20-2. The oscillator generates the RF carrier of a predetermined frequency and maintains it within required limits. The buffer amplifier isolates the oscillator from the load and thus improves the frequency stability of the transmitter. The buffer may also provide amplification of the signal and frequency multiplication. The power amplifier provides additional amplification of the RF carrier.

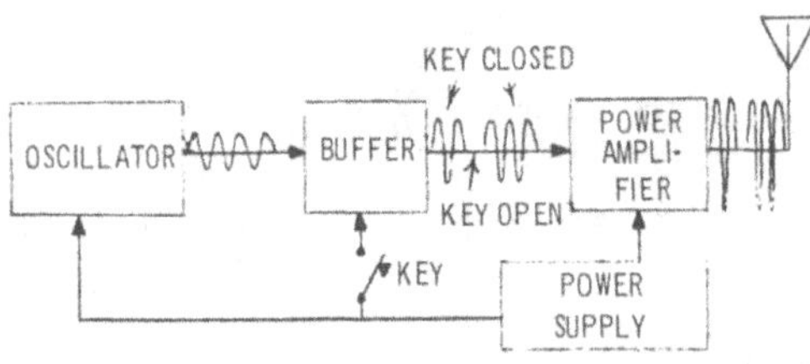

Figure 20-2.—CW transmitter block diagram.

One or more stages of amplification may be used between the buffer and the antenna. The stage which is connected to the antenna is usually called the final power amplifier (FPA), and any other stages of amplification are known as intermediate power amplifiers (IPA).

The key is used to turn the buffer on and off. When the key is closed, the RF carrier passes through the buffer stage, and when the key is open, the RF carrier is prevented from getting through.

Amplitude Modulated (AM) Transmitter

In amplitude modulation the instantaneous amplitude of the RF output signal is varied in proportion to the modulating signal. The modulating signal may consist of many frequencies of various amplitudes and phases, such as the signals comprising speech.

The block diagram of a simple AM radiotelephone transmitter is shown in figure 20-3. The oscillator, buffer, and power amplifier serve the same purpose as in the CW transmitter. The microphone converts the audio (sound) input into corresponding electrical energy. The modulator amplifies the audio signal to the amplitude necessary to fully modulate the carrier. The output of the modulator is applied to the power amplifier. The RF carrier and the modulating signal are combined in the power amplifier to produce the amplitude modulated RF carrier output for transmission. In the absence of a modulating signal, a continuous RF carrier is radiated by the antenna.

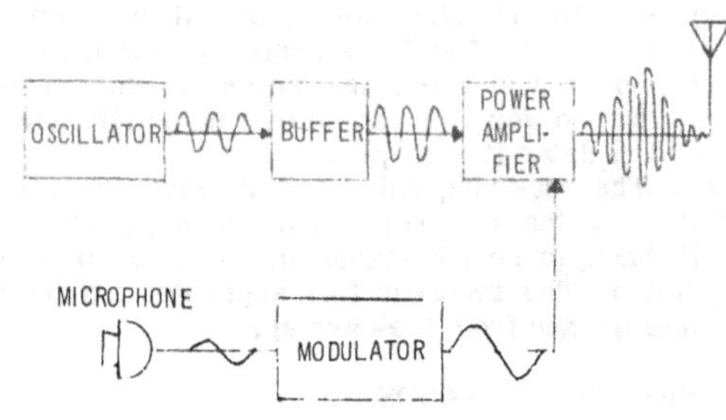

Figure 20-3.—AM radiotelephone transmitter block diagram.

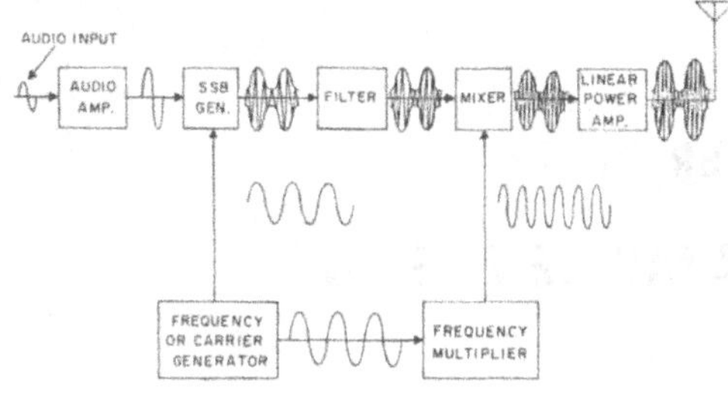

179.385

Figure 20-4.—SSB transmitter block diagram.

Single Sideband (SSB) Transmitter

A single sideband (SSB) transmitter translates audio frequency intelligence to desired radio frequencies. Unlike the amplitude modulated (AM) transmitter, usually, only one of the sidebands, either the upper or the lower sideband, is transmitted while the remaining sideband and the carrier are suppressed.

Figure 20-4 is the block diagram of an SSB transmitter. The audio amplifier increases the amplitude of the signal to a level adequate to operate the SSB generator. Usually the audio amplifier is just a voltage amplifier.

The SSB generator combines the audio input and the carrier input from the frequency generator to produce the two sidebands, and then suppresses the carrier. The two sidebands are then fed to a filter which selects the desired sideband and suppresses the other one.

The SSB generator in most cases operates at a very low frequency compared with the normal transmitted frequency. It is necessary, therefore, to convert (or translate) the sideband output from the filter to the desired frequency. This is the purpose of the mixer stage. To obtain a higher carrier frequency for the mixer stage, a second output is obtained from the frequency generator and fed to a frequency multiplier. The output from the mixer is fed to a linear power amplifier to build up the level of the signal for transmission.

Frequency Modulated (FM) Transmitter

In frequency modulation (FM), the modulating signal combines with the carrier in such a way as to cause the frequency of the resultant wave to vary in accordance with instantaneous amplitude of the modulating signal.

Figure 20-5 is the block diagram of a narrow band frequency modulation (NBFM) transmitter. The modulating signal is applied to a reactance tube causing the reactance to vary. The reactance tube is connected across the tank circuit of the oscillator. With no modulation, the oscillator generates a steady center frequency. With modulation applied, the reactance tube causes the frequency of the oscillator to vary around the center frequency in accordance with the modulating signal. The output of the oscillator is then fed to a frequency multiplier to increase the frequency and then to a power amplifier to increase the amplitude to the desired level for transmission.

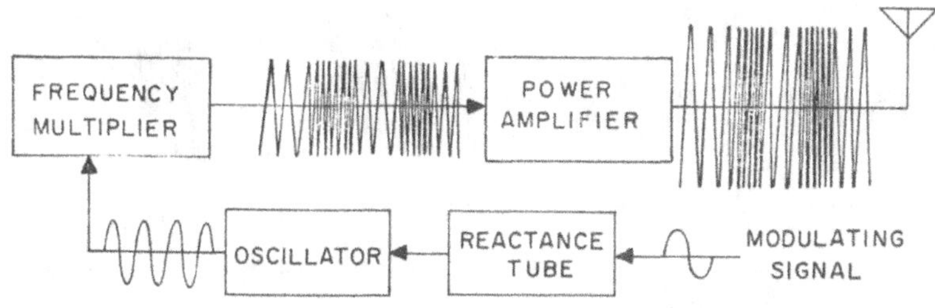

179.386

Figure 20-5.—Narrow band FM transmitter.

CHAPTER 21

RF POWER AMPLIFIERS

The simplest transmitter consists of an oscillator which generates a high frequency radio signal and an antenna system which propagates the energy. There are two principal disadvantages in connecting the oscillator directly to an antenna. The first is that the power output would be limited because there are no stages of amplification between the oscillator and the antenna. Power output is important because it determines the distance over which the transmitted signal can be picked up by a receiver. The other consideration is frequency stability. The load impedance of the oscillator, in this case the antenna, reflected into the tank circuit has a great effect upon frequency stability of the oscillator. Reflected impedance (chapter 10) contains both resistive and reactive components. The resistive component lowers the Q of the tank circuit and the reactive component alters the resonant frequency. A drift in frequency of the transmitted signal could result in a portion of the signal not being received. To overcome limitations in connecting an oscillator directly to the transmitting antenna, one or more stages of amplification are used. These stages are called RF power amplifiers, buffers, or frequency multipliers. The stage which is connected to the antenna is usually called the final power amplifier (FPA).

The other stages of amplification are known as intermediate power amplifiers (IPA). The first power amplifier, since it serves to isolate the oscillator from variations of load, is frequently called a buffer amplifier. If the frequency of the plate tank circuit of the buffer amplifier is the same as that of the oscillator driving it, the stage is a conventional type of amplifier. If the plate tank circuit of an amplifier is tuned to a harmonic of the driving signal (in order to increase the frequency of the radiated signal) applied to the grid, the stage becomes a frequency multiplier. Figure 21-1 is the block diagram of a simple transmitter.

The oscillator generates the RF at some desired frequency. The buffer amplifier isolates the oscillator from the load and amplifies the signal. The IPA and FPA provide further amplification to build up the RF to a power level sufficient for transmission. The RF is then fed to the antenna which radiates it.

The following terms are used to describe RF power amplifiers: The angle of collector or plate current is defined as the interval (measured in degrees) of the input signal during which the transistor or tube conducts. The collector or plate input power is that power delivered to

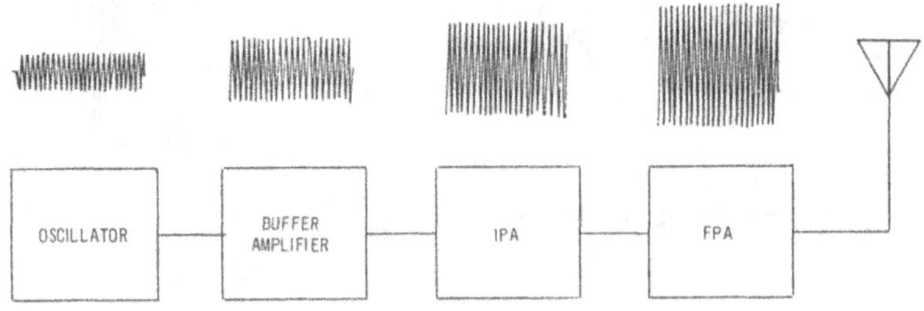

179.387

Figure 21-1.—Simple transmitter.

the transistor or tube. It is the product of the collector or plate supply voltage and the average value of collector or plate current.

The output power is that power delivered to the load. Collector or plate circuit efficiency is the ratio of the output power to the input power. Collector or plate dissipation is the factor determining collector or plate efficiency.

CLASS OF OPERATION

Amplifiers may be classified according to the conditions under which the transistor or tube operates—that is, according to the portion of the a.c. signal voltage cycle during which the collector or plate current flows as controlled by the bias on the base or grid. The four classes of amplifier operation according to bias are A, B, AB, and C, illustrated in figure 21-2. In class A operation, the base or grid is biased near the mid-point of the linear portion of the characteristic curve. The a.c. signal input causes the base or grid voltage to vary above and below the bias value.

The current variations are proportional to the voltage since the voltage swing does not go beyond the linear portion of the curve. Collector or plate current flows throughout the entire a.c. cycle. The principle characteristics of class A amplifiers are minimum distortion, low power output for a given transistor or tube (relative to class B and class C amplifiers), and relatively low collector or plate efficiency (20 to 25 percent). This type of amplifier finds wide use in various audio systems where low distortion is important. In class B operation, the base or grid is biased at or near its cutoff value. The a.c. signal drives the transistor or tube into cutoff for approximately half of the cycle. Thus, the transistor or tube conducts for about 180 degrees of the input signal cycle and is cut off during the other 180 degrees. Such amplifiers are characterized by medium power output, medium collector or plate efficiency (40 to 60 percent) and moderate power amplification.

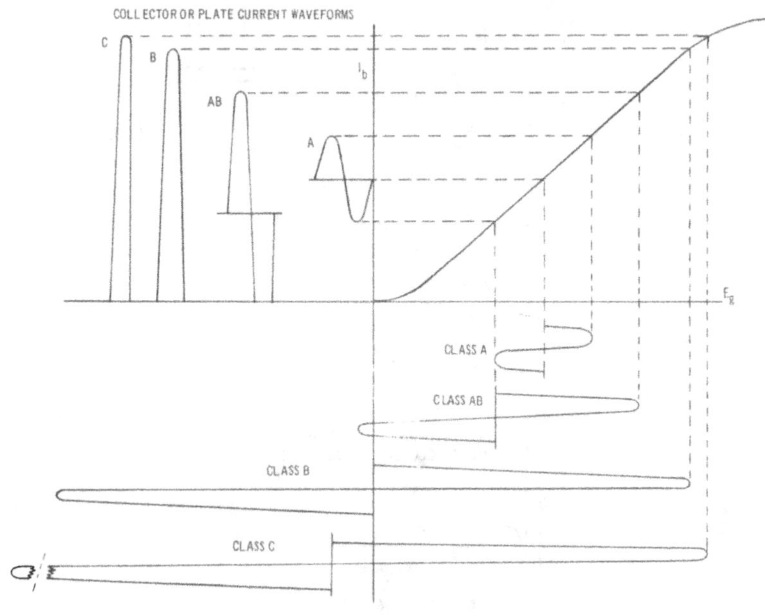

179.388
Figure 21-2.—Classes of operation.

Class AB amplifiers are biased so that the transistor or tube conducts for appreciably more than 180° but less than 360°. The power output will be higher than a class A amplifier and efficiencies will usually be between 45 and 55 percent.

In class C operation, the base or grid is biased considerably beyond cutoff with reference to the emitter or cathode, so that the transistor or tube conducts appreciably less than 180°. The transistor or tube remains in cutoff for most of each input signal cycle and current flows in the transistor or tube only when the input signal increases the base or grid voltage above cutoff. The collector or plate current, therefore, flows in pulses as shown. This class of amplifier

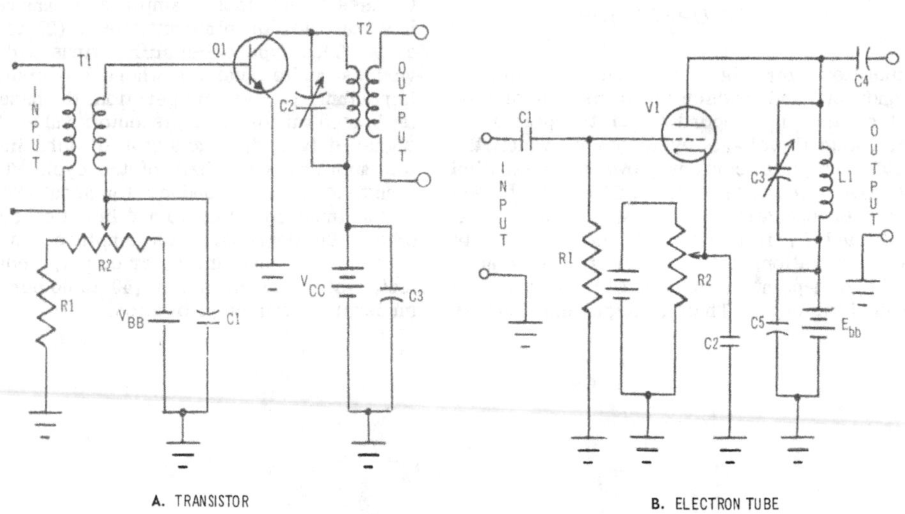

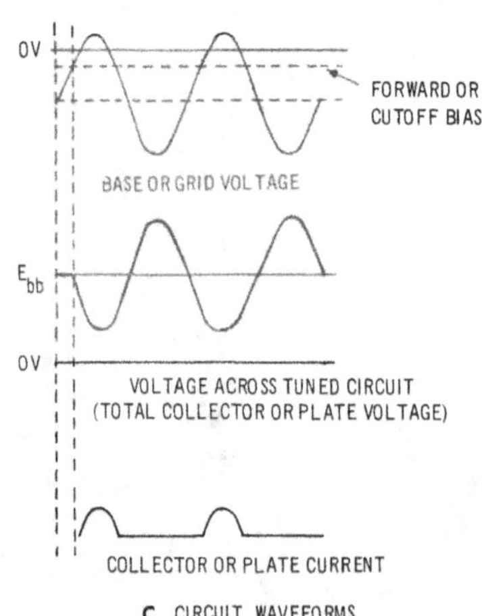

179.389

Figure 21-3.—Class C amplifier.

has relatively high collector or plate efficiency (70 to 80 percent) and high power output.

CLASS C RF AMPLIFIER

Figure 21-3A is the schematic diagram of a class C transistor RF power amplifier. Reverse bias for the emitter-base junction is supplied by V_{bb}, R1 and R2. R2 is adjusted for class C operation. Capacitors C1 and C3 are RF bypass capacitors. C2 and the primary of T2 make up the collector tank circuit.

An electron tube class C RF amplifier with resonant load, series fed, is shown in figure 21-3B. C1 couples the input signal from the preceding stage. The signal is developed across the grid resistor, R1. The bias supply is adjusted for class C operation. Capacitor C2 bypasses the RF signal around the bias supply. The plate tank circuit consists of C3 and L1. C5 bypasses the RF signal around the plate supply voltage. C4 couples the output signal to the next stage.

Assume that the frequency applied is equal to the resonant frequency of the tank circuit. The charge on the capacitor in the tank circuit equals zero volts, and the collector or plate voltage is equal to the supply voltage. When the base or grid signal reaches a value sufficient to bring the transistor into forward bias or the tube out of cutoff, collector or plate current begins to flow. Since the parallel resonant circuit offers maximum impedance at the resonant frequency, a large voltage drop will be developed across the tank circuit. The capacitor in the tank circuit charges to this value of voltage. The voltage across the tuned circuit must also be equal to the same value. Maximum voltage will be dropped across the tank circuit when the current through the transistor or tube reaches its maximum value. This is shown in the waveform diagram of figure 21-3C. The instantaneous collector or plate voltage is the algebraic sum of the supply voltage and the peak alternating voltage across the tuned circuit. When the collector or plate current starts to flow, the voltage at the collector or plate end of the tuned circuit goes in a negative direction and subtracts from the supply voltage causing the collector or plate voltage to decrease in value toward zero. When the collector or plate current decreases, the voltage across the tank begins to decrease and the capacitor starts to discharge through the inductor. When the transistor or tube is sent into reverse bias or cutoff (respectively), collector or plate current ceases and the field about the inductor collapses causing the collector or plate voltage to rise toward V_{CC} or E_{bb}. The collapsing magnetic field of the inductor maintains current in the same direction through the capacitive branch of the tank circuit charging the capacitor in the opposite direction. When the field has completely collapsed, the capacitor charge is maximum with a polarity opposite to the initial charge. When the voltage across the tuned circuit goes positive at the collector or plate end, it adds to the supply voltage causing the instantaneous collector or plate voltage to be nearly double the value of V_{CC} or E_{bb}. Once fully charged, the capacitor begins to discharge since there is no potential difference across it to sustain the charge and its discharge path will be through the inductor. The actions described above are collectively known as the "flywheel effect." This action is extremely important, because a complete cycle of output is obtained by using only a small portion of the input cycle to replace tank circuit losses. When the base or grid signal again reaches a value sufficient to bring the transistor into forward bias or the tube out of cutoff, the tank circuit receives another burst of energy due to transistor or tube conduction at that time. This action is continuous until the base or grid signal is removed, at which time the oscillations will continue, but subside in amplitude until all of the energy contained in the tank is dissipated by the normal losses in the circuit.

The collector or plate circuit efficiency may be expressed as:

$$\text{Efficiency} = \frac{P_{a.c.}}{P_{d.c.}} \text{ or } \frac{P_{out}}{P_{in}} \times 100$$

For example, with a d.c. input power of 620 watts and 451.5 watts of useful power delivered to the load, the efficiency would be:

$$\text{Efficiency} = \frac{451.5}{620} \times 100 = 72.8\%$$

This means that with a d.c. input power of 620 watts, 168.5 watts are dissipated in the transistor or tube and circuit components (mainly in the form of heat).

$$P_{dissipated} = P_{input} - P_{out} = 620 - 451.5 = 168.5 \text{ watts}$$

FREQUENCY MU TIPLICATION

In a class C amplifier, the pulse of collector or plate current is nonsinusoidal in shape and therefore has an appreciable harmonic content. Because of this harmonic content, a class C amplifier can be used to generate output power that is a harmonic of the input signal. A class C amplifier operated in this manner is called a frequency multiplier

The need for frequency multiplier circuits results from the fact that the frequency stability of most oscillators decreases as frequency increases. At the lower frequencies, relatively good oscillator stability can be achieved. Thus, to achieve optimum stability, an oscillator is operated at a low frequency, and one or more stages of multiplication are used to raise the signal to the desired operating frequency.

CONDITIONS FOR FREQUENCY MULTIPLICATION

Three important conditions must prevail in order to obtain frequency multiplication: (1) collector or plate tank circuit tuned to the desired harmonic, (2) reverse base bias or high negative grid bias, and (3) high base or grid driving voltage.

The first condition necessary to obtain frequency multiplication is to tune the collector or plate tank circuit to the desired harmonic. For example, if the input frequency is 1 MHz and the 2nd harmonic is desired, the collector or plate tank circuit would be tuned to 2 MHz. When the output tank circuit is tuned to the required harmonic frequency, the tank acts like a filter, accepts the desired harmonic and rejects all other harmonics, and frequency multiplication results.

When the collector or plate tank circuit of a class C amplifier is tuned to the same frequency as the input signal, the stage is said to be operating "straight through." That is, no multiplication is taking place. When the collector or plate tank circuit is tuned to select the 2nd harmonic, the stage is called a doubler. That is, the output frequency is twice the input frequency. The 3rd or 4th harmonic may be selected to make a tripler or a quadrupler, with the 5th harmonic being the highest ever selected.

The second condition necessary to obtain efficient frequency multiplication is a reduced angle of collector or plate current flow. This is accomplished by increasing the reverse or negative bias. Amplitude relationships are shown in figure 21-4.

In a harmonic amplifier, reducing the angle of collector or plate current flow reduces the output power and increases the efficiency of the stage.

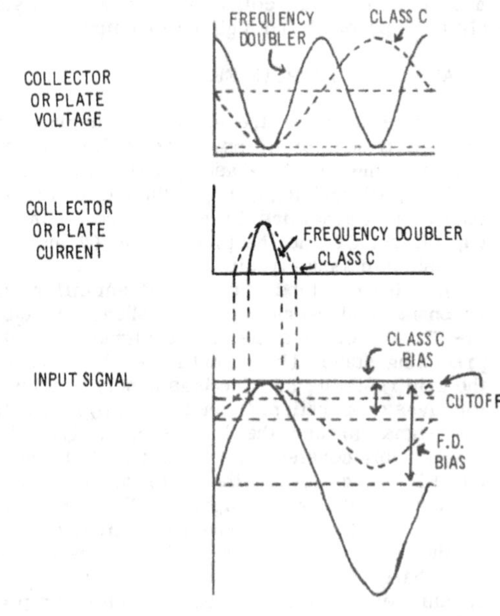

Figure 21-4.—Comparison of frequency doubler and class C voltage and current amplitudes.

The harmonic power output obtained decreases with the order of the harmonic as shown in table 21-1.

FREQUENCY MULTIPLIER OPERATION

Figure 21-5A is the schematic diagram of a transistor frequency multiplier, operating as a doubler. A basic electron tube multiplier circuit, operating as a frequency tripler, is illustrated in figure 21-5B. The frequency multiplier circuit is operated class C with the collector or plate tank resonant to twice or triple the base or grid signal frequency respectively. When the signal applied to the base or grid rises above the cutoff value of the transistor or tube, there will be a pulse of current (at the same frequency as the input signal) flowing from the emitter or cathode to the collector or plate, energizing the collector or plate tank circuit. As pulses of the collector or plate current contain appreciable energy at the second and third harmonics, and the resonant frequency of the tank circuit (determined by values of L and C) is twice or

Table 21-1. — Pulse requirements and power outputs for a variety of harmonics

HARMONIC	OPTIMUM LENGTH OF PULSE, ELECTRICAL DEGREES AT THE FUNDAMENTAL FREQUENCY	APPROXIMATE PERCENTAGE POWER OUTPUT, ASSUMING THAT NORMAL CLASS C OUTPUT IS 100%
2	90 TO 120	50 TO 65%
3	80 TO 120	30 TO 40%
4	70 TO 90	25 TO 30%
5	60 TO 72	20 TO 25%

179.391

three times the input frequency; collector or plate current will arrive at the tuned circuit during every second or third cycle of output voltage and deliver sufficient energy to the tuned circuit to sustain oscillations (by flywheel effect) during those cycles when no current flows.

When the transistor or tube goes into cutoff, energy supplied to the collector or plate tank circuit is sufficient to maintain oscillations between current pulses. The reason the tuned circuit continues to oscillate is that pulses of current always arrive at the same time during alternate or every third cycle of the doubled or tripled frequency respectively, thus energizing the tank circuit at the proper time.

LINEAR RF POWER AMPLIFIERS

An RF signal, regardless of the intelligence impressed on it, can be divided into an upper, or positive, portion and a lower, or negative, portion. Therefore, a single-ended linear amplifier is one in which the transistor or tube conducts for at least 50% (180°) of each input cycle, causing the necessary true reproduction of the input signal. In a push-pull linear amplifier, both positive and negative portions of the input signal are truely reproduced.

A linear amplifier may be operated class A or AB single-ended, or B push-pull. Class A RF power amplifiers, due to their low efficiency, are primarily used in low power applications. Class B power amplifiers provide greater power output with increased efficiency but require well regulated bias supply voltages. Class AB amplifiers represent a compromise in power and efficiency between class A and class B. Linear amplifiers are used to increase levels of power or voltage in cases such as amplitude-modulated carriers or single-sideband signals. (A subscript number is commonly added to the electron tube class designator to indicate whether or not the tube is operated in the positive grid region over part of the cycle. For example, class AB_1 indicates that the grid never goes positive with respect to the cathode, so that no grid current is drawn. The subscript "2" may be used to indicate that the grid is positive with respect to the cathode, and current flows during some parts of the input cycle. An electron tube linear amplifier may be operated class AB_1 or AB_2.)

RF LINEAR AMPLIFIER OPERATION

Figure 21-6A is the schematic of a transistor class B RF amplifier. Transistor Q1 has its base returned to ground through the secondary of T1 and is essentially cut off. The input signal is coupled to the base of Q1 by T1. The collector tank circuit is made up of C2 and the primary of T2. C1 bypasses any RF signal around the collector supply voltage.

When the input signal is applied to the base of Q1, the negative alternation will forward bias Q1. Collector current will flow delivering energy to the tank circuit for one-half cycle of operation. The positive alternation of the input signal reverse biases Q1, cutting it off. The tank circuit then flywheels for a half-cycle of operation.

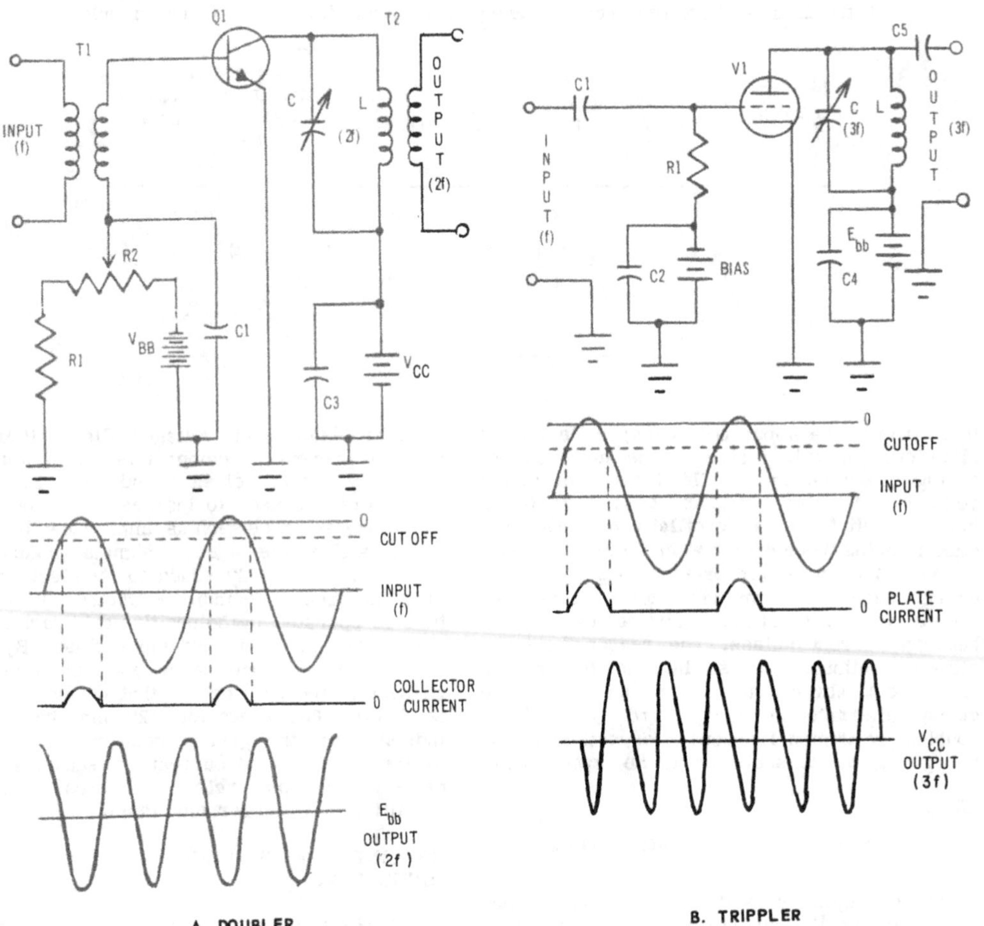

179.392

Figure 21-5.—Frequency multipliers.

Figure 21-6B is the schematic of an electron tube class B RF amplifier. The grid bias is such that the tube is approximately cut off when there is no input signal. The input signal (excitation) is coupled through C1 and developed across RFC1. C2 bypasses RF currents around the bias supply. The output tank circuit consists of C3 and L_p. RFC2 and C4 form a decoupling network to keep RF currents out of the plate supply voltage.

When the grid excitation is such that it brings the tube out of cutoff, plate current flows. When plate current flows, the tuned tank circuit offers a high impedance at the frequency of resonance, which drops the effective plate voltage. At this time the tank circuit absorbs energy from the power supply and, simultaneously, the plate is dissipating some energy in the form of heat.

As the amplitude of the grid excitation decreases toward zero and then swings negative

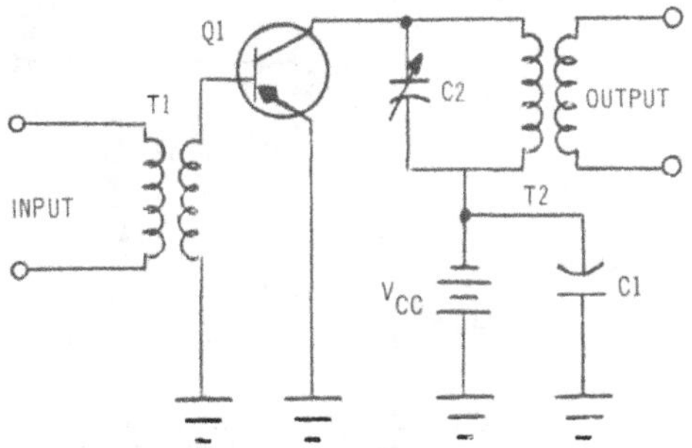

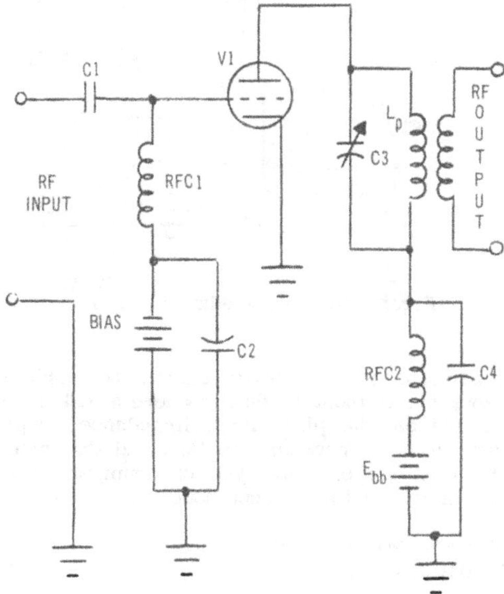

179.393
Figure 21-6.—Class B RF amplifier.

on the following half-cycle, grid voltage increases more negatively, and plate current is decreased. When plate current decreases, the voltage drop across the tank also decreases, and plate voltage rises. Since the tank circuit contains RF energy, the flywheel effect of the tank completes its half-cycle of operation during the period of zero plate current.

Figure 21-7 shows the relationship between grid voltage and plate current when modulation is applied. As long as limits of the linear portion are not exceeded (drive, input signal, is adjusted correctly) no distortion exists. This is normal class B linear operation. Should the drive be too low, the output will be undistorted, but maximum power output will not be realized, and efficiency will be reduced.

In the class B_2 amplifier, the grid is usually driven into the current region (positive with respect to the cathode) during the positive peak of the excitation cycle. Therefore, grid-leak bias cannot be used since a change in modulation would cause a change in the average grid current and result in an effective change in bias.

When grid current is drawn in amplifiers operating AB_2 and class B_2, the input impedance of the stage changes. This presents a nonlinear load to the driving source and will result in a distorted signal. To compensate for this, the grid is loaded (swamped) with a resistor which maintains a relatively constant impedance to the driver.

COUPLING

A single stage of power amplification normally is not sufficient for radio transmitters. To obtain the necessary gain, several stages must be connected together. The output of one stage then becomes the input of the next,

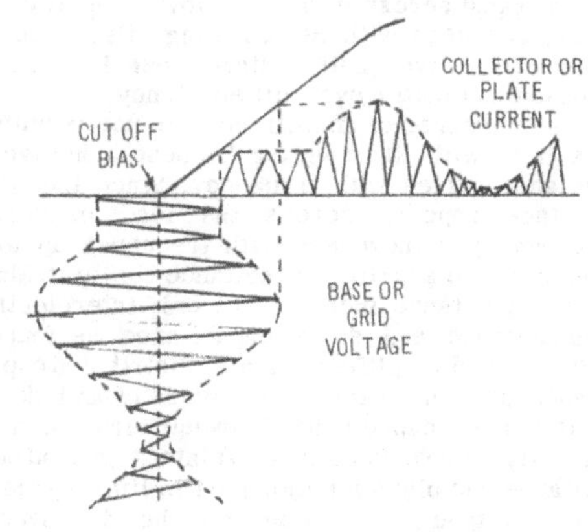

179.394
Figure 21-7.—Class B waveforms.

and this arrangement is called a cascade amplifier. There are several types of cascade amplifiers; each has certain advantages or disadvantages, and the reason for using a particular application depends on the needs of the circuit. The basic methods of coupling are: (1) resistance capacitance; (2) impedance coupling; (3) transformer coupling and (4) link coupling (a special form of transformer coupling).

RESISTANCE-CAPACITANCE COUPLING

One of the most widely used methods of connecting amplifier stages is resistance-capacitance (RC) coupling (which was covered in detail in chapter 13). Wide band voltage amplifiers are commonly coupled in this manner because of the ease with which broad frequency coverage is achieved and the economy in the cost, size, and weight of the components required. Generally, RC coupling is not used in RF power amplifiers because of excessive I^2R losses.

UNTUNED IMPEDANCE COUPLING

Impedance coupling is obtained by replacing the load resistor of a normal RC coupled amplifier with an inductance as shown in figure 21-8. To obtain as much amplification as possible, particularly at low frequencies, inductance is made as large as practicable. Because of an inductor's low d.c. resistance, less d.c. voltage is dropped across it. Thus, a lower supply source than one used with RC coupling will provide the same effective plate voltage, less I^2R (power) losses and better over-all efficiency.

The degree of amplification is not as uniform as it is with RC coupling because inductive reactance varies with frequency. Since the output voltage appears across the load impedance, voltage gain increases with frequency up to the point where shunting capacitance limits it. Shunting capacitance includes not only interelectrode capacitance and stray capacitance as found in RC coupled amplifiers but also distributed capacitance as found between the turns of an inductor. Distributed capacitance between turns of a coil greatly increases collector/plate to ground capacitance and plays a major part in limiting the use of this type of coupling at higher frequencies.

Generally, impedance coupling is not used in high power RF power amplifiers operated class C because the flywheel effect needed to faithfully reproduce a sine wave cannot be achieved. However, in those instances where a tank circuit comprises the plate load, impedance coupling can be used between the tank and the input to the next stage. This type of coupling is also known as tuned impedance coupling.

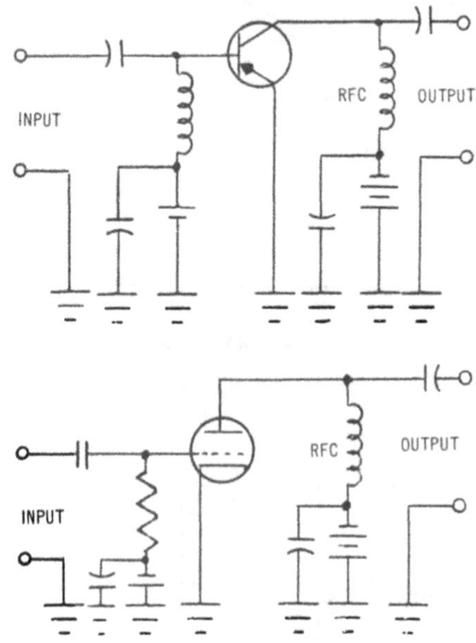

Figure 21-8.—Impedance coupling.

TUNED IMPEDANCE COUPLING

Tuned impedance coupled RF amplifiers find wide application in transmitters because of their amplification, selectivity and ability to reproduce a sine wave output. Figure 21-9 illustrates two types of tuned RF amplifiers. They differ essentially in the manner in which energy is supplied to the tank circuits.

Figure 21-9A shows a typical circuit for a triode tube using series plate feed; that is, the d.c. component of plate current flows through the d.c. supply and the tuned circuit in series. A resonant circuit is used as the load and is tuned to the frequency of the signal impressed on the grid of the tube. The resonant circuit presents a high impedance at this frequency,

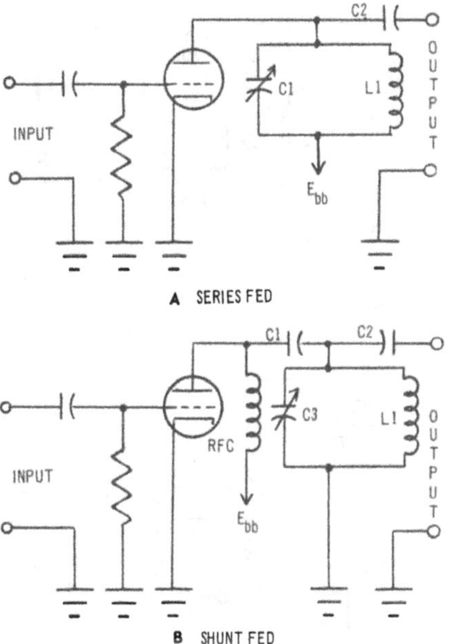

179.396
Figure 21-9.—Tuned RF amplifiers.

whereas its impedance is lower at all other frequencies. This impedance decreases rapidly as frequency shifts from resonance.

A commonly used form of class C amplifier load circuit is shown in figure 21-9B. It differs from the series feed circuit in that the plate tank circuit is connected where it shunts the plate load (RF choke). In this position, the full plate voltage signal is still developed across the tank circuit without passing the high d.c. component of plate current through it. The amount of plate current may still be changed by varying the impedance of the tank circuit. This is true because the tank circuit is in parallel with the RF choke and thereby has an effect on the total impedance offered to the operating frequency. The danger of a high value of plate voltage being present during adjustment is eliminated when using the shunt fed amplifier. However, there still is a hazard involved because of the RF voltages present. RF burns can be just as lethal as the high direct voltage. Shunt fed amplifier tanks do not require as much insulation as series fed tank.

TRANSFORMER COUPLING

Inductive or transformer coupling is a very efficient means of transferring energy from one stage to the next. Figure 21-10A illustrates a transformer coupled amplifier with a tuned primary. Two coils labeled L1 and L2 constitute an air core transformer. The plate tank is tuned to the operating frequency. When a positive voltage of sufficient amplitude is applied to the grid of the tube, plate current will flow. The output of the amplifier will be a sine wave due to the flywheel effect in the plate tank circuit.

A characteristic of a parallel resonant circuit is that it offers maximum impedance to the resonant frequency causing plate current to be minimum. Therefore, line current is minimum, but circulating current within the tank is maximum. The maximum circulating current passing through the inductor L1 will cause a magnetic field to fluctuate about it. Since the inductor L2 is in close proximity, it also will be under the influence of the same field. Therefore, a voltage will be induced into the inductor L2 through basic transformer action. The quantity of coupling between L1 and L2 may be varied

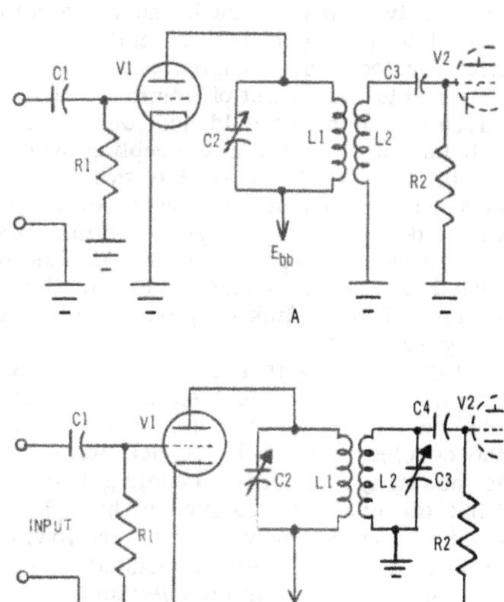

179.397
Figure 21-10.—Transformer coupled amplifiers.

by physically moving the secondary of the transformer closer to or farther away from the primary depending upon the amount of drive required by the next stage. If capacitors are placed across the primary and secondary windings of the transformer in a transformer coupled network, a double-tuned transformer coupling system is obtained (fig. 21-10B).

Coil L1 is the primary and coil L2 is the secondary of the transformer. C2 tunes L1 to resonance at the signal frequency. A large signal voltage is produced across the high impedance of the parallel resonant circuit formed by L1 and C2. The large circulating tank current in the primary of the transformer creates a magnetic field which induces a voltage in the secondary winding, L2. The voltage across the secondary winding is applied to the grid of V2.

LINK COUPLING

Link coupling is a special form of transformer coupling. The type illustrated (fig. 21-11) requires the use of two tuned circuits, one in the output circuit of the driver and another in the input circuit of the power amplifier. A low impedance RF transmission line having a coil of one or two turns at each end is used to couple the plate and grid tank circuits. The coupling links or loops are coupled to each tuned circuit at its cold end (point of minimum RF potential). Circuits which are cold near one end are called unbalanced circuits. Link coupling systems normally are used where the two stages to be coupled are separated by a considerable distance. One side of the link is grounded in cases where capacitive coupling between stages must be eliminated or where harmonic elimination is important. Typical link coupled circuits are shown in figure 21-11.

Link coupling is a very versatile interstage coupling system. It is used in transmitters when the equipment is sufficiently large to permit the coupled coils to be so positioned that there is no stray capacitive coupling between them. Link circuits are designed to have low impedance so that RF power losses are low. Coupling between links and their associated tuned circuits can be varied without complex mechanical problems.

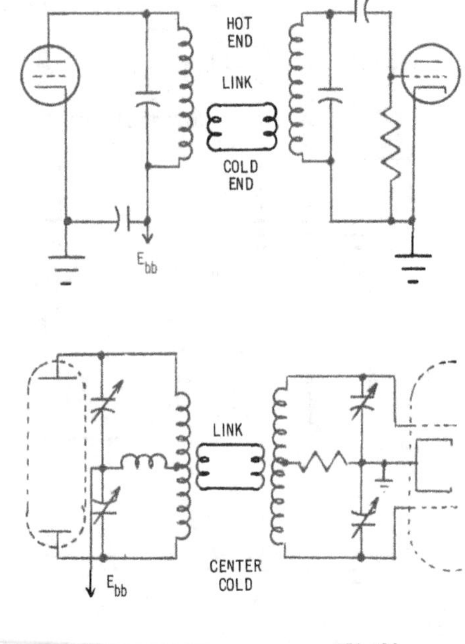

Figure 21-11.—Link coupling.

NEUTRALIZATION

Any amplifier will oscillate if sufficient energy having the same frequency and the same phase as the input signal is fed back from the output circuit to the input circuit. Feedback in the proper phase for oscillation (regenerative feedback) may take place through the collector-base capacitance of a transistor, grid-plate capacitance of a tube, or through external capacitive or inductive coupling between their output and input circuits. Although regenerative feedback is employed in oscillator circuits, it is undesirable in amplifier applications because of the resulting distortion and spurious radiation. It is possible to eliminate these oscillations by a process called neutralization. In neutralization, a network is included in the amplifier which feeds back to the input a voltage of proper amplitude and polarity to cancel the regenerative feedback.

There are several neutralization systems in use. Two of these, the collector or plate neutralization system, and the base or grid system, have the advantage of being useful over a wide frequency range. Their names are derived from the part of the circuit in which feedback voltage is developed.

COLLECTOR/PLATE NEUTRALIZATION

Collector/plate neutralization is shown in figure 21-12A/B. This is a typical transformer coupled RF amplifier to which has been added a neutralization inductance L2 closely coupled to L1 and a neutralizing capacitor C_n. L2 is connected in such a manner that the polarity of voltage at point B of this transformer is in phase opposition to the voltage at the corresponding end, point A. The center of this transformer (point C) is placed at RF ground through the low reactance of the RF bypass capacitor C4. Therefore, the RF voltages measured at points A and B with respect to ground are equal in amplitude (assuming point C is the exact center) and opposite in polarity, C_f represents the internal feedback capacitance, represented in the schematic as an external capacitor. C_n is the capacitor through which the neutralizing signal is coupled to cancel the effects of C_f.

The operation of this neutralization circuit may be explained with the aid of an equivalent circuit as in figure 21-12C. Point C of the output tank circuit is effectively the same point as the bottom of the input tank (point C') through the low reactance of C4. The low reactance of the coupling capacitor C_c (fig. 21-12B) makes the top of the input tank (point D') effectively the same as point D. Therefore, the neutralization circuit is resolved into a bridge arrangement, consisting of the input tank circuit, L1, L2, C_f, and C_n.

When the potential across the output tank is positive at the top and negative at the bottom, current will flow in the directions indicated in

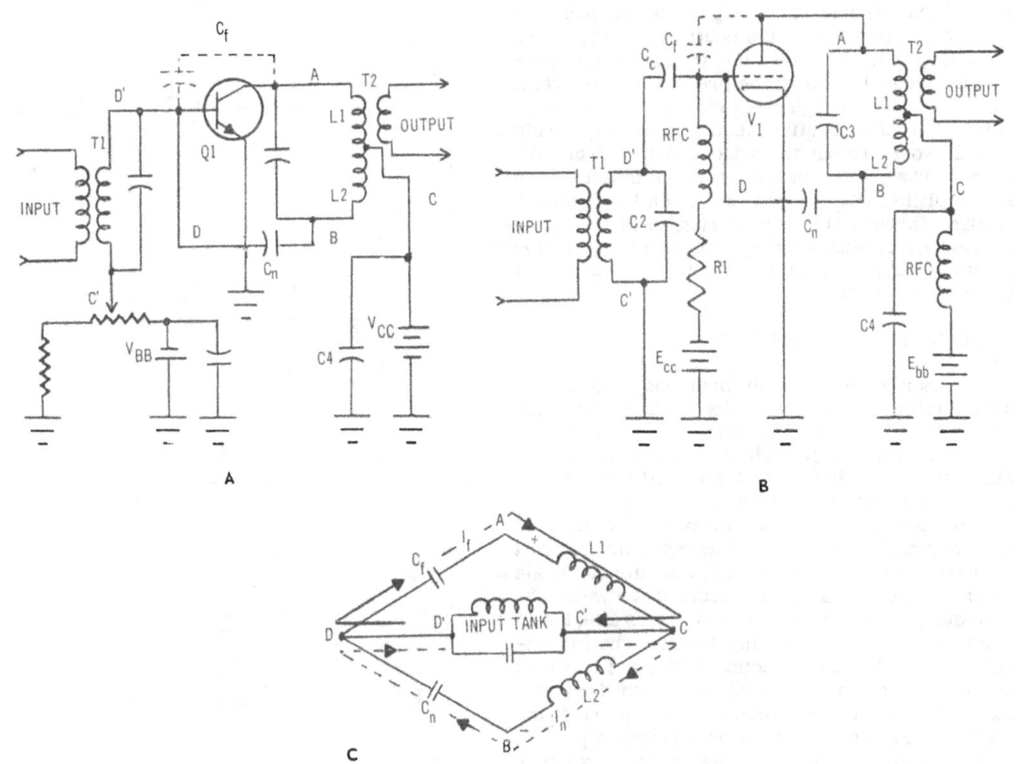

Figure 21-12.—Collector/plate neutralization.

figure 21-12C. When the output tank voltage reverses its polarity, the direction of current flow will be opposite to the direction indicated. Assuming the potentials developed across L1 and L2 are equal and the values of capacitance C_f and C_n are equal, the two currents will also be equal. When these feedback currents are equal, there will be no difference in potential between points C and D caused by feedback; therefore, current in the input tank is in series opposition and for all practical purposes, there is no current flow through the input tank circuit (in the equivalent circuit). No current flow through the input tank indicates that no energy has been fed back from the output to the input; thus, no regeneration or degeneration can take place and the circuit is considered to be neutralized. If the neutralizing capacitor C_n were smaller in value than C_f, its higher reactance would cause I_f to be the predominate current resulting in regenerative feedback current through the input tank circuit. This signal would promote circuit oscillations. If the neutralizing capacitor were larger in value than C_f, its lower reactance would cause I_n to be the predominate current. This current through the grid input tank would develop a degenerative feedback voltage. This signal would result in reduced output from the stage. Therefore, proper neutralization of an RF amplifier stage is realized when the feedback voltage through the neutralizing capacitor (C_n) cancels the feedback voltage through C_f resulting in zero energy transfer from the output circuit to the input circuit.

BASE/GRID NEUTRALIZATION

Another circuit which provides a means of neutralizing the effects of feedback capacitance is the base/grid neutralization circuit shown in figure 21-13A/B. It differs from collector or plate neutralization in that the split tank circuit is located in the input circuit.

The operation of this circuit may best be demonstrated with the aid of an equivalent circuit as shown in figure 21-13C. C_f, the interelectrode feedback capacitance, is located between the collector/plate and base/grid which is effectively point A, the top of the input tank. The low reactance of C2 places point C of the input tank at RF ground. When a signal is developed between the collector or plate (point D) and ground, the feedback capacitance (C_f) will couple a portion of this signal back to the top of the input tank (point A). Simultaneously, the neutralizing capacitor will couple an equal portion of the output

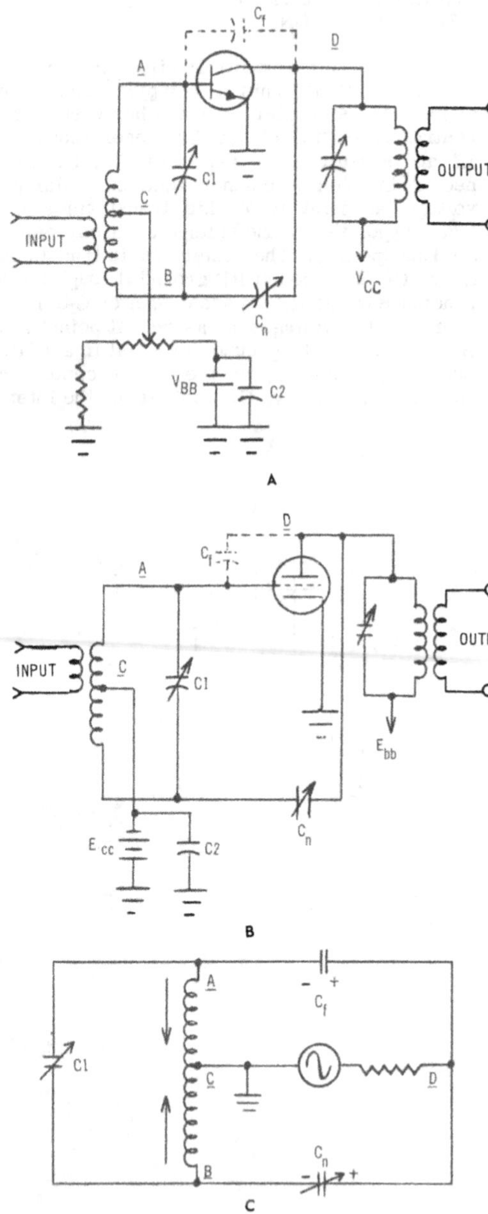

Figure 21-13.—Base/grid neutralization.

signal back to the bottom of the input tank (point B). Since both signals will have the same instantaneous amplitude and polarity, there will be no net difference of potential developed across the input tank; thus, no regeneration or degeneration can take place and the circuit is neutralized.

COIL NEUTRALIZATION

Coil neutralization differs from the systems previously described. In the other circuits, regenerative voltage fed back through the interelectrode capacitance is canceled by a degenerative voltage fed back through an external path (C_n). In this circuit, figure 21-14, the effect of interelectrode capacitance is nullified by paralleling this capacitance with an inductor of equal reactance. Since the two reactances are equal and opposite, a parallel resonant circuit is formed and there is no transfer of energy through the circuit from output to input. C1 is a blocking capacitor which prevents the supply voltage from being felt on the base or grid.

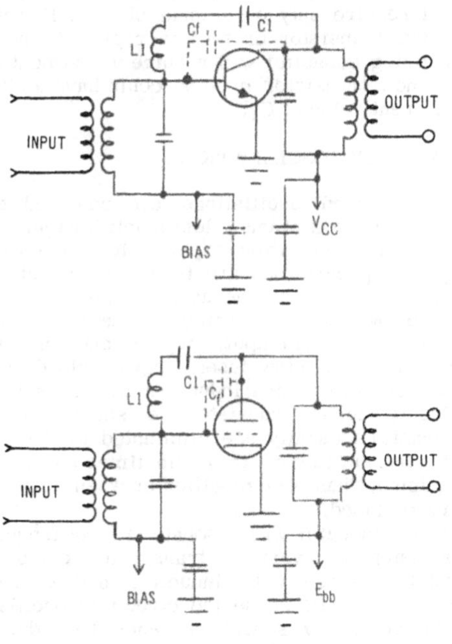

179.401

Figure 21-14.—Coil neutralization.

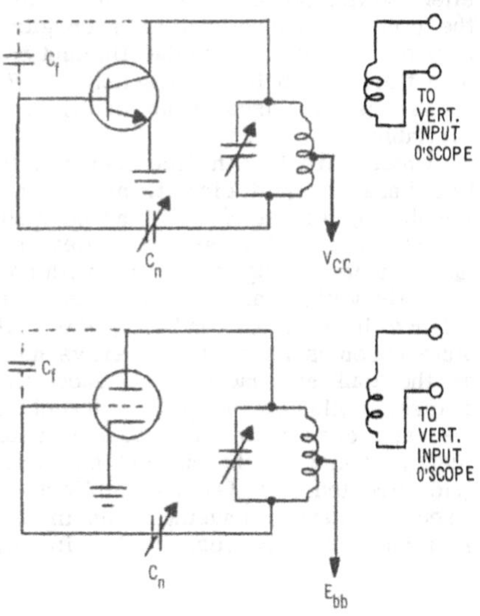

179.402

Figure 21-15.—Neutralizing procedure using an oscilloscope.

NEUTRALIZING PROCEDURES

The procedures for neutralizing are almost independent of the type of neutralizing circuit used. At the start of neutralization, the collector or plate voltage is removed from the stage to be neutralized so that any signal present in the collector or plate circuit is due to interelectrode capacity coupling. Then the master oscillator and those amplifier stages which precede the unneutralized stage are tuned. This will provide a strong signal to the base or grid of the unneutralized stage. The next step depends on the indicator used but it always results in the adjustment of the neutralizing capacitor (C_n) until there is a minimum amount of energy transferred to the collector or plate circuit.

One method places a pick up coil near the collector or plate tank. Coil leads are connected to the vertical input of an oscilloscope, as shown in figure 21-15. C_n is then adjusted so that no RF voltage appears on the scope when the tank is tuned to resonance. Under these circumstances, the RF current divides equally through C_f and C_n. The resulting RF currents in the tank flow in opposite directions and cancel the tank inductive effect so that no resonant buildup occurs between the coil and capacitors. A neon glow lamp, a loop of wire attached to the filament connections of a flashlight bulb, or a sensitive RF galvanometer may be used if an oscilloscope is not available.

When there is a milliammeter in the amplifier base or grid circuit, as shown in figure 21-16, adjustment of C_n may be made by observing the meter as the collector or plate tank is tuned through resonance with no collector or plate voltage applied. When there is an unbalance between C_f and C_n, the collector or plate becomes alternately positive and negative as the tank approaches resonance. On positive swings, collector or plate current flows. As the tank circuit is tuned to the resonant frequency, some of the current that was base or grid directed now becomes collector or plate directed, thereby causing a dip in the base or grid current. However, if C_n is adjusted to neutralize the amplifier stage, the RF current from the input stage divides equally and attempts to flow in opposite directions through the two halves of the tank coil. This cancels the inductive effect of the coil and prevents the buildup of resonant voltage in the tank. There is no rise in tank current and voltage and the collector or plate remains at zero potential. Accordingly, with C_n properly adjusted, no dip in base or grid current occurs as the collector or plate tank is tuned through the resonant frequency.

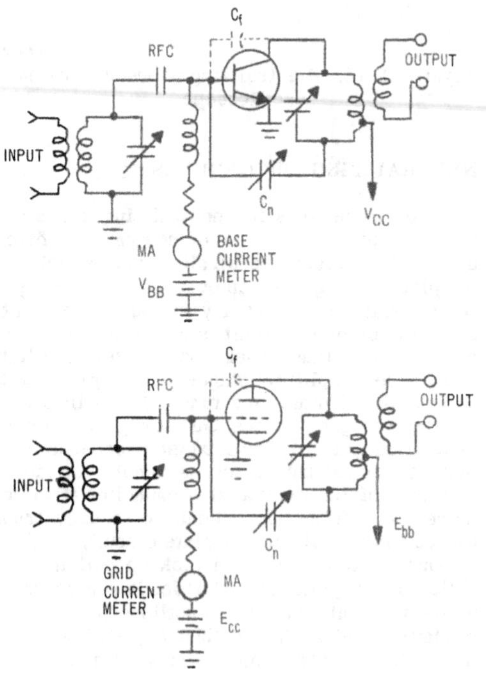

Figure 21-16.—Neutralizing procedure using current meter.

In some electron tube transmitter circuits, it is more convenient to turn off the filament voltage on the amplifier stage instead of removing plate voltage. If this is done, the process of neutralizing the amplifier is carried out in the same way, except that no current flows in the amplifier grid circuit. The absence of RF in the amplifier plate tank, as evidence of the correct adjustment of C_n, may be determined by the effect on the exciter stage or on an RF pickup coil and associated indicator, as previously mentioned.

Once a neutralizing capacitor is adjusted for use with a particular transistor or tube, it will require only occasional checks. However, if the transistor or tube is replaced, the neutralizing capacitor will require adjustment since the new transistor or tube could have a slightly different value of C_f.

PARASITIC OSCILLATIONS

Undesired oscillations are referred to as parasitic oscillations. Most amplifiers and oscillators are susceptible to parasitic troubles unless special precautions are taken to prevent them. Parasitics reduce the useful power output of a stage by absorbing some of the power which should be useful output. They may cause excessive currents that blow fuses, trip overload relays, ruin capacitors and inductors in the oscillating circuit, damage transistors, or shorten tube life. Parasitics usually are eliminated in the design of a transmitter but they sometimes appear after equipment has been modified or when components are replaced.

Any inductor will resonate at some frequency. Occasionally, various transmitter components which possess both inductive and capacitive properties will cause the circuit to oscillate at their common resonant frequency. The inductance may be that of wiring, leads of capacitors or a section of a coil or RF choke. The capacitance could be the interelectrode capacitance of a transistor or tube.

Figure 21-17A is the schematic of a simplified class C amplifier and figure 21-17B is the equivalent circuit showing one possible parasitic circuit. At frequencies well above the normal operating frequency (high-frequency parasitics), the tank circuit capacitances (C1 and C2) can be considered to be short circuits. The tank circuit inductances (L1) can be considered to be an open circuit. Under these conditions, the circuit reduces to a tuned-plate tuned-grid oscillator as shown in figure 21-17B. The tank circuit capacitances are the interelectrode capacitances of the tube, i.e., grid-to-cathode (C_{gk}) and plate-to-cathode (C_{pk}). The tank circuit inductances (L_g and L_p) are the inductances of the leads between the tube electrodes and ground through the tuning capacitors C1 and C2. Since C2 is considered to be a short circuit, the neutralizing capacitor (C_n) is now in parallel with the grid-to-plate capacitance (C_{gp}); thus, the feedback capacitance is the sum of C_n and C_{gp}.

Parasitic oscillations at frequencies lower than the normal operating frequency (low-frequency parasitics) can arise from the use of RF chokes for decoupling. Figure 21-18A is the simplified schematic of a class C push-pull amplifier and figure 21-18B is the equivalent parasitic oscillator circuit.

At frequencies much lower than the normal operating frequency, the tank circuit inductances (L1 and L2) can be considered as short circuits. This effectively places the two tubes in parallel; therefore, only one tube is shown in the equivalent circuit. The interelectrode capacitance of each tube is also effectively in parallel.

Notice that a tuned-plate tuned-grid oscillator is formed, as shown in figure 21-18B. The tank capacitances are the interelectrode capacitances of the tubes, while the tank inductances are the RF chokes (L3 and L4).

PARASITIC SUPPRESSION

High-frequency parasitics may sometimes be eliminated by changing the frequency of the plate and/or grid which form the parasitic oscillator. The plate circuit is tuned to a lower frequency than the grid circuit.

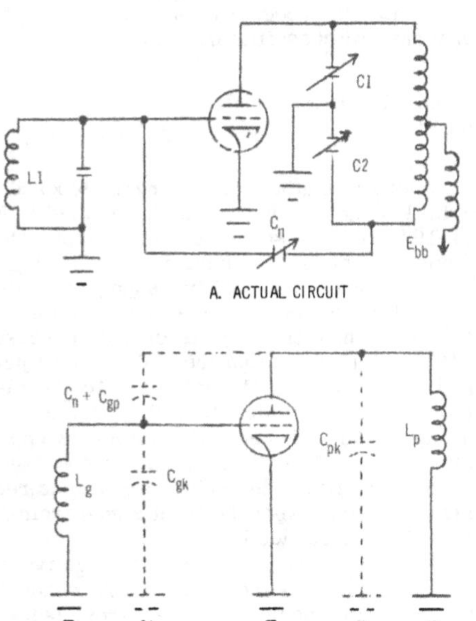

179.404
Figure 21-17.—High-frequency parasitics.

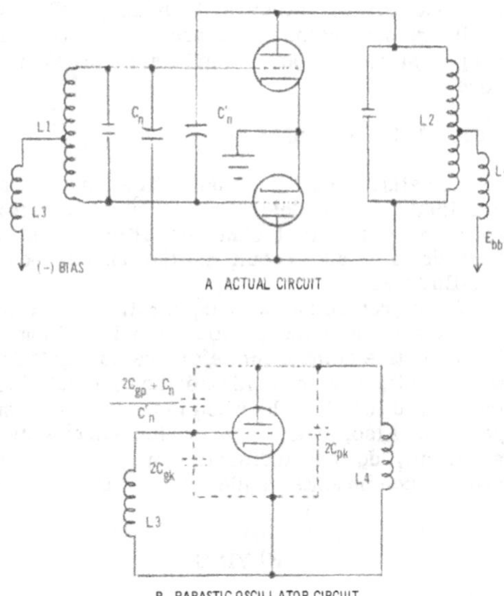

179.405
Figure 21-18.—Low-frequency parasitics.

One method of changing frequency is to make the grid lead considerably shorter than the plate lead, and/or lengthen the plate lead. This results in the inductance of the plate lead being greater than the inductance of the grid lead. The same effect may also be obtained by placing a small value RF choke, usually only a few microhenries, in the plate lead.

Another method of suppressing high-frequency parasitics is to reduce the Q of the parasitic circuit. This is done by placing low value resistors, 50 to 100 ohms, in the grid and plate leads, as close as possible to the tube terminals.

The combination of a small inductance and resistor connected in parallel and placed in the plate lead is sometimes used. Sometimes the inductance is made by winding from four to six turns of wire around the body of the resistor and then soldering the leads to the resistor leads close to the body of the resistor.

Since it is possible for low-frequency parasitics to develop due to the use of RF chokes, it is desirable to eliminate such chokes from both the grid and plate circuits of the same tube. When chokes must be employed in both circuits, they should be of such relative size that the plate circuit resonant frequency is lower than the grid circuit resonant frequency. Whenever resistors and/or inductors are used for parasitic suppression, they are called parasitic suppressors.

REPAIR TECHNIQUES

As stated earlier, parasitics are usually eliminated in the design of a transmitter. However, poor repair techniques often defeat the best designs and result in the appearance of oscillations.

When replacing a component, care should be taken to note the position and lead length of the defective component before removing it from the circuit. Try to duplicate the original placements and lead lengths when installing new components. Also, when making measurements in a circuit, do not indiscriminately push wires and/or components aside to use a test point.

KEYING

In CW radiotelegraph transmission, the carrier is turned on and off to form the dots and dashes of transmitted code characters. Turning on and off of the transmitter for this purpose is called "keying."

One method of keying is to turn the oscillator on and off. This may be accomplished by opening and closing the plate or cathode circuit with a key. The oscillator should be absolutely stable while keyed. If it is not stable, frequency shifts, causing a varying note (chirp) which makes the signal difficult to copy.

Formerly, where greatest frequency stability was required, the oscillator remained in operation continuously while the transmitter was in use. This procedure kept the oscillator tube at normal operating temperature and offered less accessibility of frequency variation when the key was used. If the oscillator is to operate continuously and the keying is to be accomplished in an amplifier stage following the oscillator, the oscillator circuit must be carefully shielded and isolated to prevent radiation of a backwave (from the oscillator). If the oscillator does not meet these conditions, a backwave may be radiated even though the stages between oscillator and antenna are cut off. Energy from the oscillator may leak to the antenna through improperly neutralized amplifiers or capacitive and/or inductive coupling between oscillator and antenna circuits. However, new techniques have virtually eliminated the necessity for continuous operation of transmitter oscillators.

BLOCKED BASE/GRID KEYING

Another method of keying is blocked base/grid keying, as shown in figure 21-19. Figure 21-19A is the schematic of a transistor RF amplifier showing blocked base keying. With the key open, the positive supply provides reverse bias to the emitter-base junction, cutting off Q1. When the key is closed, reverse bias will be removed from the emitter base junction, and Q1 will conduct. Referring to the electron tube circuit of figure 21-19B, with the key open, a high value of negative voltage is applied to the grid of V1, cutting it off. This voltage must be several times the value of grid bias required for cutoff to prevent the input signal from driving the tube into conduction.

When the key is closed, the negative voltage is removed from the grid and V1 conducts. R2 prevents the negative supply from being short circuited to ground when the key is closed. Blocked grid keying may be used in any one of several stages but presents problems in high-powered amplifiers due to the high value of voltage necessary to cut off such amplifiers.

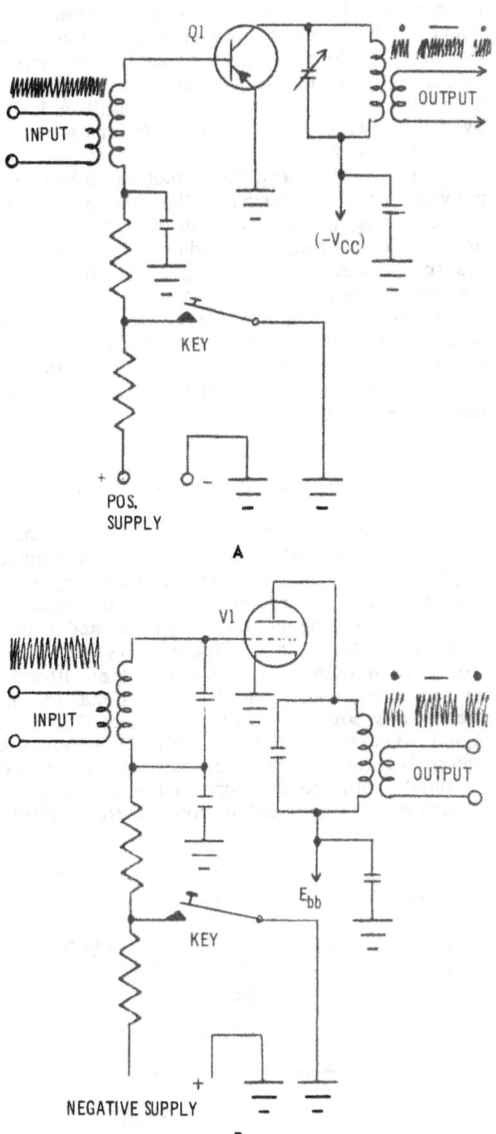

179.406
Figure 21-19.—Blocked base/grid keying.

Another disadvantage of the system is the requirement for a source of negative voltage.

CATHODE KEYING

The amplifier stages of a transmitter may be keyed by opening and closing the plate or cathode circuit. The key is rarely placed in the plate circuit due to the hazardous d.c. potentials that would be present across the key.

Figure 21-20 shows two methods of keying in the cathode circuit. Figure 21-20A shows center-tap keying of a directly heated cathode or filament, and figure 21-20B shows basic cathode keying of an indirectly heated cathode.

In the figure 21-20A, keying is accomplished by opening and closing the filament transformer center tap line to ground. With the key open the tube is cut off since there is no complete d.c. path (filament to plate). Capacitors C1 and C2 are RF bypass capacitors. They keep the RF

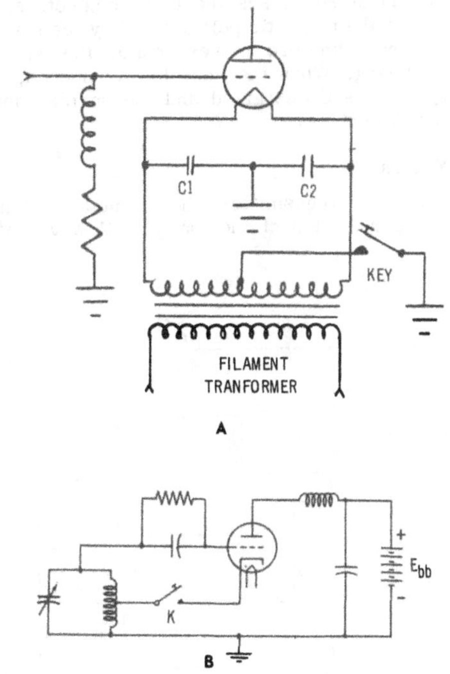

20.213
Figure 21-20.—Keying in the cathode circuit.

signal out of the filament transformer and the keying line.

In figure 21-20B, the key opens and closes the circuit. With the key open, there is no complete d.c. path (cathode to plate), neither grid nor plate current can flow, and the tube is cut off.

KEYING RELAYS

In large transmitters (75 watts or higher) the ordinary hand key would not generally accommodate the plate current without excessive arcing. In these large transmitters some local low voltage supply, such as a battery or the filament supply, is used with the hand key to open and close a circuit through the coils of a keying relay. The relay contacts, in turn, open and close the keying circuit of the amplifier tube. A schematic diagram of a typical relay-operated keying system is shown in figure 21-21. The hand key closes the circuit from the low voltage supply through coil (L) of the keying relay. The relay armature closes the relay contacts as a result of the magnetic pull exerted by the armature. The armature moves against the tension of a spring. When the hand key is opened, the relay coil is deenergized and the spring opens the relay contacts.

KEY CLICKS

Keying a transmitter should instantly start and stop radiation of the carrier. However, the sudden application and removal of power creates rapid surges of current which cause interference to be radiated. Though receivers are tuned to frequencies far removed from that of the transmitter, interference is present in the form of clicks or thumps. To prevent such interference, key click filters are used in transmitter keying systems. Two types of key click filters are shown in figure 21-22.

The capacitors and RF chokes in both circuits prevent current surges. The choke coil (L) causes a lag in current when the key is closed, and current builds up gradually instead of instantly. Capacitor C charges as the key is opened and slowly releases the energy stored in the inductor's magnetic field. Resistor R controls the rate of charge and discharge of capacitor C and also prevents sparking at the key contacts by the sudden discharge of C when the key is closed.

AUTOMATIC LEVEL CONTROL

Automatic level control (ALC) is a means of maintaining a signal at such a level that the power amplifier performs near its maximum capability without being overdriven on signal peaks. In AM systems, speech compressors and speech clipping perform this function. However, in an SSB system these methods are not as effective because the peaks of the SSB signal do not necessarily correspond with the peaks of the audio signal. Therefore, the most effective means of control is obtained by a circuit which receives its input from the envelope peaks of the power amplifier and whose output controls the amplitude

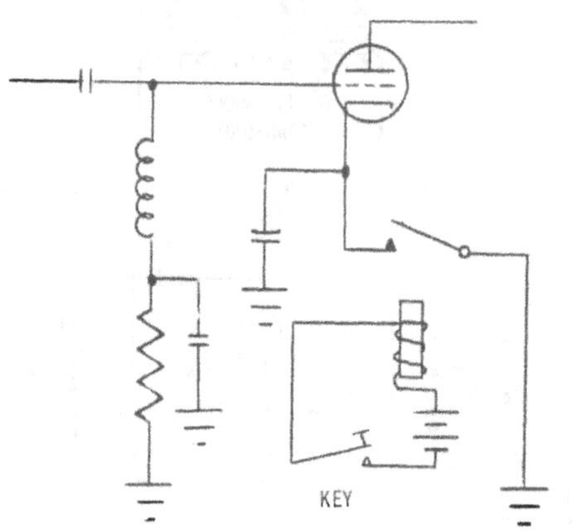

Figure 21-21.—Relay-operated keying system.

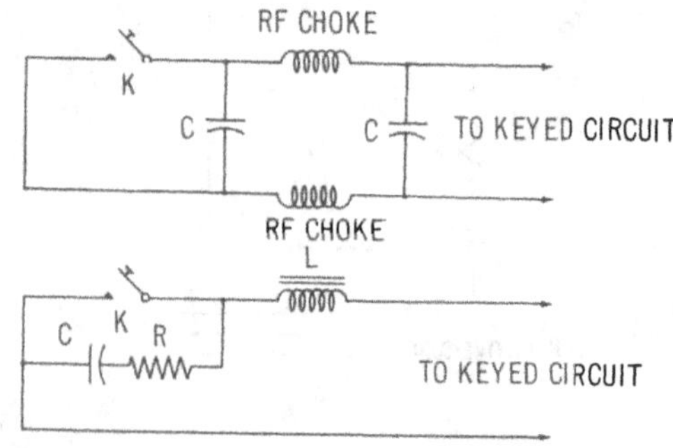

Figure 21-22.—Key click filters.

of the excitation signal. Such a circuit is an automatic level control (ALC) circuit.

Figure 21-23 is a simplified schematic of an ALC circuit. This circuit employs two variable gain stages of remote cutoff tubes operating similarly to the IF stages of a receiver with automatic gain control. The grid bias on the variable gain amplifiers is obtained from the ALC rectifier. This rectifier derives its input from the power amplifier plate circuit. The capacitor voltage divider (C1 and C2) steps down the RF voltage of the power amplifier plate. A large delay bias is used on the rectifier so that no reduction of gain takes place until the signal level is nearly at full power capability of the power amplifier. The output of the ALC rectifier passes through RC networks R1, R2, and C3 to obtain the desired attack and release times. Usually a fast attack time, about two milliseconds, is used for voice signals so that the gain is reduced rapidly to remove the overload from the power amplifier. After a signal peak passes, a release time of about one-tenth second returns the gain to normal. A meter calibrated in decibels of compression is used to adjust the gain for the desired amount of level control.

TUNING METHODS

If a class C amplifier is to operate efficiently, the plate tank circuit must be resonant at the same frequency as the grid signal. If the tuning capacitor of the plate tank circuit is variable, illustrated in figure 21-24A, the plate circuit will be either on or off resonance, depending upon the setting of the variable capacitor. Adjusting the plate tank variable capacitor to be resonant at the same frequency as the grid tank is called tuning. When a transmitter is detuned, a weak signal will be radiated and receivers tuned to the transmitter frequency may not receive the signal. When a transmitter is tuned to a given frequency, the final tank circuit is tuned to resonate at this frequency. The transmitter then radiates a signal at maximum efficiency and maximum power output.

A tank circuit in series with the plate of a class C amplifier may be compared to a rheostat (fig. 21-24B) in series with the plate. When the plate tank circuit is completely detuned, it performs inefficiently as there is low resistance in the plate circuit because tank impedance is minimum. As a result, plate voltage will

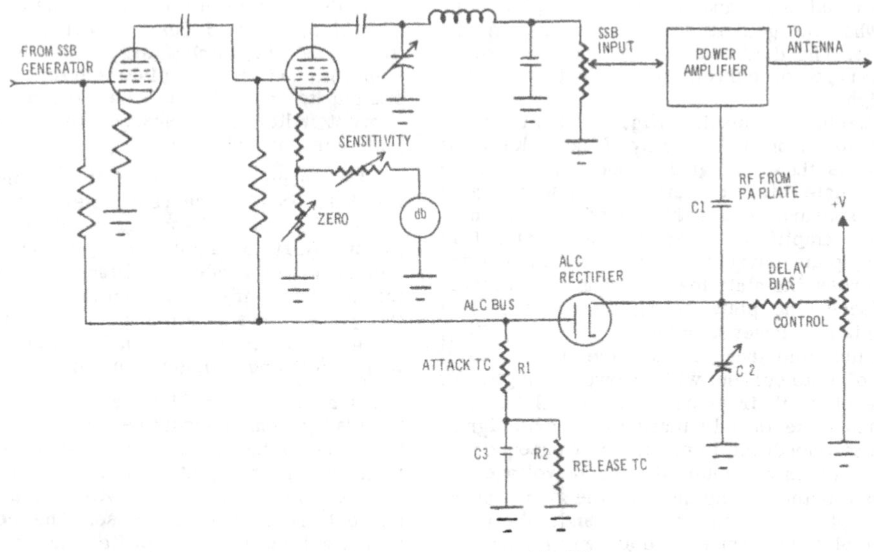

179.408
Figure 21-23.—ALC circuit.

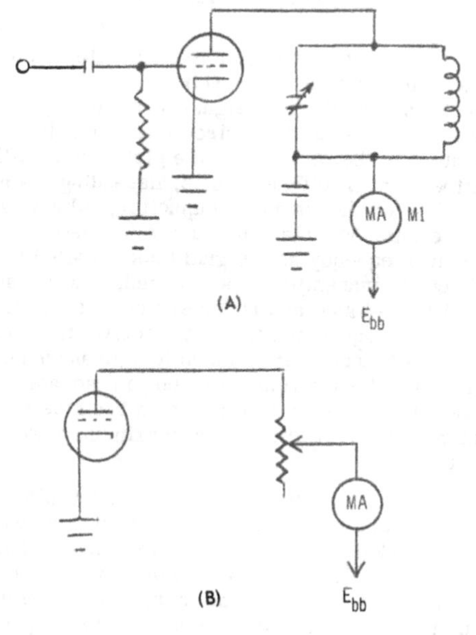

Figure 21-24.—Tuning action in a tank circuit.

almost equal E_{bb} and the pulses of plate current (when the grid is driven above cutoff) will be large. The d.c. meter (M1) which measures the average of current pulses will therefore read high.

As the tuning capacitor (fig. 21-24A) is varied so that the resonant frequency of the tank circuit approaches the grid signal frequency, impedance of the plate circuit increases. Now a signal voltage appears across this impedance. As in an ordinary amplifier, when the grid signal is positive, plate current increases and the voltage drop across the plate load impedance increases, thus decreasing plate voltage. Since the plate voltage is now lower than before (lower than E_{bb}), during the time the grid is driven above cutoff, pulses of plate current will be lower in amplitude, and therefore their average value will be less. When the plate tank is tuned to the grid signal, plate load impedance is maximum. Consequently, plate voltage is minimum. Since plate voltage is at its lowest point (during the time the grid is above cutoff), plate current pulses, and, therefore average plate current will be at their minimums. The sequence of tuning is illustrated in figure 21-25.

Minimum d.c. plate current is therefore an indication that the plate tank is tuned to the grid signal frequency. When a plate tank circuit is tuned for minimum plate current, it is said to be tuned to the dip. In addition to the plate current meter, there is another meter which indicates correct tuning of the plate circuit. This meter is in the grid circuit of the following stage and is labeled M2 in figure 21-26.

When the plate circuit is tuned to the frequency of the input signal, the voltage developed across the circuit and its output are at peaks. The larger the output from a stage, the greater is the signal on the grid of the following stage.

The grid of the following stage will draw current whenever the input signal drives the grid positive. The larger the signal input, the greater will be the flow of current from cathode to grid. Since the signal input to the grid will be greatest when the plate circuit of the previous stage is accurately tuned, the grid will draw maximum current, and milliammeter M2 (which measures the average grid current) will indicate a maximum reading. Thus, when the plate tank is accurately tuned, the plate current meter indicates a dip, and the grid current meter of the following stage simultaneously registers a rise known as a peak reading.

If the grid circuit has fixed bias or combination bias, no grid current will be drawn until the signal is fairly large. This will happen sometime after the plate current meter has started to dip. For this reason, a rising grid current indication is sharper than a decreasing plate current indication.

The normal procedure for tuning a stage which has a plate current meter and is followed by a stage which has a grid current meter, is to tune first for a minimum plate current. This indication is broader and less likely to be overlooked when varying the tuning. After observing the plate current starting to decrease, watch the grid current for a rise. The final adjustment will be for a peak in grid current.

In adjusting the FPA stage, loose coupling to the load (antenna) should be used when you begin. The stage is then tuned to resonance as indicated by a dip in the plate current meter. Coupling is then increased. This will cause the plate current reading to increase. The coupling is increased until the amplifier draws its rated plate current. Changing the coupling to the load will usually detune the tank circuit, so that it

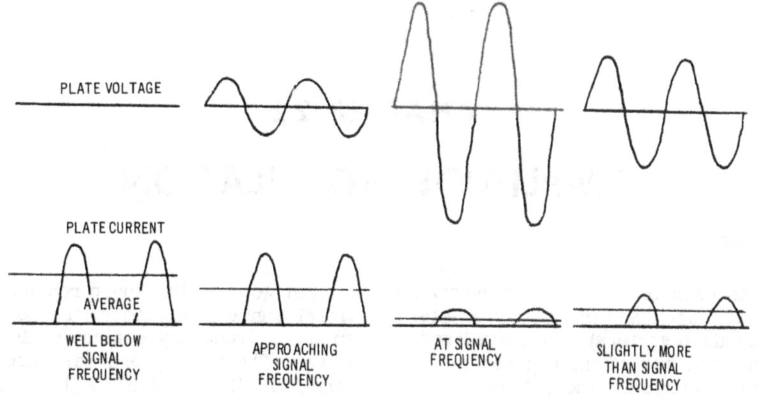

179.410
Figure 21-25.—Sequence of plate voltage and current as tuning varies.

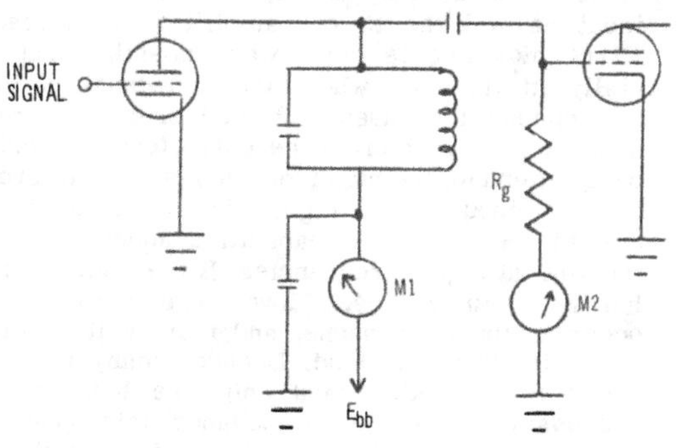

179.411
Figure 21-26.—Grid meter tuning.

will be necessary to readjust for resonance each time a change in coupling is made.

Coupling should be as loose as possible between the final tank circuit and antenna but should still maintain the desired output power. Overcoupling lowers the Q of the tank circuit and considerably increases harmonic output.

When tuning a transmitter, unauthorized frequencies or other improper signals may be produced and radiated by the antenna, causing interference. To prevent the radiation of any undesired signal, the antenna is disconnected from the transmitter and a "dummy load" is connected in its place during preliminary adjustments.

A dummy load may be an incandescent lamp or a noninductive resistor having suitable resistance and wattage rating. If one lamp, or resistor, does not have sufficient power handling capability or the proper resistance more units may be added in series, parallel, or series-parallel arrangements to meet the requirements.

Commercial dummy loads are also available. They frequently employ carbon piles or stacks of carbon discs mounted on a suitable rod to convert the RF input to heat. These in turn are surrounded by a coolant (such as oil) and metal fins to dissipate heat.

CHAPTER 22

AMPLITUDE MODULATION

The circuits used to produce high power RF signals have been discussed in previous chapters. These RF signals (carriers) can be utilized to transmit various forms of intelligence such as speech, music, teletype, video, facsimile, or telemetry.

If an RF signal is to convey intelligence, some feature of the carrier must be varied in accordance with the information to be transmitted. The process of combining the intelligence with the RF carrier to produce this variation is called modulation.

In this chapter, we will discuss the theory and operation of a variety of microphones. Additionally, we will discuss a variety of methods and techniques for modulating carriers.

MICROPHONES

A microphone is an energy converter that changes sound energy into electrical energy. The diaphragm of the microphone moves in and out in accordance with the compression and rarefaction of the atmosphere known as sound waves. The diaphragm is connected to some device that causes current flow in proportion to the instantaneous pressure delivered to it. Many devices can perform this function. The particular device used in a given application depends on characteristics desired, such as sensitivity, frequency response, impedance, power requirements, and ruggedness.

The sensitivity or efficiency of a microphone is usually expressed in terms of the electrical power level which the microphone delivers to a matched impedance load compared to the sound level being converted. The sensitivity is rated in db. It is important that the sensitivity be as high as possible. A high microphone output requires less gain in the amplifiers used in conjunction with the microphone. This provides a greater margin over thermal noise, amplifier hum. and noise pickup.

For good quality sound reproduction, the electrical signal from the microphone must correspond in frequency content to the original sound waves. The microphone response should be uniform, or flat, within its frequency range and free from the electrical or mechanical generation of new frequencies.

The impedance of a microphone is important in that it must be matched to the microphone cable between the microphone and amplifier input as well as to the amplifier input load. Exact matching is not always possible, especially in the case where the impedance of the microphone increases with an increase in frequency. A long microphone cable tends to seriously attenuate the high frequencies if the microphone impedance is high. This attenuation is due to the increased capacitive appearance of the line at higher frequencies. If the microphone has a low impedance, a lower voltage drop will occur in the microphone, and more voltage will be available at the load. Because many microphone lines used aboard ship are long, it is necessary to use low impedance microphones to preserve a sufficiently high signal voltage level over the required frequency range.

The schematic symbol used to represent a microphone in a schematic diagram is shown in figure 22-1. The schematic symbol does not identify the type of microphone used or its characteristics.

CARBON MICROPHONE

Operation of the single-button carbon microphone is based on varying the resistance of a pile of carbon granules by varying the pressure on the pile. The insulated cup, called the button, which holds the loosely piled granules, is so mounted that it is in constant contact with the thin metal diaphragm, as shown in figure 22-2A.

Sound waves striking the diaphragm vary the pressure on the button, and thus vary the pressure on the pile of carbon granules. The d.c.

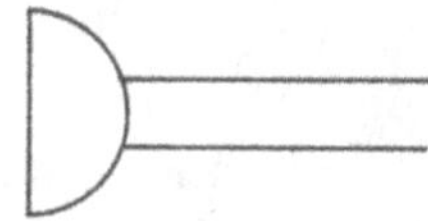

179.412
Figure 22-1.—Microphone schematic symbol.

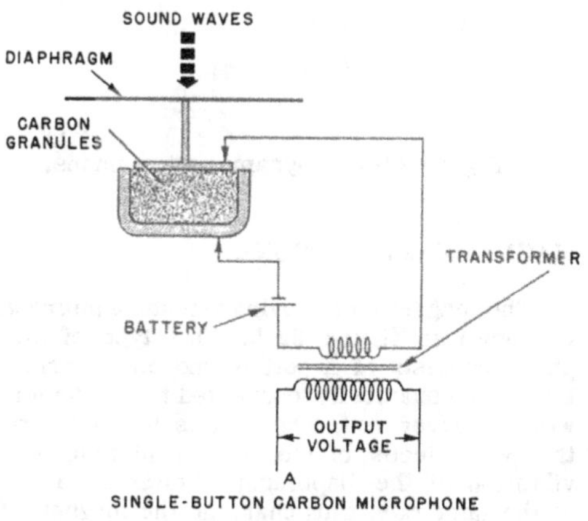

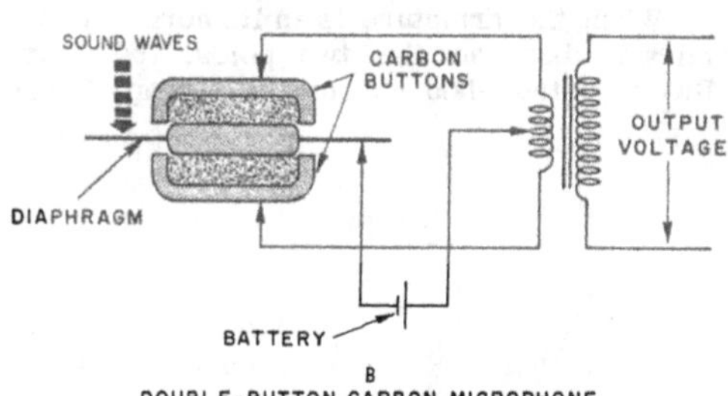

20.218
Figure 22-2.—Carbon microphones.

resistance of the carbon granule pile is varied by this pressure. This varying resistance is in series with a battery and the primary of a transformer. The changing resistance of the carbon pile produces a corresponding change in the current of the circuit. The varying current in the transformer primary produces an alternating voltage in the secondary. The transformer steps up the voltage, as well as matches the low impedance of the microphone to the high impedance of the first AF amplifier. The voltage across the secondary may be as high as 25 volts peak. The impedance of this type of microphone varies from 50 to 200 ohms.

The double-button carbon microphone is shown in figure 22-2B. Here one button is positioned on each side of the diaphragm so that an increase in pressure and resistance on one side is accompanied by a simultaneous decrease in pressure and resistance on the other. Each button is in series with the battery and one-half the transformer primary. The decreasing current in one half of the primary and the increasing current in the other half produce an output voltage in the secondary that is proportional to the sum of the primary signal components. This action is similar to that of push-pull amplifiers.

One disadvantage of carbon microphones is a constant background hiss, resulting from random changes in the resistance between individual carbon granules. Another disadvantage is the reduced sensitivity and distortion that may result from the granules packing or sticking together. Sometimes this may be cured by tapping the microphone. The carbon microphone also has a limited frequency response. Still another disadvantage is the requirement for an external voltage source.

The disadvantages, however, are offset by advantages that make its use in military applications widespread. It is light-weight, rugged, and can produce an extremely high output when its frequency response is limited to those frequencies that contribute most to intelligibility.

CRYSTAL MICROPHONE

The crystal microphone utilizes the piezoelectric property of Rochelle salt, quartz, or other crystalline materials. This means they generate a voltage (EMF) when mechanical stress (as in the pressure of a sound wave) is placed across them (fig. 22-3). Since Rochelle salt has the largest voltage output for a given mechanical stress, it is the most commonly used crystal in microphones.

A crystal microphone has a high impedance and does not require an external voltage source. It can be connected directly into the input circuit of a high gain AF amplifier. Because its output is low, however, several stages of high gain amplification are required. Crystal microphones are delicate and fragile and must be handled with care. Exposure to high temperatures (above 125°F) may permanently damage the crystal

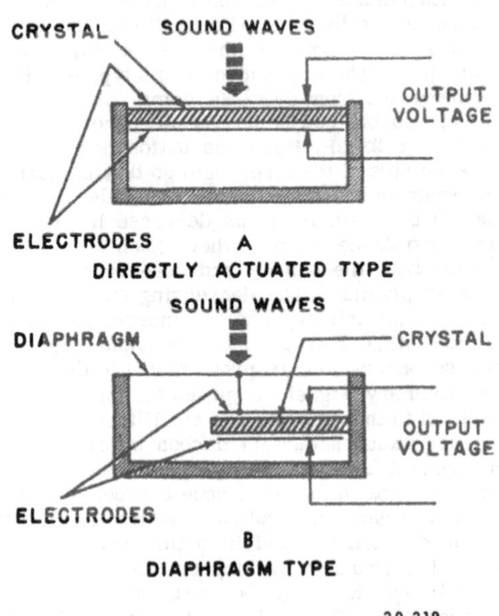

20.219
Figure 22-3.—Crystal microphone.

unit. Crystals are also soluble in water and other liquids and, therefore, must be protected from moisture and excessive humidity.

DYNAMIC MICROPHONE

The dynamic or moving coil microphone is shown in cross section in figure 22-4. A coil of fine wire is mounted on the back of the diaphragm and located in the magnetic field of a permanent magnet. When sound waves strike the diaphragm, the coil moves back and forth cutting the magnetic lines of force. This induces a voltage in the coil that is an electrical representation of the sound waves.

The sensitivity of the dynamic microphone is almost as high as the carbon type. It is light in weight and requires no external voltage. The dynamic microphone is rugged and practically immune to the effects of vibration, temperature, and moisture. The microphone has a uniform response over a frequency range that extends from 40 to 15,000 Hz. The impedance is very low (generally 50 ohms or less). A transformer is, therefore, required to match it to the input of an AF amplifier.

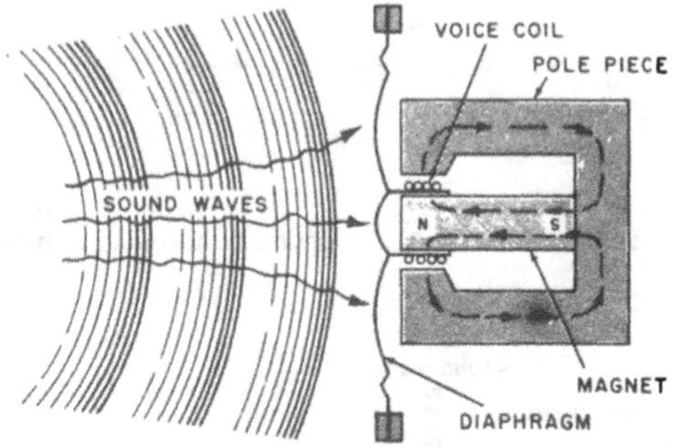

7.12
Figure 22-4.—Dynamic microphone.

MAGNETIC MICROPHONE

The magnetic or moving armature microphone is shown in figure 22-5. This type of microphone consists of a coil wound on an armature that is mechanically connected to the diaphragm with a driver rod. The coil is located between the pole pieces of the permanent magnet. Any vibration of the diaphragm vibrates the armature at the same rate, thus changing the magnetic flux in the armature and through the coil.

When the armature is in its normal position, midway between the two poles, the magnetic flux is established across the airgap. There is

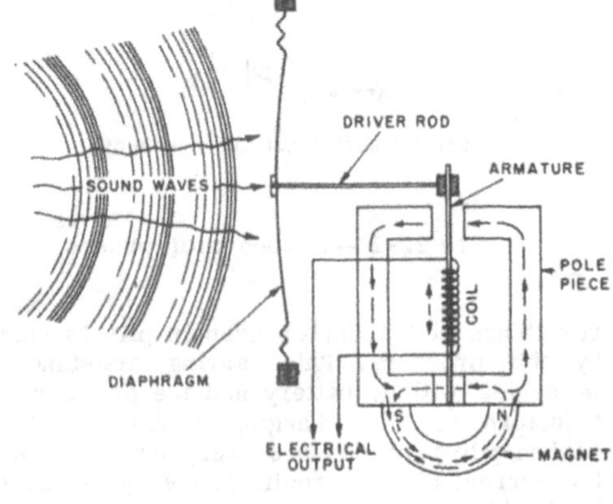

7.11
Figure 22-5.—Magnetic microphone action.

no resultant flux in the armature. When a compression wave strikes the diaphragm, the armature is deflected to the right. Although a considerable amount of the flux continues to move in the direction of the arrows, some of it now flows from the north pole of the magnet across the reduced gap at the upper right, down through the armature, and around to the south pole of the magnet.

When a rarefaction wave occurs at the diaphragm, the armature is deflected to the left. Some of the flux is now directed from the north pole of the magnet, up through the armature, through the reduced gap at the upper left, and back to the south pole.

The vibrations of the diaphragm thus cause an alternating flux in the armature which in turn induces an alternating voltage in the coil. This voltage has the same waveform as that of the sound waves striking the diaphragm.

The magnetic microphone is very similar to the dynamic microphone in terms of impedance, sensitivity and frequency response. However, it is more resistant to vibration, shock, and rough handling than other types of microphones.

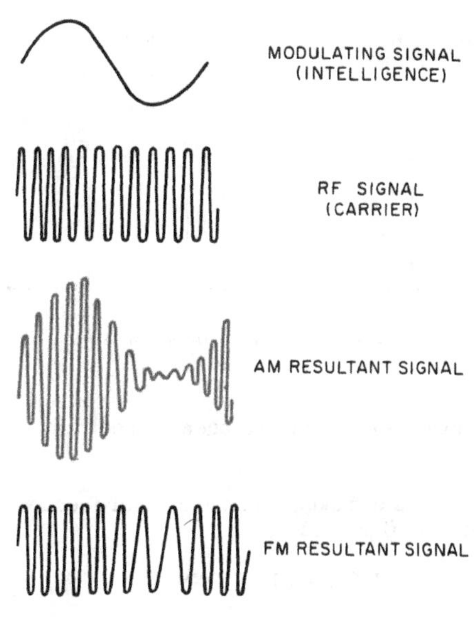

Figure 22-6.—Amplitude and frequency modulation.

MODULATION PRINCIPLES

The two basic types of modulation used to transmit intelligence are amplitude modulation and frequency modulation. Both types of modulation produce new frequencies which combine vectorially with the carrier frequency to produce a resultant waveform. In amplitude modulation, the amplitude of the resultant waveform is made to vary in accordance with the intelligence. In frequency modulation, the frequency of the resultant waveform varies with the intelligence. In either case, the intelligence is called the modulating signal. Amplitude modulation and frequency modulation are usually referred to as AM and FM. A modulating signal, carrier signal, and the resultant waveforms for amplitude and frequency modulation are shown in figure 22-6.

When the intelligence to be transmitted is superimposed on the carrier in the form of changes in the amplitude of the RF energy, the resulting modulation is called amplitude modulation or AM. The frequency of the carrier is not affected by this type of modulation.

When an RF carrier is modulated by a single audio modulating frequency, two additional frequencies are produced. One of these frequencies is the sum of the RF carrier and the audio frequency. The other is the difference between the RF carrier and the audio frequency. When the modulating signal is complex, each individual frequency component of the modulating signal produces a sum and difference frequency with the RF carrier. The sum frequencies are known as the upper sideband and the difference frequencies are known as the lower sideband. The location of the various frequencies in the case of a single modulating frequency is shown in figure 22-7.

The space which a carrier and its associated sidebands occupy in the frequency spectrum is called the bandwidth. The bandwidth is equal to twice the highest modulating frequency. In figure 22-7, the bandwidth is equal to 2 x 5 kHz or 10 kHz. A tank circuit tuned to the carrier frequency is used as the impedance across which is developed the resultant amplitude modulated waveform. The half power points of the response curve for the tuned circuit must enclose the bandwidth of the resultant signal to be transmitted. The audio modulation frequency lies outside of the response curve, and, therefore, will not be developed as part of the resultant waveform. The result of a single-tone amplitude modulation is simply the algebraic combination

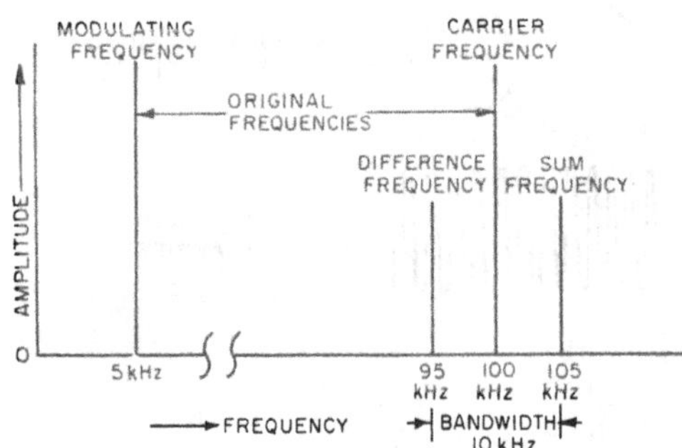

179.414

Figure 22-7.—Radio frequency spectrum.

of the two sidebands, and the carrier frequency as shown in figure 22-8.

MODULATION CALCULATIONS

The degree of modulation is the ratio of modulating voltage amplitude to carrier voltage amplitude as shown in the formula:

$$M = \frac{E_{modulation}}{E_{carrier}} = \frac{E_M}{E_C}$$

where: M = degree of modulation
E_M = the peak, peak to peak, or r.m.s. value of the modulating voltage
E_C = the carrier voltage in the same units as E_M

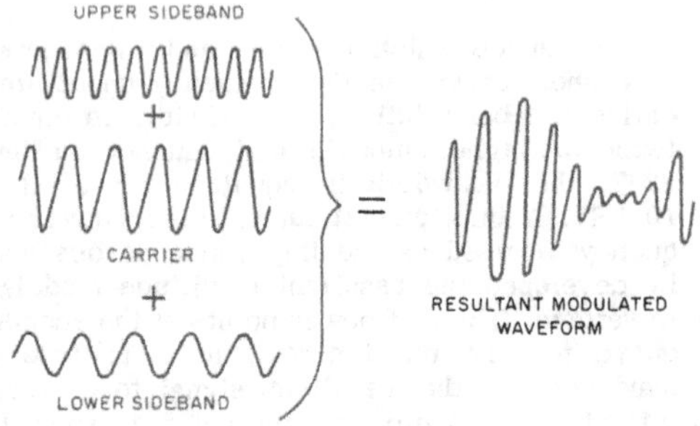

UPPER SIDEBAND
+
CARRIER
+
LOWER SIDEBAND
=
RESULTANT MODULATED WAVEFORM

179.415

Figure 22-8.—Combining of the carrier and sidebands to produce an AM wave with 100% modulation.

The degree of modulation is usually expressed as percent of modulation where 100% is the maximum permissible modulation. To determine the percentage of modulation, merely multiply the degree of modulation in the above equation by 100. From the above formula, it can be seen that the modulation voltage is equal to the carrier voltage at 100% modulation. It is desirable to operate close to 100% modulation as this will allow the greatest amount of power in the intelligence portion of the resultant modulated waveform. The intelligence is carried in the sidebands so, in effect, the maximum sideband power is developed at 100% modulation.

If the carrier is overmodulated, additional sidebands are created, extreme distortion occurs, and the bandwidth of the transmitted signal may cause interference with other transmitters.

When the modulated waveform is displayed on an oscilloscope and the degree of modulation is less than or equal to 1, the following formula can be used to calculate percent of modulation.

$$\% M = \frac{e_{max} - e_{min}}{2\, e_o} \times 100$$

where: E_{max} = the maximum voltage in the modulated waveform
e_{min} = the minimum voltage in the modulated waveform
e_o = the carrier voltage

The three voltages may be measured in peak or r.m.s. values as long as the same unit is used throughout any single calculation. The voltages used in the formula are shown in figure 22-9.

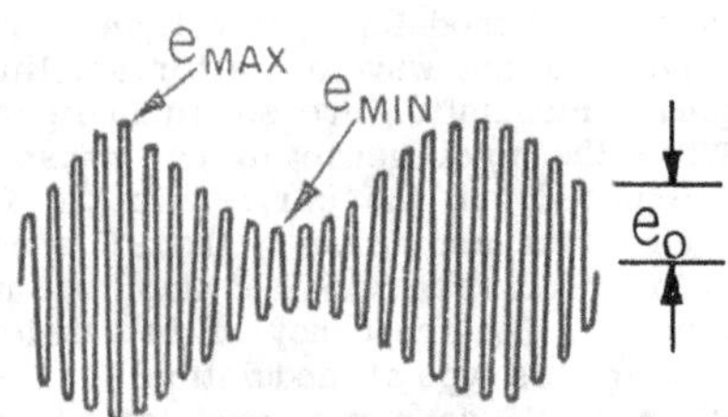

179.416

Figure 22-9.—Voltages used in computing the percentage of modulation from oscilloscope presentation.

Since e_o may be difficult to determine on the oscilloscope, the above formula may be rewritten $\% M = \frac{e_{max} - e_{min}}{e_{max} + e_{min}} \times 100$. Figure 22-9 illustrates a waveform that is less than 100% modulated.

Figure 22-10A shows the waveform for 100% modulation, while figure 22-10B shows the frequency spectrum. At a 100% modulation condition, the voltage magnitude of a single sideband is .5 the voltage amplitude of the carrier.

If the modulating signal amplitude is greater than the carrier amplitude, overmodulation will occur. The waveform and frequency spectrum when this occurs are shown in figure 22-11.

The power in an amplitude modulated wave is divided between the carrier and the sidebands. The carrier power is constant (except in cases of overmodulation), and so the sideband power is the difference between the carrier power and the total power of the modulated wave or:

$$P_{SB} = P_T - P_C$$

where: P_{SB} = power in both sidebands
P_T = total power
P_C = carrier power

The power in the sidebands can be calculated when carrier power and degree of modulation are known with the following formula:

$$P_{SB} = \frac{M^2}{2} \times P_C$$

For example, if the degree of modulation is 70% and the carrier power is 500 watts:

$$P_{SB} = \frac{.70^2}{2} \times 500$$

$$= \frac{.4900}{2} \times 500$$

$$= .2450 \times 500$$

$$= 122.5 \text{ watts}$$

Since the total sideband power is twice the power in a single sideband, the formula can also be written:

$$P_{SF} = \frac{M^2}{4} \times P_C$$

where: P_{SF} = power in one sideband

Thus, using the above values:

$$P_{SF} = \frac{.4900}{4} \times 500$$

$$= .1225 \times 500$$

$$= 61.25 \text{ watts}$$

Total power can be calculated using the following formula if carrier power and the degree of modulation are known.

$$P_T = P_C \left(1 + \frac{M^2}{2}\right)$$

Using the same values previously given:

$$P_T = 500 \left(1 + \frac{.4900}{2}\right)$$

$$= 500\ (1 + .2450)$$

$$= 500 \times 1.2450$$

$$= 622.5 \text{ watts}$$

MODULATION SYSTEMS

An RF carrier may be amplitude modulated at various locations in the RF amplifier section of the transmitter. The method of modulation refers to the electrode or element of the RF amplifier to which the modulating voltage is applied.

In collector/plate modulation, the most commonly used method, the modulating voltage is impressed on the d.c. supply voltage to the collector/plate of one of the RF amplifiers in the transmitter. The output of the modulated stage is varied by varying collector/plate supply voltage with the modulating signal.

Application of the modulating voltage to the base or grid of an RF amplifier is called base or grid modulation. Emitter/cathode modulation is a method where the modulating voltage is applied to the emitter/cathode circuit of the modulated stage.

In base/grid or emitter/cathode modulation, the modulation is accomplished by varying the bias on the RF amplifier with the modulating signal. In either method, the d.c. collector/plate supply voltage is held constant.

Pentode type power amplifiers can be modulated by applying the modulating voltage to the suppressor grid for suppressor modulation or

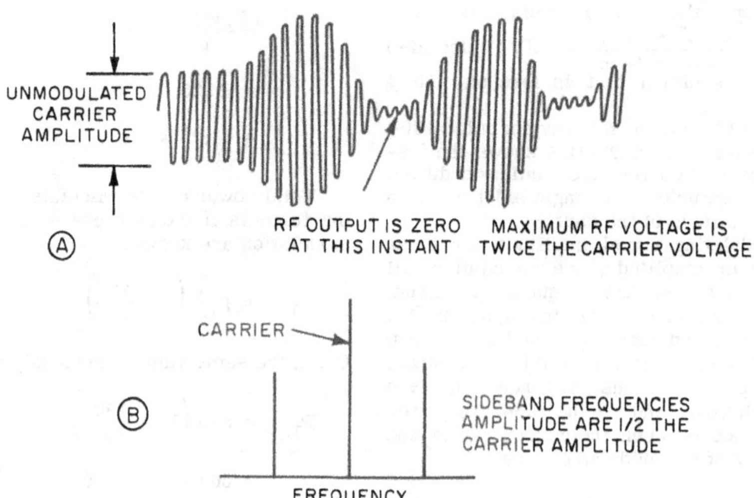

179.417
Figure 22-10.—Waveform spectrum for 100 percent modulation.

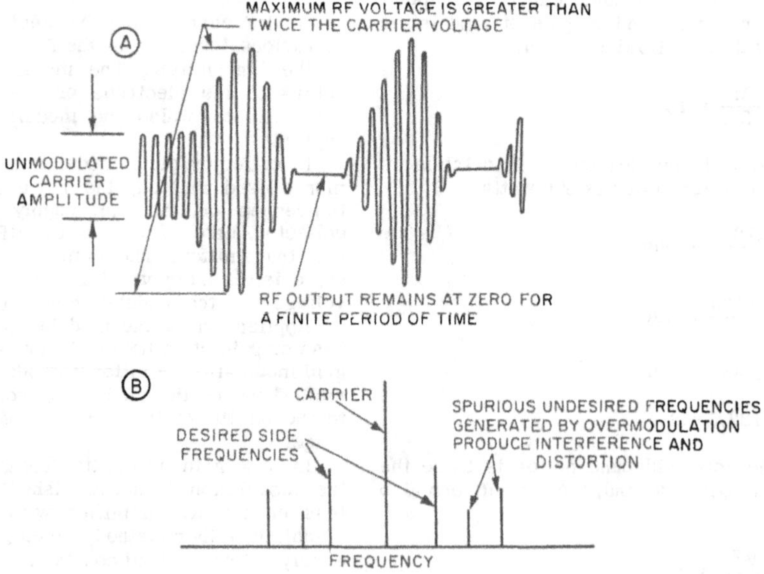

179.418
Figure 22-11.—Waveform and spectrum for overmodulation.

to the screen grid for screen grid modulation. Screen grid modulation can also be used with tetrode power amplifier tubes. In many cases, screen grid and plate modulation are used in combination.

Modulation is also identified as to level. High level modulation occurs when the modulating voltage is applied to the collector/plate circuit of the final power amplifier. With high level modulation, the final stage is operated class C and the preceding stages are also operated class C. The overall efficiency of such a transmitter is very high. A disadvantage of high level modulation is that comparatively high audio power is needed and several stages of voltage and power amplification may be required in the speech amplifier and modulator circuits.

Low level modulation takes place in the control grid or cathode of the final power amplifier. It is difficult to obtain any large degree of modulation using this method.

The modulation of a carrier signal by the modulating signal can be accomplished in any nonlinear device. However, since high power output is a requirement of the transmitter, modulation usually takes place in a transistor or electron tube amplifier.

Figure 22-12 shows a block diagram of amplitude modulation in a transistor or electron tube final power amplifier. The input to the modulator is the audio frequency intelligence from previous audiofrequency voltage amplifiers. The modulator is an audio frequency power amplifier whose output will vary some conduction controlling factor of the F.P.A. The RF input to the F.P.A. is the unmodulated RF carrier from the previous intermediate power amplifiers. Modulation can take place at various points in the F.P.A. or, for that matter, in any of the previous buffer or intermediate power amplifiers.

Collector Modulation

At present, semiconductor modulators operate over a relatively low power range compared with electron tubes. A maximum of 100 watts represents the high end at this time. As high power RF and audio transistors are developed, this power range will probably be extended and become comparable to that of the electron tube. In terms of fidelity and efficiency, the transistor is the equal of the electron tube.

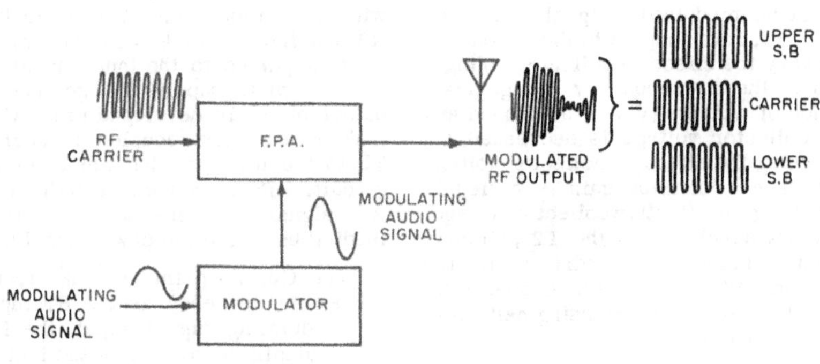

179.419
Figure 22-12.—Amplitude modulation, block diagram.

Modulation may take place in the emitter, base, or collector circuits of a transistor. Emitter and base modulation, however, are used only at low levels and at less than 100% modulation. The collector modulator will be covered in detail as this method can be used to produce modulation at either low or relatively high levels and up to a maximum of 100% modulation.

A collector modulated stage is shown in figure 22-13. The unmodulated RF input is applied to the Q1 base through the secondary of transformer T1. The base is held at a fixed forward bias by the voltage divider made up of R1 and R2. C1 bypasses R1 to prevent the feedback of RF into the power supply. Emitter resistor R3 provides thermal stabilization and is bypassed by C2 to prevent degeneration. The AF modulating signal is applied in series with the collector circuit through the secondary of transformer T3. Capacitor C6 decouples RF from the power supply and the audio circuits. C3 and the primary of T2 form the output tuned tank circuit. The modulated RF output is coupled through T2 to the next stage or antenna circuit.

The fixed negative bias applied to the Q1 base, causes the circuit to operate class A with a small signal input and class B or C with a large signal input. With only the RF applied, collector current decreases during the positive half cycles and increases during the negative half cycles. This is normal RF amplifier operation and produces the conventional RF output.

When an audio modulating signal is applied to transformer T3, it is coupled to the secondary where it will vary the effective collector voltage. When the top of the T3 secondary swings positive, the collector voltage (V_{CC}) is opposed and the effective collector voltage is decreased. On the negative swing the effective collector voltage is increased. The transistor gain is varied in step with the change in effective collector voltage. Thus, the output developed in the T2 primary, C3 tank circuit varies in accordance with the modulating signal. With class B or C operation, the tank circuit restores the missing half cycle that is lost when Q1 is cut off.

Plate Modulation

Figure 22-14 illustrates a plate modulated RF power amplifier. The V1 driver circuit is an audio frequency voltage amplifier whose output is transformer coupled to the V2 modulator stage. The modulator is an audio frequency power amplifier whose output is transformer coupled to the RF power amplifier plate circuit. The AF driver and modulator circuits must be biased class A to minimize audio modulating frequency distortion. The RF power amplifier is usually operated class C to obtain the necessary nonlinearity for good modulation and maximum efficiency.

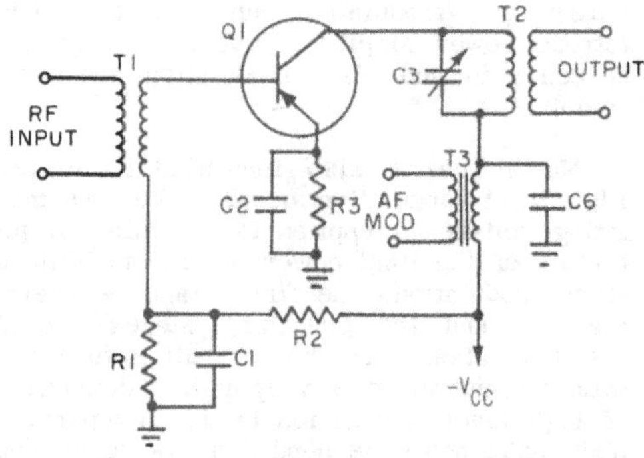

Figure 22-13.—Collector modulated transistor amplifier.

The unmodulated RF input is applied to the V3 grid through the coupling capacitor C_c and developed across RFC1 and R2. V3 is biased with a combination of grid leak bias across R2 and fixed bias E_{cc}. C2 couples the V3 plate current pulses to the tank circuit while blocking the d.c. plate supply voltage. The modulated RF output of V3 is developed in the C3, T3 primary tank circuit and coupled through transformer T3 to the following RF amplifier or the antenna circuit. RFC2 presents a high impedance to the RF signal and prevents the RF signal from feeding back into the power supply.

The C3, T3 primary tank circuit is tuned to the same frequency as the RF input to V3. With no modulating signal input, the RF output will be an amplified version of the input. Even though V3 is operated class C, the pulses of current through the tank circuit will trigger the flywheel action and the missing portions of the waveform will be restored.

Assume a polarity reversal across the transformers in the circuit. When the AF input on the V2 grid is on the negative alternation, the

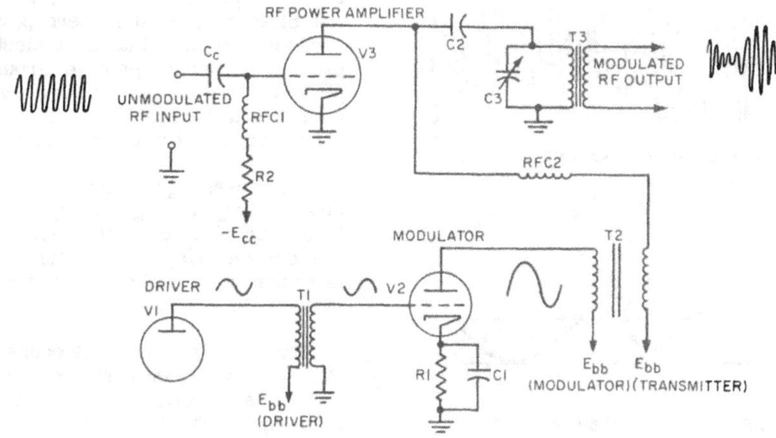

Figure 22-14.—Class C plate modulated RF amplifier.

V2 plate current will decrease, and plate voltage will increase. This results in a positive alternation at the top of the T2 primary. The voltage induced in the T2 secondary will oppose the V3 power supply voltage and the V3 plate current pulse amplitude will decrease. The signal being developed in the tank circuit will decrease also. The modulated RF in the tank circuit can follow the amplitude change of the modulating signal due to the low Q of the tank circuit. The low Q is primarily the result of heavy loading of the tank circuit by the following stage.

When the audio signal input to the V2 grid swings positive, the tube will conduct harder, and plate voltage will decrease. The voltage induced in the T2 secondary now aids the V3 plate supply voltage. The V3 plate current pulses increase in amplitude and the signal developed in the tank circuit increases in amplitude accordingly.

Figure 22-15 shows the various waveforms present in the V3 circuit. Figure 22-15A is the unmodulated RF signal applied to the grid. Figure 22-15B shows how the equivalent plate supply voltage is the sum of the fixed d.c. plate voltage and the audio modulating signal developed across the T2 secondary. Figure 22-15C is the actual instantaneous plate voltage, the amplified RF varying around the equivalent plate supply voltage (B) as a reference. Figure 22-15D shows the amplitude modulated RF across the tank circuit.

Plate/Screen Grid Modulation

High power modulation applications use plate and screen grid modulation with tetrode and pentode tubes. The plate and screen grid voltages are both varied to achieve modulation. The modulation is required to produce a power equal to one half of the RF carrier plate and screen power to produce 100% modulation. Efficiency is extremely high and the circuit will produce maximum power output for the tube used.

A simplified schematic of a basic plate and screen modulated pentode is shown in figure 22-16. V1 is operated class C with a combination of grid leak and fixed E_{cc1} bias. If the RF input signal is lost, the fixed bias provides enough protective bias to prevent tube damage.

Since screen grid voltage is usually lower than the plate voltage, the screen grid voltage is supplied through a dropping resistor (R2) or a separate power supply. The dropping resistor is the simpler method even though it does waste power. Screen current through this resistor produces sufficient drop to lower the screen grid voltage to the proper value.

The unmodulated RF input is applied to the V1 grid through a coupling capacitor and is developed across RFC1 and R_g. C1 decouples the fixed grid power supply E_{cc1}.

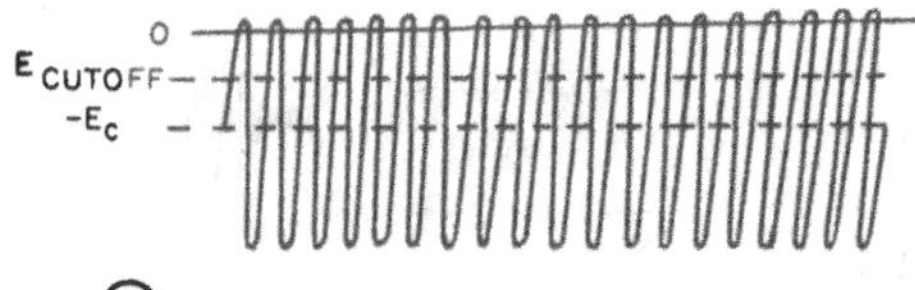

Ⓐ VOLTAGE APPLIED TO V3 CONTROL GRID

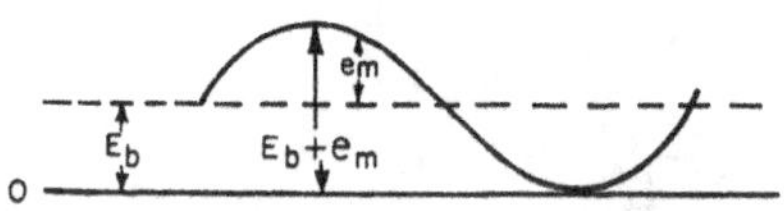

Ⓑ EQUIVALENT V3 PLATE SUPPLY VOLTAGE

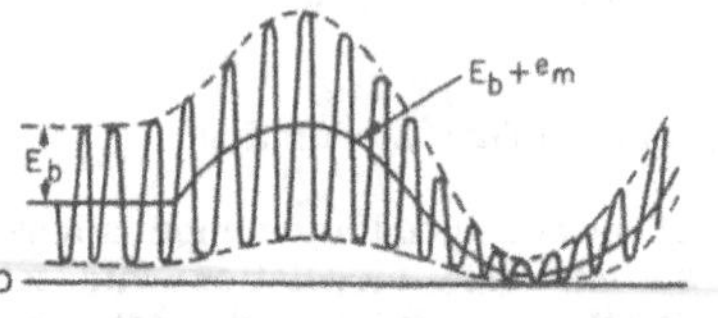

Ⓒ TOTAL V3 PLATE VOLTAGE

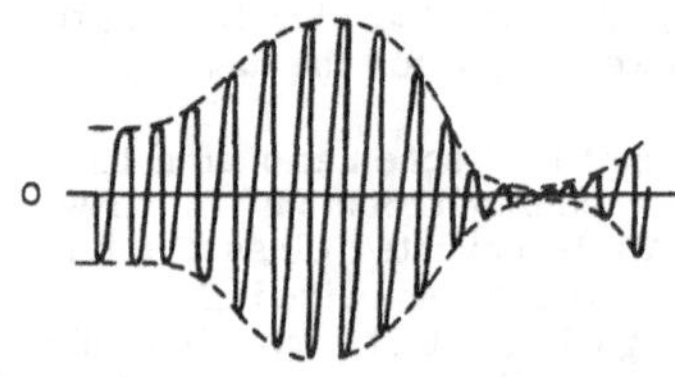

Ⓓ MODULATED RF ACROSS TANK CIRCUIT

179.422

Figure 22-15.—Voltage waveforms in modulated RF power amplifier.

The suppressor grid is shown connected to a small fixed positive bias E_{cc3} and placed at RF ground by capacitor C2. In other types of pentodes, the suppressor may be connected directly to the cathode, either internally or externally. In tetrodes, of course, the suppressor grid is not present. The positive bias, as used in the schematic, improves the shielding effect of the suppressor at low plate voltages. When the plate is almost at zero potential, the suppressor grid will intercept electron energy and provide a return path to ground. Thus, zero plate current is possible with a sharper cutoff than would be possible if the suppressor were connected directly to the cathode.

The screen grid voltage is obtained from the plate supply through dropping resistor R2 and is decoupled by capacitors C3 and C4. C3 places the screen grid at RF ground potential and C4 prevents RF feedback into the audio circuits through T1.

The secondary of T1 is connected between the E_{bb} supply and the common junction of the screen and plate supply. The audio modulating signal from the modulator is induced in the T1 secondary and produces modulation by varying the instantaneous effective plate and screen grid voltages. RFC2 isolates T1 and the audio circuits from tank circuit RF energy variations. The plate tank circuit is comprised of C5 and the T2 primary. T2 inductively couples the modulated RF to the output load.

A cycle of operation will now be analyzed in detail. With no modulating signal input, the plate and screen voltages applied are d.c. and the output is an amplified unmodulated carrier. When the signal applied to the control grid swings far enough positive to overcome the class C bias, the tube will conduct until the signal again drops the bias below cutoff. This pulse of current excites oscillations in the plate tank circuit which is tuned to the frequency of the input. The oscillations in the tank circuit restore the missing portions of the tube current pulse for a sinusoidal output.

When the applied modulating signal is going positive at the top of the secondary of T1, the induced voltage in the secondary will add to E_{bb} to increase plate and screen grid voltages. When the signal is reversed, the voltage at the secondary (top) of T1 will subtract from E_{bb} and decrease the plate and screen grid voltages. Figure 22-17 shows the modulating signal and the combined plate supply and modulation voltages. The screen grid voltage is varied similarly. Varying either plate or screen grid voltage alone will vary plate current, resulting in modulation. However, the variation is usually not great enough to produce 100% modulation in tetrode or pentode electron tubes. Changing both

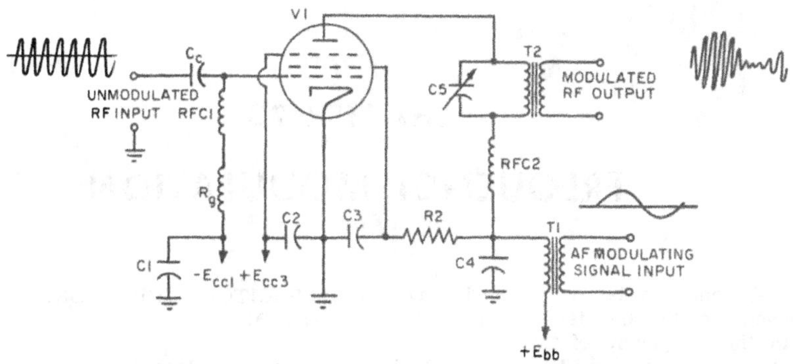

179.423
Figure 22-16.—Plate and screen modulated pentode.

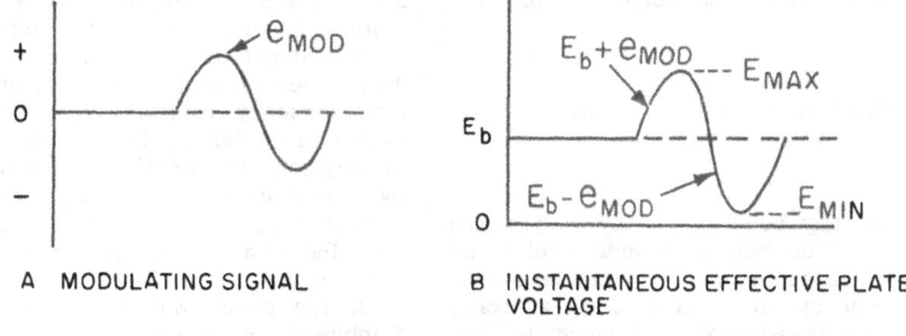

179.424
Figure 22-17.—Plate and modulation voltage combination.

plate and screen voltages simultaneously will result in a change in plate current large enough to produce 100% modulation.

When plate and screen modulation is to be obtained using a separate screen supply, the screen voltage must be varied in some manner. One method is to apply the modulating signal to the plate only, but as tube current varies a large choke in the screen circuit will develop the modulating signal on the screen grid as well. Another method would be to inductively couple the modulating signal to both the plate and screen grid circuits through a transformer with separate secondaries.

CHAPTER 23

FREQUENCY MODULATION

The intelligence to be transmitted may be superimposed on the carrier in the form of changes in the frequency of the carrier. This type of modulation is called FREQUENCY MODULATION and has certain inherent advantages over the conventional AM transmission, particularly when static-free transmission is desired. In addition to FM transmitters and receivers, frequency modulation is used for aircraft altimeters and some radar and sonar equipments.

FREQUENCY MODULATION PRINCIPLES

Intelligence may be conveyed by varying the frequency of a constant amplitude continuous radio wave. The carrier frequency can be varied a small amount on either side of its average, or assigned, value by means of an audio frequency (AF) modulating signal. The amount the carrier is varied depends on the amplitude of the modulating signal. And the rate at which the carrier frequency is varied depends on the frequency of the modulating signal. The amplitude of the RF carrier remains constant with or without modulation.

There are several systems of frequency modulation that meet these requirements. A mechanical modulator employing a capacitor microphone is the simplest system of frequency modulation. Two other systems of frequency modulation are reactance-tube and phase-angle modulation. The main difference between these two systems is that in reactance-tube modulation the RF wave is modulated at its source (the oscillator), while in phase modulation the RF wave is modulated in some stage following the oscillator. The results of each of these systems are basically the same; the FM wave created by either system can be received by the same receiver.

CAPACITOR-MICROPHONE SYSTEM

The simplest form of frequency modulation is that of a capacitor microphone, which shunts the oscillator tank circuit, LC, as shown in figure 23-1. The capacitor microphone is equivalent to an air-dielectric capacitor, one plate of which forms the diaphragm of the microphone. Sound waves striking the diaphragm compress and release it, thus causing the capacitance to vary in accordance with the spacing between the plates. This type of transmitter is not practical because the frequency range is very limited, but it is useful in explaining the principles of frequency modulation. The oscillator frequency depends on the inductance and the capacitance of the tank circuit, LC, and therefore varies in accordance with the changing capacitance of the capacitor microphone.

If the sound waves vibrate the microphone diaphragm at a low frequency, the oscillator

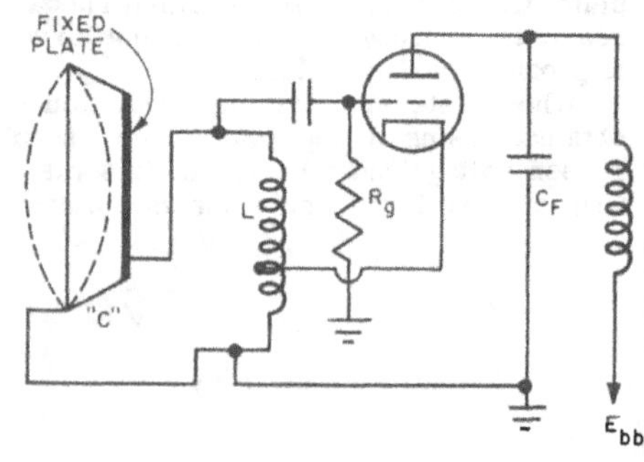

179.425

Figure 23-1.—FM transmitter modulated by a capacitor microphone.

frequency is changed only a few times per second. If the sound frequency is higher, the oscillator frequency is changed more times per second. When the sound waves have low amplitude, the extent of the oscillator frequency change from the no signal, or resting, frequency is small. A loud AF signal changes the capacitance considerably and therefore deviates the oscillator frequency to a greater degree.

Thus, the frequency of the AF signal determines the number of times per second (RATE OF DEVIATION) that the oscillator tank frequency changes. The amplitude of the AF signal determines the extent of the tank circuit frequency change (DEGREE OF DEVIATION). A graph representing the changing tank circuit frequency at an audio rate is shown in figure 23-2.

F_r represents the resting frequency of the tank circuit when there is no modulating signal present, and the diaphragm of the microphone is motionless (in this case f_r is equal to 100 MHz). With a sinusoidal sound source of 500 Hz, the amount of deviation on either side of the rest frequency is 25 kHz and the rate of deviation is 500 Hz. On the graph this is represented by curve A.

When the amplitude of the 500 Hz sound source is increased to B (curve B on the graph), the microphone diaphragm moves farther, but still crosses its resting point at the same rate or the same number of times per second. In this case the rate of deviation remains constant but the amount of deviation increases to 75 kHz. The oscillator frequency now changes from 99.925 MHz to 100.075 MHz.

It is also possible to keep the amount of deviation constant but vary the rate of deviation. In a practical FM transmitter the two independent variables, rate and amount of deviation, are continually changing. This occurs because the amplitude and frequency of the modulating signal continually changing.

Figure 23-3 shows the actual tank circuit output voltage compared, with respect to time, to the modulating signal as might be seen on an oscilloscope. The horizontal axis represents a linear time base. The degree of compression or rarefaction is dependent on the amplitude of the audio signal. The number of compressions or rarefactions per second would be determined by the audio frequency. Thus, it is seen how RF energy can carry intelligence in its frequency changes.

ESTABLISHING FREQUENCY MODULATION

Previously (in chapter 5) it was shown how a d.c. control voltage could change the frequency of an oscillator by use of a varactor. A more practical method of obtaining FM is by use of a frequency modulated oscillator, which will be controlled by an audio signal rather than by a d.c. control voltage. Figure 23-4 contains a block diagram of a frequency modulated oscillator, showing the input and output waveforms.

A typical frequency modulated oscillator is shown in figure 23-5. The frequency modulation is established by use of a reactance modulator. The

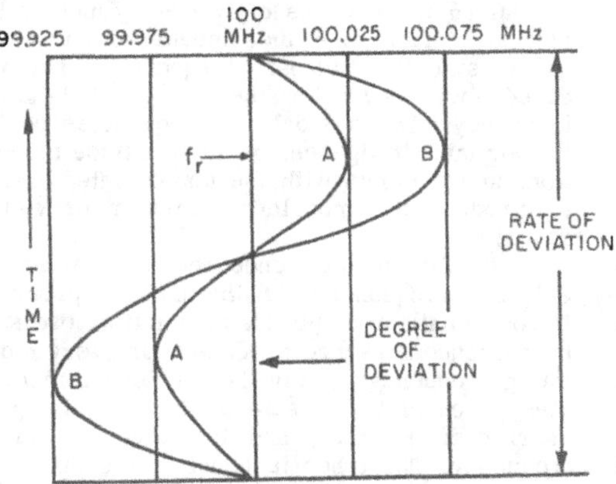

179.426

Figure 23-2.—Deviation changes in FM.

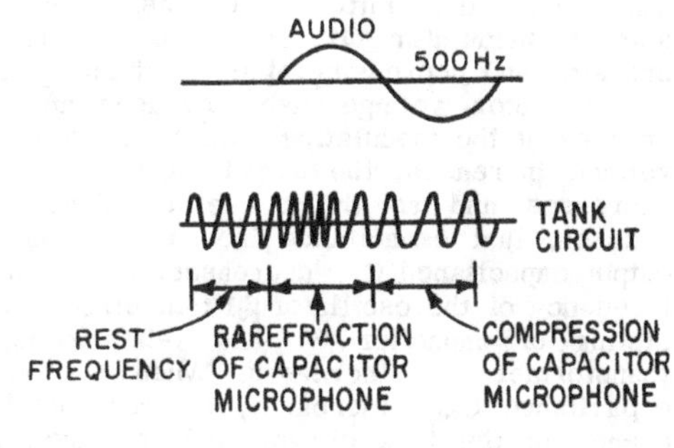

179.427

Figure 23-3.–FM waveforms of a capacitor microphone modulator.

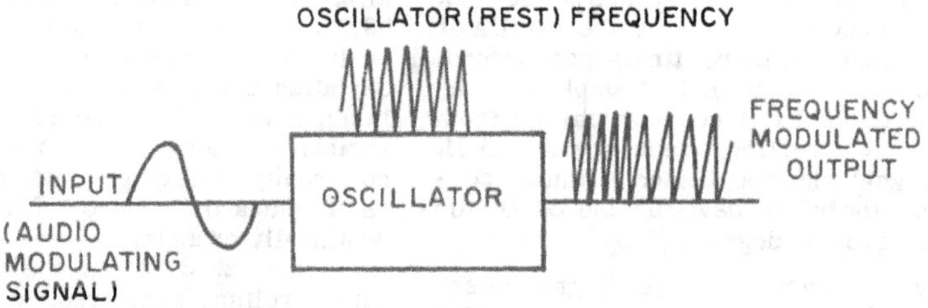

OSCILLATOR (REST) FREQUENCY
INPUT (AUDIO MODULATING SIGNAL)
OSCILLATOR
FREQUENCY MODULATED OUTPUT

179.428
Figure 23-4.—Block diagram showing the basic process of frequency modulation.

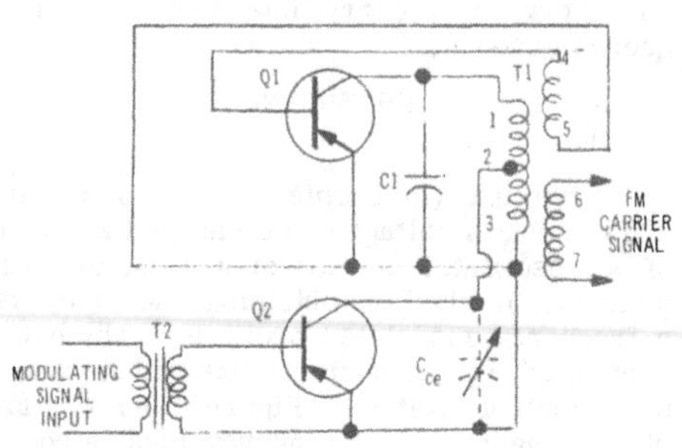

179.429
Figure 23-5.—Oscillator circuit, frequency modulated by reactance modulation.

modulating signal, coupled through transformer T2, varies the emitter-base bias of the reactance modulator Q2. Since the bias is increasing and decreasing at the modulating rate, the collector voltage also increases and decreases at the modulating rate. As the collector voltage increases, the output capacitance C_{ce} decreases and as the collector voltage decreases, that capacitance increases. When the output capacitance C_{ce} decreases, the resonant frequency of the oscillator Q1 tank circuit (consisting of capacitor C1 and 1-3 winding of transformer T1) increases. When the output capacitance C_{ce} increases, the resonant frequency of the tank circuit will decrease. The resonant frequency of the oscillator tank circuit is therefore increasing and decreasing at the modulating rate. Thus, the frequency of the signal generated by the oscillator is increasing and decreasing at the modulating rate. The output of the oscillator is therefore a frequency modulated carrier signal.

Transistor Q1 provides the oscillator signal. Capacitor C1 and winding 1-3 of transformer T1 form a parallel resonant circuit at the oscillator frequency. Winding 4-5 of transformer T1 provides the required feedback and winding 6-7 couples the oscillator signal to the following stage. Transformer T2 couples the modulating signal to the reactance modulator Q2. The reactance of the output capacitance C_{ce} across winding 2-3 of transformer T1 varies the resonant frequency of the oscillator tank circuit.

FM SIDEBANDS

During the process of frequency modulation, just as during amplitude modulation, new frequencies, called sideband frequencies, are produced above and below the unmodulated carrier frequency. These sideband frequencies contain the signal intelligence, as in amplitude modulation, and combine with the unmodulated carrier to produce the modulation carrier previously described.

A significant difference between AM and FM sideband frequencies is the number produced. If you recall, in amplitude modulation, two sideband frequencies are produced for every modulating frequency. One of these sideband frequencies is equal to the sum of the modulating and carrier frequencies, and is above the carrier frequency. The other is equal to the difference between the modulating and carrier frequencies, and is below the carrier frequency. In FM, each modulating frequency produces a similar pair of sum and difference sideband frequencies.

However, in addition to the basic pair, a theoretically infinite number of additional sideband frequencies are produced. These additional frequencies are equal to whole number multiples of the basic pair.

For example, if a 100 MHz carrier is frequency modulated by a single frequency of 10 kHz, the basic pair of sideband frequencies will be 100.01 MHz and 99.99 MHz. The additional frequencies will be 100.02 & 99.98 MHz, 100.03 and 99.97 MHz, 100.04 and 99.96 MHz and so on (fig. 23-6). Although theoretically these sideband frequencies extend outward from the carrier indefinitely, only a limited number of them contain sufficient power to be significant. Even so, this limited number is always far greater than the number produced by comparable amplitude modulation.

The number of significant sideband frequencies produced in any particular case depends on the amplitude and frequency of the modulating signal. The larger the amplitude or the lower the frequency of the modulating signal, the greater is the number of significant sideband frequencies. The exact number can be found using a ratio called the MODULATION INDEX. You will learn about this shortly.

You recall that in amplitude modulation, the amplitude of the sideband frequencies, or the power contained in them, was independent of the amplitude of the unmodulated carrier, and depended only on the amplitude, or power, of the modulating signal. In FM, the situation is different. The sidebands derive their power from the carrier, which means that the unmodulated carrier component of an FM wave has less power, or smaller amplitude, after modulation than it does before modulation. The amount of power removed from the carrier and placed in the sidebands depends on the modulating frequencies and the maximum deviation of the carrier. It is possible, under certain conditions, for the carrier power to be zero, with all the power contained in the sidebands. This, of course, is quite desirable, since the carrier itself contains no intelligence.

The amplitude of the individual sideband frequencies depend on the modulation index, explained later. The pattern of the individual amplitudes is highly irregular. There is no continuous increase in amplitude as frequencies go further from the carrier, nor is there a continuous decrease. However, in all cases, there is a point relatively distant from the carrier where the amplitudes of the sideband frequencies drop below 1 percent of the amplitude of the unmodulated carrier. Past this point, the sidebands are insignificant, and can be ignored.

FM BANDWIDTH

In the chapter on AM, you learned that the term bandwidth meant the entire range of frequencies in a modulated wave. Because of the many sideband frequencies contained in a FM wave, bandwidth when applied to FM is more restrictive. It includes only the significant frequencies. The bandwidth of an FM wave is the frequency range between the extreme upper and the extreme lower sideband frequencies whose amplitudes are 1 percent or more of the unmodulated carrier amplitude. Since these sideband frequencies are multiples of the modulating frequency, you can see that the bandwidth of an FM wave can be many times greater than that of an AM wave.

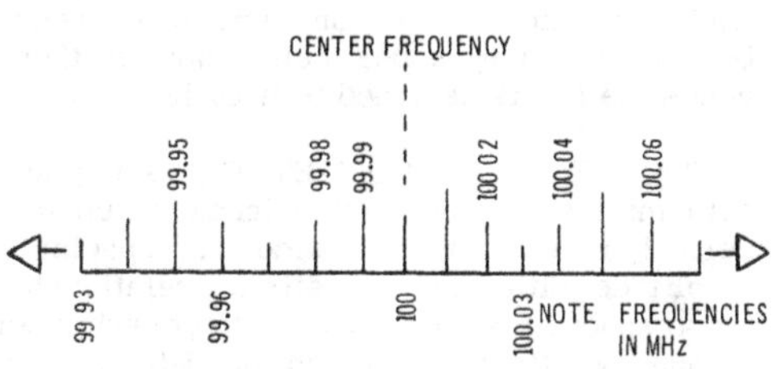

179.430
Figure 23-6.—FM sidebands.

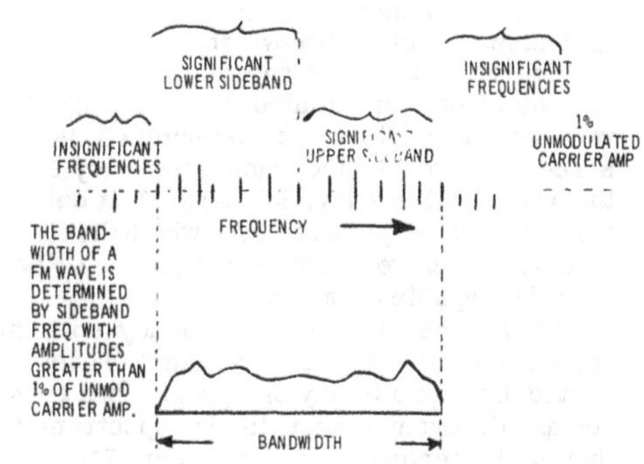

179.431
Figure 23-7.—Spectrum of modulated wave (wideband FM).

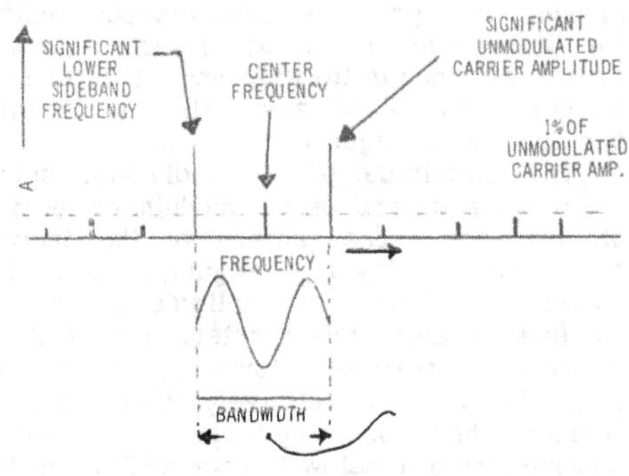

Figure 23-8.—Spectrum of modulated wave (narrow-band FM).

When an FM wave has a very wide bandwidth, it is called WIDEBAND FM, and requires the use of carrier frequencies much higher than those used for AM carrying similar intelligence. These high carrier frequencies are necessary so that a maximum number of FM waves can be transmitted by radio without interferring with each other. In wideband FM, many of the sideband frequencies are included in the bandwidth, as shown in figure 23-7.

It is possible, by limiting the maximum deviation of the FM carrier, to produce FM having the same bandwidth as an AM wave carrying the same intelligence. This is called NARROW-BAND FM (fig. 23-8). While this process causes some distortion of the intelligence, it allows carrier frequencies to be used that are lower than some required for wideband FM.

You should understand at this point that the bandwidth of a modulated wave is important for two reasons: first, it determines how much space or room in the radio frequency spectrum the wave will occupy; and second, it determines the range of frequencies over which the electronic circuits used to receive and process the wave must be capable of operating.

As far as the radio frequency spectrum is concerned, all of the modulated waves transmitted by radio in any one geographic area must occupy different places in the spectrum or else they will interfere with each other. For example, the lower frequencies of a modulated wave with a 100 MHz carrier and an 8 kHz bandwidth would overlap and interfere with the upper frequencies of a 99 MHz carrier with an 8 kHz bandwidth. You can see that interference between radio waves can be avoided either by reducing bandwidths or by moving carrier frequencies farther apart.

If bandwidths are made too narrow, though, distortion of the intelligence carried by the wave will result, since many of the sidebands, which contain the intelligence, will be eliminated. On the other hand, if carrier frequencies are too far apart, a very limited number of radio waves would completely fill the radio spectrum.

MODULATION INDEX

In FM, the modulation index is the ratio of the carrier's frequency deviation to the modulating frequency. Hence:

$$M = \frac{F_d}{F_m}$$

where: M = modulation index
F_d = frequency deviation of the carrier
F_m = frequency of the highest modulating signal

A carrier's frequency deviation is the amount of "swing" or deviation, in kilohertz or megahertz, caused by modulation with audio voltages.

For example, a maximum deviation of 75 kHz on the frequency bands up to 108 MHz can be interpreted in terms of percent by referring to this carrier swing as a limit and saying; that 75 kHz is 100 percent modulation when the carrier is allowed to swing to this maximum.

Overmodulating, or overdeviating, will occur if the maximum carrier swing is exceeded. Transmission is not particularly affected, but audio distortion may result; moreover, the bandwidth or slice of the communication frequency band occupied by a particular carrier then exceeds the limits assigned to it by law.

The BESSEL ZERO METHOD is a system of determining the modulation index which is concerned more with the sideband spacing than amplitude differences. Definite relationship of these spacings with respect to the carrier center frequency offers a method of determining the index.

Specifically, the carrier sideband components vary cyclically in amplitude the farther they progress from the center frequency or carrier, which disappears completely for certain values of M, such as 2.405, 5.52, 8.654, etc... At these points none of the transmitted power is contained in the sidebands. This fact and the distribution of energy of other sideband frequencies are disclosed by a mathematical series known as Bessel's functions—hence the name Bessel zero. The technique of measurement calls for increasing the amplitude of the modulating signal until the carrier of the FM wave disappears. The first disappearance point occurs where 2.405 is substituted for M in the previously given equation. The frequency deviation can then be calculated:

$$F_d = 2.405\ F_m$$

To measure and adjust an FM wave, the carrier being analyzed must first be monitored by a receiver so that disappearance of the received sidebands can be detected as the amplitude of the modulating signal is increased.

For purposes of discussion, suppose we desire to limit a transmitter to a maximum frequency deviation of 25 kHz. The modulating frequency that will produce carrier disappearance at an index of 2.405 equals 10,395 hertz (F_m equals F_d divided by M and F_d equals 25 kHz and M equals 2.405). An audio signal of this frequency when applied to the FM modulator and gradually increased in amplitude, will eventually extinguish the carrier. In the audio measurement setup this arrangement produces a steady lessening of the beat note until it becomes inaudible. It is this audio amplitude, then, that produces 25 kHz deviation, and this value must be constantly maintained through constant carrier operation for maximum efficiency and confinement within prescribed limits.

If a large frequency deviation is to be established, we repeat the above computation and find that F_m equals 75/2.405 or 31.2 kHz. Since this frequency is too high for audio stages we utilize the second carrier disappearance point, 5.52, which leads to the relationship:

$$F_m = \frac{75}{5.52} = 13.583 \text{ kHz}$$

and provides a frequency within the audio bandpass of most FM transmitters.

Using this special form of mathematics, tables like the one in table 23-1 show the relationship between the modulation index, and the bandwidth and number of significant sideband frequencies of an FM wave. As an example of how to use the table, consider a modulating frequency of 1 kHz that causes a maximum carrier deviation of 7 kHz. The modulation index is:

$$\frac{7 \text{ kHz}}{1 \text{ kHz}} = 7$$

So there are 11 significant sidebands in the wave, and the bandwidth is 22 x 1 kHz, or 22 kHz.

Table 23-1.—Modulation index, number of sidebands and bandwidths

MODULATION INDEX	NUMBER OF SIDEBAND FREQUENCIES ABOVE AND BELOW REST FREQUENCY	BANDWIDTH F-MODULATING FREQUENCY
0.5	2	4 X F
1	3	6 X F
2	4	8 X F
3	6	12 X F
4	7	14 X F
5	8	16 X F
6	9	18 X F
7	11	22 X F
8	12	24 X F

179.433

PERCENT OF MODULATION

To explain 100 percent modulation in an FM system, it is desirable to first review the same conditions for an AM wave. As has been stated, 100 percent modulation (AM) exists when the amplitude of the envelope varies between zero and twice its normal unmodulated value. There is a corresponding increase in power of 50 percent. The amount of power increase depends upon the degree of modulation, and because the degree of modulation is constantly varying under dynamic conditions, the tubes cannot be operated at maximum efficiency continuously.

In frequency modulation, 100 percent modulation has a different meaning. The AF signal varies only the frequency of the oscillator. Therefore, the tubes operate at maximum efficiency continuously and the FM signal has a constant power input at the transmitting antenna, regardless of the degree of modulation. A modulation of 100 percent simply means that the

carrier is deviated in frequency by the full permissible amount. For example, an 88 MHz FM station has 100 percent modulation when its audio signal deviates the carrier 75 kHz above and 75 kHz below the 88 MHz carrier, when this value is assumed to be the maximum permissible frequency swing, as is the case for commercial stations under the FCC. For 50 percent modulation, the frequency would be deviated 37.5 kHz above and below the resting frequency.

CHAPTER 24

RECEIVER AND TRANSMITTER TROUBLESHOOTING

The technician who approaches all troubleshooting problems in a logical manner locates troubles faster than a technician who tries in a haphazard or aimless way (easter egging). It is impossible to anticipate all possible troubles that might occur in the many complex systems aboard ships and aircraft. A service manual provides most of the information required to work through a troubleshooting problem. However, there are six basic steps that are very useful when troubleshooting. These are: Symptom Analysis; Equipment Inspection; Signal Tracing or Substitution; Voltage and Resistance Measurements; Transistor and Tube Testing; and Basic Tests.

SYMPTOM ANALYSIS is a study of trouble symptoms used to isolate the trouble to as few components of an equipment as possible. In a radio receiver, low sound output, distortion, or hum would be trouble symptoms.

An EQUIPMENT INSPECTION is an inspection using only the senses; that is, sight, smell, hearing, and touch to locate the cause of trouble in an equipment. A few things that might be uncovered during an equipment inspection are, for example, a charred resistor, arcing, open fuse, power cord not connected, or equipment not in the "ON" position.

SIGNAL TRACING is tracing the path of a normal signal which is passing through the equipment in order to isolate the section of the equipment where the trouble has developed. SIGNAL SUBSTITUTION is injecting a test signal at various points in the equipment and observing the resulting output of the equipment in order to isolate the trouble.

VOLTAGE AND RESISTANCE MEASUREMENTS are measurements of the operating voltages and the resistance (with power off) at various points in a single stage of an equipment after the trouble has been isolated to that stage. Because voltage and resistance measurements require only a V.T.V.M. or a multimeter and are easy to perform, some technicians have a tendency to try to isolate a trouble almost entirely by voltage and resistance measurements instead of the above troubleshooting procedure. Considering the hundreds, and sometimes thousands, of measurements that today's complex equipment would require, this is not always practical from a time standpoint.

TRANSISTOR TESTING may be done by substitution or with a transistor tester. When testing transistors, remember, certain precautions must be observed or the transistor might be permanently damaged. These precautions were described in chapter 6 of this manual. TUBE TESTING is one of the most misused troubleshooting procedures. Quite often, a technician will test tubes as the first step in troubleshooting. In a 5 or 6 tube receiver, this may not be a waste of time. However, most Navy equipment contains many more than 5 or 6 tubes. A complete radar or communications system contains between 30 and 1500 tubes; a computer might contain many thousands of tubes. It would be completely impractical from a time standpoint to attempt to test all of these tubes.

BASIC TESTS provide the most reliable indication of how well the equipment is working. Basic tests are measurements and observations taken to determine agreement or management with previously established references or standards for any portion of an electronic equipment for the purpose of treating and correcting malfunctioning.

Technical diagrams also play an important part in efficient troubleshooting. Functional Block Diagrams show the stage or stages function (by name and waveforms). Servicing Block Diagrams include voltages and test points. Wiring Diagrams show interconnecting cables between physically separated units. The information contained on wiring diagrams aids in finding the best place to connect test equipment in order to obtain voltages and waveforms shown on the servicing diagrams. A Complete Schematic, as the name implies, shows the entire circuitry of a particular equipment and may or may not show voltages or waveforms. Stage Schematics deal with one specific stage rather than the whole system.

RECEIVER TROUBLESHOOTING

The following illustrates the use of the preceding steps, in efficient troubleshooting of a superheterodyne radio receiver. Trouble symptoms indicate no output from the speaker when the volume control is adjusted to mid point and the tuner is rotated through its range. Symptom Analysis of these symptoms would indicate that the trouble could be anywhere within the receiver. To isolate the trouble in this instance would require proceeding to the next step.

The next step would be "Equipment Inspection." Deenergize and remove the radio receiver from its case, or pull it out on slides if rack mounted, and use the second troubleshooting step "Equipment Inspection." Visually inspect all connections, and check for odors. The odor of varnish is a sign of an over-heated resistor while the odor of rotten eggs is that of a shorted selenium rectifier. Look for melted tar on transformers. With the power on, look for tubes glowing red and look and listen for arcing. If nothing is found during this inspection proceed to the next troubleshooting procedure "Signal Injection and Signal Substitution."

To perform this step, more information is required such as:

1. How many stages does this receiver contain?
2. What are the functions of these stages?
3. What path or paths does the signal travel?

With the aid of the Functional Block Diagram these questions can be answered. The Functional Block Diagram of a receiver is shown in figure 24-1.

From the Functional Block Diagram of the receiver, it can be seen this receiver has eleven stages and that the signal travels from left to right. Also, it indicates what each stage does. For example, V4 is the 1st Intermediate Frequency Amplifier. Information contained on the Functional Block Diagram gives an overall understanding of the receiver's makeup.

The need now arises for location of test points, circuit voltages and waveforms. This specific information is found on the Servicing Block Diagram shown in figure 24-2.

The Servicing Block Diagram, figure 24-2, illustrates test points (TP1, TP2, etc.), waveforms found at test points (such as those found in the IF Section), sectional division of the receiver (indicated by the dotted lines), and power supply voltages.

Since Signal Substitution will be utilized, only a function generator will be required. The signal source (function generator) is turned on and allowed to warm up. The warm up period permits the generator's circuits to stabilize. Also, at this time, the volume control on the receiver is rotated to its mid-range position. A quick voltage check of the receiver's power supply will ensure that proper operating voltages are available.

After the normal warm-up period, logical troubleshooting can begin. As seen in figure

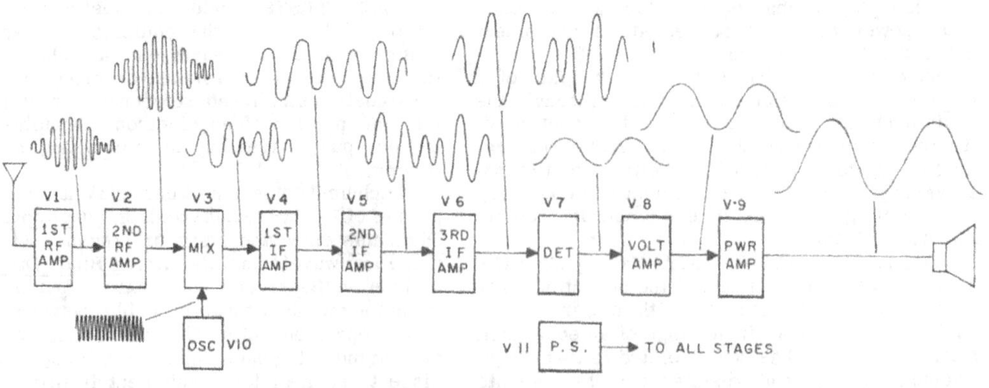

179.434

Figure 24-1.—Functional block diagram.

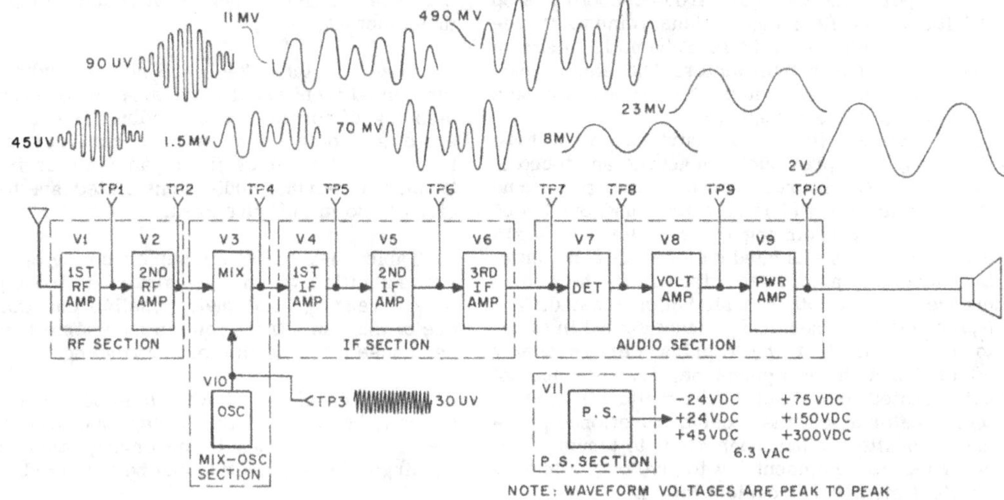

179.435

Figure 24-2.—Servicing block diagram.

24-2, ten of the eleven stages are associated with the signal. Since it is very unlikely that more than one stage is not operating, it is logical to start in the middle of the signal stages. The signal source is adjusted for a modulated output (type of signal source is dictated by point of insertion) and set to the receiver's intermediate frequency. This frequency can be found in the receiver's technical manual. When the signal source is connected to the mid-point test point (TP5) an output should be heard from the speaker.

If the proper output is heard from the speaker, it would logically indicate stages V5, V6, V7, V8 and V9 were operating properly and that the trouble was in one of the other five stages. The signal source would then be connected to the mid-point of the five remaining stages, and this procedure would be repeated until the trouble is located.

If no output was heard when the signal source was connected to test point TP5 the trouble must be in the remaining stages. The source signal is then connected to the mid-point of these five stages (V5, V6, V7, V8 and V9), which is test point TP8. Note: Test point TP7 could have been used. The signal source must now be adjusted to an audio frequency since TP8 is located in the audio section of the receiver.

A sound output at this time would indicate that the non-operating stage is one of the previous stages (V5, V6, or V7). If there is no output, then the signal source is connected to TP9. No output from the speaker would indicate stage V9 is inoperative. An output would indicate that stage V8 is not functioning.

For the purpose of explanation it is assumed an output is heard when the signal source is connected to TP9. As stated in the above paragraph, an output would indicate that stage V8 was not functioning.

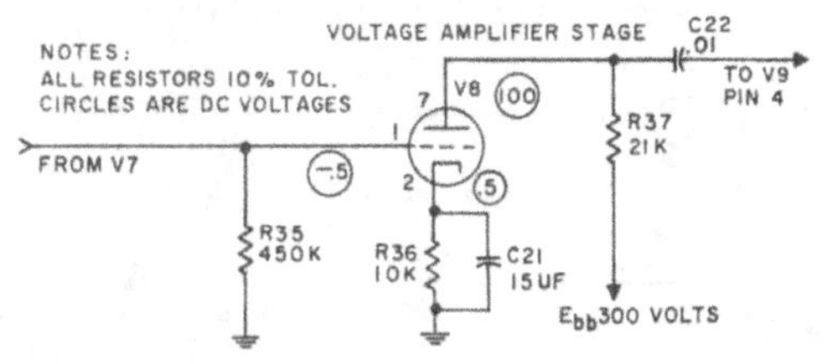

179.436

Figure 24-3.—Stage schematic diagram.

To perform the next troubleshooting step ''Voltage and Resistance Measurements,'' requires the use of a Stage Schematic Diagram (fig. 24-3), and a multimeter. The stage schematic diagram indicates the proper voltages and resistances for stage V8.

V8's plate voltage is measured and found to be zero. E_{bb} was previously measured and found to be 300 volts. The receiver is deenergized, and the resistance of R37 is measured and found to be infinite. After replacing the resistor with a 21k ohm resistor as indicated on the stage schematic, the tube and associated circuit is checked to ensure that no other malfunction existed. The equipment was then performance tested in order to determine that the repairs were properly effected and the equipment performance had not deteriorated as a result of the malfunction. Once this is determined, the receiver functioned properly. No attempt has been made to troubleshoot any specific equipment but to give only a logical efficient method of troubleshooting.

TRANSMITTER TROUBLESHOOTING

Visual checks and signal tracing, which were employed to troubleshoot receivers, are not used as such when troubleshooting Navy transmitters. Because of the high voltages present in transmitters, it is not safe to work with them while they are energized. Therefore, an alternate method is provided by means of meters mounted on the transmitter's front panel. These front panel meters are permanently wired (or may be switched) into various transmitter circuits, and most troubles can be isolated using only these meters.

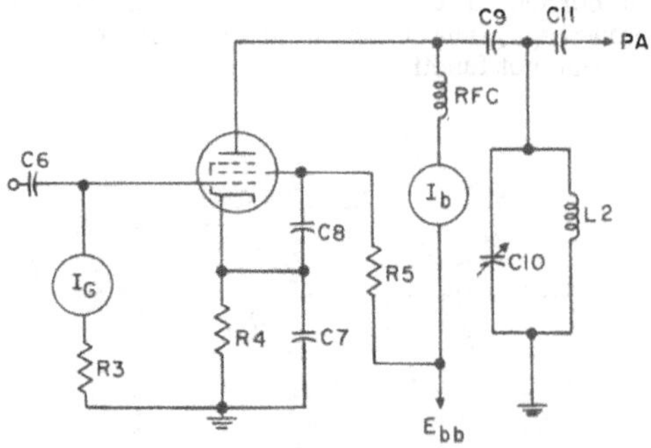

179.437

Figure 24-4.—Transmitter amplifier stage.

Listed in table 24-1 are normal indications for typical front panel meters. A thorough understanding of normal meter readings is necessary, since any determination of a malfunction may first be indicated by front panel meter indications. The normal indications listed are for the circuit shown in figure 24-4.

Table 24-2 shows a troubleshooting chart for a transmitter which lists abnormal front panel meter readings and their possible causes. Any one of the items in the column of possible troubles can cause the abnormal meter reading.

The user of the chart should be aware of all the possible troubles that could cause an abnormal meter reading. Then by observing other meter readings, the malfunction can be localized.

SAFETY PRECAUTIONS

Some transmitter tubes in the Navy operate with very high plate voltages and it is, therefore,

Table 24-1.—Normal indications for front panel meters

METER	NORMAL INDICATIONS	IMMEDIATE CONCLUSIONS
D.C. GRID CURRENT	CURRENT RISES TO NORMAL VALUE WHEN PLATE CIRCUIT OF PREVIOUS STAGE IS TUNED	1. RF SIGNAL PRESENT ON THE GRID 2. COUPLING CIRCUIT OK 3. D.C. GRID CIRCUIT COMPLETE TO GROUND 4. CATHODE EMITTING
D.C. PLATE CURRENT	CURRENT DIPS TO NORMAL VALUE WHEN PLATE CIRCUIT IS TUNED	1. RF PRESENT IN PLATE CIRCUIT 2. D.C. PLATE CIRCUIT COMPLETE TO POWER SUPPLY 3. TUBE OPERATING NORMALLY

179.438

Table 24-2.—Transmitter troubleshooting chart

METER	SYMPTOMS	POSSIBLE TROUBLE
D.C. GRID CURRENT	ZERO READING	1. NO GRID DRIVE 2. NO RF IN PLATE OF PREVIOUS STAGE 3. BAD TUBE
	LOW READING	1. WEAK GRID DRIVE 2. MISTUNED PLATE CIRCUIT OF PREVIOUS STAGE 3. BAD TUBE
D.C. GRID CURRENT	HIGH READING	1. OPEN SCREEN DROPPING RESISTOR (NO SCREEN VOLTAGE) 2. NO PLATE VOLTAGE
D.C. PLATE CURRENT	ZERO READING	1. BAD TUBE 2. NO GRID DRIVE (FIXED BIAS, BELOW CUTOFF) 3. OPEN SCREEN DROPPING RESISTOR (FIXED BIAS, BELOW CUTOFF) 4. NO PLATE VOLTAGE 5. OPEN FILAMENT RETURN TO GROUND
	LOW READING	1. WEAK TUBE 2. OPEN GRID LEAK RESISTOR
	HIGH READING	1. NO GRID DRIVE (NO FIXED BIAS) 2. MISTUNED PLATE CIRCUIT

179.439

dangerous to reach inside a transmitter to measure voltages or to change a part while the power is on. This may seem obvious to you but unless you are aware of the danger, an accident is possible.

Navy transmitters are designed to protect you from accidents that could happen through carelessness. They have access doors through which you can remove bad tubes. If you open one of these doors when the transmitter is on, the power should turn off automatically. The doors operate switches, called INTERLOCKS, which are connected between the on-off switch and the circuits of the transmitter. When an access door is open, the interlock is opened and no power is applied to the transmitter. Thus, opening the access door shuts down the transmitter. Because of the important role played by an interlock, unauthorized shorting out of an interlock is dangerous and forbidden.

There is still a possibility of a serious shock—even when you do turn the power off before reaching into a transmitter—a fault in the circuit may have prevented a capacitor from discharging and your body may provide a discharge path.

As a safeguard against this, a shorting bar, consisting of a metal rod with a wooden handle, is used. The metal end is connected with a copper braid to the ground of the transmitter. Before putting your hand inside, probe about with the shorting rod, touching every point which might conceivably be charged. If nothing is charged, this procedure costs you about 30 seconds; if something is charged, the shorting rod discharges it and the procedure saves you from injury or a severe jolt.

CHAPTER 25

SINGLE SIDEBAND

It has been previously shown that an amplitude modulated (AM) radio frequency signal may be considered to consist of a carrier, an upper sideband, and a lower sideband. The frequency spectrum analyzer presentation of an RF carrier, amplitude modulated by a single audio frequency, is illustrated in figure 25-1. It should be recalled that if a more complex waveform such as speech, was used to modulate the carrier the sidebands would be composed of many frequencies. A single audio modulating frequency will be used for simplicity of explanation.

The intelligence, carried by an amplitude modulated signal, is contained in both sidebands. The amplitude of the intelligence is represented by the relative amplitude of either sideband, and the frequency of the intelligence is represented by the frequency difference between the carrier and either sideband. Since the intelligence contained in one sideband is a duplicate of the intelligence contained in the other sideband, only one sideband is required for communication. The other sideband may be eliminated by the use of filtering. As the carrier is totally unnecessary for the transmission of intelligence, it too may be filtered out. Such communication depends, however, upon the reinsertion of the carrier at the receiver in order to acquire the proper demodulated frequencies. This system is properly referred to as Single Sideband Suppressed Carrier (SSBSC) communiction.

A frequency spectrum analyzer presentation of a single sideband signal is illustrated in figure 25-2. In the presentation shown, the upper sideband has been chosen for transmission. In practice, either sideband may be transmitted.

SSB ADVANTAGES AND DISADVANTAGES

A comparison of figures 25-1 and 25-2 indicates one advantage of SSB over AM. An SSB signal occupies a relatively small portion of the frequency spectrum in comparison to the bandwidth of the AM signal. Therefore, many SSB signals may be present in a small portion of the frequency spectrum, without interference with each other.

SSB transmitters are rated in terms of peak envelope power (PEP). PEP is the average power of the transmitter divided by the fraction of each second that an output is actually produced. Normally, it is considered that the carrier power of an AM transmitter must be twice the PEP of an SSB transmitter for comparable operation. This is somewhat misleading in that the PEP of an SSB transmitter is based on a duty cycle of .5. (Duty cycle is the fraction of each second that a device is actually producing an output.)

The output stages of an SSB transmitter may be operated class AB or class B. They are frequently said to be more efficient than the final power amplifier of an AM transmitter operating class C. This is because it is assumed that the power output stages of an SSB transmitter are actually operating 50% of the time (.5 duty cycle), or only when modulation is occurring.

The power comparison between SSB and AM, given in the preceding paragraphs, is based on ideal propagation conditions. However, an SSB signal is more consistent than an AM signal under conditions of adverse propagation.

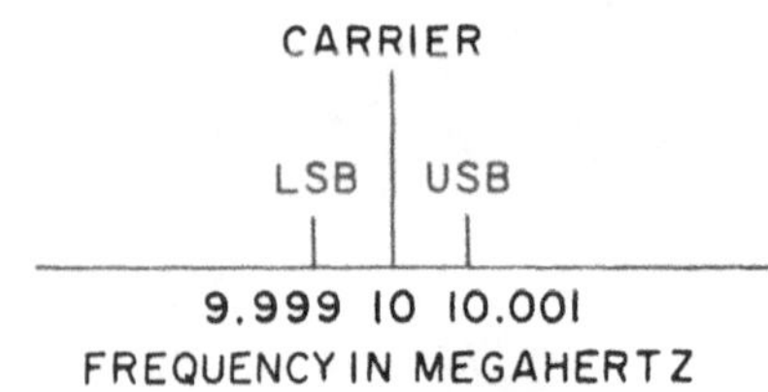

179.440

Figure 25-1.—AM spectrum analyzer presentation.

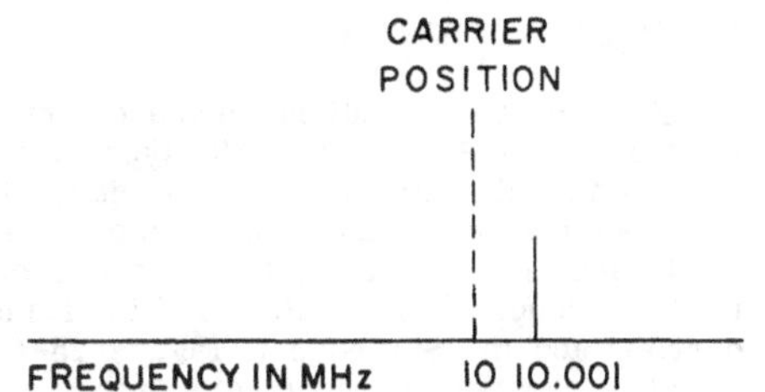

179.441
Figure 25-2.—SSB spectrum analyzer presentation.

SSB communication systems do have several disadvantages. The process of producing an SSB signal is somewhat more complicated than simple amplitude modulation. Secondly, frequency stability is much more critical in SSB communication. Additionally, while there is not the annoyance of heterodyning from adjacent signals, a weak SSB signal may be completely masked or hidden from the receiving station by a stronger signal. Also, a carrier of proper frequency and amplitude must be reinserted at the receiver.

SSB RECEIVERS

Figure 25-3 illustrates the block diagram of a basic SSB receiver. A basic SSB receiver is not significantly different from a conventional superheterodyne AM receiver. However, a special type of detector, and a carrier reinsertion oscillator, must be used. The carrier reinsertion oscillator must insert a carrier in the detector circuit at a frequency which corresponds almost exactly with the relative position of the carrier in the original spectrum.

The filters used in SSB receivers serve several purposes. As was previously stated, many SSB signals may exist in a small portion of the frequency spectrum. Filters supply the selectivity necessary to adequately receive only one of the many signals which may be present. They may also select USB or LSB operation, as well as reject noise and other interference.

The oscillators in a SSB receiver must be extremely stable. In some types of SSB data transmission a frequency stability of plus or minus 2 hertz is required. For simple voice communication a deviation of plus or minus 50 hertz may be tolerable.

SSB receivers may employ additional circuits which enhance frequency stability, improve image rejection, or provide automatic gain control (AGC). However, the circuits contained in the basic receiver of figure 25-3 will be found in all single sideband receivers.

CARRIER REINSERTION

The need for extreme frequency stability may be understood if one considers the fact that a small deviation in local oscillator frequency, from the correct value, will cause the IF produced by the mixer to be displaced from its correct value. In AM reception this is not too damaging, since the carrier and sidebands are all present and will all be displaced an equal amount. Therefore, the relative positions of carrier and sidebands will be retained. However, in SSB reception, there is no carrier present in the incoming signal.

The carrier reinsertion oscillator frequency will be set to correspond to the IF frequency of the carrier—if a carrier were present. The carrier reinsertion frequency is determined by the local oscillator frequency. For example,

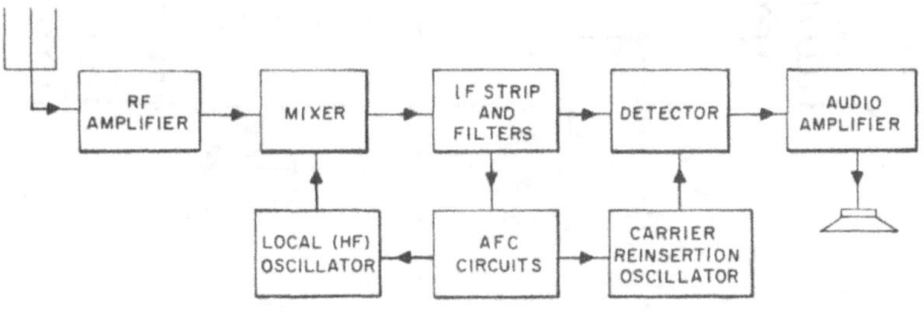

179.442
Figure 25-3.—Basic SSB receiver.

assume that a transmitter, with a suppressed carrier frequency of 3 MHz, is emanating a USB signal. Also assume that the intelligence consists of a 1-kHz tone. The transmitted frequency will be 3,001 kHz. If the receiver has a 500-kHz IF, the correct local oscillator frequency should be 3,500 kHz. The IF output of the mixer would be 499 kHz. If a carrier were present, it would correspond to an IF frequency of 500 kHz. Therefore, the carrier reinsertion oscillator frequency should be 500 kHz, in order to preserve the frequency relationship of the carrier and sideband at 1 kHz.

If the local oscillator frequency drifts to 3500.5 kHz, then the IF output of the mixer will become 499.5 kHz. However, the carrier reinsertion oscillator would still be operating at 500 kHz. This will result in an incorrect audio output of .5 kHz. If the intelligence transmitted was a complex signal, such as speech, it would be unintelligible, due to the displacement of the side frequencies caused by the local oscillator deviation.

Even with the correct local oscillator frequency, distortion may occur due to a shift in carrier reinsertion oscillator frequency. This would also cause a displacement in the relative positions of the carrier and sideband.

The carrier reinsertion oscillator may be any type of stable oscillator. It is tuned to the correct frequency of the IF bandpass, as was previously described. Usually a crystal oscillator will be used, as this type of oscillator provides good stability.

PRODUCT DETECTOR

Although a conventional diode detector may be used for SSB reception, the type of detector most commonly employed is the product detector. It is so called because under ideal operating conditions, its output amplitude is proportional to the product of the amplitude of the reinserted carrier and the SSB signal. Figure 25-4 illustrates a transistor produce detector.

Transistors Q1 and Q2 form a balanced mixer circuit. The bias for these transistors is obtained from the voltage divider formed by R1 and R2 and is applied to the bases of Q1 and Q2 through the secondary of transformer T1. The emitter operating voltage is applied to Q1 and Q2 through emitter resistors R6 and R7.

The IF signal is applied to the base of each transistor, 180° out of phase, by transformer T1. The carrier reinsertion oscillator signal is coupled, in phase, to the emitters of Q1 and Q2, through capacitors C2 and C3 respectively. Resistors R4 and R5 provide isolation between the emitters of Q1 and Q2.

The IF signal and the reinserted carrier are heterodyned in the transistors. The resulting output consists of the sum and difference of the original input frequencies, as well as the two original frequencies. Capacitors C4 and C5 bypass the sum and original frequencies to ground. Since the circuit is balanced, the outputs from transistors Q1 and Q2 (developed across transformer T2) are 180° out of phase with each other.

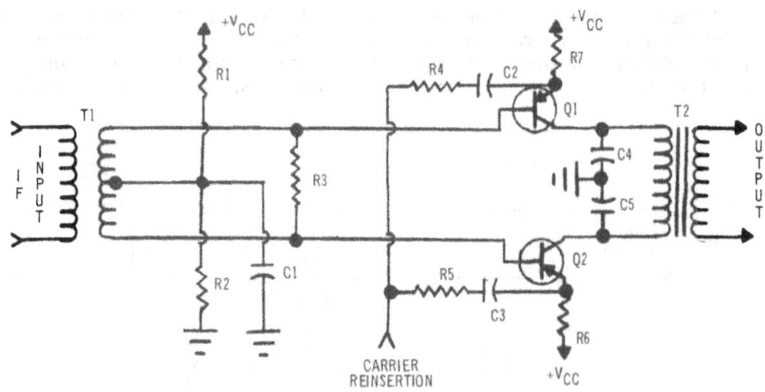

Figure 25-4.—Transistor product detector.

This results in additional cancelling of the reinserted carrier, because it was applied in phase to the emitters of the transistors. Transformer T2 has an audio frequency response that will attenuate any of the RF signals not previously cancelled. The difference between the two input signals is the desired intelligence, and this difference is developed across transformer T2.

Figure 25-5 illustrates an electron tube product detector. The sideband signal from the IF amplifier is applied to the control grid of V1 through transformer T1. The reinserted carrier is applied to the suppressor grid. The two input signals are heterodyned within the tube, and the output will contain frequencies equal to the sum and difference of the input signals, as well as the original input frequencies. All frequency components, with the exception of the difference frequency, will be attenuated by the low pass filter made up of C5, L1, and C6. The difference frequency, which is the desired audio signal, will be developed across R4.

It should be noted, that in both detectors the reinserted carrier amplitude must be appreciably greater than the sideband IF amplitude. A carrier too small in relative amplitude would result in distortion of the output signal.

BASIC SINGLE SIDEBAND TRANSMITTER

A functional block diagram of a basic SSB transmitter is shown in figure 25-6. The audio amplifier is of conventional design. Audio filtering is not required because the highly selective filtering which takes place in the SSB generator attenuates the unnecessary frequencies below 300 Hz and above 3 kHz. The input signal may be any desired intelligence signal and may cover all or any part of the frequency range between .1 and 6 kHz. The upper limit of the input audio signal is determined by the channel bandwidth and upper cutoff frequency of the filter in the SSB generator. The lower limit of the input audio signal is determined by the lower cutoff frequency of the filter in the SSB generator.

The SSB generator produces the SSB signal at an intermediate frequency (IF). To produce the SSB a double sideband (DSB) signal is generated and passed through a highly selective filter to reject one of the sidebands. The signal is generated at a fixed IF frequency because highly selective circuits are required. The highly selective filter requirements for the filter method of SSB generation are met by either crystal or mechanical filters.

The generated SSB signal at a fixed IF is passed through mixers and amplifiers where it is converted to the transmitted RF. Two stage conversion is shown with the second conversion frequency being a multiple of the first conversion frequency. The frequency conversions required to produce the RF frequencies produce sum and difference frequencies as well as higher order mixing products inherent in mixing circuits. However, the undesired difference frequency or sum frequency, along with the higher order mixing products, is attenuated by interstage tuned circuits.

Because a SSB system without a pilot carrier demands an extremely stable frequency system,

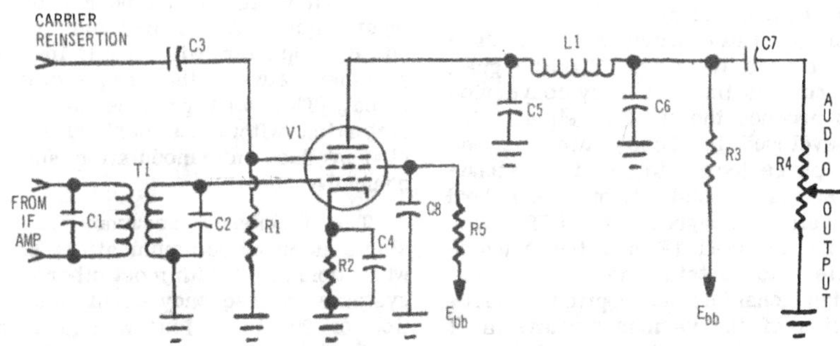

179.444

Figure 25-5.—Electron tube product detector.

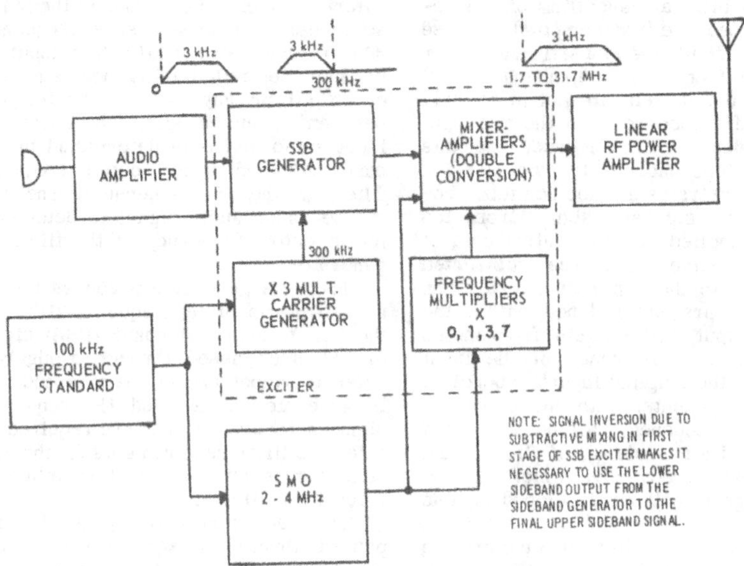

179.445

Figure 25-6.—Basic block diagram of an SSB transmitting system.

the frequency standard and stabilized master oscillator (SMO) are extremely important. The standard frequency is obtained from a crystal oscillator with the crystal housed in an oven. Since the stability of the crystal frequency depends upon the stability of the oven temperature, stable temperature control of the oven is necessary. This thermal control of the oven is obtained by using heat-sensitive semiconductors in a bridge network. Any variation in the oven temperature is indicated and corrected by an unbalance in the control bridge.

The carrier generator provides the IF carrier used to produce the fixed IF SSB signal, and the SMO provides the necessary conversion frequencies to produce the RF SSB signal. The frequencies developed in these units are derived from or phase locked to the single standard frequency so that the stability of the standard frequency prevails throughout the SSB system. Choice of the fixed IF and the conversion frequencies to obtain the RF is an important design consideration. Optimum operating frequencies of the various circuits must be considered as well as the control of the undesirable mixing products. The use of harmonically related conversion frequencies in the mixer permits the frequency range to be covered with a single 2- to 4-MHz oscillator, a very practical range for obtaining high oscillator stability. The 300-kHz fixed IF is the optimum operating frequency for the filter required in the SSB generator.

The SSB exciter output drives a linear power amplifier to produce the high power RF signal. A linear power amplifier is required for SSB transmission, because it is essential that the plate output RF signal be a replica of the grid input signal. Any non-linear operation of the power amplifier will result in intermodulation (mixing) between the frequencies of the input signal. This will produce not only undesirable distortion within the desired channel but will also produce intermodulation outputs in adjacent channels.

The frequency accuracy requirements for single sideband communications are very precise when compared with most other communications systems. A frequency error in carrier reinsertion of 20 Hz or less will give good voice reproduction. Errors of only 50 Hz result in noticeable distortion; this is considered the maximum error allowable.

There are significant frequency errors introduced by the propagation medium and by doppler shifts due to relative motion between transmitter and receiver in aircraft communications. In high frequency sky wave transmission, the doppler shift caused by the motion of the ionosphere introduces a frequency shift of several hertz. Doppler shift due to relative motion amounts to one part in 10^6 for every 670 miles per hour (mach 1) difference in velocity between the transmitting and receiving station. At a carrier frequency of 20 MHz, communicating from a jet aircraft to ground, the frequency shift will be approximately 20 Hz. Inasmuch as this represents approximately half of the maximum error, the error introduced by the transmitting and receiving equipment, due primarily to oscillator drift, must be comparatively small. This dictates a design goal in the vicinity of $\pm 1/2$ part in 10^6 in both ground and aircraft installations.

Present day trends demand that communications be established on prearranged frequencies without searching a portion of the spectrum in order to obtain communications, and therefore, the figure of ± 0.1 part in 10^6 presents the required absolute accuracy rather than short term stability. Most military and some commercial applications demand that operation be obtained on any of the seven thousand SSB voice channels in the HF band. A channel frequency generator having an absolute accuracy of ± 0.1 part in 10^6 and providing either continuous coverage or channelized coverage in steps no greater than 4 kHz is required in many SSB systems.

FREQUENCY SYNTHESIZER

In order to meet the requirements stipulated in the preceding paragraph a device known as a frequency synthesizer has been developed. By general definition a frequency synthesizer is a circuit in which a signal frequency is produced by heterodyning and otherwise combining frequencies not necessarily harmonically related to each other or the frequency produced.

Throughout the years many variations of frequency synthesizers have been developed. It is our intent here to select and explain the operation of a frequency synthesizer that best represents current operational units.

In the most current frequency control circuits a frequency synthesizer is used to provide a reference signal to control the frequency of a variable frequency master oscillator. Such a circuit has come to be known as a stabilized master oscillator (SMO). The output signal from the SMO is further processed with mixers and frequency multipliers (see fig. 25-7).

Figure 25-7 is a functional block diagram of a SSB transmitter that emphasizes the action of the RF frequency multipliers and mixers in the process of frequency snythesis. Note that figure 25-7 is similar to figure 25-6. However, it contains a more detailed display of the mixer and multiplier sections. To explain the action of the mixer-multiplier section we will assume an arbitrary signal frequency from the SMO of 2 MHz and a 300-kHz signal from the SSB generator. Switch S1 will be in position 1 as shown. The 2-MHz signal from the SMO is fed through a buffer amplifier, which serves to isolate the SMO from any load variations, to the subtractive mixer and multiplier stages. Also at the input to the subtractive mixer is the 300-kHz signal from the SSB generator. The subtractive mixer mixes the 2-MHz and 300-kHz signals, and their difference frequency, 1.7 MHz, is coupled to position 1 of S1-A through an amplifier. With S1 in position 1 the 1.7-MHz signal bypasses the additive mixer and is fed directly to the linear RF amplifier. Notice that in position 1 of S1, no output is selected from the multiplier stages.

Next, we will maintain the same signal frequencies from the SMO and SSB generator, but switch S1 is in position 2. The output from the subtractive mixer will remain at 1.7 MHz, since its inputs did not change, and again will be fed to the input of the additive mixer. However, in position 2 the additive mixer receives a second input from position 2 of S1-C, which is connected directly to the output from the buffer amplifier. In this case (S1 in position 2) the additive mixer receives, as its second input, the same signal frequency as the subtractive mixer (in this case 2 MHz). The additive mixer will mix the 1.7-MHz and 2-MHz signals, and their sum (3.7 MHz) will be coupled to the linear RF amplifier.

Again maintaining the same signal frequencies from the SMO and SSB generator but placing switch S1 in position 3, the first input to the additive mixer will remain 1.7 MHz since the signal frequencies to the subtractive mixer did not change. However, the second input to the additive mixer is now being received from the X3 multiplier. The output from the X3 multiplier will be the third harmonic of its input, in this case 6 MHz. At the inputs to the additive mixer we then have a 1.7-MHz and 6-MHz signal. The output from the additive mixer will be the sum

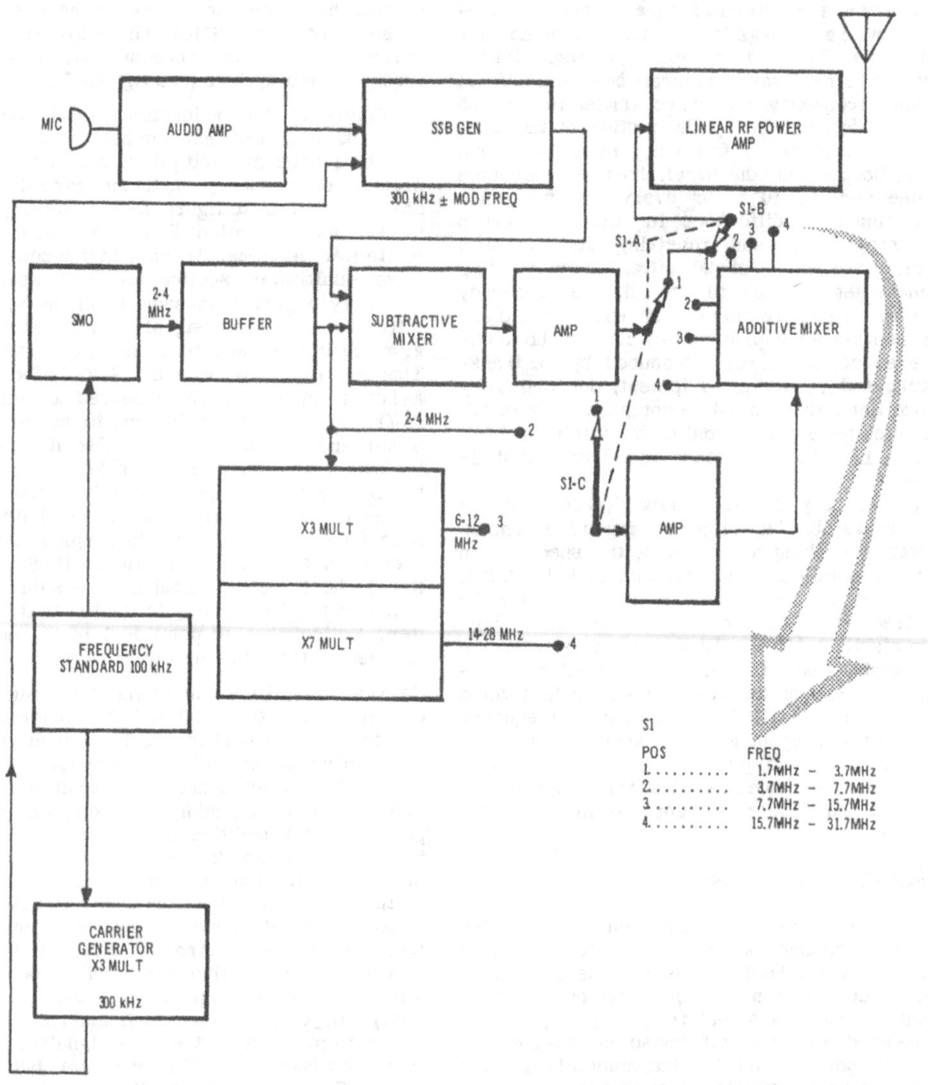

179.446

Figure 25-7.—Functional block single sideband transmitter.

of its inputs, in this case 7.7 MHz, which is fed to the linear RF amplifier.

Placing S1 in position 4 and again maintaining the same signal frequencies from the SMO and SSB generator, the first input to the additive mixer will remain 1.7 MHz since the frequency of the signal inputs to the subtractive mixer remained the same. The second input to the additive mixer will be from the X7 multiplier. The X7 multiplier is receiving the same 2 MHz signal being fed to the subtractive mixer. The output from the X7 multiplier will be the 7th harmonic of its input, 14 MHz in this case. The output from the additive mixer will be the sum of these two inputs, 15.7 MHz in this case, which is fed to the linear RF amplifier.

It should now be clear that with one signal frequency from the SMO, four output frequencies can be obtained from the transmitter. The output from the SMO is variable from 2 to 4 MHz in increments of 500 Hz (4000 possible frequencies). Since for each possible frequency from the SMO you can obtain 4 output frequencies, the total channel capabilities of this transmitter would be 16,000 different frequencies.

STABILIZED MASTER OSCILLATOR

The master oscillator circuit is a conventional series fed Hartley oscillator composed (see fig. 25-8) of V2, R4, R5, R6, R7, R8, C6, C7, C8, C9, CR10, L1, and CR3. CR10 is a Varicap, a semiconductor diode with inherent capacitance which is variable and dependent upon the degree of reverse bias. The degree of reverse bias of CR10, and, therefore, its capacitance, depends on its anode potential which is directly coupled from the output of the phase discriminator circuit and its reverse bias which is fixed by zener diode CR3. The capacitance of CR10 along with C6, C7 and L1 make up the tank circuit of the oscillator and determine its frequency of operation. Since CR10's capacitance can be controlled by the phase discriminator, it holds that the operating frequency of the oscillator can be controlled by the phase discriminator. The other components in the oscillator serve the following functions: R8—load resistor for CR3, C8 and R4—grid coupling and grid leak bias, R7 along with R6—plate load resistors, R7 along with R5—screen load resistors, C9—RF bypass capacitor, C11—coupling and d.c. blocking. C7 while serving as part of the tank capacitance serves as a d.c. blocking capacitor for CR10's bias potential.

The action of CR10 is as follows: CR10's reverse bias, as fixed by CR3, is such that its anode potential will never become great enough to overcome it, so CR10 will never conduct. As the output from the phase discriminator becomes more positive it decreases the reverse bias of CR10 and increases its capacitance. This increase in capacitance will lower the oscillating tank's frequency.

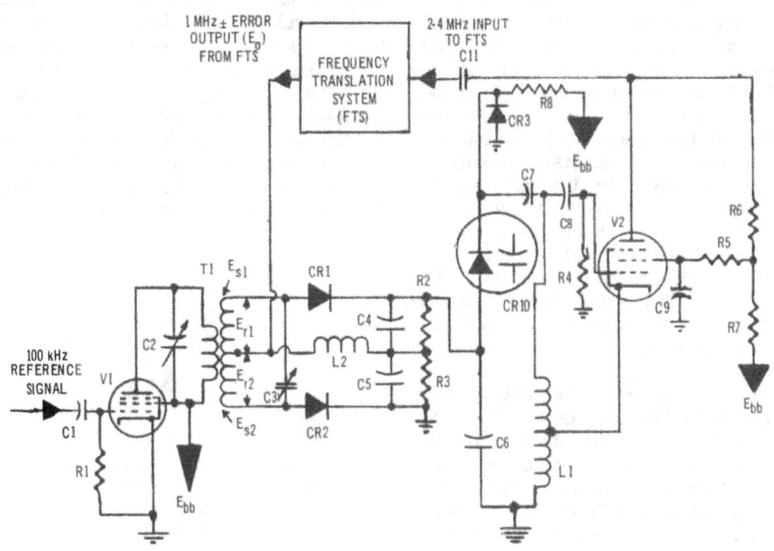

Figure 25-8.—Stabilized master oscillator.

In order for the output from the phase discriminator to become more positive, the oscillator's tank frequency had to originally increase. When the oscillator's tank frequency increased it caused the phase discriminator's output to become more positive which in turn decreased the tank frequency to its original value.

Magnetostriction Oscillator

The magnetostriction oscillator is usually used at audio or very low radio frequencies to provide a highly stabilized sine wave output. It is used in preference to the crystal type of oscillator at the very low frequencies because of its simplicity and economy due to the lack of suitable quartz crystals.

MAGNETOSTRICTION EFFECT.—The magnetostriction effect is similar to the piezoelectric effect found in crystal oscillators. Instead of using electric charges, however, it operates by the effect of a changing magnetic field. When an iron-alloy is placed within a magnetic field, there is a change in length due to the strain placed on the rod by the magnetic field. This compressional strain, in effect, squeezes the rod and makes it longer; when the field is removed, the rod returns to nearly its former length. Similarly, when a rod located within a magnetic field changes its length, it also induces a change in the magnetic field, increasing or decreasing the field. The induced change is dependent upon the original direction of the magnetic field and the polarization of the metal rod and its composition. When the change in length of the bar is performed at the resonant (mechanical) frequency of the bar, and the induced change is properly phased to enchance the field that produces the strain, mechanical oscillations are set up at the fundamental frequency of the bar. When clamped in a fixed position at the middle, the bar will vibrate with a flexural motion similar to a tuning fork.

The metal composition of the bar determines its efficiency and effectiveness as a resonator. Temperature effects will cause changes in the bar length and thus affect the frequency of operation; hence, for extreme stability, the alloy must have a small temperature coefficient or temperature control must be used. Operation in this respect is similar to the operation of crystal oscillators.

CIRCUIT OPERATION.—The basic circuit of the magnetostriction oscillator is shown in figure 25-9. Grid bias is obtained conventionally through grid-leak operation ($C_g - R_g$), and series plate feed is used. The coils are wound so as to produce the same flux at the plate end of the bar for an increasing plate current or an increasing grid current. This condition, which is opposite to that of the normal feedback oscillator, produces degenerative feedback, rather than regenerative feedback. In addition, these plate and grid coils have little or no coupling. Thus, the oscillator will normally not oscillate at all, or oscillate only feebly, as a result of feedback between the interelectrode capacitances. Capacitor C1 is the tuning capacitor, and is connected so as to tune both the plate and grid coils (either one alone may be tuned if desired). For a single frequency or frequencies within a narrow range of operation, C1 may be fixed and different lengths of rod may be inserted in the coils (similar to crystal plug-in operation) for the different frequencies.

Assume that plate voltage is applied and that the plate current is producing a steady strain on the rod. If a noise pulse occurs and produces an increase of flux in the plate coil, a compressional wave will be started at the plate end of the bar and will travel to the left toward the grid end. This compressional wave due to magnetostriction will travel through the bar in a manner similar to the propagation of sound waves through a metal bar. When the compressional wave reaches the grid coil, L_g, the lengthening of the bar induces a positive voltage in the grid coil, which is applied to the grid and causes the plate current to increase. The increased plate current induces a stronger field around the plate coil, inducing another compressional wave into the

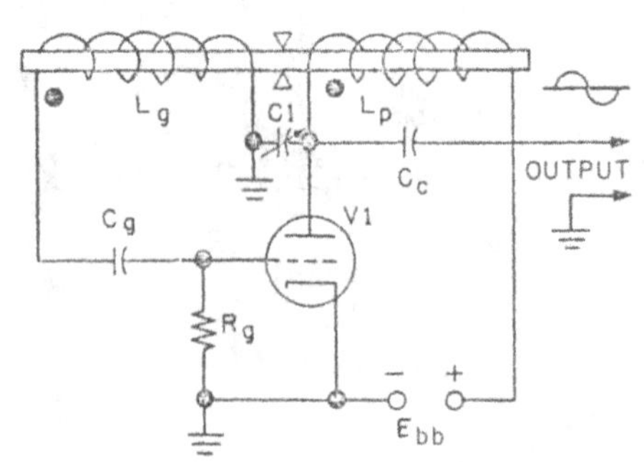

Figure 25-9.—Magnetostriction oscillator.

bar. The compressional waves in the bar are reflected from the grid end and travel back to the plate end, where they again are reflected back toward the grid end. When the compressional wave which is reflected from the grid or left end of the bar reaches the right or plate end, a voltage is induced in plate coil L_p by the lengthening of the bar. This voltage, by induction, creates a stronger field around the plate coil. Consequently, the motion of the plate end of the bar is further reinforced, causing it to vibrate more strongly. As a result, another compressional wave is started and the cycle repeats. When the induced and reflected waves are in phase, the grid-reflected wave arriving at the plate end will always reinforce the induced wave at the plate end, producing a stronger oscillation. Therefore, the length of time it takes for the wave to travel from one end of the bar to the other and return will determine the phasing of the reflected and induced waves. For each length of bar, there will be a specific time required for a wave to travel to the end and return; if the length is made equivalent to an electrical half-wave length, each wave will reinforce the other. When the frequency of the tuned circuit (as adjusted by C1) is approximately the same as the resonant frequency of the bar, maximum reinforcement will occur and maximum mechanical vibration will be produced. With the bar initially unpolarized, operation will occur at the second harmonic. Therefore, the bar is usually permanently magnetized and inserted into the coils so that the field of the plate coil increases the polarization. Operation then takes place at the fundamental frequency.

PHASE DISCRIMINATOR

To follow the phase discriminator operation, it will first be shown without any input from the master oscillator. The 100 kHz reference signal from the frequency standard (fig. 25-7) is coupled through network C1-R1 (fig. 25-8) to the grid of V1. V1 is an amplifier-multiplier. The plate tank circuit of V1, C2 and the primary of T1, is tuned to the 10th harmonic of the input signal (1 MHz). This 1-MHz signal will be inductively coupled to the tuned secondary (secondary T1 and C3) of T1. The 1-MHz signal across the secondary of T1 will be used as the reference for the phase discriminator circuitry. The signal voltage across the secondary of T1, E_{s1} in respect to E_{s2}, will have a mutual phase displacement of 180 degrees. This is shown graphically and vectorially in figure 25-10A. Diodes CR1 and CR2 will conduct alternately but equally through R2 and R3, and in this case will be equal and essentially d.c. due to the filtering action of components C4, C5, and L2. The potential from the top of R2 to ground will be zero volts d.c. (disregarding the slight positive potential due to the bias voltage applied to the SMO) since the ground side of R3 is at the same d.c. potential as the top of R2. With this zero volt potential applied to the anode of CR10, its reverse bias is at its reference potential (a value at which the capacitance is such as to cause the tank circuit to oscillate at a frequency which will cause a 1 MHz output from the frequency translation system).

Now consider the effect of applying the second input to the phase discriminator (the output from the master oscillator) to the center tap of T1's secondary, through the frequency translation system. This signal will always be equal to the reference signal (1 MHz) with the system in a stable condition. This 1 MHz signal in a stable condition will equal the reference signal frequency, but be 90 degrees out of phase. Since it is applied to the center tap of T1 it will be exactly 90 degrees out of phase with both E_{s1} and E_{s2}. The signal potential from the center tap to either E_{s1} or E_{s2} will be composite signals, equal to the sum of the reference signal and the output from the translation system. These composite signals are labeled E_{r1} and E_{r2}, in figure 25-10. This is shown graphically and vectorially in figure 25-10B. E_{r1} is the anode potential for CR1 and E_{r2} is the anode potential for CR2. As can be seen by the vector and graphic representation of figure 25-10A the vector sums E_{s1} and E_o (E_o = oscillator signal applied to the center tap of T1 through the translation system), and E_{s2} and E_o are equal. Since equal voltages are applied to the diodes CR1 and CR2 the circuit will function in the same manner as it did with the reference signal only applied.

In order to observe how the phase discriminator stabilizes the master oscillator, the operation of the circuits will be explained in a condition where the oscillator attempts to drift off frequency. If the oscillator drifts higher in frequency it will cause the signal output from the frequency translation system (E_o) to decrease to a frequency lower than 1 MHz. Now, with a frequency applied to the center tap of T1 that is lower in frequency than the reference signals (E_{s1} and E_{s2}), the vector sums of E_{s1} $+E_o$ and E_{s2} $+ E_o$ will change. E_o will become

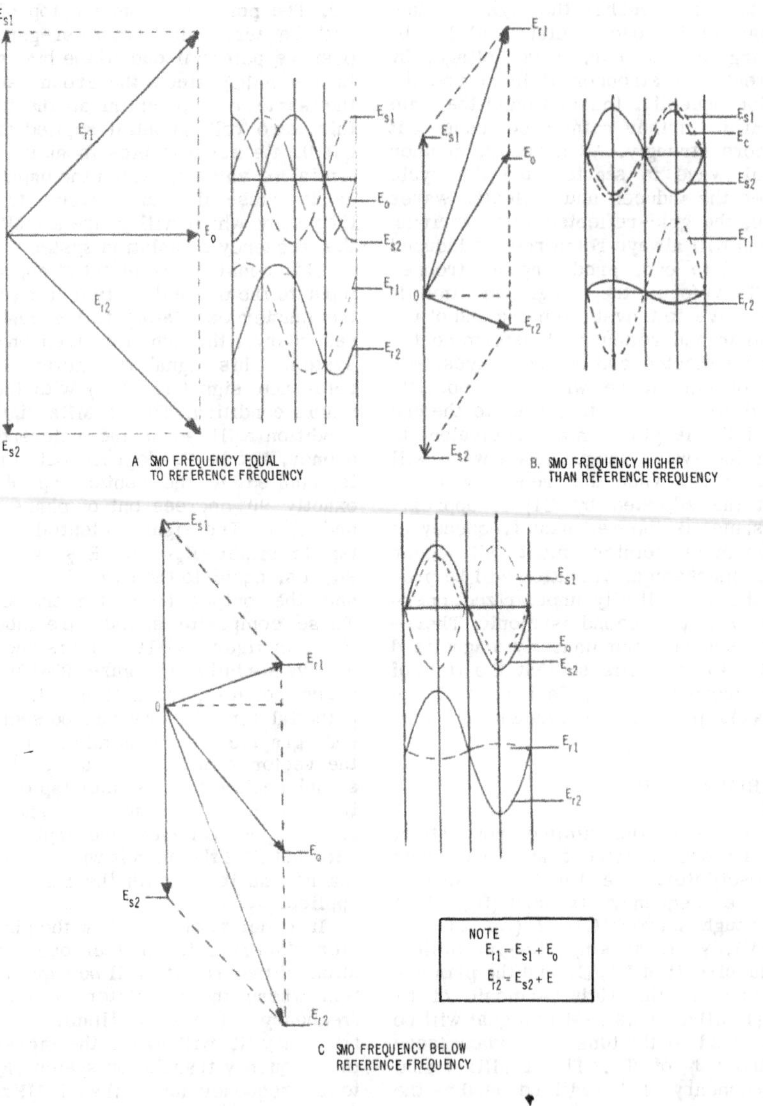

179.448

Figure 25-10. — Phase discriminator action.

more nearly in phase with E_{s1} and further out of phase with E_{s2}. This is shown vectorially and graphically in figure 25-10B. As can be seen from figure 25-10B the amplitude of E_{r2} will decrease. CR1 will now have a greater anode potential than CR2, so the potential across R2 will be greater than that across R3. The potential from the top of R2 to ground will now be positive in respect to ground. This positive potential will be felt at the anode of CR10, decreasing its reverse bias, increasing its capacitance, and thus lowering the frequency of the oscillator tank. When the frequency of the oscillator tank is again such as to cause a 1 MHz signal output from the frequency translation system, the phase discriminator will be balanced and the oscillator stabilized.

If the oscillator drifts lower in frequency it will cause the signal output from the frequency translation system (E_o) to increase to a frequency higher than 1 MHz. Now E_o will become more nearly in phase with E_{s2} and further out of phase with E_{s1}. This is shown vectorially and graphically in figure 25-10C. As can be seen from figure 25-10D the amplitude of E_{r1} ($E_{s1} + E_o = E_{r1}$) will decrease and the amplitude of E_{r2} ($E_{s2} + E_o = E_{r2}$) will increase. CR2 will now have a greater anode potential than CR1, so the potential across R3 will be greater than that across R2. The ground side of R3 will now be more positive than the top of R2 so the top of R2 will now be negative in respect to ground. This negative potential will be felt at the anode of CR10, increasing its reverse bias, decreasing its capacitance, and thus raising the frequency of the oscillator tank. When the frequency of the oscillator tank is again such as to cause a 1 MHz signal output from the frequency translation system, the discriminator will be balanced and the oscillator stabilized.

FREQUENCY TRANSLATION SYSTEM

How the frequency of the master oscillator can be varied and still remain stabilized will now be discussed. In the following explanation refer to the overall block diagram of the stabilized master oscillator circuit (SMO) that includes the block functions of the frequency translation system, figure 25-11.

The frequency translation system is a means of varying the frequency of the master oscillator throughout a range of 2-4 MHz in increments of 500 Hz. It must always be kept in mind that the output from the translation system being fed back to the phase discriminator must be 1 MHz in order for the phase discriminator to stabilize the master oscillator. If a frequency generated within the translation system is changed, the master oscillator frequency must change to maintain a 1-MHz output from the translation system. Note that the output from the master oscillator is fed through an X8 multiplier in the translation system. This is necessary due to the fact that a change of 8 Hz in the translation system causes a change of only 1 Hz in the master oscillator.

When the master oscillator is generating a 2-MHz signal oscillator number 1 must be set at 19.5 MHz and oscillator number 2 at 2.5 MHz. The output of the master oscillator (which is 2 MHz) is fed through the X8 multiplier. The output from the X8 multiplier is then 16 MHz. This 16-MHz signal along with the 19.5-MHz signal from the oscillator number 1 is fed to mixer number 1. The output from mixer number 1, as selected by the bandpass filter, will be the difference frequency of these two inputs (3.5 MHz). This 3.5 MHz along with the 2.5-MHz signal from oscillator number 2 are the inputs to mixer number 2. The output from mixer number 2, as selected by the bandpass filter, will be the difference frequency of these two inputs (1 MHz). (The outputs from both mixers shown in figure 25-10 will contain primarily the input frequencies plus their sum and difference. However, the difference frequency will be selected by the mixer's respective bandpass filters.) Since a 1-MHz signal out of mixer number 2 is being fed to the phase discriminator, the conditions have been satisfied to allow the master oscillator to operate at 2 MHz.

In order to increase the frequency of the master oscillator by 500 Hz, the frequency of oscillator number 2 must be changed to 2.496 MHz (2.5 MHz minus 4 kHz). The output from mixer number 2 at this instant, as selected by the bandpass filter, will be the difference between oscillator number 2 (2.496 MHz) and the output from mixer number 1 which is still 3.5 MHz. This difference frequency, 1.004 MHz, appears as the input to the phase discriminator. The phase discriminator sees a frequency that is higher than its reference frequency. The output from the phase discriminator will become more negative causing an increase in the master oscillator frequency (this can be related to a low frequency drift as previously discussed).

An increase of 4 kHz above 1 MHz at the input to the phase discriminator will cause the

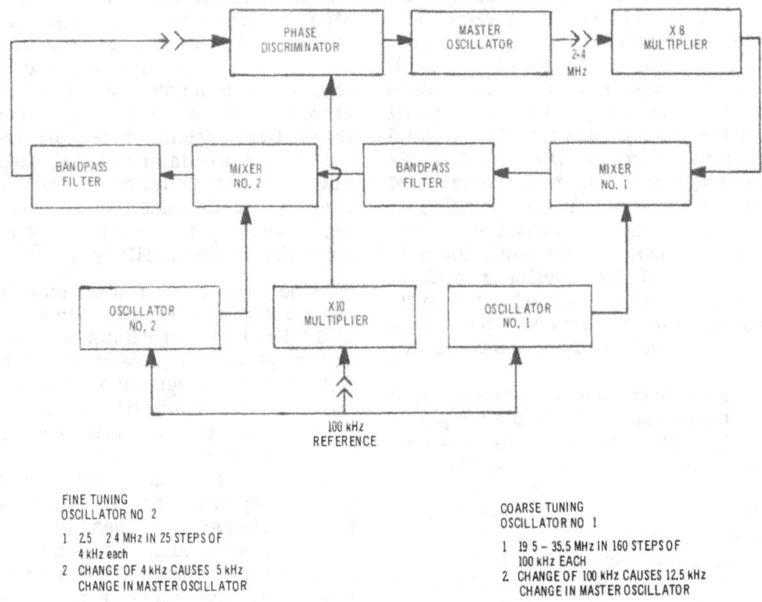

179.449

Figure 25-11.—Block diagram of stabilized master oscillator including frequency translation system.

output from the phase discriminator to go sufficiently negative to increase the master oscillator frequency 500 Hz. The output from the master oscillator is now 2,000.5 kHz. This 2,000.5-kHz signal is fed to the X8 multiplier. The output of the X8 multiplier will be 16.004 MHz. This 16.004-MHz signal will be fed to mixer number 1 along with the 19.5-MHz output from oscillator number 1. The output from mixer number 1, as selected by the bandpass filter, will be the difference frequency of these two inputs (3.496 MHz). This 3.496-MHz signal along with the output from oscillator number 2, which is now 2.496 MHz, is fed to mixer number 2. The output from mixer number 2, as selected by the bandpass filter, is the difference frequency of these two inputs (1 MHz). This 1-MHz signal is the same frequency as the reference signal to the phase discriminator, so the master oscillator will now be stabilized at 2,000.5 kHz. The operating frequency of the oscillator may be increased in this manner, in increments of 500 Hz, up to 2,012.5 kHz (25 increments). Then oscillator number 1, coarse tuning, must be utilized.

The loop analysis for varying the frequency of the master oscillator utilizing oscillator number 1, coarse tuning, is the same as for fine tuning with oscillator number 2 except that oscillator number 2 is held constant. A change of 100 kHz in oscillator number 1 will cause a change of 12.5 kHz in the master oscillator. Oscillator number 1 has the ability to vary the master oscillator 2 MHz (from 2-4 MHz) in 160 increments. To show the action of the translation system using oscillator number 1, set up the same conditions as previously. The master oscillator will be operating at 2 MHz so oscillator number 1 must be at 19.5 MHz and oscillator number 2 must be at 2.5 MHz. Next, change the output frequency of oscillator number 1 to 19.6 MHz. This 19.6-MHz signal is fed to mixer number 1 along with the output from the X8 multiplier which is 16 MHz since at this instant the master oscillator is still at 2 MHz. The output from mixer number 1, as selected by the bandpass filter, will be the difference frequency of these two inputs (3.6 MHz). This 3.6-MHz signal along with the output from oscillator number 2, which is at 2.5 MHz, will be fed to mixer number 2, as selected by the bandpass filter. The difference frequency of these two inputs (1.1 MHz) is fed to the input of the phase discriminator. The phase discriminator sees a signal that is 100-kHz higher than its reference signal of 1 MHz. The output from the phase discriminator will go negative causing an increase in the master oscillator frequency (this again can be related to a low frequency drift as previously discussed). The output from the phase discriminator will increase the master oscillator frequency 12.5 kHz. The output from the master oscillator is now 2,012.5 kHz. This 2,012.5-kHz signal is fed to the X8 multiplier. The output from the X8 multiplier will be 16.1 MHz. This 16.1-MHz signal will be fed to mixer number 1 along with the 19.6-MHz signal from oscillator number 1. The output from mixer number 1, as selected by the bandpass filter, will be the difference frequency of these two inputs (3.5 MHz). This 3.5-MHz signal along with the output from oscillator number 2 which is 2.5 MHz, is fed to mixer number 2. The output from mixer number 2, as selected by the bandpass filter, will be the difference frequency of these two inputs (1 MHz). This 1-MHz signal is the same frequency as the reference signal to the phase discriminator so the master oscillator will now be stabilized at 2,012.5 kHz.

FILTER TYPE SSB TRANSMITTER

Figure 25-12 shows the fundamental block diagram of a basic filter type single sideband transmitter. The RF carrier oscillator provides one of the inputs to the balanced modulator. The RF carrier oscillator generally operates in the 100- to 500-kHz range. It must be extremely accurate and possess a high degree of frequency stability. To meet these requirements a crystal controlled oscillator is used (the crystal is usually placed in a temperature controlled oven). Since quartz crystal is quite fragile, the oscillator must operate with a relatively low voltage. The resultant low power output necessitates low level modulation in the SSB transmitter.

The other input to the balanced modulator is the audio modulating signal. The function of the balanced modulator is to produce a double-sideband, amplitude-modulated signal and suppress the RF carrier. The output consists of only the upper and lower sidebands of the amplitude modulated carrier. The audio input frequency will be rejected by the limited bandpass of the modulator output circuit. The balanced modulator may utilize transistors, electron tubes or diodes. An example of each type will be covered in detail later in this chapter.

The double sideband output from the low frequency balanced modulator is next applied to a bandpass filter. This allows the desired sideband frequencies to pass to the following stage and rejects the undesired sideband. The design of a filter sharp enough to select the desired sideband and reject the undesired sideband would be difficult if the modulation process were performed at a high radio frequency. This is because the percentage of separation between the two sidebands would be relatively small. For greater ease of filter design, the modulation is accomplished at relatively low radio frequencies (100 to 500 kHz). The filter may be of the LC, crystal, or mechanical type. Since the principles of LC and basic crystal filters have been covered in previous chapters, only the mechanical and crystal lattice type filters will be covered in this chapter.

The RF carrier oscillator, balanced modulator, and sideband filter make up the single sideband exciter section of the transmitter. The output of the SSB exciter is a low power, low frequency single sideband signal.

The circuits following the SSB exciter are similar to those used in AM systems. Since frequency multipliers and class C power amplifiers would impair the frequency relationship between the sideband components, balanced mixers and linear power amplifiers are used to increase the frequency to the desired final value.

The HF mixer must be balanced if the HF oscillator is too close to the desired sideband output. In this case the balanced mixer operates very similarly to the balanced modulator and cancels the oscillator input. The sum and difference frequencies, and the sideband input from the exciter are separated enough in frequency so that the output tank in the mixer can be tuned to pass the desired frequencies and reject the others.

Although only one frequency translating step is shown in figure 25-12, any number of these steps may be used to reach the desired output frequency. An important fact to remember is that the frequency of the signal radiated from the antenna is not the frequency of the final HF oscillator, but the sum or difference (whichever is selected) of the oscillator and the sideband input to the final mixer. Regardless of the number of translating steps used, the frequency of the final HF oscillator is always chosen so

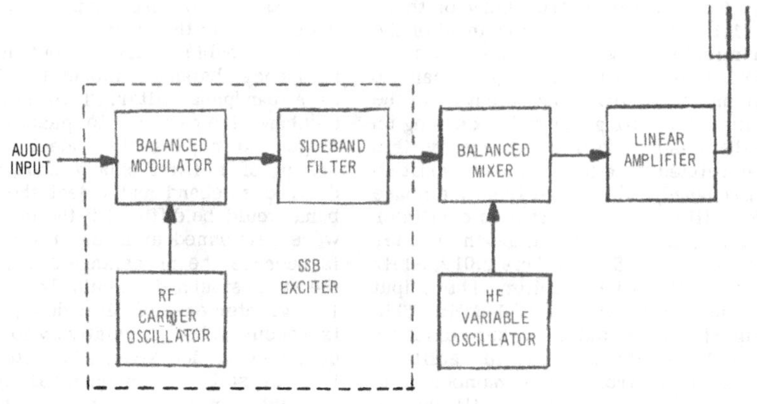

Figure 25-12.—Block diagram of basic filter type SSB transmitter.

as to place the single sideband output signal in the desired frequency range for transmission.

The linear amplifier is used to meet the low distortion and high power gain requirements of the output circuit. The linear amplifier is required to amplify the signal containing a modulation envelope with minimum distortion. Basically, a linear amplifier is one in which the output signal is a direct function of, and proportional to the input signal. In order to accomplish this, the output amplifiers in an SSB transmitter are generally class AB or class B push-pull circuits.

SSB TRANSISTOR BALANCED MODULATOR

The balanced modulator provides a double sideband output and suppresses the low frequency RF carrier. A transistor balanced modulator is shown in figure 25-13. The RF carrier input amplitude is 8 to 10 times larger than the modulating signal amplitude in order to reduce distortion. When both the RF and modulating signal are applied simultaneously, the sum and difference sidebands are developed across Q1 and Q2. Since the RF is applied to the bases in phase, the RF component of the output will cancel by the push-pull action across T2. T2 also presents a very low impedance to the modulating signal component, and this signal will not appear in the output. The generated sidebands will be out of phase with each other on the collectors because the modulating signal is out of phase on the bases. The sidebands will be coupled across the transformer as the output.

Audio transformer T1, acting as a phase inverter, couples the modulating signal to the bases. R1 and R2 provide the correct bias. Capacitor C1 decouples the AF signal and puts the T1 secondary center tap at AF ground potential. Capacitor C2 couples the 100-kHz RF from the carrier oscillator to the wiper arm on the carrier balance potentiometer R3. R3 adjusts the amplitude of the RF carrier components for equal outputs from Q1 and Q2 so the RF will be cancelled. C3 and C4 act as d.c. blocking capacitors while coupling the RF to the transistor bases. The transistors Q1 and Q2 are the nonlinear devices used to heterodyne the RF carrier and audio modulating signals. R4, bypassed by C5, is used for emitter stabilization. C6 decouples the power supply and places the center tap of T2 primary at RF ground potential.

To analyze the operation of the circuit assume first that only the RF carrier input is applied. The RF carrier from the 100-kHz carrier oscillator is coupled through C2 to the R3 wiper arm. R3 is adjusted so that the effective RF signal will be the same on both bases. When the RF is on its negative half cycle the forward bias on both transistors is increased, and both conduct harder. The voltage drop across both halves of T2 is positive going, thus equal and opposite voltages are developed. Since equal currents are flowing in opposite directions, the fields

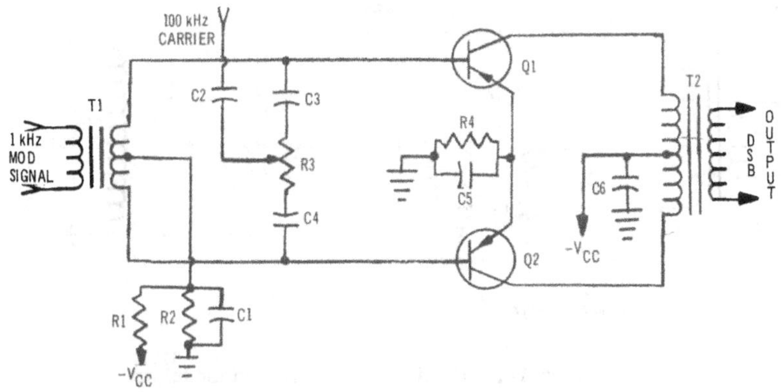

179.451
Figure 25-13.—Transistor balanced modulator.

produced cancel and no signal is coupled to the secondary. On the positive half cycle of the RF, the forward bias on Q1 and Q2 is decreased. Collector currents decrease and the voltage drop across each half of T2 is negative going. The currents are still flowing in opposite directions with equal magnitudes, and again no signal is developed in the secondary. If R3 is properly adjusted the carrier will be completely suppressed.

When the audio modulating signal and the RF carrier signal are applied simultaneously, the RF at the bases will be of the same instantaneous polarity, while the AF will be of opposite instantaneous polarity. This is due to the center tapped transformer T1 acting as a phase inverter for the AF. Capacitors C3 and C4 are of such value that they present a high impedance to the audio frequencies and thus isolate the bases as far as the audio signals are concerned. With the two signals present at each base, modulation occurs, and the upper and lower sidebands are produced.

Because the modulating signals are out of phase on the bases of Q1 and Q2 the sideband signals across T2 will be out of phase also. Push-pull action occurs and the sidebands are coupled into the secondary of T2. The audio is not developed because T2 has a very low reactance to audio frequencies. The RF carrier is suppressed as described earlier, so only the upper and lower sidebands will be present in the output of the balanced modulator.

SSB ELECTRON TUBE BALANCED MODULATOR

A balanced modulator utilizing electron tubes rather than transistors is shown in figure 25-14. This circuit produces amplitude modulated sidebands and suppresses the RF carrier in a fashion similar to that of the transistor balanced modulator.

The tubes are connected for a push-pull output. Since the RF carrier is applied to the control grids in phase it will be suppressed in the output. The RF carrier is usually 8 to 10 times larger in amplitude than the modulating signal to keep distortion at a minimum. Transformer T1 couples the audio modulating signal to the tube grids. R1 and R2 provide grid bias from E_c and effectively centertap T1. The modulating signal is therefore applied to the grids at opposite instantaneous polarities. The RF carrier is coupled through C1 to the carrier balance potentiometer, R3. R3 is adjusted to balance the RF carrier on the control grids so that the carrier will be completely suppressed in the output. Capacitors C2 and C3 couple the RF to the grids while providing AF modulating signal isolation between the grids. Resistor R4, bypassed by C4, provides common cathode bias for both tubes. The center tapped transformer T2 provides a push-pull plate load, and C5 and C6 bypass any higher order harmonics to ground. Resistor R5 is a plate voltage dropping resistor

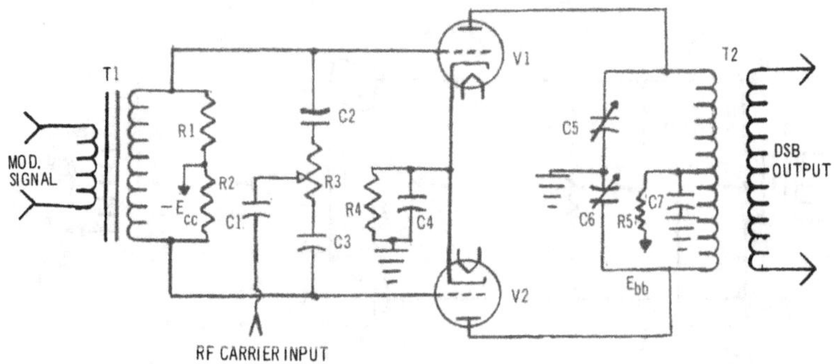

179.452
Figure 25-14.—Electron tube balanced modulator.

decoupled by C7 which places the center of T2 at RF ground.

To examine the operation of the circuit the RF carrier will first be applied alone. The RF carrier produced by the carrier oscillator is coupled through C1 and is the same instantaneous polarity at opposite ends of R3. The RF is coupled through C2 and C3 to the control grids of V1 and V2. On the positive half cycle of the RF carrier, both tubes increase conduction and the voltage drop across each half of T2 will be negative going. The equal and opposing voltages developed in the T2 primary will cancel and no output will be coupled to the secondary. On the negative half cycle of the input, both tubes decrease in conduction, the voltage at the ends of the T2 primary will be positive going, and the same cancellation takes place. If R3 is properly adjusted, the RF current in the plate circuits will be equal in amplitude and of opposite direction, resulting in carrier suppression.

When the AF modulation signal is also supplied, modulation will take place across the tubes biased for nonlinear operation and sidebands will be produced. The audio modulation signal is applied to the grids at opposite instantaneous polarities by the phase splitting action of the effectively center tapped T1 secondary. Since the audio signal is out of phase at the grids, the sidebands developed in the plate circuit will also be out of phase. Push-pull action occurs across the T2 primary, and the sideband frequencies are coupled to the T2 secondary. The RF carrier is suppressed as described earlier, and the original modulating frequency is not developed due to the low reactance of T2 to this frequency. Therefore, only the upper and lower sideband frequencies will be present in the output of this circuit.

SSB DIODE BALANCED RING MODULATOR

A balanced modulator used for the development of a double sideband output with suppressed carrier can be constructed using diodes. A balanced ring modulator with diodes is shown in figure 25-15.

To analyze the operation of the balanced ring modulator, the circuit current paths with the RF carrier alone will be presented first. Figure 25-16A and B illustrate the current paths for

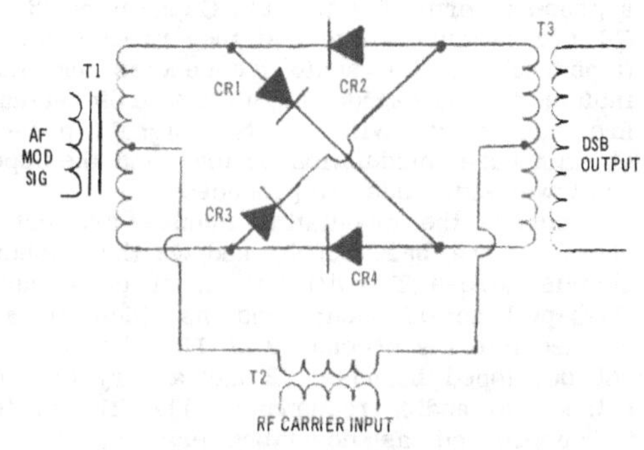

179.453
Figure 25-15.—Balanced ring modulator.

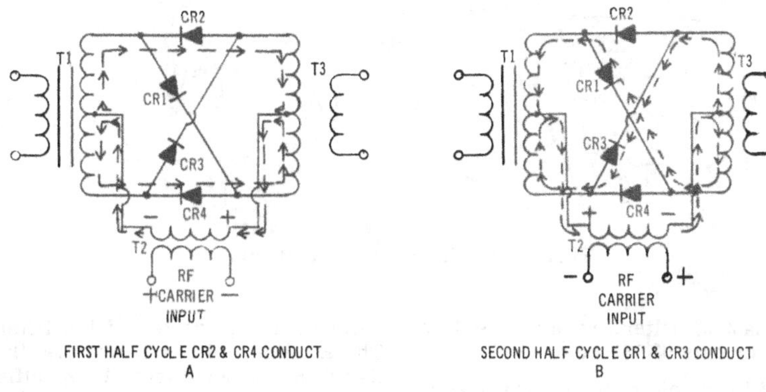

179.454
Figure 25-16.—Current paths of balanced ring modulator.

each polarity of the RF carrier input. During the first half cycle CR2 and CR4 are forward biased while CR1 and CR3 are reverse biased. Current flow is from the negative side of the T2 secondary to the center tap of the T1 secondary where current divides into equal quantities flowing in opposite directions. One path is through the upper half of T1, CR2, and the upper half of T3 to the center tap. The other path is through the lower half of T1, CR4, and the lower half of T3 to the center tap. The currents through T3 primary are equal and opposite. The RF is therefore cancelled and no output is developed. The currents combine at the T3 center tap and return to the positive side of T2 secondary.

During the second half-cycle CR1 and CR3 are forward biased while CR2 and CR4 are reverse biased. Again the currents divide in the T1 and T3 transformers, and the RF carrier is suppressed in the output.

When an audio signal is applied simultaneously with the RF carrier, modulation will occur across the nonlinear diodes, and sidebands will be produced. Since the RF carrier is approximately 10 times the amplitude of the AF modulating signal, the carrier will determine which diodes will conduct.

When the RF and AF signal are both applied, the current flows in the same directions as shown (fig. 25-16) due to the RF carrier being greater in amplitude. However, the amount of current in each half of T1 and T3 will now be unequal. During the first half cycle the current through CR2 and the upper half of T1 and T3 will increase while the current through CR4 and the lower half of T1 and T3 will decrease. This is due to the audio modulating signal further increasing the forward bias on CR2 while decreasing the forward bias on CR4. As far as the audio sideband frequencies are concerned, T3 presents enough impedance to develop the sideband frequencies and couple them to the output. The impedance is very low for the audio signal frequency, therefore, the audio signal frequency will not appear in the output. The circuit is still balanced for the RF carrier and this will be suppressed as described previously.

During the second half cycle the circuit is again unbalanced by the AF modulating signal input which causes CR1 to conduct less and CR3 to conduct harder. The sideband frequencies will be produced and coupled to the T3 secondary while the AF and RF will not appear in the output.

MECHANICAL FILTERS

SSB transmitters and receivers require very selective bandpass filters in the region of 100 kHz to 600 kHz. The filters used must have very steep skirt characteristics and a flat passband characteristic.

In previous chapters, basic crystal filters and filters made up of inductors, resistors, and capacitors have been discussed. A third type of

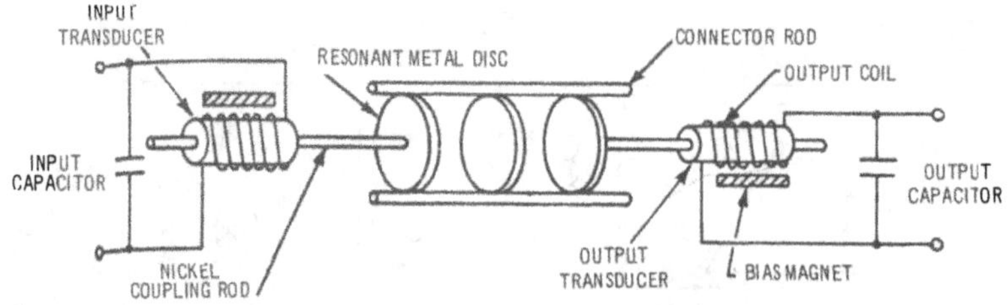

179.455
Figure 25-17.—Mechanical filter.

filter, the mechanical filter, is also used to satisfy the requirements of SSB equipment.

Mechanical filters offer many advantages over LC and crystal filters. They are small, have excellent rejection characteristics, and are comparatively rugged. The Q obtainable with a mechanical filter is much greater than that obtained with an LC filter.

Basic Mechanical Filter

Figure 25-17 illustrates a basic mechanical filter. The filter consists of an input transducer, bias magnets, resonant metal disks, nickel coupling rods, external capacitors, and an output transducer.

The input transducer converts electrical energy to mechanical energy by utilizing the magnetostrictive effect previously described. The mechanical oscillations of the input transducer are transmitted to the coupling rods and metal disks. Each disk acts as a mechanical series resonant circuit. The disks are designed to resonate at the center of the filter's passband. The number of disks determines the skirt selectivity. Skirt selectivity is specified as "shape factor", which is the ratio of the passband 60 db below peak, to the passband 6 db below peak. Practical manufacturing presently limits the number of disks to eight. A six-disk filter has a shape factor of approximately 2.2, a seven disk filter 1.85, and an eight disk filter 1.5.

The passband of the filter is primarily determined by the area of the coupling rods. Passband may be increased by using more or larger coupling rods. Mechanical filters, with bandwidths as narrow as .5 kHz and as wide as 35 kHz are practical in the 100-kHz to 600-kHz range.

A terminal disk vibrates the output transducer rod, which induces, by means of "generator action," a current in the output transducer coil. The external capacitors are used to form parallel resonant circuits with the input and output transducer coils, at the filter frequency. The equivalent circuit for the filter in figure 25-17 is shown in figure 25-18. C1

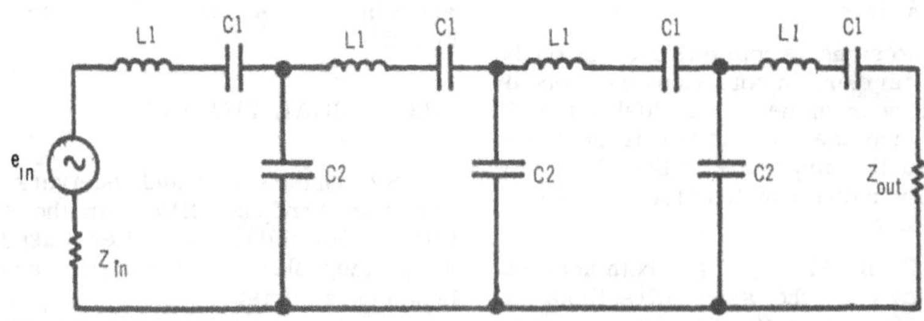

179.456
Figure 25-18.— Equivalent circuit of mechanical filter shown in figure 25-17.

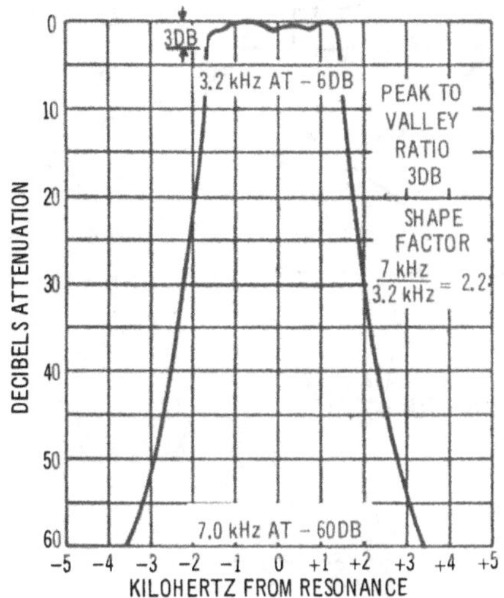

179.457

Figure 25-19.—Mechanical filter response curve.

and L1 represent a resonant metal disk; C2 represents the coupling rods.

Figure 25-19 illustrates a mechanical filter frequency response curve. Although an ideal filter would have a flat "peak" or passband, practical limitations prevent the ideal from being obtained. The term "ripple amplitude" or "peak-to-valley ratio" is used to specify the peak characteristic of the filter. The peak-to-valley ratio is the ratio of the maximum to minimum output level across the useful frequency range of the filter. Mechanical filters, with a peak-to-valley ratio of 1 db, may be produced with accurate adjustment of filter elements.

Mechanical filters, other than the disk type, are presently in use. All mechanical filters are similar in that they employ mechanical resonance. Mechanical filters differ in that they employ various methods of mechanical oscillation to achieve their purpose.

CRYSTAL LATTICE FILTER

Another type of filter commonly used in SSB equipment is the crystal lattice filter. This type of filter makes use of extremely high Q quartz crystal resonators. Lattice filters often consist of six or more crystals. This type of device offers an economical method of selective filtering.

Figure 25-20 illustrates a single section crystal lattice filter and its equivalent electrical circuit. The series crystals, Y1 and Y2, are a matched pair, as are the shunt or lattice crystals, Y3 and Y4. The resonant frequency of Y3 and Y4 is higher than the resonant frequency of Y1 and Y2. The passband of the filter is primarily determined by the difference in frequency between the two sets of crystals.

The operation of this filter may be better visualized if the circuit of figure 25-20 is redrawn to form a bridge. This is accomplished in figure 25-21. It may be observed that these circuits are electrically identical by referring to the lettered points in the two figures.

The bridge will be balanced when the effective reactances of the crystals are equal in amplitude and of the same type (i.e., inductive or capacitive). Under this condition there is zero output. The bridge is at its unbalanced limit when the effective reactances are equal in amplitude but opposite in type. Under such a condition the output is maximum.

A graph of the effective reactances of the two sets of crystals versus frequency is shown in figure 25-22. Note that the parallel resonant point of the series crystals corresponds to the series resonant point of the lattice crystals. Figure 25-22 shows that the conditions previously given for an unbalanced bridge are satisfied between points f_{ss} and f_{p1}. This frequency range is the passband of the filter. At $f_{\infty 1}$ and $f_{\infty 2}$ the conditions for a balanced bridge are satisfied. At these frequencies output will be zero, and attenuation will be maximum. Below $f_{\infty 1}$ and above $f_{\infty 2}$ attenuation, though not maximum, will be very high, and output very low.

A typical response curve for a crystal lattice filter is depicted in figure 25-23. It should be noted that the bridge balance and/or unbalance may not be absolute. This, of course, would result in an imperfect response curve. To improve the response, additional filtering sections are cascaded to form a multisection filter.

FREQUENCY STANDARDS

Due to the required accuracy for single sideband communications, it is necessary to

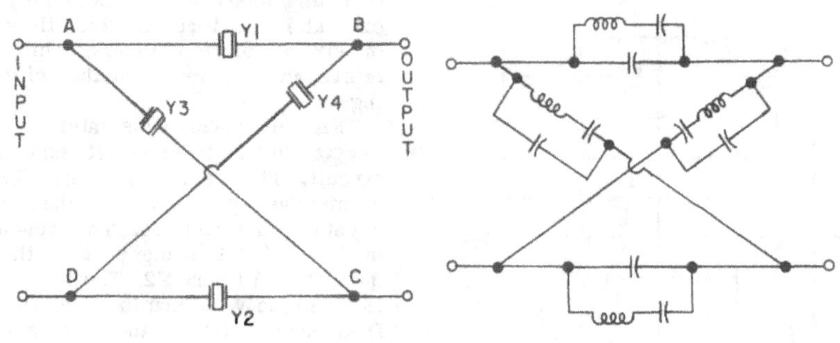

179.458
Figure 25-20.—Crystal lattice filter.

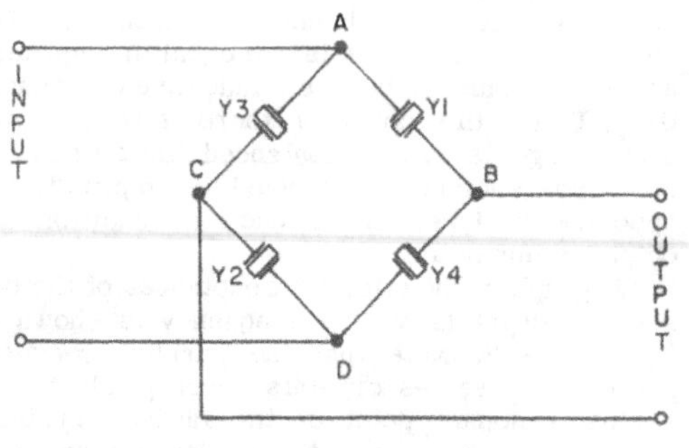

179.459
Figure 25-21.—Crystal lattice filter (redrawn).

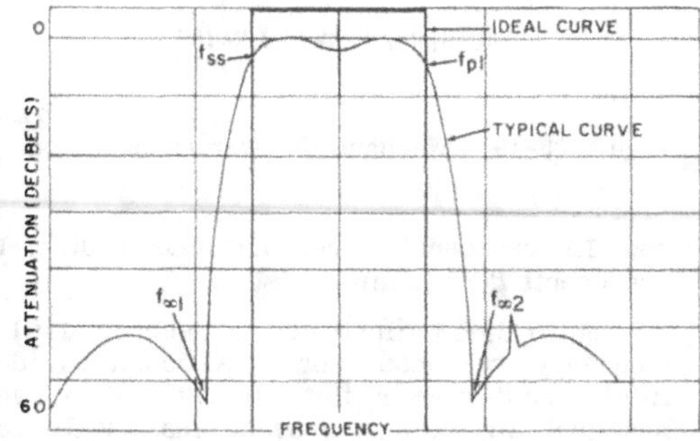

179.461
Figure 25-23.—Crystal lattice filter response curve.

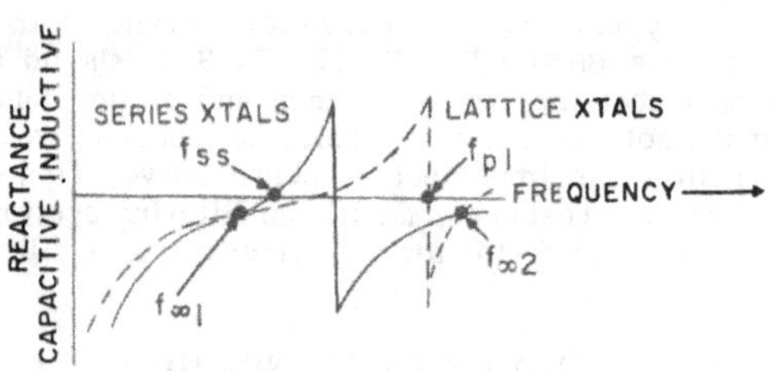

179.460
Figure 25-22.—Effective crystal reactances.

standardize the frequencies of the oscillators used. A frequency standard is a universally agreed upon unit of measurement of a recurring event. Once a standard is agreed upon, it can be used for comparison with any other frequency. Since frequency is the reciprocal measurement of time, it can be only as accurate as the time unit used.

The primary time standard must use measurement of a recurring event which does not change as time goes on. Until recent years, most primary time standards were based on astronomical phenomena, such as the time of rotation of the Earth on its axis or the revolution of the Earth around the Sun. Now, the best primary time standards employ the invariance of time required for certain atomic transitions to occur. These can be duplicated in any laboratory where accuracy is required. However, it is more convenient to transmit such a standard for general use so that oscillators can be synchronized to it. Such oscillators, as those used in a SSB system, then become secondary frequency standards.

These secondary frequency standards can then be used to accurately calibrate the frequency of oscillators used within a system. The primary standard most conveniently used to set a local secondary frequency standard is the U.S. National Frequency Standard, which is transmitted as time ticks by the National Bureau of Standards (NBS). NBS presently maintains two stations: WWV at Fort Collins, Colorado, which transmits on 2.5, 5, 10, 15 and 20 MHz; and WWVH at Kauai, Hawaii, which transmits on 2.5, 5, 10, and 15 MHz. A time tick is transmitted each second, and time announcements are sent in English every minute.

There are many situations in which it is necessary to accurately determine the output frequency of an oscillator or transmitter. This can be accomplished by any number of methods.

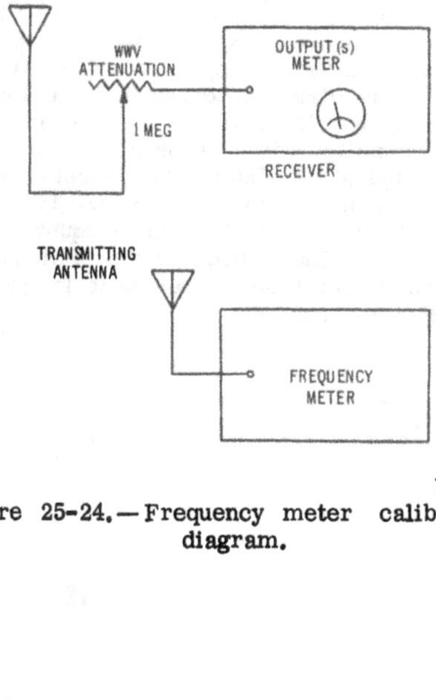

Figure 25-24.—Frequency meter calibration diagram.

The type of result desired, approximate or extremely accurate, will determine the method.

FREQUENCY METER

If a frequency meter, a secondary standard, were to be considered for measuring a frequency, it must first be determined whether it is accurately calibrated. The following method may be used to determine the accuracy of a frequency meter. The frequency meter is warmed up for at least six hours to permit stabilization of its output. Meanwhile, a receiver is tuned to a WWV frequency, and the RF gain of the receiver is adjusted to give a signal strength indication of about S5 on the receiver's sensitivity meter (see fig. 25-24). The frequency meter is then tuned to the same WWV frequency and the position of the transmitting portion of the frequency meter is changed so as to raise the signal strength at the receiver to about S8. The sensitivity meter should now periodically swing indicating a beat, or difference frequency, between the WWV signal and the frequency meter. No swing indicates a zero beat, or the two are on the exact same frequency. The frequency meter's instruction book is consulted for the proper procedure used to adjust for a zero beat, at which point the frequency meter is accurately calibrated. It can then be used to measure the unknown output frequency of a transmitter.

CALIBRATED RECEIVER

Another method of determining a transmitter's frequency uses a calibrated receiver. This method is used if no specific frequency measuring equipment is available. Since a well designed receiver is usually accurate to better than 0.04 percent, its dial setting is a fairly accurate indication of the received, and therefore transmitted, frequency. Accuracy is best found by using an unmodulated transmission. The receiver is tuned to the transmitted frequency, with the beat frequency oscillator turned off, to get maximum deflection of the tuning indicator. An even more accurate method can be used if a stable beat frequency oscillator is available: inject a standard signal into the receiver, and adjust the best frequency oscillator until zero beat is obtained. Then, without disturbing the beat frequency oscillator setting, the receiver can be tuned to the unknown frequency by listening for the zero beat. The

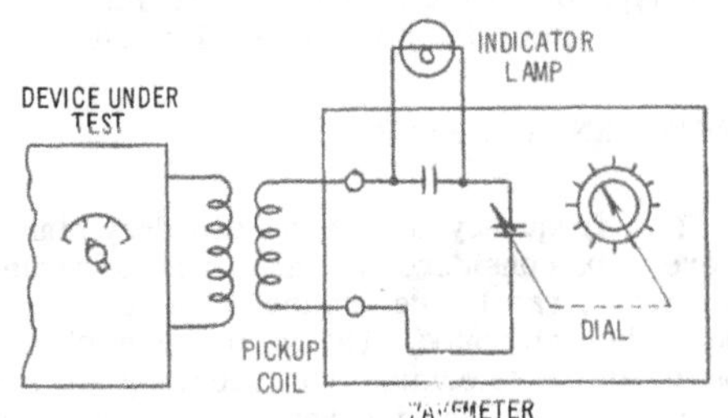

179.463

Figure 25-25.—Absorption wavemeter circuit.

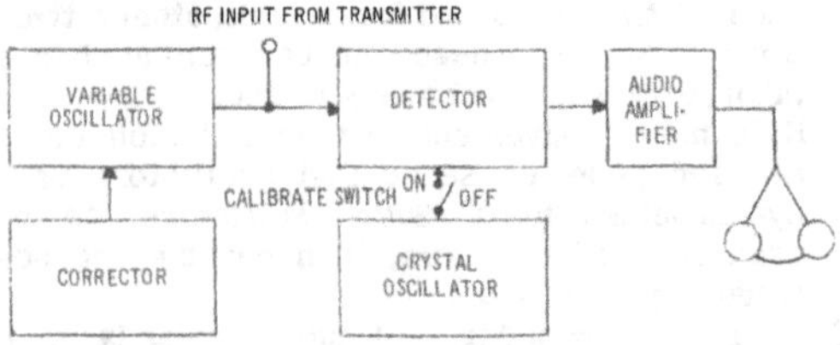

RF INPUT FROM TRANSMITTER
VARIABLE OSCILLATOR
DETECTOR
AUDIO AMPLI-FIER
CALIBRATE SWITCH ON / OFF
CORRECTOR
CRYSTAL OSCILLATOR

179.464

Figure 25-26.—Heterodyne frequency meter block diagram.

setting of the calibrated tuning dial on the receiver is then the unknown frequency. To use this method with a nearby transmitter, it may be necessary to disconnect the receiver's antenna or turn off the transmitter's power amplifier in order to prevent overloading the receiver.

ABSORPTION WAVEMETER

The next method of determining a transmitter's frequency is by use of an absorption wavemeter. It is not as accurate as other methods; but it may be employed when approximate frequency measurement is needed or when simple detection of the presence or absence of RF energy at some point is necessary. The absorption wavemeter consists of a pickup coil, a fixed capacitor, a lamp and a variable capacitor with a calibrated dial, as in figure 25-25. When the wavemeter's components are at resonance, maximum current flows in the loop giving the indicating lamp maximum brilliance. The calibrated dial setting is converted to frequency by means of a chart or graph in the instruction manual. If the lamp glows very bright the wavemeter should be coupled more loosely to the oscillator by moving it further away. For greatest accuracy the wavemeter should be coupled so that its indicator lamp provides only a faint glow at its maximum brilliance.

HETERODYNE FREQUENCY METER

The next method to be considered for determining a transmitter's frequency uses the Heterodyne Frequency Meter. This method should be used whenever high accuracy is required. The Heterodyne Frequency Meter consists of a stable variable frequency oscillator, a crystal oscillator, a detector and an audio amplifier, as in figure 25-26. The detector measures the difference frequency between the variable oscillator and the accurate standard crystal (or one of its harmonics). The resulting frequency difference is in the audio range, so that the beat frequency is audible in headphones after amplification. The variable oscillator is tuned by means of a corrector knob to get a zero beat; then the variable oscillator is known to be at the same frequency as the crystal. The RF input from the transmitter will now be connected in place of the crystal oscillator. The variable oscillator will be tuned to obtain a zero beat; the variable oscillator is now at the same frequency as the transmitter. The setting of the calibrated dial is read directly or converted to frequency by means of a chart.

CHAPTER 26

CW AND FM RECEPTION

This chapter will discuss the conversion of received CW and FM signals into usable intelligence. You will probably note that the receivers are similar to those discussed previously for AM reception. Even so, some repetition will be necessary, although mostly the differences between AM, CW, and FM receivers will be stressed.

CONTINUOUS WAVE (CW) TRANSMISSION

An example of continuous wave transmission is the interruption of a constant amplitude, constant frequency RF carrier to form the dots and dashes of the international Morse Code. This type of transmission was the first type of radio communication used and is still used for reliable long range communication.

Figure 26-1 shows the wave envelope of a CW signal representing the letter "R." Since there is only one frequency present, the carrier frequency, the CW signal occupies a very narrow portion of the frequency spectrum. This narrow bandwidth gives CW transmission certain advantages over AM or SSB transmission, both of which require a wider bandwidth. A comparison of the relative amount of the frequency spectrum used by each of the aforementioned types of transmission is shown in figure 26-2.

Due to its narrow bandwidth, there can be several CW signals in the frequency occupied by one AM voice transmission. Also, CW transmission is negligibly affected by fading and atmospheric disturbances. Again, due to its narrow bandwidth, CW transmission is not easily distorted by adjacent channel signals.

A disadvantage of CW transmission is that the speed of transmission is limited by the skill of the operator in either transmitting or receiving the intelligence.

RECEPTION AND DEMODULATION OF CW TRANSMISSION

Figure 26-3 is a block diagram of a superheterodyne receiver capable of receiving and demodulating a CW signal. The action of the RF amplifier, mixer, local oscillator, and IF strip is the same for both the CW signal and a conventional AM signal, but the CW signal reaches the detector as a single frequency signal with no sideband components. In order to produce an AF output it must be heterodyned (beat) with an RF signal of the proper frequency. This separate signal is obtained from a tunable oscillator known as a beat frequency oscillator (BFO). For example, if the intermediate frequency is 465 kHz and the BFO is tuned to 464 kHz or 466 kHz, the output of the detector will be 1 kHz. Figure 26-4 illustrates a typical transistor and electron tube BFO. The transistor type is a modified Colpitts (described in chapter 15). The electron tube type is a Hartley oscillator. Its operation is also described in chapter 15. The BFO stage should be shielded to prevent its own output from being radiated and combined with desired signals ahead of the detector.

If AGC voltage is to be used, it should be obtained from a separate diode, rather than the detector; otherwise, the output of the BFO would develop an AGC voltage even with no received signal present. One way of obtaining a proper AGC voltage when using a BFO, is to couple the

179.465

Figure 26-1.—Wave envelope of a CW signal representing the letter "R."

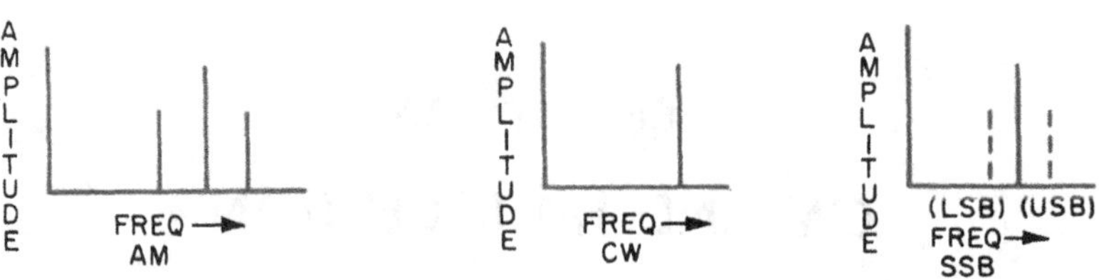

Figure 26-2.—Comparison of the relative amounts of the frequency spectrum used by AM and CW transmission.

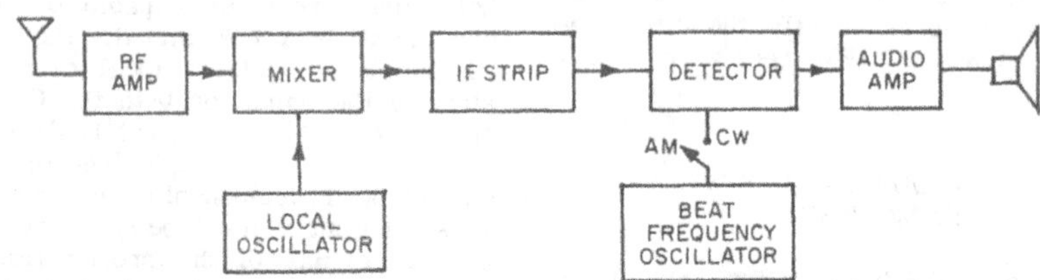

Figure 26-3.—Block diagram of a superheterodyne receiver capable of receiving and demodulating a CW signal.

output of an IF amplifier stage ahead of the detector to a separate AGC diode.

CRYSTAL FILTER

Often, in CW reception, the channel is crowded, and considerable noise is present. To properly copy the desired CW signal, maximum selectivity is necessary. A quartz crystal, used as a selective filter in the IF section of a communications receiver, is one of the most effective methods of achieving maximum selectivity.

One possible circuit arrangement is shown in figure 26-5. The crystal is in one arm of a bridge circuit. The secondary of the input transformer (L2) is balanced to ground through the center tap connection. The phasing capacitor, C4, is another arm of the bridge circuit. The crystal acts as a high "Q" series resonant circuit and allows signals within the immediate vicinity of resonance to pass through the crystal to the output coil, L3. The desired signal appears between the center tap of L3 and ground.

The capacity between the crystal holder plates (C3) may bypass unwanted signals around the crystal; therefore, some method must be provided to balance out this capacitance. In the above circuit balancing is accomplished by taking a voltage 180° out of phase with the instantaneous voltage across the crystal and applying it via C4 in such a way as to neutralize the undesired signal voltage. The balanced input circuit in this case is obtained by the use of a center tapped inductor. The tap on L3 permits the proper impedance match.

FM RECEIVERS

The TRF and superheterodyne receivers that have been described in the preceding chapters are designed to receive RF signals that vary in amplitude according to the audio modulation at the transmitter. The amplitude of the RF signal is increased by one or more RF amplifier stages, and the modulation component is reproduced by the detector. Each of the tuned circuits preceding the detector is designed to pass only a relatively narrow band of frequencies containing the necessary upper and lower sideband

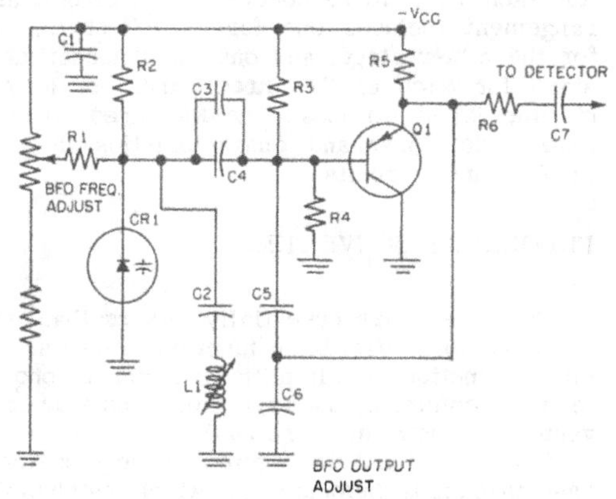

A TRANSISTOR

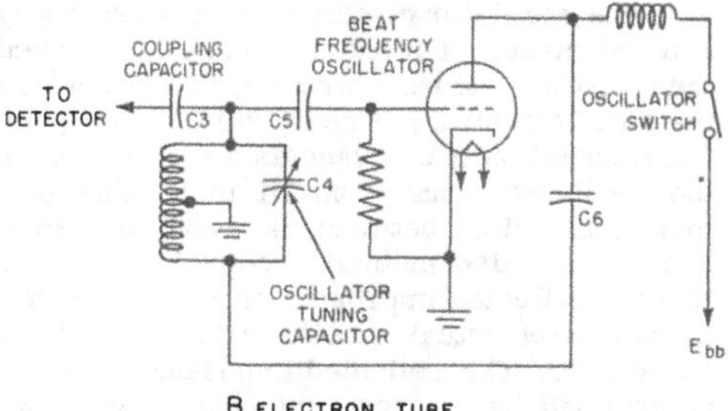

B ELECTRON TUBE

20.271

Figure 26-4.—Typical transistor and electron tube BFO.

frequencies associated with the amplitude modulated carrier.

FM receivers are supplied RF signals that vary in frequency according to the information being transmitted. The amount of variation or deviation from the center, or resting, frequency at a given instant depends on the amplitude of the impressed audio signal. The rate at which the variations from the center frequency occur depends on the frequency of the impressed audio signal. The function of the FM receiver is basically the same as that of the superheterodyne AM receiver; that is, the amplitude of the incoming RF signal is increased in the RF stages, then the frequency is reduced in the mixer stage to the intermediate frequency and amplified in the IF amplifier stage. Finally, the amplitude is clipped in the limiter stage and the modulation component is reproduced by the detector, or discriminator, as it is called in the FM receiver.

There are a few major differences between AM and FM receivers. The greatest difference is the method of detection. Also, the tuned circuits of the FM receivers have a wider passband and the last IF stage is especially adapted for limiting the amplitude of the incoming signal. However, in both systems the audio amplifiers and reproducers are similar. A superheterodyne receiver designed for FM reception is shown in figure 26-6 in block diagram form.

The function of the FM antenna is to provide maximum signal voltage to the receiver input. FM antennas are cut to the required length in order to receive a signal of sufficient amplitude to drive the first RF amplifier. If a single

20.266

Figure 26-5.—Crystal filter used in the IF section of a superheterodyne receiver.

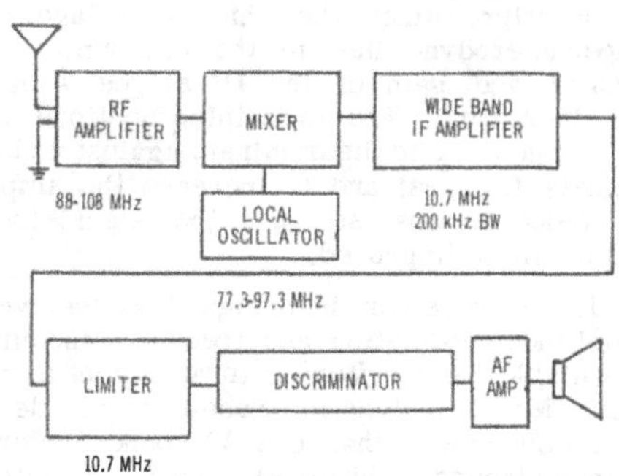

13.67

Figure 26-6.—FM receiver block diagram.

frequency is to be received, the antenna may be designed for maximum response at that frequency. If, however, a band of frequencies is to be received, the antenna length will represent a compromise. Usually the length is chosen so that it will be at resonance at the geometric center of the band. The geometric center is equal to $\frac{\lambda_1 + \lambda_2}{2}$, where λ_1 and λ_2 are the wavelengths at the two ends of the band. There are many types of FM antennas, but probably the simplest is the half wave dipole. The length of the half wave dipole is measured in feet according to the following formula:

$$\text{Approximate length in feet} = \frac{468}{F_c}$$

F_c equals the frequency, in megahertz, at the center of the band. Because the resistance at the center of a half wave dipole equals approximately 72 ohms, the transmission line connecting the receiver and the antenna should have a characteristic impedance of 72 ohms in order to operate as a non-resonant transmission line with no standing waves. More detail will be given on this in chapter 28. The transmission line feeds the signal to the receiver. If the input impedance of the preselector (RF amplifier) stage is not approximately 72 ohms, the signal is fed via an input matching transformer.

The RF amplifier or the preselector stage performs the same function in the FM receiver as it does in the AM version, that is to increase the sensitivity of the receiver. Such an increase in sensitivity is often a practical necessity in fringe areas. However, the gain of the IF stages is much greater, perhaps 10 times that of the preselector, since the chief advantage of the superheterodyne lies in the uniformity of response and gain of the IF stages within the receiver band. The principle functions of the RF stages are to discriminate against undesired signals (images) and to increase the amplitude of weak signals so that the signal-to-noise ratio will be improved.

If the receiver is designed to receive both amplitude modulation and frequency modulation, a suitable band switching arrangement is necessary. Many combination receivers are designed to receive more than one AM band. Under such circumstances additional tuned circuits are needed. Thus, if two AM bands and one FM band are used and one RF stage is used ahead of the mixer, three tuned circuits are needed for each band to be covered. This circuit arrangement includes one for the RF stage, one for the mixer stage, and one for the oscillator stage for each of the three bands, or a total of nine tuned circuits. The FM tuned circuits have wider passband characteristics than do the AM tuned circuits.

FREQUENCY CONVERTER

Converters are essentially mixers that perform the local oscillator function as well as the mixer function. A simplified version of one of the most commonly used types of transistor converters is shown in figure 26-7.

The RF signal is applied to the transistor base through transformer T1, which matches the impedance of impedance of the antenna or transmission line to the input impedance of the transistor. The oscillator circuit consists essentially of a tuned circuit (C2 and L1) and L2 connected in the collector and base circuits, respectively, and the tuned circuit made up of coil L3 and capacitor C3. Oscillation takes place as a result of coupling the feedback signal from L1 to L2 through the mutual induction between the two coils. Both L1 and L2 are also mutually coupled to coil L3, so that the reflected impedance from L3 determines the degree of mutual inductance that exists between L1 and L2. The reflected impedance, in turn, is determined by the resonant frequency of L3 and C3. At the resonant frequency the mutual inductance, and, therefore, coupling, between L1 and L2 is maximum. At all other frequencies, the feedback is negligible. Thus, variable capacitor

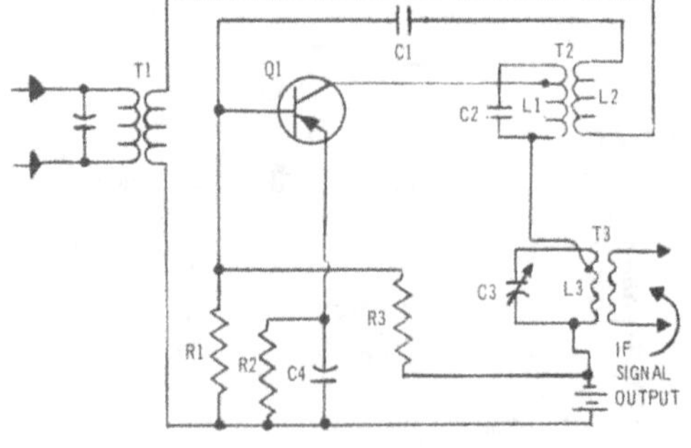

179.468

Figure 26-7.—Transistor FM receiver converter.

C3 determines the frequency at which the collector-to-base feedback takes place. It, therefore, serves as the collector tuning capacitor.

As a result of the oscillator feedback, both the oscillator signal and the RF signal are applied in series between the base and the emitter. The circuit is biased for nonlinear operation, so the two signals are heterodyned in the output. The difference frequency is selected by the tuned collector tank and coupled to the IF amplifier stage.

The most widely used type of tube converter is the pentagrid converter. These converters use special multigrid tubes that are basically the same as those used for pentagrid mixers. In the typical pentagrid converter circuit shown in figure 26-8, the RF signal is applied to the third grid. The oscillator signal is generated by the cathode, the first grid, and the second grid, which together with their associated components function as a series fed Hartley oscillator. The oscillator causes variations in the electron flow within the tube, as does the RF signal. Heterodyning therefore takes place the same as in mixer circuits. Actually, except for the fact that the oscillator is part of the circuit, the operation of the pentagrid converter is basically the same as that of a mixer.

It must be noted that most modern receivers are employing separate oscillators and mixers to perform the heterodyning action. This is done because of the inherent instability of the oscillator portion of the converter, due to interaction between the RF signal and the oscillator signal.

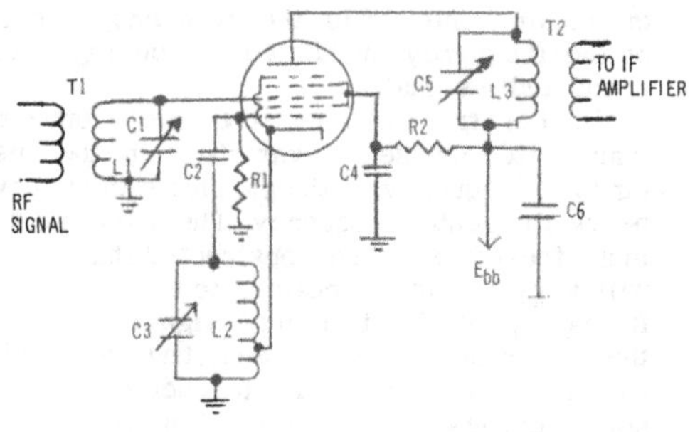

Figure 26-8.—Pentagrid converter circuit.

IF AMPLIFIER

The IF amplifier in an FM receiver is usually tuned to a center frequency. It generally employs double tuned transformers. The passband is from 150 to 200 kHz. The last one or two stages function as a limiter. The gain of each IF stage, due to the wide bandwidth required is less than that of a narrow band IF AM amplifier; therefore, more IF stages are needed for FM reception.

The optimum IF is affected by such factors as image response, response to signals at the same frequency as the intermediate frequency, response to beat signals produced by two stations separated in frequency by the IF value, and response to harmonic frequencies.

Two stations separated in frequency by the IF value will, if sufficiently powerful, produce a beat frequency that will pass through the receiver. This type of interference may be eliminated if the intermediate frequency chosen is greater than the entire FM bandwidth. It may be minimized by adequate discrimination in the preselector stage.

Harmonics of the local oscillator may combine with harmonics produced when a strong incoming signal overloads the input stage to produce the intermediate frequency. Harmonics produced at the input may be reduced by increasing the selectivity of the tuned circuits and using variable mu tubes that do not overload easily. The production of harmonics by the local oscillator may be reduced by maintaining a satisfactory high circuit Q and by reducing its loading.

LIMITER

The limiter in an FM receiver removes amplitude variations and passes on to the discriminator an FM signal of constant amplitude. As the FM signal leaves the transmitting antenna it is varying in frequency according to an audio modulating signal, but it has essentially a constant amplitude. As the signal travels between the transmitting and receiving antennas, natural and man made noises are combined with it to produce variations in amplitude of the modulated signal. Other variations are caused by fading. Fading might be caused, for instance, by movement of the ship. Still other amplitude variations are caused by the receiver itself because of the lack of uniform response of the tuned circuits.

All of these undesired variations in the amplitude of the FM signal are amplified as the signal passes through the successive stages of the receiver up to the input to the limiter. This condition in which both frequency modulation (desired) and amplitude modulation (undesired) are present at the same time is illustrated in figure 26-9A. Figure 26-9B illustrates the waveform after limiting action has occurred. This limiting action is performed by the last IF amplifier stage and is commonly called a saturation-cutoff limiter. Figure 26-10 shows both the transistor and electron tube versions of this circuit.

The transistor is biased in such a manner as to cause one portion of the input signal to drive the transistor into saturation while the other portion will drive it into cutoff. Hence, the output waveform will be limited to a certain amplitude regardless of greater amplitude variations at the input. Since transistors have a forward bias voltage at which runaway occurs, saturation limiting must be approached with care, lest excessive current flow damage the transistor. An emitter resistor is commonly used in order to assure stability of operation.

The same circuit finds its counterpart in electron tubes. The electron tube is biased so that the input signal will, on the positive cycle, drive the tube into saturation, and so that the negative cycle drives the tube into cutoff. In the first case the constant current at saturation limits the amplitude of the negative excursion of the plate voltage waveform, and in the second

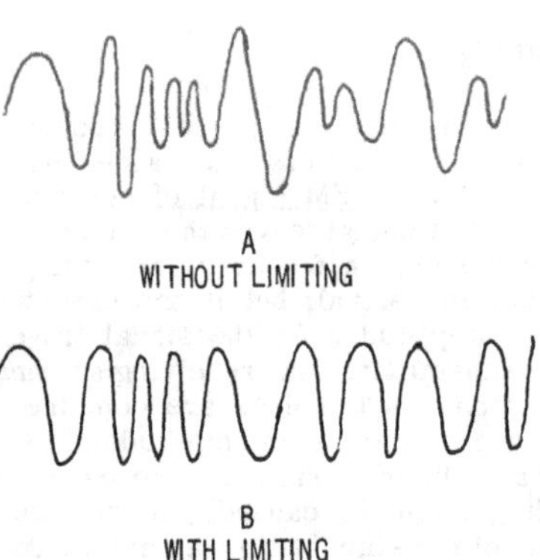

Figure 26-9.—FM signals.

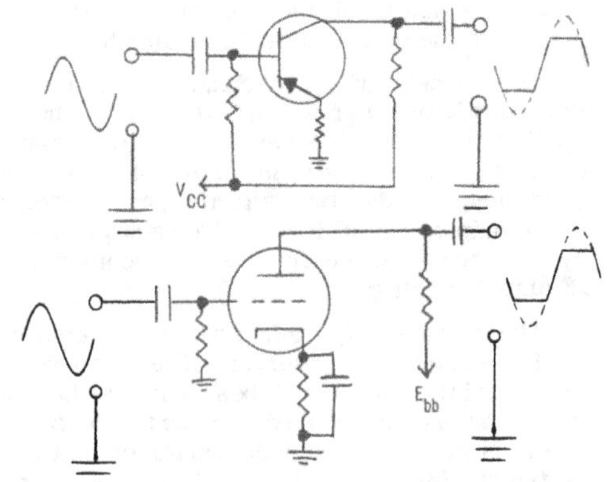

Figure 26-10.—Saturation-cutoff limiter, transistor and electron tube.

case, the cutoff state limits the positive excursion of the plate voltage. Normally, the plate voltage supply is rather small to insure that the amplitude will be limited to the proper value. At saturation, the grid circuit must be able to handle the large current that will flow due to the positive voltage on the grid.

DEMODULATION OF FM WAVES

Another major difference between the AM receiver and the FM receiver is the method of detection of the signal. The detector in an AM receiver interprets the amplitude variations of the amplitude modulated RF energy in terms of the audio signal. In the FM receiver, the discriminator interprets the frequency variations of the frequency modulated RF energy in terms of the audio signal.

In FM transmission, the intelligence to be transmitted causes a variation in the instantaneous frequency of the carrier either above or below the center frequency. The detecting device must therefore be so constructed that its output will vary linearly according to the instantaneous frequency of the incoming signal. Also, since the detecting device is sensitive to amplitude variation produced by interference or by receiver nonlinearities; a special limiting device, called a limiter, must precede the FM detector.

Several types of FM detectors have been developed and are in use, but perhaps two of the most common types are the Foster-Seeley discriminator and the ratio detector. The discriminator requires a limiter, which in turn requires considerable amplification ahead of its input.

Foster Seeley Discriminator Circuit

The detection of a change of frequency by the discriminator is accomplished by a transformer, which, due to special connections, converts the frequency change into a voltage or varying amplitude change. The voltage or varying amplitude is detected in the manner employed by any amplitude modulation detector. Figure 26-11 shows a typical Foster-Seeley discriminator.

The two coils L1 and L2 are the primary and the secondary, respectively, of the IF transformer. The secondary coil is tuned to the correct resting frequency of the incoming carrier by capacitor C3. CR1 and CR2 are the detecting diodes. The filters R1C4 and R2C5 remove the RF component from the circuit, leaving only the audio component at the output. The presence of L3, tertiary winding of the transformer, and its connection to the primary at the top and to the secondary at the center presents an unusual feature in the discriminator circuit. L3 has a high reactance to radio frequencies. Since it is connected across the primary, the primary voltage appears across it at all times. The connection of each diode to this coil (through L2) causes primary voltage to appear at the anodes of the semiconductor diodes with the same phase shift in each case. This voltage causes currents in the opposite directions in the resistors R1 and R2, resulting in zero output at the resting frequency. It must be noted that the voltage drop across L3 will depend on its reactance, and, therefore, the reference voltage will depend on the frequency.

The energy developed in the output tank of the last IF stage (used as a limiter) is coupled to the discriminator in two fashions; one is through mutual inductance between the primary L1 and the secondary L2, and a second is through capacitor C2 to L3. The diodes will thus find two voltages applied to them and the resultant voltage will determine their degree of conduction.

At the center frequency, with a positive swing in the primary, we have a negative swing in the secondary, from top to bottom, as shown in uncircled polarities in figure 26-12. Across L3 we have a voltage drop which puts a positive potential on the anodes of CR1 and CR2. CR1 is thus cut off while CR2 is allowed to conduct charging C5 as shown in figure 26-12. With the following negative swing in the primary, we find the circled polarities in the circuit and we see that CR2 is cut off, while CR1 is allowed to conduct slightly. On successive alternations CR2 will not conduct as long as C5 is charged and CR1 will conduct until C4 charges up to the same value as C5. The output then will be zero, since the charges on the capacitors are series opposing.

Above the center frequency, we find that the X_L will increase both across L2 and L3 while the X_C of C3 is decreasing along with the X_C of C2. Thus, the voltage from the top of the primary across L3 will be greater since less is dropped across C2. Also, this voltage

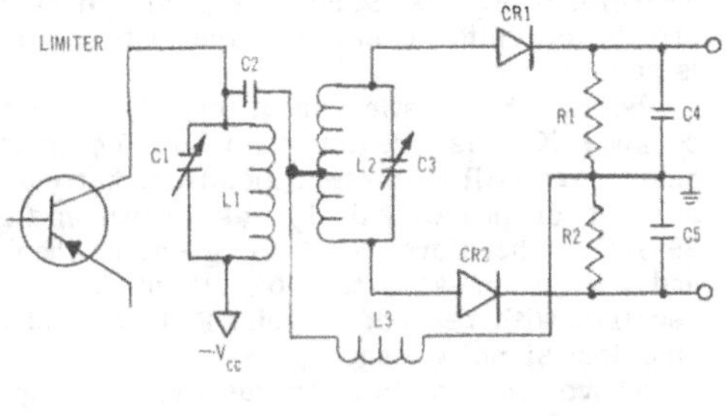

179.471

Figure 26-11.—Foster-Seeley discriminator.

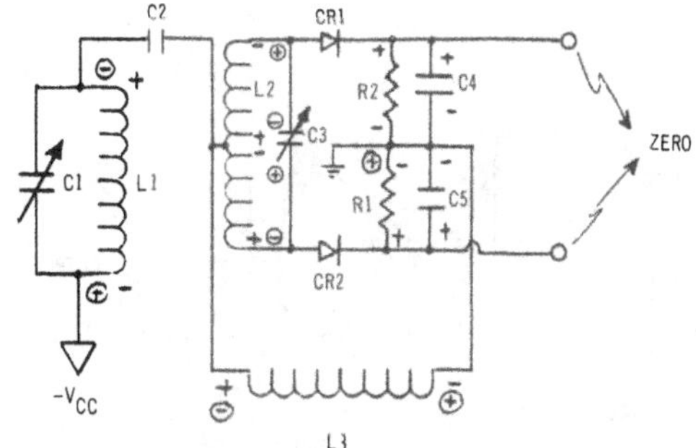

179.472

Figure 26-12.—Discriminator with the center frequency applied.

will appear at the anodes of the diodes before the induced voltage. With the primary swinging positive, the uncircled polarities will be found in the circuit of figure 26-13. CR1 conducts slightly on this alternation along with CR2. On the negative alternation CR1 conducts again while CR2 is cut off. Overall, the output will be positive, since C4 has a greater charge than C5.

With a signal that is below the carrier frequency, the coupling capacitor, C2, will drop more voltage, since its reactance will increase, and L3 will drop less voltage; thus, the reference voltage decreases with respect to the potential applied to the diodes. On the positive swing, CR2 will conduct much more than CR1, charging up C5; on the negative swing, neither diode conducts, and the overall charge at the output is negative. In short, we can examine this circuit and say that there are in effect three sources in the circuit, as shown in figure 26-14. At the center frequency all sources are equal; above the resting frequency source 1 is greater, therefore, a positive output results; below the resting frequency source 2 is the greatest and, therefore, a negative output results. All this depends on the amplitude of the voltage dropped across L3.

EXPLAINING THE DISCRIMINATOR VECTORIALLY. — The discriminator can also be explained by using vectors as follows. The input voltage, e_1, is applied across the input tuned circuit. The current i_1, lags e_1 by 90 degrees. The mutually induced voltage, e_2, lags i_1 by 90 degrees. Thus, e_2 is 180° out of phase with e_1, as shown in part (1) of figure 26-15B.

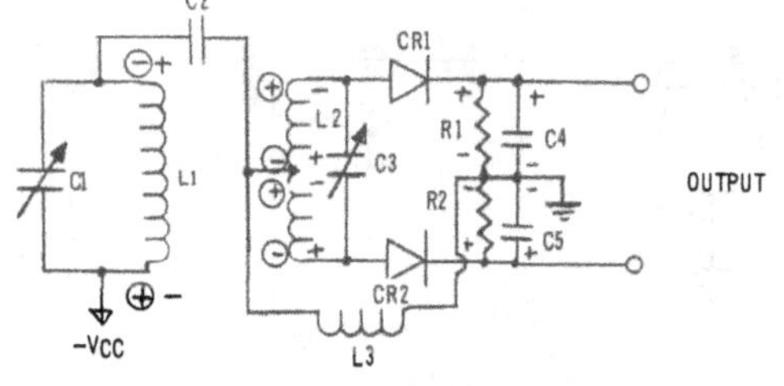

179.473

Figure 26-13.—Discriminator with above carrier signal applied.

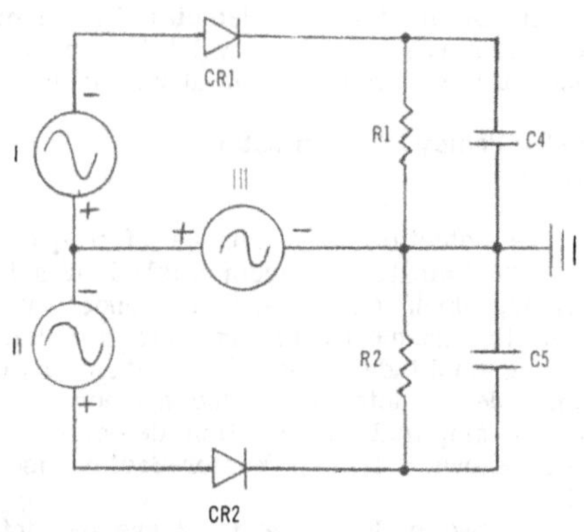

179.474

Figure 26-14.—Discriminator equivalent diagram.

Inductor L4 is shunted across the input tuned circuit via C2 and C5, which have negligible reactance at the resonant frequency. Thus, e_1 is also applied across L4.

Assume first that the incoming signal is at the resting frequency. The induced current, i_2, is in phase with e_2, as shown in parts (2) and (3) of figure 26-15B. The voltages, e_3 and e_4 are the iX_L drops across L2 and L3 respectively. From figure 26-15A, and part (3) of figure 26-15B, it may be seen that e_6, the the voltage applied to CR1 is the vectorial sum of e_1 and e_3; and e_7, the voltage applied to CR2, is the vectorial sum of e_1 and e_4. The rectified output voltage of CR1 is e_8 and that of CR2 is e_9. The output voltage, e_{10}, is the algebraic sum of e_8 and e_9. In part (3) of figure 26-15B, e_6 and e_7 are equal because the incoming signal is at the resting frequency. Therefore, e_8 is equal to e_9; and since they are in opposite directions, the output voltage is zero.

Below the resting frequency i_2 leads e_2 because X_C is greater than X_L. Voltages e_3 and e_4 are still in phase opposition, but each is 90° out of phase with i_2, as shown in figure 26-15C. Therefore, e_7 is greater than e_6, and e_9 is greater than e_8. Point A becomes negative with respect to point B, thus producing an output signal voltage.

Above the resting frequency, i_2 lags e_2 because X_L is greater than X_C. Voltages e_3 and e_4 bear the same phase relation with each other and with i_2 as they did for each of the

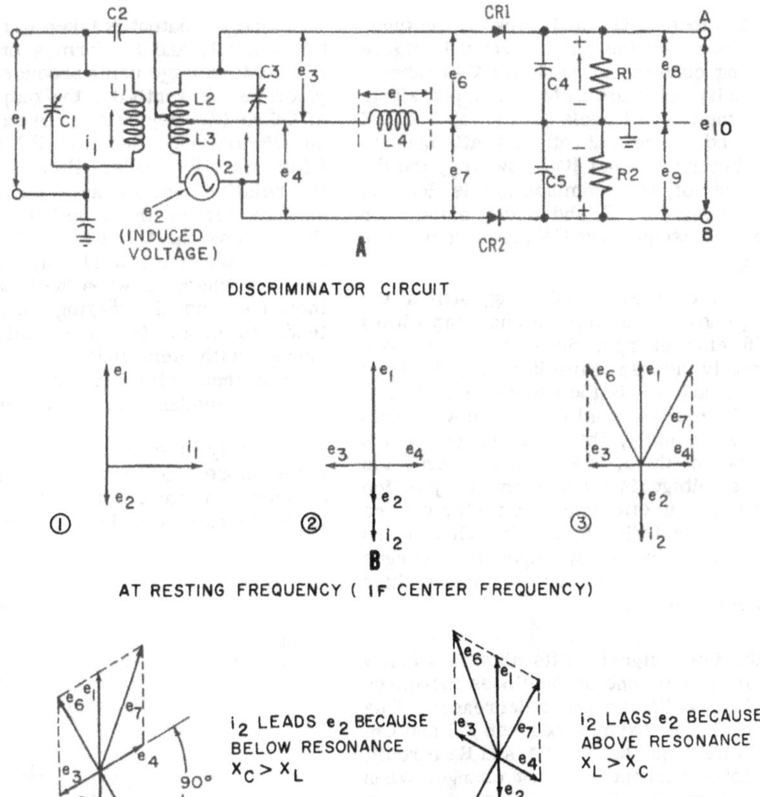

20.199

Figure 26-15.—Discriminator circuit and vector diagram.

above conditions. However, from figure 26-15D, e_8 is now greater than e_7. Therefore, e_8 is greater than e_9, and point A becomes positive with respect to point B, thus producing the other half of the audio signal voltage waveform.

Ratio Detector

A commonly used FM detector that is relatively insensitive to amplitude variations of the input signal and, therefore, does not usually have to be preceded by a limiter, is the ratio detector. Unlike the Foster-Seeley discriminator, the diodes in the ratio detector are connected in series with respect to the tuned input circuit, L2C3.

At the FM center frequency, both diodes have equal applied voltages, so they conduct equally. Conduction takes place on the half cycles of the input signal when the top of the input transformer secondary L2 is negative. The conduction path is through diode CR1, resistors

R1 and R2, diode CR2, and back to the tuned circuit as shown by the solid arrows in figure 26-16. During conduction, capacitor C4 charges, with a polarity as shown, to the input signal voltage. During the half cycles of the input signal when the diodes are cut off, C4 attempts to discharge through R1 and R2. However, the RC time constant of the combination is long in relation to both the IF and audio modulation frequencies. Consequently, C4 loses very little of its charge.

While the diodes are conducting, still at the center frequency of the input signal, capacitors C5 and C6 also charge. Since they are connected directly across capacitor C4, the total charge on C5 and C6 is the same as that on C4. Both C5 and C6 have equal values, so when both diodes conduct equally, the total charge divides equally between them. As you can see, the audio output voltage is taken from the junction of C5 and C6, and is effectively the center voltage that exists across both C5 and C6. This means that at the center frequency, since the charges on C5 and C6 are equal, the voltage at their junction is effectively zero.

When the input signal shifts above or below the center frequency, one of the diodes increases in conduction, while the other decreases. This changes the charge balance between C5 and C6, but has no effect on the C4, R1 and R2 circuit, since the total current does not change. When the charge on one of the capacitors is greater than the other, the voltage at their junction changes above or below zero according to the charges.

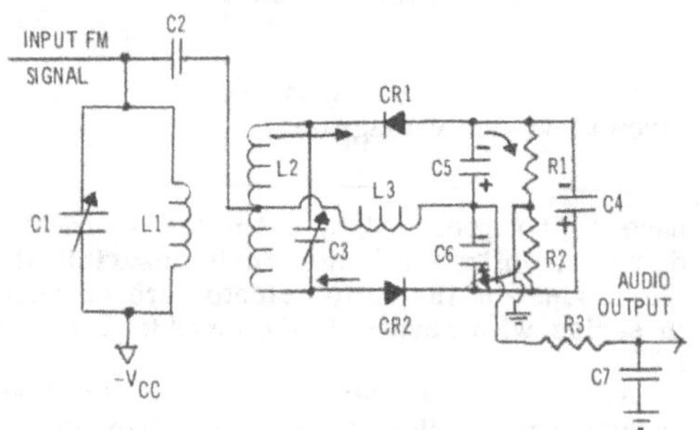

179.475
Figure 26-16.—Ratio detector.

Since the output is taken in relation to ground, C5, C6, R1, and R2 form a bridge circuit, with the audio voltage being produced between the two junctions. Essentially, the output at the junction of C5 and C6 depends on the ratio of the charges on C5 and C6. Resistor R3 and capacitor C7 filter the IF carrier. However, no matter how the relative charges on C5 and C6 vary, the total voltage across them both is still the same as that across the C4, R1, and R2 circuit. Capacitor C4 keeps the total voltage effectively constant by charging when the total voltage tends to increase, and discharging when the total voltage tends to drop. The total voltage only tends to change with amplitude variation. So, by maintaining that voltage steady, C4 counteracts the effect of sudden changes relatively in amplitude.

Actually, the overall operation of C4 is a little more complex. For example, when the carrier amplitude rises and the output voltage tends to rise, C4 charges and causes a higher

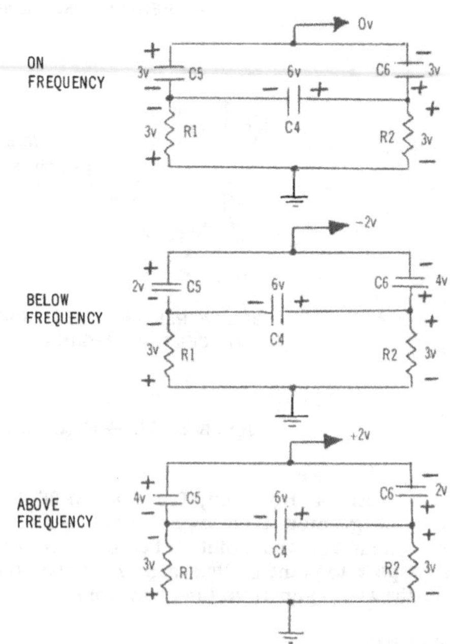

179.476
Figure 26-17.—Ratio detector equivalent circuits.

average current to flow through the entire circuit, including the transformer. This loads the transformer down, lowering its Q. As a result, the phase-shift sensitivity of the transformer is reduced. This reduces the discriminator action of the circuit, to lower the output, to help compensate for the amplitude increase. The opposite happens when the amplitude of the carrier drops.

In summary then, since R1 and R2 are equal, the voltage drop across one will equal the voltage drop across the other, and the amplitude of this drop will be determined by the charge on C4. However, the charges on C5 and C6 vary according to the FM signal. With this bridge circuit, the capacitor frequency of the charges oppose the resistor voltage drops, and the result is the output. The algebraic sum of each branch (capacitive or resistive) is always zero, since C5 and C6 vary oppositely. This is illustrated in figure 26-17.

EXPLAINING THE RATIO DETECTOR VECTORIALLY.—The induced voltage, e_2 is 180° out of phase with e_1, as indicated by the vector diagrams in figure 26-18. At resonance (fig. 26-18B), i_2 is in phase with e_2, as in all tuned circuits. Voltages e_3 and e_4, developed by the i_2X_{L2} and i_2X_{L3} voltage drops respectively, are 180° out of phase with each other and 90° out of phase with i_2. This out-of-phase relation holds true at resonance as well as below or above resonance.

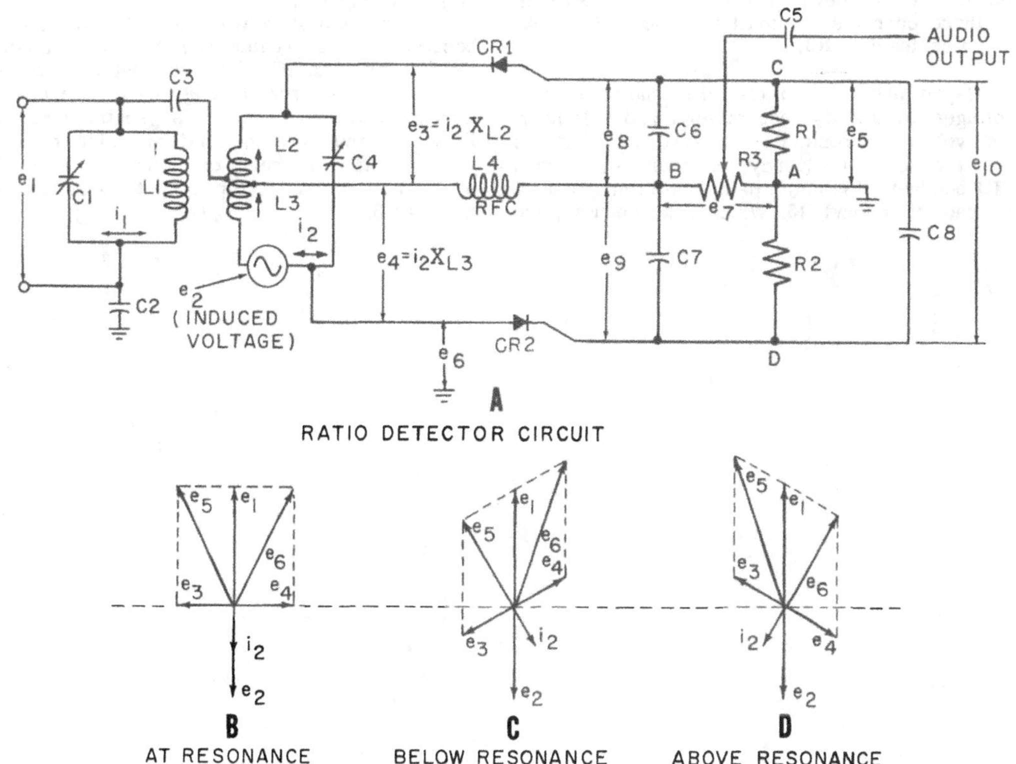

Figure 26-18.—Ratio detector and vector diagrams.

From figure 26-18B, e_5, the a.c. voltage applied across CR1, is the vector sum of e_1 (coupled through C3 between the center tap and ground) and e_3. Also, e_6, the voltage applied across CR2 is the vector sum of e_1 and e_4. At resonance, e_5 and e_6 are equal.

The angle of current flow through CR1 and CR2 may be of the order of 40° or less. In figure 26-18B, vector e_5 leads e_6 by approximately 50°. Thus, CR1 will conduct for 40° and be cut off for 10° of the input cycle before CR2 begins conduction. The current path of CR1 is from C to A, B, L4, L2, CR1 and back to C. The current path of CR2 is from D to CR2, L3, L4, B, A, and back to D. Hence, RF rectified current flows via CR1 from A to B and via CR2 from B to A. Without capacitors C6 and C7, the RF voltage developed across R3 as a result of these currents is small because of RF choke L4 in series with R3.

Capacitors C6 and C7 will charge to d.c. voltages e_8 and e_9 respectively, and will keep the voltage across R3 at zero volts at the rest (resonant) frequency. In other words when CR1 conducts, the major path for current is via C6 instead of R1 and R3. When CR2 conducts, the major path for current is via C7 instead of R2 and R3. During the interval when neither diode is conducting, C6 and C7 will discharge only slightly. Thus, the output voltage e_7 across R3 is zero at the resonant frequency because the CR1 and CR2 currents are equal.

Below resonance (fig. 26-18C) i_2 leads e_2 because X_C is greater than X_L. Therefore, e_6, the vector sum of e_1 and e_4, is greater than e_5, the vector sum of e_1 and e_3. The current paths are the same as previously stated but because e_6 is greater than e_5 the current through CR2 is greater than through CR1. Thus, the current through R3 from B to A (CR2 conducting) will be greater than from A to B (CR1 conducting). Point B, therefore, swings negative with respect to point A to develop the audio output voltage e_7.

Above resonance (fig. 26-18D), i_2 lags e_2 because X_L is greater than X_C. Therefore, e_5 is greater than e_6. The current through CR1 exceeds the current through CR2. The current through R3 from A to B is greater than from B to A and the potential at point B swings positive with respect to point A. Thus, the other half of the audio output cycle appears across R3.

CHAPTER 27

SIX-STEP TROUBLESHOOTING

As a result of the tremendous expansion of the electronics industry to meet the needs of World War II and the postwar demand for more powerful and accurate instruments of destruction and detection, there exists today a great number of electronic equipments. All of these require preventive and corrective maintenance to keep them in continuous operation. Troubleshooting, along with all of its necessary tests, is the most important part of this maintenance.

"Troubleshooting" is a term we in the electronics field hear continually. But, what does it mean? Troubleshooting is sometimes misinterpreted to mean simply fixing an equipment when it fails. This is only part of the "big picture." In addition, the troubleshooter must be able to evaluate equipment performance by comparing his theoretical knowledge with the present indications of performance characteristics. This evaluation must be made both before and after repairs are accomplished for reasons which will become apparent as you progress through this chapter.

Performance data, along with other general information for various electronic equipments, are available to aid you in making an intelligent comparison of the operation characteristics of specific Navy equipment.

The functioning of any shore or shipboard naval installation is almost entirely dependent upon the proper continuous operation of many different types of electronic equipments. This requirement clearly supports the fact that the primary job of the technician is the maintenance of these equipments.

The term "maintenance" refers to all actions which a person performs on an actual piece of equipment or machinery to retain the equipment in a serviceable condition or to restore it to serviceability. This involves inspection, testing, servicing, repair, rebuilding, etc. Proper maintenance of equipment cannot be performed by an untrained person, but must be performed by a person who is thoroughly familiar with the equipment. This familiarity requires a knowledge of how the equipment performs its task—the "theory of operation."

A logical or systematic approach to troubleshooting is paramount in a technician's overall knowledge of electronics. Many man-hours have been lost because of the time consuming "hit-or-miss" methods of trouble analysis. The trouble analysis procedure presented in this chapter has been developed to give you a path to follow toward the ultimate goal of effective equipment maintenance and optimum operational capability. If you can grasp the concept and basic importance of the suggested troubleshooting steps explained in the following pages, you will gain the ability to troubleshoot any electronic equipment regardless of its complexity or purpose.

THE LOGICAL APPROACH

Before entering into a discussion of the details of the primary subject matter (troubleshooting) it is necessary to establish the basic element upon which satisfactory trouble analysis is based. This basic element, so very often overlooked in almost every endeavor, is a logical approach. According to up-to-date terminology, the definition of logic is: "the system or principles of reasoning applicable to any branch of knowledge or study." Examining the definition and its relation to our subject, it would be well to underline the words principles of reasoning. In a broad sense, reasoning is, in itself, logic.

When one examines the complex nature of most of today's electronic systems, whether military or commercial, it should be apparent that the personnel who are assigned to keeping the equipment operational must have specific training. These technicians are far from superhuman beings in understanding and maintaining such devices. So what is the secret of their capability? It is simply and basically the fact that they have learned to think logically.

When you have assimilated the fundamental theories related to basic electronic circuitry, you will be well prepared to learn how they may be combined to form a complete system designed to do a specific task. Armed with this knowledge and a logical approach, you can functionally divide any electronic (or for that matter, non-electronic) equipment and test it in an orderly and professional fashion. This procedure will save you valuable man-hours which are completely wasted when a haphazard technique is employed.

THE SIX-STEP PROCEDURE

At this point, it should be apparent that a standardized approach toward electronic equipment troubleshooting and maintenance procedures will ultimately save many hours of needless equipment "down time" and costly repairs caused by improper maintenance techniques. Another equally important point is to keep electronic equipment in a constant state of operational readiness equivalent to the design standards specified by the equipment manufacturer.

The SIX-STEP Procedure is as follows:

1. Symptom recognition
2. Symptom elaboration
3. Listing of probable faulty functions
4. Localizing the faulty function
5. Localizing trouble to the circuit
6. Failure analysis

STEP 1—SYMPTOM RECOGNITION

This is the first step in our logical approach to trouble analysis. In order to repair an equipment you must first determine whether it is functioning correctly or incorrectly. All electronic equipments are designed to do a specific job or group of jobs according to the requirements established by the Navy and the equipment manufacturer. This demands that a certain type of performance be obtainable at all times. If it were impossible to know when the equipment is performing poorly, it would be equally impossible to maintain the equipment in tip-top shape. For this reason the recognition of trouble symptoms is the first step in troubleshooting an equipment.

A trouble symptom is a sign or indicator of some disorder or malfunction in an electronic equipment. Symptom recognition is the act of identifying such a sign when it appears.

When you have a fever or a headache, you know that there is a disorder somewhere in your body. When you hear a loud "knocking" sound in the engine of your car, you know that some part of the engine is not performing properly. Similarly, when you observe that the sound from a receiver is distorted, you know that there is a fault somewhere in the receiver or its supporting equipment.

Normal and Abnormal Performance

Since a trouble symptom is a manifestation of an undesirable change in equipment performance, we must have some standard of normal performance to serve as a guide. By comparing the present performance with the normal, you can recognize that a trouble symptom exists and make a decision as to just "what" the symptom is.

Your normal body temperature is 98.6° Fahrenheit. A change above or below this temperature is an abnormal condition—a trouble symptom. If you determine that your body temperature is 102°F., by comparing this with the normal you can say that the symptom is an excess temperature of 3.4°F. Thus, you have exactly defined the symptom.

The normal television picture is a clear, properly contrasted representation of an actual scene. It should be centered within the vertical and horizontal boundaries of the screen. If the picture suddenly begins to "roll" vertically, you should recognize this as a trouble symptom because it does not correspond to the normal performance which is expected.

The normal sound from a superheterodyne receiver is a clearly understandable reproduction of the message sender's voice. If it sounds as though the sender is talking from the bottom of a barrel filled with water, the receiver operator knows that this distortion is a trouble symptom.

Performance Evaluation

During the process of doing their assigned job, most electronic equipments yield information which an operator or technician can either see or hear. The senses of hearing and sight, therefore, allow you to recognize the symptoms of normal and abnormal equipment performance. The display of information may be the sole job of the equipment or it may be a secondary job to permit performance evaluation.

Electrical information, to be presented as sound, must be applied to a loudspeaker or a headset. A visual display results when the information is applied to a cathode-ray tube or to an indicating meter which is built into the equipment control panel and which can be viewed by the operator. Pilot lights also provide a visual indication of equipment operation.

As an example of how these various displays can be used to evaluate or "monitor" equipment performance, consider the plate current meter or "tuning meter" which monitors the plate current in the final stage of a broadcast transmitter. When the transmitter is tuned to the proper frequency, there should be a "dip" in the plate current meter. Tuning the transmitter corresponds to adjusting the capacitance of the parallel inductor-capacitor tank circuit so that the condition of parallel resonance exists at the desired frequency. The parallel resonance condition results in maximum impedance to plate current flow—hence, a very low value of plate current. (The plate current meter would "dip" at this point.)

In the detuned condition, the plate current may be quite high, as shown in part A of figure 27-1. As the tuning control is adjusted to approach the proper frequency, the current will abruptly decrease, as shown in part B. The lowest reading (part C) will occur at the correct frequency if the equipment is performing normally.

Knowledge of the normal equipment displays will enable you to recognize an abnormal display, which provides the trouble symptoms we are concerned with in our first troubleshooting step.

Equipment Failure

Electronic equipment failure is the simplest type of trouble symptom to recognize. Equipment failure means that either the entire equipment or some part of the equipment is not functioning and will, therefore, show no performance display. The absence of sound from a superheterodyne receiver when all controls are in their proper positions indicates a complete or partial failure. Similarly, the absence of a visible trace or picture on the screen of a cathode-ray tube when all controls are properly set points to some form of equipment failure. If you have observed the plate current reading on the tuning meter of a broadcast transmitter and the reading suddenly dropped to zero, you have observed an equipment failure.

Degraded Performance

Even if the audible and visual information described in the previous section is present, the equipment may not be performing normally. Whenever the equipment is doing its job, but is presenting the operator with information that does not correspond with the design specifications, the performance is said to be degraded. Such performance must be corrected just as quickly as an equipment failure. This performance may range from a nearly perfect operating condition to the condition of barely operating.

If you were sick but still went to your regular duty assignment, the chances are your performance would be degraded for the period of your illness. You would still be performing your job, but not as well.

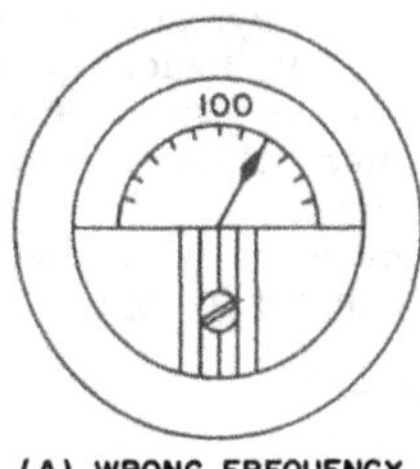

(A) WRONG FREQUENCY

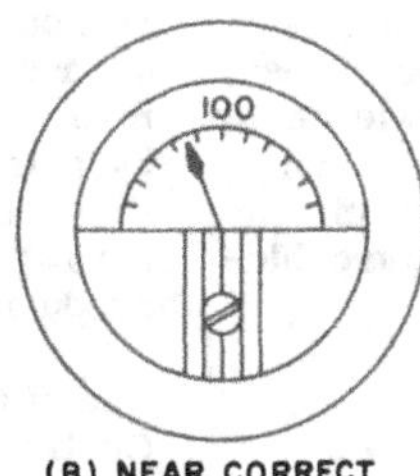

(B) NEAR CORRECT FREQUENCY

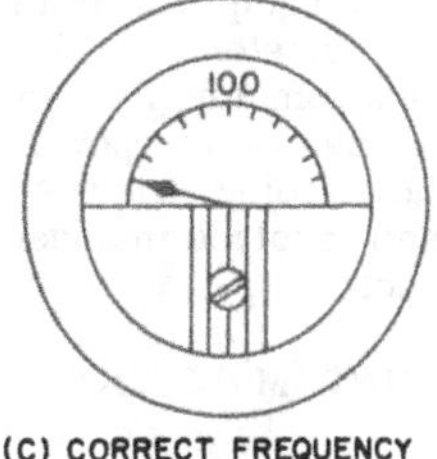

(C) CORRECT FREQUENCY (LOWEST READING)

179.477

Figure 27-1.—Plate current indications during transmitter tuning process.

All of the trouble symptoms presented in this chapter, with the exception of those given in the section, Equipment Failure, have been symptoms of degraded performance. These include distortion in the sound from a superheterodyne receiver, the vertical "rolling" of a television picture, an incorrect meter reading, and an undesirable cathode-ray tube display. A reduction in radar range capability, as indicated when a search radar set which is designed to spot targets 50 miles away will not pick up targets beyond 30 miles, is another example of a degraded performance symptom.

Again, all of these indications are audible and/or visual. They can be recognized by comparing the normal equipment displays with the present operating displays.

Know Your Equipment

Before you can decide whether a piece of electronic equipment is doing its job and doing it properly, you must have a complete and thorough knowledge of its normal operating characteristics. Remember that any electronic equipment, no matter how complex, is built by using the basic electronic circuits and devices you have already studied. These components are combined in such a manner that the desired performance is produced. Therefore, a knowledge of electronic fundamentals will allow you to analyze the performance of any equipment.

This basic knowledge can be supplemented by consulting the equipment handbooks, instruction books, and maintenance directives associated with each piece of equipment.

The information you need to evaluate equipment performance is usually provided by audible means or visual displays on the equipment control panel(s). However, unless you can apply your knowledge of the equipment in interpreting these displays, their existence will be meaningless. Unless you can employ your own knowledge to recognize or discredit reported trouble symptoms, you are going to waste a lot of time getting yourself involved in unnecessary troubleshooting projects.

STEP 2—SYMPTOM ELABORATION

As a second step, the obvious or not so obvious symptom should be further defined. Most electronic devices or systems have operational controls, additional indicating instruments other than the main indicating device, or other built-in aids for evaluating performance. These should be utilized at this point to see whether they will effect the symptom under observation or provide additional data that further defines the symptom.

Breaking out the test equipment and equipment diagrams and proceeding headlong into testing procedures on just the original recognition of a trouble symptom is an unrealistic approach. Unless you completely define a trouble symptom first, you can quickly and easily be led astray. The result, as before, would be loss of time, unnecessary expenditure of energy, and perhaps even a total dead-end approach. This step is the "I need more information" step in our systematic approach.

Symptom elaboration is the process of obtaining a more detailed description of the trouble symptom. Recognizing that the fluorescent screen of a cathode-ray tube is not lighted is not sufficient information for you to decide exactly what could be causing the trouble. This symptom could mean that the cathode-ray tube is burned out, that there is some disorder in the internal circuitry associated with this tube, the intensity control is turned down too low, or even that the equipment is not turned on. Think of all the time you may waste if you tear into the equipment and begin testing procedures when all you may need to do is flip the "on-off" switch to "on", adjust the intensity control, or just plug in the main power cord.

Similarly, recognizing an undesirable hum in a superheterodyne receiver as a trouble symptom could lead you in several directions if you do not obtain a more detailed description of the symptom. This receiver hum may be due to poor filtering action in the power supply, heater-cathode leakage, a.c. line voltage interference, or other internal and/or external faults.

It should be apparent by now that the primary reason for placing symptom elaboration as the second step in our logical procedure is that many similar trouble symptoms can be caused by a large number of equipment faults. In order to proceed efficiently, it is necessary to make a valid decision as to which fault(s) is probably producing the specific symptom in question.

Use of Operating Controls

Operating controls are considered to be all front panel switches, variable circuit elements, or mechanical linkages connected to internal circuit components which can be adjusted without going inside the equipment enclosure. These are the controls which the equipment operator must "operate" in order to supply power to the equipment circuits, to tune or adjust the performance characteristics, or to select a particular type of performance.

By their very nature, operating controls must produce some sort of change in the circuit conditions. This change will indirectly alter current or voltage values by the direct variation of resistance, inductance, and/or capacitance elements in the equipment circuitry. The information displays associated with the equipment—front panel meters and other indicating devices—will enable you to "see" the changes which take place when the controls are operated.

Control manipulation can cause detrimental effects in equipment performance, as well as the desirable effects for which they are primarily intended. Manipulating controls in an improper order or allowing voltage and current values to exceed maximum design specifications may have resulted in the damage which brought about the original trouble symptom. Unless you observe the proper precautions while investigating the symptom, even more damage to the equipment can result from the improper use of operating controls.

Every electronic circuit component has definite maximum current and voltage limits below which it must be operated in order to prevent "burn out" or insulation breakdown. The meters placed on the front panel of electronic equipments serve as an aid in determining voltage and current values at crucial points in the equipment circuitry. Operating controls should never be adjusted so that these meters indicate values above the maximum ratings.

Precautions for Specific Equipment

In addition to exceeding maximum ratings and manipulating controls in an improper sequence, there are certain other precautions associated with specific types of equipment. For example, an intensity control on an oscilloscope should never be adjusted to produce an excessively bright spot on the fluorescent screen. A bright spot indicates a high current which may burn the screen coating and decrease the life of the tube. Also, for the same reason, you should never permit a sharply focused spot to remain stationary for any length of time.

Another precaution concerns the adjustment of a range selector switch on any type of indicating meter. If the switch is carelessly positioned to a range below the value of the quantity being measured, the needle will strike its upper mechanical limit. This may bend the needle and result in inaccurate (offset) readings.

It would be impossible within the scope of this chapter to list all the precautions associated with the various equipments. The examples above are included to increase your recognition of the importance of these precautions.

A knowledge of the circuit changes that take place when you adjust a control will enable you to think ahead of each step and to anticipate any damage which the adjustment might produce. Do not reach to make an adjustment in haste or panic. Remember that any damage resulting from the improper use of operating controls will have to be repaired. You can save a lot of unnecessary troubleshooting time by exercising care when you are at the controls.

Further Defining the Symptom

The first step of our procedure (symptom) recognition), required adequate knowledge of equipment operation before you could be "aware" of a trouble symptom. This knowledge is to be assumed throughout the remaining steps of the logical troubleshooting procedure. It cannot be over-emphasized that knowledge of how an equipment works and a systematic approach to troubleshooting have equal importance, and that possessing one of these factors alone is not sufficient.

The purpose of symptom elaboration is to enable you to fully understand what the symptoms are and what they truly indicate. This elaboration is required in order to gain further insight into the problem.

Incorrect Control Settings

Incorrect operating control settings will produce an apparent trouble symptom. We use the word apparent because the equipment may be operating perfectly, but, because of the incorrect setting, the information display will not correspond with the expected performance. An incorrect setting may be brought about by an accidental movement of the control, as well as careless misadjustment. The discovery of such an incorrect setting permits sufficient "elaboration" of the trouble symptom to "fix" it immediately, thereby ending your troubleshooting project if you can verify that the incorrect setting was the only cause of the trouble symptom.

Assume that you are checking the voltage across the load resistor of an audio amplifier stage in a superheterodyne receiver with an oscilloscope. The waveform should be 100 volts from the positive peak to the negative peak, and you are trying to verify this amplitude in order to evaluate the amplifier stage performance. You intend to set the vertical sensitivity switch to 50 volts per centimeter and expect to see a consequent display waveform similar to that shown in figure 27-2A. However, in haste you accidentally set the vertical sensitivity switch to 10 volts per centimeter. As a result of this carelessness, the display you see is actually similar to that of part B in the figure, and at first glance, you assume the amplitude is 500 volts since you think the switch is set at 50 volts per centimeter. Certainly, the first thought to enter your mind is that the amplifier is not functioning properly.

At this point your knowledge of amplifier operation should be applied. Figure 27-3 is the circuit diagram for the amplifier you are checking. Immediately you should realize that, since the supply voltage for the amplifier is only 150 volts, it is an operational inaccuracy for 500 volts to exist across the load. The amplifier shown in the figure cannot produce an output voltage larger than its own plate supply voltage.

The next logical assumption is that the oscilloscope is in error. Since it is the vertical dimension which is apparently in error, this should immediately direct your attention to the vertical sensitivity control. Once you discover the error in control setting, you will realize that the display actually represents 100 volts peak-to-peak. Therefore, there is no real trouble symptom.

This application of equipment knowledge and logic should have enabled you to discover the error in less time that it took you to read this section. This represents a considerable savings over the time that would have been consumed if you had torn into the amplifier or oscilloscope looking for a nonexistent fault.

Aggravating the Trouble Symptom

If all controls are set at their correct positions but the symptom persists, it is still possible that an operating control is responsible for the trouble symptom. However, in this case the trouble would have to fall in the general area of component failure. If a control is faulty, this may be immediately apparent—especially when it is a mechanical failure. However, additional information may be required to determine when a control has failed electronically since the trouble symptom produced may also point to other electronic failures.

Have we wasted the time involved in checking the controls if they are all in their correct positions? Definitely not. First of all, the time involved will be only a matter of seconds, or minutes at the most. Secondly, there is very logical reason for checking and manipulating the controls, even if all settings are correct. This is to gain information which will define the trouble symptom still further and aid you in proceeding with your trouble analysis.

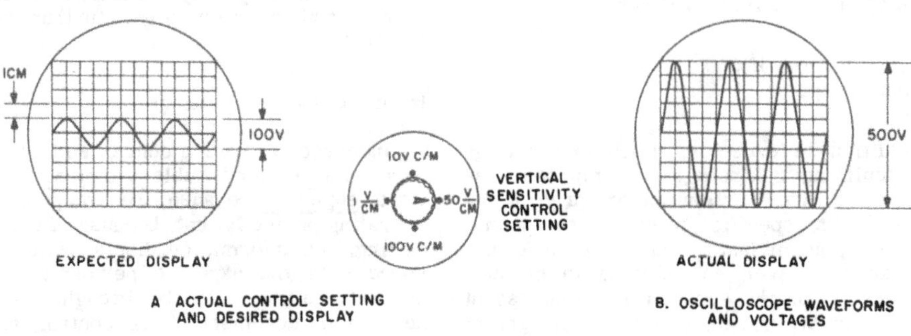

Figure 27-2.—Oscilloscope waveforms and voltages.

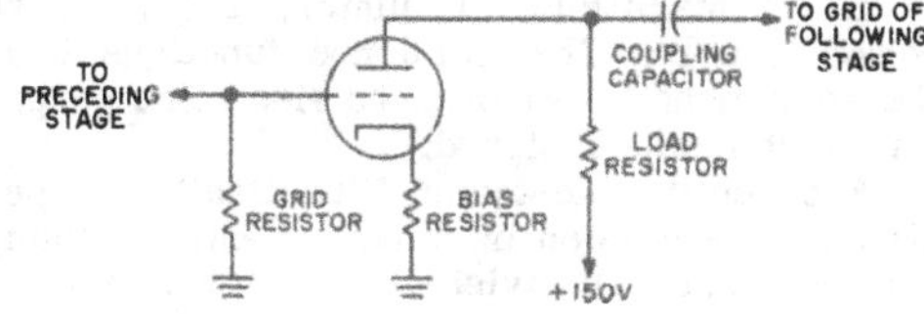

179.479

Figure 27-3.—Typical audio amplifier stage.

The specific intent is to aggravate the trouble symptom, if possible. By observing the changes this aggravation produces in the trouble symptom, you will be able to make a valid estimate as to just what is probably causing the panel meter readings and the displays produced by other front-panel devices, such as cathode-ray tube screens and indicator lights.

As an example, consider the frequency range switch on a broadcast transmitter. This control is a multi-position switch with each position connecting a different RF coil in parallel with the main tuning capacitor of the oscillator tank circuit. The value of each coil is such that it will cause the oscillator to vary over a different range of frequencies as the tuning capacitor is varied. If a weak transmission symptom is reported, it is logical to try other frequency ranges by manipulating the selector switch. If normal transmission is achieved for any of the range positions, the fault probably lies either in the switch itself or in a few (perhaps only one) of the tuning coils. Such an observation would provide a quick location of the trouble area.

Further Aggravating the Trouble Symptom

Similarly in a receiver, if a mode selector switch can be changed from AM operation to FM operation, it is logical to check the receiver in both positions. If the symptom persists only in the AM mode, the circuitry associated with the FM mode can be eliminated as a probable cause later in our troubleshooting procedure.

If a broadcast transmitter uses plate modulation to add the information signal (voice) to the RF signal, the degree of modulation will be controllable by a front panel knob. The modulating signal is applied to the plate circuit of the RF amplifier through a coupling transformer. Between the input of this transformer and the microphone that gathers the voice information is an audio amplifier. The modulation control may vary the gain of this amplifier; hence, any additional undesirable changes in the trouble symptom produced by varying this control would point to faults in the audio units which precede the modulation transformer.

The examples above represent only a small portion of the various controls associated with a transmitter or receiver. The controls and instrumentation associated with every type of equipment are specifically incorporated to provide information about that particular device. Therefore, it is necessary to understand the operation of the equipment in order to appreciate the aid which can be obtained by manipulating controls.

Data Recording and its Purpose

Symptom elaboration cannot be fully accomplished unless the observed displays can be completely evaluated. This means that the indications must be evaluated in relation to one another, as well as in relation to the overall operation of the equipment. The easiest method for accomplishing this evaluation is to have all data handy for reference by recording the information as it is obtained.

This will enable you to sit back a moment and "think" the information over before jumping to a conclusion as to where the trouble lies. It will also enable you to check the equipment manual and compare the information with detailed descriptions if this is necessary—a particularly useful technique for someone just becoming familiar with troubleshooting. Finally, by recording all control positions and the associated meter and indicator information, you can quickly reproduce the information and check to see that it is correct, as well as put the equipment in exactly the operating condition that you wish to test. Thus, the recording of information will enable you to save time and become a more efficient troubleshooter.

Whenever the adjustment of a control has no effect upon the symptom, this fact should also be recorded. This information may later prove to be just as important as any changes a control may produce in the trouble symptom. This procedure may seem unnecessary as this portion of the text is read, but it will definitely pay off in systematic trouble analysis. This fact will become more obvious as we probe deeper into the equipment under test.

Gaining further information about a trouble symptom by manipulating the operating controls and instruments will help you identify the probable faulty function required in the next step. This procedure will give you an estimate of where the trouble lies and will permit you to eventually classify the problem down to the exact item responsible.

If the trouble is cleared up by manipulating the controls, the trouble analysis may stop at this point. However, by using your knowledge of the equipment involved, you should find the reason why the specific control adjustment removed the apparent malfunction. This action is necessary to assure yourself, as well as the operator, that there are no additional faulty items which will produce the same trouble later.

In manipulating controls, you must be aware of the circuit area in which the control is located. Only those controls that will logically affect the indicated symptom should be adjusted. When adjusting controls, use extreme caution—a misadjustment may cause additional circuit damage.

Whether or not you will proceed from step 2 to step 3 (listing of probable faulty functions) or to step 5 (localizing trouble to the circuit) will depend on the number of units in the equipment and/or the complexity of a single-unit equipment, as previously described.

STEP 3—LISTING OF PROBABLE FAULTY FUNCTIONS

The performance of the third step is dependent upon the information gathered in the two previous steps. Step 1, remember, was "symptom recognition", that is, becoming aware of the fact that an equipment is performing its operational function in an abnormal manner. Step 2, symptom elaboration, allows you to use the operating controls and front-panel indicators to obtain as much information about the abnormality as you possibly can.

Step 3, listing of probable faulty functions, is applicable to equipments that contain more than one functional area, or unit. It allows you to mentally select the functional unit (or units) which probably contains the malfunction, as indicated by the information obtained in steps 1 and 2. The selection is made by stopping to think "Where can the trouble logically be in order to produce the information I have gathered?"

The term, function, is used here to denote an electronic operation performed by a specific area (or unit) of an equipment. A transceiver, for example, may include the following functional areas: transmitter, modulator, receiver, and power supply. The combined functions cause the equipment to perform the electronic purpose for which it was designed.

Frequently, the terms "function" (an operational sub-division of an equipment) and "unit" (its physical sub-division) are synonymous. A functional unit may be located in one or more physical locations. For example some components of a receiver, in a transceiver set, may be located in the transmitter compartment. Normally, the physical location, such as a drawer containing a receiver, is referred to as a "unit." A functional unit consists of all the components that are required for the unit to perform its function, whether these components are packaged in an individual drawer or in two drawers. Within this text the terms function, unit, and functional unit will be used interchangeably, although in some equipments one or more circuits of a given function may have been built into a unit other than that bearing the title of the function.

You cannot converse with the set the way a doctor converses with his patient and ask where it "hurts." You must determine this directly by surveying the information you have gathered and by using your knowledge of how the set works electronically. Here you will be aided by the technical manual's description of equipment operation.

Selection Logic

Faulty unit or function selection requires the use of logic similar to that employed by a medical doctor, auto mechanic, or other "technical doctor" when he searches for the cause of an illness or malfunction.

Assume that you are continually plagued with headache and you finally go to a doctor. If the doctor elaborates on the symptom by checking your eyes, ears, nose, and throat, taking your temperature, and listening to your heartbeat, but then promptly sends you to the operating room to have your foot amputated, you would certainly question his diagnosis. Instead of taking such an illogical step, the doctor will decide on the basis of his examination, whether the most probable trouble is poor eyesight, a sinus infection, or some other logical disorder. Only after making such a decision will the doctor prescribe a possible remedy.

Similarly, the technician who accomplishes the first two steps of our six-step procedure and then picks just any test or repair procedure in an attempt to correct the trouble is indeed a poor troubleshooter. He must first survey the information he has gathered; then, using his knowledge of equipment operation along with the aids provided in the applicable technical manuals, he must make a technically sound decision as to what is probably causing the recorded symptoms.

The millions of cells and thousands of parts in the human body could provide a considerable complication to the medical doctor when he is making his diagnosis if he had to check each part or each cell separately to find the exact cause of the illness. Instead, he divides the body into functional groups, each containing many associated parts. He then associates the symptoms of the illness with the normal performance of the functional groups. Any indication of abnormal performance provides him with a clue to the exact cause of the illness.

The abnormal performance indications you noted in steps 1 and 2 should also give you clues as to the probable location of an electronic malfunction. Electronic equipment can have as many as 10,000 circuits, or 70,000 individual parts. The probability of finding the faulty part by methodically checking each of the 70,000 parts in turn is highly remote. The size of the task can be reduced by a factor of seven by checking the outputs of each circuit rather than checking each part separately.

However, 10,000 tests is still a job of considerable magnitude. By dividing the 10,000 circuits into their normal groupings of electronic functional units — seven, a dozen, or two dozen — you can reduce the job to a practical number of tests. Whether the equipment contains thousands, hundreds, or just a few circuits, logical reasoning dictates that the troubleshooting problem can be resolved more quickly and accurately by reducing the total circuits into a small number of groups.

Assume that we've divided the 10,000 circuits into 12 functional units. Locating the faulty unit might require 12 output tests unless you were lucky enough to find it before all units were tested. This still represents a departure from our basic logic. Why should the doctor amputate your foot if you had a sinus headache? Why should you test the turntable of a radio-television-phonograph console set if the picture on the TV tube is bad? You can predict that the trouble lies in the television receiver unit and confine your tests to that unit.

The Functional Block Diagram

Naval electronic equipments and sets are subdivided into functional units. Each functional unit is generally contained within a single case or box or, in some instance, within drawers arranged in a rack comprising the overall set. The term functional is applied to these units because each one accomplishes a specific electronic function. The units are interconnected so that the individual functions will be performed in the proper sequence to accomplish the overall operational function of the set.

The equipment functional block diagram is an overall symbolic representation of the functional units within the equipment, as well as the signal flow paths between them.

Figure 27-4 is a typical functional block diagram. This particular diagram is for an AM transceiver set composed of six separate units. Each unit performs an electronic function and conforms to the input-conversion-output (ICO) concept universally applied to all electronic circuits, units, and sets. Briefly, the function of each unit is:

1. The sound pickup unit (microphone) changes (converts) the sound information (input) to be transmitted into an electrical signal (output) of audio frequency (AF).

2. The modulator unit amplifies the AF signal and applies it to the transmitter in such a manner as to cause the amplitude of the RF carrier signal to vary at an audio rate.

3. The transmitter unit provides the RF signal, as well as the proper "boost" for the power in the AM signal, to achieve the desired transmission range.

4. The antenna assembly unit converts the electrical AM signal into electromagnetic energy suitable for transmission through the atmosphere. When the transceiver is serving as a receiver,

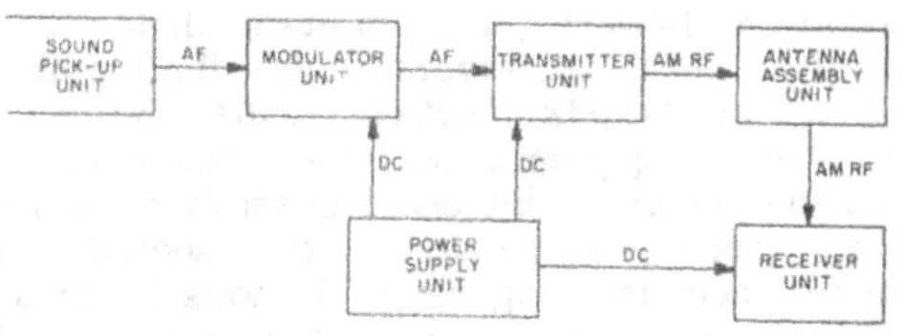

179.480

Figure 27-4. — Functional block diagram for AM transceiver set.

this unit converts the electromagnetic energy transmitted from another location into electrical signals to be applied to the receiver unit.

5. The receiver unit converts the AM signal received from another location into sound.

6. The power supply unit converts the line voltage into a low-value a.c. voltage suitable for heating the filaments of the tubes within the set and into a d.c. voltage suitable for operating the various units.

There is no indication in the equipment functional block diagram as to how each function is accomplished. Thus, each functional unit may consist of a variety of circuits or stages, each performing its own electronic function. For example, the transmitter unit may contain an RF oscillator stage, a voltage amplifier stage, and several power amplifier circuits.

Notice that the connecting lines between the various functional blocks represent important signal flow connections, but that the diagram does not necessarily indicate where these connections can be found in the actual equipment circuitry.

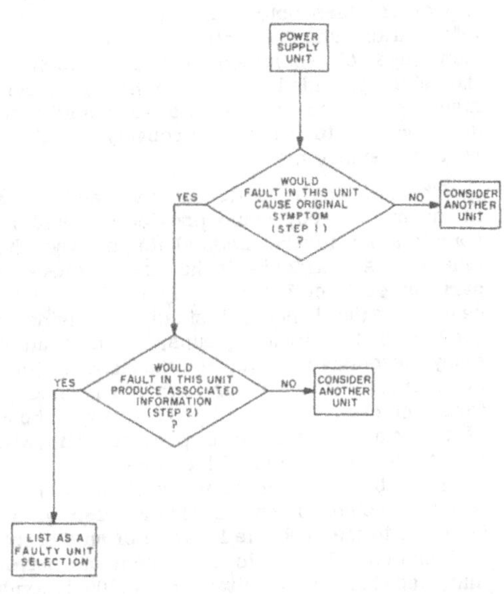

Figure 27-5.—Considering power supply as a faulty unit selection.

Formulating a Faulty Unit Selector

As explained previously, making faulty unit selections requires that you reach a decision as to the possible equipment area(s) which could probably produce the trouble symptom and associated information. At this point in our six-step procedure, the trouble area will be restricted to a functional unit of the equipment. Thus, the functional block diagram is indispensable at this point.

Assume that you have found no reception as the trouble symptom for the transceiver whose functional block diagram is illustrated in figure 27-4. Manipulation of the receiver volume and tuning controls has no effect upon the no reception condition. However, the power-on light and the dial lights of the receiver unit are all illuminated.

Out of the six functional units shown, only the power supply unit, the antenna assembly unit, and the receiver unit could possibly be at fault since these are the only units associated with signal reception. Figure 27-5 shows the thought process involved in formulating a valid faulty unit selection. The answers to the questions you must ask will be obtained from your knowledge of how the equipment operates and/or from your use of the technical diagrams in the technical manual. Most of them should come from a study of the functional block diagram.

First we consider the power supply unit and ask ourselves, "Would a failure or abnormal performance on the part of the power supply cause the original trouble symptom?" If the answer is "no," we can go on to consider another unit. If the answer is "yes," (as it is for the symptom given above), we ask ourselves, "Would a failure or abnormal performance on the part of the power supply produce the associated information obtained during symptom elaboration?" For this example the answer would be "yes," because the fact that the dial lights and power-on light are illuminated does not prove that the proper operating voltages are being produced. This is true because these lights are in the filament voltage circuit only. Therefore, we list the power supply unit as a faulty unit selection, and we also note that the portion of this unit responsible for providing filament voltages is probably okay.

Next we consider the antenna assembly unit and receiver unit (separately) in the same manner. For the no-reception symptom and associated information given above, both of these units must also be listed as faulty unit selections. A break in the antenna lead could easily cause our trouble. Similarly, many different faults in the receiver unit could be the cause of no reception.

The condition described above represents the maximum number of technically accurate selections—every functional unit associated with receiving an external signal may be at fault. The number of selections can be reduced if the second step—symptom elaboration—yields more information about the trouble symptom.

The Exception to the Rule

There are some equipments which do not require a functional block diagram or a faulty unit selection process. These equipments are relatively simple devices and consist of only one functional unit.

Figure 27-6 shows the circuit for a multirange ohmmeter used to make resistance checks. It consists of a current indicating meter, a battery, and resistors of known value connected so that the unknown resistance can be compared with one of the known resistors. When the test leads are open, no current flows through the meter, and the meter is mechanically set to indicate an infinite resistance. When the test leads are shorted together, the meter is electrically adjusted to give a zero reading.

When the test leads are placed across an unknown resistance, this resistance is in series with the battery and meter. The meter is both in series and in parallel with the known resistors. Thus, the current through the meter will be some intermediate value between those values which produced zero reading and infinite reading. The actual value of this current will be determined by the ratio of the fixed and unknown resistors. An appropriately calibrated scale will allow the meter to indicate the value of the resistor being measured.

Equipments of this type represent almost an extreme in simplicity. For such equipments not only is a functional block diagram unnecessary, but, troubleshooting step 3 (listing probably faulty functions) and step 4 (localizing the faulty function) can be omitted entirely from our six-step procedure.

Now you should have a very good idea of how to go about making a faulty unit selection—the decision as to which equipment functional units are probably causing the trouble. These selections must, of course, be logical and must be technically substantiated by the information

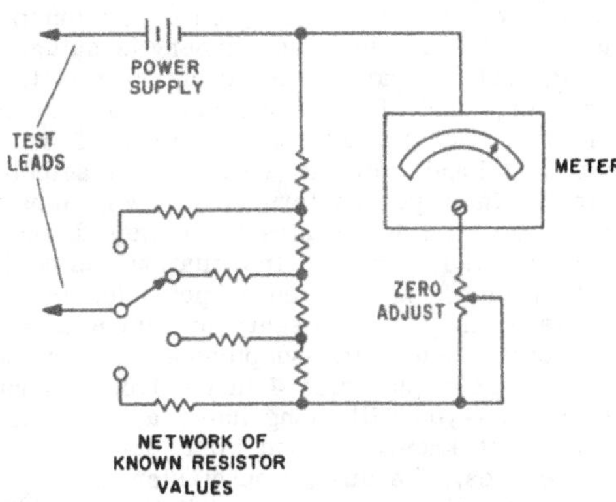

Figure 27-6.—Simple ohmmeter circuit.

you obtained during symptom elaboration, the relationships between signal paths and functional units, and your operational knowledge of the equipment.

One of the reasons for performing this step is to save time. This is accomplished by making technically accurate selections of units which could contain the malfunction. Doing this eliminates the necessity of making illogical checks of all units. However, it must be understood that each unit so selected is only a probable source of the trouble even though its selection is based on technically valid evidence. The next step explains a time-saving and logical method of locating the unit that is most probably faulty.

STEP 4—LOCALIZING THE FAULTY FUNCTION

The first three steps in our systematic approach to troubleshooting have dealt with the examination of "apparent" and "not so apparent" equipment performance deficiencies, as well as a logical selection of the probable faulty functional units. Up to this point no test equipment other than the controls and indicating devices physically built into the equipment has been utilized. No dust covers or equipment drawers have been removed to provide access to any of the parts or internal adjustments. After evaluating the symptom information, you have made mental decisions as to the most probable areas in which the malfunction could occur.

Localizing the faulty function means that you will have to determine which of the functional units of the multi-unit equipment is actually at fault. This is accomplished by systematically checking each faulty functional unit selection until the actual faulty unit is found. If none of the functional units in your list of selections display improper performance, it will be necessary for you to backtrack to step 3 and re-evaluate the symptom information, as well as obtain more information if possible. In some cases it may be necessary to return to step 2 and obtain additional symptom-elaboration data.

At this point—step 4 in the troubleshooting sequence—you will bring into play your factual equipment knowledge and your skill in testing procedures. The utilization of standard or specailized test instruments and the interpretation of the test data will be very important throughout this and the remaining troubleshooting steps.

Testing a Faulty Functional Unit Selection

The intent of the fourth step is to determine the functional unit of an electronic equipment which is responsible for the indicated failure. The selection of any unit as the probable cause should always be based upon equipment knowledge and basic electronic principles. In step 3 it was pointed out that there may be only a few possibilities or there may be many possibilities for faulty functional unit selections. The number of selections is entirely dependent upon the type of equipment and the information gathered in steps 1 and 2 of the troubleshooting procedure.

The use of a logical approach in choosing the first faulty functional unit selection to be tested is of prime importance. The need for a logical approach was pointed out at the beginning of this chapter. When learning the operation of an equipment or when troubleshooting an equipment, this continuous use of logic must be followed. A logical approach is dependent upon equipment knowledge and an understanding of the situation.

Factors to Consider

The simultaneous elimination of several functional units as the cause of the trouble symptom should be an important factor in deciding which faulty unit selection to test first. This requires an examination of the functional block diagram to see whether satisfactory test results from any one of the selections could also eliminate other units listed as probable causes.

Test point accessibility is also an important factor affecting the technician's logic in choosing a faulty functional unit selection for further examination. A test point, as described in some instruction books, is a special jack located at some accessible spot on the equipment, such as the front panel or chassis. The jack is electrically connected (directly or by means of a switch) to some important operating potential or signal voltage. Actually, any point where wires join or where components are connected can serve as a test point.

Another factor to consider is past experience and history of repeated failures. Past experience with similar equipment and related trouble symptoms, as well as the probability of unit failure based upon records of repeated failures, should have some bearing upon the choice of a first test point. However, the selection should be based mainly upon a logical conclusion formed from data obtained in previous steps, without undue emphasis upon past experience and history of the equipment.

In summary, then, the factors to be considered in selecting the first test point, listed in their usual order of importance, are as follows:

1. The functional unit that will give the best information for simultaneously eliminating other units, based upon the data obtained in steps 1, 2, and 3 of our procedure, provided that a certain unit is not obviously the cause.

2. Accessibility of test points—for example, a test point might be avoided as a first choice merely because the equipment must be disassembled to obtain access to this test point.

3. Past experience and history of repeated failures, provided that this factor is carefully weighed in the light of data obtained in steps 1, 2, and 3.

Test Results and Conclusions

Now that you have mastered the process of choosing the first faulty functional unit selection to test, you might ask "Where do I go from here?" The answer, of course, depends upon the results of your first step.

There can be only two results—a satisfactory indication or an unsatisfactory indication. The latter may be in the form of no indication or a degraded indication. In any event, the result obtained should lead you to the next logical test.

Analyzing the Tests

Once the units have been isolated, what happens if the last check doesn't pinpoint the faulty unit? In this case you have either made an error in making one of your checks or the results of the check were misunderstood, leading you down the wrong path. This points up the importance of writing down your results. If you have the information written down, it is not difficult to look back and determine where you went astray.

Further Elaboration

If a final check shows the suspected units to be satisfactory, it will be necessary to reevaluate the information obtained from the previous checks. The question now is how far back should you go?

You could ignore all the information and start over at the beginning with step 1, Symptom Recognition; however, this should not be necessary because the fact that there is a trouble should have been pretty well established when the trouble was first reported. Returning to step 2, Symptom Elaboration, would allow you to reevaluate the meter readings or other indications that were present when the operating controls were manipulated. A return to step 3, Listing of Probable Faulty Functions would permit a review of the list of faulty units previously prepared to insure that a possible faulty unit was not overlooked.

Trouble Verification

Having isolated the trouble to the actual functional unit at fault, it is now necessary to reconsider whether a fault in this unit could logically produce the trouble symptom and fits the associated information obtained during symptom elaboration. To do this you will have to use the functional block diagram again.

In order to locate the faulty functional unit, we proceeded from symptom information to actual location. To verify the located functional unit we will proceed in the reverse direction. We will ask the question, "What trouble symptoms would this faulty unit produce?" Thus, equipment knowledge is very important.

If you will refer to figure 27-4, you will be able to follow the logic in the example to be presented. Assume that the fault lies in the antenna assembly unit. It does not automatically switch over to the "receiver function" as it should. What trouble symptom would this failure produce?

First of all, we know that the symptom should occur only in the units associated with the receiver function. This would include the receiver unit. The modulator and transmitter units should be performing properly. The receiver should provide all normal responses, noise in the speaker and the ability to vary the noise with the operating controls. However, no signal would be present.

If the original trouble symptom and the associated data collected during symptom elaboration fit the above expectations, we have verified the faulty functional unit.

Step 4 (localizing the faulty function) has been concerned with the testing of an equipment on a limited basis; that is, it has considered only those tests that are necessary to isolate a faulty functional unit. A logical application of equipment knowledge and symptom analysis, coupled with the three factors—simultaneous elimination of several functional units, test point accessibility, and past experience and history of repeated failures, enabled you to take the list of faulty functional unit selections made in step 3 and pick the most logical one for the first test. This same logic was then applied to the systematic selection of all subsequent test points. At each point, a new bit of information enabled you to narrow the trouble area until the faulty functional unit was located.

The completion of this step as presented in this text should leave no doubt as to which functional unit is at fault. However, as a final check of your work, you have been made aware of the necessity to verify the isolated faulty functional unit by "back-tracking" and matching the theoretical trouble symptoms with those actually present.

STEP 5—LOCALIZING TROUBLE TO THE CIRCUIT

Steps 1 and 2—symptom recognition and symptom elaboration—of the six-step troubleshooting procedure provide you with initial diagnostic information. This information, ascertained from the operation of controls on the equipment, provides visual indications, such as meter readings or scope presentations, so that the effects of the trouble can be further evaluated. Step 3—listing of probable faulty functions—applies this information and your equipment knowledge so that you can select the functional units in a multi-unit set which are most probably causing the trouble. Actual tests were performed in step 4—localizing the faulty function—so that you could isolate the faulty portion of the equipment.

Radar and communication equipment normally consists of multi-unit sets and, each step in the logical process must be considered in the proper sequence. However, some equipment such as ohmmeters consists of a single functional unit. In this case steps 3 and 4 can be eliminated. After completing step 2, you would proceed directly to step 5—localizing trouble to the circuit.

In step 5—localizing trouble to the circuit—you will do rather extensive testing in order to isolate the trouble to a specific circuit within the faulty functional unit. To accomplish this, you will first be concerned with isolating a group of circuits within a functional unit, arranged according to a common electronic sub-function. Once the faulty circuit group has been located, you can perform the tests which will isolate the faulty circuit(s).

This procedure adheres to the same reasoning we have used throughout our six-step troubleshooting procedure—the continuous "narrowing down" of the trouble area by making logical decisions and performing logical tests. Such a process reduces the number of tests which must be performed. This reduction of tests not only saves time but also minimizes the possibility of error.

To gain a better understanding of this successive functional division, refer to figure 27-7. First, there is the equipment or set which is designed to perform an overall operational function. We see that steps 1 and 2 of our troubleshooting procedure are associated with this functional classification. The set is then divided into functional units, each designed to perform a major electronic function vital to the overall operational function. Steps 3 and 4 are associated with this category. When there is only one functional unit, steps 3 and 4 are skipped.

The next division—the circuit group—is a convenient subdivision of the functional unit. The circuits and stages in the circuit group perform an electronic sub-function vital to the task assigned to the functional units. Our first concern in step 5 is to determine which of these groups is at fault. After this is done, we can go deeper into the equipment to isolate the final equipment division—the individual circuit.

The Correct Approach

Before you continue the troubleshooting procedures into step 5, you should pause and assimilate all of the data obtained at this point which may aid you in performing the next step. After completing step 4, you know that all of the inputs to the faulty function are correct and that one or more of the outputs is incorrect or nonexistent. The incorrect output waveform(s) obtained in step 4 should be analyzed to obtain any information which may indicate possible trouble areas within the functional unit. It is important to remember that the original symptoms and clues obtained in the first two steps should not be discarded merely because steps 3 and 4 are completed. This information will be helpful throughout the troubleshooting procedure, and should be reviewed, together with all clues discovered in subsequent steps, before continuing to the next step.

Step 5 should be a continuation of the narrowing-down process, and the ICO principle should be employed in each part of this step. Each functional unit has a separate function within an equipment, and within each functional unit there may be two or more groups of circuits, each with a sub-function. This means that the input to each group (sub-function) is converted, and the output emerges in a different form. An understanding of the conversions which occur within a functional unit makes it possible to logically select possible trouble areas within the unit. Testing is then performed to isolate the defective circuit group. The same principles are applied to the circuit group to locate the faulty circuit within the group.

Servicing Block Diagrams

The purpose of the servicing block diagram is to provide you with a pictorial guide for use in step 5. Figure 27-8 is such a diagram for the receiver unit of a transceiver set. There will also be a servicing block diagram for every other unit in the transceiver set—modulator, transmitter, and power supply. Occasionally, the entire equipment will be represented by one service block diagram.

The use of a servicing block diagram is facilitated by the fact that all circuits within the functional unit are enclosed in heavy dashed lines, while circuits comprising a circuit group within the function are enclosed with light dashed lines. Within each dashed enclosure is the name

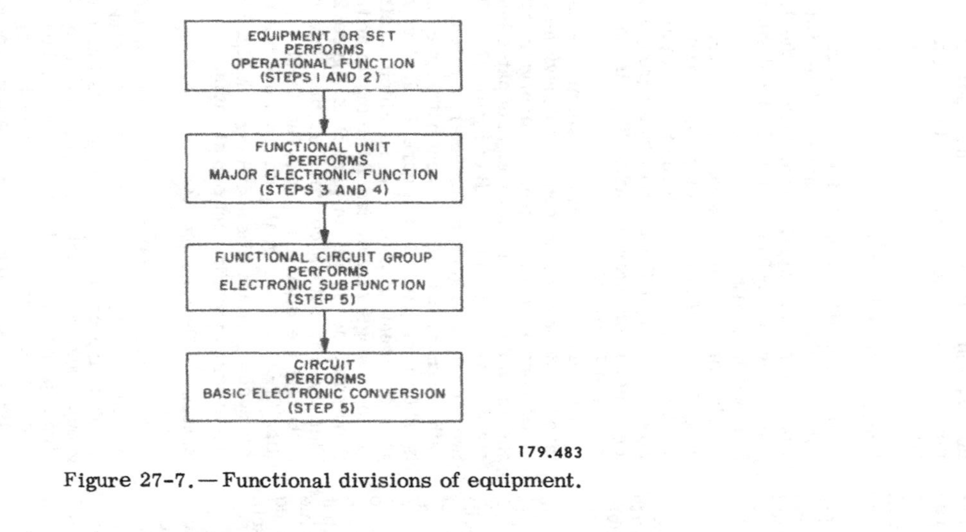

179.483

Figure 27-7.—Functional divisions of equipment.

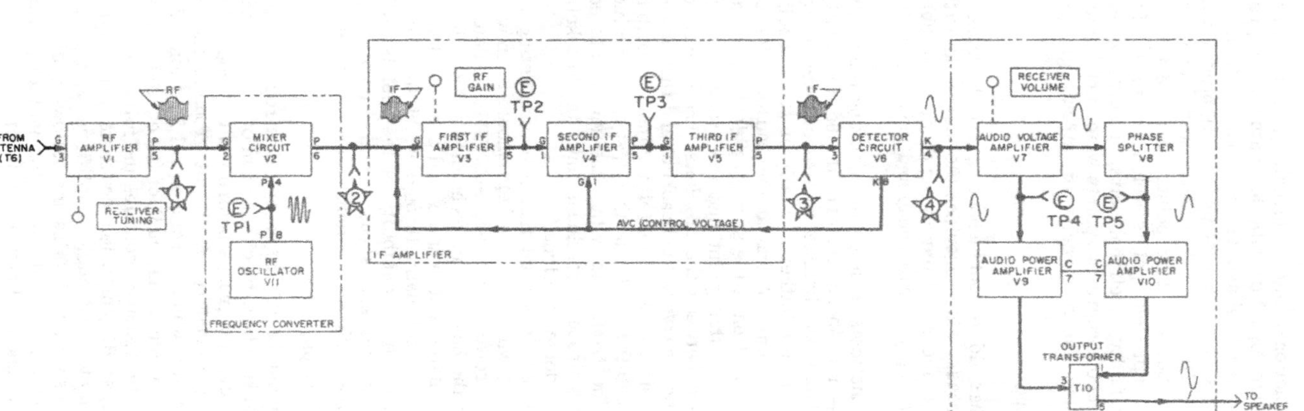

179.484

Figure 27-8.—Simplified servicing block diagram for receiver unit (transceiver set).

of the functional unit or circuit group it represents. Main signal or data flow paths are represented by heavy lines, and secondary signal or data paths are represented by lighter lines.

Notice in figure 27-8, that waveforms are given at several test points. Star test point symbols represent points which are useful for isolating faulty functional units, and circled test point symbols represent points which are helpful in locating faulty circuits or circuit groups.

Bracketing

Another aid to troubleshooting is the "bracketing" process, which provides the technician with a physical means of narrowing down the trouble area to a faulty circuit group and then to a faulty circuit.

Once the tests in step 4—localizing the faulty function—have been performed and the faulty unit isolated, the bracketing process begins by placing brackets (either mentally or with a pencil) at the good input(s) and at the bad output(s) of the faulty function is the servicing block diagram. You know at this point that the trouble exists somewhere between the brackets. The idea is to make a test between the brackets and then move the brackets one at a time (either input or output bracket) and then make another test to determine whether the trouble is within the new bracketed area. This process continues until the brackets isolate the defective circuit.

The most important factor in bracketing is determining where the brackets should be moved in this narrowing-down process. This is determined on the basis of the technician's deductions from the analysis of systems and previous tests, the type of circuit paths through which the signal flows, and the accessibility of test points. All moves of the brackets should be aimed at isolating the trouble with a minimum number of tests.

Circuit Groups

You must be able to recognize circuit groups and to subdivide a block diagram of a functional unit into circuit groups before you can apply the bracketing procedure. A circuit group is one or more circuits which form a signal functional division of a functional unit of an equipment. A typical radio receiver contains the following circuit groups: RF amplifier, converter, IF amplifiers, detector and audio amplifiers. A typical radio transmitter contains the following circuit groups: master oscillator, intermediate power amplifiers, and final power amplifier.

You can see that the circuit groups named above perform subfunctions in a receiver or transmitter, and that their combined operations perform the complete function of the unit they constitute.

Types of Signal Paths

The signals associated with a circuit group normally flow in one or more of four different types of signal paths. These include the linear path, convergent-divergent path, feedback path, and switching path.

The linear path is a series of circuits arranged so that the output of one circuit feeds the input of the following circuit. Thus, the signal proceeds straight through the circuit group without any return or branch paths. This is shown in part A of figure 27-9.

The convergent-divergent path may be any of three kinds: divergent, convergent, and the combined convergent-divergent. A divergent path is one in which two or more signals paths leave a circuit, as shown in part B of figure 27-9. When two or more signal paths enter a circuit, the path is known as a convergent path. An example is shown in part C of figure 27-9. A convergent-divergent path is one in which a circuit group or single circuit has multi-inputs and multi-outputs, as shown in part D of figure 27-9. This type is not as common as the convergent path and the divergent path. The feedback path (fig. 27-9E) is a signal path from one circuit to a preceding circuit. The switching path (fig. 27-9F) has a switch for different signal paths.

Bracketing Procedures

You have been introduced to several aids to be used in step 5: the servicing block diagram, bracketing procedures, dividing units into circuit groups, and recognizing the four basic types of signal paths. Now it is time to show how these aids are employed in this troubleshooting step.

Before beginning the bracketing procedure, the data obtained in the previous steps should be reviewed and evaluated, and the servicing block diagram should be employed to provide a pictorial guide of the signal paths through the faulty functional unit. The functional unit is divided into circuit groups, and the subfunction of each circuit group is considered to determine which could cause the observed symptoms. The bracketing procedure begins by placing opening brackets at the good inputs, and closing brackets at the bad outputs, of the functional unit on the servicing block diagram.

When tests have indicated a list of possible defective circuit groups, a bracket is moved to

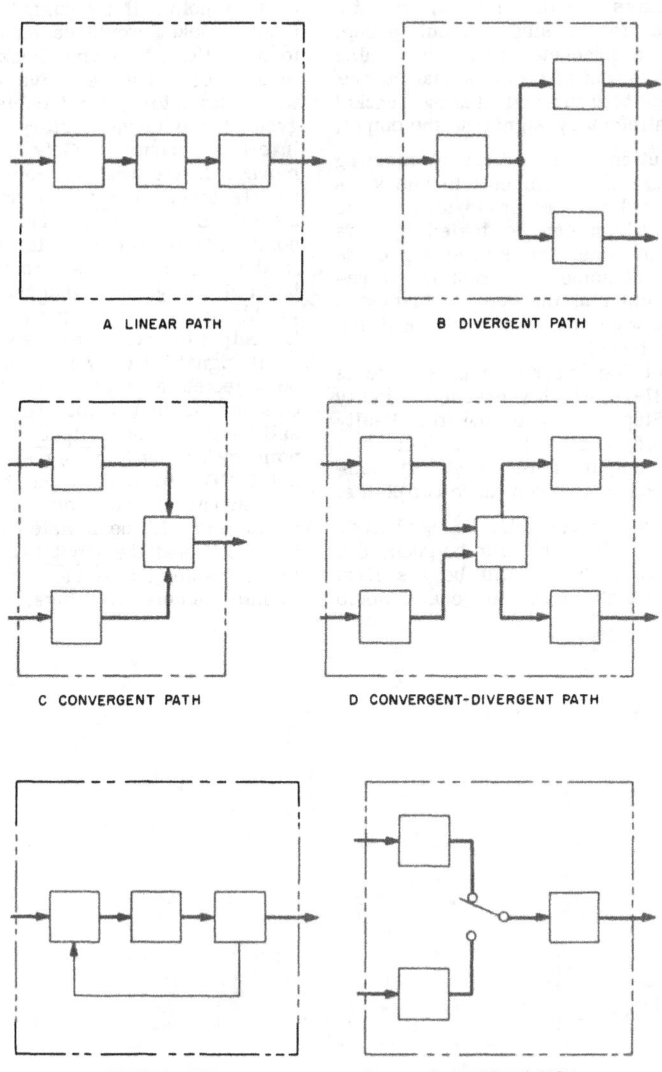

Figure 27-9.—Types of signal paths.

the input or output of one of these circuit groups, and further tests are made to verify whether the test points within the circuit group are good or bad. The brackets are then moved, one at a time, until they enclose a single circuit group. A circuit group is bracketed when an opening bracket indicates a satisfactory signal at the input of the circuit group and closing bracket indicates an unsatisfactory signal at the output.

Signal substitution is a method of injecting an artificial signal into a circuit to check its performance. A radio receiver is an example of an equipment which can be tested by this method. If a signal generator is used to provide the proper signal at some test point in the receiver, a good output at the speaker indicates that the area between the test point and the speaker is free of trouble.

An example of the bracketing procedure is shown in figure 27-10, which represents a faulty functional unit. Step 4—Localizing the Faulty Function—indicated a good input at test point ''A'' and a faulty output at test point ''E'' indicating a trouble located between these two points.

Signal tracing would normally be used since the signal at point ''A'' is known to be good. The signal at test point ''C'' would be the first logical place to check since this check would eliminate half of the circuit. (This will be discussed later in the text.) The signal at test point ''C'' can be checked by placing a meter or scope at this point. If the signal is satisfactory, the input bracket would be moved from point ''A'' to point ''C''. The next check at point ''D'' would then isolate the defective circuit. If the test was satisfactory, the trouble would be in circuit group IV; if unsatisfactory, the trouble would be in circuit group III. Only the input bracket was moved. If the test at point ''C'' had been unsatisfactory, the output bracket would have been moved to this point. The next check at test point ''B'' would then isolate the defective circuit. If the signal is satisfactory, the trouble would be in circuit group II; if unsatisfactory, the trouble would be in circuit group I. In this case, only the output bracket was moved.

If signal injection is used, a test signal would be injected at test point ''C'', and the output checked at test point ''E''. If the results are satisfactory, the output bracket is moved to point ''C'', and a signal is injected at test point ''B'' to isolate the defective circuit. If the signal is not correct at point ''E'', the trouble should be isolated between points ''C'' and ''E'' and the input bracket moved to point ''C''. Injecting a signal at point ''D'' would then isolate the defective stage.

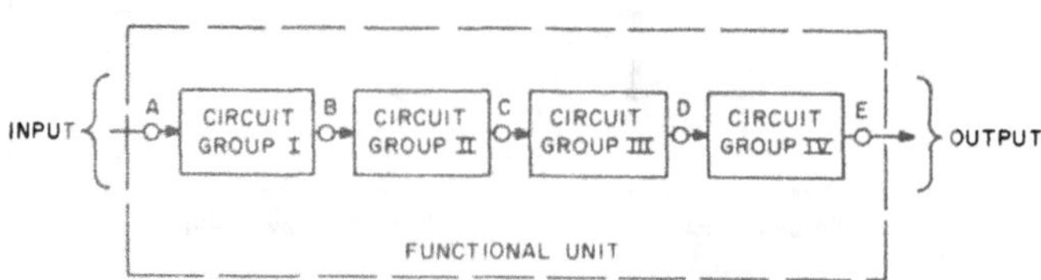

179.486
Figure 27-10.—Functional unit with input and output brackets.

The first method of bracketing to be considered will be the method used for linear circuit arrangements. The best method of troubleshooting this type of circuit path is the half-split method. Assume that you have brackets at the input and output of a number of circuits or circuit groups in which the signal path through all circuits is linear (fig. 27-11A). Unless the symptoms point to one circuit in particular which might be the trouble source, the most logical place to move a bracket is to a convenient test point near the center of the bracketed area (point C fig. 27-11A). If a test indicates that the signal is good at this point, an input bracket should be left there (fig. 27-11B). The brackets will then surround the second half of the linear circuit path, and the other half will be eliminated from the trouble area. If an incorrect signal is found at the test point, an output bracket placed at this point will show that the trouble exists in the first half of the linear circuit path (fig. 27-11C). This process should be repeated with the area now enclosed with brackets until the brackets surround only one circuit. With the half-split method, a defective circuit can be located with a minimum of tests. By testing the circuits in sequence, the trouble may take many tests before being located.

It is unusual to find a complete functional unit with only one linear signal path; however, the half-split method can be applied to any part of a unit which contains a linear path. Convergent-divergent signal paths require a different technique. This type of signal path is not as easily recognized as linear paths; however, if you follow the definitions previously given, you should have little difficulty with this part of the procedure.

The next method of bracketing to be considered is applied to feedback signal paths. Before describing this method, a short discussion of the principles of feedback circuits will be necessary. As stated previously, a feedback signal path is one in which a signal is removed from some point in a circuit chain and applied to a point preceding its source. Since this feedback signal is combined with the original signal, it will tend to either increase or decrease the signal amplitude. If the feedback signal arrives back at the main signal path in phase with that signal, it is called regenerative feedback and will increase the gain of the circuit chain. When the feedback signal is out of phase with the main signal, it is called degenerative feedback and will decrease the circuit gain.

Regenerative feedback is used extensively in oscillator circuits. This type of feedback makes it possible for an oscillator to generate an a.c. signal. Sometimes regenerative feedback is used to increase the gain of an amplifier; however, its tendency to cause oscillations limits its use in amplifiers. Degenerative feedback is often employed in amplifiers to decrease distortion and increase the bandwidth of a circuit. Another application of degenerative feedback is in automatic-volume-control (AVC) systems. In this case the feedback has the function of keeping the

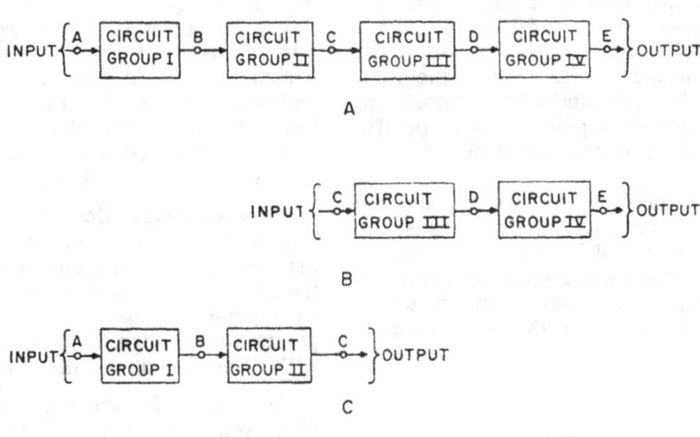

Figure 27-11.—Half-split method of bracketing.

output of a receiver essentially constant despite varying input signal strength.

When troubleshooting circuits with feedback paths, it is important to consider the type of feedback and the function of feedback in the circuit. Since a regenerative feedback path results in an increased circuit gain, a trouble in the feedback path results in a decreased output signal. Conversely, a trouble in a degenerative feedback path results in an increased output signal.

Another method of troubleshooting circuits containing feedback loops is to disable the feedback loop. This may be accomplished by disconnecting the feedback path (opening the loop) or by shorting the feedback signal to ground. The first method is sometimes inconvenient, and the second method should be used only when it has been determined that shorting the signal to ground will not cause damage to the circuit.

Disabling the AVC loop is convenient in the case of a receiver with an AVC ON-OFF switch. If a trouble is located in the circuit containing the AVC loop, the switch can be used to determine whether the trouble still exists without AVC.

The last type of signal path is the switching path. You have seen how electronic equipments are composed of various circuit chains interconnected to perform a desired task. Control of these circuits is usually accomplished by the use of switches placed directly in the circuits, or by remote switching relays.

In order to isolate faulty circuits along a switching branch, we initially test the final signal output for the branch following the switch. When the switch is a multiple-contact type, each contact may be connected to a different circuit branch. In this case, it may be necessary to place the switch in each position and check the final output of the branch associated with that position. If the symptoms and data point to one specific branch, it may not be necessary to check every switch position.

Once this test has been performed and the trouble is isolated to one or more branches, the suspected branches should be checked to locate the faulty branch. The next step is to apply the half-split, convergent-divergent, or feedback method, as required, to isolate the faulty circuit.

Summary

From the discussion of bracketing procedures and applications presented, it should now be apparent that the procedure to be used is dependent upon the signal paths and circuit arrangement involved in the equipment being analyzed. It should also be apparent that, in general, two or three — or even all — of the bracketing techniques may have to be used in combination, since most equipments are complex enough to contain several circuit arrangements.

The overall procedure to be used in step 5, however, should always follow a certain pattern. Before the narrowing down process begins, you should analyze the information previously obtained and determine whether you have sufficient data to begin the bracketing procedure. Brackets are originally placed at the correct input(s) and incorrect output(s) of the faulty functional section, and the ICO analogy is applied to determine whether the symptoms point to one or more circuit groups in particular.

The subfunctions performed by the various circuit groups and the signal paths interconnecting these groups should be considered in this selection of possible faulty circuit groups. The inputs and/or outputs of these circuit groups should be checked, the most probable faulty subfunctions being checked first.

The accessibility of test points is an important consideration when checking these circuit groups. As each test is performed, an input bracket if the signal is satisfactory, and an output or closing bracket if the signal is unsatisfactory, is moved to this point.

When the brackets enclose a signal circuit group, the ICO analogy and the signal paths within the circuit groups are considered to determine where to make the next test. Each test should attempt to eliminate as many circuits from the bracketed area as possible, unless the symptoms indicate that one or more circuits are more likely to be at fault than the rest of the circuit group. This process is continued until the brackets are at the input and output of a signal circuit.

The servicing block diagram is an invaluable aid in each step of the bracketing procedure. It provides you with a pictorial guide which indicates the main and secondary signal paths, shows the waveforms to be expected at important test points, and divides the function into circuit groups with titles to indicate their subfunctions.

In step 6, failure analysis, which is the next step, you will learn how to continue the troubleshooting procedure to locate the faulty part in the circuit isolated in step 5.

STEP 6—FAILURE ANALYSIS

The recognition, verifications, and descriptive information obtained in steps 1 and 2—symptom recognition and symptom elaboration—enabled you to make a logical and valid estimate concerning the selection of the faulty functional unit in step 3. In step 4—localizing the faulty function—you performed simple input-output tests. Step 5 carried you deeper into the circuits comprising the equipment being tested. This step—localizing trouble to the circuit—required that you perform extensive tests as prescribed by the bracketing procedure suited for the particular circuit arrangement involved. This bracketing procedure enabled you to find the malfunctioning circuit or stage within the faulty functional unit.

The final step in our six-step troubleshooting procedure—failure analysis—will require that you test certain branches of the faulty circuit in order to determine where the faulty part lies. These branches are the interconnected networks associated with each element of the transistor, electron tube, or other active device in the faulty circuit.

Step 6—failure analysis—places you in the position necessary to replace or repair faulty circuit components so that the equipment can be returned to optimum serviceability. However, locating the faulty part does not complete step 6. You will also be concerned with determining the cause of the failure. It is quite possible that still another failure occurred and, unless all faults are corrected, the trouble will reoccur at a later date. The final step in failure analysis requires that certain records be maintained. These records will aid you or some other technician in the future. They may also point out consistent failures which could be caused by a design error. When this step has been finished satisfactorily you can perform whatever repairs are necessary.

Schematic Diagrams

Schematic diagrams illustrate the detailed circuit arrangements of electronic parts (represented symbolically) which make up the complete circuits within the equipment or unit. These diagrams show what is inside the blocks on a servicing block diagram and provide the final picture of an electronic equipment. Figure 27-12, shows the schematic diagram of the receiver unit for a transceiver set. The receiver unit differs slightly from the one illustrated by the functional diagram in figure 27-4. The frequency conversion function is accomplished in a single tube—there are no separate mixer and RF oscillator circuits. Only one IF amplifier circuit is used.

Note that the values of the circuit parts are listed. Each part is also given a reference designation for identification purposes. These diagrams will be very helpful in making tests not shown directly on the servicing block diagram, as well as determining which branch of an isolated faulty circuit needs to be repaired. For example, to check the bias resistor of the phase splitter tube, V6, you could place the multimeter probes on pin 7 and the junction of R19 and R21. The value, as shown, should be 1000 ohms.

Voltage and Resistance Charts

Once the faulty circuit has been isolated, the voltages and resistances of the various circuit branches must be measured, in order to determine which components within the circuit are at fault. The measurement results must be compared to voltage and resistance charts or tables in order to evaluate them. This information may appear on the apron of its associated fold-out schematic diagram, or it may be on a separate page(s) in the manual. The normal voltage and resistance reading to ground (or other point of significance) for each tube socket pin is given. Also listed are the conditions necessary to observe the gain reading, such as control settings and equipment connections.

Types of Circuit Trouble

In step 1, we recognized that an equipment was not functioning from the audible and visual indications of performance. Equipment failure results in a complete lack of performance information which would normally be seen or heard by the equipment operator. The second type of abnormal performance we discussed was "degraded performance." For this type the visual and audible indications are present, but they are not up to the specifications provided for normal operation.

Regardless of the type of trouble symptom, the actual fault can eventually be traced to one or more of the circuit parts—resistors, capacitors, etc—within the equipment. The actual fault may also be classified by the degree of malfunction. The complete failure or abnormal performance of a part, of course, falls in line with

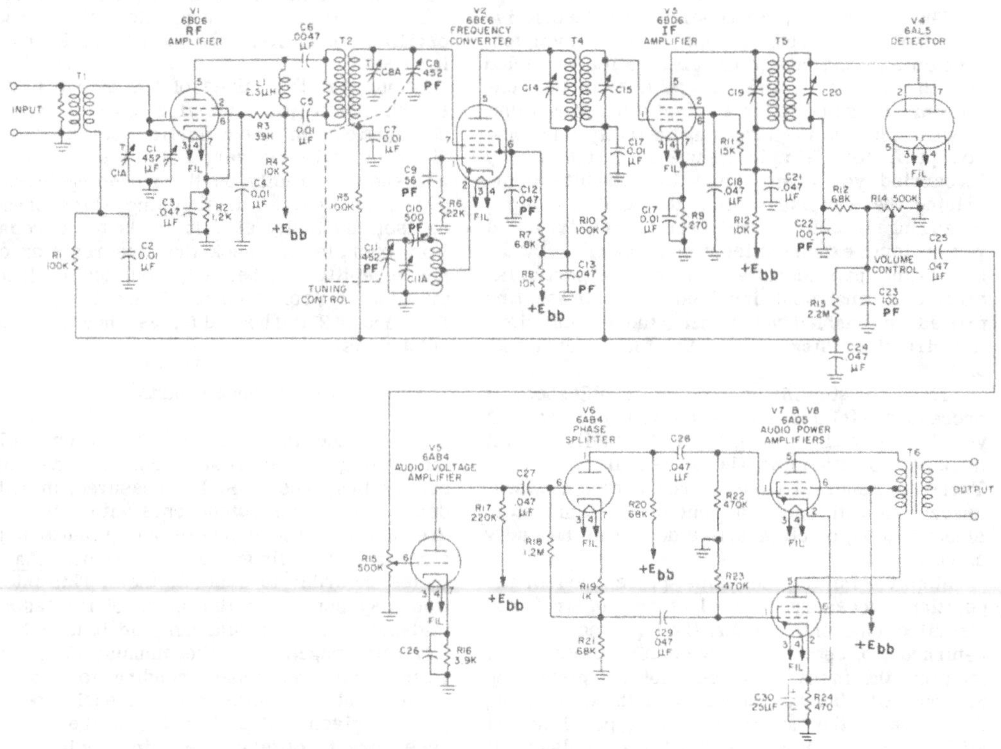

179.488

Figure 27-12.—Schematic diagram of receiver unit, transceiver set.

the previous use of these terms. These types of faults are easily discovered.

There is a third degree of part malfunction which is not always so obvious. This is the intermittent part malfunction. Intermittent, by definition, refers to something which alternately ceases and begins again. This same definition applies to electronic part malfunctions. The part operates normally for a period of time, then fails completely or operates on a degraded level for a while, and then returns to normal operation. The cyclic nature of this malfunction is an aid in determining that it exists. However, it is often difficult to locate the actual faulty part. This is true because of the fact that while you are testing the circuit in which this part lies, it may be operating normally. Thus you will pass it by as satisfactory, only to be faced with trouble again as soon as the cycle of operation completes itself.

Isolation of Faulty Parts

The first step in isolating a faulty part within a circuit is to apply the same ICO method used in previous steps. The output signal should be analyzed to aid in making a valid selection of the parts or branch of the circuit which may cause the defective output. The voltage, duration, and/or shape of the output waveform may be indications of possible open or shorted parts or out-of-tolerance values. This step performs two functions: it reduces to a minimum the number of test readings required, and it helps determine whether the faulty part, when located, is the sole cause of the malfunction.

The second step in isolating a faulty part is a visual inspection of the parts and leads in the circuit. Often this inspection will reveal burned or broken parts or defective connections. Open filaments in electron tubes may also be spotted in this check.

Voltage measurements at transistor leads or at the pins of electron tubes can be compared with the normal voltages listed in available voltage charts to provide valuable aid in locating the trouble. This check will often help isolate the trouble to a single branch of a circuit. A separate circuit branch is generally associated with each pin connection of the transistor or electron tube. Resistance checks at the same points are also useful in locating the trouble. Suspected parts can often be checked by a resistance measurement.

When a part is suspected of being defective, a good part may be substituted for it in the circuit. You must keep in mind, however, that an undetermined fault in the circuit may also damage the substituted part. Another factor to consider before performing this step is that some circuits are critical and substituting parts (especially transistors or electron tubes) may alter the circuit parameters.

In some equipments the circuits are specifically designed for the substitution process. For example, the plug-in circuit module is being employed in many new electronic equipments. It contains all of the necessary parts (resistors, capacitors, and inductors) for a circuit branch or even the entire circuit. Once a trouble has been bracketed to a module, substitution of the module is the only method of correcting the fault.

Systematic Checks

Probable deductions should always be checked first. Next because of the safety practice of setting a voltmeter to its highest scale before making measurements, the points having the highest voltages should be checked. (Transistor collector and electron tube plate and screen grid.) Then the elements having smaller voltages should be checked in the descending order of their applied voltage, that is, the transistor emitter and base or the electron tube cathode and control grid.

Voltage, resistance, and waveform readings are seldom identical to those listed in the manual. The most important question concerning voltage checks is, "How close is good enough?" In answering this question, there are many factors to consider. The tolerances of the resistors, which greatly affect the voltage readings in a circuit, may be 20%, 10%, or 5%; in some critical circuits, precision parts are used. The tolerances marked or color-coded on the parts are, therefore, one important factor. Transistors and electron tubes have a fairly wide range of characteristics and will thus cause variations in voltage readings. The accuracy of the test instruments must also be considered. Most voltmeters have accuracies of 5 to 10 percent, while precision meters are more accurate.

For proper operation, critical circuits require voltage readings within the values specified in the manufactures technical manual; however, most circuits will operate satisfactorily if the voltages are slightly off. Important factors to consider are the symptoms and the output signal. If no output signal is produced at all, you should expect a fairly large variation of voltages in the trouble area. A trouble which results in a circuit performance just out of tolerance, however, may cause only a slight change in circuit voltages.

Locating the Faulty Part

The voltage and/or resistance checks discussed previously indicate which branch within a circuit is at fault. We must now isolate the trouble to a particular part (or parts) within the branch.

One procedure for accomplishing this is to move the test probe to the different points where two or more parts are joined together electrically and measure the voltage or resistance with respect to ground. Generally, however, the correct values (particularly voltage) will be difficult to determine from these points on a schematic diagram and may not be available elsewhere. Thus, we shall reserve this procedure for making resistance checks to locate shorts and openings in the branch. A better check to use when voltage readings are not normal is a systematic check of the value of each resistor, capacitor, and/or inductor in the branch. The instruments required for these measurements include impedance bridges, Q-meters, etc.

Review of Previous Data

A review of all the symptom and test information obtained thus far will help isolate other faulty parts, whether the malfunction of these parts is due to the isolated malfunction or to some entirely unrelated cause (multiple malfunction).

In order to determine if there is an indication of multiple malfunction, we will start with the part we have isolated and ask the question, "What effect does the malfunction of this part have on the operational function of the equipment?" If the isolated malfunction can produce all of the normal and abnormal symptoms and indications we have accumulated, we can logically assume that it is the sole item at fault. If not, we must utilize our knowledge of electronics and of the equipment itself to determine what other malfunction(s) must also occur to provide all of the symptoms and test data.

Common Malfunction Causes

The review of accumulated test information as discussed in the preceding section is indispensable to the recognition and isolation of multiple malfunction or malfunction directly caused by the isolated malfunction. How about the isolated malfunction itself? What could have caused the trouble?

Consider the transistor amplifier circuit shown in figure 27-13. Assume that our troubleshooting procedures have isolated the transistor as the cause of trouble—it is burned out. What could cause this? Excessive current can destroy the transistor by causing internal shorts or by altering the characteristics of the semiconductor material, which may be very temperature-sensitive. Thus, the problem reduces to a matter of determining how excessive current can be produced.

Excessive current could be caused by an excessively large input signal, which would overdrive the transistor. Such an occurrence would indicate a fault somewhere in the circuitry preceding the input connection. Power surges (intermittent excessive outputs) from the power supply could also cause the burn-out. In fact, power supply surges are a common cause of transistor (and electron tube) burn-out.

It is advisable to check for the conditions just mentioned before placing a new transistor in the circuit. Bias stabilization circuits are generally included (as in figure 27-13) to reduce the effect of excessive bias currents.

Some other malfunctions, along with their common causes, include:

1. Burned-out cathode resistors caused by shorts in electron tube elements.

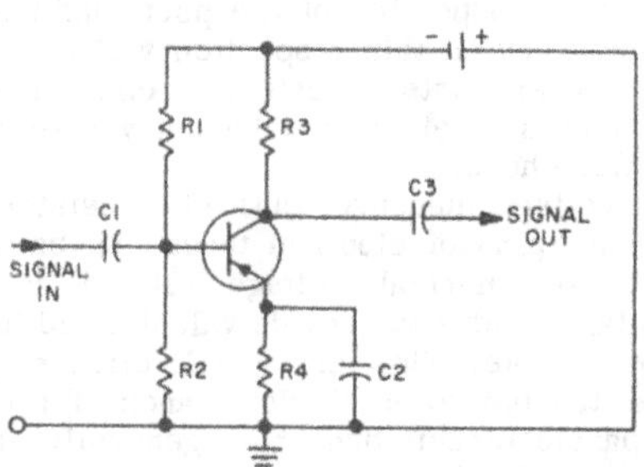

Figure 27-13.—Transistor amplifier circuit.

2. Power supply overload caused by a short-circuit in some portion of the voltage distribution network.

3. Burned-out transformer in shunt feed system caused by shorted blocking capacitor.

4. Burned-out fuses caused by power-supply surges or shorts in filtering (power) networks.

In general, a degraded component characteristic can be traced to an operating condition which caused the maximum ratings of the component to be exceeded. The condition may be temporary and accidental, or it may be deeply rooted in the circuitry itself.

Bad tubes account for over 60 percent of all equipment malfunctions. For this reason, the possibility of such a fault should be uppermost in your mind when you have reached this point in our six-step troubleshooting procedure.

As tubes age they undergo certain inherent changes. For example, there is a change in transconductance which generally lowers the gain of the tube. Also, the heater-to-cathode leakage current increases with age. This current is responsible for hum and other undesirable "coupling" effects between the heater and cathode. The period of time required for the leakage increase is shortened if the tube is frequently operated at a higher-than-normal temperature.

Other tube troubles include grid current variations, gas leakage, and improper usage (operation at excessive values). A consideration of these factors can locate multiple malfunctions, as well as prevent a future failure of similar nature.

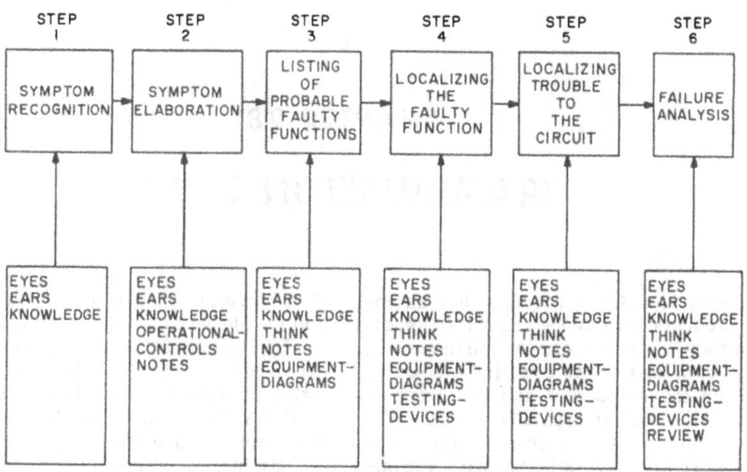

Figure 27-14.—The six step procedure.

This final step—Failure Analysis—has brought you to the end of our six-step troubleshooting procedure. You have "narrowed down" the trouble area with each successive step until the malfunctioning part has been isolated. You have reviewed your procedure to ensure that multiple malfunctions do not exist and to verify the cause of the malfunction. You have also made the necessary records of your actions.

By replacing the faulty part and rechecking equipment operation, you can return the equipment to operation with the knowledge that you have completed your troubleshooting duties.

Although not directly connected with troubleshooting procedures previously outlined, the technician should reorder any parts used in the repair of the faulty equipment. Proper logistic support will enable the technician to return the equipment to an operating status once the trouble has been located.

The six-step troubleshooting procedure is summarized in figure 27-14.

CHAPTER 28

TRANSMISSION LINES

Practical considerations require that an antenna be located some distance from its associated transmitter or receiver. A means must, therefore, be provided to transfer the electrical energy between the equipment and the antenna. Basically, an RF transmission line electrically connects an antenna and a transmitter or a receiver. It must do so efficiently with a minimum loss of power or signal.

TYPES OF TRANSMISSION LINES

There are five general types of transmission lines—the parallel two-wire line, the twisted pair, the shielded pair, the concentric (coaxial) line and the waveguides. The use of a particular type of line depends, among other things, on the frequency and the power to be transmitted and on the type of installation.

PARALLEL TWO-WIRE LINE

One type of transmission line consists of two parallel conductors that are maintained at a fixed distance by means of insulating spacers or spreaders that are placed at suitable intervals. This type of line is shown in figure 28-1A. The line has the assets of ease of construction, economy, and efficiency. In practical applications two-wire transmission lines (with individual insulators rather than spacers) are used for power lines, rural telephone lines, and telegraph lines. This type of transmission line is also used as the connecting link between an antenna and transmitter or an antenna and receiver.

In practice, such lines used in radio work are generally spaced from 2 to 6 inches apart at frequencies of 14 MHz and below. The maximum spacing for frequencies of 18 MHz and above is 4 inches. In any case, in order to effect the best cancellation of radiation, it is necessary that the wires be separated by only a small fraction of a wavelength. For best results, the separation should be less than 0.01 wavelength.

The principal disadvantage of the parallel-wire transmission line is that it has relatively high radiation loss and therefore cannot be used in the vicinity of metallic objects, especially when high frequencies are used, because of the greatly increased loss which results.

Uniform spacing of a two-wire transmission line may be assured if the wires are imbedded in a solid low-loss dielectric throughout the length of the line, as indicated in figure 28-1B. This type of line is often called a two-wire ribbon type. The ribbon type is commonly made with two characteristic impedance values, 300 ohms and 75 ohms. The 300-ohm line is about one-half inch wide and is made of stranded wire. Because the wires are imbedded in only a thin ribbon of polyethylene, the dielectric is partly air and partly polyethylene. Moisture or dirt will change the characteristic impedance of the line. This effect becomes more serious if the line is not terminated in its characteristic impedance.

The wires of the 75-ohm line are closer together, and the field between the wires is confined largely to the dielectric. Weather and dirt affect this line less than they affect the 300-ohm

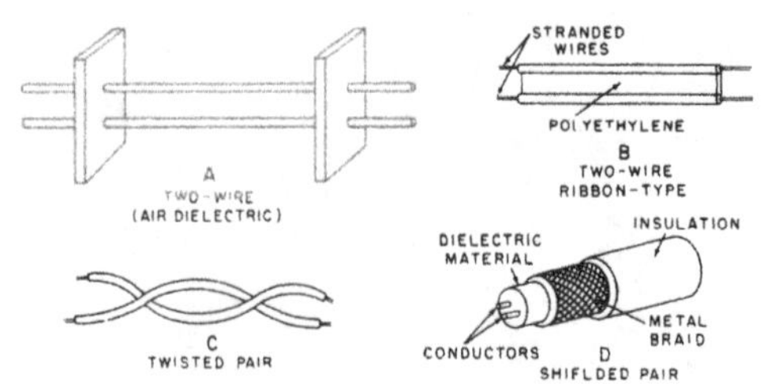

Figure 28-1.—Four types of transmission lines.

line. The ribbon type of line is widely used to connect television receivers to antennas.

TWISTED PAIR

The twisted pair is shown in figure 28-1C. As the name implies, it consists of two insulated wires twisted to form a flexible line without the use of spacers. It is used as an untuned line (on a tuned line the insulation might be broken down by arc-over at voltage loops) for low frequency transmission. It is not used for the higher frequencies because of the high losses occurring in the rubber insulation. When the line is wet, the losses increase greatly. The characteristic impedance of such lines is about 100 ohms, depending on the type of cord used.

SHIELDED PAIR

The shielded pair (shown in fig. 28-1D) consists of two parallel conductors separated from each other and surrounded by a solid dielectric. The conductors are contained within a copper-braid tubing that acts as a shield. This assembly is covered with a rubber or flexible composition coating to protect the line against moisture and friction. Outwardly, it looks much like an ordinary power cord for an electric motor.

The principal advantage of the shielded pair is that the two conductors are balanced to ground — that is, the capacitance between each conductor and ground is uniform along the entire length of the line and the wires are shielded against pickup of stray fields. This balance is effected by the grounded shield that surrounds the conductors at a uniform spacing throughout their length.

If radiation from an unshielded line is to be prevented, the current flow in each conductor must be equal in amplitude in order to set up equal and opposite magnetic fields that are thereby canceled out. This condition may be obtained only if the line is clear of all obstructions, and the distance between the wires is small. If, however, the line runs near some grounded or conducting surface, one of the two conductors will be nearer that obstruction than the other. A certain amount of capacitance exists between each of the two conductors and the conducting surface over the length of the line, depending upon the size of the obstruction. This capacitance acts as a parallel conducting path for each half of the line, causing a division of current between each conductor. Since one conductor may be nearer the obstruction than the other, the current flow will accordingly be increased, resulting in an inequality of current flow in the two conductors and therefore incomplete-cancellation of radiation. The shielded pair, therefore, eliminates such losses to a considerable degree by maintaining balanced capacitances to ground.

COAXIAL

Coaxial lines, or coaxial cables as they are called, are the most widely used type of RF transmission line. They consist of an outer conductor and an inner conductor held in place exactly at the center of the outer conductor.

Several types of coxial cable have come into wide use for feeding power to an antenna system. Figures 28-2A and B illustrate the construction of flexible and rigid coaxial cables.

The power lost in a properly terminated coaxial line is the sum of the effective resistance losses along the length of the cable and the dielectric losses between the two conductors. The dielectric losses between the two conductors are the greater, since they are largely due to the spacers (which have a greater than unity dielectric constant), that support the inner conductor.

Figure 28-2 shows that one of the conductors is placed inside the other. Since the outside conductor completely shields the inner one, no radiation loss takes place. The conductors may both be tubes, one within the other; the line may consist of a solid wire within a tube, or it may consist of a stranded or solid inner conductor with the outer conductor made of one or two wraps of copper shielding braid.

In the type of cable most popular for military use, the outer conductor consists of a braid of copper wire, and the inner conductor is supported within the outer by means of a semi-solid dielectric with exceedingly low-loss characteristics called polyethylene. The Army-Navy designation for one size of this cable, which is suitable for power levels up to one kilowatt at frequencies as high as 30 MHz, is RG-8/U. The outside diameter of this type of cable is approximately one half inch. Its characteristic impedance is 52 ohms.

When using solid electric coaxial cables, it is necessary that precautions be taken to ensure that moisture cannot enter the line. If the better grade of connectors manufactured for the line are employed as terminations, this condition is automatically satisfied. If connectors are not used, it is necessary that some type of moisture-proof

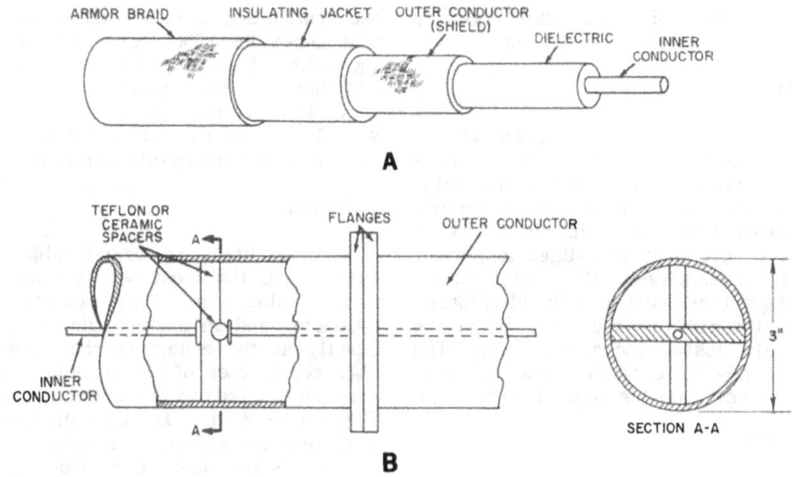

67.82(67C)
Figure 28-2.—Coaxial cables: (A) Flexible; (B) Rigid.

sealing compound be applied to the end of the cable where it would be exposed to the weather.

The chief advantage of the coaxial line is its ability to keep down radiation losses, very little electric or magnetic fields extend outside the outer conductor. Therefore, nearby metallic objects cause minimum loss, and the coaxial cable may be run up air ducts or elevator shafts, inside bulkheads, or through metal conduit. Insulation troubles can be forgotten. The coaxial cable may be buried in the ground or suspended above ground.

WAVEGUIDES

Figure 28-3 illustrates circular and rectangular waveguides. The behavior of a waveguide is similar in many ways to that of a transmission line. Thus, waves traveling along a guide have a phase velocity and are attenuated. When a wave reaches the end of a guide, it is reflected unless the load impedance is carefully adjusted to absorb the wave; also an irregularity in a waveguide produces reflection just as an irregularity in a transmission lines does. Again, reflected waves can be eliminated by the use of an impedance-matching system, exactly as with a transmission line. Finally, when both incident and reflected waves are simultaneously present in a waveguide, the result is a standing wave pattern.

In some respects, waveguides and transmission lines are unlike in their behavior. The most striking difference is that a particular mode will propagate down a waveguide with low attenuation only if the wavelength of the wave is less than some critical value determined by the dimensions and the geometry of the guide. If the wavelength is greater than this cutoff value, the waves in the waveguide die out very rapidly in amplitude even when the walls of the guides are of material having infinite conductivity. Waveguides are discussed in detail in Volume 2 of this training manual.

DISTRIBUTED IMPEDANCE CONSTANTS

At radio frequencies, the capacitance and inductance of any conductor become significant.

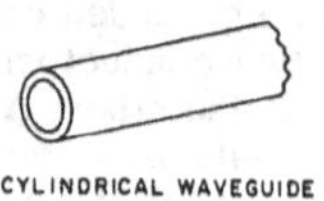

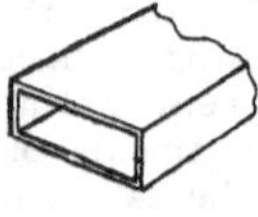

179.491
Figure 28-3.—Waveguides.

This creates special problems and considerations in the use of these conductors as efficient transmission lines. (This is especially true of transmission lines which are relatively long.)

The amount of inductance and capacitance depends on the length of the transmission line, the size of the conducting wires, the spacing between the wires, and the dielectric material separating the wires. Transmission lines also have resistance, which is dependent upon the conductor diameter, length, and conductor material.

Figure 28-4 illustrates how these constants are distributed along a line. The resistance is distributed uniformly along the entire length of the line and is usually measured in ohms per unit length. This is represented by R1 and R2. There is also a certain quantity of leakage resistance between the wires represented by R3. This resistance is in shunt with the input and output ends and is the result of current leakage between conductors through the dielectric.

The wires forming a transmission line also possess distributed inductance. This inductance can be seen in the action of the magnetic fields set up by current flow. For example, if the current attempts to drop to zero suddenly, the collapsing magnetic fields sustain the current for a time, which is a measure of the distributed inductance. This inductance, L1 and L2, is effectively in series with the line. It is measured in microhenries per unit length.

Finally, there is an electric field set up between the wires forming a transmission line. The intensity of this electric field is a measure of the distributed capacitance, which is expressed in picofarads per unit length, and acts as a capacitance, C, shunted across the line.

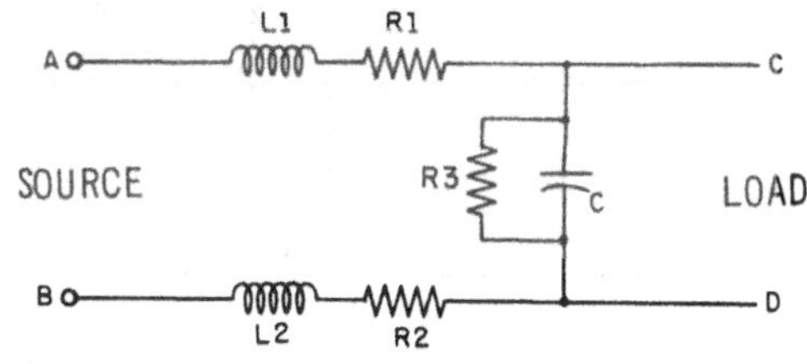

Figure 28-4.—Equivalent circuit of a two-wire transmission line of unit length.

CHARACTERISTIC IMPEDANCE

In addition to having the distributed constants, a transmission line has a "characteristic impedance." Essentially, the characteristic impedance, Z_o, of a transmission line is the impedance it would have if it were infinite in length. This impedance is considered to be purely resistive, and is constant for a given transmission line. The characteristic impedance is important in determining how well energy is transferred from the source to the load. For the infinitely long line, all of the energy sent out on the line would appear to be dissipated in some remote load.

In any circuit, such as the one shown in figure 28-4, some current will flow if a voltage is applied across terminals A and B. The ratio of the voltage to the current is the impedance, Z, ($Z = \frac{E}{I}$). The impedance presented to the input terminals is much more than the simple resistance of the wires in series with the impedance of the load. The effects of series inductance and shunt capacitance distributed along the line are appreciable and constitute the principle components of the equivalent network, shown in figure 28-5. The formula for the characteristic impedance as a function of the L and C of a unit length of transmission line may be determined from the simplified T network in figure 28-5.

The conductor resistance R_s, and the insulation leakage conductance $\frac{1}{R_{SH}}$, of figure 28-4, are low and considered negligible; hence, they are not shown in the simplified network. The distributed inductance of the line is divided equally in two parts on the horizontal arms of the T. The distributed capacitance is lumped in one value in the center leg of the T. The line is terminated in a resistive load having a value equal to that of the characteristic impedance of the line (Z_o).

Therefore, since $Z_o = \sqrt{\frac{L}{C}}$, the characteristic impedance depends on the distributed inductance and capacitance of the line. By increasing the spacing between the wires of a transmission line, the following effect is noted. The capacitance of the line is lowered due to the increased distance between the plates of the capacitor (in this case the plates are the two wires). This increased distance causes a weaker electric field to exist between the two wires. The inductance of the line remains essentially unchanged and is relatively independent of the spacing between the wires. Thus, the effect of increasing the spacing of the wires is to increase the

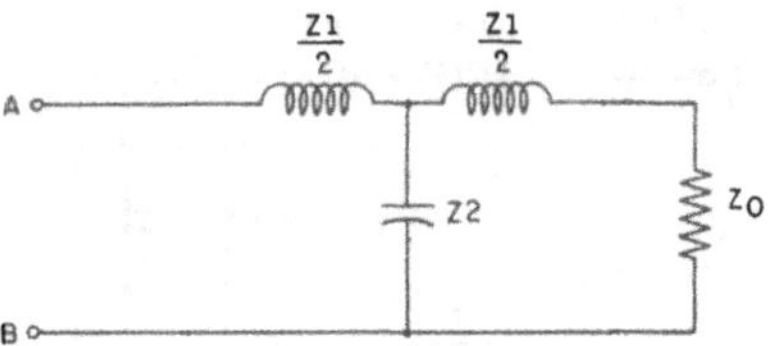

20.224

Figure 28-5.—Simplified equivalent circuit terminated in its characteristic impedance.

characteristic impedance, because the $\frac{L}{C}$ ratio is increased.

Similarly, a reduction in the diameter of the wires also increases the characteristic impedance. The reduction in the size of the wire affects the capacitance more than the inductance, as the effect is equivalent to decreasing the size of the plates in a capacitor. Any change in the dielectric material between the two wires also changes the characteristic impedance. Thus, if a change in dielectric material increases the capacitance between the wires, the characteristic impedance is reduced.

Whenever a transmission line is terminated with a purely resistive load equal to the characteristic impedance of the line, the line behaves like an infinitely long line. All of the energy from the source is transferred down the line to the load and dissipated in the form of heat. This is the condition desired when the transmission line is connected between the transmitter or receiver and an antenna.

In the transmitter-antenna system, the transmitter acts as the source and the antenna as the load. The antenna dissipates the energy in the form of an electromagnetic field rather than as heat. When the antenna is connected to a receiver, the antenna acts as the source and the receiver input impedance is the load.

WAVE MOTION IN AN INFINITE LINE

Since the desired impedance match between the antenna and transmitter or receiver caused the line to act as if it were infinitely long, a detailed analysis of the infinite line will be presented in this section.

One wavelength is the distance traveled by a wave during a period of one cycle. It can easily be determined by the formula $\lambda = \frac{V}{f}$ where λ = the distance in meters from the crest of one wave to the crest of the next; f = frequency in hertz and V = 300×10^6 meters per second (the velocity of the radio wave). Actually, the radio wave velocity is 299.8×10^6 meters per second but can be rounded off to 300×10^6 for general usage.

Traveling waves exist on the line because it takes a certain time to propagate RF energy down a line. Figure 28-6 shows sine waves of voltage and current that travel at high speed along a two wire transmission line of infinite length. Because this line is of infinite length, no reflections occur; therefore, the voltage and current are in phase with each other, everywhere along the line. Because of line losses, the curves diminish in amplitude as the waves progress along the line. If a voltage is impressed on a line such as this, an electric field will be established between the wires. Likewise, current will flow in the wires, and a magnetic field will be established around each wire. These two fields constitute an electromagnetic wave that travels down the wire at a velocity approximately equal to that of light.

Actually, figure 28-6 illustrates what would happen if the voltage and the current could be stopped for an instant in time. Assume that at a given instant the voltage at the generator terminals is zero. An instant later one terminal

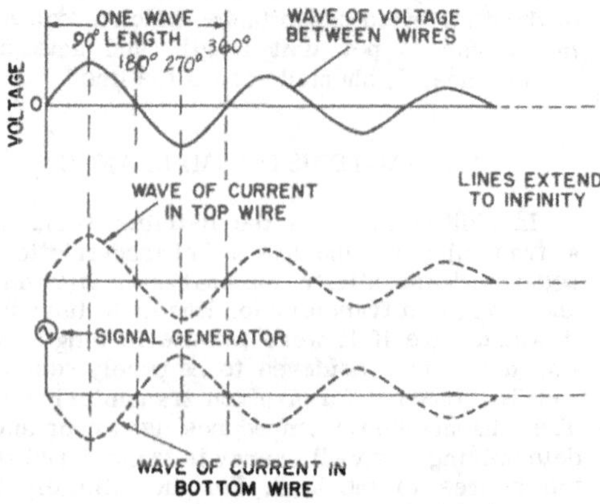

20.225

Figure 28-6.—Traveling waves of current and voltage on a line of infinite length.

becomes more positive and the other becomes more negative. The electric field between the wires increases in strength; the current and the magnetic field increase proportionately. The perpendicular distance from any point along the wire to the current curve indicates the relative magnitude and direction of the current at that point. The perpendicular distance from any point along the voltage axis to the voltage curve represents the relative magnitude and polarity of the voltage across the line at the corresponding location.

At 90° in the electrical cycle the electric and magnetic fields are at their maximum, and from 90° to 180° they decrease in amplitude to zero. At 180° the voltage at the generator terminals reverses polarity, and the electric field between the wires reverses polarity. Similarly, the current reverses direction, which causes the magnetic field to reverse direction. The fields increase in strength from 180° to 270°, and then decrease in strength to zero at 360°. These electric and magnetic impulses do not return to the generator once they start down a line of infinite length.

The characteristics of a theoretically infinite line may be summarized as follows:

1. The voltage and the current are in phase throughout the line.

2. The ratio of the voltage to the current is constant over the entire line and is known as the characteristic impedance.

3. The input impedance is equal to the characteristic impedance.

4. Since the voltage and current are in phase, the line operates at maximum efficiency.

5. Any length of line can be made to appear infinite in length if it is terminated in a load equal to its characteristic impedance.

NONRESONANT LINES

A nonresonant transmission line can be defined as a line that has no standing waves of current and voltage. Such a line is either infinitely long or is terminated in its characteristic impedance. Because there are no reflections, all of the energy passed along the line is absorbed at the load except for the small quantity of energy dissipated by the line. The voltage and current waves are traveling waves that move in phase with each other from the source to the load.

On lines carrying radio frequencies, the characteristic impedance is almost always purely resistive. Therefore, it is customary to say that a non-resonant line is terminated in a resistive load equal to the characteristic impedance of the non-resonant line. The nonresonant line is used for maximum efficiency when transferring power from a source to a load.

RESONANT LINES

A resonant transmission line is one that has standing waves of current and voltage. The line is of finite length, is not terminated in its characteristic impedance, and, therefore, reflections are present.

A resonant transmission line, like a tuned circuit, is resonant at some particular frequency. A resonant transmission line will present to its source of energy a high or low resistive impedance at multiples of a quarter wavelength. Whether the impedance is high or low at these points depends on whether the line is short or open-circuited at the output end. At points that are not exact multiples of a quarter wavelength, the line acts as a capacitor or an inductor.

Resonant transmission lines possess many characteristics of a resonant circuit. Some of these characteristics are: resonant rise of voltage across reactive circuit elements, low impedance across the resonant circuit (series resonance), resonant rise of current in the reactive circuit elements, and high impedance across the resonant circuit (parallel resonance).

As there are many applications for resonant lines, an analysis using two extreme cases of impedance mismatch will be discussed below. Applications for resonant lines will follow later in this chapter and subsequent chapters.

RESONANCE IN OPEN-END TRANSMISSION LINES

The impedances a generator sees at various distances from the output end are shown in the impedance curves (fig. 28-7). The curves above the letters (R, X_L, X_C) of various heights indicate the relative magnitude of impedance presented to the generator for the various lengths of lines indicated. The letters themselves indicate the type of impedance offered at the corresponding inputs. The circuit symbols above the various transmission lines indicate the equivalent electrical circuits for a transmission line of that particular length measured from the output

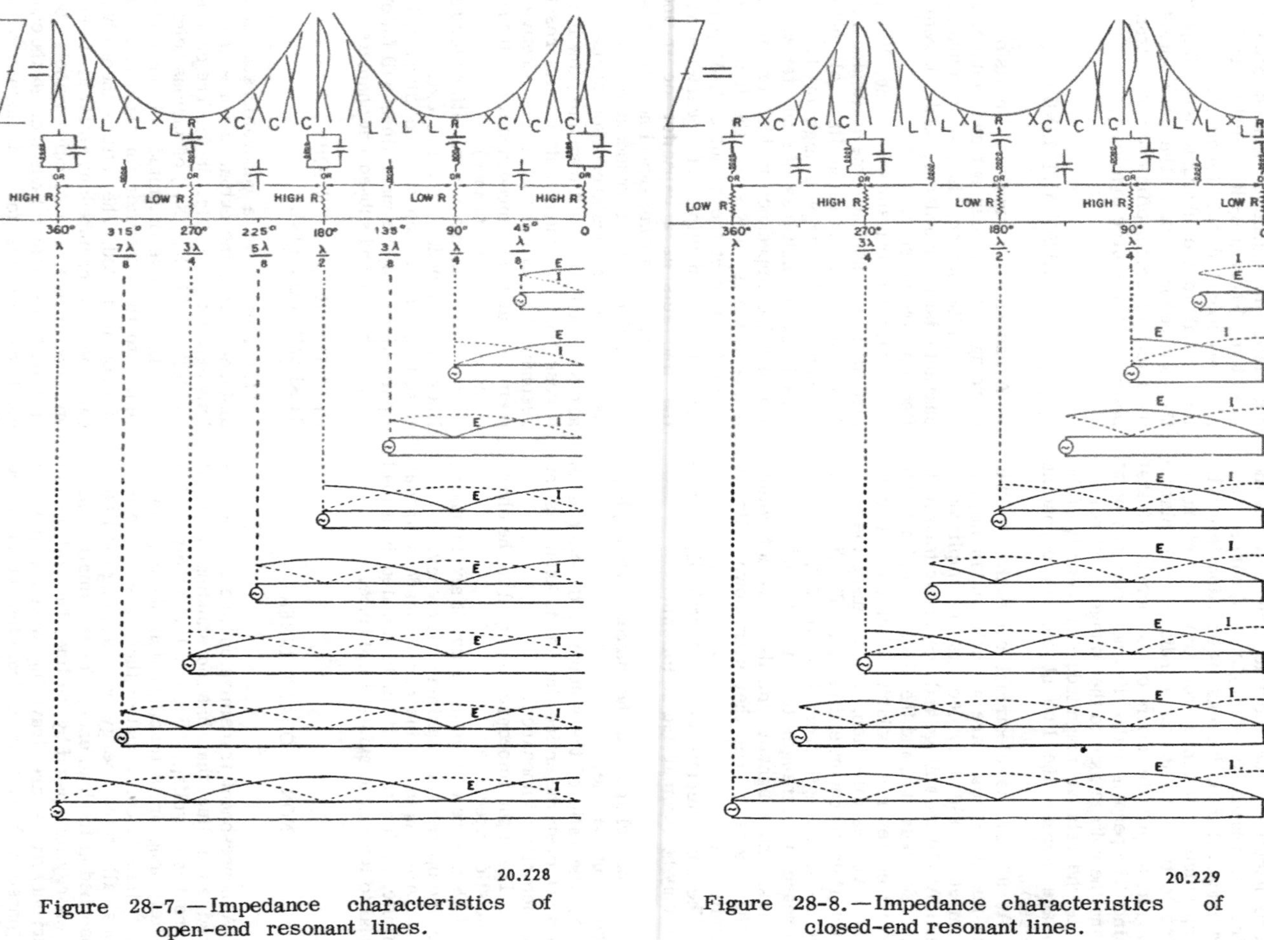

20.228
Figure 28-7.—Impedance characteristics of open-end resonant lines.

20.229
Figure 28-8.—Impedance characteristics of closed-end resonant lines.

end. The curves of effective E and I, whose ratio E/I is the impedance Z, are shown above each line.

At all odd quarter wavelength points ($\frac{\lambda}{4}$, $\frac{3\lambda}{4}$, $\frac{5\lambda}{4}$, etc.), measured from the output end, the current is a maximum and the impedance is a minimum. In addition, there is a resonant rise of voltage from the odd quarter-wavelength points toward the open end. Thus, at all odd quarter-wavelength points, the open-end transmission line acts like a series resonant circuit. The impedance is therefore very low. The generator sees a very low impedance as a load.

At all even quarter-wavelength points ($\frac{\lambda}{2}$, $\frac{3\lambda}{2}$, etc.) the voltage is maximum and the impedance is maximum. Thus, at even quarter wavelengths, the line acts as a parallel resonant circuit.

In addition to acting as series or parallel LC resonant circuits, resonant open end lines also may act as nearly pure capacitances or inductances. This occurs when the lengths of the lines are not an exact multiple of the fundamental quarter-wavelength corresponding to the frequency of the generator at the input terminals. Figure 28-7 shows that an open end line less than a quarter-wavelength long acts like a capacitance, between $\frac{\lambda}{4}$ and $\frac{\lambda}{2}$ as an inductance, between $\frac{\lambda}{2}$ and $\frac{3\lambda}{4}$ as a capacitance, between $\frac{3\lambda}{4}$ and λ as an inductance, and so forth.

RESONANCE IN CLOSED-END LINES

At odd quarter-wavelengths from the shorted (closed-end) of the line, (fig. 28-8), the voltage is high, the current low and the impedance high. Thus, a shorted end transmission line of odd quarter-wavelength acts as a parallel resonant circuit.

At even quarter-wavelength points from the shorted end, the voltage is minimum, the current is maximum and the impedance is minimum. Thus, a shorted transmission line of even quarter-wavelengths acts as a series resonant circuit.

Resonant closed-end lines may also act as purely capacitive or inductive when the lengths of the lines are not exact multiples of the fundamental quarter-wavelength corresponding to the frequency of the applied voltage at the input terminals.

APPLICATIONS OF RESONANT LINES

When a quarter-wave line is shorted at the output end and is excited to resonance at the other end by the correct frequency, there are standing waves of current and voltage on the line. At the short circuit, the voltage is zero while the current is at a maximum. At the input end the current is nearly zero and the voltage is at a maximum. This makes the input impedance very high and appears as an insulator at the line's two open terminals.

Figure 28-9 shows a quarter-wave section of line acting as a standoff insulator for a two wire transmission line. For the particular frequency that makes the section a quarter-wavelength, it acts as a highly efficient insulator. At terminals A and B there is a high voltage and a low current. Because $Z = \frac{E}{I}$, the impedance between A and B must be very high. The insulator obtains a negligible amount of energy from the line to make up any losses caused by the circulating current. If the frequency varies too widely from the value for which the section is designed, the section rapidly becomes a poor insulator and begins to act as a capacitor or inductor across the line.

The impedance of a quarter-wave section of transmission line shorted at one end varies widely over its length (fig. 28-10A). At the shorted end the current is high and the voltage is low. Because $Z = \frac{E}{I}$, the impedance at the shorted end is low. At the open end, the conditions are reversed, and the impedance is high. For example, a 300-ohm line may be matched to a 70-ohm line without the production of standing waves on either of the two lines that are

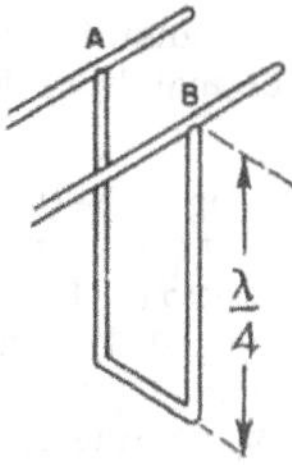

20.237
Figure 28-9.—Quarter-wave sections as insulator for a two-wire transmission line.

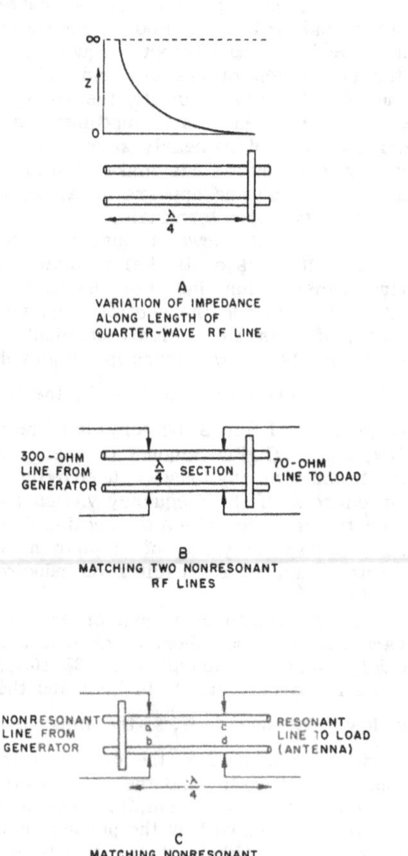

20.238
Figure 28-10.—Transmission line as an impedance matching device.

being matched (fig. 28-10B). Energy from the 300-ohm line sets up standing waves on the quarter-wave section. The connection between the 300-ohm line and the quarter-wave matching section is made at a point where the impedance of the quarter wave line is 300 ohms. When this adjustment is made, the SWR on the 300-ohm line should be at a minimum, essentially unity. The 70-ohm line is similarly adjusted to bring about an impedance match near the shorted end.

The quarter-wave line may also be used to match a nonresonant line to a resonant line, as shown in figure 28-10C. In order to be nonresonant, a line must be terminated in its characteristic impedance, and the terminating impedance should be approximately a pure resistance. The impedance of a shorted quarter-wave resonant section is zero at the shorting bar and increases along the line toward the open end. The shorting bar is adjusted to make a voltage maximum appear at cd; and the contacts, ab, between the nonresonant line and the quarter-wave section, are adjusted for the best match.

Quarter-wave lines may also be used as filters. The characteristic of a quarter-wave line allows it to be used as an efficient filter or suppressor of even harmonics.

If a transmitter is operating on a frequency of 5 MHz and it is found that the transmitter is causing excessive interference on 10 and 20 MHz, a resonant transmission line may be used as a harmonic suppressor.

A quarter-wave line shorted at one end offers a high impedance at the unshorted end to the fundamental frequency. At a frequency twice the fundamental (second harmonic) such a line is a half-wave line. At a frequency four times the fundamental it becomes a full wave line and offers a very low impedance at its input end. Therefore, the radiation of even harmonics from the transmitting antenna can be eliminated almost completely by the circuit shown in figure 28-11.

The resonant filter line, ab, as shown, is a quarter-wave in length at the fundamental frequency, and offers almost infinite impedance at this frequency; in other words, it acts as an insulator. At the second harmonic the line, ab, is a half-wave line and offers zero impedance at the antenna, thus shorting this frequency to ground.

A quarter-wave filter (b, in figure 28-12) that is open at the output end can be inserted in

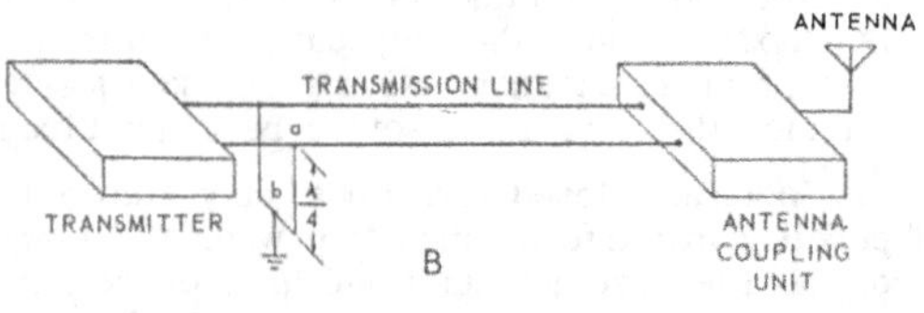

20.240
Figure 28-11.—Shorted-end quarter-wave filter.

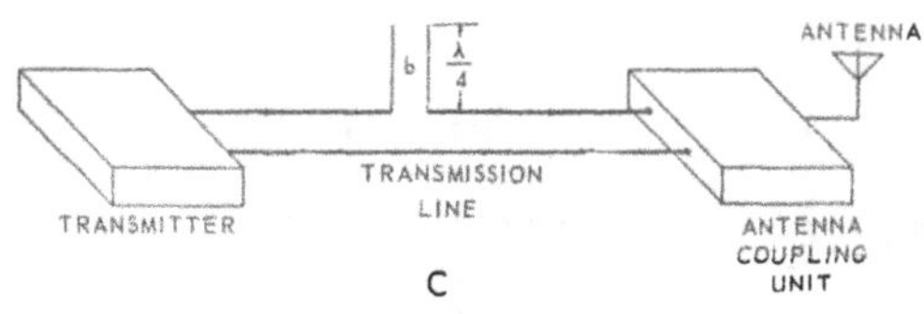

20.240
Figure 28-12.—Open-end quarter-wave filter.

series with the transmission line. Being open at its output end it offers a low impedance at the input end to the fundamental frequency. At the fundamental and odd harmonics the impedance at b is so low that it may be considered a continuous line, as if a short were placed across the base of the quarter-wave line. Thus, the quarter-wave open filter line, b, in figure 28-12, passes the fundamental and odd harmonics along the line to the antenna-coupling unit. At even harmonics however, the length of the open line (at b) becomes a half wave, or some multiples of a half-wave, so that line b offers high impedance to the even harmonics and blocks their passage to the antenna coupling unit.

CREATION OF STANDING WAVES (VOLTAGE AND CURRENT)

If a transmission line is infinitely long, or if it is terminated in its characteristic impedance, reflections do not occur. However, if there is an abrupt impedance change in the line, such as an open or short circuit, complete reflection will occur. An impedance change of less magnitude will cause some reflection. The amount of reflection will depend on the mismatch between the termination impedance and the characteristic impedance. We will limit our discussion to the two extreme cases, that is to open- and closed-end lines.

(The wave leaving the source is called the incident wave and, when reflection occurs, the wave returning to the source is called the reflected wave.)

OPEN-END LINES

In an open-end line, the impedance at the output end can be considered as practically infinite because no load is attached. When energy is applied at the generator end, the first surge consists of waves of current and voltage that sweep down the line in phase with each other. The initial current and voltage waves must travel down the line in phase because the characteristics of the line are the same as those of a line that is truly infinite while the initial wave is traveling toward the output end. The in-phase condition of these waves can be changed only when they encounter a difference in the impedance between the two wires of the line, and then reflections occur. Thus, when the wave of current reaches the open-circuited output end of the line the current must collapse to zero. When the current wave collapses, the magnetic field that was set up by it also collapses. This collapsing magnetic field cuts the conductors near the output end and induces an additional voltage across the line. This voltage acts as a reverse generator and sets up new current and voltage waves that travel back along the line toward the input end.

An open-end transmission line, one wavelength long and having no attenuation is shown in figure 28-13A. In figure 28-13B Part 1, we have assumed that the generator voltage vector has gone through at least two complete revolutions so that the voltage wave has had time to travel down the line and return to the generator end. The waveforms are stopped in time, in this figure, at the instant that the generator vector is at the zero position.

It may be observed that the initial voltage wave is reflected at the output end of the line in phase with the voltage wave that would have continued along the line in the original direction of travel if the line had been longer. For example, the dotted waveform extending slightly beyond the end of the line in figure 28-13B (1), indicates that the incident waves would have started going negative. Therefore, the reflected wave will start back in a negative direction. At the instant in time being considered here, the incident and the reflected wave add vectorially to give a zero resultant wave.

Next, consider the voltage variation across the line that occurs with respect to time at certain points along the line. At point a of figure 28-13B (2) the incident and the reflected wave combine to produce a voltage loop, (maximum voltage); the same is true also for points c and e. Also at point a of figure 28-13B (4) the incident and reflected wave combine to produce a voltage loop, (maximum voltage); the same is also true of points c and e of figure 28-13B (4). (2) and (4) of figure 28-13B are 180° out of phase. Figure 28-13B (1) and (3) have the incident and reflected wave 180° out of

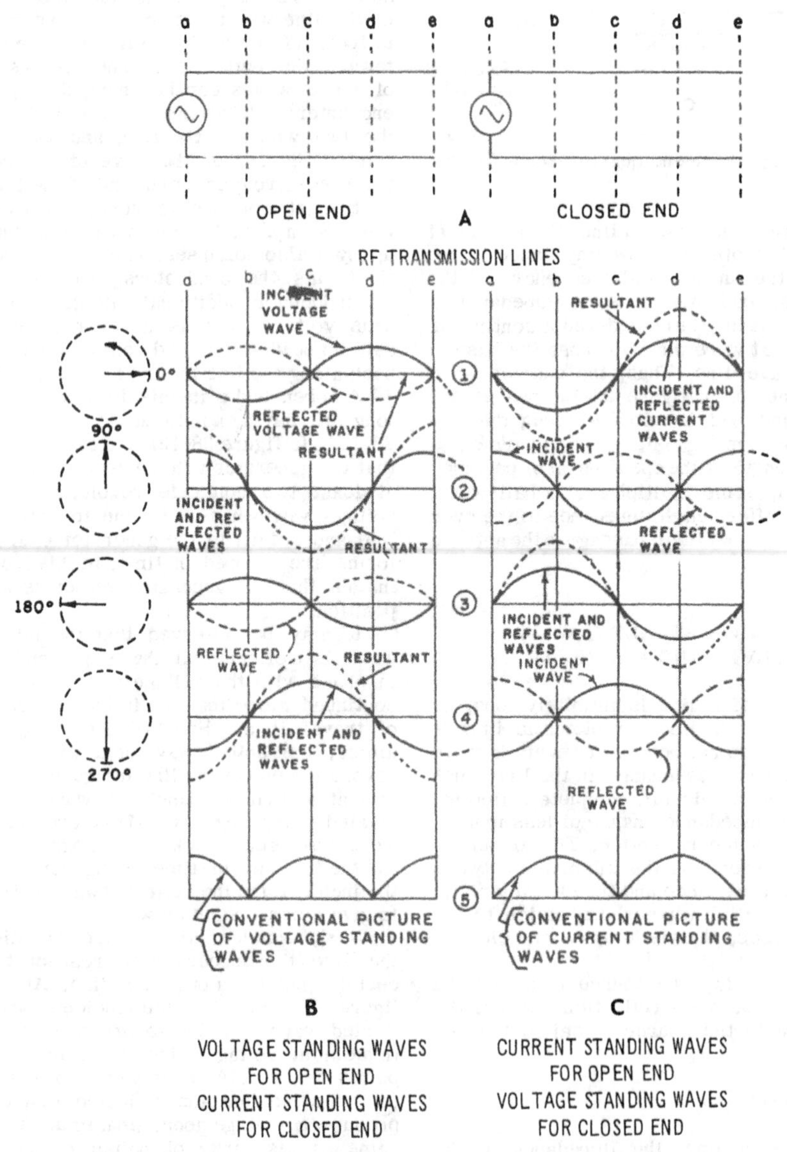

Figure 28-13.—Voltage current waveforms for open-end and closed-end transmission lines.

phase so at all points along the line the voltages will be zero.

Also consider the voltage variations across the line that occur with respect to time at certain locations along the line. At point a the voltage is zero (fig. 28-13B (1)); then maximum in one direction (fig. 28-13B (1) point b); then zero again (fig. 28-13B (1) point c); then maximum in the other direction (fig. 28-13B (1) point d); and back to zero again, (fig. 28-13B (1) point e). A suitable voltage indicating device located at a, c, or e will indicate voltage loops (points of maximum voltage); and at b and d the device will indicate voltage nodes (points of zero voltage).

Standing waves of voltage are shown in (5) of figure 28-13B. This curve represents effective values of voltage at the various points along the line. These values are actaully the effective values of the sinusoidal voltage variations occurring across the line at points where the measurements are being made. Thus, at point c, the voltage will be zero at one instant of time (fig. 28-13B (1)), then it will build up to a maximum with one polarity (fig. 28-13B (2)), then it will become zero (fig. 28-13B (3)), and finally it will build up to a maximum with the opposite polarity (fig. 28-13B (4)).

Standing waves of current are shown in figure 28-13C. They are occurring simultaneously with the voltage waves on the transmission line, but they are shown here separately in order to simplify the figure. The initial current wave is reflected at the output end of the line 180° out of phase with the current wave that would have continued along the line in the original direction of travel if the line had been extended. (See dotted extensions). In other words, the reflected current reverses direction at the open end of the line and is shown 180° out of phase with the incident wave except when the incident and reflected waves have zero values at the end of the line. This reversal is opposite to the condition for voltage, because the reflected voltage wave has the same polarity that the incident wave would have had if it continued down the extended line in the original direction of travel. (See the dotted line extensions.)

Because the incident and the reflected current waves are 180° out of phase at the open end of the line, they cancel at this point, and the resultant current at the open end is always zero. The rotating vectors at the left of the figures indicate the generation of sine waves of both voltage and current. At point b (fig. 28-13C (1)) the incident and the reflected current waveforms combine to produce a current loop (maximum current); the same is also true at point d. At a, c, and e the incident and the reflected waveforms combine to produce current nodes (points of zero current).

In figure 28-13C (2), the current vector has rotated 90 degrees. Combining the incident and the reflected waves at this instant gives a resultant current waveform that has an amplitude of zero throughout the entire length of the line. In figure 28-13C (3), the current vector has completed 180° of its cycle. Again combining the incident and the reflected waves at this instant gives a resultant current that has a maximum amplitude at points b and d although the direction of the current is opposite to that in figure 28-13C (1) at these points. In figure 28-13C (4), the resultant current is again zero all along the line.

Figure 28-13C (5) is a plot of the effective values of current at the various points along the line. These current values are actually the effective values of the resultant current variations through the cycle at the respective points where the measurements are made. For example, at point b, the current is maximum in one direction at one instant of time (fig. 28-13C (1)) then it becomes zero (fig. 28-13C (2)), then it builds up to a maximum in the opposite direction (fig. 28-13C (3)), and finally returns to zero fig. 28-13C (4)). The effective value of this current variation is plotted at b in figure 28-13C (5).

CLOSED-END LINES

In a closed-end line the voltage and current waveforms exchange places with respect to their locations in an open-end line. Thus, in figure 28-13A, if the line is closed at the end, figure 28-13B becomes current standing waves, and figure 28-13C becomes voltage standing waves. At the short circuited end of the line (point e), the current varies from zero to a maximum in one direction and back to zero and a maximum in the other direction. The effective value of current will, therefore, be maximum at point e (a current loop). The voltage (E = I x R) across the short circuit approaches zero because the resistance of the short circuit is negligible.

WAVELENGTH MEASUREMENTS

There are a number of ways in which the length of a standing wave can be ascertained. Figure 28-14 illustrates the Lecher-wire system,

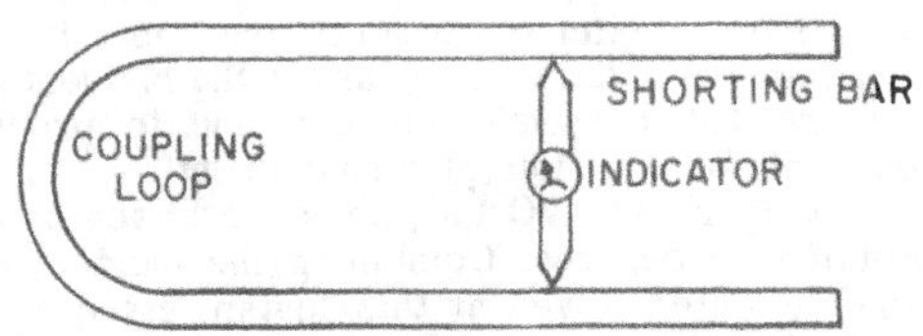

179.492

Figure 28-14.—Lecher-wire system for ultra-high frequency measurements.

which determines either current peaks or current nodes in the transmission line. The length of the Lecher wire is not important provided it is at least a wavelength long.

In the Lecher wire method, standing waves are set up on an open two wire parallel transmission line. Since the distance between successive peaks (or nodes) is equal to a half wavelength, a direct measurement of this distance will enable the frequency to be determined.

There should be loose coupling between the line and the source of current so that the normal frequency of the source is unchanged during the measurement. Some standing waves are produced when the far end of a line is open or short-circuited. Either of these terminations may be used. By sliding a current indicating device, preferable a thermocouple ammeter, along the Lecher-wire frame, points of maximum current (loops) will be disclosed by maximum deflection of the meter. Points of minimum current (nodes) will be indicated by minimum deflection. The distance between successive maxima or minima is always equal to a half-wavelength. This distance can be readily measured with a suitably calibrated scale. With the measurement in meters, the frequency in megahertz will be 150 divided by the distance in meters.

There are a number of precautions to observe. The measurement itself must be made with extreme care, as the accuracy of the result is limited by the physical measurement. The looser the coupling, the sharper will be the indications of maxima or minima. It is desirable for air to be the dielectric separating the conductors of the line. These conductors should be made of copper tubing or of wires stretched tautly between convenient supports. Satisfactory spacing between the conductors is approximately from one to one-and-a-half inches. The contacts of the meter should be maintained at right angles to the line. A test system can be employed more conveniently and accurately if built solidly and in a permanent fashion. It has been found from experience that somewhat greater accuracy is possible by measuring the distance between nodes rather than peaks. If the foregoing precautions are taken, an accuracy as high as 0.1 percent is possible.

An alternate method uses the resonant properties of the short-circuited transmission line to measure wavelength (fig. 28-15). Here, the current indications are found by moving a shorting bar along the length of Lecher wire. Once again, the coupling between the source and the line must be as loose as possible. Also, loosely coupled to the source is some indicating device, such as a wavemeter. As the shorting bar is moved toward the far end of the line, it will reach a resonant point at which a pronounced dip in the meter indication will take place. With the shorting bar in such a position, the line is tuned to the frequency of the source and so absorbs maximum energy from it. At a half-wavelength from this first resonant point, the line will exhibit the same characteristics. The distance between these points must be measured accurately.

The ratio of the value of current or voltage at a loop to the value of current or voltage at a node is called the standing wave ratio (SWR). This SWR depends upon the ratio of the resistance of the load connected to the output end of the line to the characteristic impedance of the line itself. SWR $= \frac{Z_o}{Z_r}$ or $\frac{Z_r}{Z_o}$ where Z_o is the characteristic impedance of the line and Z_r is the terminating resistance.

The standing wave ratio provides a direct indication of the degree of mismatch along a transmission line of a feed system.

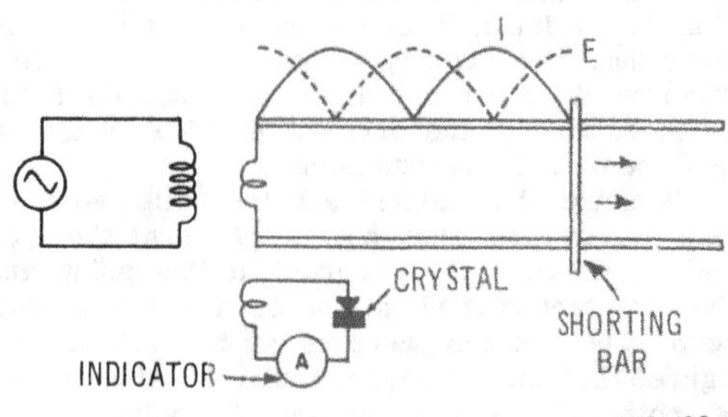

179.493

Figure 28-15.—Short circuited Lecher line.

CHAPTER 29

WAVE PROPAGATION AND ANTENNAS

This chapter discusses the basic theory involved in the propagation of radio waves, basic antenna theory, and some of the commonly used types of antennas.

WAVE PROPAGATION

Radio waves propagated into space are considered to be a radiant form of energy, similar to light and heat. They travel at approximately the speed of light. Conventional concepts of how the waves are radiated impose a severe strain on the average person's imagination.

The theory of wave progagation as presented in this text, although greatly simplified, has found general acceptance. Our major concern is in showing how to make antennas operate efficiently, both for transmission and reception, under various conditions.

The radio frequency spectrum, which extends from .01 megahertz (very low frequency) to 30,000 megahertz (super high frequency), is shown in table 1-1 (chapter 1) and explained in chapter 9.

VARIOUS COMPONENTS OF A PROPAGATED ELECTROMAGNETIC WAVE

When a radio wave leaves a vertical antenna the field pattern of the wave resembles a huge doughnut lying on the ground with the antenna in the hole at the center, as seen in figure 29-1. Part of the wave moves outward in contact with the ground to form the GROUND WAVE, and the rest of the wave moves upward and outward to form the SKYWAVE, as shown in figure 29-2. The ground and sky portions of the radio wave are responsible for two different methods of carrying the messages from transmitter to receiver. The ground wave is used both for short-range communications at high frequencies with low power and for long-range communication at low frequencies and with very high power. Daytime reception from most commercial stations is carried by the ground wave.

The sky wave is used for long-range high frequency daylight communication. At night, the sky wave provides a means for long-range contacts at somewhat lower frequencies.

GROUND WAVE

The ground wave is responsible for most of the daytime broadcast reception. As it passes over and through the ground, this wave induces a voltage in the earth, setting up eddy currents. The energy used to establish these currents is absorbed from the ground wave, thereby weakening it as it moves away from the transmitting antenna. Increasing the frequency rapidly increases the attenuation so that ground wave transmission is limited to relatively low frequencies. Shore-base transmitters are able to transmit long-range ground wave transmissions

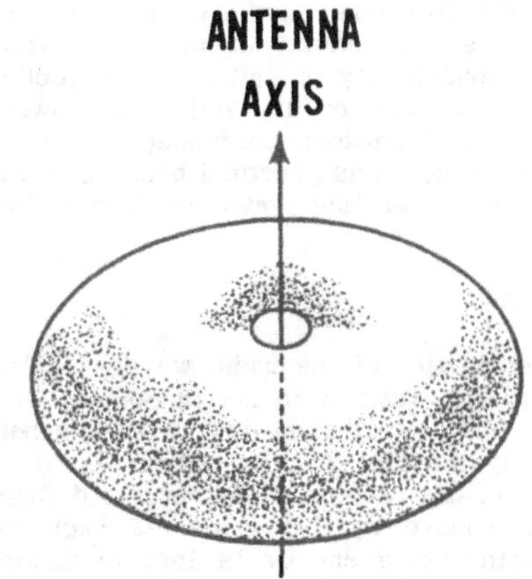

Figure 29-1.—Vertical antenna field pattern.

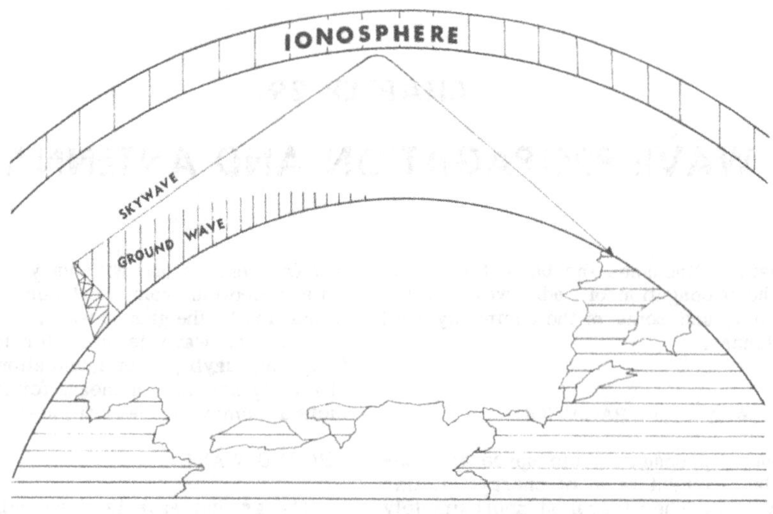

179.495

Figure 29-2.—Ground and sky waves.

by using frequencies between 18 and 300 kHz with extremely high power.

Since the electrical properties of the earth along which the surface wave travels are relatively constant, the signal strength from a given station at a given point is nearly constant. This holds essentially true in all localities except those having distinct rainy and dry seasons. There the difference in the amount of moisture causes the conductivity of the soil to change.

The conductivity of salt water is 5,000 times as great as that of dry soil. High power, low frequency transmitters are placed as close to the edge of the ocean as practical because of the superiority of surface wave conduction by salt water.

SKY WAVE

That portion of the radio wave which moves upward and outward is not in contact with the ground and is called the SKY WAVE. It behaves similarly to the ground wave. Some of the energy of the sky wave is refracted (bent) by the ionosphere so that it comes back toward the earth. Some energy is lost in dissipation to particles of the atmospheric layers. A receiver located in the vicinity of the returning sky wave will receive strong signals even though several hundred miles beyond the range of the ground wave.

IONOSPHERE

The ionosphere is found in the rarefied atmosphere approximately 40 to 50 miles above the earth. It differs from other atmospheric parts in that it contains a much higher number of positive and negative ions. The negative ions are believed to be atoms whose energy levels have been raised to a high level by solar bombardment of ultra-violet and particle radiations. The rotation of the earth on its axis, the annual course of the earth around the sun, and the development of sun spots all affect the number of ions present in the ionosphere, and these in turn affect the quality and distance of electronic transmissions.

The ionosphere is constantly changing. Some of the ions are returning to their normal energy level, while other atoms are being raised to a higher energy level. The rate of variation between high and low level of energy depends upon the amount of air present, and the strength of radiation from the sun, as well as the propagation relation to the earth's magnetic field.

At altitudes above 350 miles, the particles of air are far too sparse to permit large-scale energy transfer. Below about 40 miles altitude, only a few ions are present because the rate of return to a normal energy level is high. Ultraviolet radiations from the sun are absorbed in passage through the upper layers of the ionosphere so that below an elevation of 40 miles, too few ions exist to affect, materially, sky wave communication.

Densities of ionization at different heights make the ionosphere appear to have layers. Actually, there is thought to be no sharp dividing line between layers, but for the purpose of discussion, such a demarcation is indicated.

The ionized atmosphere at an altitude of between 40 and 50 miles is called the "D" layer. Its ionization is low and it has little effect on the propagation of radio waves except for the absorption of energy from the electronic waves as they pass through it. The "D" layer is present only during the day. Its presence greatly reduces the field intensity of transmissions that must pass through daylight zones.

The band of atmosphere at altitudes between 50 and 90 miles contains the so called "E" layer. It is a well defined band with greatest density at an altitude of about 70 miles. This layer is strongest during daylight hours, and is also present, but much weaker, at night. The maximum density of the "E" layer appears at about noon local time.

The ionization of the "E" layer at the middle of the day is sometimes sufficiently intense to refract frequencies up to 20 MHz back to the earth. This action is of great importance to daylight transmissions for distances up to 1,500 miles.

The "F" layer extends approximately from the 90 mile level to the upper limits of the ionosphere. At night only one "F" layer is present, but during the day, especially when the sun is high, this layer often separates into two parts, F_1 and F_2, as shown in figure 29-3. As a rule, the F_2 layer is at its greatest density during early afternoon hours, but there are many notable exceptions of maximum F_2 density existing several hours later. Shortly after sunset, the F_1 and F_2 layers recombine into a single F layer.

In addition to the layers of ionized atmosphere that appear regularly, erratic patches of ionized atmosphere occur at "E" layer heights in the manner that clouds appear in the sky. These patches are referred to as Sporadic-E ionizations. They are often present in sufficient number and intensity to enable good VHF transmissions over distances not normally possible.

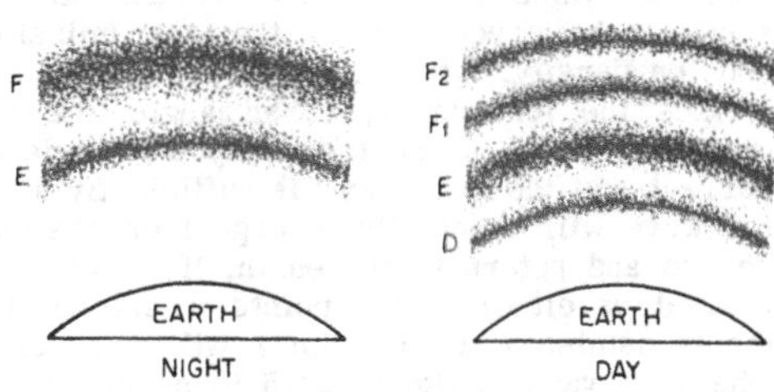

Figure 29-3.—E and F layers of the ionosphere.

Sometimes sporadic ionizations appear in considerable strength at varying altitudes and actually prove harmful to electronic transmissions.

EFFECT OF IONOSPHERE ON THE SKY WAVE

The ionosphere has many characteristics. Some waves penetrate and pass entirely through it into space, never to return. Other waves penetrate but bend. Generally, the ionosphere acts as a conductor, and absorbs energy in varying amounts from the electronic wave. The ionosphere also acts as an electronic mirror and refracts (bends) the sky wave back to the earth, as illustrated in figure 29-4. Here the ionosphere does by refraction what water does to a beam of light.

The ability of the ionosphere to return an electronic wave to the earth depends upon the

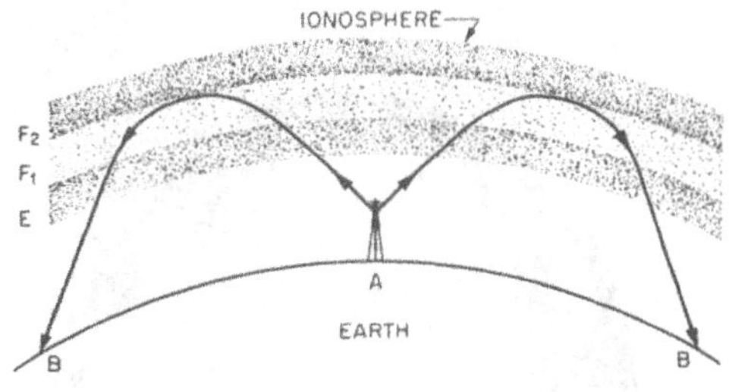

Figure 29-4.—Refraction of the sky waves by the ionosphere.

angle at which the sky wave strikes the ionosphere, the frequency of the transmissions, and ion density. When the wave from an antenna strikes the ionosphere, at an angle, the wave begins to bend. If the frequency and angle are correct and the ionosphere is sufficiently dense, the wave will eventually emerge from the ionosphere and return to the earth. If a receiver is located at either of the points B, figure 29-4, the transmission from point A will be received. The sky wave in figure 29-5 is assumed to be composed of rays that emanate from the antenna in three distinct groups that are identified according to the angle of elevation. The angle at which the group 1 rays strike the ionosphere is too nearly vertical for the rays to be returned to earth. The rays are bent out of line, but pass completely through the ionosphere and are lost. A currently popular theory on propagation is explained as follows.

The angle made by the group 2 rays is called the CRITICAL ANGLE for that frequency. Any ray that leaves the antenna at an angle greater than this angle (θ) will penetrate the ionosphere.

Group 3 rays strike the ionosphere at the smallest angle that will be refracted and still return to the earth. At any smaller angle the rays will be refracted but will not return to the earth.

As the frequency increases, the initial angle decreases. Low frequency fields can be projected straight upward and will be returned to the earth. The highest frequency that can be sent directly upward and still be refracted back to the earth is called the CRITICAL FREQUENCY. At sufficiently high frequencies, regardless of the angle at which the rays strike the ionosphere, the wave will not be returned to the earth. The critical frequency is not constant but varies from one locality to another, with the time of day, with the season of the year, and with the sunspot cycle.

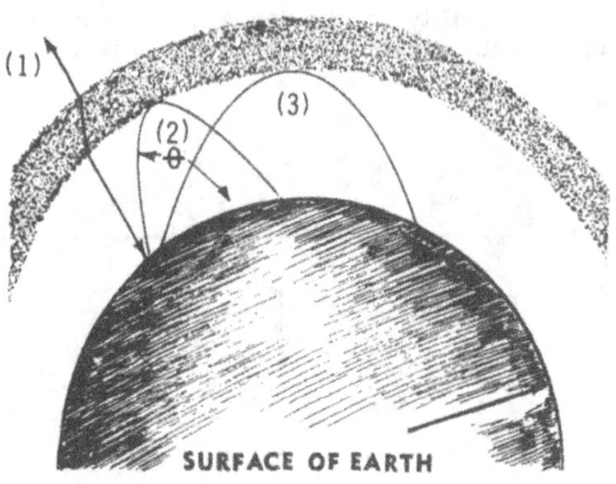

13.30
Figure 29-5.— Effect of the angle of departure on the area of reception.

Because of this variation in the critical frequency, nomograms and frequency tables are issued that predict the maximum usable frequency (MUF) for every hour of the day for every locality in which transmissions are made. Nomograms and frequency tables are prepared from data obtained experimentally from stations scattered all over the world. All this information is pooled, and the results are tabulated in the form of long range predictions that remove some of the guesswork from transmissions.

SKIP ZONE

Between the point where the ground wave is completely dissipated and the point where the first sky wave returns, no signals will be heard. This area is called the SKIP ZONE, and is illustrated in figure 29-6. The skip zone for the lower high frequencies (3 to 9 MHz) will be greater at night than during the day. For instance, signals transmitted on 6 MHz from Philadelphia, Pa., can be heard from 1300 hours to 1700 hours in Pittsburgh, Pa., but not in Chicago, Ill.; transmissions from the same station may be heard from 1730 hours to 2200 hours in Chicago, but not in Pittsburgh. However,

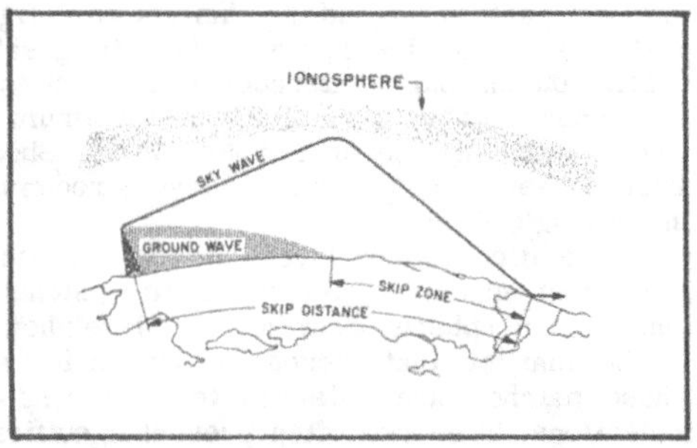

179.496
Figure 29-6.—Skip zone.

skip effects can be minimized by utilizing lower frequencies for nighttime communication. As a general rule, it can be said that as the frequency decreases, the skip distance decreases.

MULTIPLE-HOP TRANSMISSION

The sky wave path of a signal propagated at two different vertical angles is illustrated in figure 29-7. When the vertical angle is Ø1, the signal is returned to the earth at point A, reflected back to the ionosphere and reappears at point B. If the same signal is transmitted at a lower vertical radiation angle Ø2, it can reach point B in a single hop. The signal transmitted at angle Ø1 will suffer more ionospheric and ground absorption losses than that signal transmitted at angle Ø2. In general, single hop transmissions result in greater field intensities at a distant point than multiple hop transmissions. By inspection of figure 29-7, it is evident that longer distances can be covered by multiple hop transmissions as the vertical radiation angle is decreased. The higher medium frequencies, 9 to 30 MHz, are generally utilized for long distance transmissions, and in order to minimize the number of reflections of the signal in arriving at a distant point, lower vertical radiation angles are used. However, there is a limit to the improvement obtained by low angle radiation, because absorption and other factors make operation on vertical radiation angles below 3 degrees impractical.

FADING

When a received signal varies in intensity over a relatively short period of time, the effect is known as fading, which is one of the most troublesome problems encountered in electronic reception.

There are several conditions which can produce fading. One type of fading is prevalent in areas where sky waves are relied upon for transmission. Figure 29-8 shows two sky waves traveling paths of different lengths, thereby varying about the same point out of phase and thus producing a weakening of the signal. For instance, if a portion of the transmitted wave front arrived at a distant point via the E layer and another via the F layer, a complete cancellation of signal voltages would occur if the waves arrived 180 degrees out of phase and with equal amplitude. Usually, one signal is weaker than the other, therefore, a usable signal is obtained.

One method of overcoming fading is to place two antennas a wavelength apart, feed two separate receivers, and combine the audio outputs. This is known as diversity reception.

EFFECT OF ATMOSPHERE ON HF TRANSMISSION

Unusual ranges of VHF and UHF contacts are caused by abnormal atmospheric conditions a few miles above the earth. Normally, the warmest air is found near the surface of the water. The air gradually becomes cooler as the altitude

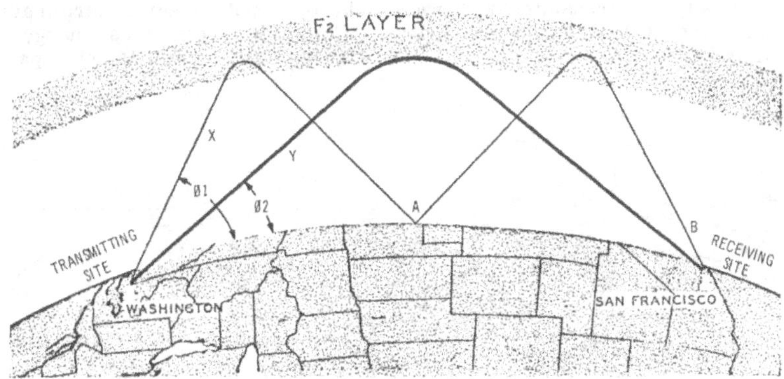

179.497

Figure 29-7.—Path of a sky-wave signal propagated at two different vertical angles.

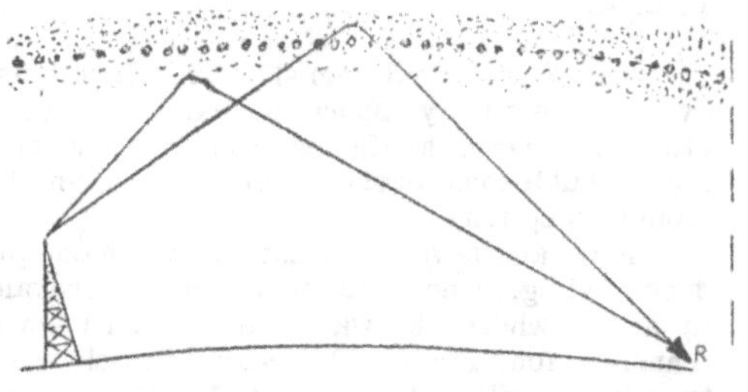

179.498

Figure 29-8.—Fading caused by arrival of two sky waves at the same point (R) out of phase.

increases. Sometimes, unusual situations develop where warm layers of air are found above cooler layers. The condition is known as TEMPERATURE INVERSION.

When a temperature inversion exists, the amount of refraction (index of refraction) is different for the particles trapped within the boundaries from those outside them. These differences form channels or ducts that will conduct the radio waves many miles beyond the assumed normal range.

Sometimes these ducts are in contact with the water and may extend a few hundred feet into the air. At other times the duct will start at an elevation of between 500 and 1,000 feet and extend an additional 500 to 1,000 feet in the air.

If an antenna extends into the duct or if the wave enters a duct after leaving an antenna, the transmission may be conducted a long distance. An example of this type of transmission of radio waves in ducts formed by temperature inversions is shown in figure 29-9.

With certain exceptions, ducts are formed over water where the following conditions are observed aboard ship:

1. A wind is blowing from the land.
2. There is a stratum of quiet air.
3. High pressure, clear skies and little wind.
4. Cool breeze over warm open ocean.
5. Smoke, haze, or dust fails to rise.
6. Received signal is fading rapidly.

DIFFRACTION

The diffraction effect of an obstacle in the path of a radio wave is shown in figure 29-10. The resultant wave fronts are distorted, and the wave front is extended downward. Instead of being shielded, an antenna erected at point A would receive some energy from the transmitting antenna.

ANTENNAS

An antenna can be defined as a conductor or system of conductors used either for radiating electromagnetic energy into space or for collecting electromagnetic energy from space. In effect, electrical energy from the transmitter is converted into electromagnetic energy by the antenna and radiated into space. On the reception end, electromagnetic energy is converted into electrical energy by the antenna and fed into the receiver.

Fortunately, separate antennas seldom are required for transmission and reception of radio energy. Any antenna transfers energy from space to its input receiver terminals with the same efficiency it transfers energy from the output transmitter terminals into space, assuming, of

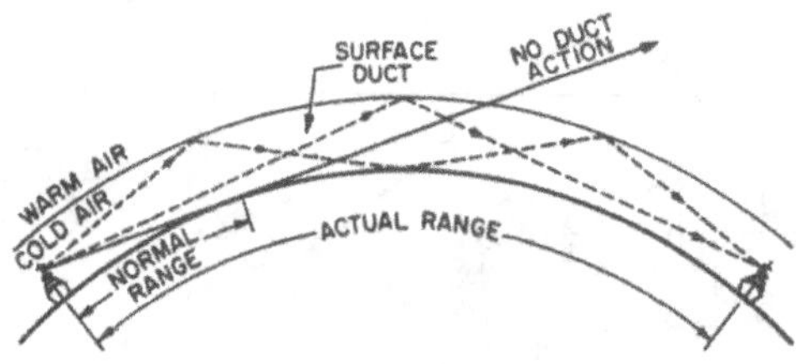

20.255

Figure 29-9.—Duct effect in high-frequency transmission.

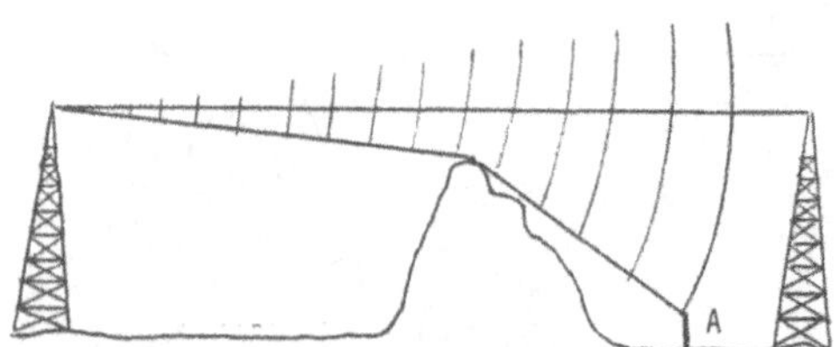

179.499

Figure 29-10.—Diffraction of transmitted energy by sharp edge of a mountain.

course, that the same frequency is used in both cases. This property of interchangeability of the same antenna for transmitting and receiving operations is known as antenna reciprocity. Antenna reciprocity is possible mainly because antenna characteristics are essentially the same regardless of whether an antenna is sending or receiving electromagnetic energy.

ELECTROMAGNETIC ENERGY IN SPACE

Because the half-wave dipole antenna is the fundamental element of an antenna system, it will be used as a starting point for discussing the radiation of electromagnetic energy into space. A dipole antenna can be developed from an open-end transmission line one quarter wavelength long. Figure 29-11 shows this quarter wave section and the standing waves of current and voltage developed on the line. The current flow, as indicated, is flowing in opposite directions in the two conductors. The magnetic fields, due to these currents, cancel. The electrostatic fields, too, are in opposite directions.

Efficient radiation requires a very heavy field density; therefore, the fields must reinforce each other. Suppose the ends of the two conductors are spread apart. The direction of current flow doesn't change, but the magnetic field is being opened into space, and less cancellation occurs. The electrostatic field is opened out instead of being concentrated in the small space between the two conductors. If we move the two conductors until they are in line with each other as in figure 29-12 the current flow in the two conductors is now in the same direction. The standing waves are exactly as they were before.

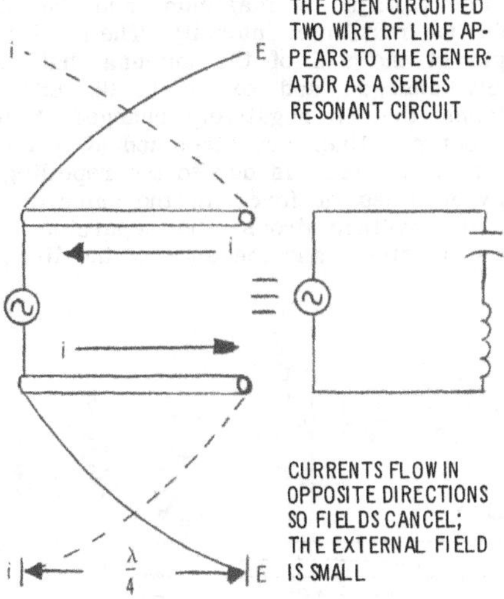
THE OPEN CIRCUITED TWO WIRE RF LINE APPEARS TO THE GENERATOR AS A SERIES RESONANT CIRCUIT
E
i
i
i
CURRENTS FLOW IN OPPOSITE DIRECTIONS SO FIELDS CANCEL; THE EXTERNAL FIELD IS SMALL
i
$\frac{\lambda}{4}$
E

179.500

Figure 29-11.—Open-end quarter-wave section of RF transmission line.

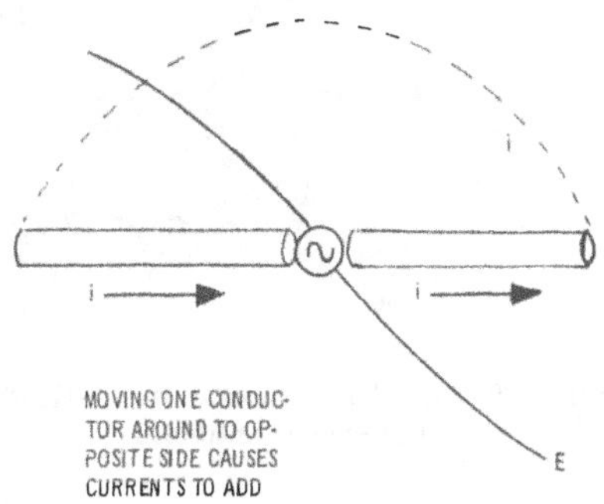
i
i
i
MOVING ONE CONDUCTOR AROUND TO OPPOSITE SIDE CAUSES CURRENTS TO ADD
E

179.501

Figure 29-12.—Effect of rearranging the conductors.

The magnetic (H) field about the wire is shown in figure 29-13. Since maximum change of current takes place at the center of the dipole, the magnetic field is greatest at the center. Figure 29-14 illustrates the electrostatic (E) field about the conductor when the difference in potential between the ends is the

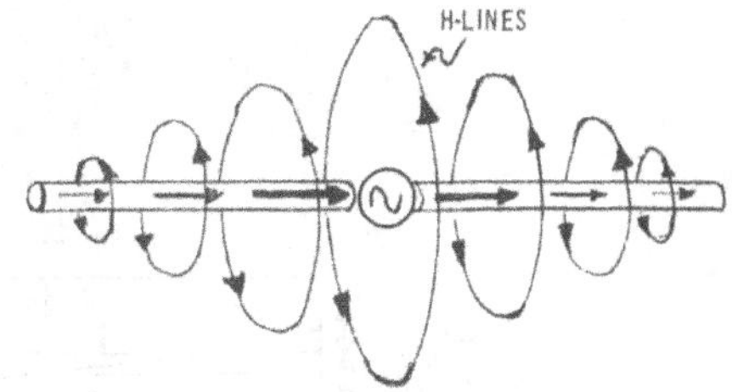
H-LINES

179.502

Figure 29-13.—Distribution of H-field around half-wave antenna.

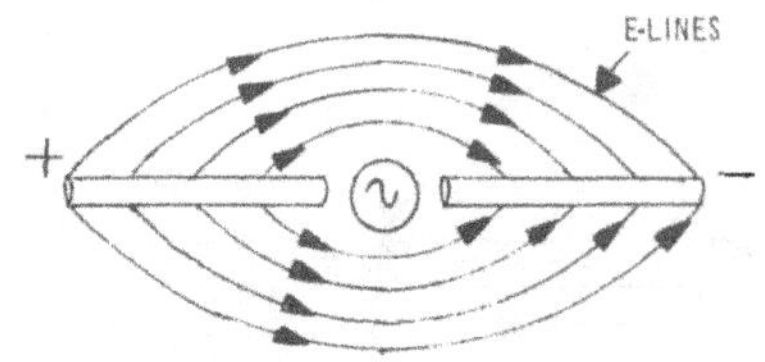

179.503

Figure 29-14.—Distribution of E-field around half-wave antenna.

greatest. This does not occur at the same time as the maximum magnetic field. Since the voltage and current on the dipole are 90° out of phase, the electrostatic and magnetic fields surrounding the antenna are also 90° out of phase.

Figure 29-15, shows the relationship between the variations in field intensities. When maximum current is flowing, there is no difference in potential between the ends of the conductors. The magnetic field about the conductor is maximum at this time (fig. 29-15A). As the current continues to flow, a charge is built up on the conductor and current decreases. The charge built up increases, and the electrostatic field which accompanies it also builds up. The current flow and its accompanying magnetic field decrease. When the charge has built up to maximum, the current flow has decreased to zero. The electrostatic (E) field is at its maximum intensity (fig. 29-15B). When the current starts to flow in the opposite direction through the conductor, the magnetic field builds up as the electric field decreases. Figure 29-15C shows the fields when current is maximum in the direction opposite to figure 29-15A.

Figure 29-15D illustrates the electric field at a maximum when the preceding current flow has stopped momentarily and the magnetic field is zero. In part E of figure 29-15, there is no charge, and the flow of current is again maximum. Notice that the magnetic and the electrostatic lines of force are at right angles to each other. In addition, these fields reach their maximum intensity a quarter of a cycle, or 90 electrical degrees apart.

The fields associated with the energy stored in the resonant circuit (antenna) are called the induction fields. These fields decrease with the square of the distance from the antenna. They are only local in effect and play no major part in the transmission of electromagnetic energy. The fields set up in the transfer of energy through space are known as the radiation fields, although these fields decrease with distance from the antenna, this decrease is much less rapid than the decrease encountered with the induction fields, and the decrease is linear. It is the radiation fields which are responsible for electromagnetic radiation.

MECHANICS OF RADIATION

Figure 29-16 gives a simple picture of an E-field detaching itself from an antenna. In part A the voltage is maximum and the electric field has maximum intensity. The lines of force begin at the end of the antenna that is positively charged and extend to the end of the antenna that is negatively charged. Note that the outer E lines are stretched away from the inner ones. This is due to the repelling force between lines of force in the same direction. As the voltage drops, the separated charges come together, and the ends of the lines move

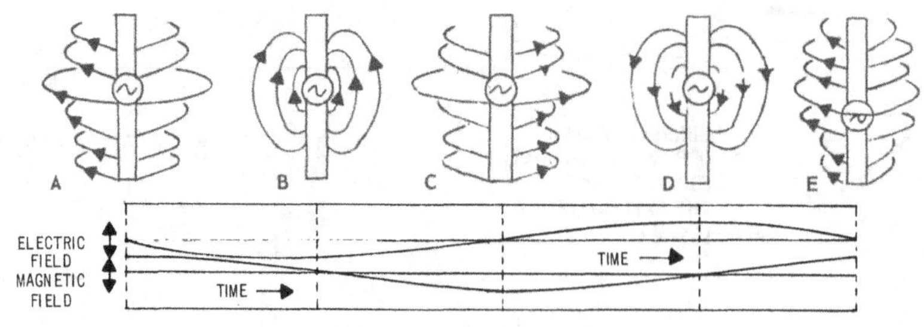

179.504

Figure 29-15.—Variations in field intensities.

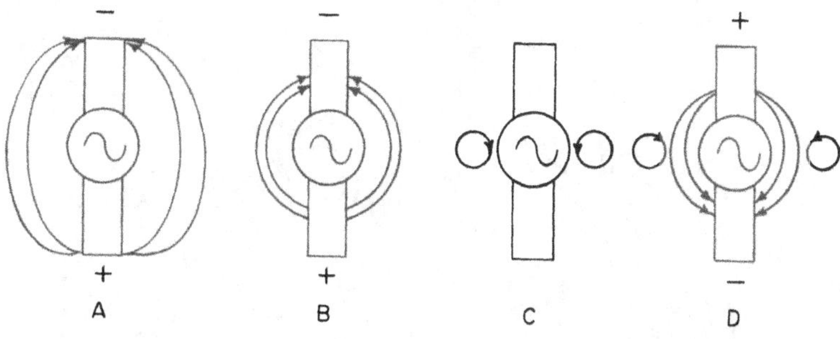

Figure 29-16.—Radiation from half-wave antenna.

toward the center of the dipole, part B. But, since lines of force in the same direction repel each other, the centers of the lines are still being held out.

As the voltage approaches zero part C, some of the lines collapse back into the dipole, but in others the ends of the line come together and form a complete loop. At this point the voltage on the antenna is zero. As the charge starts to build up in the opposite direction, part D, electric lines of force again begin at the positive end of the antenna and stretch to the negative end of the antenna. These lines of force, being in the same direction as the sides of the closed loop next to the antenna, repel these closed loops and force them out into space at the speed of light. As they travel they generate a magnetic field in phase with them.

Figure 29-17 shows the manner in which radiation fields are propagated from the antenna. The E-field and the H-field are shown as separate sets of flux lines about the antenna. The electric flux lines are closed loops on each side of the antenna. The magnetic flux lines are closed circular loops which have their axis around the antenna. They are represented as dots and crosses. The sine waves, that are labeled "curves of radial variation of flux density," indicate the relative field strength at various distances and angles from the antenna.

In the direction perpendicular to the antenna, both the H-field and the E-field are strong, for this is the direction where both fields originally formed. In the direction parallel to the antenna, no H-field forms at all and only a very small E-field. The flux density, therefore, is small in these two directions.

EFFECTS OF LENGTH ON ANTENNA IMPEDANCE

An antenna of the correct length acts as a resonant circuit and presents pure resistance to the excitation circuit at a specific frequency (fig. 29-18A). An antenna having other than the correct length displays both resistance and reactance to the excitation circuit. An antenna slightly longer than a half-wave, for example, acts as an inductive circuit (fig. 29-18B). This is understandable if you think of the antenna in terms of a quarter-wave RF line. When the antenna is excited at the center, it is equivalent to a quarter-wave RF line looking at it from the generator end (fig. 29-18A). The inductive reactance is exactly equal to the capacive reactance and the circuit appears purely resistive to the source. When the antenna is longer than a half wave, it behaves like an RF line that is slightly longer than a quarterwave. According to the equivalent lumped circuit (fig. 29-18B) the inductance is not entirely balanced by the capacitance and thus appears inductive to the source. The remedy for correcting the length is either cutting the conductor shorter, or tuning out the inductive reactance by adding capacitance in series.

If the antenna is physically shorter than its resonant length, the input impedance appears as a capacitive reactance. The two-wire line shorter than a quarter-wavelength displays capacitance at its input terminals (fig. 29-18C). The correction in this case is either to add inductance in series with the antenna to bring it back to resonance or to add physical length to the antenna.

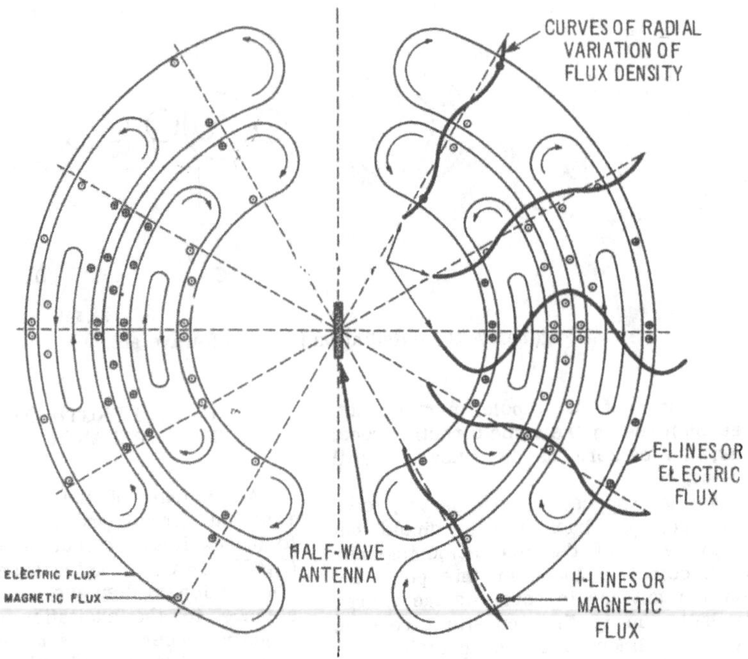

Figure 29-17.— Fields in space around a half-wave antenna.

INTRODUCTION OF ENERGY TO THE ANTENNA

Energy may be fed to an antenna in a variety of ways; the most common are shown in figure 29-19. When the excitation energy is introduced to the antenna at the point of high circulating currents, the antenna must be current fed. When the energy is introduced at the point of maximum voltage, the antenna is voltage-fed. In the case of half-wave antennas isolated in free space, the impedance is 73 ohms at the center and 2,500 ohms at the ends. The intermediate points have intermediate values of impedance.

POLARIZATION OF AN ELECTROMAGNETIC WAVE

The radiated energy from an antenna is in the form of an expanding sphere. A small section of this sphere is called a wave front; it is perpendicular to the direction of travel of the energy. All energy on this surface is in phase. Usually all points on the wave front are at equal distances from the antenna. The farther from the antenna, the less spherical the wave appears. At a considerable distance the wave front can be considered as a plane surface at right angles to the direction of propagation.

The radiation field is made up of magnetic and electric lines of force which are always at right angles to each other. Most electromagnetic fields in space are said to be linearly polarized. The direction of polarization is the direction of the electric vector. That is, if the electric lines of force are horizontal, the wave is said to be horizontally polarized (fig. 29-20A), and if the E lines are vertical the wave is said to be vertically polarized.

As the electric field is parallel to the axis of the dipole, the antenna is in the plane of polarization. The horizontally-placed antenna produces a horizontally polarized wave, and the

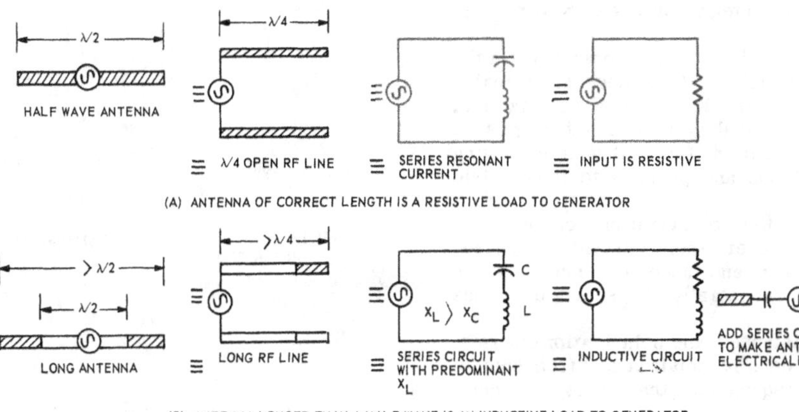

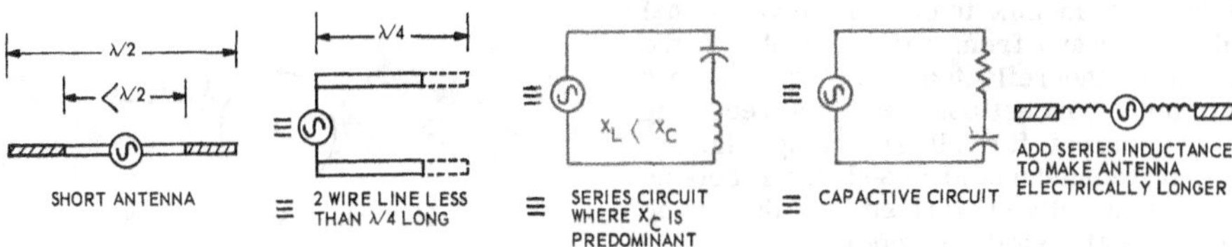

(C) ANTENNA SHORTER THAN A HALF WAVE IS A CAPACTIVE LOAD TO GENERATOR

179.507

Figure 29-18.— Effect of length on antenna impedance.

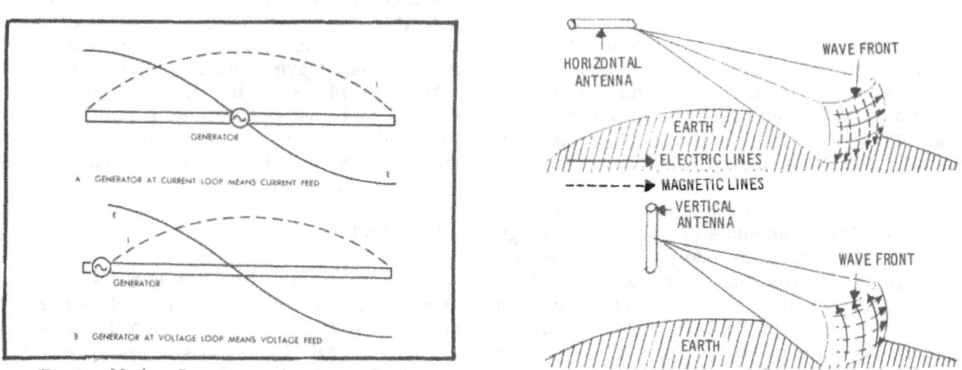

Figure 28-6 - Current and voltage feed.

20.245

Figure 29-19.—Current and voltage fed antennas.

25.92

Figure 29-20.—Horizontal and vertical polarization.

vertically-placed antenna produces a vertically polarized wave.

For maximum absorption of energy from the electromagnetic fields, it is necessary that a dipole be located in the plane of polarization. This places the conductor at right angles to the magnetic lines of force that are moving through the antenna and parallel to the electric lines.

In general, the polarization of a wave does not change over short distances. Therefore, transmitting and receiving antennas are oriented alike, especially if short distances separate them.

Over long distances, the polarization changes. The change is usually small at low frequencies, but at high frequencies the change is quite rapid.

With radar transmissions, a received signal is a reflected wave from some object. As the polarization of the reflected signal varies with the type of object, no set position of the receiving antenna is correct for all returning signals. Where separate antennas are used, the receiving antenna is generally polarized in the same direction as the transmitting antenna.

When the transmitting antenna is close to the ground insofar as propagation is concerned, vertically polarized waves cause a greater signal strength along the earth's surface. On the other hand, antennas high above ground should be horizontally polarized to get the greatest signal strength possible to the earth's surface.

MARCONI ANTENNA

A quarter-wave grounded vertical antenna is called a Marconi antenna. Figure 29-21 illustrates the quarter-wave grounded antenna. Note the amplitude of the standing waves of current and voltage on this type of antenna. Note also the similarity to the half wave dipole when the image is included.

The ground is a fairly good conductor for medium and low frequencies and acts as a large mirror for the radiated energy. This results in the ground reflecting a large amount of energy that is radiated downward from an antenna mounted over it. It is just as though a mirror image of the antenna is produced; the image being located the same distance below the surface of the ground as the actual antenna is located above it. Even in the high frequency range and higher, many ground reflections occur, especially if the antenna is erected over highly

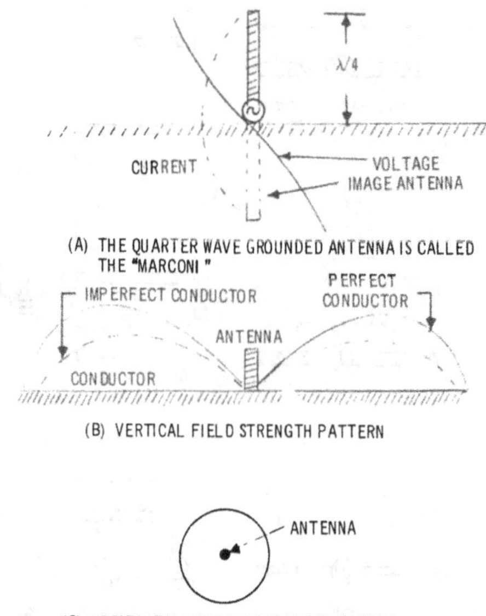

Figure 29-21.—Quarter-wave grounded antenna.

conducting earth, salt water or a grounded screen.

Utilizing this characteristic of the ground, an antenna only a quarter-wavelength long, can be made into the equivalent of a half wave antenna. If such an antenna is erected vertically and its lower end is connected electrically to the ground, the quarter-wave antenna behaves like a half-wave antenna. Here, the ground takes the place of the missing quarter-wavelength, and the reflections supply that part of the radiated energy that normally would be supplied by the lower half of an ungrounded half-wave antenna.

WHIP ANTENNA

The vertical whip antenna is most often used in the frequency range between 1.8 MHz and 30 MHz. Long whips introduce mechanical problems, but the average whip length used with Naval communications equipment is approximately 35 feet long. A whip antenna represents a resonant quarter-wave antenna at a frequency near 6 MHz. When the whip length is such that it is one quarter-wavelength at the operating frequency, the antenna will appear as a pure resistance at the point where energy is fed to the base of the antenna.

Any antenna displays a resistive component to an input signal. That portion of an antenna's total input impedance which causes the radiator (antenna) to consume power is frequently called its RADIATION RESISTANCE.

When the input impedance of an antenna is primarily comprised of radiation resistance, the efficiency of the antenna is high. However, at frequencies above or below antenna resonance, inductive or capacitance reactances dominate the input impedance, and antenna efficiency is low. When such reactances are unavoidable, as frequently occurs in whip antennas, opposite reactances are used to cancel each other and improve antenna efficiency.

SLEEVE ANTENNA

A sleeve antenna is a broad band antenna. It is similar in appearance to the whip antenna but has a large diameter sleeve at its base (fig. 29-22). Critical dimensions are determined by the lowest frequency to be used by the antenna.

The large diameter sleeve may be welded to the deck, and the sleeve diameter, in feet, should be equal to 30 divided by the frequency in megahertz. The sleeve section length is found by dividing 75 by the frequency in megahertz. The same method is used to find the diameter and the length of the whip section.

As figure 29-22 illustrates, the impedance at the base of the antenna is matched to the 50 ohm transmission line impedance by a quarter-wave transformer section. The impedance of this section is found by multiplying the antenna base impedance by the transmission line impedance, then taking the square root of the result. The impedance of the whip section is 238 ohms, and the transmission line impedance is 50 ohms. Multiplying these values and taking the square root of the product gives an impedance of 109 ohms for the quarter-wave section.

When used for the higher frequencies, sleeve antennas have several desirable characteristics. They are superior to single-wire or regular whip antennas in both vertical pattern and impedance characteristics when fed against ground. The impedance characteristic of a sleeve antenna is such that satisfactory standing wave ratios (3 to 1, or less) can be obtained over a wider frequency range, and radiation at low angles is much greater than a regular whip.

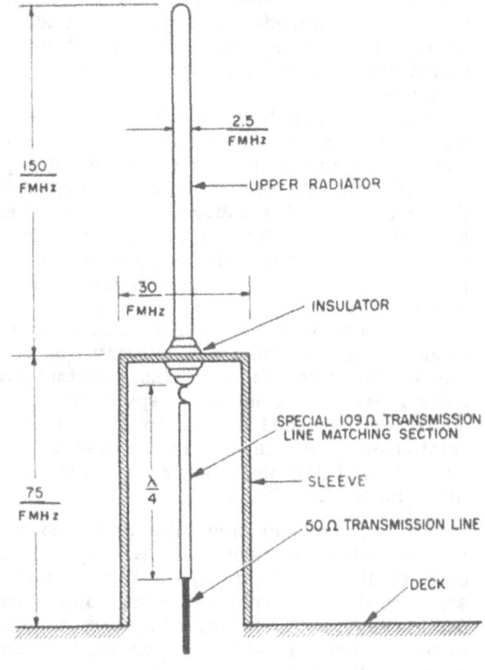

Figure 29-22.—Sleeve antenna.

HERTZ ANTENNA

Any antenna that is one-half wavelength long, or any even or odd multiple thereof, is a Hertz antenna and may be mounted either vertically or horizontally. A distinguishing feature of all Hertz antennas is that they need not be connected conductively to ground. The frequency spectrum covered in shipboard installations is 3.5 MHz to 70 MHz.

Many broad-band UHF antenna have been developed for use aboard ship. They are used for transmitting or receiving vertically polarized waves in the 220 to 400 MHz range.

The radiation pattern is similar to that of a dipole. However, the pattern may be affected by the external supporting structure and the proximity of metallic masses.

FOLDED DIPOLE

In multi-element arrays the input impedance drops. In these arrays special matching devices are needed for matching this low impedance to the higher impedance of most RF lines. One method for making the match is to use the folded dipole. This is most often used in radar equipment.

A folded dipole is a full wavelength conductor that is folded to form a half-wave element. A better description is that it consists of a pair of half-wave elements connected together at the ends. In it the voltage at the ends of each element must be the same. In operation, the field from the driven element induces a current in the second element. This current is the same as the current in the driven element.

An ordinary dipole with a given current (I) produces a certain field intensity in space. Due to this field, there is also a certain power density per square meter in space. This power density is produced by the input power P. The relationship between the input resistances, the current, and the input power is expressed by the equation, $R = P/I^2$.

Another consideration with the folded dipole is that when the same current (I) exists in each of the two sections, the field strength in space is doubled. This causes the power density per square meter to increase four times. In turn, the input power must be four times as great. (In this case,

$$R = \frac{4P}{I^2}\Big)$$

As long as each section of a dipole has the same diameter, the input resistance is four times that of the simple (half-wave) dipole. Increasing the diameter of one section makes the increase in impedance still greater. While the input impedance to the driven element of a parasitic array drops to about a fourth of the value of the coaxial cable impedance, the use of a folded dipole increases the impedance by about four times. In this way a good impedance match is affected.

Parasitic arrays are made up of non-driven (free) elements that absorb energy from the driven element and in turn re-radiate it in such a manner that the gain in one direction is increased while the gain in the other direction is greatly reduced. Longer parasitic elements that are placed parallel to the driven antenna and in the direction of minimum radiation are called reflectors (fig. 29-23A). Short parasitic elements that are placed parallel to the driven element and in the direction of maximum radiation are called directors (fig. 29-23B).

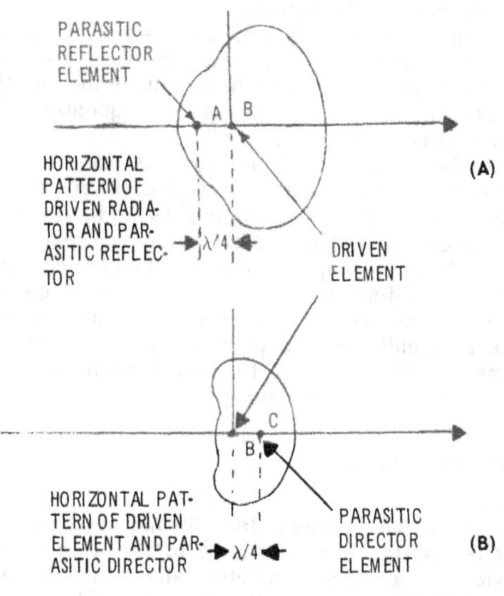

Figure 29-23.—Single parasitic arrays.

Antenna A (figure 29-23A) is slightly longer (tuned to a lower frequency) than a half wavelength and is not connected to a power source. If it is placed parallel to and slightly less than a quarter-wavelength from the driven half-wave antenna, B, it will act as a parasitic reflector. Power is absorbed from the driven element and re-radiated in such a phase and manner that the combined electric fields add in one direction, (from reflector toward the driven element) and subtract or cancel in the opposite direction. This action and interaction produce the pattern shown in figure 29-23A.

When the field produced by the driven element cuts the reflector, it induces a voltage in the reflector in such a direction as to produce a field opposite in polarity to the inducing field. If the reflector is properly designed, the induced field is comparable in magnitude to the inducing field. Therefore, only a small quantity of energy can travel behind the reflector because the two fields cancel when they are of opposite polarity.

The time required for the field to travel from the driven element to the reflector and back to the driven element corresponds to the time for one half cycle or 180°. Since fields undergo a phase shift of 180° when they are reflected, the field returning to the driven element arrives at this element exactly in phase with the next radiated field and hence adds to it. Thus, the effective radiation is in a direction away from the reflector.

A parasitic element shorter (tuned to a higher frequency) than a half wavelength placed parallel to and slightly less than a quarter wavelength from a half wave driven element is called a parasitic director. It absorbs power and radiates it with such a phase relation that the fields of the two elements add in the direction of the director as shown in figure 29-23B.

A wide variety of field patterns can be obtained even with a single directional system containing only one parasitic element. Two variables can change this pattern—the length or tuning of the parasitic element and the spacing between the parasitic element and the driven element.

INVERTED L MILITARY ANTENNA

A bent antenna is a grounded antenna so constructed that a portion of it is mounted horizontally. Such an antenna takes the form of an inverted L. In this type of antenna, a fairly long horizontal portion, or flattop, is used, and the vertical down lead, which forms an important part of the radiating system, is connected to one end of the flattop. The length of the antenna is measured from the far end of the flattop to the point at which the downlead is connected to the transmitter.

Figure 29-24A shows an inverted L, single wire vertical antenna. The length of the straight vertical portion is one quarter-wavelength; the length of the horizontal portion is also a quarter-wavelength. Thus, as can be seen, the current loop is at the topmost part of the straight vertical portion, with resultant increased radiation efficiency. However, considerable energy leaves the antenna at relatively high radiation angles, in a direction opposite the free end, due to current flow in the horizontal section.

Figure 29-24C shows an inverted L antenna with a multiple wire flattop arrangement. The efficiency of this inverted L is somewhat better than that of the single-wire type. Although it has good ground wave propagation characteristics, it also radiates considerable energy at a high radiation angle. Since a high voltage loop exists

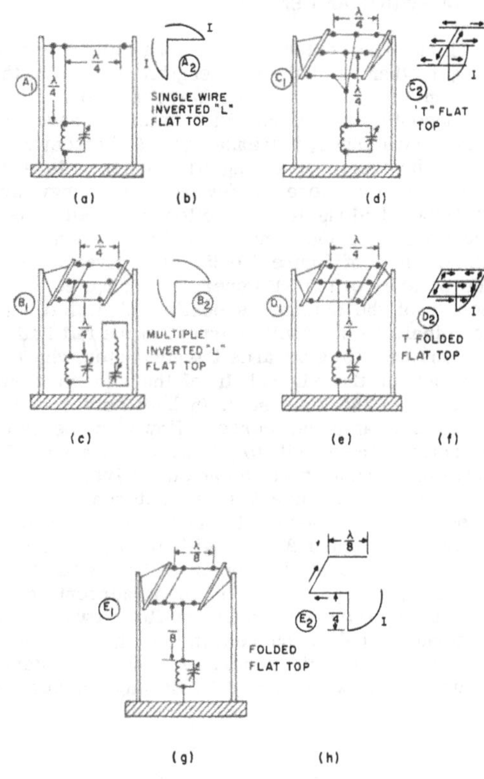

Figure 29-24.—Flat-top vertical antennas, show- the coupling circuits and the current distribution.

at the input end, a parallel tuned coupling arrangement is suitable.

Figure 29-24C shows the method for obtaining some cancellation of the fields in the vertical plane due to the out-of-phase currents in the flattop conductors.

Figures 29-24D and E show the arrangement of horizontal conductors to effect a maximum cancellation of the fields due to current flow in the flattop. The folded top arrangement offers the best possibilities for minimizing high angle radiation, especially in the low frequency operating range.

LONG-WIRE ANTENNAS

Long-wire antennas are long single wires, longer than a half-wavelength, in which the current in adjacent half wave sections flows in opposite directions. Such antennas have two basic advantages, increased gain and directivity.

If the length of a long-wire antenna is such that two or more half-waves of energy are distributed along it, it is often referred to as a harmonic antenna. Consider the half-wave antenna shown in A of figure 29-25. At a given instant, the polarity of the RF generator connected to the center of the antenna is positive at its left hand terminal. As a result, current in the left half of the antenna flows towards the generator, whereas current in the right half of the antenna flows away from the generator. In both halves of the half wave antenna, current flows in the same direction, from left to right, as shown by the wave of current above the antenna wire.

Now let's assume that the antenna first discussed is increased until it is 2 half-wavelengths, as in B of figure 29-25. With the RF generator still connected at the center and with the same instantaneous polarities as in A, current in the left side of the antenna must flow towards the generator, and the current in the right side must flow away from the generator. Since the antenna is now 2 half-wavelengths, 2 half-waves of current can be accommodated on the antenna and the current polarity is the same in both halves of the antenna. It is important to note that this is not a true long-wire or harmonically operated antenna since there is no reversal of current flow in adjacent half-wave sections. Instead this arrangement is simply 2 halfwave antennas operating in phase at their fundamental frequency. Such an arrangement is called a driven collinear array and has characteristics quite different from those to be discussed for the true harmonically operated or long-wire antenna.

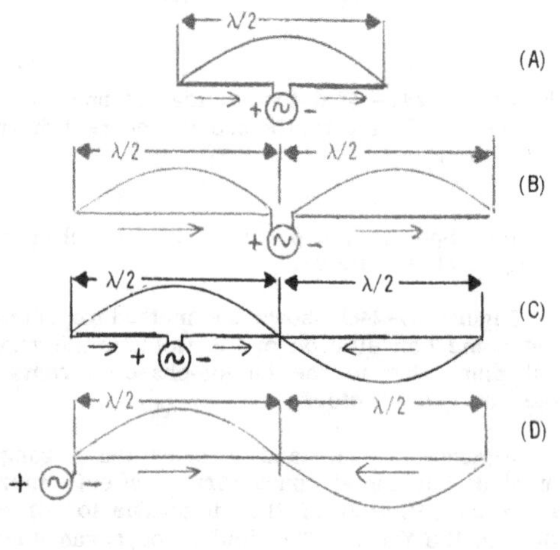

Figure 29-25.—Harmonic and nonharmonic antennas.

The antenna in B can be converted into a true long-wire, harmonically operated antenna simply by moving the generator to a current loop as shown in C. With the RF generator polarity as shown, current flows from left to right in the half-wave section of the antenna. The direction of current flow is then reversed in the second half-wave section. If the generator is moved to the extreme end of the antenna as shown in D, the antenna is also a long-wire antenna, and the current distribution on the antenna is exactly the same as in C. The harmonically operated antenna, therefore, must be fed either at a current loop or at its end for proper operation.

As the length of the antenna is increased, it is natural to expect a change in the radiation pattern produced by the antenna. A long wire antenna can be considered to be made up of a number of half wave sections fed 180° out of phase and spaced a half-wavelength apart. As a result, there is no longer zero radiation off the ends of the antenna, but considerable radiation occurs in the direction of the long wire as a result of the combined fields produced by the individual half-wave sections. In addition, radiation also occurs broadside to the long wire. Consequently, the resultant maximum radiation is neither completely at right angles to the long wire nor completely along the line of the long wire. Instead, the maximum radiation occurs at some acute angle in respect to the wire, the exact angle being determined by the length of the antenna.

As shown in figure 29-26, as the length of a long wire antenna is increased, the following characteristic changes occur: First, the gain of the antenna increases considerably compared with that of the basic half wave antenna, especially when the long wire is many wavelengths. Second, the direction along which maximum radiation occurs makes a smaller angle with respect to the wire itself. Consequently, as the

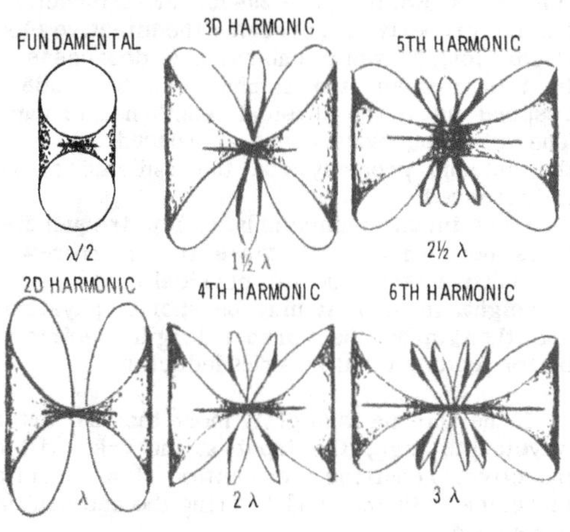

Figure 29-26.—Radiation patterns of harmonic antennas.

antenna is made longer, its major lobe of radiation lies closer to the direction of the wire itself. Third, more minor lobes are produced as the antenna length is increased.

CYCLIC VARIATIONS OF REACTANCE AND RESISTANCE AT CENTER OF ANTENNA

When the frequency applied to an antenna is no longer the resonant frequency, a considerable reactive component is present. Consider a half-wave antenna that is fed at the center. As the frequency of the RF generator is increased above the resonant frequency of the antenna, the antenna becomes inductive. This inductive reactance reaches a maximum value and then begins to fall off as the frequency continues to rise. When the frequency is such that the antenna length is slightly less than a full wavelength, the reactance is zero, but resistance is apparently very high. An increase in frequency above this value causes the antenna to behave as a capacitive reactance.

As the frequency is raised still higher, the capacitive reactance reaches a maximum value and then falls off toward zero. When the frequency is such that the antenna is slightly less than 1 1/2 wavelengths, the reactance is again zero, but resistance is apparently very low. An increase in frequency above this causes the antenna once more to behave as an inductive reactance. As a result of these variations, the input impedance of the antenna also has a cyclic variation as the frequency is raised from one harmonic to another.

FEEDING LONG-WIRE ANTENNAS

Both resonant and non-resonant lines can be used to feed long-wire antennas. The same general principles apply here as in the half-wave antenna. Since a point on the antenna which is a current loop becomes a current node on the next higher harmonic, a current-fed antenna behaves as a true long wire only at odd harmonics of the original frequency. Therefore, for operation on all harmonics, end feeding is preferred. However, with end feeding, unbalanced transmission line currents result and a nonsymmetrical radiation pattern is produced. The intensity of the lobes in the direction of the feeder end of the long wire is reduced, and the intensity of the lobes in the direction away from the feeder end is increased. When matching sections of line are used with non-resonant feeders, it must be realized that these operate over only a narrow band of frequencies.

An end-fed long wire antenna with a resonant feeder line is shown in figure 29-27A. Operation on all harmonic frequencies is possible with this arrangement, provided the tuning unit at the input end of the resonant line has sufficient range to match the input impedance to the transmitter. Arrangements for using nonresonant lines are shown in B and C. In both quarter-wave matching sections are used to match the nonresonant line to the long-wire antenna. In B the feeder is tapped on the matching section at a point where an impedance match occurs.

In C the feeder is connected to a Q-matching section, the characteristic impedance of which is made equal to the square root of the product of the radiation resistance of the long-wire antenna and the impedance of the nonresonant line.

The antenna may be excited either by a direct connection or by radiation (induction fields) from nearby antennas. In addition to being excited by a nearby antenna, an antenna may also be excited by radiation fields from a distant transmitting antenna. This takes place in any receiving antenna. In a receiving antenna the current that is conducted to the receiver is the result of

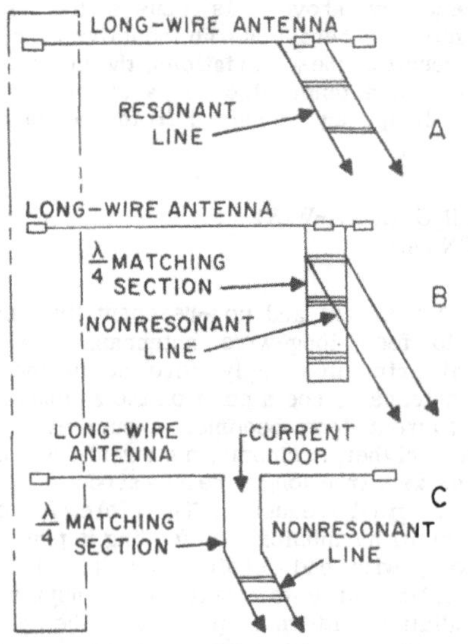

179.512

Figure 29-27.—Feeding long-wire antennas.

current induced by radiation fields generated by the transmitting antenna.

ANTENNA TUNING

Aboard ship, antennas used for communications at the medium frequencies are not usually of the proper length to give optimum performance at the operating frequency. This condition exists because these antennas are all of standard size and shape or are installed in whatever space may be available for them and because they are each operated at more than one frequency. All equipment must be able to operate at any frequency within its tuning range. In this case, then, it is necessary to employ some means at the transmitter to adjust the antenna for reasonable efficiency at any frequency regardless of the physical dimensions or arrangement of any antennas that might be available.

As each transmitter is usually associated with only one antenna, which is of fixed length, the adjustment of the effective length of the antenna must be made by electrical means. This process is called "antenna tuning," and is accomplished by adding either inductance or capacitance to the antenna at the point where it is fed from the transmitter or transmission line, as shown in figure 29-28. Added inductance has the property of increasing the effective electrical length, while capacitance decreases it. In this manner the antenna can be made to respond as if the physical length had changed. The standing waves are increased by tuning the antenna properly and the radiated energy is increased.

Sometimes, particularly at low frequencies, it is not practical to make the quarter-wave grounded antenna the full physical quarter-wave in height. Instead, it may be shorter physically and then made the correct length electrically by top loading it with a series inductor.

If the antenna is slightly more than a quarter-wavelength high, the input at the base will be inductive, requiring the addition of a capacitor in series with the feed to bring the antenna into resonance.

EFFECTS OF PRACTICAL GROUNDS

Up to this point, all of the effects produced have been the result of a ground which has a uniform high conductivity. In practice, the nature of the ground over which the antenna is erected is subject to considerable variation. For example, the antenna may be erected over a ground which has high or low conductivity, and over a ground with uniform or non-uniform conductivity in the vicinity of the antenna. All of these characteristics

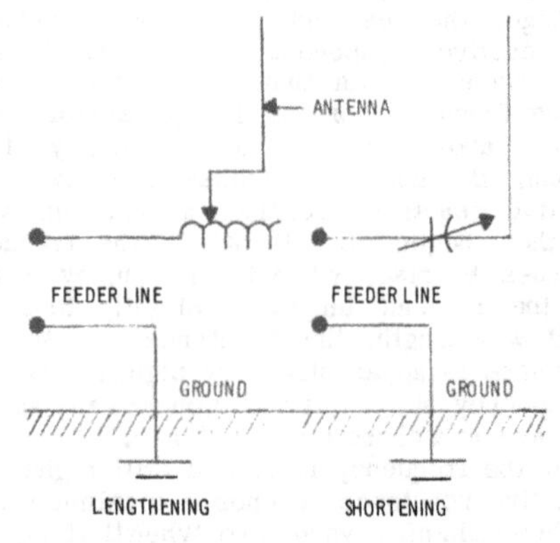

179.513

Figure 29-28.—Methods of correcting the electrical length of a grounded antenna.

of the ground have an effect on the radiation patterns and resistance of the antenna.

Unless the ground behaves as a nearly perfect conductor, the amplitude of the ground-reflected wave will be much less than the amplitude of the wave before reflection. A portion of the wave which ordinarily would be reflected is absorbed by the resistance of the ground. Such absorption by the ground constitutes ground losses. This means that the value of the ground reflection factor will be reduced considerably.

The most common ground system used with vertical grounded antennas at fixed stations is the radial ground. This consists of a number of bare conductors arranged radially and connected. The conductors, which may be from a tenth to a half-wavelength or more, are buried a short distance beneath the surface of the earth. Sometimes one radial ground system serves two vertical antennas which operate over different frequency ranges.

With a less elaborate ground system, a number of ground rods can be used. These rods usually are made of galvanized iron, steel, or copper plated steel in lengths up to 8 feet. One end of the rod is pointed so that it can be driven easily into the earth. The other end frequently is fitted with some type of clamp so that the ground lead can be attached. Some ground rods are supplied with a length of ground lead already attached. If possible, the rods should be planted in a moist section of ground or in a depression which will collect moisture. Ground resistance can be reduced considerably by treating the soil directly, with coal dust for instance. A trench about a foot deep is dug around each ground rod and filled with some common rock salt. The trench then is flooded with water, after which it is covered with earth. To remain effective this treatment is renewed every few years.

When an actual ground connection cannot be used because of high resistance of the soil or because a large buried ground system is not practical, a counter-poise may replace the usual direct ground connection in which current actually flows to and from the antenna through the ground itself. The counterpoise consists of a structure made of wire erected a short distance off the ground and insulated from the ground. The size of the counterpoise should be at least equal to and preferably longer than the size of the antenna.

The counterpoise operates by virtue of its capacitance to ground. Because of this capacitance, the ground currents which flow normally and usually are collected by conduction, now are collected in the form of charge and discharge currents. The end of the antenna which normally is connected directly to ground now is connected to ground through the large capacitance formed by the counterpoise. If the counterpoise is not well insulated from ground, the effect is much the same as that of a leaky capacitor. Leakage currents flow between the counterpoise and ground so that a poorly insulated counterpoise introduces more losses than no counterpoise at all.

FIELD INTENSITY

Field intensity or field strength as it is sometimes called, is the effective value of the electric field intensity in microvolts or milliwatts per meter produced at a point by radio waves from a particular station. Unless otherwise specified, the measurement is assumed to be in the direction of maximum field intensity.

Seldom are the actual operational characteristics of an antenna exactly the same as those determined on the basis of theoretical considerations. In order to determine these differences, various measurements must be made after the antenna is installed and while it is being test operated. Often, on the basis of these measurements, changes are made in the designs or installation of the antenna to improve the radiation pattern.

It is very important to know the direction and intensity of the power being radiated from an antenna; and, to determine these values, measurements of the field intensity are made at various distances and directions from and around the antenna.

In order to determine the field strength, or field intensity, it is desirable to use some type of standard antenna as a basis for all comparisons. The standard antenna is a wire 1 meter long. The magnitude of the signal voltage (in microvolts) induced into this antenna is called the Absolute Field Strength, and is measured in microvolts per meter. Other types of antennas are commonly used, but they are calibrated against a standard one meter antenna to obtain an absolute field-strength measurement. Rod antennas are used to measure ground-wave signal intensities; when the signal is reflected from the ionosphere, a loop antenna is commonly used.

The calibrated antenna picks up the radio wave, and feeds the induced voltage to a sensitive receiver that is well shielded against extraneous signals. An indicating voltmeter is connected across the output of the receiver.

When the receiver is properly calibrated, the absolute value of the field strength in microvolts per meter ($\mu v/m$) is indicated on the voltmeter.

If the meter in the available equipment is not calibrated in $\mu v/m$, but a signal generator, (covering the desired frequency range and whose output is calibrated in $\mu v/m$) is available, then then the calibrated field-strength readings may be made without difficulty. The antenna is connected to the receiver and the meter reading noted. The antenna is then disconnected and the calibrated signal generator connected to the receiver input. The output of the calibrated signal generator is adjusted until the receiver meter reads the same as it did when the antenna was connected. The output of the signal generator is then equal to the output of the standard meter antenna. If the antenna is not 1 meter long, an antenna correction factor is used. The field strength is $\mu v/m$ is computed as:

$$\text{field strength} = \frac{\text{output of calibrated signal generator}}{\text{antenna correction factor}}$$

The power of a radiated wave, such as a light wave, falls off as the square of the distance between the source and point of measurement. This same law holds here for the field intensity of electromagnetic waves when the power intercepted in a unit area is considered. However, since electric power expressed in terms of the voltage present is proportional to E^2 (because $P = E^2/R$), then the square of the voltage falls off as the square of the distance, or voltage itself falls off as the distance.

Absolute field-intensity measurements are not difficult to make. However, the necessary equipment is relatively complex, bulky, and must be calibrated very carefully. Often, all that is necessary to know is Relative Field Strength, and simple field-strength meters and a pick-up antenna are all that is necessary to make the measurements. The pick-up antenna should be polarized the same as the antenna whose field intensity is being measured.

FIELD-STRENGTH METER

In the discussion of antenna field strength, the use of a receiving device that gives a relative indication of field strength was mentioned. The use of some kind of field-strength measuring device is essential if the antenna system is to be adjusted to produce maximum radiation. Basically, the field strength meter consists of a pick-up antenna, a tuned input circuit, a crystal rectifier, and some kind of indicating device such as a microammeter. The field strength meter will only give relative measurements, but is very useful in the VHF range, and for determining the beam pattern when using directional antennas.

When using the field-strength meter, care must be taken to make the field-pattern checks at least several wave lengths away from the antenna, and at heights corresponding with the desired angle of directivity.

Figure 29-29 shows a field-strength measuring device especially suitable for measurements in the VHF range. The input or antenna coil shown, which is of the variometer type, may be changed for different frequency ranges. The amount of energy introduced into the tuned circuit is determined by the amount of coupling between the primary and the secondary. Capacitor C1 is tuned for maximum indication of the meter at one particular location of the pick-up antenna. Resistors R1 and R2 make the response of the crystal more linear with variations in radiated

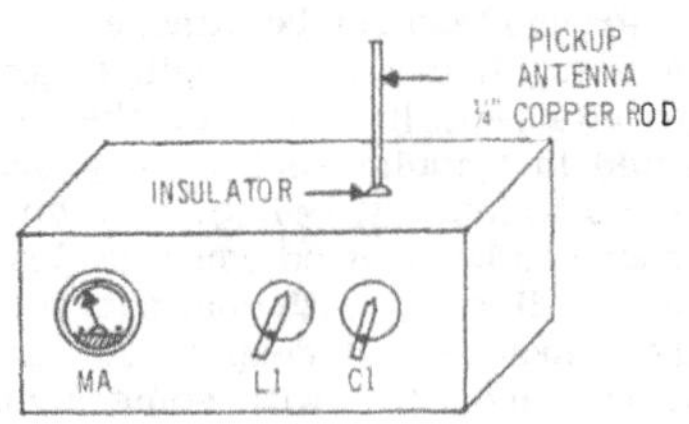

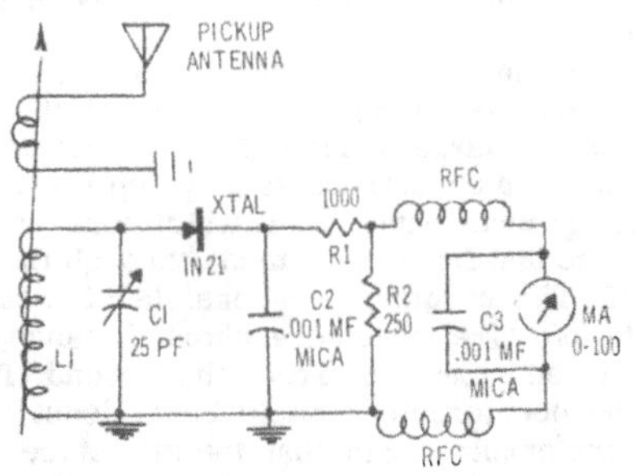

179.514

Figure 29-29.—Field strength measurement device.

power, and also lessen the loading effect of the meter on the tuned circuit.

ANTENNA INSULATORS

The strain or suspension insulator is used to support the wire antenna as well as to isolate it from the ship's structure. This type insulator is made of hard rubber, isolantite, or ceramic and may be made in rectangular or cylindrical shapes with holes at each end.

The stand off type insulator is used for supporting a transmission line away from a wall or siding. This insulator has a base in which holes are drilled for mounting and a top clamping arrangement through which the transmission line is fed and secured.

Lead-through insulators serve the purpose of providing a good insulating surface for feeding transmission lines through a wall. These are usually shaped like a bowl and are made of porcelain or ceramic.

INSULATION RESISTANCE

For proper operation of an antenna it is important that the insulation between the antenna and the ship's structure be in excellent condition. A megger is the test instrument used to determine the insulation resistance of an antenna. If the megger is applied to an antenna between the radiating element and its insulation, the reading will indicate the condition of the insulation.

Generally, if the reading is more than 100 Megohms, the insulation is acting as a good dielectric and the antenna should operate well from this standpoint. If the reading falls between 55-100 Megohms, it indicates dirt or a minor crack in the insulation. Below 35 Megohms indicates insulation breakdown and excessive loss of power.

It should be noted that some antennas operate properly with insulation readings of about 5 megohms and others are designed to show a low resistance (possibly a direct short) to d.c. Therefore, always consult the technical manual for an antenna when in doubt as to its insulation resistance.

Antennas are an important part of the communications system and require a certain amount of maintenance. Almost all of this maintenance deals with care of the insulators. Insulators should not be cracked, chipped or damaged in any way. They must be kept free from dirt, salt, soot, or other deposits.

All insulators, transmission line end-seals, and wave-guide windows should be cleaned with fresh water or with a dishwashing compound in hot water. They should then be rinsed with fresh water and dried with a clean rag. Application of a thin coating of Dow-Corning Compound No. 4 will prevent buildup of salt or dirt deposits.

Paint, varnish, shellac, or grease should not be applied to any portion of the insulating materials or the antenna bus.

SAFETY PRECAUTIONS

All personnel should be cautioned not to go closer than 1 foot of an exposed radio antenna unless it is first determined from competent authority that the antenna is not energized. Unauthorized personnel must not tamper with or attempt an inspection of any electronic equipment as this equipment carries dangerous voltages.

Before any work may be done aloft authorization must be obtained from the CDO or OOD. Before sending men aloft, the CDO or OOD directs the communication watch officer and the radar watch officer to secure the proper transmitters in order to render this area safe. He also notifies the engineer duty officer that men will be working in a prescribed area aloft in order that the engineer duty officer may take the necessary precautions to prevent the boiler safety valves from lifting. Until he has received a report from the communication watch officer that the proper transmitters are secured, the CDO or OOD permits no man to go aloft. (If antennas must be energized, men shall not be permitted to go aloft except by means of ladders and platforms rendered safe by grounded hand rails or similar structure.)

After the work has been completed, a report is made to the CDO or OOD, and his authorization must be obtained before equipment is again energized.

Radar and other antennas which rotate or swing through horizontal or vertical arcs may cause men working aloft to fall. Therefore, the motor switches which control the direction of these antennas should be locked open and tagged before men are permitted to ascend or go within reach of them.

The use of electronic equipment in the frequency range of 30 megahertz and below will cause voltages to be induced in the standing rigging and other portions of a ship's structure

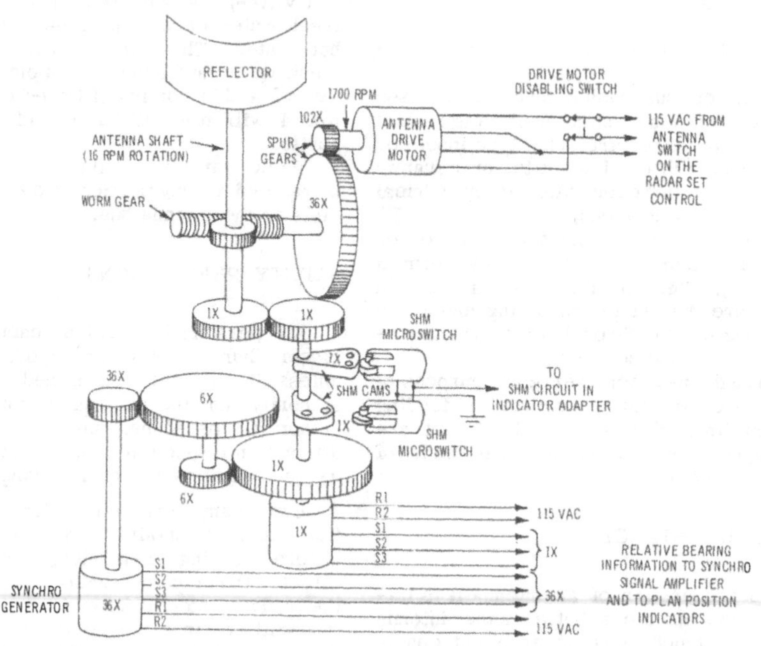

179.515

Figure 29-30.—Antenna assembly, antenna drive, and synchro system.

that, under certain conditions, must be considered hazardous. Operation in the higher frequencies (above 30 MHz) is not considered hazardous in this respect because of the remote location of antennas from the normal working areas in a ship. The voltage, or resonant circuits set up in a ship's structure or rigging will cause shock to personnel or produce open sparks when contact is made or broken, when the circuit is opened, or when metallic objects are in contact with the structure.

During the handling of ammunition, volatile liquids, or gases (particularly fueling operations involving delivery of gasoline from hoses, spouts, cans, or any place where vapors are present), high frequency transmitters are secured.

ANTENNA PEDESTAL

The antenna pedestal is the support for the radar antenna and houses the motors, synchros, and switches that run and control the antenna. These components and devices require preventative maintenance, which if properly carried out will greatly reduce the need for corrective maintenance.

A typical radar antenna pedestal contains the following components: antenna drive motor and associated gear trains, synchro generators, terminals, boards for incoming and outgoing power leads, ship's heading marker assembly, disabling switch, oil reservoirs with level indicating device, ship heater with thermal switch, and a rotary joint for the waveguide.

The ship's heading marker is a device that gives a flash on the radar indicator to indicate the direction in which the ship is moving. This is accomplished by SHM microswitches in the antenna assembly. These switches, are shown in figure 29-30. Complete instructions for setting these switches can be found in the instruction book for the equipment.

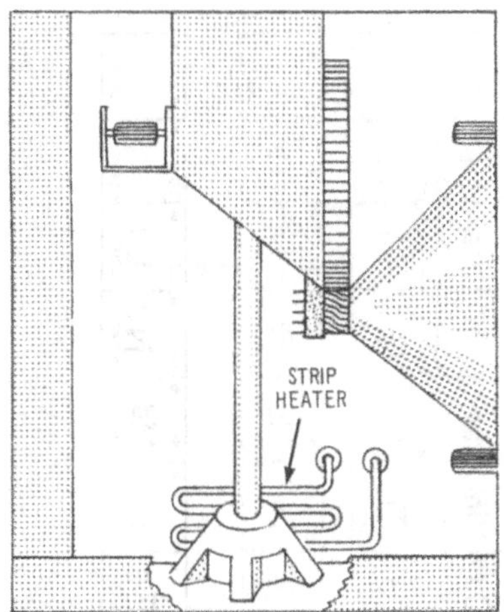

179.516

Figure 29-31.—Antenna assembly, synchro section of antenna gear box.

The strip heater shown in figure 29-31 has a dual purpose. By application of heat to the antenna pedestal, moisture is driven out of the enclosure. When the ship is operating in cold climates, the oil is kept heated preventing unnecessary strain on the drive motor and gears.

The rotary joint allows electrical connection of antenna to waveguide without mechanical connection. When upper and lower joint sections are properly aligned with each other, the rotary joint acts as a good conductor of the RF energy while being mechanically separated.

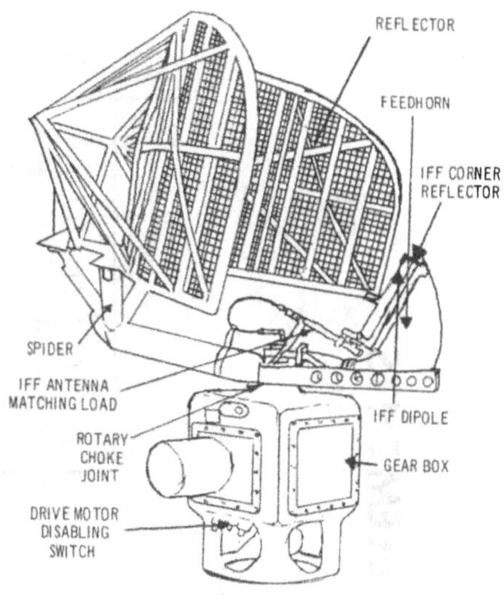

20.291

Figure 29-32.—Antenna assembly.

The antenna disabling switch is located conveniently at the base of the antenna pedestal. It disables the antenna by removing power to the antenna drive motor, thereby preventing the antenna from being rotated. Disabling of the antenna is necessary to prevent the antenna from being placed into operation from a remote station, causing serious injury to personnel working on the antenna or the antenna pedestal. The location of the disabling switch is shown in figure 29-32.

APPENDIX I

PERIODIC TABLE OF THE ELEMENTS

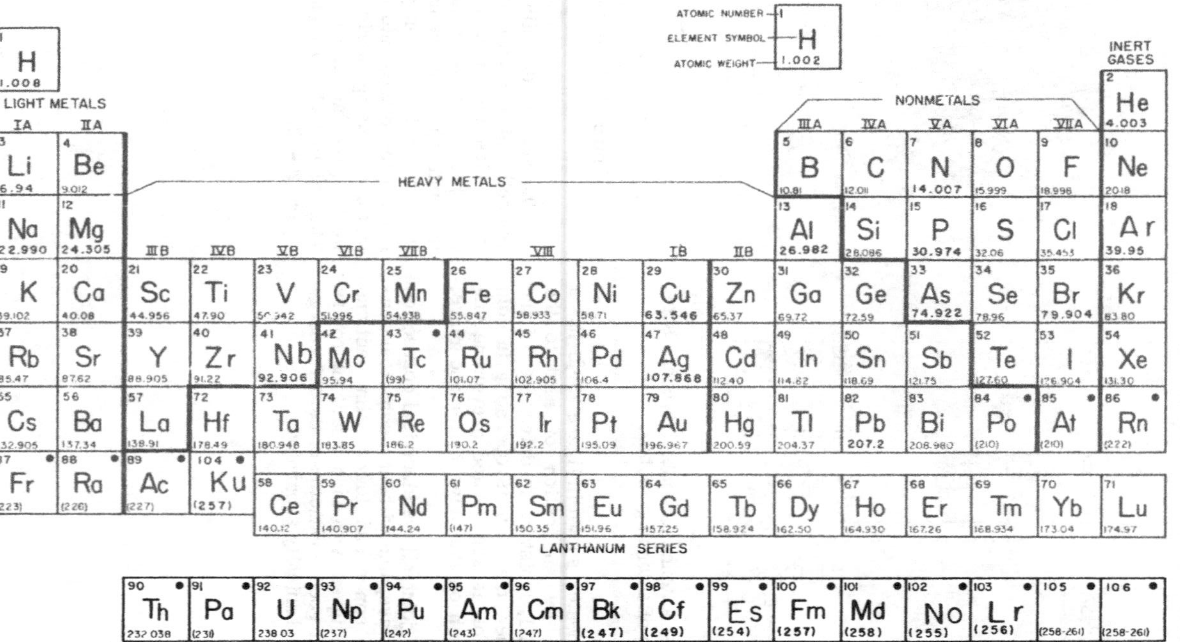

SEE NEXT PAGE FOR INTERPRETATION OF SYMBOLS

Figure A1-1. – Periodic table of the elements. 5.36(179)A

Symbol	Name	Atomic Number	Atomic Weight
Ac	Actinium	89	1(227)
Ag	Silver	47	107.868
Al	Aluminum	13	26.982
Am	Americium	95	(243)
Ar	Argon	18	39.95
As	Arsenic	33	74.922
At	Astatine	85	(210)
Au	Gold	79	196.967
B	Boron	5	10.81
Ba	Barium	56	137.34
Be	Beryllium	4	9.012
Bi	Bismuth	83	208.980
Bk	Berkelium	97	(247)
Br	Bromine	35	79.904
C	Carbon	6	12.011
Ca	Calcium	20	40.08
Cd	Cadmium	48	112.40
Ce	Cerium	58	140.12
Cf	Californium	98	(249)
Cl	Chlorine	17	35.453
Cm	Curium	96	(247)
Co	Cobalt	27	58.933
Cr	Chromium	24	51.996
Cs	Cesium	55	132.905
Cu	Copper	29	63.546
Dy	Dysprosium	66	162.50
Es	Einsteinium	99	(254)
Er	Erbium	68	167.26
Eu	Europium	63	151.96
F	Fluorine	9	18.998
Fe	Iron	26	55.847
Fm	Fermium	100	(257)
Fr	Francium	87	(223)
Ga	Gallium	31	69.72
Gd	Gadolinium	64	157.25
Ge	Germanium	32	72.59
H	Hydrogen	1	1.008
He	Helium	2	4.003
Hf	Hafnium	72	178.49
Hg	Mercury	80	200.59
Ho	Holmium	67	164.930
I	Iodine	53	126.904
In	Indium	49	114.82
Ir	Iridium	77	192.2
K	Potassium	19	39.102
Kr	Krypton	36	83.80
*Ku	Kurchatovium	104	(257)
La	Lanthanum	57	138.91
Li	Lithium	3	6.94
Lr	Lawrencium	103	(256)
Lu	Lutetium	71	174.97
Md	Mendelevium	101	(258)
Mg	Magnesium	12	24.305

Figure A1-1.—Periodic table of the elements—continued. 5.36(179)B

Symbol	Name	Atomic Number	Atomic Weight
Mn	Manganese	25	54.938
Mo	Molybdenum	42	95.94
N	Nitrogen	7	14.007
Na	Sodium	11	22.990
Nb	Niobium	41	92.906
Nd	Neodymium	60	144.24
Ne	Neon	10	20.18
Ni	Nickel	28	58.71
No	Nobelium	102	(255)
Np	Neptunium	93	(237)
O	Oxygen	8	15.999
Os	Osmium	76	190.2
P	Phosphorus	15	30.974
Pa	Protactinium	91	(231)
Pb	Lead	82	207.2
Pd	Palladium	46	106.4
Pm	Promethium	61	(147)
Po	Polonium	84	(210)
Pr	Praseodymium	59	140.907
Pt	Platinum	78	195.09
Pu	Plutonium	94	(242)
Ra	Radium	88	(226)
Rb	Rubidium	37	85.47
Re	Rhenium	75	186.2
Rh	Rhodium	45	102.905
Rn	Radon	86	(222)
Ru	Ruthenium	44	101.07
S	Sulfur	16	32.06
Sb	Antimony	51	121.75
Sc	Scandium	21	44.956
Se	Selenium	34	78.96
Si	Silicon	14	28.086
Sm	Samarium	62	150.35
Sn	Tin	50	118.69
Sr	Strontium	38	87.62
Ta	Tantalum	73	180.948
Tb	Terbium	65	158.924
Tc	Technetium	43	(99)
Te	Tellurium	52	127.60
Th	Thorium	90	232.038
Ti	Titanium	22	47.90
Tl	Thallium	81	204.37
Tm	Thulium	69	158.934
U	Uranium	92	238.03
V	Vanadium	23	50.942
W	Tungsten	74	183.85
Xe	Xenon	54	131.30
Y	Yttrium	39	88.905
Yb	Ytterbium	70	173.04
Zn	Zinc	30	65.37
Zr	Zirconium	40	91.22

* Note: Element proposed but not confirmed.

Figure A1-1.—Periodic table of the elements—continued. 5.36(179)B

APPENDIX II

TRIGONOMETRIC FUNCTIONS

0°–14.9°

Degs.	Function	0.0°	0.1°	0.2°	0.3°	0.4°	0.5°	0.6°	0.7°	0.8°	0.9°
0	sin	0.0000	0.0017	0.0035	0.0052	0.0070	0.0087	0.0105	0.0122	0.0140	0.0157
	cos	1.0000	1.0000	1.0000	1.0000	1.0000	1.0000	0.9999	0.9999	0.9999	0.9999
	tan	0.0000	0.0017	0.0035	0.0052	0.0070	0.0087	0.0105	0.0122	0.0140	0.0157
1	sin	0.0175	0.0192	0.0209	0.0227	0.0244	0.0262	0.0279	0.0297	0.0314	0.0332
	cos	0.9998	0.9998	0.9998	0.9997	0.9997	0.9997	0.9996	0.9996	0.9995	0.9995
	tan	0.0175	0.0192	0.0209	0.0227	0.0244	0.0262	0.0279	0.0297	0.0314	0.0332
2	sin	0.0349	0.0366	0.0384	0.0401	0.0419	0.0436	0.0454	0.0471	0.0488	0.0506
	cos	0.9994	0.9993	0.9993	0.9992	0.9991	0.9990	0.9990	0.9989	0.9988	0.9987
	tan	0.0349	0.0367	0.0384	0.0402	0.0419	0.0437	0.0454	0.0472	0.0489	0.0507
3	sin	0.0523	0.0541	0.0558	0.0576	0.0593	0.0610	0.0628	0.0645	0.0663	0.0680
	cos	0.9986	0.9985	0.9984	0.9983	0.9982	0.9981	0.9980	0.9979	0.9978	0.9977
	tan	0.0524	0.0542	0.0559	0.0577	0.0594	0.0612	0.0629	0.0647	0.0664	0.0682
4	sin	0.0698	0.0715	0.0732	0.0750	0.0767	0.0785	0.0802	0.0819	0.0837	0.0854
	cos	0.9976	0.9974	0.9973	0.9972	0.9971	0.9969	0.9968	0.9966	0.9965	0.9963
	tan	0.0699	0.0717	0.0734	0.0752	0.0769	0.0787	0.0805	0.0822	0.0840	0.0857
5	sin	0.0872	0.0889	0.0906	0.0924	0.0941	0.0958	0.0976	0.0993	0.1011	0.1028
	cos	0.9962	0.9960	0.9959	0.9957	0.9956	0.9954	0.9952	0.9951	0.9949	0.9947
	tan	0.0875	0.0802	0.0910	0.0928	0.0945	0.0963	0.0981	0.0998	0.1016	0.1033
6	sin	0.1045	0.1063	0.1080	0.1097	0.1115	0.1132	0.1149	0.1167	0.1184	0.1201
	cos	0.0945	0.9943	0.9942	0.9940	0.9938	0.9936	0.9934	0.9932	0.9930	0.9928
	tan	0.1051	0.1069	0.1086	0.1104	0.1122	0.1139	0.1157	0.1175	0.1192	0.1210
7	sin	0.1219	0.1236	0.1253	0.1271	0.1288	0.1305	0.1323	0.1340	0.1357	0.1374
	cos	0.9925	0.9923	0.9921	0.9919	0.9917	0.9914	0.9912	0.9910	0.9907	0.9905
	tan	0.1228	0.1246	0.1263	0.1281	0.1209	0.1317	0.1334	0.1352	0.1370	0.1388
8	sin	0.1392	0.1409	0.1426	0.1444	0.1461	0.1478	0.1495	0.1513	0.1530	0.1547
	cos	0.9903	0.9900	0.9898	0.9895	0.9893	0.9890	0.9888	0.9885	0.9882	0.9880
	tan	0.1405	0.1423	0.1441	0.1459	0.1477	0.1495	0.1512	0.1530	0.1548	0.1566
9	sin	0.1564	0.1582	0.1599	0.1616	0.1633	0.1650	0.1668	0.1685	0.1702	0.1719
	cos	0.9377	0.9874	0.9871	0.9869	0.9866	0.9863	0.9860	0.0857	0.9854	0.9851
	tan	0.1584	0.1602	0.1620	0.1638	0.1655	0.1673	0.1691	0.1709	0.1727	0.1745
10	sin	0.1736	0.1754	0.1771	0.1788	0.1805	0.1822	0.1840	0.1857	0.1874	0.1891
	cos	0.9848	0.9845	0.9842	0.9839	0.9836	0.9833	0.9829	0.9826	0.9823	0.9520
	tan	0.1763	0.1781	0.1799	0.1817	0.1835	0.1853	0.1871	0.1890	0.1908	0.1926
11	sin	0.1908	0.1925	0.1942	0.1959	0.1977	0.1994	0.2011	0.2028	0.2045	0.2062
	cos	0.9816	0.9813	0.9810	0.9806	0.9803	0.9799	0.9796	0.9792	0.9789	0.9785
	tan	0.1944	0.1962	0.1980	0.1998	0.2016	0.2035	0.2053	0.2071	0.2089	0.2107
12	sin	0.2079	0.2096	0.2113	0.2130	0.2147	0.2164	0.2181	0.2198	0.2215	0.2232
	cos	0.9781	0.9778	0.9774	0.9770	0.9767	0.9763	0.9759	0.9755	0.9751	0.9748
	tan	0.2126	0.2144	0.2162	0.2180	0.2199	0.2217	0.2235	0.2254	0.2272	0.2290
13	sin	0.2250	0.2267	0.2284	0.2300	0.2318	0.2334	0.2351	0.2368	0.2385	0.2402
	cos	0.9744	0.9740	0.9736	0.9732	0.9728	0.9724	0.9720	0.9715	0.9711	0.9707
	tan	0.2309	0.2327	0.2345	0.2364	0.2382	0.2401	0.2419	0.2438	0.2456	0.2475
14	sin	0.2419	0.2436	0.2453	0.2470	0.2487	0.2504	0.2521	0.2538	0.2554	0.2571
	cos	0.9703	0.9699	0.9694	0.9690	0.9686	0.9681	0.9677	0.9673	0.9668	0.9664
	tan	0.2493	0.2512	0.2530	0.2549	0.2568	0.2586	0.2605	0.2623	0.2642	0.2661
Degs.	**Function**	**0′**	**6′**	**12′**	**18′**	**24′**	**30′**	**36′**	**42′**	**48′**	**54′**

Figure A2-1.—Natural Sines, Cosines, and Tangents. 45.45.2(179)A

15°–29.9°

Degs.	Function	0.0°	0.1°	0.2°	0.3°	0.4°	0.5°	0.6°	0.7°	0.8°	0.9°
15	sin	0.2588	0.2605	0.2622	0.2639	0.2656	0.2672	0.2689	0.2706	0.2723	0.2740
	cos	0.9659	0.9655	0.9650	0.9646	0.9641	0.9636	0.9632	0.9627	0.9622	0.9617
	tan	0.2679	0.2698	0.2717	0.2736	0.2754	0.2773	0.2792	0.2811	0.2830	0.2849
16	sin	0.2756	0.2773	0.2790	0.2807	0.2823	0.2840	0.2857	0.2874	0.2890	0.2907
	cos	0.9613	0.9608	0.9603	0.9598	0.9593	0.9588	0.9583	0.9578	0.9573	0.9568
	tan	0.2867	0.2886	0.2905	0.2924	0.2943	0.2962	0.2981	0.3000	0.3019	0.3038
17	sin	0.2924	0.2940	0.2957	0.2974	0.2990	0.3007	0.3024	0.3040	0.3057	0.3074
	cos	0.9563	0.9558	0.9553	0.9548	0.9542	0.9537	0.9532	0.9527	0.9521	0.9516
	tan	0.3057	0.3076	0.3096	0.3115	0.3134	0.3153	0.3172	0.3191	0.3211	0.3230
18	sin	0.3090	0.3107	0.3123	0.3140	0.3156	0.3173	0.3190	0.3206	0.3223	0.3239
	cos	0.9511	0.9505	0.9500	0.9494	0.9489	0.9483	0.9478	0.9472	0.9466	0.9461
	tan	0.3249	0.3269	0.3288	0.3307	0.3327	0.3346	0.3365	0.3385	0.3404	0.3424
19	sin	0.3256	0.3272	0.3289	0.3305	0.3322	0.3338	0.3355	0.3371	0.3387	0.3404
	cos	0.9455	0.9449	0.9444	0.9438	0.9432	0.9426	0.9421	0.9415	0.9409	0.9403
	tan	0.3443	0.3463	0.3482	0.3502	0.3522	0.3541	0.3561	0.3581	0.3600	0.3620
20	sin	0.3420	0.3437	0.3453	0.3469	0.3486	0.3502	0.3518	0.3535	0.3551	0.3567
	cos	0.9397	0.9391	0.9385	0.9379	0.9373	0.9367	0.9361	0.9354	0.9348	0.9342
	tan	0.3640	0.3659	0.3679	0.3699	0.3719	0.3739	0.3759	0.3779	0.3799	0.3819
21	sin	0.3584	0.3600	0.3616	0.3633	0.3649	0.3665	0.3681	0.3697	0.3714	0.3730
	cos	0.9336	0.9330	0.9323	0.9317	0.9311	0.9304	0.9298	0.9291	0.9285	0.9278
	tan	0.3839	0.3859	0.3879	0.3899	0.3919	0.3939	0.3959	0.3979	0.4000	0.4020
22	sin	0.3746	0.3762	0.3778	0.3795	0.3811	0.3827	0.3843	0.3859	0.3875	0.3891
	cos	0.9272	0.9265	0.9259	0.9252	0.9245	0.9239	0.9232	0.9225	0.9219	0.9212
	tan	0.4040	0.4061	0.4081	0.4101	0.4122	0.4142	0.4163	0.4183	0.4204	0.4224
23	sin	0.3907	0.3923	0.3939	0.3955	0.3971	0.3987	0.4003	0.4019	0.4035	0.4051
	cos	0.9205	0.9198	0.9191	0.9184	0.9178	0.9171	0.9164	0.9157	0.9150	0.9143
	tan	0.4245	0.4265	0.4286	0.4307	0.4327	0.4348	0.4369	0.4390	0.4411	0.4431
24	sin	0.4067	0.4083	0.4099	0.4115	0.4131	0.4147	0.4163	0.4179	0.4195	0.4210
	cos	0.9135	0.9128	0.9121	0.9114	0.9107	0.9100	0.9092	0.9085	0.9078	0.9070
	tan	0.4452	0.4473	0.4494	0.4515	0.4536	0.4557	0.4578	0.4599	0.4621	0.4642
25	sin	0.4226	0.4242	0.4258	0.4274	0.4289	0.4305	0.4321	0.4337	0.4352	0.4368
	cos	0.9063	0.90[illegible]6	0.9048	0.9041	0.9033	0.9026	0.9018	0.9011	0.9003	0.8996
	tan	0.4663	0.4684	0.4706	0.4727	0.4748	0.4770	0.4791	0.4813	0.4834	0.4856
26	sin	0.4284	0.4399	0.4415	0.4431	0.4446	0.4462	0.4478	0.4493	0.4509	0.4524
	cos	0.8988	0.8980	0.8973	0.8965	0.8957	0.8949	0.8942	0.8934	0.8926	0.8918
	tan	0.4877	0.4899	0.4921	0.4942	0.4964	0.4986	0.5008	0.5029	0.5051	0.5073
27	sin	0.4540	0.4555	0.4571	0.4586	0.4602	0.4617	0.4633	0.4648	0.4664	0.4679
	cos	0.8910	0.8902	0.8894	0.8886	0.8878	0.8870	0.8862	0.8854	0.8846	0.8838
	tan	0.5095	0.5117	0.5139	0.5161	0.5184	0.5206	0.5228	0.5250	0.5272	0.5295
28	sin	0.4695	0.4710	0.4726	0.4741	0.4756	0.4772	0.4787	0.4802	0.4818	0.4833
	cos	0.8829	0.8821	0.8813	0.8805	0.8796	0.8788	0.8780	0.8771	0.8763	0.8755
	tan	0.5317	0.5340	0.5362	0.5384	0.5407	0.5430	0.5452	0.5475	0.5498	0.5520
29	sin	0.4848	0.4863	0.4879	0.4894	0.4909	0.4924	0.4939	0.4955	0.4970	0.4985
	cos	0.8746	0.8738	0.8729	0.8721	0.8712	0.8704	0.8695	0.8686	0.8678	0.8669
	tan	0.5543	0.5566	0.5589	0.5612	0.5635	0.5658	0.5681	0.5704	0.5727	0.5750
Degs.	Function	0′	6′	12′	18′	24′	30′	36′	42′	48′	54′

Figure A2-1.—Natural Sines, Cosines, and Tangents—continued. 45.45.2(179)B

30°-44.9°

Degs.	Function	0.0°	0.1°	0.2°	0.3°	0.4°	0.5°	0.6°	0.7°	0.8°	0.9°
30	sin	0.5000	0.5015	0.5030	0.5045	0.5060	0.5075	0.5090	0.5105	0.5120	0.5135
	cos	0.8660	0.8652	0.8643	0.8634	0.8625	0.8616	0.8607	0.8599	0.8590	0.8581
	tan	0.5774	0.5797	0.5820	0.5844	0.5867	0.5890	0.5914	0.5938	0.5961	0.5985
31	sin	0.5150	0.5165	0.5180	0.5195	0.5210	0.5225	0.5240	0.5255	0.5270	0.5284
	cos	0.8572	0.8563	0.8554	0.8545	0.8536	0.8526	0.8517	0.8508	0.8499	0.8490
	tan	0.6009	0.6032	0.6056	0.6080	0.6104	0.6128	0.6152	0.6176	0.6200	0.6224
32	sin	0.5299	0.5314	0.5329	0.5344	0.5358	0.5373	0.5388	0.5402	0.5417	0.5432
	cos	0.8480	0.8471	0.8462	0.8453	0.8443	0.8434	0.8425	0.8415	0.8406	0.8396
	tan	0.6249	0.6273	0.6297	0.6322	0.6346	0.6371	0.6395	0.6420	0.6445	0.6469
33	sin	0.5446	0.5461	0.5476	0.5490	0.5505	0.5519	0.5534	0.5548	0.5563	0.5577
	cos	0.8387	0.8377	0.8368	0.8358	0.8348	0.8339	0.8329	0.8320	0.8310	0.8300
	tan	0.6494	0.6519	0.6544	0.6569	0.6594	0.6619	0.6644	0.6669	0.6694	0.6720
34	sin	0.5592	0.5606	0.5621	0.5635	0.5650	0.5664	0.5678	0.5693	0.5707	0.5721
	cos	0.8290	0.8281	0.8271	0.8261	0.8251	0.8241	0.8231	0.8221	0.8211	0.8202
	tan	0.6745	0.6771	0.6796	0.6822	0.6847	0.6873	0.6899	0.6924	0.6950	0.6976
35	sin	0.5736	0.5750	0.5764	0.5779	0.5793	0.5807	0.5821	0.5835	0.5850	0.5864
	cos	0.8192	0.8181	0.8171	0.8161	0.8151	0.8141	0.8131	0.8121	0.8111	0.8100
	tan	0.7002	0.7028	0.7054	0.7080	0.7107	0.7133	0.7159	0.7186	0.7212	0.7239
36	sin	0.5878	0.5892	0.5906	0.5920	0.5934	0.5948	0.5962	0.5976	0.5990	0.6004
	cos	0.8090	0.8080	0.8070	0.8059	0.8049	0.8039	0.8028	0.8018	0.8007	0.7997
	tan	0.7265	0.7292	0.7319	0.7346	0.7373	0.7400	0.7427	0.7454	0.7481	0.7508
37	sin	0.6018	0.6032	0.6046	0.6060	0.6074	0.6088	0.6101	0.6115	0.6129	0.6143
	cos	0.7986	0.7976	0.7965	0.7955	0.7944	0.7934	0.7923	0.7912	0.7902	0.7891
	tan	0.7536	0.7563	0.7590	0.7618	0.7646	0.7673	0.7701	0.7729	0.7757	0.7785
38	sin	0.6157	0.6170	0.6184	0.6198	0.6211	0.6225	0.6239	0.6252	0.6266	0.6280
	cos	0.7880	0.7869	0.7859	0.7848	0.7837	0.7826	0.7815	0.7804	0.7793	0.7782
	tan	0.7813	0.7841	0.7869	0.7898	0.7926	0.7954	0.7983	0.8012	0.8040	0.8069
39	sin	0.6293	0.6307	0.6320	0.6334	0.6347	0.6361	0.6374	0.6388	0.6401	0.6414
	cos	0.7771	0.7760	0.7749	0.7738	0.7727	0.7716	0.7705	0.7694	0.7683	0.7672
	tan	0.8098	0.8127	0.8156	0.8185	0.8214	0.8243	0.8273	0.8302	0.8332	0.8361
40	sin	0.6428	0.6441	0.6455	0.6468	0.6481	0.6494	0.6508	0.6521	0.6534	0.6547
	cos	0.7660	0.7649	0.7638	0.7627	0.7615	0.7604	0.7593	0.7581	0.7570	0.7559
	tan	0.8391	0.8421	0.8451	0.8481	0.8511	0.8541	0.8571	0.8601	0.8632	0.8662
41	sin	0.6561	0.6574	0.6587	0.6600	0.6613	0.6626	0.6639	0.6652	0.6665	0.6678
	cos	0.7547	0.7536	0.7524	0.7513	0.7501	0.7490	0.7478	0.7466	0.7455	0.7443
	tan	0.8693	0.8724	0.8754	0.8785	0.8816	0.8847	0.8878	0.8910	0.8941	0.8972
42	sin	0.6691	0.6704	0.6717	0.6730	0.6743	0.6756	0.6769	0.6782	0.6794	0.6807
	cos	0.7431	0.7420	0.7408	0.7396	0.7385	0.7373	0.7361	0.7349	0.7337	0.7325
	tan	0.9004	0.9036	0.9067	0.9099	0.9131	0.9163	0.9195	0.9228	0.9260	0.9293
43	sin	0.6820	0.6833	0.6845	0.6858	0.6871	0.6884	0.6896	0.6909	0.6921	0.6934
	cos	0.7314	0.7302	0.7290	0.7278	0.7266	0.7254	0.7242	0.7230	0.7218	0.7206
	tan	0.9325	0.9358	0.9391	0.9424	0.9457	0.9490	0.9523	0.9556	0.9590	0.9623
44	sin	0.6947	0.6959	0.6972	0.6984	0.6997	0.7009	0.7022	0.7034	0.7046	0.7059
	cos	0.7193	0.7181	0.7169	0.7157	0.7145	0.7133	0.7120	0.7108	0.7096	0.7083
	tan	0.9657	0.9691	0.9725	0.9759	0.9793	0.9827	0.9861	0.9896	0.9930	0.9965
Degs.	**Function**	**0′**	**6′**	**12′**	**18′**	**24′**	**30′**	**36′**	**42′**	**48′**	**54′**

Figure A2-1.—Natural Sines, Cosines, and Tangents—continued. 45.45.2(179)C

45°–59.9°

Degs.	Function	0.0°	0.1°	0.2°	0.3°	0.4°	0.5°	0.6°	0.7°	0.8°	0.9°
45	sin	0.7071	0.7083	0.7096	0.7108	0.7120	0.7133	0.7145	0.7157	0.7169	0.7181
	cos	0.7071	0.7059	0.7046	0.7034	0.7022	0.7009	0.6997	0.6984	0.6972	0.6959
	tan	1.0000	1.0035	1.0070	1.0105	1.0141	1.0176	1.0212	1.0247	1.0283	1.0319
46	sin	0.7193	0.7206	0.7218	0.7230	0.7242	0.7254	0.7266	0.7278	0.7290	0.7302
	cos	0.6947	0.6934	0.6921	0.6909	0.6896	0.6884	0.6871	0.6858	0.6845	0.6833
	tan	1.0355	1.0392	1.0428	1.0464	1.0501	1.0538	1.0575	1.0612	1.0649	1.0686
47	sin	0.7314	0.7325	0.7337	0.7349	0.7361	0.7373	0.7385	0.7396	0.7408	0.7420
	cos	0.6820	0.6807	0.6794	0.6782	0.6769	0.6756	0.6743	0.6730	0.6717	0.6704
	tan	1.0724	1.0761	1.0799	1.0837	1.0875	1.0913	1.0951	1.0990	1.1028	1.1067
48	sin	0.7431	0.7443	0.7455	0.7466	0.7478	0.7490	0.7501	0.7513	0.7524	0.7536
	cos	0.6691	0.6678	0.6665	0.6652	0.6639	0.6626	0.6613	0.6600	0.6587	0.6574
	tan	1.1106	1.1145	1.1184	1.1224	1.1263	1.1303	1.1343	1.1383	1.1423	1.1463
49	sin	0.7547	0.7559	0.7570	0.7581	0.7593	0.7604	0.7615	0.7627	0.7638	0.7649
	cos	0.6561	0.6547	0.6534	0.6521	0.6508	0.6494	0.6481	0.6468	0.6455	0.6441
	tan	1.1504	1.1544	1.1585	1.1626	1.1667	1.1708	1.1750	1.1792	1.1833	1.1875
50	sin	0.7660	0.7672	0.7683	0.7694	0.7705	0.7716	0.7727	0.7738	0.7749	0.7760
	cos	0.6428	0.6414	0.6401	0.6388	0.6374	0.6361	0.6347	0.6334	0.6320	0.6307
	tan	1.1918	1.1960	1.2002	1.2045	1.2088	1.2131	1.2174	1.2218	1.2261	1.2305
51	sin	0.7771	0.7782	0.7793	0.7804	0.7815	0.7826	0.7837	0.7848	0.7859	0.7869
	cos	0.6293	0.6280	0.6266	0.6252	0.6239	0.6225	0.6211	0.6198	0.6184	0.6170
	tan	1.2349	1.2393	1.2437	1.2482	1.2527	1.2572	1.2617	1.2662	1.2708	1.2753
52	sin	0.7880	0.7891	0.7902	0.7912	0.7923	0.7934	0.7944	0.7955	0.7965	0.7976
	cos	0.6157	0.6143	0.6129	0.6115	0.6101	0.6088	0.6074	0.6060	0.6046	0.6032
	tan	1.2799	1.2846	1.2892	1.2938	1.2985	1.3032	1.3079	1.3127	1.3175	1.3222
53	sin	0.7986	0.7997	0.8007	0.8018	0.8028	0.8039	0.8049	0.8059	0.8070	0.8080
	cos	0.6018	0.6004	0.5990	0.5976	0.5962	0.5948	0.5934	0.5920	0.5906	0.5892
	tan	1.3270	1.3319	1.3367	1.3416	1.3465	1.3514	1.3564	1.3613	1.3663	1.3713
54	sin	0.8090	0.8100	0.8111	0.8121	0.8131	0.8141	0.8151	0.8161	0.8171	0.8181
	cos	0.5878	0.5864	0.5850	0.5835	0.5821	0.5807	0.5793	0.5779	0.5764	0.5750
	tan	1.3764	1.3814	1.3865	1.3916	1.3968	1.4019	1.4071	1.4124	1.4176	1.4229
55	sin	0.8192	0.8202	0.8211	0.8221	0.8231	0.8241	0.8251	0.8261	0.8271	0.8281
	cos	0.5736	0.5721	0.5707	0.5693	0.5678	0.5664	0.5650	0.5635	0.5621	0.5606
	tan	1.4281	1.4335	1.4388	1.4442	1.4496	1.4550	1.4605	1.4659	1.4715	1.4770
56	sin	0.8290	0.8300	0.8310	0.8320	0.8329	0.8339	0.8348	0.8358	0.8368	0.8377
	cos	0.5592	0.5577	0.5563	0.5548	0.5534	0.5519	0.5505	0.5490	0.5476	0.5461
	tan	1.4826	1.4882	1.4938	1.4994	1.5051	1.5108	1.5166	1.5224	1.5282	1.5340
57	sin	0.8387	0.8396	0.8406	0.8415	0.8425	0.8434	0.8443	0.8453	0.8462	0.8471
	cos	0.5446	0.5432	0.5417	0.5402	0.5388	0.5373	0.5358	0.5344	0.5329	0.5314
	tan	1.5399	1.5458	1.5517	1.5577	1.5637	1.5697	1.5757	1.5818	1.5880	1.5941
58	sin	0.8480	0.8490	0.8499	0.8508	0.8517	0.8526	0.8536	0.8545	0.8551	0.8563
	cos	0.5299	0.5284	0.5270	0.5255	0.5240	0.5225	0.5210	0.5195	0.5180	0.5165
	tan	1.6003	1.6066	1.6128	1.6191	1.6255	1.6319	1.6383	1.6447	1.6512	1.6577
59	sin	0.8572	0.8581	0.8590	0.8599	0.8607	0.8616	0.8625	0.8634	0.8643	0.8652
	cos	0.5150	0.5135	0.5120	0.5105	0.5090	0.5075	0.5060	0.5045	0.5030	0.5015
	tan	1.6643	1.6709	1.6775	1.6842	1.6909	1.6977	1.7045	1.7113	1.7182	1.7251
Degs.	**Function**	**0′**	**6′**	**12′**	**18′**	**24′**	**30′**	**36′**	**42′**	**43′**	**54′**

Figure A2-1.—Natural Sines, Cosines, and Tangents—continued. 45.45.2(179)D

60°–74.9°

Degs.	Function	0.0°	0.1°	0.2°	0.3°	0.4°	0.5°	0.6°	0.7°	0.8°	0.9°
60	sin	0.8660	0.8669	0.8678	0.8686	0.8695	0.8704	0.8712	0.8721	0.8729	0.8738
	cos	0.5000	0.4985	0.4970	0.4955	0.4939	0.4924	0.4909	0.4894	0.4879	0.4863
	tan	1.7321	1.7391	1.7461	1.7532	1.7603	1.7675	1.7747	1.7820	1.7893	1.7966
61	sin	0.8746	0.8755	0.8763	0.8771	0.8780	0.8788	0.8796	0.8805	0.8813	0.8821
	cos	0.4848	0.4833	0.4818	0.4802	0.4787	0.4772	0.4756	0.4741	0.4726	0.4710
	tan	1.8040	1.8115	1.8190	1.8265	1.8341	1.8418	1.8495	1.8572	1.8650	1.8728
62	sin	0.8829	0.8838	0.8846	0.8854	0.8862	0.8870	0.8878	0.8886	0.8894	0.8902
	cos	0.4695	0.4679	0.4664	0.4648	0.4633	0.4617	0.4602	0.4586	0.4571	0.4555
	tan	1.8807	1.8887	1.8967	1.9047	1.9128	1.9210	1.9292	1.9375	1.9458	1.9542
63	sin	0.8910	0.8918	0.8926	0.8934	0.8942	0.8949	0.8957	0.8965	0.8973	0.8980
	cos	0.4540	0.4524	0.4509	0.4493	0.4478	0.4462	0.4446	0.4431	0.4415	0.4399
	tan	1.9626	1.9711	1.9797	1.9883	1.9970	2.0057	2.0145	2.0233	2.0323	2.0413
64	sin	0.8988	0.8996	0.9003	0.9011	0.9018	0.9026	0.9033	0.9041	0.9048	0.9056
	cos	0.4384	0.4368	0.4352	0.4337	0.4321	0.4305	0.4289	0.4274	0.4258	0.4242
	tan	2.0503	2.0594	2.0686	2.0778	2.0872	2.0965	2.1060	2.1155	2.1251	2.1348
65	sin	0.9063	0.9070	0.9078	0.9085	0.9092	0.9100	0.9107	0.9114	0.9121	0.9128
	cos	0.4226	0.4210	0.4195	0.4179	0.4163	0.4147	0.4131	0.4115	0.4099	0.4083
	tan	2.1445	2.1543	2.1642	2.1742	2.1842	2.1943	2.2045	2.2148	2.2251	2.2355
66	sin	0.9135	0.9143	0.9150	0.9157	0.9164	0.9171	0.9178	0.9184	0.9191	0.9198
	cos	0.4067	0.4051	0.4035	0.4019	0.4003	0.3987	0.3971	0.3955	0.3039	0.3923
	tan	2.2460	2.2566	2.2673	2.2781	2.2889	2.2998	2.3109	2.3220	2.3332	2.3445
67	sin	0.9205	0.9212	0.9219	0.9225	0.9232	0.9239	0.9245	0.9252	0.9259	0.9265
	cos	0.3907	0.3891	0.3875	0.3859	0.3843	0.3827	0.3811	0.3795	0.3778	0.3762
	tan	2.3559	2.3673	2.3789	2.3906	2.4023	2.4142	2.4262	2.4383	2.4504	2.4627
68	sin	0.9272	0.9278	0.9285	0.9291	0.9298	0.9304	0.9311	0.9317	0.9323	0.9330
	cos	0.3746	0.3730	0.3714	0.3697	0.3681	0.3665	0.3649	0.3633	0.3616	0.3600
	tan	2.4751	2.4876	2.5002	2.5129	2.5257	2.5386	2.5517	2.5649	2.5782	2.5916
69	sin	0.9336	0.9342	0.9348	0.9354	0.9361	0.9367	0.9373	0.9379	0.9385	0.9391
	cos	0.3584	0.3567	0.3551	0.3535	0.3518	0.3502	0.3486	0.3469	0.3453	0.3437
	tan	2.6051	2.6187	2.6325	2.6464	2.6605	2.6746	2.6889	2.7034	2.7179	2.7326
70	sin	0.9397	0.9403	0.9409	0.9415	0.9421	0.9426	0.9432	0.9438	0.9444	0.9449
	cos	0.3420	0.3404	0.3387	0.3371	0.3355	0.3338	0.3322	0.3305	0.3289	0.3272
	tan	2.7475	2.7625	2.7776	2.7929	2.8083	2.8239	2.8397	2.8556	2.8716	2.8878
71	sin	0.9455	0.9461	0.9466	0.9472	0.9478	0.9483	0.9489	0.9494	0.9500	0.9505
	cos	0.3256	0.3239	0.3223	0.3206	0.3190	0.3173	0.3156	0.3140	0.3123	0.3107
	tan	2.9042	2.9208	2.9375	2.9544	2.9714	2.9887	3.0061	3.0237	3.0415	3.0595
72	sin	0.9511	0.9516	0.9521	0.9527	0.9532	0.9537	0.9542	0.9548	0.9553	0.9558
	cos	0.3090	0.3074	0.3057	0.3040	0.3024	0.3007	0.2990	0.2974	0.2957	0.2940
	tan	3.0777	3.0961	3.1146	3.1334	3.1524	3.1716	3.1910	3.2106	3.2305	3.2506
73	sin	0.9563	0.9568	0.9573	0.9578	0.9583	0.9588	0.9593	0.9598	0.9603	0.9608
	cos	0.2924	0.2907	0.2890	0.2874	0.2857	0.2840	0.2823	0.2807	0.2790	0.2773
	tan	3.2709	3.2914	3.3122	3.3332	3.3544	3.3759	3.3977	3.4197	3.4420	3.4646
74	sin	0.9613	0.9617	0.9622	0.9627	0.9632	0.9636	0.9641	0.9646	0.9650	0.9655
	cos	0.2756	0.2740	0.2723	0.2706	0.2689	0.2672	0.2656	0.2639	0.2622	0.2605
	tan	3.4874	3.5105	3.5339	3.5576	3.5816	3.6059	3.6305	3.6554	3.6806	3.7062
Degs.	Function	0′	6′	12′	18′	24′	30′	36′	42′	48′	54′

Figure A2-1. — Natural Sines, Cosines, and Tangents — continued. 45.45.2(179)E

75°–89.9°

Degs.	Function	0.0°	0.1°	0.2°	0.3°	0.4°	0.5°	0.6°	0.7°	0.8°	0.9°
75	sin	0.9659	0.9664	0.9668	0.9673	0.9677	0.9681	0.9686	0.9690	0.9694	0.9699
	cos	0.2588	0.2571	0.2554	0.2538	0.2521	0.2504	0.2487	0.2470	0.2453	0.2436
	tan	3.7321	3.7583	3.7848	3.8118	3.8391	3.8667	3.8947	3.9232	3.9520	3.9812
76	sin	0.9703	0.9707	0.9711	0.9715	0.9720	0.9724	0.9728	0.9732	0.9736	0.9740
	cos	0.2419	0.2402	0.2385	0.2368	0.2351	0.2334	0.2317	0.2300	0.2284	0.2267
	tan	4.0108	4.0408	4.0713	4.1022	4.1335	4.1653	4.1976	4.2303	4.2635	4.2972
77	sin	0.9744	0.9748	0.9751	0.9755	0.9759	0.9763	0.9767	0.9770	0.9774	0.9778
	cos	0.2250	0.2232	0.2215	0.2198	0.2181	0.2164	0.2147	0.2130	0.2113	0.2096
	tan	4.3315	4.3662	4.4015	4.4374	4.4737	4.5107	4.5483	4.5864	4.6252	4.6646
78	sin	0.9781	0.9785	0.9789	0.9792	0.9796	0.9799	0.9803	0.9806	0.9810	0.9813
	cos	0.2079	0.2062	0.2045	0.2328	0.2011	0.1994	0.1977	0.1959	0.1942	0.1925
	tan	4.7046	4.7453	4.7867	4.8288	4.8716	4.9152	4.9594	5.0045	5.0504	5.0970
79	sin	0.9816	0.9820	0.9823	0.9826	0.9829	0.9833	0.9836	0.9839	0.9842	0.9845
	cos	0.1908	0.1891	0.1874	0.1857	0.1840	0.1822	0.1805	0.1788	0.1771	0.1754
	tan	5.1446	5.1929	5.2422	5.2924	5.3435	5.3955	5.4486	5.5026	5.5578	5.6140
80	sin	0.9848	0.9851	0.9854	0.9857	0.9860	0.9863	0.9866	0.9869	0.9871	0.9874
	cos	0.1736	0.1719	0.1702	0.1685	0.1668	0.1650	0.1633	0.1616	0.1599	0.1582
	tan	5.6713	5.7297	5.7894	5.8502	5.0124	5.9758	6.0405	6.1066	6.1742	6.2432
81	sin	0.9877	0.9880	0.9882	0.9885	0.9888	0.9890	0.9893	0.9895	0.9898	0.9900
	cos	0.1564	0.1547	0.1530	0.1513	0.1495	0.1478	0.1461	0.1444	0.1426	0.1409
	tan	6.3138	6.3859	6.4596	6.5350	6.6122	6.6912	6.7720	6.8548	6.9395	7.0264
82	sin	0.9903	0.9905	0.9907	0.9910	0.9912	0.9914	0.9917	0.9919	0.9921	0.9923
	cos	0.1392	0.1374	0.1357	0.1340	0.1323	0.1305	0.1288	0.1271	0.1253	0.1236
	tan	7.1154	7.2066	7.3002	7.3962	7.4947	7.5958	7.6996	7.8062	7.9158	8.0285
83	sin	0.9925	0.9928	0.9930	0.9932	0.9934	0.9936	0.9938	0.9940	0.9942	0.9943
	cos	0.1219	0.1201	0.1184	0.1167	0.1149	0.1132	0.1115	0.1097	0.1080	0.1063
	tan	8.1443	8.2636	8.3863	8.5126	8.6427	8.7769	8.9152	9.0579	9.2052	9.3572
84	sin	0.9945	0.9947	0.9949	0.9951	0.9952	0.9954	0.9956	0.9957	0.9959	0.9960
	cos	0.1045	0.1028	0.1011	0.0993	0.0976	0.0958	0.0941	0.0924	0.0906	0.0889
	tan	9.5144	9.6768	9.8448	10.02	10.20	10.39	10.58	10.78	10.99	11.20
85	sin	0.9962	0.9963	0.9965	0.9966	0.9968	0.9969	0.9971	0.9972	0.9973	0.9974
	cos	0.0872	0.0854	0.0837	0.0819	0.0802	0.0785	0.0767	0.0750	0.0732	0.0715
	tan	11.43	11.66	11.91	12.16	12.43	12.71	13.00	13.30	13.62	13.95
86	sin	0.9976	0.9977	0.9978	0.9979	0.9980	0.9981	0.9982	0.9983	0.9984	0.9985
	cos	0.0698	0.0680	0.0663	0.0645	0.0628	0.0610	0.0593	0.0576	0.0558	0.0541
	tan	14.30	14.67	15.06	15.46	15.89	16.35	16.83	17.34	17.89	18.46
87	sin	0.9986	0.9987	0.9988	0.9989	0.9990	0.9990	0.9991	0.9992	0.9993	0.9993
	cos	0.0523	0.0506	0.0488	0.0471	0.0454	0.0436	0.0419	0.0401	0.0384	0.0366
	tan	19.08	19.74	20.45	21.20	22.02	22.90	23.86	24.90	26.03	27.27
88	sin	0.9994	0.9995	0.9995	0.9996	0.9996	0.9997	0.9997	0.9997	0.9998	0.9998
	cos	0.0349	0.0332	0.0314	0.0297	0.0279	0.0262	0.0244	0.0227	0.0209	0.0192
	tan	28.64	30.14	31.82	33.69	35.80	38.19	40.92	44.07	47.74	52.08
89	sin	0.9998	0.9999	0.9999	0.9999	0.9999	1.000	1.000	1.000	1.000	1.000
	cos	0.0175	0.0157	0.0140	0.0122	0.0105	0.0087	0.0070	0.0052	0.0035	0.0017
	tan	57.29	63.66	71.62	81.85	95.49	114.6	143.2	191.0	286.5	573.0
Degs.	**Function**	**0′**	**6′**	**12′**	**18′**	**24′**	**30′**	**36′**	**42′**	**48′**	**54′**

Figure A2-1.—Natural Sines, Cosines, and Tangents—continued. 45.45.2(179)F

APPENDIX III

TABLE OF LOGARITHMS

N	0	1	2	3	4	5	6	7	8	9
0	....	0000	3010	4771	6021	6990	7782	8451	9031	9542
1	0000	0414	0792	1139	1461	1761	2041	2304	2553	2788
2	3010	3222	3424	3617	3802	3979	4150	4314	4472	4624
3	4771	4914	5051	5185	5315	5441	5563	5682	5798	5911
4	6021	6128	6232	6335	6435	6532	6628	6721	6812	6902
5	6990	7076	7160	7243	7324	7404	7482	7559	7634	7709
6	7782	7853	7924	7993	8062	8129	8195	8261	8325	8388
7	8451	8513	8573	8633	8692	8751	8808	8865	8921	8976
8	9031	9085	9138	9191	9243	9294	9345	9395	9445	9494
9	9542	9590	9638	9685	9731	9777	9823	9868	9912	9956
10	0000	0043	0086	0128	0170	0212	0253	0294	0334	0374
11	0414	0453	0492	0531	0569	0607	0645	0682	0719	0755
12	0792	0828	0864	0899	0934	0969	1004	1038	1072	1106
13	1139	1173	1206	1239	1271	1303	1335	1367	1399	1430
14	1461	1492	1523	1553	1584	1614	1644	1673	1703	1732
15	1761	1790	1818	1847	1875	1903	1931	1959	1987	2014
16	2041	2068	2095	2122	2148	2175	2201	2227	2253	2279
17	2304	2330	2355	2380	2405	2430	2455	2480	2504	2529
18	2553	2577	2601	2625	2648	2672	2695	2718	2742	2765
19	2788	2810	2833	2856	2878	2900	2923	2945	2967	2989
20	3010	3032	3054	3075	3096	3118	3139	3160	3181	3201
21	3222	3243	3263	3284	3304	3324	3345	3365	3385	3404
22	3424	3444	3464	3483	3502	3522	3541	3560	3579	3598
23	3617	3636	3655	3674	3692	3711	3729	3747	3766	3784
24	3802	3820	3838	3856	3874	3892	3909	3927	3945	3962
25	3979	3997	4014	4031	4048	4065	4082	4099	4116	4133
26	4150	4166	4183	4200	4216	4232	4249	4265	4281	4298
27	4314	4330	4346	4362	4378	4393	4409	4425	4440	4456
28	4472	4487	4502	4518	4533	4548	4564	4579	4594	4609
29	4624	4639	4654	4669	4683	4698	4713	4728	4742	4757
30	4771	4786	4800	4814	4829	4843	4857	4871	4886	4900
31	4914	4928	4942	4955	4969	4983	4997	5011	5024	5038
32	5051	5065	5079	5092	5105	5119	5132	5145	5159	5172
33	5185	5198	5211	5224	5237	5250	5263	5276	5289	5302
34	5315	5328	5340	5353	5366	5378	5391	5403	5416	5428
35	5441	5453	5465	5478	5490	5502	5514	5527	5539	5551
36	5563	5575	5587	5599	5611	5623	5635	5647	5658	5670
37	5682	5694	5705	5717	5729	5740	5752	5763	5775	5786
38	5798	5809	5821	5832	5843	5855	5866	5877	5888	5899
39	5911	5922	5933	5944	5955	5966	5977	5988	5999	6010
40	6021	6031	6042	6053	6064	6075	6085	6096	6107	6117
41	6128	6138	6149	6160	6170	6180	6191	6201	6212	6222
42	6232	6243	6253	6263	6274	6284	6294	6304	6314	6325
43	6335	6345	6355	6365	6375	6385	6395	6405	6415	6425
44	6435	6444	6454	6464	6474	6484	6493	6503	6513	6522
45	6532	6542	6551	6561	6571	6580	6590	6599	6609	6618
46	6628	6637	6646	6656	6665	6675	6684	6693	6702	6712
47	6721	6730	6739	6749	6758	6767	6776	6785	6794	6803
48	6812	6821	6830	6839	6848	6857	6866	6875	6884	6893
49	6902	6911	6920	6928	6937	6946	6955	6964	6972	6981
50	6990	6998	7007	7016	7024	7033	7042	7050	7059	7067
N	0	1	2	3	4	5	6	7	8	9

Figure A3-1.—Table of Logarithms. 45.45.1(179)A

N	0	1	2	3	4	5	6	7	8	9
50	6990	6998	7007	7016	7024	7033	7042	7050	7059	7067
51	7076	7084	7093	7101	7110	7118	7126	7135	7143	7152
52	7160	7168	7177	7185	7193	7202	7210	7218	7226	7235
53	7243	7251	7259	7267	7275	7284	7292	7300	7308	7316
54	7324	7332	7340	7348	7356	7364	7372	7380	7388	7396
55	7404	7412	7419	7427	7435	7443	7451	7459	7466	7474
56	7482	7490	7497	7505	7513	7520	7528	7536	7543	7551
57	7559	7566	7574	7582	7589	7597	7604	7612	7619	7627
58	7634	7642	7649	7657	7664	7672	7679	7686	7694	7701
59	7709	7716	7723	7731	7738	7745	7752	7760	7767	7774
60	7782	7789	7796	7803	7810	7818	7825	7832	7839	7846
61	7853	7860	7868	7875	7882	7889	7896	7903	7910	7917
62	7924	7931	7938	7945	7952	7959	7966	7973	7980	7987
63	7993	8000	8007	8014	8021	8028	8035	8041	8048	8055
64	8062	8069	8075	8082	8089	8096	8102	8109	8116	8122
65	8129	8136	8142	8149	8156	8162	8169	8176	8182	8189
66	8195	8202	8209	8215	8222	8228	8235	8241	8248	8254
67	8261	8267	8274	8280	8287	8293	8299	8306	8312	8319
68	8325	8331	8338	8344	8351	8357	8363	8370	8376	8382
69	8388	8395	8401	8407	8414	8420	8426	8432	8439	8445
70	8451	8457	8463	8470	8476	8482	8488	8494	8500	8506
71	8513	8519	8525	8531	8537	8543	8549	8555	8561	8567
72	8573	8579	8585	8591	8597	8603	8609	8615	8621	8627
73	8633	8639	8645	8651	8657	8663	8669	8675	8681	8686
74	8692	8698	8704	8710	8716	8722	8727	8733	8739	8745
75	8751	8756	8762	8768	8774	8779	8785	8791	8797	8802
76	8808	8814	8820	8825	8831	8837	8842	8848	8854	8859
77	8865	8871	8876	8882	8887	8893	8899	8904	8910	8915
78	8921	8927	8932	8938	8943	8949	8954	8960	8965	8971
79	8976	8982	8987	8993	8998	9004	9009	9015	9020	9025
80	9031	9036	9042	9047	9053	9058	9063	9069	9074	9079
81	9085	9090	9096	9101	9106	9112	9117	9122	9128	9133
82	9138	9143	9149	9154	9159	9165	9170	9175	9180	9186
83	9191	9196	9201	9206	9212	9217	9222	9227	9232	9238
84	9243	9248	9253	9258	9263	9269	9274	9279	9284	9289
85	9294	9299	9304	9309	9315	9320	9325	9330	9335	9340
86	9345	9350	9355	9360	9365	9370	9375	9380	9385	9390
87	9395	9400	9405	9410	9415	9420	9425	9430	9435	9440
88	9445	9450	9455	9460	9465	9469	9474	9479	9484	9489
89	9494	9499	9504	9509	9513	9518	9523	9528	9533	9538
90	9542	9547	9552	9557	9562	9566	9571	9576	9581	9586
91	9590	9595	9600	9605	9609	9614	9619	9624	9628	9633
92	9638	9643	9647	9652	9657	9661	9666	9671	9675	9680
93	9685	9689	9694	9699	9703	9708	9713	9717	9722	9727
94	9731	9736	9741	9745	9750	9754	9759	9763	9768	9773
95	9777	9782	9786	9791	9795	9800	9805	9809	9814	9818
96	9823	9827	9832	9836	9841	9845	9850	9854	9859	9863
97	9868	9872	9877	9881	9886	9890	9894	9899	9903	9908
98	9912	9917	9921	9926	9930	9934	9939	9943	9948	9952
99	9956	9961	9965	9969	9974	9978	9983	9987	9991	9996
100	0000	0004	0009	0013	0017	0022	0026	0030	0035	0039
N	0	1	2	3	4	5	6	7	8	9

Figure A3-1. — Table of Logarithms — continued. 45.45.1(179)B

APPENDIX IVa

SEMICONDUCTOR LETTER SYMBOLS

Letter symbols used in solid state circuits are those proposed as standard for industry, or are special symbols not included as standard. Semiconductor symbols consist of a basic letter with subscripts, either alphabetical or numerical, or both, in accordance with the following rules:

1. A capital (upper case) letter designates external circuit parameters and components, large-signal device parameters, and maximum (peak), average (d.c.), or root-mean-square values of current, voltage, and power (I, V, P, etc.)
2. Instantaneous values of current, voltage, and power, which vary with time, and small-signal values are represented by the lower case (small) letter of the proper symbol (i, v, p, i_e V_{eb}, etc.).
3. D.c. values, instantaneous total values, and large-signal values, are indicated by capital subscripts (i_C, I_C, v_{EB}, V_{EB}, P_C, etc.).
4. Alternating component values are indicated by using lower case subscripts: note the examples i_c, I_c, v_{eb}, V_{eb}, P_c, p_c.
5. When it is necessary to distinguish between maximum, average, or root-mean-square values, maximum or average values may be represented by addition of a subscript m or av; examples are i_{cm}, I_{CM}, I_{cav}, i_{CAV}.
6. For electrical quantities, the first subscript designates the electrode at which the measurement is made.
7. For device parameters, the first subscript designates the element of the four-pole matrix; examples are I or i for input, O or o for output, F or f for forward transfer, and R or r for reverse transfer.
8. The second subscript normally designates the reference electrode.
9. Supply voltages are indicated by repeating the associated device electrode subscript, in which case, the reference terminal is then designated by the third subscript; note the cases V_{EE}, V_{CC}, V_{EEB}, V_{CCB}.
10. In devices having more than one terminal of the same type (say two bases), the terminal subscripts are modified by adding a number following the subscript and placed on the same line, for example, V_{B1-B2}.
11. In multiple-unit devices the terminal subscripts are modified by a number preceding the electrode subscript; note the example, V_{1B-2B}.

Semiconductor symbols change, and new symbols are developed to cover new devices as the art changes; an alphabetical list of the complex symbols is presented below for easy reference.

The list is divided into six sections. These sections are signal and rectifier diodes, zener diodes, thyristors and SCR's, transistors, unijunction transistors, and field effect transistors.

SIGNAL AND RECTIFIER DIODES

PRV	Peak Reverse Voltage
I_o	Average Rectifier Forward Current
I_r	Average Reverse Current
I_{surge}	Peak Surge Current
V_F	Average Forward Voltage Drop
V_R	D.c. Blocking Voltage

ZENER DIODES

I_F	Forward current
I_Z	Zener current
I_{ZK}	Zener current near breakdown knee
I_{ZM}	Maximum D.c. zener current (limited by power dissipation)

Figure A4a-1.—Semiconductor Letter Symbols.

I_{ZT} Zener test current

V_f Forward voltage

V_Z Nominal zener voltage

Z_Z Zener impedance

Z_{ZK} Zener impedance near breakdown knee

Z_{ZT} Zener impedance at zener test current

I_R Reverse current

V_R Reverse test voltage

THYRISTORS AND SCRs

I_f Forward current, r.m.s. value of forward anode current during the "on" state.

$I_{FM(pulse)}$ Repetitive pulse current. Repetitive peak forward anode current after application of gate signal for specified pulse conditions.

$I_{FM(surge)}$ Peak forward surge current. The maximum forward current having a single forward cycle in a 60 Hz single-phase resistive load system.

I_{FOM} Peak forward blocking current, gate open. The maximum current through the thyristor when the device is in the "off" state for a stated anode-to-cathode voltage (anode positive) and junction temperature with the gate open.

I_{FXM} Peak forward blocking current. Same as I_{FOM} except that the gate terminal is returned to the cathode through a stated impedance and/or bias voltage.

I_{GFM} Peak forward gate current. The maximum instantaneous value of current which may flow between gate and cathode.

I_{GT} Gate trigger current (continuous d.c.). The minimum d.c. gate current required to cause switching from the "off" state at a stated condition.

I_{HO} Holding current. That value of forward anode current below which the controlled rectifier switches from the conducting state to the forward blocking condition with the gate open, at stated conditions.

I_{HX} Holding current (gate connected). The value of forward anode current below which the controlled rectifier switches from the conducting state to the forward blocking condition with the gate terminal returned to the cathode terminal through specified impedance and/or bias voltage.

$P_{F(AV)}$ Average forward power. Average value of power dissipation between anode and cathode.

P_{GFM} Peak gate power. The maximum instantaneous value of gate power dissipation permitted.

I_{ROM} Peak reverse blocking current. The maximum current through the thyristor when the device is in the reverse blocking state (anode negative) for a stated anode-to-cathode voltage and junction temperature with the gate open.

I_{RXM} Peak reverse blocking current. Same as I_{ROM} except that the gate terminal is returned to the cathode through a stated impedance and/or bias voltage.

$P_{GF(AV)}$ Average forward gate power. The value of maximum allowable gate power dissipation averaged over a full cycle.

V_F Forward "on" voltage. The voltage measured between anode and cathode during the "on" condition for specified conditions of anode and temperature.

$V_{F(on)}$ Dynamic forward "on" voltage. The voltage measured between anode and cathode at a specified time after turn-on function has been initiated at stated conditions.

Figure A4a-1.—Semiconductor Letter Symbols—continued.

V_{FOM}	Peak forward blocking voltage, gate open. The peak repetitive forward voltage which may be applied to the thyristor between anode and cathode (anode positive) with the gate open at stated conditions.
V_{FXM}	Peak forward blocking voltage. Same as V_{FOM} except that the gate terminal is returned to the cathode through a stated impedance and/or voltage.
V_{GFM}	Peak forward gate voltage. The maximum instantaneous voltage between the gate terminal and the cathode terminal resulting from the flow of forward gate current.
V_{GRM}	Peak reverse gate voltage. The maximum instantaneous voltage which may be applied between the gate terminal and the cathode terminal when the junction between the gate region and the adjacent cathode region is reverse biased.
V_{GT}	Gate trigger voltage (continuous d.c.). The d.c. voltage between the gate and the cathode required to produce the d.c. gate trigger current.
$V_{ROM(rep)}$	Peak reverse blocking voltage, gate open. The maximum allowable value of reverse voltage (repetitive or continuous d.c.) which can be applied between anode and cathode (anode negative) with the gate open for stated conditions.
V_{RXM}	Peak reverse blocking voltage. Same as V_{ROM} except that the gate terminal is returned to the cathode through a stated impedance and/or bias voltage.

TRANSISTORS

A_G	Available gain
A_P	Power gain
A_I	Current gain
B or b	Base electrode
BV_{BCO}	D.c. base-to-collector breakdown voltage, base reverse-biased with respect to collector, emitter open.
BV_{BEO}	D.c. base-to-emitter breakdown voltage, base reverse-biased with respect to emitter, collector open.
BV_{CBO}	D.c. collector-to-base breakdown voltage, collector reverse-biased with respect to base, emitter open.
BV_{CEO}	D.c. collector-to-emitter breakdown voltage, collector reverse-biased with respect to emitter, base open.
BV_{EBO}	D.c. emitter-to-base breakdown voltage, emitter reverse-biased with respect to base, collector open.
BV_{ECO}	D.c. emitter-to-collector breakdown voltage, emitter reverse-biased with respect to collector, base open.
C or c	Collector electrode
C_c	Collector junction capacitance
C_e	Emitter junction capacitance
C_{ib}, C_{ic}, C_{ie}	Input capacitance for common base, collector, and emitter, respectively.
C_{ob}, C_{oc}, C_{oe}	Output terminal capacitance, a.c. input open, for common base, collector and emitter, respectively.
D	Distortion
E or e	Emitter electrode
$f_{\alpha b}$, $f_{\alpha c}$, $f_{\alpha e}$	Alpha cutoff frequency for common base, collector, and emitter, respectively.
f_{co}	Cutoff frequency
f_{max}	Maximum frequency of oscillation
GC (CB), GC (CC), GE (CE)	Grounded (or common) base, collector, and emitter, respectively.
G_b, G_c, G_e	Power gain for common base, collector, and emitter, respectively.

Figure A4a-1.—Semiconductor Letter Symbols—continued.

h	Hybrid parameter
h_{fe}, h_{fb}, h_{fc}	Small signal forward current transfer ratio, a.c. output shorted, common emitter, common base, common collector, respectively.
h_{ib}	Small-signal input impedance, a.c. output shorted, common base.
h_{ob}	Small-signal output admittance, a.c. input open, common base.
I	Direct current (d.c.).
I_B, I_C, I_E	D.c. current for base, collector, and emitter, respectively.
I_{CBO}	D.c. collector current, collector reverse-biased with respect to base, emitter-to-base open.
I_{CES}	D.c. collector current, collector reverse-biased with respect to emitter, base shorted to emitter.
I_{EBO}	D.c. emitter current, emitter reverse-biased with respect to base, collector-to-base open.
NF	Noise Figure
P_D	Total average power dissipation of all electrodes of a semiconductor device.
P_G	Power gain
P_{Go}	Over-all power gain
P_{in}	Input power
P_{out}	Output power
r'b	Equivalent base resistance, high frequencies
T_j	Junction temperature
T_{stg}	Storage temperature
t_f	Fall time, from 90 percent to 10 percent of pulse (switching applications).
t_r	Rise time, from 10 percent to 90 percent pulse (switching applications).
t_s	Storage time (switching applications).
V_{BE}	Base-to-emitter d.c. voltage
V_{CE}	Collector-to-base d.c. voltage
V_{CE}	Collector-to-emitter d.c. voltage
V_{CEO}	D.c. collector-to-emitter voltage with collector junction reverse-biased, zero base current.
V_{CER}	Similar to V_{CEO}, except with a resistor (of value R) between base and emitter.
V_{CES}	Similar to V_{CEO}, except with base shorted to emitter.
V_{CEV}	D.c. collector-to-emitter voltage, used when only voltage bias is used.
V_{CEX}	D.c. collector-to-emitter voltage, base-emitter back biased.
V_{EB}	Emitter-to-base d.c. voltage
V_{pt}	Punch-through voltage

UNIJUNCTION TRANSISTORS

I_E	Emitter current
I_{EO}	Emitter reverse current. Measured between emitter and base-two at a specified voltage, and base-one open-circuited.
I_p	Peak point emitter current. The maximum emitter current that can flow without allowing the UJT to go into the negative resistance region.
I_V	Valley point emitter. The current flowing in the emitter when the device is biased to the valley point.
r_{BB}	Interbase resistance. Resistance between base-two and base-one measured at a specified interbase voltage.
V_{B2B1}	Voltage between base-two and base-one. Positive at base-two.

Figure A4a-1.—Semiconductor Letter Symbols—continued.

V_p — Peak point emitter voltage. The maximum voltage seen at the emitter before the UJT goes into the negative resistance region.

V_D — Forward voltage drop of the emitter junction.

V_{EB1} — Emitter to base-one voltage

$V_{EB1(SAT)}$ — Emitter saturation voltage. Forward voltage drop from emitter to base-one at a specified emitter current (larger than I_V) and specified interbase voltage.

V_v — Valley point emitter voltage. The voltage at which the valley point occurs with a specified V_{B2B1}.

V_{OB1} — Base-one peak pulse voltage. The peak voltage measured across a resistor in series with base-one when the UJT is operated as a relaxation oscillator in a specified circuit.

α_{rBB} — Interbase resistance temperature coefficient. Variation of resistance between B2 and B1 over the specified temperature range and measured at the specific interbase voltage and temperature with emitter open circuited.

$I_{B2(mod)}$ — Interbase modulation current. B2 current modulation due to firing. Measured at a specified interbase voltage, emitter and temperature.

FIELD EFFECT TRANSISTORS

I_D — Drain current

I_{DGO} — Maximum leakage from drain to gate with source open

I_{DSS} — Drain current with gate connected to source

I_G — Gate current

I_{GSS} — Maximum gate current (leakage) with drain connected to source

$V_{(BR)DGO}$ — Drain to gate, source open

V_D — D.c. drain voltage

$V_{(BR)DGS}$ — Drain to gate, source connected to drain

$V_{(BR)DS}$ — Drain to source, gate connection not specified

$V_{(BR)DSX}$ — Drain to source, gate biased to cutoff or beyond

$V_{(BR)GS}$ — Gate to source, drain connection not specified

$V_{(BR)GSS}$ — Gate to source, drain connected to source

$V_{(BR)GD}$ — Gate to drain, source connection not specified

$V_{(BR)GDS}$ — Gate to drain, source connected to drain

V_G — D.c. gate voltage

$V_{G1S(OFF)}$ — Gate 1-source cutoff voltage (with gate 2 connected to source)

$V_{G2S(OFF)}$ — Gate 2-source cutoff voltage (with gate 1 connected to source)

$V_{GS(OFF)}$ — Cutoff

Figure A4a-1.—Semiconductor Letter Symbols—continued.

APPENDIX IVb

ELECTRON TUBE LETTER SYMBOLS

A number of letter symbols which are used as a form of shorthand notation in technical literature when designating electron-tube operating conditions are explained and listed below.

1. Maximum, average, and root-mean-square values are represented by capital (upper case) letters, for example: I, E, P.

2. Where needed to distinguish between values in item 1 above, the maximum value may be represented by the subscript "m", for example: E_m, I_m, P_m.

3. Average values may be represented by the subscript "av," for example: E_{av}, I_{av}, P_{av}. (When items 2 and 3 above are used, then item 1 indicates r-m-s, or effective, values.)

4. Instantaneous values of current, voltage, and power which vary with time are represented by the small (lower case) letter of the proper symbol, for example: i, e, p.

5. External resistance, impedance, etc, in the circuit external to an electron tube electrode may be represented by the upper case symbol with the proper electrode subscripts, for example: R_g, R_{sc}, Z_g, Z_{sc}.

6. Values of resistance, impedance, etc, inherent within the electron tube are represented by the lower case symbol with the proper electrode subscripts, for example: r_g, Z_g, r_p, Z_p, C_{gp}.

7. The symbols "g" and "p" are used as subscripts to identify a.c. values of electrode currents and voltages, for example: e_g, e_p, i_g, i_p.

8. The total instantaneous values of electrode currents and voltages (d.c. plus a.c. components) are indicated by the lower case symbol and the subscripts "b" for plate and "c" for grid, for example: i_b, e_c, i_c, e_b.

9. No-signal or static currents and voltages are indicated by upper case symbol and lower case subscripts "b" for plate and "c" for grid, for example: E_c, I_b, E_b, I_c.

10. R.m.s. and maximum values of a varying component are indicated by the upper case letter and the subscripts "g" and "p", for example: E_g, I_p, E_p, I_g.

11. Average values of current and voltage for the with-signal condition are indicated by adding the subscript "s" to the proper symbol and subscript, for example: I_{bs}, E_{bs}.

12. Supply voltages are indicated by the upper case symbol and double subscript "bb" for plate, "cc" for grid, "ff" for filament, for example: E_{ff} , E_{cc}, E_{bb}.

An alphabetical list of electron tube symbols follows for easy reference.

C_{gk}	Grid-cathode capacitance
C_{gp}	Grid-plate capacitance
C_{pk}	Plate-cathode capacitance
E_b	Plate voltage, d.c. value
E_c	Grid voltage, d.c. value
E_{cc}	Grid bias supply voltage
E_{co}	Negative tube cutoff voltage
E_f	Filament voltage
E_{ff}	Filament supply voltage
E_k	D.c. cathode voltage
e_b	Instantaneous plate voltage
e_c	Instantaneous grid voltage
e_g	A.c. component of grid voltage
e_p	A.c. component of plate voltage

Figure A4b-1.—Electron tube letter symbols.

I_b, I_o	D.c. plate current
I_c	D.c. grid current
I_f	Filament current
I_k	Cathode current
i_b	Instantaneous plate current
i_c	Instantaneous grid current
i_g	A.c. component of grid current
i_p	A.c. component of plate current
P_g	Grid dissipation power
P_o	Output power
P_p	Plate dissipation power
R_g	Grid resistance
R_L	Load resistance
R_k	Cathode resistance
R_p	Plate resistance, d.c.
R_{sc}	Screen resistance
r_L	A.c. load resistance
r_p	Plate resistance, a.c.
t_k	Cathode heating time

Figure A4b-1.— Electron tube letter symbols—continued.

APPENDIX V

JOINT ELECTRONICS TYPE DESIGNATION (AN) SYSTEM

THE AN NOMENCLATURE WAS DESIGNED SO THAT A COMMON DESIGNATION COULD BE USED FOR ARMY, NAVY, AND AIR FORCE EQUIPMENT. THE SYSTEM INDICATOR AN DOES NOT MEAN THAT THE ARMY, NAVY, AND AIR FORCE USE THE EQUIPMENT, BUT MEANS THAT THE TYPE NUMBER WAS ASSIGNED IN THE AN SYSTEM.

AN NOMENCLATURE IS ASSIGNED TO COMPLETE SETS OF EQUIPMENT AND MAJOR COMPONENTS OF MILITARY DESIGN, GROUPS OF ARTICLES OF EITHER COMMERCIAL OR MILITARY DESIGN WHICH ARE GROUPED FOR MILITARY PURPOSES, MAJOR ARTICLES OF MILITARY DESIGN WHICH ARE NOT PART OF OR USED WITH A SET; AND COMMERCIAL ARTICLES WHEN NOMENCLATURE WILL NOT FACILITATE MILITARY IDENTIFICATION AND/OR PROCEDURES.

AN NOMENCLATURE IS NOT ASSIGNED TO ARTICLES CATALOGED COMMERCIALLY EXCEPT AS STATED ABOVE, MINOR COMPONENTS OF MILITARY DESIGN FOR WHICH OTHER ADEQUATE MEANS OF IDENTIFICATION ARE AVAILABLE, SMALL PARTS SUCH AS CAPACITORS AND RESISTORS; AND ARTICLES HAVING OTHER ADEQUATE IDENTIFICATION IN JOINT MILITARY SPECIFICATIONS. NOMENCLATURE ASSIGNMENTS REMAIN UNCHANGED REGARDLESS OF LATER CHANGES IN INSTALLATION AND/OR APPLICATION.

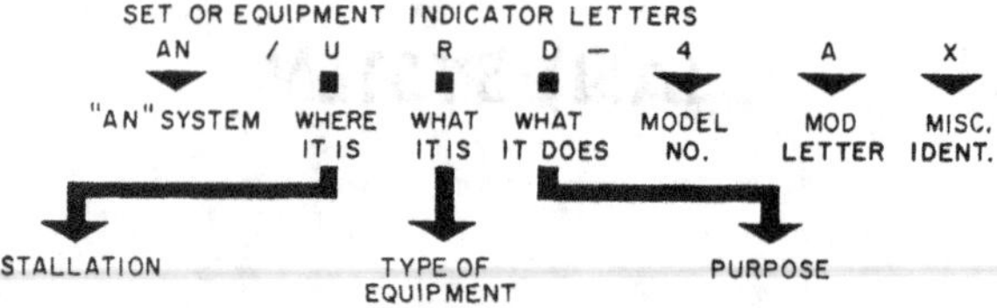

A--AIRBORNE (INSTALLED AND OPERATED IN AIRCRAFT).
B--UNDERWATER MOBILE, SUBMARINE.
C--AIR TRANSPORTABLE (INACTIVATED, DO NOT USE).
D--PILOTLESS CARRIER.
F--FIXED
G--GROUND, GENERAL GROUND USE (INCLUDES TWO OR MORE GROUND-TYPE INSTALLATIONS)
K--AMPHIBIOUS.
M--GROUND, MOBILE (INSTALLED AS OPERATING UNIT IN A VEHICLE WHICH HAS NO FUNCTION OTHER THAN TRANSPORTING THE EQUIPMENT).
P--PACK OR PORTABLE (ANIMAL OR MAN).
S--WATER SURFACE CRAFT
T--GROUND, TRANSPORTABLE
U--GENERAL UTILITY (INCLUDES TWO OR MORE GENERAL INSTALLATION CLASSES, AIRBORNE, SHIPBOARD, AND GROUND).
V--GROUND, VEHICULAR (INSTALLED IN VEHICLE DESIGNED FOR FUNCTIONS OTHER THAN CARRYING ELECTRONIC EQUIPMENT, ETC., SUCH AS TANKS)
W--WATER SURFACE AND UNDERWATER.

A--INVISIBLE LIGHT, HEAT RADIATION
B--PIGEON
C--CARRIER.
D--RADIAC.
E--NUPAC
F--PHOTOGRAPHIC.[1]
G--TELEGRAPH OR TELETYPE
I--INTERPHONE AND PUBLIC ADDRESS.
J--ELECTROMECHANICAL OR INERTIAL WIRE COVERED
K--TELEMETERING
L--COUNTERMEASURES
M--METEOROLOGICAL.
N--SOUND IN AIR.
P--RADAR
Q--SONAR AND UNDERWATER SOUND.
R--RADIO
S--SPECIAL TYPES, MAGNETIC, ETC., OR COMBINATIONS OF TYPES.
T--TELEPHONE (WIRE)
V--VISUAL AND VISIBLE LIGHT
W--ARMAMENT (PECULIAR TO ARMAMENT, NOT OTHERWISE COVERED).
X--FACSIMILE OR TELEVISION
Y--DATA PROCESSING

A--AUXILIARY ASSEMBLIES (NOT COMPLETE OPERATING SETS USED WITH OR PART OF TWO OR MORE SETS OR SETS SERIES).
B--BOMBING.
C--COMMUNICATIONS (RECEIVING AND TRANSMITTING).
D--DIRECTION FINDER, RECONNAISSANCE, AND/OR SURVEILLANCE
E--EJECTION AND/OR RELEASE
G--FIRE-CONTROL OR SEARCHLIGHT DIRECTING.
H--RECORDING AND/OR REPRODUCING (GRAPHIC METEOROLOGICAL AND SOUND)
K--COMPUTING.
L--SEARCHLIGHT CONTROL (INACTIVATED, USE G).
M--MAINTENANCE AND TEST ASSEMBLIES (INCLUDING TOOLS).
N--NAVIGATIONAL AIDS (INCLUDING ALTIMETERS, BEACONS, COMPASSES, RACONS, DEPTH SOUNDING, APPROACH, AND LANDING).
P--REPRODUCING (INACTIVATED, DO NOT USE)
Q--SPECIAL, OR COMBINATION OF PURPOSES
R--RECEIVING, PASSIVE DETECTING
S--DETECTING AND/OR RANGE AND BEARING, SEARCH.
T--TRANSMITTING.
W--AUTOMATIC FLIGHT OR REMOTE CONTROL
X--IDENTIFICATION AND RECOGNITION.

[1]NOT FOR US USE EXCEPT FOR ASSIGNING SUFFIX LETTERS TO PREVIOUSLY NOMENCLATURED ITEMS

20.484
Figure A5-1.—AN system.

APPENDIX VI

ELECTRONICS COLOR CODING

COLOR	1ST DIGIT	2ND DIGIT	MULTIPLIER	TOLERANCE (percent)
Black	0	0	1	
Brown	1	1	10	
Red	2	2	100	
Orange	3	3	1,000	
Yellow	4	4	10,000	
Green	5	5	100,000	
Blue	6	6	1,000,000	
Violet	7	7	10,000,000	
Gray	8	8	100,000,000	
White	9	9	1,000,000,000	
Gold			.1	5
Silver			.01	10
No color				20

A

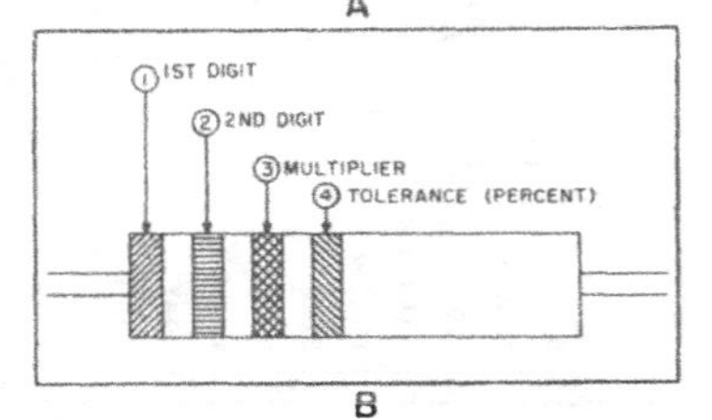

B

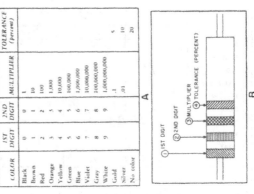

20.373:.374
Figure A6-1.—Resistor color code.

DIRECTION OF READING DOTS
TYPE
1ST DIGIT
2ND DIGIT
MULTIPLIER
TOLERANCE
CHARACTERISTIC OR CLASS

TYPE	COLOR	1ST DIGIT	2ND DIGIT	MULTIPLIER	TOLERANCE (PERCENT)	CHARACTERISTIC OR CLASS
JAN, MICA	BLACK	0	0	1.0		APPLIES TO
	BROWN	1	1	10	± 1	TEMPERATURE
	RED	2	2	100	± 2	COEFFICIENT
	ORANGE	3	3	1,000	± 3	OR METHODS
	YELLOW	4	4	10,000	± 4	OF TESTING
	GREEN	5	5	100,000	± 5	
	BLUE	6	6	1,000,000	± 6	
	VIOLET	7	7	10,000,000	± 7	
	GRAY	8	8	100,000,000	± 8	
EIA, MICA	WHITE	9	9	1,000,000,000	± 9	
	GOLD			.1		
MOLDED PAPER	SILVER			.01	±10	
	BODY				±20	

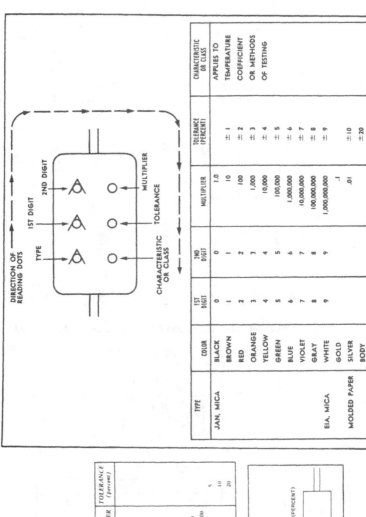

20.376
Figure A6-2.—6-Dot color code for mica and molded paper capacitors.

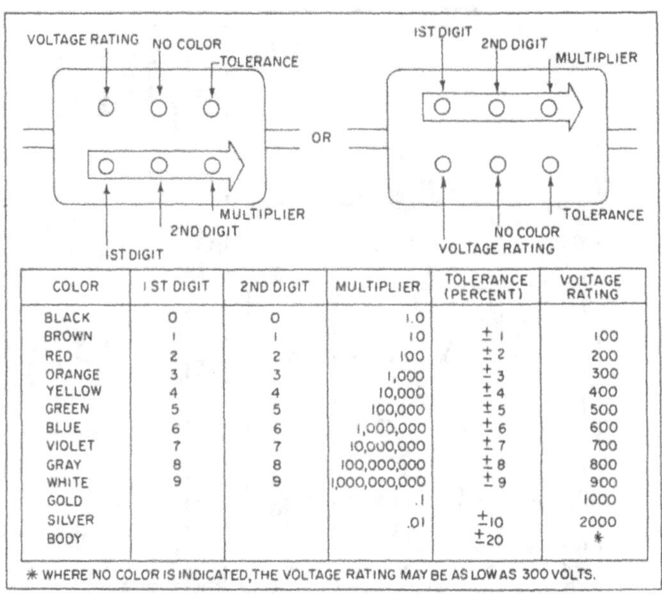

COLOR	1ST DIGIT	2ND DIGIT	MULTIPLIER	TOLERANCE (PERCENT)	VOLTAGE RATING
BLACK	0	0	1.0		
BROWN	1	1	10	±1	100
RED	2	2	100	±2	200
ORANGE	3	3	1,000	±3	300
YELLOW	4	4	10,000	±4	400
GREEN	5	5	100,000	±5	500
BLUE	6	6	1,000,000	±6	600
VIOLET	7	7	10,000,000	±7	700
GRAY	8	8	100,000,000	±8	800
WHITE	9	9	1,000,000,000	±9	900
GOLD			.1		1000
SILVER			.01	±10	2000
BODY				±20	*

* WHERE NO COLOR IS INDICATED,THE VOLTAGE RATING MAY BE AS LOW AS 300 VOLTS.

Figure A6-3.—5-Dot color code for capacitors (dielectric not specified). 20.486

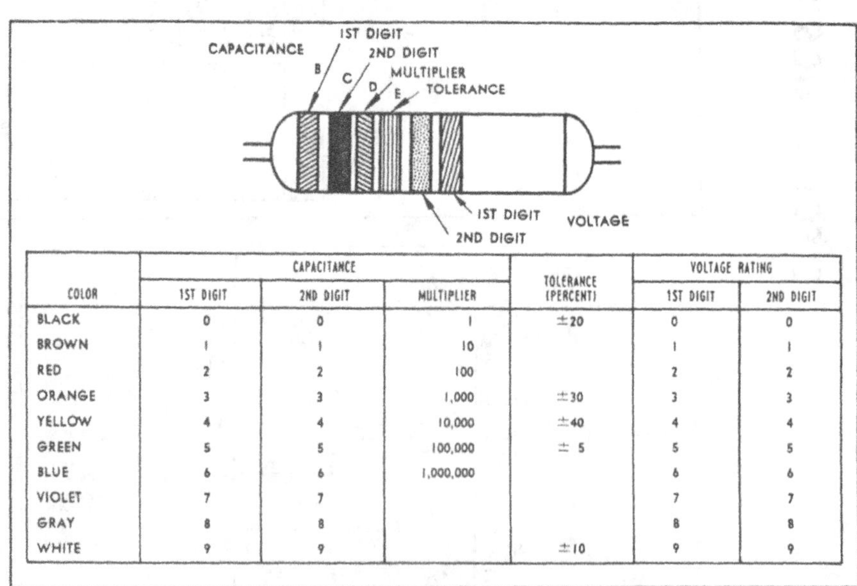

COLOR	CAPACITANCE			TOLERANCE (PERCENT)	VOLTAGE RATING	
	1ST DIGIT	2ND DIGIT	MULTIPLIER		1ST DIGIT	2ND DIGIT
BLACK	0	0	1	±20	0	0
BROWN	1	1	10		1	1
RED	2	2	100		2	2
ORANGE	3	3	1,000	±30	3	3
YELLOW	4	4	10,000	±40	4	4
GREEN	5	5	100,000	± 5	5	5
BLUE	6	6	1,000,000		6	6
VIOLET	7	7			7	7
GRAY	8	8			8	8
WHITE	9	9		±10	9	9

Figure A6-4.—6-Band color code for tubular paper dielectric capacitors. 20.487

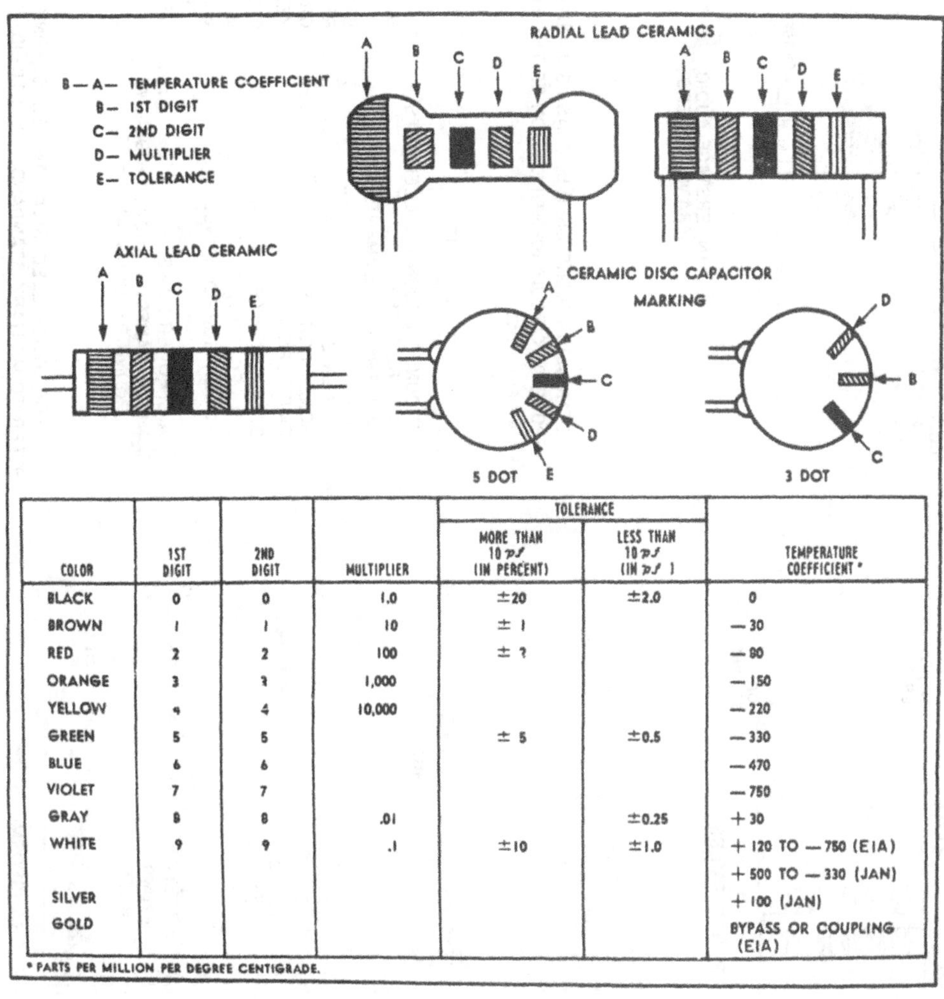

COLOR	1ST DIGIT	2ND DIGIT	MULTIPLIER	TOLERANCE: MORE THAN 10 pf (IN PERCENT)	TOLERANCE: LESS THAN 10 pf (IN pf)	TEMPERATURE COEFFICIENT *
BLACK	0	0	1.0	±20	±2.0	0
BROWN	1	1	10	± 1		—30
RED	2	2	100	± 2		—80
ORANGE	3	3	1,000			—150
YELLOW	4	4	10,000			—220
GREEN	5	5		± 5	±0.5	—330
BLUE	6	6				—470
VIOLET	7	7				—750
GRAY	8	8	.01		±0.25	+30
WHITE	9	9	.1	±10	±1.0	+120 TO —750 (EIA)
						+500 TO —330 (JAN)
SILVER						+100 (JAN)
GOLD						BYPASS OR COUPLING (EIA)

* PARTS PER MILLION PER DEGREE CENTIGRADE.

20.488
Figure A6-5.—Color code for ceramic capacitors having different configurations.

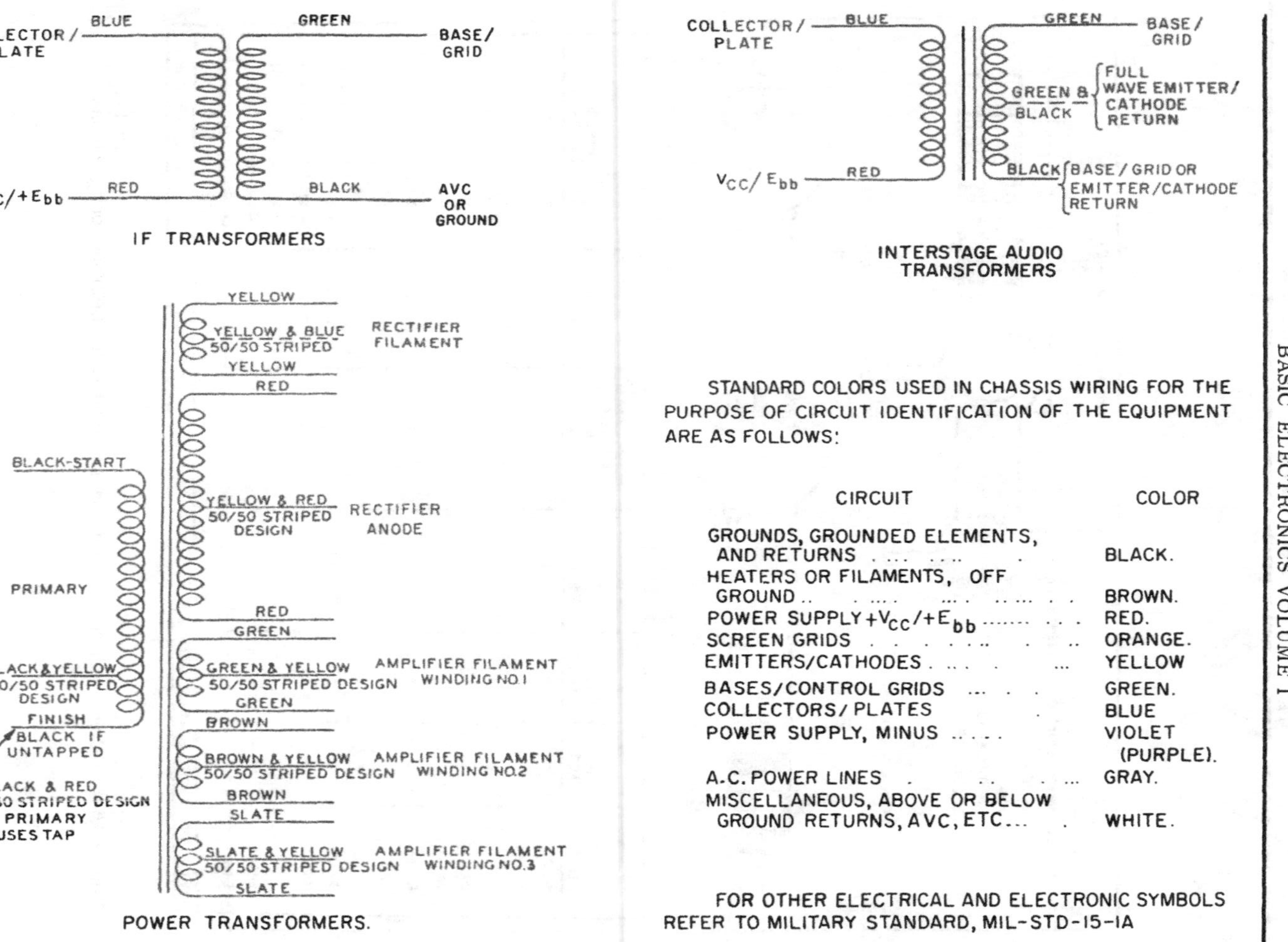

Figure A6-6. – Color code for transformers.

APPENDIX VII

ELECTRONICS SYMBOLS

APPENDIX VII
ELECTRONICS SYMBOLS

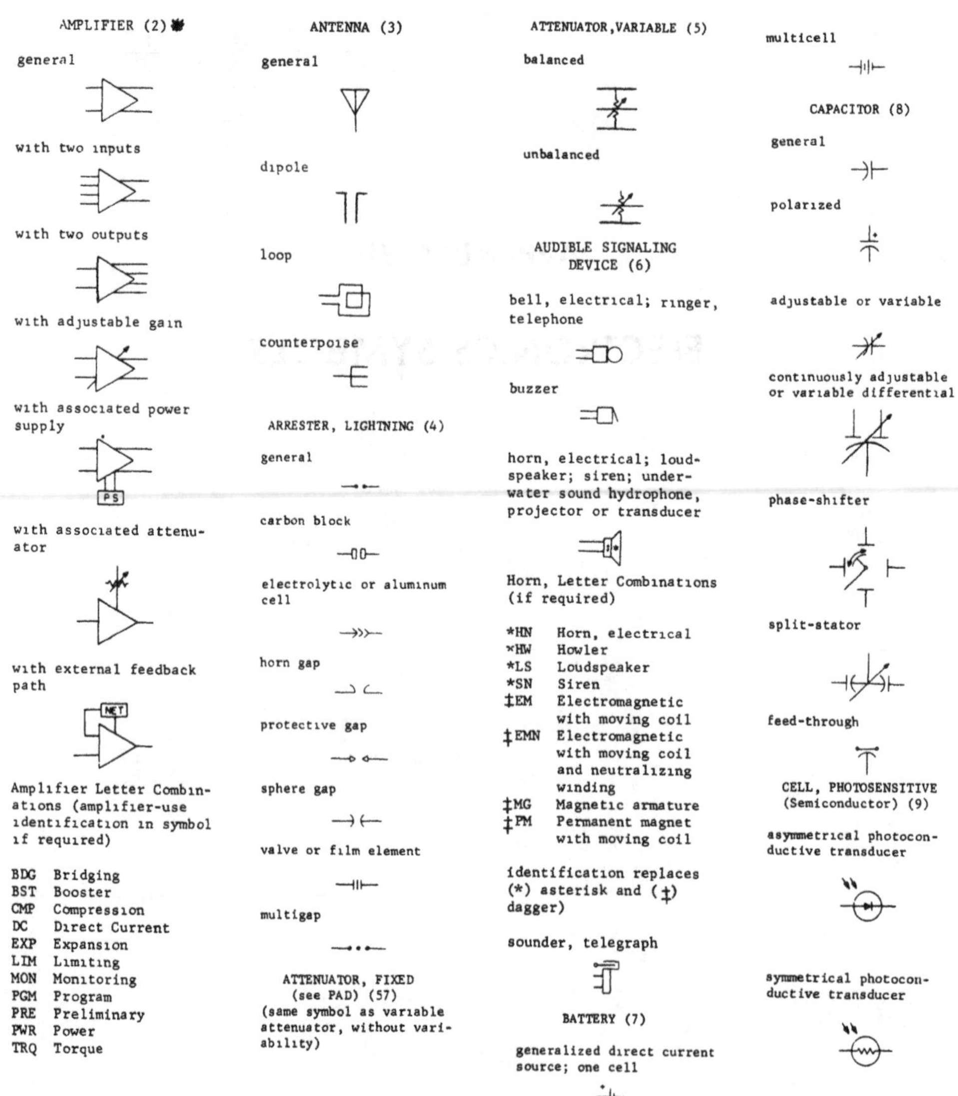

*** NUMBER IN PARENTHESES INDICATES LOCATION OF SYMBOL IN MIL-STD PUBLICATION**

Figure A7-1. — Electronics symbols. 13.5(179)A

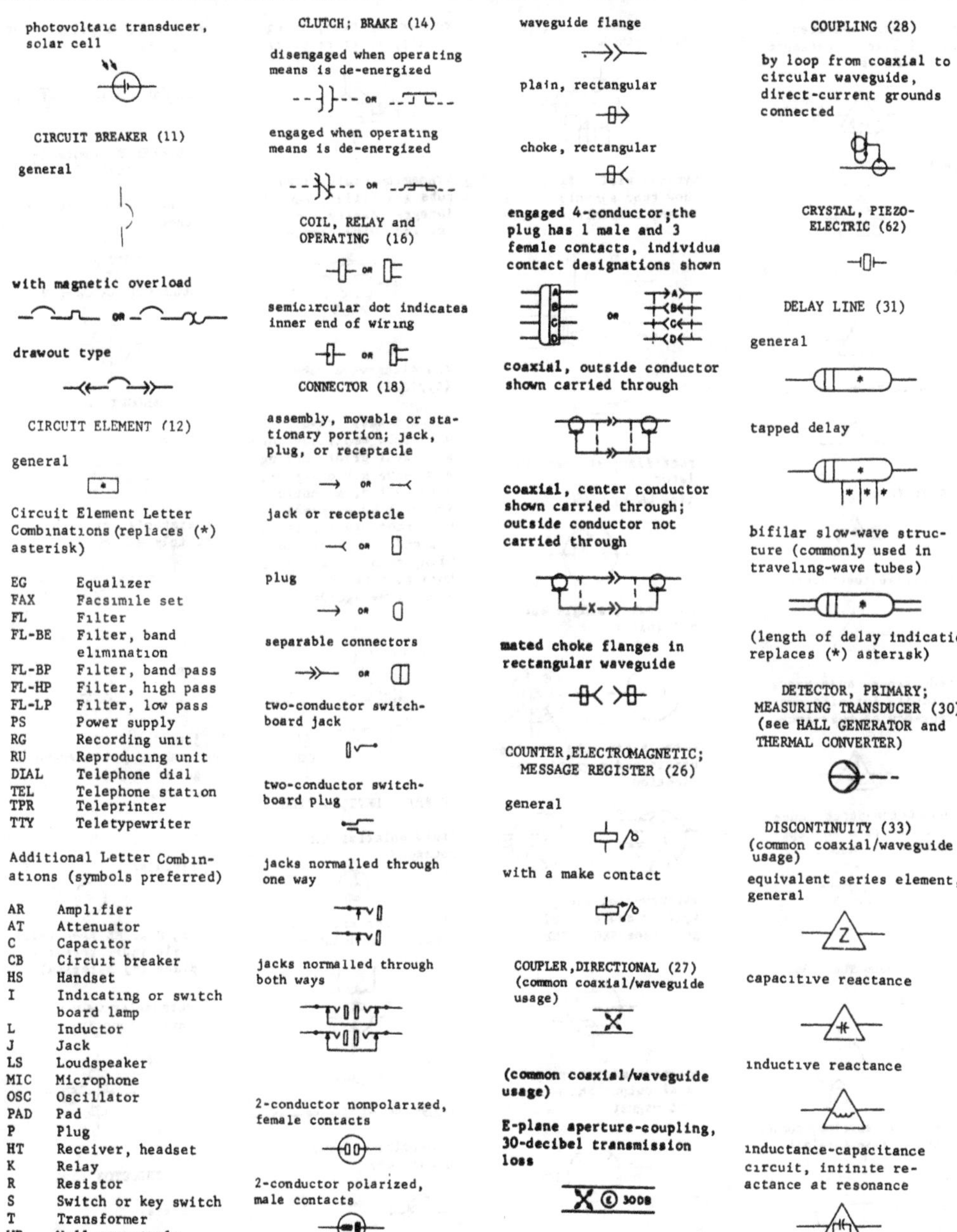
photovoltaic transducer, solar cell
CIRCUIT BREAKER (11)
general
with magnetic overload
OR
drawout type
CIRCUIT ELEMENT (12)
general
Circuit Element Letter Combinations (replaces (*) asterisk)
EG Equalizer
FAX Facsimile set
FL Filter
FL-BE Filter, band elimination
FL-BP Filter, band pass
FL-HP Filter, high pass
FL-LP Filter, low pass
PS Power supply
RG Recording unit
RU Reproducing unit
DIAL Telephone dial
TEL Telephone station
TPR Teleprinter
TTY Teletypewriter
Additional Letter Combinations (symbols preferred)
AR Amplifier
AT Attenuator
C Capacitor
CB Circuit breaker
HS Handset
I Indicating or switch board lamp
L Inductor
J Jack
LS Loudspeaker
MIC Microphone
OSC Oscillator
PAD Pad
P Plug
HT Receiver, headset
K Relay
R Resistor
S Switch or key switch
T Transformer
WR Wall receptacle
CLUTCH; BRAKE (14)
disengaged when operating means is de-energized
OR
engaged when operating means is de-energized
OR
COIL, RELAY and OPERATING (16)
OR
semicircular dot indicates inner end of wiring
OR
CONNECTOR (18)
assembly, movable or stationary portion; jack, plug, or receptacle
OR
jack or receptacle
OR
plug
OR
separable connectors
OR
two-conductor switchboard jack
two-conductor switchboard plug
jacks normalled through one way
jacks normalled through both ways
2-conductor nonpolarized, female contacts
2-conductor polarized, male contacts
waveguide flange
plain, rectangular
choke, rectangular
engaged 4-conductor; the plug has 1 male and 3 female contacts, individua contact designations shown
A
B
C
D
OR
coaxial, outside conductor shown carried through
coaxial, center conductor shown carried through; outside conductor not carried through
X
mated choke flanges in rectangular waveguide
COUNTER, ELECTROMAGNETIC; MESSAGE REGISTER (26)
general
with a make contact
COUPLER, DIRECTIONAL (27) (common coaxial/waveguide usage)
X
(common coaxial/waveguide usage)
E-plane aperture-coupling, 30-decibel transmission loss
X
E
30DB
COUPLING (28)
by loop from coaxial to circular waveguide, direct-current grounds connected
CRYSTAL, PIEZO-ELECTRIC (62)
DELAY LINE (31)
general
*
tapped delay
*
bifilar slow-wave structure (commonly used in traveling-wave tubes)
*
(length of delay indication replaces (*) asterisk)
DETECTOR, PRIMARY; MEASURING TRANSDUCER (30) (see HALL GENERATOR and THERMAL CONVERTER)
DISCONTINUITY (33) (common coaxial/waveguide usage)
equivalent series element, general
Z
capacitive reactance
inductive reactance
inductance-capacitance circuit, infinite reactance at resonance

Figure A7-1. — Electronics symbols — continued. 13.5(179)B

inductance-capacitance circuit, zero reactance at resonance

resistance

equivalent shunt element, general

capacitive susceptance

conductance

inductive susceptance

inductance-capacitance circuit, infinite susceptance at resonance

inductance-capacitance circuit, zero susceptance at resonance

ELECTRON TUBE (34)

triode

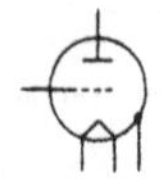

pentode, envelope connected to base terminal

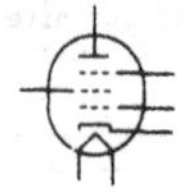

twin triode, equipotential cathode

typical wiring figure to show tube symbols placed in any convenient position

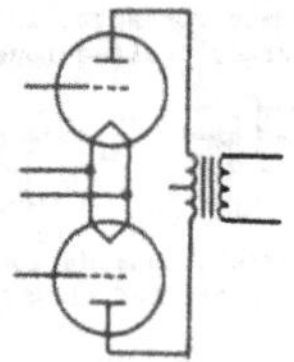

rectifier; voltage regulator
(see LAMP, GLOW)

phototube, single and multiplier

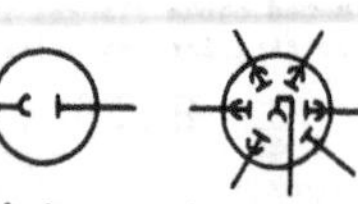

cathode-ray tube, electrostatic and magnetic deflection

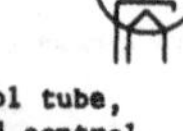

mercury-pool tube, ignitor and control grid (see RECTIFIER)

resonant magnetron, coaxial output and permanent magnet

reflex klystron, integral cavity, aperture coupled

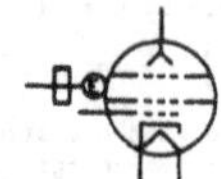

transmit-receive (TR) tube gas filled, tunable integral cavity, aperture coupled, with starter

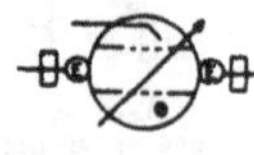

traveling-wave tube (typical)

forward-wave traveling-wave-tube amplifier shown with four grids, having slow-wave structure with attenuation, magnetic focusing by external permanent magnet, rf input and rf output coupling each E-plane aperture to external rectangular waveguide

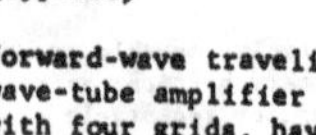

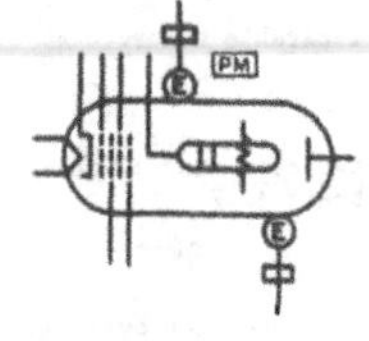

FERRITE DEVICES (100)

field polarization rotator

field polarization amplitude modulator

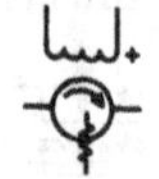

FUSE (36)

high-voltage primary cutout, dry

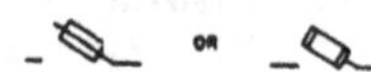

high-voltage primary cutout, oil

GOVERNOR (Contact-making) (37)

contacts shown here as closed

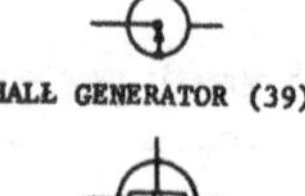

HALL GENERATOR (39)

HANDSET (40)

general

operator's set with push-to talk switch

HYBRID (41)

general

junction
(common coaxial/waveguide usage)

circular

(E, H or HE transverse field indicators replace (*) asterisk)

rectangular waveguide and coaxial coupling

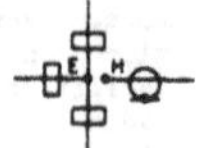

INDUCTOR (42)

general

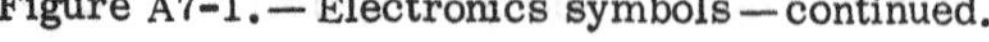

Figure A7-1. — Electronics symbols — continued. 13.5(179)C

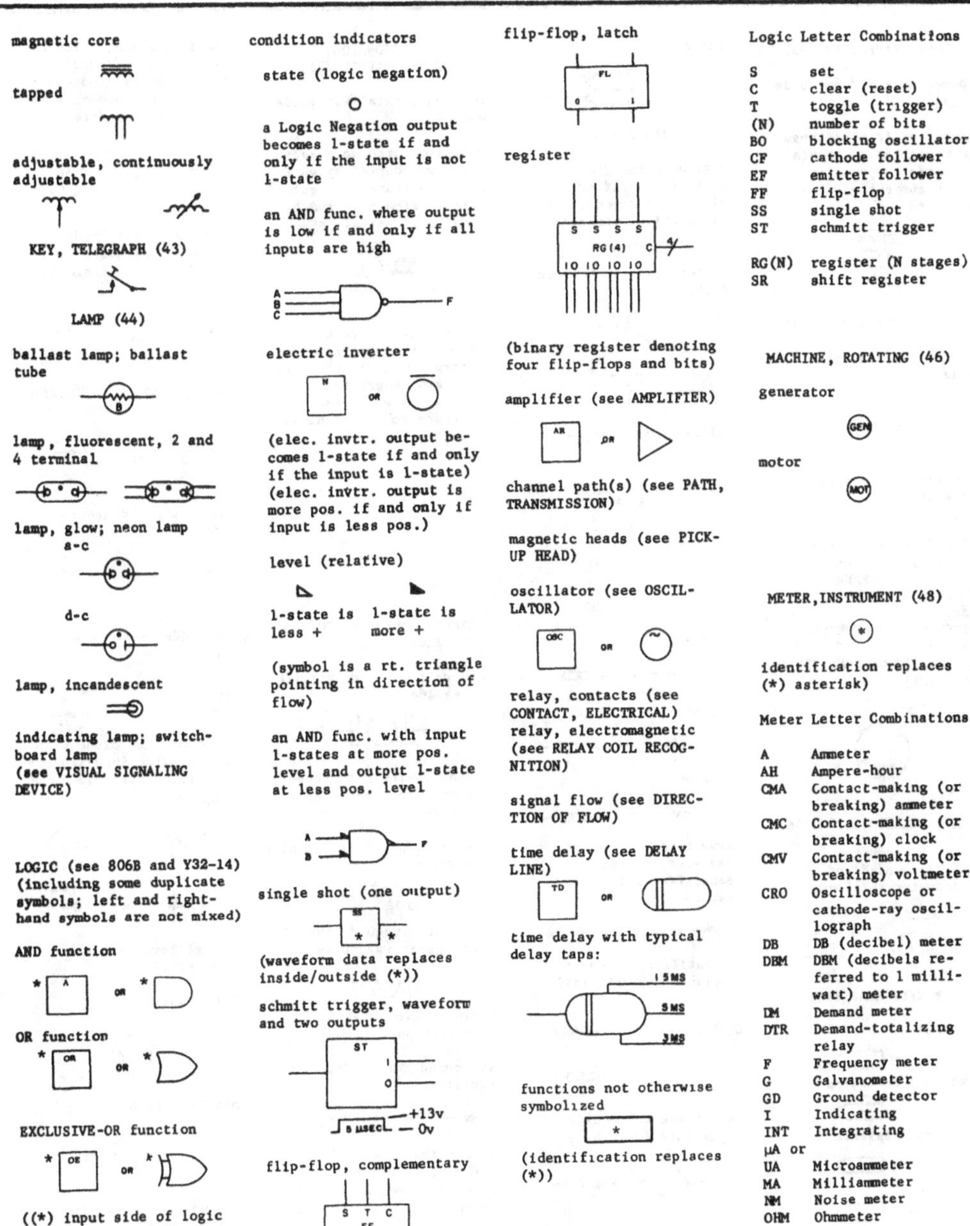
magnetic core
tapped
adjustable, continuously adjustable
KEY, TELEGRAPH (43)
LAMP (44)
ballast lamp; ballast tube
B
lamp, fluorescent, 2 and 4 terminal
lamp, glow; neon lamp
a-c
d-c
lamp, incandescent
indicating lamp; switch-board lamp
(see VISUAL SIGNALING DEVICE)
LOGIC (see 806B and Y32-14) (including some duplicate symbols; left and right-hand symbols are not mixed)
AND function
A
OR
OR function
OR
OR
EXCLUSIVE-OR function
OE
OR
((*) input side of logic symbols in general)
condition indicators
state (logic negation)
a Logic Negation output becomes 1-state if and only if the input is not 1-state
an AND func. where output is low if and only if all inputs are high
A
B
C
F
electric inverter
N
OR
(elec. invtr. output becomes 1-state if and only if the input is 1-state)
(elec. invtr. output is more pos. if and only if input is less pos.)
level (relative)
1-state is less +
1-state is more +
(symbol is a rt. triangle pointing in direction of flow)
an AND func. with input 1-states at more pos. level and output 1-state at less pos. level
A
B
F
single shot (one output)
SS
(waveform data replaces inside/outside (*))
schmitt trigger, waveform and two outputs
ST
1
0
+13v
0v
8 µSEC
flip-flop, complementary
S T C
FF
1 0
flip-flop, latch
FL
0 1
register
S S S S
RG(4)
C
4
10 10 10 10
(binary register denoting four flip-flops and bits)
amplifier (see AMPLIFIER)
AR
OR
channel path(s) (see PATH, TRANSMISSION)
magnetic heads (see PICK-UP HEAD)
oscillator (see OSCILLATOR)
OSC
OR
relay, contacts (see CONTACT, ELECTRICAL)
relay, electromagnetic (see RELAY COIL RECOGNITION)
signal flow (see DIRECTION OF FLOW)
time delay (see DELAY LINE)
TD
OR
time delay with typical delay taps:
15MS
5MS
3MS
functions not otherwise symbolized
*
(identification replaces (*))
Logic Letter Combinations
S set
C clear (reset)
T toggle (trigger)
(N) number of bits
BO blocking oscillator
CF cathode follower
EF emitter follower
FF flip-flop
SS single shot
ST schmitt trigger
RG(N) register (N stages)
SR shift register
MACHINE, ROTATING (46)
generator
GEN
motor
MOT
METER, INSTRUMENT (48)
identification replaces (*) asterisk)
Meter Letter Combinations
A Ammeter
AH Ampere-hour
CMA Contact-making (or breaking) ammeter
CMC Contact-making (or breaking) clock
CMV Contact-making (or breaking) voltmeter
CRO Oscilloscope or cathode-ray oscillograph
DB DB (decibel) meter
DBM DBM (decibels referred to 1 milliwatt) meter
DM Demand meter
DTR Demand-totalizing relay
F Frequency meter
G Galvanometer
GD Ground detector
I Indicating
INT Integrating
µA or UA Microammeter
MA Milliammeter
NM Noise meter
OHM Ohmmeter
OP Oil pressure

Figure A7-1. — Electronics symbols — continued. 13.5(179)D

MODE TRANSDUCER (53)

(common coaxial/waveguide usage)

transducer from rectangular waveguide to coaxial with mode suppression, direct-current grounds connected

MOTION, MECHANICAL (54)

rotation applied to a resistor

(identification replaces (*) asterisk)

NUCLEAR-RADIATION DETECTOR, gas filled; IONIZATION CHAMBER; PROPORTIONAL COUNTER TUBE; GEIGER-MULLER COUNTER TUBE (50) (see RADIATION-SENSITIVITY INDICATOR)

PATH, TRANSMISSION (58)

cable; 2-conductor, shield grounded and 5-conductor shielded

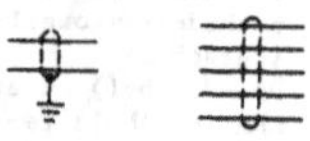

PICKUP HEAD (61)

general

writing; recording

reading; playback

erasing

writing, reading, and erasing

stereo

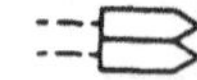

RECTIFIER (65)

semiconductor diode; metallic rectifier; electrolytic rectifier; asymmetrical varistor

mercury-pool tube power rectifier

fullwave bridge-type

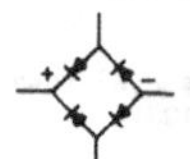

RESISTOR (68)

general

tapped

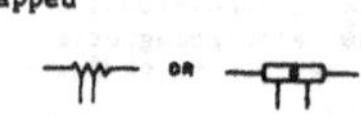

heating

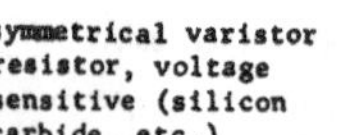

symmetrical varistor resistor, voltage sensitive (silicon carbide, etc.)

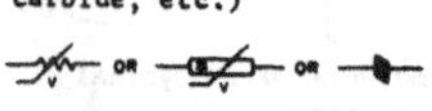

(identification marks replace (*) asterisk)

with adjustable contact

adjustable or continuously adjustable (variable)

(identification replaces (*) asterisk)

RESONATOR, TUNED CAVITY (71)

(common coaxial/waveguide usage)

resonator with mode suppression coupled by an E-plane aperture to a guided transmission path and by a loop to a coaxial path

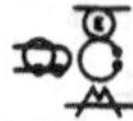

tunable resonator with direct-current ground connected to an electron device and adjustably coupled by an E-plane aperture to a rectangular waveguide

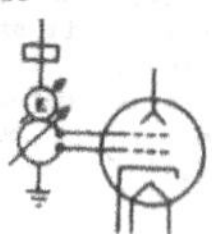

ROTARY JOINT, RF (COUPLER) (72)

general; with rectangular waveguide

(transmission path recognition symbol replaces (*) asterisk)

coaxial type in rectangular waveguide

circular waveguide type in rectangular waveguide

SEMICONDUCTOR DEVICE (73) (Two Terminal, diode)

semiconductor diode; rectifier

capacitive diode (also Varicap, Varactor, reactance diode, parametric diode)

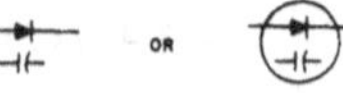

breakdown diode, unidirectional (also backward diode, avalanche diode, voltage regulator diode, Zener diode, voltage reference diode)

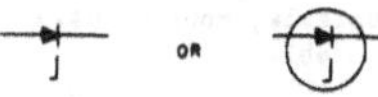

breakdown diode, bidirectional and backward diode (also bipolar voltage limiter)

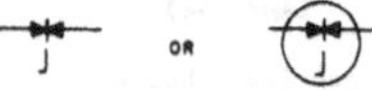

tunnel diode (also Esaki diode)

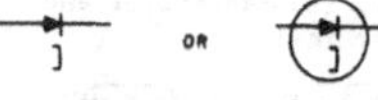

temperature-dependent diode

photodiode (also solar cell)

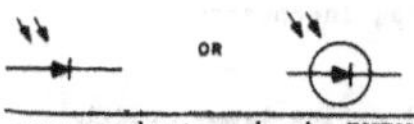

semiconductor diode, PNPN switch (also Shockley diode, four-layer diode and **SCR**).

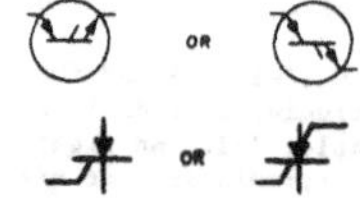

(Multi-Terminal, transistor, etc.)

PNP transistor

NPN transistor

unijunction transistor, N-type base

Figure A7-1.—Electronics symbols—continued. 13.5(179)E

unijunction transistor, P-type base

field-effect transistor, N-type base

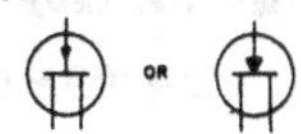

field-effect transistor, P-type base

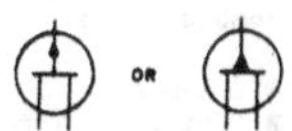

semiconductor triode, PNPN-type switch

semiconductor triode, NPNP-type switch

NPN transistor, transverse-biased base

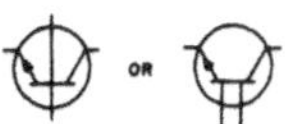

PNIP transistor, ohmic connection to the intrinsic region

NPIN transistor, ohmic connection to the intrinsic region

PNIN transistor, ohmic connection to the intrinsic region

NPIP transistor, ohmic connection to the intrinsic region

SQUIB (75)

explosive

igniter

sensing link; fusible link operated

SWITCH (76)

push button, circuit closing (make)

push button, circuit opening (break)

nonlocking; momentary circuit closing (make)

nonlocking; momentary circuit opening (break)

transfer

locking, circuit closing (make)

locking, circuit opening (break)

transfer, 3-position

wafer

(example shown: 3-pole 3-circuit with 2 non-shorting and 1 shorting moving contacts)

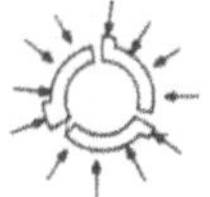

safety interlock, circuit opening and closing

2-pole field-discharge knife, with terminals and discharge resistor

(identification replaces (*) asterisk)

SYNCHRO (78)

Synchro Letter Combinations

CDX	Control-differential transmitter
CT	Control transformer
CX	Control transmitter
TDR	Torque-differential receiver
TDX	Torque-differential transmitter
TR	Torque receiver
TX	Torque transmitter
RS	Resolver
B	Outer winding rotatable in bearings

THERMAL ELEMENT (83)

actuating device

thermal cutout; flasher

thermal relay

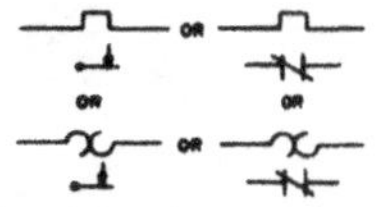

thermostat (operates on rising temperature), contact)

thermostat, make contact

thermostat, integral heater and transfer contacts

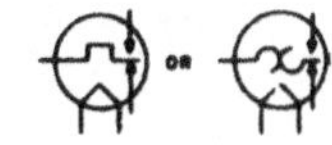

THERMISTOR; THERMAL RESISTOR (84)

with integral heater

THERMOCOUPLE (85)

temperature-measuring

current-measuring, integral heater connected

current-measuring, integral heater insulated

temperature-measuring, semiconductor

current-measuring, semiconductor

TRANSFORMER (86)

general

magnetic-core

one winding with adjustable inductance

separately adjustable inductance

adjustable mutual inductor, constant-current

Figure A7-1. — Electronics symbols — continued. 13.5(179)F

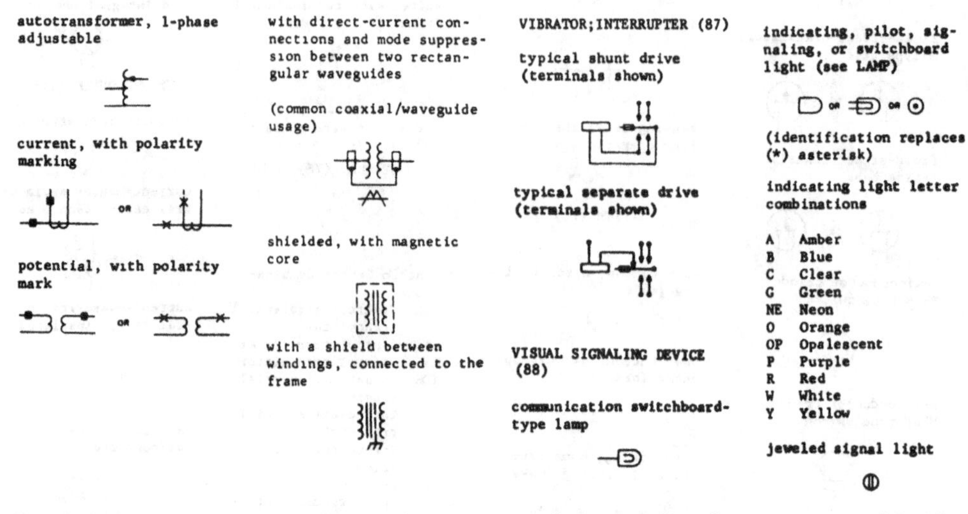
autotransformer, 1-phase adjustable
current, with polarity marking
OR
potential, with polarity mark
OR
with direct-current connections and mode suppression between two rectangular waveguides
(common coaxial/waveguide usage)
shielded, with magnetic core
with a shield between windings, connected to the frame
VIBRATOR; INTERRUPTER (87)
typical shunt drive (terminals shown)
typical separate drive (terminals shown)
VISUAL SIGNALING DEVICE (88)
communication switchboard-type lamp
indicating, pilot, signaling, or switchboard light (see LAMP)
OR
OR
(identification replaces (*) asterisk)
indicating light letter combinations
A Amber
B Blue
C Clear
G Green
NE Neon
O Orange
OP Opalescent
P Purple
R Red
W White
Y Yellow
jeweled signal light

Figure A7-1.—Electronics symbols—continued. 13.5(179)G

INDEX

D

E

F

U

V

W

Z